SOLID-PHASE SYNTHESIS AND COMBINATORIAL TECHNOLOGIES

SOLID-PHASE SYNTHESIS AND COMBINATORIAL TECHNOLOGIES

Pierfausto Seneci
GlaxoWellcome Medicines Research Centre

WILEY-INTERSCIENCE
A JOHN WILEY & SONS, INC., PUBLICATION
New York • Chichester • Weinheim • Brisbane • Singapore • Toronto

This book is printed on acid-free paper. ♾

Published simultaneously in Canada.

For ordering and customer service, call 1-800-CALL-WILEY.

Library of Congress Cataloging-in-Publication Data:

Seneci, Pierfausto, 1960–
Solid-phase synthesis and combinatorial technologies / by Pierfausto Secenci
p. cm.
Includes index.
ISBN 0-471-33195-3 (cloth: alk. paper)
1. Solid-phase synthesis. 2. Combinatorial chemistry. I. Title
QD262.S465 2000
547'.2—dc21 99-086954

Printed in the United States of America.

10 9 8 7 6 5 4 3 2 1

To Dora and Lorenzo, for their continuous support and for their indefatigable patience, which allows my being a father, a husband, and a writer.

CONTENTS

PREFACE

During the last decade, the emergence of the so-called high-throughput screening (HTS) technique in pharmaceutical research has allowed biologists to design and set up assays, aimed toward the identification of active compounds, that can test large numbers of compounds in a short time period. Great advances in automation, information science, data management, and related disciplines have contributed to create a typical environment where every biological laboratory requires tens of thousands of compounds to be tested on several assays in a week.

The challenge represented by this phenomenon for the medicinal chemists immediately appeared too demanding if only classical organic synthesis was to be used to provide biologists with the large number of compounds they needed. The assembly of large chemical collections derived from either commercial sources or from the proprietary chemical stores of major pharmaceutical companies was used to partially fulfil these needs; however, the limitations of these collections in terms of chemical diversity and as sources of positives on different biological targets were immediately apparent.

A new discipline capable of significantly increasing the throughput of chemical synthesis in terms of diversity and numbers of biologically relevant compounds has emerged to fulfill the HTS needs. This discipline, called combinatorial chemistry, officially dates to a few key papers that appeared in the mid-1980s and since then has experienced an enormous growth and has steadily attracted the interest of many researchers, at first only in pharmaceutical research but more recently in many other disciplines.

Since the beginning, combinatorial chemistry has strongly depended on the techniques of solid-phase synthesis (SPS). For this reason a detailed presentation of solid-phase (SP) chemistry in which the differences compared to classical organic chemistry in solution are highlighted appears at the beginning of this book.

The strong involvement of many other disciplines together with chemistry in the combinatorial arena prompts the more correct definition of combinatorial technologies, in which organic, inorganic, and analytical chemistry meet automation, statistical sciences, science information, data management, and various biological disciplines with the aim of producing and testing large number of high-quality compounds for one or more specific activity. The result is the extreme acceleration of the process of discovery of active entities, and examples of the use of combinatorial technologies in various applications will be presented. The main goal of this book is to provide the reader with the general concepts related to the core aspects of combinatorial technologies, and just to mention less well established approaches that have yet to prove their worth. An integrated description of related combinatorial disciplines will provide the reader with a general multidisciplinary overview of combinatorial technologies and will hopefully be of help in understanding their enormous and still partially unexploited potential. The strengths and weaknesses of each significant approach that has appeared in the literature will be analyzed and critically discussed. Significant new trends will be presented and their possible future impact on combinatorial technologies will be analyzed.

Combinatorial technologies have been historically associated with pharmaceutical applications, and several chapters of this book are centered around this application. Several other emerging applications, though are exhaustively treated: entire chapters are dedicated to combinatorial libraries from biological sources and to inorganic or polymeric combinatorial libraries, and several sections illustrate the use of synthetic organic libraries in catalysis research, in molecular recognition, and in agricultural research among others.

This book is aimed at three main groups of readers. First, experienced combinatorial chemists and scientists will find coverage of the most recent combinatorial approaches and a detailed multidisciplinary bibliography, including more than 1700 relevant papers, reviews, books, abstracts, and patents, as well as coverage of every relevant aspect of combinatorial technologies. Second, experienced chemists who are approaching SPS and combinatorial technologies for the first time and who wish to enhance their knowledge of the area will discover the basic concepts of SPS and combinatorial chemistry and a critical evaluation of their applications to specific strategies and disciplines. This book, though, is mainly aimed at chemistry students at both graduate and postgraduate advanced course level who will find the concepts of SPS, combinatorial chemistry, and related combinatorial technologies presented in a clear and exhaustive format. The first five chapters are conceived to provide the material for a basic course on solid-phase synthesis and combinatorial technologies, with focus both on explaining the theoretical fundamentals of these disciplines and making them more obvious through many examples; appropriate citations allow the expansion of any subject according to the reader's interests. The following six chapters are dedicated to an expert treatment of combinatorial technologies which hopefully covers most, if not all of the combinatorial hot topics; each of these chapters, either alone or together with several others, can represent the material for advanced, postgraduate courses. A large number of relevant examples will be thoroughly described in Chapters 6 to 11 to clarify each of the theoretical sections. The accessibility of the original papers will allow both the students to follow up an intriguing subject and their professors to expand a specific subject and to make it even more suitable for an advanced, postgraduate course.

The goal to reach such a wide and diverse audience may be overambitious, but a good blend of basic principles and detailed information about combinatorial technologies should really be useful for many workers, or future workers, in the field. The homogeneous organization of the book should also be instrumental for the reader and for the student to receive a balanced but complete overview of this new and exciting discipline and hopefully attract the attention of new talented scientists, or soon-to-be scientists, who will eventually contribute to the future development of combinatorial technologies.

Pierfausto Seneci
Verona, Italy

SOLID-PHASE SYNTHESIS AND COMBINATORIAL TECHNOLOGIES

1 Solid-Phase Synthesis: Basic Principles

Chemical reactions can be divided into two main classes: those that take place in a single phase and are therefore said to be homogeneous and those that occur at the interface of two phases and are described as heterogeneous. Typical examples of heterogeneous reactions are catalytic hydrogenations, where the substrate in solution is reduced by means of an insoluble catalyst and the reaction takes place at the interface between the solid and the solution.

In this and the following two chapters we will cover a specific family of heterogeneous reactions in which a reagent is coupled to a solid support via a chemical functionality that is present on the solid support. A multistep synthesis on the solid phase (SP) then transforms the bound intermediate into the target molecule that eventually is cleaved from the support. This technique, commonly referred to as *solid-phase synthesis* (SPS), was introduced in the 1960s by Merrifield (1) for the efficient synthesis of polypeptides and has rapidly become a standard technology for the preparation of oligopeptides and oligonucleotides. The advent of *combinatorial chemistry* has led to an increased awareness and use of SPS for the preparation of small organic molecules via classical organic reactions (2–8) that have been adapted to the SP.

We will now describe the main differences between SPS and organic synthesis in solution, focusing on (i) the types of *solid supports* available for SPS, (ii) the *linkers* used to anchor compounds to those supports, (iii) *monitoring reactions* in SPS, and (iv) *estimation of the purity and yield* of SP reactions.

1.1 SOLID SUPPORTS

1.1.1 Hydrophobic Polystyrene Resins

The most common solid supports in SPS are hydrophobic polystyrene (PS) resin beads (9), which are representatives of the class of so-called *gelatinous solid supports*. They consist of PS cross-linked with 1–2% divinylbenzene (DVB) and are described schematically in Fig. 1.1, which shows the appearance of a hydroxymethyl-grafted PS resin.

On the left in Fig. 1.1, at low magnification the regular spherical shape of the bead is very apparent. It should be pointed out that this is the common representation of a resin in equations describing a chemical reactions carried out in the SP. On the right,

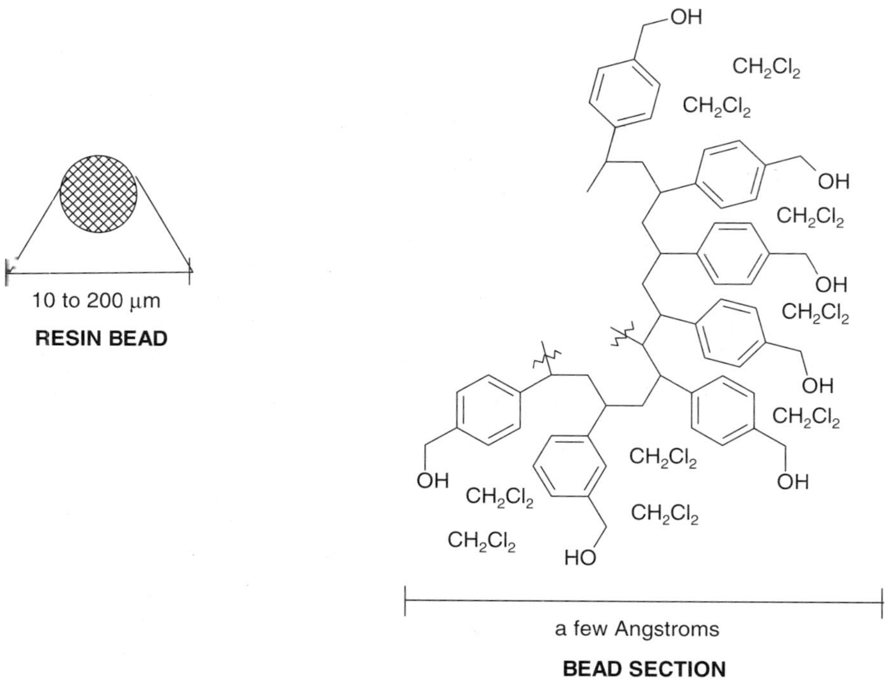

Figure 1.1 Representation of a resin bead (low magnification, left) and the inner structure of a bead (higher magnification, right).

the complex nature of the solid support is shown, with many different hydroxymethylated chains distributed three dimensionally throughout the whole resin bead and solvated by the surrounding solvent molecules.

The solvation of gelatinous resins strongly influences the reactivity of the bead sites. These supports swell in most solvents and behave as gels. The reaction with reagents in solution is facilitated because access to the inner reaction sites of the bead becomes easy. Swelling depends heavily on the solvent (which at best must dissolve the resin bead) and on the percentage of cross-linking of the PS support. While hydrophobic PS resins swell properly in apolar solvents (from 3 to 8 times their starting volume), their swelling is poor in polar protic solvents such as alcohols and water. This precludes any reaction with reagents dissolved in these solvents since only around 1–2% of the reaction sites of a bead are located on its surface. The cross-linking of these resins is normally between 1 and 2%, which gives a reasonable compromise between a good ability to swell resulting from a low level of cross-linking and the stability of the beads (low levels of cross-linking result in beads that are very fragile).

The PS resin beads commonly used in SPS have particle sizes between 90 and 200 μm. This makes them big enough to have large numbers of reaction sites on a single bead (typically in the range of hundreds of picomoles) but small enough to be resistant to physical shock during handling. Their loading, defined as the number of sites per resin gram, is typically in the range of 1 mmol/g. They are cheap, when compared to other supports, and are commercially available with many different functional groups for a wide variety of loading chemistries. Examples are the chloromethylated- (Merrifield), aminomethylated-, carboxy, hydroxymethyl, formyl, and bromo resins (Fig. 1.2). All of these functionalized supports can be prepared from underivatized PS resin by simple chemical transformations.

1.1.2 Hybrid Hydrophilic Polystyrene Resins

A major drawback of the hydrophobic PS resins is their poor swelling in polar protic solvents, as mentioned previously. The grafting of hydrophilic monofunctional or bifunctional polyethylene glycol (PEG) chains to the PS resin to produce a hybrid support (10, 11) provides a solution to this problem that allows the use of hydrophilic solvents with these resins. The swelling properties of hydrophobic PS resins, monofunctional PEG-grafted (Tentagel, TG), and bifunctional PEG-grafted (Argogel, AG) PS resins are compared in Table 1.1. The structures of the two hybrid PEG–PS resins are shown schematically in Fig. 1.3.

These newer resins have had a great impact on SPS. The higher degree of flexibility of the terminal PEG chains produces a real "solutionlike" microenvironment and allows on-bead reaction monitoring and structure determination by gel-phase ^{13}C nuclear magnetic resonance (NMR) or magic angle spinning (MAS) ^{1}H-NMR spectroscopy. Many examples of the MAS–NMR spectra of compounds bound to a hydrophobic PS or to a hybrid PEG–PS resin clearly show the better signal resolution obtained with PEG–PS. The use of NMR in SPS will be described in more detail in Sections 1.3.4 and 1.4.6.

The grafting of hydrophilic groups onto PEG–PS resins decreases their loading, which typically drops to around 0.2–0.3 mmol/g for TG or 0.4–0.6 mmol/g for high-loading TG or AG. The PEG–PS-based resins have several disadvantages, and although they are commercially available with various grafted functionalities, they are significantly more expensive than hydrophobic PS supports. The introduction of PEG chains on the solid support sometimes has negative effects on the quality of the

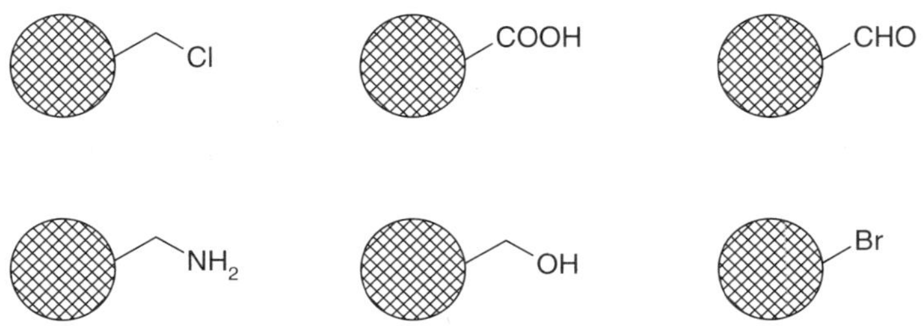

Figure 1.2 Selection of commerically available functional groups grafted onto PS resins.

TABLE 1.1 Swelling of PS-Based Resins in Common SP Solvents[a]

Resins/Solvents	THF[b]	DMF	DCM	MeOH	Water
Hydrophobic PS	8.8	5.6	8.3	1.6	—
TG–PS	5.8	5.6	5.1	4.2	4.2
AG–PS	5.7	5.8	7.2	4.1	3.2

[a]*Note*: THF = tetrahydrofuran; DMF = dimethyl formamide; DCM = dichloromethane.
[b]Resin swelling measured as volume increase in respect to the dry bead.

chemistry, which may be due to the leakage of PEG chains due to the use of aggressive reagents such as strong electrophiles or to the use of Lewis acids, which can form complexes with the PEG chains.

The general tendency is to use PEG–PS resins for detailed SP studies, where reaction monitoring by NMR is necessary or when polar protic solvents are required. When large quantities of supports must be used and the SPS strategy has been thoroughly validated, hydrophobic PS resins are normally used.

Other more hydrophilic, non-PS-based solid supports were reported with suitable properties for specific SP protocols. Radical polymerization of acrylamide-substituted PEG chains yields the so-called PEGA support (12), which is ideal for aqueous solvents and allows the inner penetration of macromolecular reagents (13; see also Section 7.2.3) but has a limited applicability for SP organic synthesis due to the reactivity of its numerous amide bonds. Polytetrahydrofuran (PTHF)-cross-linked PS resin is obtained from suspension copolymerization (14) of PTHF-linked styrenes with styrene and 4-vinylbenzyl chloride; its swelling properties and stability to commonly encountered organic reaction conditions is extremely encouraging, as confirmed by the compatibility with harsh reagents such as *n*-Buli (15). Newer, more stable and promising non-PS-based solid supports were recently reported and validated in demanding organic reaction conditions (16–19); they will soon represent further alternative support options for the bench chemist involved in SPS.

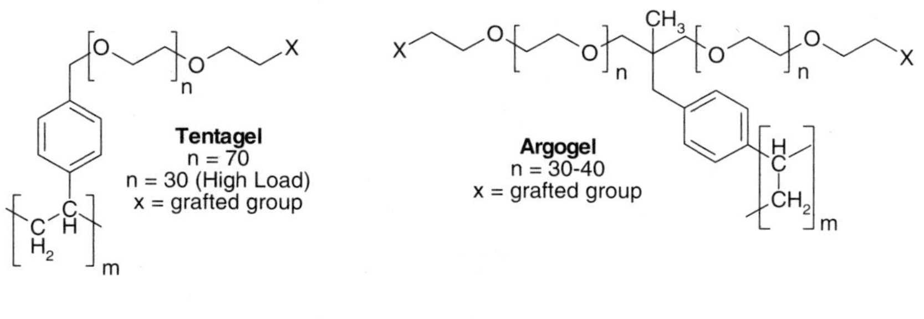

Figure 1.3 Structure of monofunctional hybrid PEG–PS resins (Tentagel, left) and of bifunctional hybrid PEG–PS resins (Argogel, right).

1.1.3 Macroporous Nonswelling Resins

A different class of supports contains macroscopic pores embedded in an extremely rigid structure that does not allow any swelling of the matrix. The SPS takes place on a gellike support that is polymerized into the permanent pores. These *macroporous nonswelling resins* are typically used for continuous-flow oligonucleotide and peptide synthesis (20, 21). The most popular are POLYHYPE resins, made of polyamide- (PA-) containing 10–50% cross-linked PS (22); PA-containing Kieselguhr, which is made from a silica-based support with a porous structure (23); and fully inorganic controlled-pore glass (CPG) supports, made of highly porous pure silica (24), which are very common in oligonucleotide SPS and will be detailed in Section 2.2. A representation of a macroporous Kieselguhr support is shown in Fig. 1.4.

A recent addition to the family of macroporous supports was designed for the SPS of organic compounds (25) and has been used as an alternative to gelatinous supports. This resin, called AP–PS, consists of a highly cross-linked macroporous PS framework and has a loading of around 0.6–1 mmol/g. It does not swell appreciably in a wide range of solvents, allowing many different experimental conditions to be used. It is now commercially available with a number of different grafted functionalities, and it is in the same price range as the PEG–PS resins.

Two characteristics of macroporous supports are worth mentioning. First, the transfer of specific reaction conditions from solution to the macroporous support should be easier because the influence of swelling in different solvents and diffusion rates, typical for low-cross-linked PS resins, is not relevant. Second, this support can be washed more easily than the classical PS resins. A study of the retention of biphenyl in hydrophobic PS and AP–PS resins treated with solutions of biphenyl in methylene chloride (26) shows how after two identical wash cycles AP–PS retains 0.01% of biphenyl while hydrophobic PS retains 2.63% of the same impurity. This is due to the

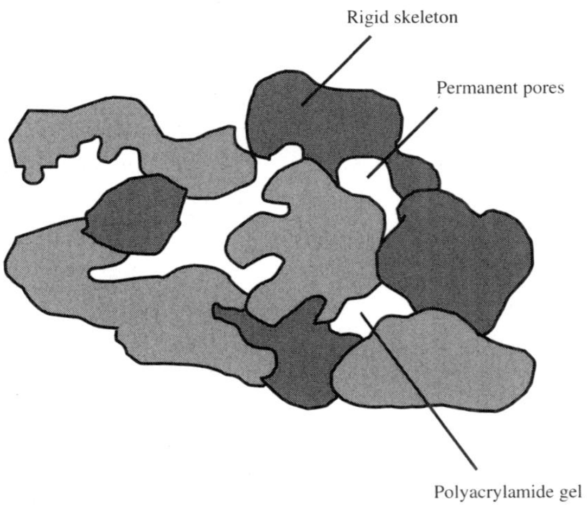

Figure 1.4 Schematic representation of a macroporous SP support.

easier accessibility for solvents of non-resin-bound impurities on macroporous resins. Disadvantages of macroporous supports are their extreme rigidity, which makes them more suitable to continuous-flow synthesis, even though they are claimed to exhibit a similar stability to physical shock as gelatinous resins (26), and the impossibility of performing on-bead NMR spectra on the rigid solid-like support.

1.1.4 Miscellaneous Solid Supports

The commercial availability of PS-based supports in nonbead format such as *pins* (27, 28), *crowns* (29, 30), *microtubes* (31), and recently *discs* (32) has also had a notable impact on the field of SPS. These devices allow the attachment of significantly higher amounts of compound onto a single support unit compared to resin beads. They are similar in nature to the PS resin supports, but their different morphology requires different handling procedures than resin beads; we will discuss some of their specific properties in the following chapters as related to specific examples.

While gellike and macroporous resins cover the vast majority of SPS, the use of other supports has also been explored. *Cellulose* (33, 34) in the form of paper sheets has been employed for multiple simultaneous SPS of peptides with a relatively low loading of 0.5–0.6 μmol/cm^2. *Cotton* (35) has been used for the same application with a loading of around 0.1 mmol/g. *Glass* (36) was among the first supports used for the synthesis of large numbers of peptides, due to its chemical inertness and solidity. Various *polymeric membranes* (37, 38) were also used to prepare peptides on SP. Several of these supports will be mentioned also as related to combinatorial library synthesis (see Section 6.4.1 and 6.4.2).

The class of so-called *soluble supports*, including PEGs, non-cross-linked PS, and the recently introduced high-loading dendrimers, will be covered in Section 8.5.

1.1.5 SPS: Reaction Kinetics and Work-Up Procedures

The use of a solid support for carrying out organic chemistry has a profound influence on some of the reaction parameters (9, 35); above all, there is a strong effect on the rate of reaction. When a reaction takes place in a homogeneous solution, the reactants can freely interact and the reaction rates depend on classical parameters (e.g., concentration and temperature). When a resin-bound reagent is involved, the reaction takes place in a heterogeneous medium and the reaction kinetics are also dependent upon the rate of diffusion of the reagent in solution into and out of the resin beads. Reaction rates are generally slower in SP than in solution-phase chemistry, and reagents supported on gelatinous resins exhibit different reactivities depending on the swelling properties of the resin in the solvent used for the reaction. Reaction rates are also highly influenced by the nature of the support (40). PEG–PS and hydrophobic PS resins show different reaction kinetics in different reactions, as expected, but it is not possible to predict their behavior due to the many factors influencing the reaction kinetics in each experimental condition (41).

Macroporous resins are not influenced by swelling, but their reactivity is comparable, and sometimes lower, than that of gelatinous resins.

Site–site interactions are an important parameter in SP reactions. When inter- and intramolecular reactions are possible in solution, their relative occurrence will depend on the number of encounters/collisions between two different molecules (intermolecular) and between two groups in the same molecule (intramolecular); the complete freedom of movement for each molecule in solution will not prevent a priori any intermolecular coupling. When the intramolecular process is disfavored (e.g. with macrocyclization of peptides), the undesired linear polypeptide will always be present in the crude reaction products even using the most careful reaction conditions (Fig. 1.5, top). The support significantly constrains the freedom of each supported molecule, thus favoring intra- versus intermolecular reactions; the above-mentioned macrocyclization of peptides usually produces pure cyclic peptides in high yields (Fig. 1.5, bottom). It must be remembered, though, that the good solvation of beads brings them toward a solutionlike environment, thus lessening the constraints for each loading site to cross-react with another site. Site isolation is generally observed in SPS, and a high cross-linking and a high backbone rigidity for the support will favor it; a low level of cross-linking and a highly flexible support, as the PEG–PS resins, may rather cause significant site–site interactions. Examples of both negligible and significant site–site interactions during SP reactions have been reported; several recent, excellent reviews present in details the current knowledge regarding this issue (9, 42, 43). The extreme flexibility of two PS-based supports bearing respectively maleimide and anthracene has recently allowed an inter-bead Diels-Alder reaction to give stable, covalently linked bead aggregates (42); the potential of such a solid–solid interface chemistry needs to be fully assessed.

Another major difference is related to the work-up procedure for the reaction. When a reaction is carried out on SP, the reaction product remains attached to resin while all of the excess reagents, the catalysts, and the impurities remain in solution. A typical work-up procedure involves simple filtration of the resin followed by repeated washings with fresh solvents in which the resin has a good swelling and the reagents/impurities are soluble. A good degree of swelling helps the solvent to access the adsorbed impurities, which must be soluble in the same solvent to allow them to be removed from the beads. It is quite common to use a number of different solvents sequentially in the washing cycle in order to remove soluble reagents/impurities with different physicochemical properties. This work-up procedure is amenable to automation and, as we will see in the following chapters, is one of the features that make SPS so appealing for combinatorial technologies.

The tedious purification of intermediates in classical solution synthesis produces pure compounds, which are then carried forward to the subsequent steps of the synthetic scheme. When a side product of the SP reaction remains attached to the resin, it becomes impossible to separate it from the desired intermediate/target molecule, thus irreversibly affecting the quality of the synthesis and the purity of the final product. This and other factors, which will be discussed in the following chapters, are considered during the so-called *chemistry assessment* phase of the synthesis in which the chemistry is transferred from solution to the solid phase and the reactions are optimized. In general, significant differences between a reaction in solution and the same reaction on a solid support are the rule rather than the exception. Several reviews related to the use of solids supports in SPS and/or in combinatorial chemistry have recently appeared (44–51); their content largely expands what has been covered in this Section and may be useful for more experienced and interested readers.

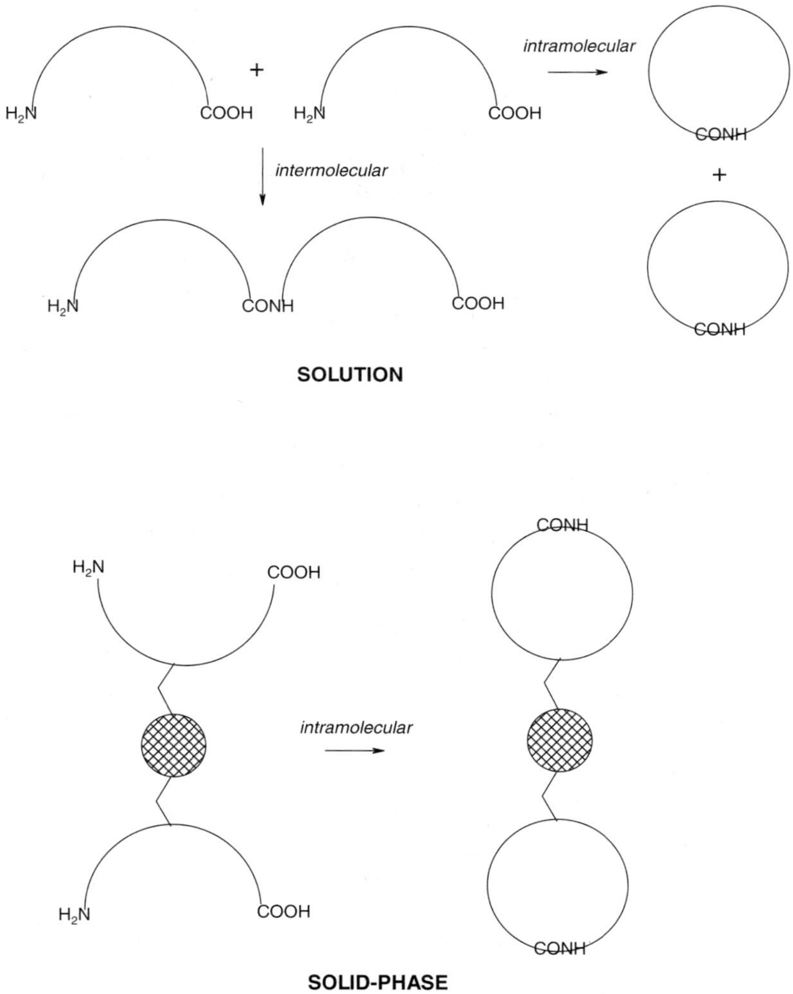

Figure 1.5 Site–site interactions in solution phase and on SP: intra- versus intermolecular reactions.

1.2 LINKERS

1.2.1 Properties of a Solid-Phase Linker

Successful SPS produces a final resin-bound target molecule that is released into solution by breaking a bond between the resin and a functional group in the final compound. Two examples are shown in Fig. 1.6. On the left, the basic hydrolysis of an ester bond releases a carboxylic acid into solution and simultaneously re-forms the original hydroxy PS resin. On the right, the acidic hydrolysis of an acetal function provides the starting aldehyde resin and a diol compound.

Often, the intermediates to be attached to the resin do not possess a suitable functional group. Moreover, the bond linking the solid support and the compound must be stable to all of the reaction conditions to be employed during the preparation of the final compounds. Even marginal sensitivity to one of the reagents used could result in the release of intermediates into solution and create free sites on the beads during the synthesis. This would decrease the amount of final compound produced and could also lead to the formation of resin-bound side products. Finally, the bond linking the substrate to the resin must be sensitive to a cleavage reaction condition that allows the release of the final compound in solution without degradation.

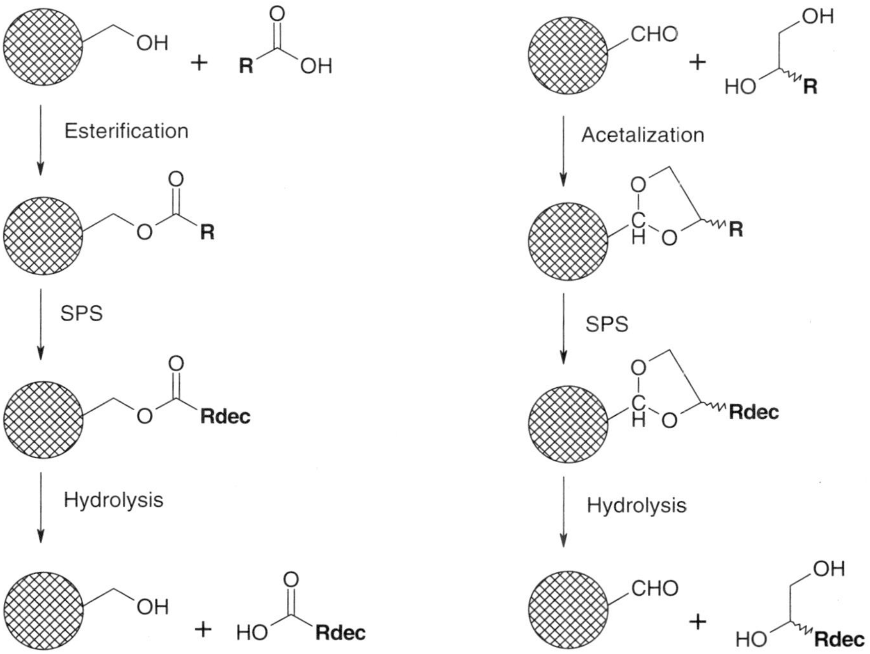

Figure 1.6 SP attachment and cleavage of an acid onto a hydroxymethyl support (left) and of a diol onto an aldehydic support (right).

A direct bond between the precursor to the final compound and a commercially available PS resin hardly satisfies all of these requirements but the problem can be circumvented by the use of a so-called *linker*, which is a chemical structure inserted between the solid support and the compound to be prepared. The linker is stable to all of the reaction conditions used during the SPS scheme but is labile under well-defined conditions. The proper selection of the linker strongly influences both the quality and the success of the SPS strategy.

The use of SPS in combinatorial chemistry has given rise to the need to transpose many different chemistries onto SP and, as a result, has driven many efforts toward the design and preparation of new SP linkers. The ideal linker must be easy to prepare and be stable to the reaction conditions used during elaboration, but at the same time, it should be highly sensitive to one, or at most a small number, of specific cleavage reagents/conditions. It should not release troublesome by-products during the final cleavage and, ideally, should release different products into solution when cleaved under different conditions. It should also allow the selection of suitable protecting groups, either commercially available for a specific building block or easily prepared, to protect reactive functionalities during the SPS and to deprotect them selectively at different stages of the SPS without affecting the linker stability.

Many families of SP linkers are currently used in SPS and can be classified according to conditions used for their cleavage. The most commonly used are commercially available already anchored onto PS resins. We will describe the most frequently encountered families of linkers such as *acid*- and *base-labile* linkers, *photolabile* linkers, *safety-catch* linkers, and *traceless* linkers. Several linkers could be attributed to more than one category, and the classification used will be motivated in the text. The so-called *cyclative cleavage*, where the last SPS step simultaneously cyclizes and cleaves the final compound from the solid support, will also be covered. An extensive bibliography will be provided, especially regarding recent examples of linkers with appealing features to recover pure compounds in high yields at the end of a SP synthesis; the reader should also refer to recent, exhaustive reviews related to SP linkers (52–56).

1.2.2 Acid-Labile Linkers

This is the most widely used class of SP linkers. Historically, the SPS of peptides (see Section 2.1) was developed using building blocks protected with acid-labile groups, thus allowing a convenient simultaneous cleavage and deprotection in the final step of the synthesis. Four commercially available acid-labile linkers are depicted in Fig. 1.7 in resin- and compound-bound forms. The preferred cleavage conditions for each linker are also provided.

The Wang linker **1.1** (57) is a very popular choice that is based on a *p*-alkoxybenzyl alcohol moiety. It is typically cleaved by trifluoroacetic acid (TFA)–DCM 1/1 in 30 min at room temperature (rt) to produce carboxylic acids or alcohols. Two slight modifications to this linker have produced the super-acid-sensitive resin (SASRIN) **1.2** (58), in which an *o*-methoxy group increases the acid sensitivity allowing cleavage with 1% TFA–DCM, and the hypersensitive acid-labile (HAL) linker **1.3** (59), where a second *o*-methoxy group further increases the acid sensitivity allowing cleavage with

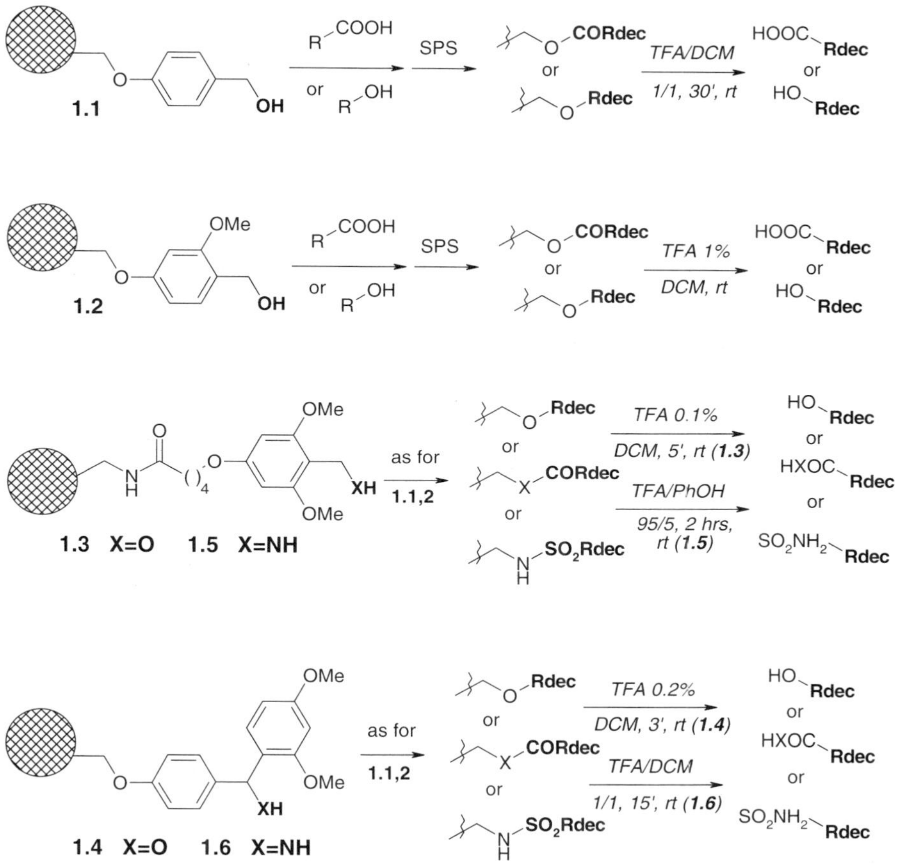

Figure 1.7 Acid-labile, commerically available SP linkers **1.1–1.6**.

either 0.1% TFA–DCM in 5 min at rt, or 10% AcOH/DCM. Another popular choice is the Rink ester linker **1.4** (60), which contains a bis-benzylic hydroxyl function and can be cleaved with 10% AcOH–DCM for 1.5 h at rt, or 0.2% TFA–DCM in 3 min at rt. Both the HAL and the Rink linker exist in the corresponding NH2 versions, called PAL **1.5** [5-(4-(g-FMOC) aminomethyl-3,5-dimethoxy phenoxy) valeric acid] (61) and Rink amide **1.6** (62), respectively, which allow the formation and final release of amides or sulfonamides in acidic conditions. All of these linkers are reasonably stable to nonacidic conditions.

Seven noncommercial acid-labile linkers have been reported recently in the literature and are shown in Figs. 1.8. (**1.7–1.10**) and 1.9 (**1.11–1.13**). The THP (tetrahydropyran) linker **1.7** (63), which is easily grafted onto Merrifield resin, has been used to support primary alcohols, secondary alcohols, hydroxylamines, and carboxylic acids. It is stable to strong nucleophiles and basic conditions and can be cleaved by

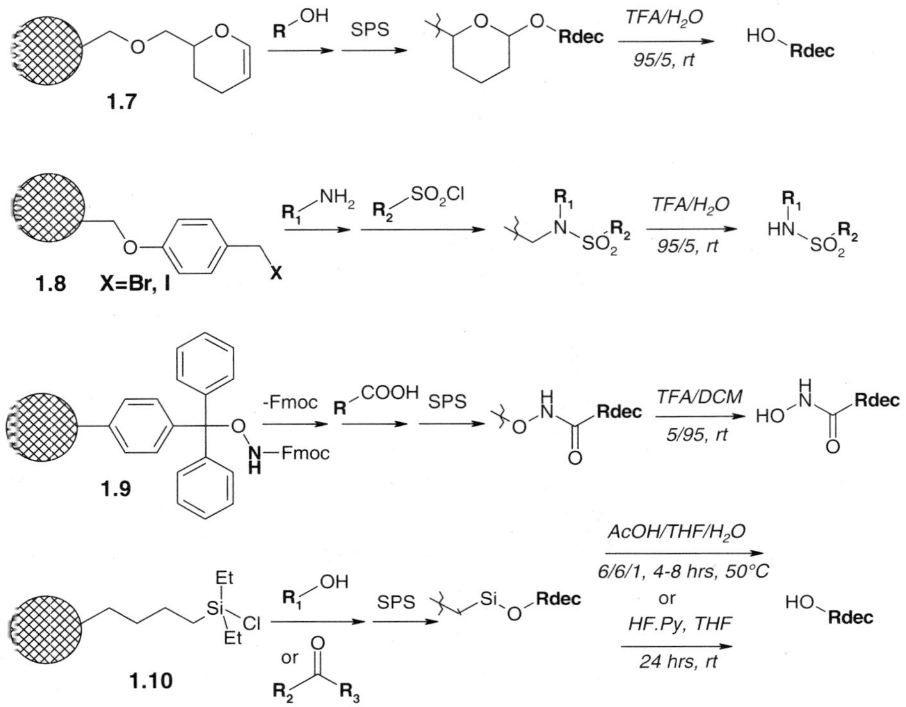

Figure 1.8 Acid-labile SP linkers **1.7–1.10**.

95% TFA–water at rt or by PPTS (pyridinium p-toluensulfonate) in DCE (dichloromethane)–*n*-BuOH 1/1 at 60 °C for 16 h to release hydroxyl-containing compounds.

The halide linkers **1.8** (64), which can be prepared from Wang resin, have been used to support amines and to release sulfonamides using 95% TFA–DCM (R = H) or 5% TFA–DCM (R = OMe).

The *N*-Fmoc-amino-oxy-2-chlorotrityl linker **1.9** (65) is derived from the commercially available, extremely popular 2-chlorotrityl chloride PS resin and has been used to prepare functionalized hydroxamic acids or peptidyl hydroxamic acids. The final compounds can be removed from the resin using 5% TFA–DCM.

The silicon-based linker **1.10** (66), which can be readily prepared from commercially available PS–diethylsilane resin, has been successfully used to perform multistep SPS to give alcohols that could be cleaved with AcOH–THF–H_2O in 4–8 h at 50 °C (48) or with HF.Py (hydrofluoric acid/pyridine complex) in THF at rt (67).

The indole linker **1.11** (68), easily prepared from aminomethyl PS resin and N-carboxyalkylated indole-3-carboxaldehyde, was used to support amines and to transform them on SP, obtaining, by release with TFA–DCM 1/1 in 30 min, a variety of compounds, including amides, sulfonamides, guanidines, ureas, and carbamates.

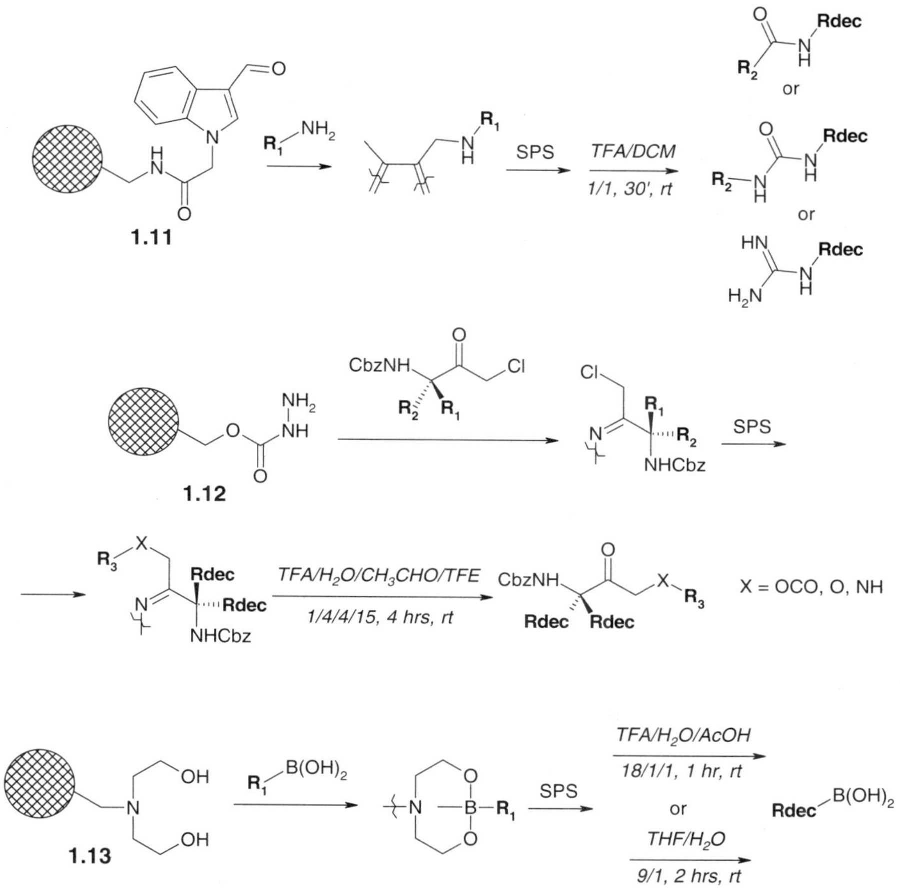

Figure 1.9 Acid-labile SP linkers **1.11–1.13**.

The carbazate linker **1.12** (69), obtained from hydroxy Argogel™ resin activated with CDI and reacted with anhydrous hydrazine, was used to support ketone-based protease inhibitors and to release them after SPS using TFA/H_2O/CH_3CHO/TFE 1/4/4/15 in 4 h at rt.

The diethanolamine linker **1.13** (70), obtained from aminomethyl PS resin and ethylene oxide, has been used to support boronic acids which, after SP transformations, were released with THF/H_2O/AcOH 18/1/1 in 1 h at rt or, when acid-labile boronic acids were involved, with THF/H_2O 9/1 in 2 h at rt.

This brief, incomplete survey should have provided a flavor of the many functional groups that can be hooked onto and released from a solid support using acid-labile linkers. Acids, alcohols, phenols, amines, hydroxylamines, and halides have been successfully attached to a variety of resins through the judicious choice of the linker that can be cleaved after elaboration to release acids, amides, sulfonamides, hydroxamic acids, and aromatics. The acidic cleavage conditions can be modulated according to the sensitivity of the final compounds, and the cleavage reagents are easily removed from the sample either by evaporation or, when necessary, by simple extraction. This is a significant advantage, particularly when many final compounds are simultaneously released and must be obtained in good purity with minimal purification, as is the case in combinatorial technologies.

1.2.3 Base- or Nucleophile-Labile Linkers

The wide use of acid-labile linkers and protecting groups in peptide SPS has reduced efforts toward the development of base-labile linkers. The commonly used SP Fmoc peptide coupling protocols require Fmoc deprotection under basic conditions during the synthesis, thus ruling out base-labile linkers. However, base-labile linkers are popular in oligonucleotide SPS and will be described in Section 2.2. Other examples of base- or nucleophile-labile linkers are shown in Fig. 1.10.

The fluorene linker **1.14** (71), which is easily attached to an aminomethyl PS resin, has been used to support C-terminal Boc-protected amino acids during oligopeptide synthesis. The cleavage of this linker requires the use of 20% piperidine, or better 20% morpholine at rt over 2 h.

The commercially available oxime linker **1.15** (72) has been used in peptide synthesis and for the SPS of small organic molecules such as indoles. Cleavage is carried out with either hydrazine (0.5 *M* hydrazine in $CHCl_3$–MeOH 2/1 for 10 min at rt) or with aliphatic amines or amino esters (DCM at rt for 12 h) to produce hydrazides and amides, respectively.

The acetyldimedone linker **1.16** (73) is another example of a linker attached to an aminomethyl PS resin and has been used to support amines and amino acids during SPS. The products are released containing free amino groups. The cleavage conditions are typically 2% hydrazine–DMF for 5 min at rt.

The silicon linker **1.17** (74), prepared by Grignard reaction of a 4-bromoarylsilane with commercial formyl PS resin, was used to support acids, alcohols, and amines either as such via its acyl imidazole derivative; cleavage with TBAF (tert-butylammonium fluoride) in DMF for 3 h at 60 °C, or with CsF in DMF for 18–24 h at 90 °C validated the release of the three different functionalities.

The *p*-thiophenol linker **1.18** (75), prepared from aminomethyl PS resin and 3-(4-thiophenyl)-propionic acid, was used to support a chloropyridazine and to release after SPS decorated aminopyridazines by treatment with primary or secondary amines and anilines for 24–48 h at 90 °C.

1.2.4 Photolabile Linkers

Photolabile linkers use light to break the bond between the elaborated intermediate and the linker, thus releasing pure compound from the SPS into solution without interference from potentially troublesome side products. This advantage has encour-

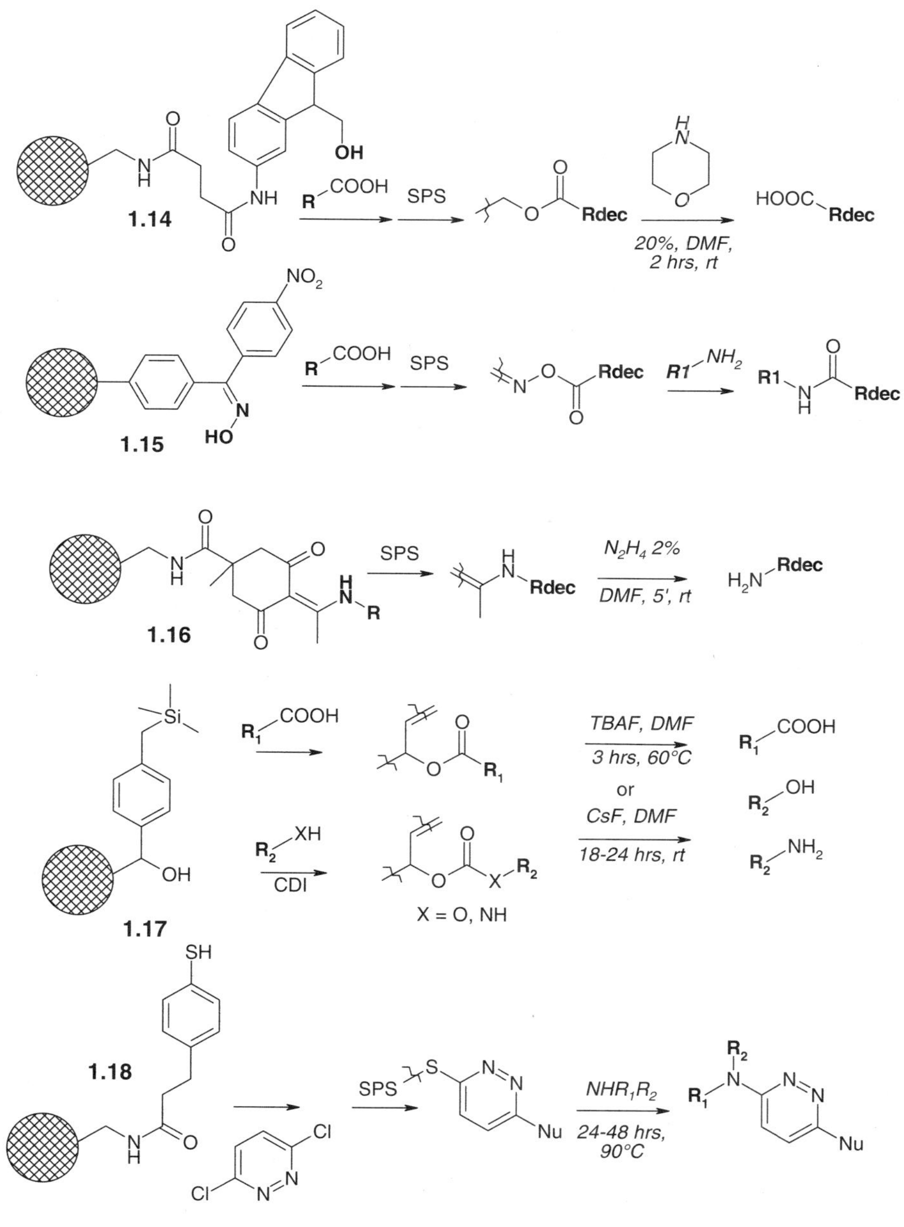

Figure 1.10 Base- or nucleophile-labile SP linkers **1.14–1.18**.

aged the development of photolabile linkers for SPS, and some examples are shown in Fig. 1.11.

The original *o*-nitrobenzyl bromide linker **1.19** (76) attached to aminomethyl PS resin has been used for both peptide and small organic molecule SPS. Cleavage by photolysis at 350 nm under anaerobic conditions gives carboxylic acids; the insertion of an α-methyl allowed an easier cleavage (77). The linker **1.19** has also been prepared

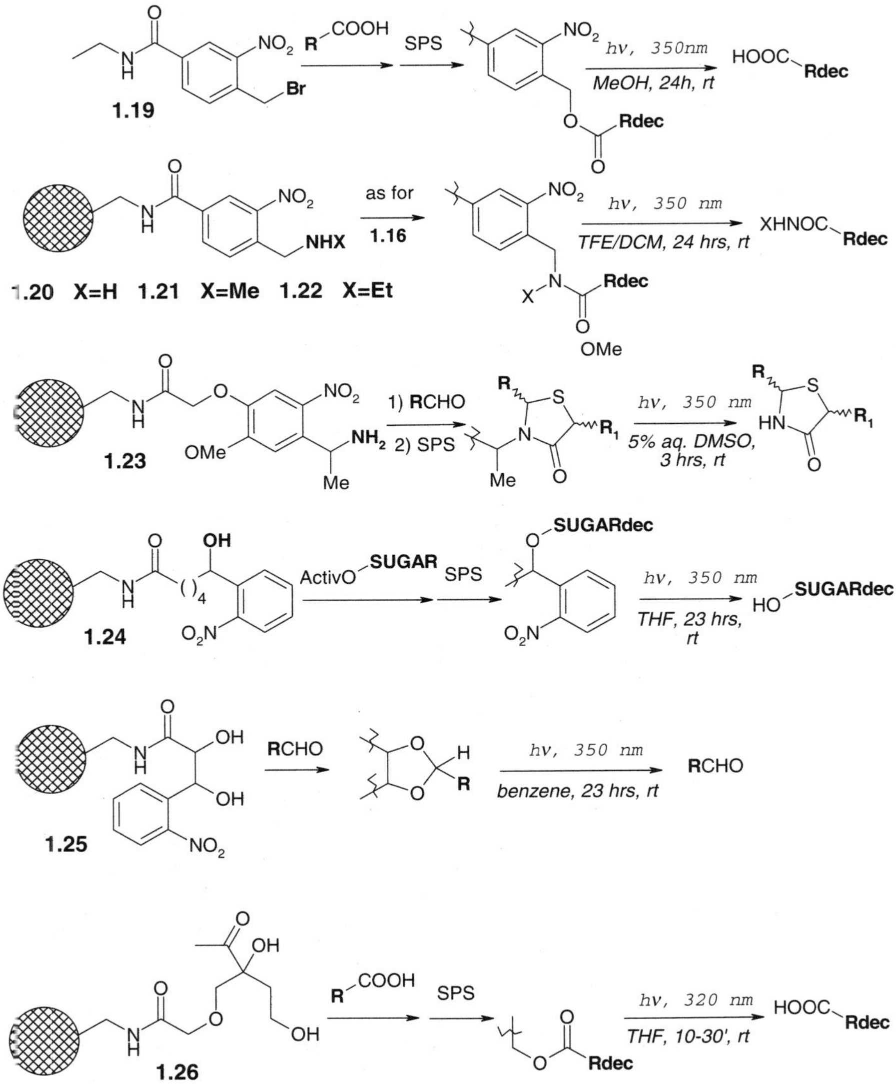

Figure 1.11 Photolabile SP linkers **1.19–1.26**.

in the NH_2 (**1.20**), NHMe (**1.21**), or NHEt (**1.22**) forms, which allow the synthesis of primary or secondary amides or C-terminal peptide amides (78, 79).

Another aminomethyl PS resin supported *o*-nitrobenzyl photolabile linker **1.23** (80) has been employed for the synthesis of heterocycles such as thiazolidinones. Attachment to the resin is through an acid or aldehyde group and photolytic cleavage is performed in 5% dimethyl sulfoxide (DMSO)–aqueous buffer to facilitate the biological testing of the final compound. A photolabile hydroxyl linker **1.24** (81), again supported on aminomethyl PS resin, has been used to synthesize carbohydrate and peptide derivatives in SP. The easily prepared diol linker **1.25** (82) allowed the attachment of aldehydes as acetals and their photo induced release using standard conditions.

A different, pivaloylglycol-based photolabile linker **1.26** (83) was prepared with a complex, eight-step procedure from dihydroxyacetone dimer and aminomethyl Tentagel resin; the linker was used to support and to further elaborate carboxylic acids. The photolytic release at 320 nm of elaborated acids compared favorably with more assessed *o*-nitrobenzyl linkers.

1.2.5 SAFETY-CATCH LINKERS

Some SP linkers are totally stable during the synthetic sequence and only become labile after a process known as *activation*, which increases the lability of the linker toward well-defined cleavage conditions. These linkers, known as safety-catch (SC) linkers, are very popular and allow the support and release of many different functionalities. Some examples that rely on different methods of activation are collected in Fig. 1.12 (**1.27–1.30**) and 1.13 (**1.31–1.34**).

The Kenner sulfonamide-based SC linker **1.27** was supported on PS resin (84) allowing the attachment of carboxylic acids or amino acids to the sulfonamide function. After synthetic elaboration, treatment with diazomethane produces the *N*-methylacylsulfonamide, which can be cleaved with nucleophiles such as 0.5 *N* NH_3–dioxane or hydrazine–MeOH, 0.5 *N* NaOH, releasing amides, hydrazides, or carboxylic acids, respectively. A modification using iodoacetonitrile produces the more labile *N*-cyanomethyl derivative, which can be cleaved completely with stoichiometric amounts of amines to release the corresponding amides into solution.

The phenol-sulfide SC linker **1.28** (85) attached to Merrifield resin has been used to support peptides, with the exclusion of S-containing aminoacids. Oxidation of the sulfur atom with hydrogen peroxide increases the reactivity of the linker towards amines and allows the facile cleavage of the ester bond to release the final compounds into solution as amides.

The acid-labile SC linker **1.29** (86) bonded to aminomethyl PS resin, has been used to support acids and C-terminal Boc protected aminoacids. Reductive acidolysis (typically $SiCl_4$–thioanisole–anisole/TFA at rt over 3 h) reduces the sulfoxide in the activation step and releases the free acid into solution because of the enhanced reactivity of the sulfide towards acid.

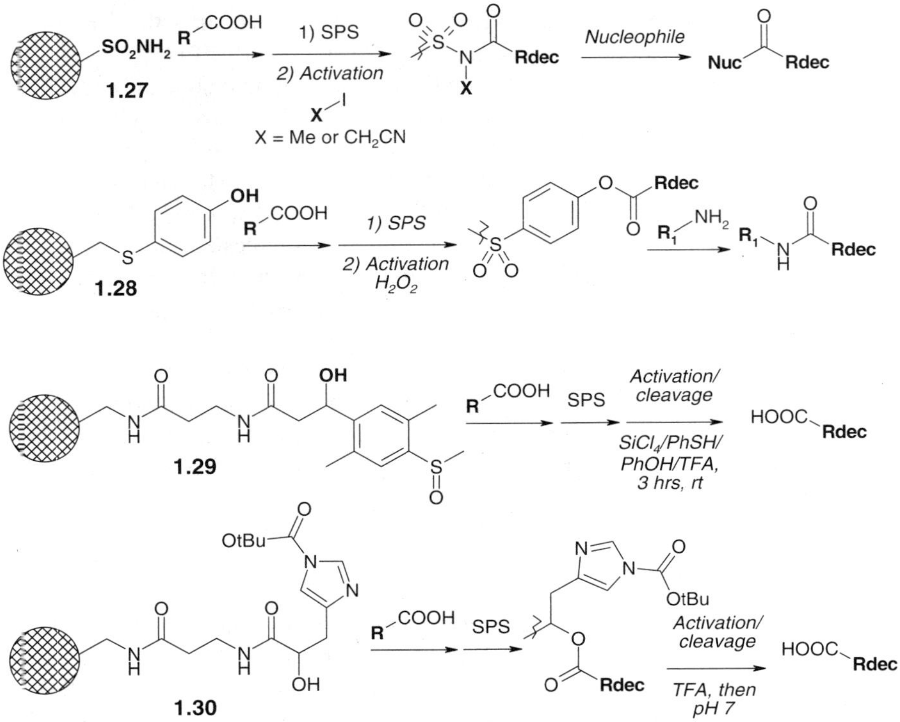

Figure 1.12 Safety-catch (SC) SP linkers **1.27–1.30**.

The imidazole hydroxyl SC linker **1.30** (87) linked to hydrophilic PS resins has been employed for loading acids and C-terminal Fmoc-protected amino acids. Activation and deprotection of functional groups is performed by TFA treatment, which produces the imidazole TFA salt. The imidazole ring intramolecularly attacks the ester bond upon neutralization at pH 7 with aqueous buffer and releases the compound as the free acid into solution.

The thioketal-containing hydroxyl linker **1.31** (88) was prepared in three steps from Merrifield resin and used to support carboxylic acids. The stable resin-bound intermediate is activated via desulfurization, with either $Hg(ClO_4)_2$ or HIO_4, and the resulting linker is photolyzed in standard conditions to give the pure, released acid.

The sulfide-based linker **1.32** (89), obtained from commercial thio-PEG–PS resin and chloropyrimidine, is activated to nucleophilic substitution via oxidation with perbenzoic acid after multistep SP transformations; treatment with amines then releases pure 2-aminopyrimidines in solution. Other nucleophiles should be suitable for the modular release of this and other heterocyclic S-supported nuclei.

The acetal linker **1.33** (90), obtained from suitably protected aminophenol (three steps from 2-nitro-5-methoxytoluene) and hydroxy PS resin, was activated by acetal hydrolysis to give acylindole derivatives which could be cleaved and diversified to give

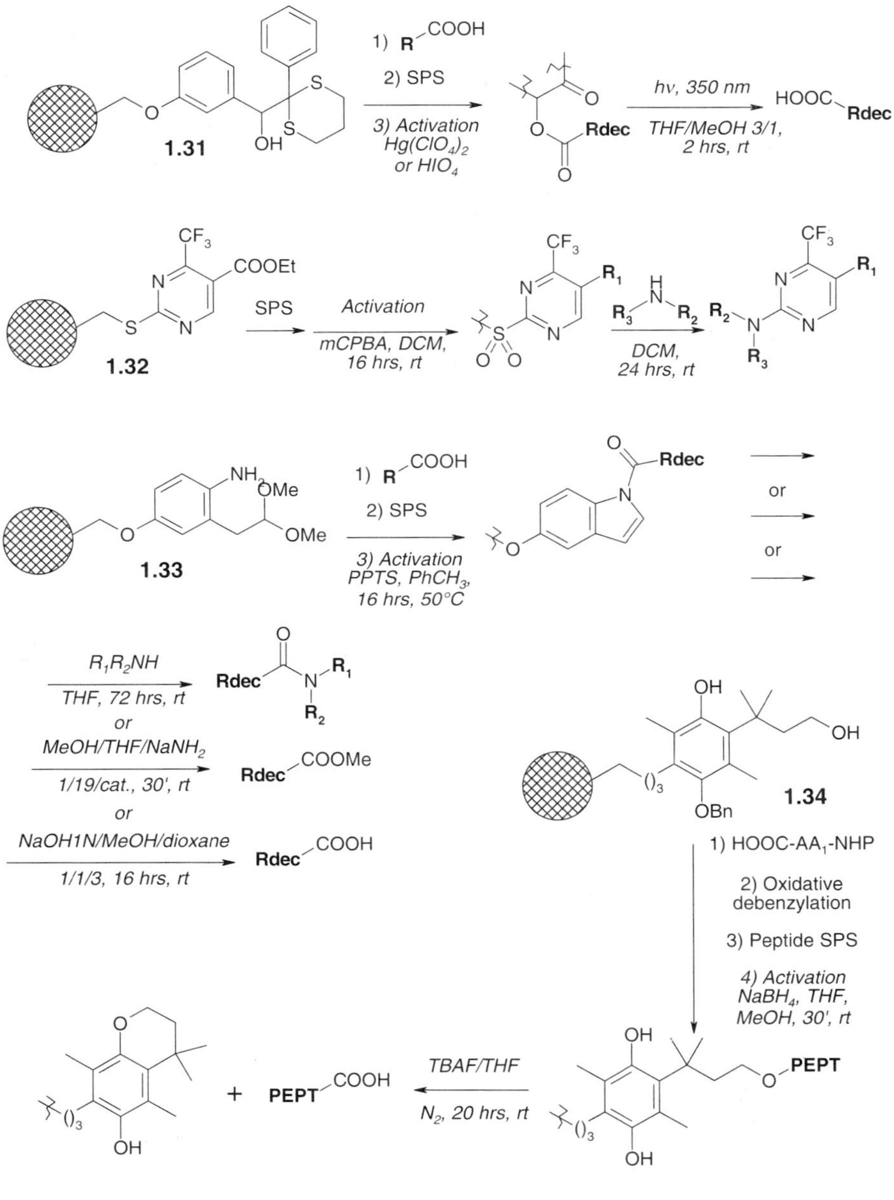

Figure 1.13 Safety-catch (SC) SP linkers **1.31–1.34**.

amides (amine, THF, 2 h, rt), esters (MeOH/THF 1/19, catalytic sodium amide, 30 min, rt) or acids (aqueous NaOH, MeOH/dioxane, 16 h, rt).

The redox-sensitive linker **1.34** (91), obtained in several steps from Merrifield resin and a lactone precursor, was charged with a N-protected aminoacid, treated with NBS to debenzylate and oxidize the linker to quinone, and submitted to SPS. The quinone linker was reductively activated to dihydroquinone with $NaBH_4$ in THF/MeOH for 30 min at rt, then cleaved by treatment with anhydrous TBAF in THF for 20 h at rt to provide the free acidic peptide via intramolecular cyclization of the linker moiety.

1.2.6 Traceless Linkers

A major drawback to all of the examples encountered so far is the necessity of having a functional group in the final compound through which it is attached to the resin during the elaboration phase. This is true both for compounds that are directly attached to the resin and for compounds attached via a linker. In the latter case, the linker often leaves a residue on the cleaved compound; for example, the Rink amine linker leaves a terminal carboxamide residue. The search for so-called traceless linkers (TLs), that is, linkers that do not leave a residual functional group deriving from the cleavage reaction, has recently become a major area of interest in SP linker chemistry. These linkers are usually substituted with a hydrogen atom during the cleavage, but some alternative quenchers have also been used (*vide infra*). Some examples are reported in Figs. 1.14 (**1.35–1.38**), 1.15 (**1.39–1.43**) and 1.16 (**1.44–1.46**); additional references can be found in a recent review on TLs (92).

The silicon-based TL **1.35** (Fig. 1.14) (93) on aminomethyl PS resin has been used in the SP preparation of 1,4-benzodiazepines by *decoration* of the phenyl ring. Cleavage is effected by protodesilylation using the somewhat harsh anhydrous HF, which releases the unsubstituted phenyl. A more labile TL **1.36** (94) obtained by substituting Ge for Si can be cleaved with TFA–Me_2S–water in the ratio 85/10/5. Many other silyl-based TLs have been reported (52–56, 92); cleavage conditions where H is replaced by I (ICl) or Br (Br_2–pyridine) have been validated.

The phosphorus-based TL **1.37** (95), which can be prepared from the commercially available triphenylphosphine resin, can be cleaved to give various chemical functionalities depending on the cleavage conditions used. For example, a methyl group can be generated under strongly basic conditions, an alkene can be formed under Wittig reaction conditions, or alternatively, indoles can be obtained via the modified Madelung synthesis.

The hydroxymethyl resin-supported regenerated Michael (REM) linker **1.38** (96), has been used to support secondary amines during elaboration followed by release as tertiary amines by classical Hoffman elimination [DIEA/(diisopropyl ethylamine)–DMF at rt over 18 h] of the resin-bound quaternary ammonium salt.

The quinodimethane linker **1.39** (97) was easily prepared from hydroxymethyl PS resin and anthranilic acid; this linker can be used to perform hetero-Diels–Alder reactions to form condensed six-member heterocycles. Cleavage with Lewis acid–nucleophile cocktails in DCM for 16–24 h at rt produces unsubstituted or alkyl-substituted heterocycles in good yields.

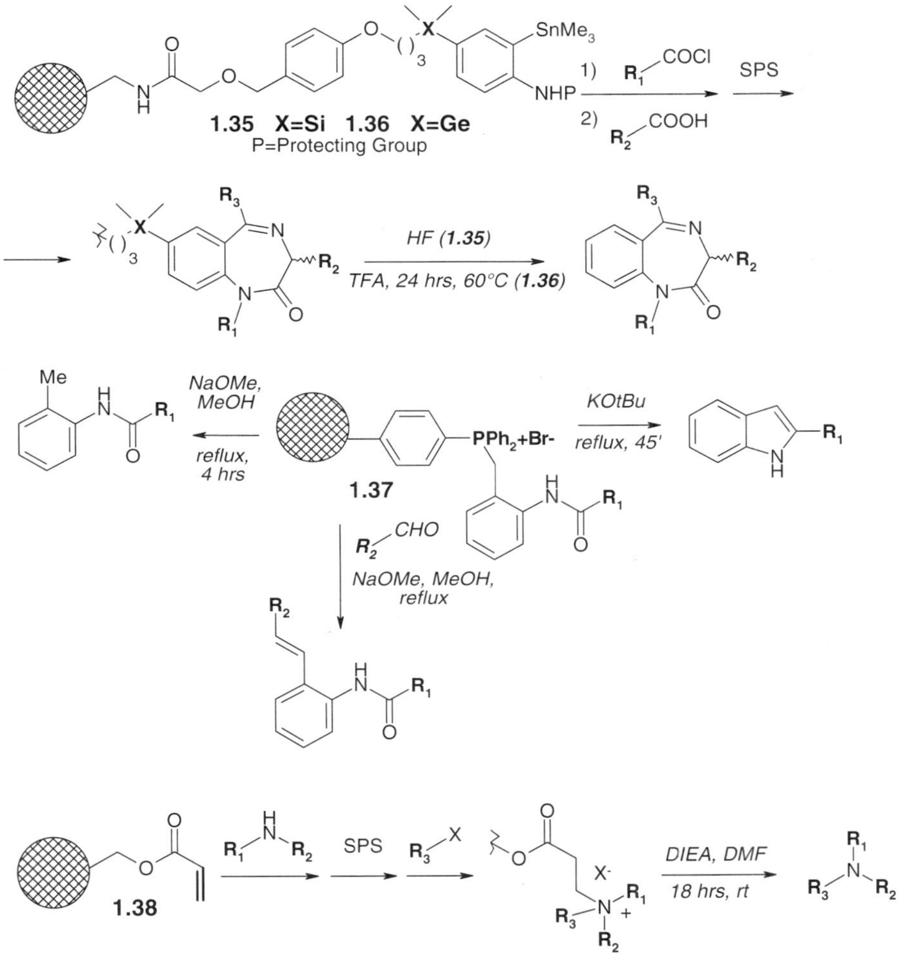

Figure 1.14 Traceless SP linkers **1.35–1.38**.

The chromium carbonyl linkers **1.40** (98) and **1.41** (99) were prepared from commercial triphenylphospine resin and respectively from pre-formed *p*-arene chromium carbenes and Fischer chromium amino carbenes. Their SP elaboration is followed by cleavage with pyridine at reflux for 2 h (**1.40**) and with iodine in DCM for 1 h at rt (**1.41**); both linkers produce the desired compounds in good yields. A similar cobalt carbonyl linker **1.42** (100) was prepared as a mixture of mono- (**1.42a**) and bis- (**1.42b**) phosphine complex, either from pre-formed alkyne complexes on triphenylphosphine resin or by direct alkyne loading on the bisphosphine cobalt complex; traceless cleavage was obtained after SP transformations by aerial oxidation (DCM, O_2, hν, 72 h, rt) and modified alkynes were released with good yields and

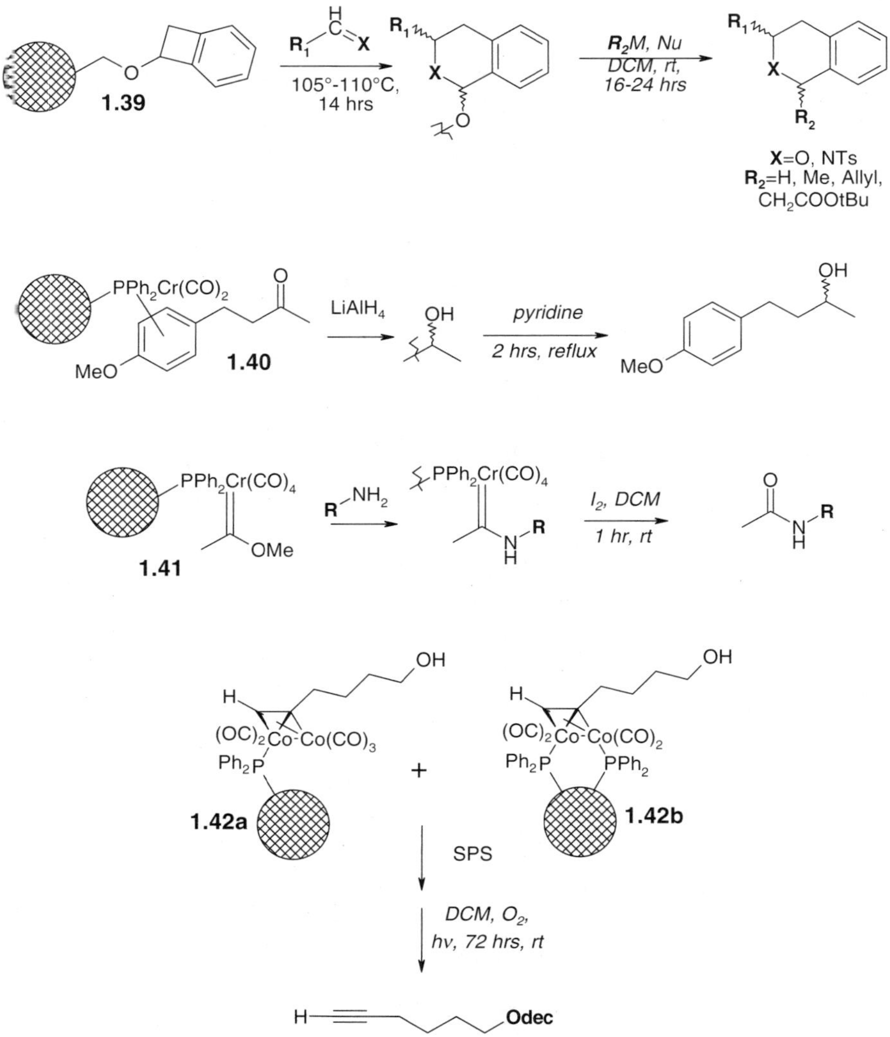

Figure 1.15 Traceless SP linkers **1.39–1.42**.

purities. A significant use of chromium and cobalt carbonyl complexes as TLs is to be expected for many solid-phase chemistry applications.

The carboxyl-based TL **1.43** (101) was easily prepared from hydroxymethyl PS resin and a trisubstituted aromatic compound; its SP functionalization on the amide carbonyl or on the chlorine atom is followed by cleavage with TMSI (trimethyl silyl iodide) for 72 h at 75 °C to obtain simultaneously ester hydrolysis and decarboxylation to 2-unsubstituted quinazolines. An expansion to other heterocyclic systems is easily foreseeable.

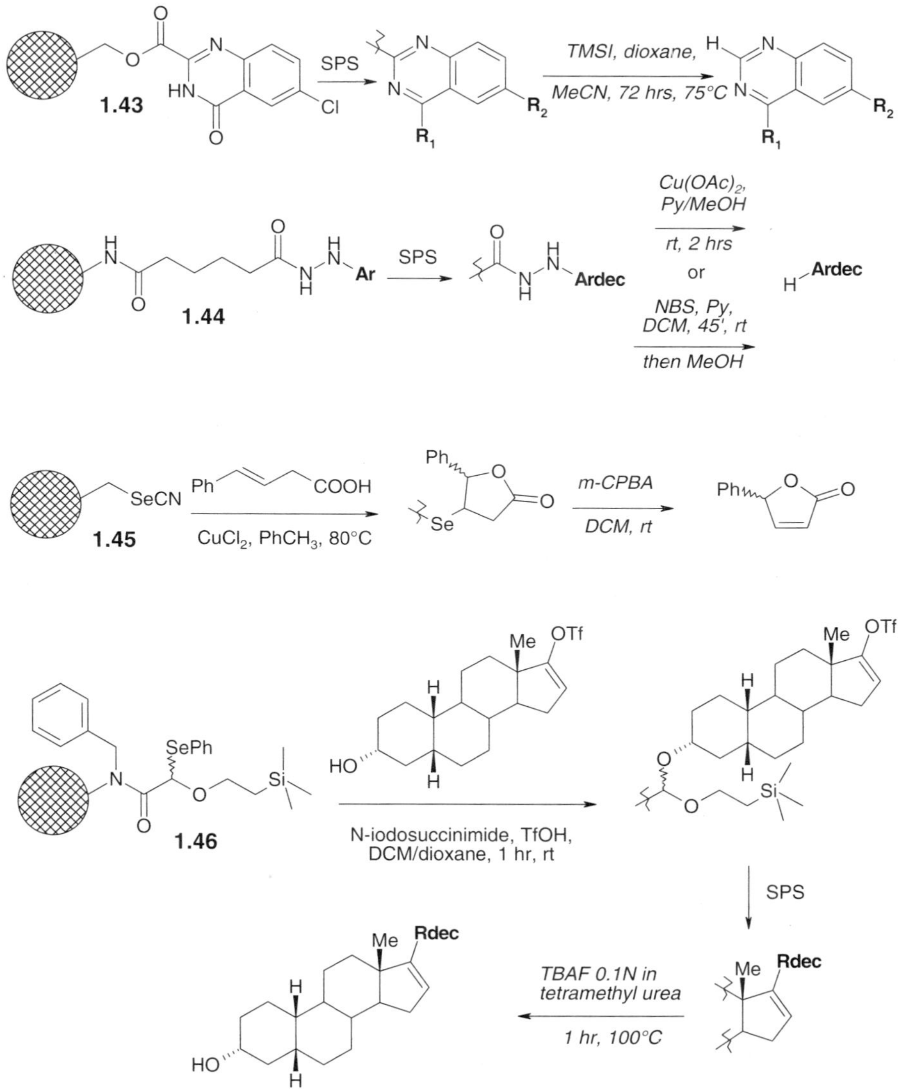

Figure 1.16 Traceless SP linkers **1.43–1.46**.

The oxidation-labile TL **1.44** (102) was readily prepared from aminomethyl PS resin and used to decorate the aryl moiety with C–C coupling reactions (e.g. Heck, Stille, Suzuki, and Sonogashira protocols). Oxidation of the hydrazide moiety to acyl diazene [copper acetate or *N*-bromosuccinimide (NBS)] and nucleophilic cleavage (methanol or *n*-propylamine) produced the desired aryl products.

The selenium linker **1.45** (103), obtained from Merrifield resin and potassium selenocyanate, was treated according to oxyselenylation conditions to give a supported selenolactone. Oxidative deselenylation (*m*-CPBA, DCM, rt) produced the unsaturated lactone in good yield and purity. An expansion of this chemistry, including several SP transformations prior to the cleavage, is mentioned in the original paper (103).

The selenium-silicon linker **1.46** (104), obtained from hydroxy PS resin and methyl glyoxylate through a simple five-step scheme, was coupled with a steroid scaffold to give an acetal then transformed on SP and finally cleaved with TBAF/tetramethyl urea, 1 h, 100 °C to give decorated alcohols *via* unstable hemiacetals with good yields and purities.

1.2.7. Cyclative Cleavage

A promising SP method of cleavage that is becoming more common is the so-called cyclative cleavage (CC). SPS produces an advanced open intermediate that undergoes cyclization with concomitant release of the final cyclized product. Only the desired product is released in the CC step, while any side product or remaining intermediate cannot cyclize and thus remains bound to the solid phase; this significantly eases the purification–work-up procedures to obtain the pure recovered compound from SPS. Several examples of CC are shown in Fig. 1.17 (**1.47–1.50**) and 1.18 (**1.51–1.54**).

In SP preparation of oxazolidinones, the carbamate SP linker **1.38** attached to a hydroxymethyl SP resin (105) was alkylated with a tosyl-epoxide and cyclized/cleaved at the end of the synthesis by aminolysis of the epoxide with spontaneous cyclization of the amino-alcohol to release the pure oxazolidinone into solution.

The allyl alcohol linker **1.48** (106), bound to an aldehyde PS resin, has been elaborated to the pentenoic acid derivative shown. This was cyclized with cleavage using ring-closing metathesis (RCM) (Ru catalyst in DCE at 80 °C for 16 h) to give pure Freidinger lactam in solution.

The commercially available, supported amino acid **1.49** was elaborated to complex linear intermediates that can easily be cyclatively cleaved to produce tetramic acids; the cleavage conditions included aqueous NaOEt (107), tetrabutyl ammonium hydroxide (108), and methanolic KOH (109). The procedures afforded the desired heterocycles in good yields and purities.

The TMS-exomethylene linker **1.50** (110), obtained from carboxyethyl PS resin and a suitable TMS alcohol, was reacted with an N-acylimine as in the imino-Sakurai protocol and then the SP transformations were performed; the intermediates were cyclatively cleaved with $Pd(acac)_2$ and dppe in refluxing THF to give highly functionalized pyrrolidines.

Commercially available (4-(4-formyl-3-methoxyphenoxy)butyryl PS resin was loaded with primary amines to give the resin-bound secondary amines then transformed to the highly functionalized resin-bound secondary amines then transformed to the highly functionalized resin-bound anilines **1.51** (111) that were cyclatively cleaved with AcOH overnight at 80 °C to give tri-decorated imidazoles with good yields and purities. A similar approach provided the highly functionalized quinoxali-

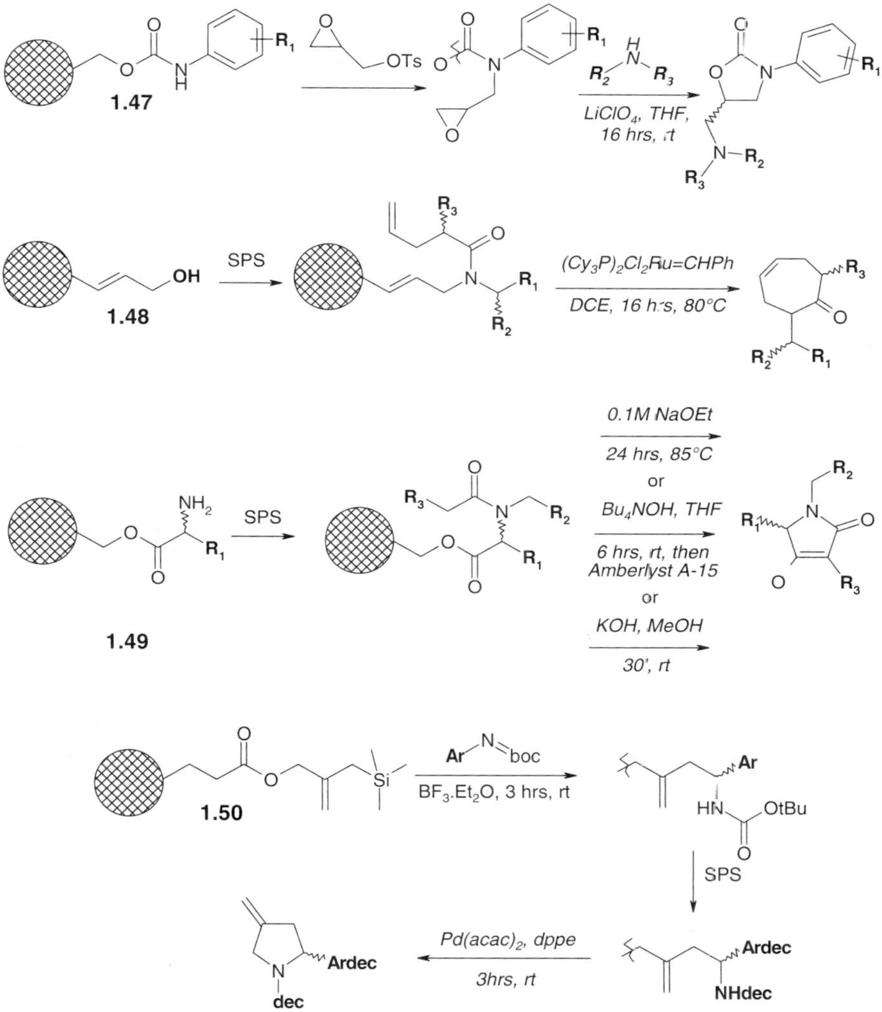

Figure 1.17 Examples of CC on SP: **1.47–1.50**.

nones **1.52** (112), which were cyclized/cleaved with TFA/DCM/Et_3SiH 10/30/1, 1 h, rt followed by heating at 50 °C in HCl, 4N, dioxane/MEOH 2/1, 3 h.

Even large linear precursors can be effectively cleaved and cyclized, releasing only the desired macrocycles; two examples (**1.41** and **1.42**) reported by Nicolaou et al. (113, 114) are shown to highlight the complexity of SP products that can be obtained and the usefulness of CC as a release protocol even for polyfunctional sensitive molecules.

A large number of additional examples of CC approaches to a wide panel of isolated and condensed ring systems can be accessed through a recent review (115).

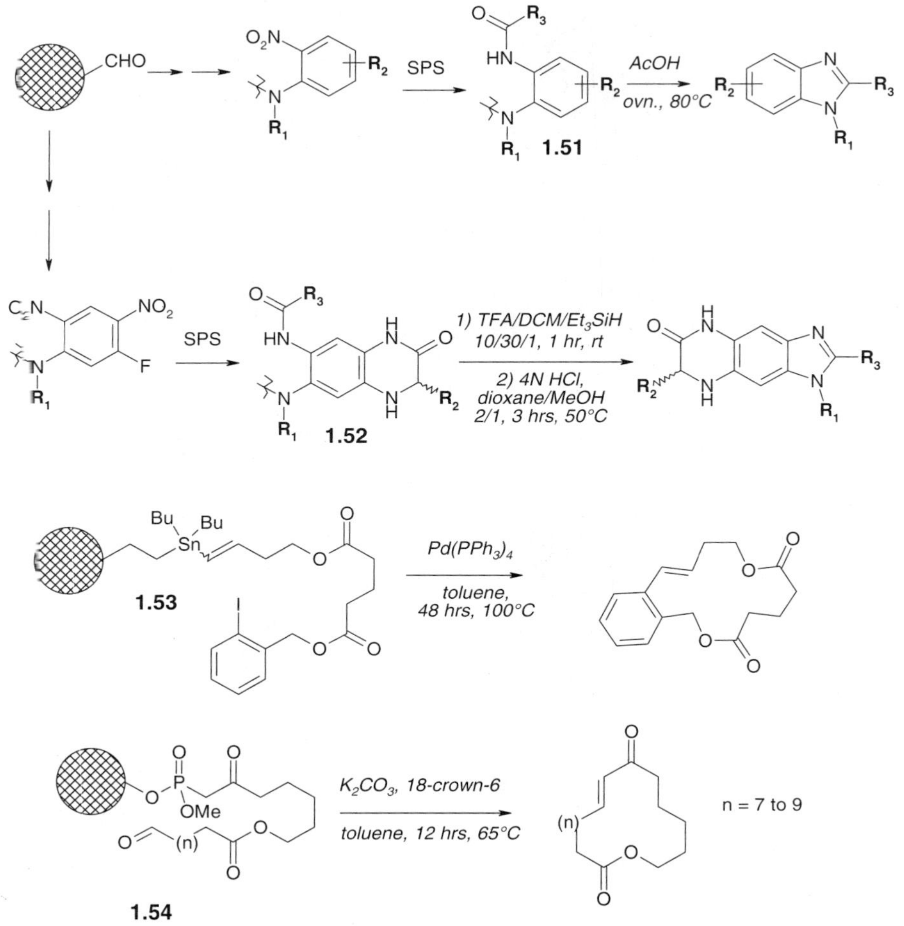

Figure 1.18 Examples of CC on SP: **1.51–1.54**.

1.3 REACTION MONITORING IN SOLID-PHASE SYNTHESIS

1.3.1 General Considerations

A key component of a successful organic synthesis is the *constant monitoring* of the reaction, which allows optimization of the yield of the target molecule while minimizing the products of side reactions. This monitoring is readily performed by either chromatographic analysis [thin-layer (TLC), high-performance liquid chromatography (HPLC), or gas (GC)] or spectroscopy. The use of excess reagents in SPS (typically 3 to 5 equivalents) precludes the monitoring of reactions by following the disappearance of reagents in solution and therefore alternative methods, specifically designed for studying reactions and determination of the structure in the solid phase, have been developed to guide the efforts of the SP chemist (116, 117).

Destructive methods, where the analytical sample is consumed by the analysis, and *nondestructive* methods will be presented and their qualitative or quantitative nature will be discussed. They will be divided into *off-bead* methods, where the resin-bound reaction product(s) are cleaved from the support with subsequent analysis of the cleavage solution, and *on-bead* methods, where single or multiple beads are analyzed directly.

1.3.2 Off-Bead Methods

The cleavage of resin-bound materials and their full analytical characterization in solution are used as the most accurate way to monitor the outcome of a reaction carried out in the SP. The methods used are those of classical organic chemistry and will not be commented on further. The reaction products can be weighed and an accurate structure determination can be obtained. There are, however, some limitations to the usefulness of off-bead methods for reaction monitoring in SPS.

First, the resin beads cleaved after each step of a multistep SPS are lost and the gravimetric yield determination requires a significant amount of compound. This may lead to a notable waste of precious materials and to a significant reduction of the target compound prepared.

The cleavage reaction may take hours, which prevents rapid monitoring of the reaction or necessitates complicated sets of parallel experiments that are quenched at different times. The reagents used for the cleavage may pollute the cleavage solution, thus requiring some purification steps prior to the analytical determination. Some reactive resin-bound intermediates may be sensitive to the cleavage conditions, thus leading to a misinterpretation of the reaction outcome.

The use of fast, reliable, sensitive on-bead methods circumvents the drawbacks to off-bead analysis outlined above. The modification of common analytical techniques has provided the SP chemist with valuable and often preferred alternatives to off-bead methods for SPS reaction monitoring.

1.3.3 On-Bead Methods: Colorimetric/Fluorescence Detection

Colored reagents to follow the appearance or the disappearance of a functional group have been widely used to monitor reactions in classical organic chemistry, particularly in TLC analysis. This technique has been successfully adapted to SPS; for example, ninhydrin (118), bromophenol blue (119), nitrophenyl isothiocyanate-*O*-trityl (120), picric acid (121), and malachite green isothiocyanate (122) have all been used to show the presence or the absence of free resin-bound amines. The presence of free resin-bound thiol groups can also be detected (123).

In the commonly used Kaiser test, a few milligrams of resin beads is withdrawn from the reaction, thoroughly washed with a range of solvents, and treated with stock ethanolic solutions of ninhydrin and phenol followed by a solution of KCN in pyridine at 100 °C for 10 min. If the beads turn deep blue in color, there are free primary amine groups on the support, while if the beads maintain their original color, no free primary amines are present. The method is strictly qualitative because beads containing between 10 and 100% of free amine groups give the same result. The quantitation of derivatized sites may be performed (124), but the procedure is complex and the risk of obtaining irreproducible results is high. The sensitivity of the test is at best around 4–5%, that is, less than 4–5% of sites bearing free amines can be overlooked using colorimetric detection. Nevertheless, the extreme rapidity of colorimetric methods and their simplicity make them useful for monitoring SPS, giving the best compromise for a qualitative evaluation of the reaction course.

Two recently reported colorimetric methods are particularly promising. The first is specific for resin-bound amines, including prolines and sterically hindered primary amines (125); the test relies on an accessible derivative of commercial Disperse Red 1 (three synthetic steps), is extremely sensitive (up to 2% of free amines on bead are spotted by a deep red color) and fast (10 min heating at 70 °C). The second is specific for hydroxyl groups *via* their tosylation, but also for any group which an undergo nucleophilic substitution (126); the test relies on a nucleophilic displacement with *p*-nitrobenzylpyridine and formation of an internal salt after treatment with piperidine, is extremely sensitive (up to 2% of free hydroxyls on bead are spotted by a pink/violet color) and fast (1 min heating with a heat gun).

The use of such methods for the detection and monitoring of other resin-bound functional groups will become common as more and more staining protocols will be transferred from solution synthesis to SPS. Adaptation of these reagents to the presence of the solid support, for example, by choosing a solvent where the beads swell properly, may sometimes be necessary.

1.3.4 On-Bead Methods: Nuclear Magnetic Resonance Spectroscopy

The use of NMR techniques to monitor reactions in solution has never been very popular because of the time required to prepare an aliquot of the solution and to record a meaningful spectrum. Faster, even if less information-rich, methods have been traditionally used.

The use of NMR in SPS is further complicated by two factors. First, the spectra of solid samples show broad NMR lines due to restricted molecular motion. This may be partially alleviated by a good swelling of the beads in a suitable deuterated solvent. Second, the heterogeneity of the resin slurry produces microenvironments with different magnetic susceptibilities that cannot be shimmed in the same way as homogeneous samples where the magnetic susceptibility is uniform. Again, this leads to broadening of the NMR signals. Nevertheless, solvent in which the resin swells properly allows the recording of an SP NMR spectrum using the so-called *gel-phase NMR technique* (127).

This method has found applications in SPS, particularly for those nuclei that are absent in the matrix and whose appearance can be easily monitored even with broad NMR signals; examples have been reported for ^{19}F-NMR (128–130), ^{31}P-NMR (131), and isotopically enriched ^{15}N-NMR (132). While examples of ^{2}H and ^{13}C gel-phase

NMR spectra exist, the technique is limited by line broadening, the long acquisition times, and the low abundance of ^{13}C and ^{2}H atoms in the matrix.

The use of a ^{13}C-enriched building block anchored to a resin makes the gel-phase spectrum selective for the appearance of the new ^{13}C signal, and the enrichment allows much shorter acquisition times (133, 134); a "real-time" kinetic was reported for the alkylation of amines with ^{13}C-enriched bromoacetic acid (135). An example from our laboratories (136) shows the formation of a cyanohydrin on SP is monitored by ^{13}C-enriched gel-phase NMR using ^{13}C-benzaldehyde. The appearance of the cyanohydrin signal (63.2 ppm) and its increase at different reaction times is easily monitored by comparison with the constant signals of the solvent (deuterated benzene, 133–126 ppm, Fig. 1.19, spectra **A–D**). A major drawback of this technique is the cost and the limited availability of ^{13}C-enriched building blocks, which currently severely limits its application.

While specific applications of gel-phase NMR have been useful for SPS reaction monitoring, the great potential of SPS NMR is in the determination of structure and the measurement of purity and yields, especially through the use of magic angle spinning–high-resolution (MAS–HR) NMR techniques. This important topic will be addressed in Section 1.4.6.

1.3.5 On-Bead Methods: Mass Spectrometry

The use of mass spectrometry (MS) techniques to monitor SP reactions has recently become possible through the use of matrix-assisted laser desorption/ionization–time of flight (MALDI–TOF) spectrometry (137) after in situ cleavage of a small number of resin beads (138–140). Although the compound is cleaved from the resin, the cleavage happens directly onto the center of the MALDI target and the method can be considered on-bead.

In a specific example (141), a small aliquot (10–100 beads) from an attempted Pd(0)-mediated deprotection of an allyl ester on Rink resin was placed on the MALDI–TOF sample plate and exposed to TFA vapor at rt for 20 min. The TFA was removed, the matrix and an internal standard were added, and the sample was analyzed. The incomplete removal of the allyl group under classical conditions was observed, and only repeated deprotection protocols with Pd(0) afforded the pure deprotected carboxylic acid.

The suitability of this technique for acid-labile (142) and photocleavable linkers (143) has been demonstrated. Other cleavage conditions that do not produce residues (e.g., cleavage with gaseous ammonia) could also be used in theory. TOF–secondary ion mass spectrometry (TOF–SIMS) (144) has also been validated to monitor SP peptide synthesis (145) and could in future increase the versatility of MS monitoring of SP reactions.

The requirement of an expensive MALDI MS spectrometer and the limited usefulness for compounds with molecular weight (MW) < 600 (presence of intense matrix signals) are serious limitations to the application of this method for the monitoring of SPS reactions. However, the extreme sensitivity (a single resin bead is usually enough)

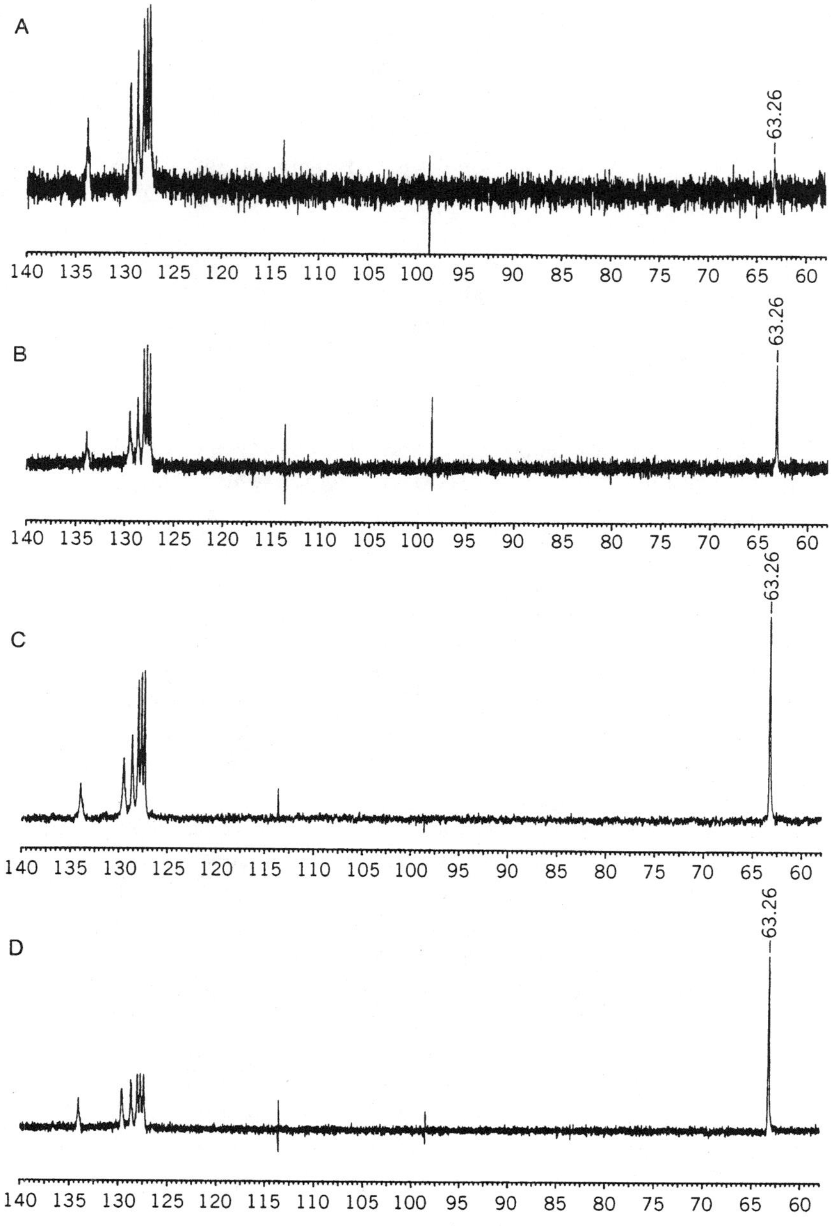

Figure 1.19 Gel-phase ^{13}C-NMR reaction monitoring of a cyanohydrin formation (63.2 ppm). Spectrum A: 10 min. Spectrum B: 1 h. Spectrum C: 2.30 h. Spectrum D: 4 h.

and reliability of the method should help to increase its popularity in the near future, at least for the analysis of ionizable compounds.

A new reliable, sensitive, and fast MS monitoring approach, based on previous MS quantitation/encoding studies (146, 147), has been recently reported (148). The use of a carbamate-based photolabile linker **1.55** containing two cleavage sites developed in-house allowed the release of the target molecule via acid cleavage (site 1, Fig. 1.20); the so-called *analytical construct* also allowed the photolytic release for analytical determinations of a larger fragment containing the target molecule, the acid linker, and a free amine (site 2, Fig. 1.20). The amine acted as a mass sensitizer, increased the MS signal of the reaction product, and via an isotopic labeling protocol (146) allowed to discard the MS signals of unbound materials; a single bead was largely enough to determine the quality of the reaction and to spot elusive side products (147, 148). Additional analytical constructs were reported by the same researchers (149); the quantitation of mass-based analytical constructs is currently under study by several groups and holds the promise to accelerate and make more reliable the on-bead reaction monitoring in SPS.

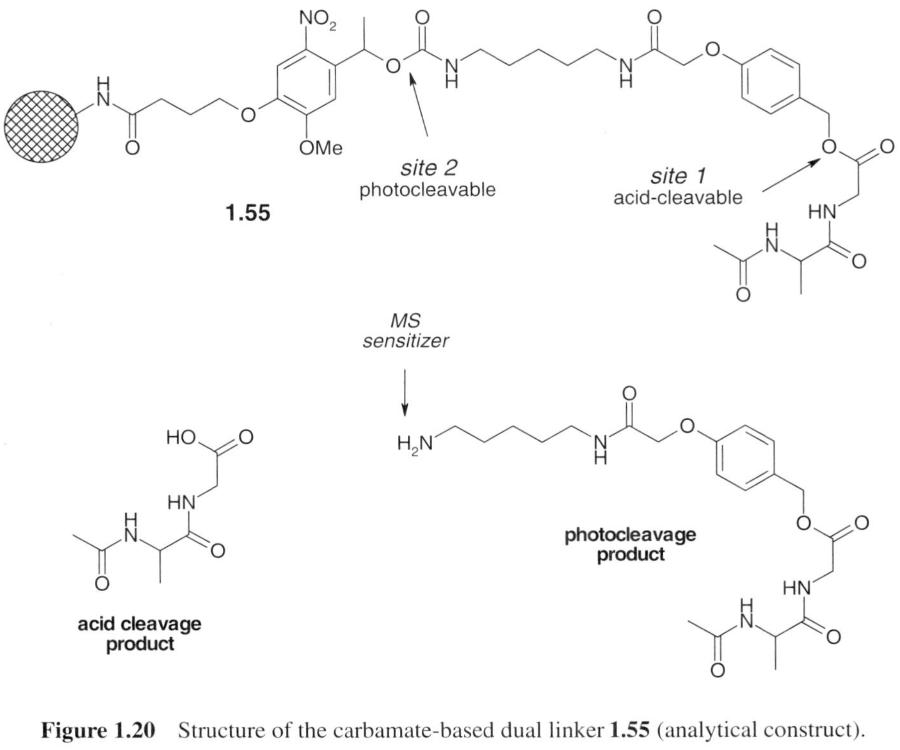

Figure 1.20 Structure of the carbamate-based dual linker **1.55** (analytical construct).

1.3.6 On-Bead Methods: Infrared Spectroscopy

The use of infrared (IR) spectroscopy as a reaction monitoring technique for SPS has become more frequent over the last few years with the introduction of technologies specifically aimed toward SP reactions. Even so, a number of reports describing the use of classical IR by thorough mixing of a few milligrams of grounded resin beads in KBr pellets have appeared (150, 151).

Single-bead Fourier transform IR (FTIR) spectroscopy is well established as a sensitive and reliable method to monitor SPS (152). Significant improvements have been obtained by flattening the bead in a manual pellet maker, thus reducing and making more homogeneous the pathlength of the incident beam compared to a spherical resin bead (153). For example, the transformation of a resin-bound aldehyde into a dansylhydrazone on hydrophobic PS and on Tentagel resin was monitored using this technique (154). The increase of the $N(Me)_2$ and the N–H hydrazone stretches at 2790 and 3218 cm^{-1}, respectively, and the decreases of the C–H stretch at 2728 cm^{-1} and of the carbonyl band at 1700 or 1723 cm^{-1} were easily monitored over time. The beads were pretreated simply by washing with DMF, THF, and DCM, followed by vacuum drying for 15 min. Other examples applied to different chemical transformations have also been reported (40, 155, 156); FTIR spectroscopy has also been used to evaluate the partitioning of reagents into PS beads and the influence of diffusion on SP reaction kinetics (157) or to evaluate the mechanism and kinetics of SP reactions (158).

Some related but more specialized SPS IR techniques have also been described. Attenuated total reflection (ATR) microspectroscopy is extremely sensitive (femtomolar quantities are routinely measured), detecting only surface-bound materials with an average penetration of about 2 μm (159), and has been used to monitor the SP synthesis of a polysubstituted piperazine (160) and of β-amino substituted piperidinols (161). Photoacoustic FTIR (PA-FTIR) provides spectra where only the absorption component of the IR radiation is measured, thus canceling interference due to light scattering and nonhomogeneity of the resin (162). Diffuse reflectance infrared Fourier transform spectroscopy (DRIFTS) combines an extremely short analysis time of around 30 sec with negligible sample preparation (163, 164). Even more promising appears near-infrared (NIR) spectroscopy, which is also nondestructive and does not require any sample preparation; both a classical example where a single reaction was monitored (165) and the multiple monitoring and detection of NIR images (166) for several SP samples were reported, showing a great potential and flexibility for such a technique in SPS.

Recently several reports highlighting the usefulness of IR and Raman spectroscopy for SP synthesis, especially during the SP chemical assessment, were published; a multistep SP scheme was monitored successfully (167) and a new encoding technique (see Section 7.4) was reported (168). Both IR (169) and Raman (170) spectroscopy applications for SP synthesis were recently reviewed.

In summary, the use of IR either as a simple (KBr pellets) or sophisticated (single-bead techniques) reaction monitoring system for SPS has become very important. The technology behind the methods outlined above is constantly evolving, and these techniques will eventually form a complementary network for monitoring reactions in SP chemistry. An interesting comparison (171) between single-bead IR and MAS NMR techniques as tools to assess a novel SP protocol has recently highlighted the complementarity of the two, which should as much as possible be used in conjunction to maximize the quality and the reliability of the analytical monitoring studies.

The further development of these methods will allow chemists to select an appropriate method on the basis of the specific SP synthetic problem in hand and the instrumentation available in their laboratory. Eventually, analytical monitoring will become a routine element of SPS, as it is in classical solution-phase organic chemistry, and one of the major bottlenecks for the further development of SPS will disappear.

1.4 PURITY AND YIELD DETERMINATION IN SOLID-PHASE SYNTHESIS

1.4.1 General Considerations

The determination of the yield of a reaction carried out in the SP and the structure and purity of the product is an essential component of the process of SPS. The same analytical methods that we examined in the previous section for reaction monitoring can be used, but their usefulness for qualitative analysis may vary, as we will describe in this section.

The problems associated with high-throughput analysis of SP reaction products, automation of purification techniques, and automation of methods for structure determination in SPS are typically encountered in combinatorial chemistry and will be dealt with more thoroughly in Chapters 6–8.

1.4.2 Off-Bead Methods

The cleavage and characterization of intermediates or final compounds, as for off-bead reaction monitoring (Section 1.3.2), are the most reliable methods to quantify the outcome of an SPS. The final target molecule on SP has to be cleaved into solution so that it can be characterized by classical off-bead methods. The synthetic intermediates obtained after each SPS step can also be characterized off-bead when the cleavage conditions do not affect their structure and when the amount of resin beads lost during the characterization process is negligible.

Examples of off-bead yield estimation that do not require cleavage include reactions where new species are stoichiometrically formed in solution, and their quantification provides an indirect, but accurate, yield estimation of the SPS step. The classical Fmoc deprotection of amines in peptide synthesis is a widely used example. In a typical experimental procedure, the beads are treated with a 20% DMF solution of piperidine at rt for 20 min and the solution is recovered together with the resin washings. The solution is brought to a constant volume (typically 10 mL) by addition of DMF, and the quantitation is carried out by reading the UV absorbance of the piperidine–

dibenzylfulvene adduct in solution at 301 nm against a blank solution of piperidine in DMF. The loading can be determined from the equation

$$L = AV/7.8W$$

where L is the Fmoc loading (mmol/g), A is the absorbance value at 301 nm, V is the solution volume in liters, 7.8 is a constant, and W is the weight of the resin in milligrams. The Fmoc loading compared with the known initial resin loading gives the yield of the deprotection. (Note: The UV readings are generally performed in triplicate and an average reading value is used in the equation.)

The outcome of an SPS can be monitored indirectly by reacting the functional group formed with an excess of a fluorescent dye and then monitoring the fluorescence decrease in the supernatant. A carefully planned and executed example (154) used dansylhydrazine to quantify the amount of carbonyl compounds on SP by forming the dansylhydrazone (DMF, at rt for 30 min). A panel of different commercially available (PS–CHO and TentaGel–CHO) and derivatized resins containing ketones or aldehydes were tested for their loading, and the results were checked with single-bead FTIR (see Section 1.3.6). The method requires the destruction of a few milligrams of resin but does not require cleavage and provides reliable results that are not affected by the presence of other functional groups on the support or by the nature of the solid support itself. The disappearance of dansylhydrazine from the supernatant is measured by subtraction of the fluorescence value at the end of the reaction (excess of reagent in solution) from the value at $t = 0$. This indirect measure of the loading sites of the support allows also a fast and reliable reaction monitoring. The same group reported recently sensible and reliable methods to quantify resin-bound hydroxylic groups *via* reaction with 9-anthroylnitrile and carboxylic groups *via* reaction with 1-pirenyldiazomethane (156); both methods could also be used qualitatively in an easy colorimetric test. The use of fluorescent dyes with other chemical functionalities should extend the usefulness of this method in the near future to the quantification of SPS reaction products for different resin-bound chemical groups.

There is a strong need for accurate quantitative and information-rich analytical methods for the on-bead estimation of purity and yield. In the transfer of the chemistry from solution to the SP, it is crucial to determine the yields of all the SPS steps and to monitor the yield and variations in the purity of products obtained by modifying selected reaction parameters. This can be done better, and with reduced amounts of materials, using nondestructive on-bead methods, which provide results that are not biased by the method of cleavage, thus showing if specific cleavage conditions are affecting a target molecule by allowing a comparison of the data obtained through on-bead and off-bead analyses.

1.4.3 On-Bead Methods: Colorimetric/Fluorescence Detection

These methods are typically qualitative rather than quantitative (see Section 1.3.3), and therefore their use for the determination of purity and yield in SPS has not been reported. The indirect use of fluorescent dyes for the estimation of yield using off-bead methods has already been discussed in the previous section.

1.4.4 On-Bead Methods: Infrared Spectroscopy

Infrared spectroscopy may be considered to be one of the analytical techniques best suited to the rapid monitoring of the progress of chemical reactions in the SP, as has been discussed in Section 1.3.6. There are some intrinsic limitations to the use of this technique in the determination of purity and yields. The difficulty in quantifying reaction yields by following the simple appearance or disappearance of IR bands is worsened by the broader bands often obtained on SP, especially when using KBr pellets. Small quantities of side products cannot be detected easily due to the reduced intensity (or absence) of IR-specific bands. While some attempts have provided quite accurate estimations of the yield in a few cases (40, 154–156), other analytical techniques appear to be more suited for the quantitative determination of yield and purity of SP reactions.

1.4.5 On-Bead Methods: Mass Spectrometry

The use of MALDI–TOF (see Section 1.3.5) in SPS purity and yield estimation takes advantage of the extreme sensitivity and reliability of the technique, which has other important advantages.

The extremely mild ionization process (137) produces no fragmentation ions, thus allowing the identification of each observed peak as corresponding to a specific reaction product. This is important for monitoring the complete disappearance of the starting material and the appearance of the expected reaction product but, most importantly, for detecting small amounts of side products through their molecular ions. Even impure samples, or mixtures, can be analyzed without affecting the reliability, rapidity, and sensitivity of the technique, which is typically in the femtomolar range. The usual drawbacks, such as the need for a cleavage step prior to analysis of the sample, do not significantly affect the usefulness of MALDI–TOF when used for a rapid estimation of the purity of the reaction products and for the monitoring conditions during the assessment of the chemistry.

The MALDI–TOF monitored SP preparation of lysobactin, a natural cyclopeptide antibiotic, on PS resin bearing the Rink amine linker **1.6** has been reported (141). The SPS scheme is shown in Fig. 1.21. The classical peptide coupling steps, the allyl deprotection (see Section 1.3.5), and the macrocyclization step were all monitored by MALDI–TOF, and the purity of the products was also determined using this technique. The presence of small impurities in compounds **1.56–1.61** was easily detected, and the reaction conditions for the key deprotection of **1.59** and cyclization of **1.60** were rapidly optimized. A total yield of 15% was obtained after HPLC purification of released **1.62** (lysobactin).

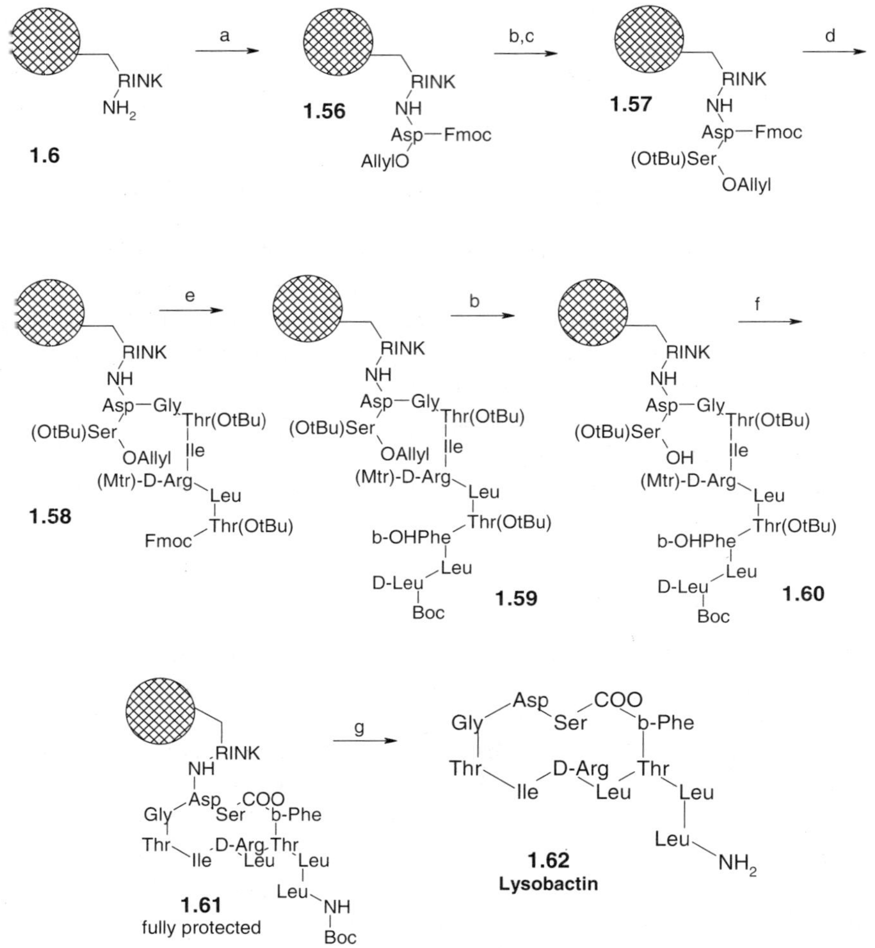

a: DIC, HOBt, Fmoc-Asp(OH)-OAllyl; b: Pd(0); c: PyBOP, Ser(OtBu)-OAllyl; d: repeated DIC/HOBt amino acid couplings and Fmoc deprotections; e: repeated EEDQ amino acid couplings and Fmoc deprotections; f: DIC, HOBt, DMAP; g: TFA/TIS/H_2O 95/2.5/2.5.

Figure 1.21 SPS of **1.62** (lysobaotin).

1.4.6 On-Bead Methods: Nuclear Magnetic Resonance Spectroscopy

The use of gel-phase NMR for monitoring SPS reactions has been described previously. The application of this method to the determination of the purity and yield of a product is not recommended for the reasons already discussed in Section 1.3.4. Another SPS NMR technique, MAS–HR NMR, is more suited to this purpose.

MAS NMR comes from an old observation (172) that spinning a heterogeneous NMR sample at an angle of 54.736° reduces the line broadening of solid/swollen polymer samples. Recent realization and use (173) of a HR NMR probe of reduced volume (NanoProbe) that holds all of the sample in the active region of the receiver coil allows the acquisition of SP MAS–HR ^{1}H-NMR spectra with line widths as small as 1 Hz using a 500-MHz spectrometer (174). A few milligrams of resin beads is typically transferred into a Nano.NMR Probe cell, and 40 μL of deuterated solvent is added. Further technical details on MAS–HR NMR can be found in a recent review (116).

A fundamental study (175) has shown that the quality of a MAS–HR spectrum is influenced primarily by the nature of the polymer matrix. Hybrid hydrophilic PS resins generally give well-resolved spectra, and the use of techniques such as presaturation of the PEG signals allows the recording of spectra of comparable quality to their solution state counterparts. The high mobility of the long and flexible tethered PEG chains is responsible for the high quality and the narrow signal widths. The use of solvents where both the matrix and the resin-bound compound have a good swelling/solubility further improves the quality of the spectra. Hydrophobic PS resins with short tethers such as the Wang linker generally produce poor-quality spectra, which may be improved by the use of appropriate solvents to swell the resin. Nontethered PS resins give poor-quality spectra under any experimental conditions, and the technique cannot be used to quantify the outcome of SPS on these resins.

Examples of MAS–HR applications to SPS reports can be accessed through recent reviews (116, 176, 177); among the most recent, significant papers, Riedl et al. (178) have brilliantly verified the feasibility and determined the scope of the asymmetric dihydroxylation of resin-bound alkenes, both in terms of reaction yields and of enantiomeric excess.

An emerging trend in the field of SPS is to use high-cost hybrid PS resins for the assessment of the chemistry and to reveal potential problems associated with a given synthetic strategy in the SP. Once this information is acquired, further preparation of the same target molecule on SP and possible exploitation of SPS to make combinatorial libraries can be carried out on cheaper hydrophobic PS resins, for which recent technological improvements (179) are significantly increasing the quality of MAS–NMR spectra. An example of the spectrum on TG–PS recorded in our laboratories (180) is shown in Fig. 1.22, pointing out both the high resolution and the ease of attribution for signals of complex molecules.

Despite the high cost of the equipment required and the time taken for sample preparation and spectra acquisition, MAS–HR NMR provides invaluable structural information about the species present in a reaction. Only a few milligrams of resin beads are required and they can be recovered as the technique is nondestructive. The complementarity of the technique with other analytical methods is clear: MALDI–TOF cannot discriminate among compounds with the same MW and depends on the ionization properties of the resin-bound compound, while FTIR depends on the presence of selected functional groups in the molecule. MAS–HR NMR can be used independently from the nature of the performed reaction and the functional groups formed or lost during the SPS step. Additionally, two-dimensional MAS techniques such as 2D-COSY (correlated spectroscopy) and TOCSY (total correlated spectroscopy) (171) or 2D-SECSY (spin echo correlation spectroscopy) (181) can provide more detailed information that may be useful in specific cases.

a	7,74	d
b	7.56	d
c	7.47	s
d	7.36	m
e	7.27	m
f	6.93	bs
g	5.49	m
h,i	4.33	m
l	4.17	m
m	4.0	m
n	3.86	s
o	3.06	m
p	2.32	m
q	2.13	m
r	2.07	m
s,t	1.54	m
u	1.46	d
v	1.25	m

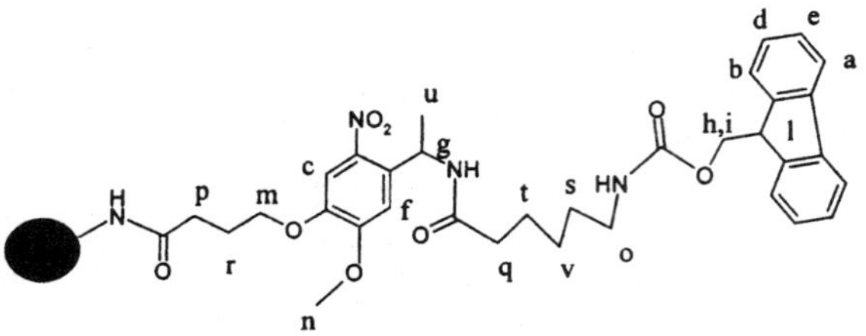

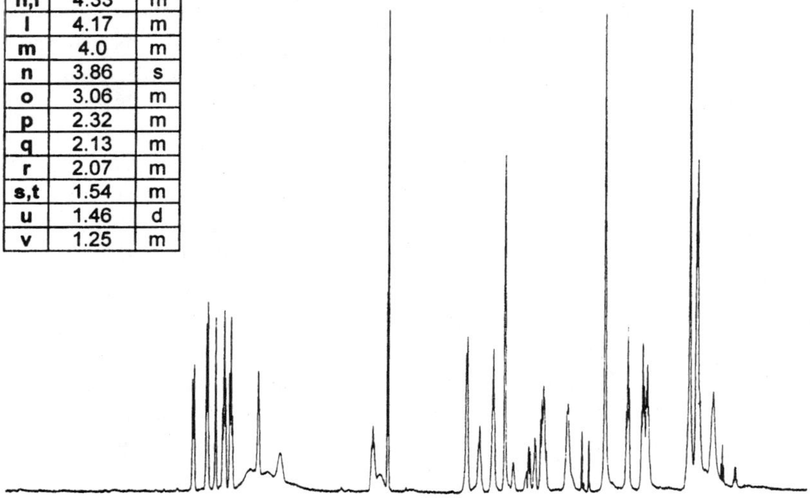

Figure 1.22 MAS–HR ^{1}H-NMR spectrum of *N*-Fmoc-6-amino hexanoic acid linked via amino photolinker to Argogel AG–PS resin.

It seems likely that the use of MAS NMR–related analytical techniques for quantitative SPS analysis will become more widespread in the future. Recent experiments have demonstrated the technical feasibility of improving the sensitivity (182) using also macrobeads (183); the integration of this with other techniques such as MALDI–TOF and the complete off-bead characterization intermediates and final products will provide fast, reliable, and complete information about the structure, purity, and yield of any product obtained through SPS. This will have a major impact on the assessment phase of the chemistry for SPS, which will, in turn, have a profound influence on combinatorial chemistry. These and related issues will be analyzed in depth in Chapters 6–8.

REFERENCES

1. Merrifield, R. B., *J. Am. Chem. Soc.* **85**, 2149–2154 (1963).
2. Fruechtel, J. S. and Jung, G., *Angew. Chem., Int. Ed. Engl.* **35**, 17–42 (1996).

3. Hermkens, P. H. H., Ottenheijm, H. C. J. and Rees, D. C., *Tetrahedron* **52**, 4527–4554 (1996).
4. Hermkens, P. H. H., Ottenheijm, H. C. J. and Rees, D. C., *Tetrahedron* **53**, 5643–5678 (1997).
5. Booth, S., Hermkens, P. H. H., Ottenheijm, H. C. J. and Rees, D. C., *Tetrahedron* **54**, 15385–15443 (1998).
6. Hall, S. E., *Annu. Rep. Combi. Chem. Mol. Div.* **2**, 15–26 (1999).
7. Andres, C. J., Denhart, D. J., Desphande, M. S. and Gillman, *Comb. Chem. High Throughput Screen.* **2**, 191–210 (1999).
8. Kingsbury, C. L., Mehrman, S. J. and Takacs, J. M., *Curr. Org. Chem.* **3**, 497–555 (1999).
9. Hodge, P., *Chem. Soc. Rev.* **26**, 417–424 (1997).
10. Bayer, E., *Angew. Chem., Int. Ed. Engl.* **30**, 113–129 (1991).
11. Rapp, W. E., in *Combinatorial Chemistry: Synthesis and Application*, S. R. Wilson and A. W. Czarnik, (Eds.) Wiley, New York, 1997, pp. 65–93.
12. Auzanneau, F.-I., Meldal, M. and Bock, K., *J Pept. Sci.* **1**, 31–44 (1995).
13. Leon, S., Quarrell, R. and Lowe, G., *Bioorg. Med. Chem. Lett.* **8**, 2997–3002 (1998).
14. Toy, P. H. and Janda, K. D., *Tetrahedron Lett.* **40**, 6329–6332 (1999).
15. Garibay, P., Toy, P. H., Hoeg-Jensen, T. and Janda, K. D., *Synlett* 1438–1440 (1999).
16. Kempe, M. and Barany, G., *J. Am. Chem. Soc.* **118**, 7083–7093 (1996).
17. Rademann, J., Grotli, M., Meldal, M. and Bock, K., *J. Am. Chem. Soc.* **121**, 5459–5466 (1999).
18. Kumar, K. S. and Rajasekharan Pillai, V. N., *Tetrahedron* **55**, 10437–10446 (1999).
19. Barrett, A. G. M., Cramp, S. M. and Roberts, R. S., *Org. Lett.* **1**, 1083–1086 (1999).
20. Jansson, A. M., *J Chem. Soc., Perkin Trans. I* 1001–1006 (1996).
21. Birch-Hirschfeld, E., Weiss, R., Witkowski, W. and Reinert, H., *J. Prakt. Chem.* **327**, 337–349 (1985).
22. Small, P. W. and Sherrington, D. C., *J. Chem. Soc. Chem. Commun.* 1589–1591 (1989).
23. Atherton, E., Brown, E., Sheppard, R. C. and Rosevear, A., *J. Chem. Soc. Chem. Commun.* 1151–1152 (1981).
24. Adinolfi, M., Barone, G., De Napoli, L., Iadonisi, A. and Piccialli, G., *Tetrahedron Lett.* **39**, 1953–1956 (1998).
25. Labadie, J. W., Porco, J. A. Jr. and Gooding, O. W., WO Patent 059727226 A2, Argonaut Technologies, San Leandro, CA, July 31, 1997.
26. ArgoPore Resins, *Technical Information*; www.argotech.com.
27. Geysen, H. M., Meloen, R. H. and Barteling, S. J., *Proc. Natl Acad. Sci. USA* **81**, 3998–4002 (1984).
28. Takahashi, T., Ebata, S. and Doi, T., *Tetrahedron Lett.* **39**, 1369–1372 (1998).
29. Ede, N. J., James, I. W., Krywult, B. M., Griffiths, R. M., Eagle, S. N. Gubbins, B., Leitch, J. A., Sampson, W. R. and Bray, A. M., *Lett. Pept. Sci.* **6**, 157–163 (1999).
30. Smith, J. M., Gard, J., Cummings, W., Kaniszai, A. and Krchnak, V., *J. Comb. Chem.* **1**, 368–370 (1999).
31. Zhao, C., Shi, S., Mir, D., Hurst, D., Li, R., Xiao, X.-y., Lillig, J. and Czarnik, A. W., *J Comb. Chem.* **1**, 91–95 (1999).

32. Hird, N., Morrison, M. G. J. T., Sherrington, D. C. and Trillow, J. C., *Tetrahedron* **55**, 9575–9584 (1999).

33. Frank, R. and Doering, R., *Tetrahedron* **44**, 6031–6040 (1988).

34. Dostmann, W. R. G., Nickl, C., Thiel, S., Tsigelny, I., Frank, R. and Tegge, W. J., *Pharmacol. Ther.* **82**, 373–387 (1999).

35. Eichler, J., Bienert, M., Stierandova, A. and Lebl, M., *Pept. Res.* **4**, 296–307 (1991).

36. Fodor, S. P. A., Read, J. L., Pirrung, M. C., Stryer, L. Lu, A. T. and Solas, D., *Science* **251**, 767–773 (1991).

37. Berg, R. H., Almdal, K., Pedersen, W. B., Holm, A., Tam, J. P. and Merrifield, R. B., *J. Am. Chem. Soc.* **111**, 8024–8026 (1989).

38. Daniels, S. B., Bematowicz, M. S., Coull, J. M. and Koster, H., *Tetrahedron Lett.* **30**, 4345–4348 (1989).

39. Sherrington, D. C., *Chem. Commun.* 2275–2286 (1998).

40. Yan, B., Fell, J. B. and Kumaravel, G., *J. Org. Chem.* **61**, 7467–7472 (1996).

41. Li, W. and Yan, B., *J. Org. Chem.* **63**, 4092–4097 (1998).

42. Thomas, I. P., Ramsden, J. A., Kovacs, T. Z. and Brown, J. M., *Chem. Commun.* 1507–1508 (1999).

43. Yan, B., *Comb. Chem. High Throughput Screening* **1**, 215–229 (1998).

44. Gordon, K. and Balasubramanian, S., *Curr. Opin. Drug Discovery Dev.* **2**, 342–349 (1999).

45. Terrett, N., *Combinatorial Chemistry*. Oxford University Press, Oxford, UK, 1999, pp. 33–54.

46. Lebl, M., *Biopolymers* **47**, 397–404 (1999).

47. Sucholeiki, I., *Annu. Rep. Combi. Chem. Mol. Div.* **2**, 9–14 (1999).

48. Bergbreiter, D. E., *Med. Res. Rev.* **19**, 439–450 (1999).

49. Mendonca, A. J. and Xiao, X.-Y., *Med. Res. Rev.* **19**, 451–462 (1999).

50. Hudson, D., *J. Comb. Chem.* **1**, 333–360 (1999).

51. Hudson, D., *J. Comb. Chem.* **1**, 403–457 (1999).

52. Backes, B. J. and Ellman, J. A., *Curr. Opin. Chem. Biol.* **1**, 86–93 (1997).

53. Bunin, B. A., *The Combinatorial Index*. Academic Press, San Diego, CA, 1998, pp. 9–76.

54. Blackbum, C., *Biopolymers* **47**, 311–351 (1999).

55. James, I. W., *Tetrahedron* **55**, 4855–4946 (1999).

56. Gordon, K. and Balasubramanian, S., *J. Chem. Technol. Biotechnol.* **74**, 835–851 (1999).

57. Wang, S.-S., *J. Am. Chem. Soc.* **95**, 1328–1333 (1973).

58. Katritzky, A. R., Toader, D., Watson, K. and Kiely, J. S., *Tetrahedron Lett.* **38**, 7849–7850 (1997).

59. Albericio, F. and Barany, G., *Tetrahedron Lett.* **32**, 1015–1018 (1991).

60. Rink, H., *Tetrahedron Lett.* **28**, 3787–3790 (1987).

61. Albericio, F., Kneib-Cordonier, N., Biancalana, S., Gera, L., Masada, R. I., Hudson, D. and Barany, G., *J. Org. Chem.* **55**, 3730–3743 (1990).

62. Brown, E. G. and Nuss, J. M., *Tetrahedron Lett.* **38**, 8457–8460 (1997).

63. Thompson, L. A. and Ellman, J. A., *Tetrahedron Lett.* **35**, 9333–9336 (1994).

64. Ngu, K. and Patel, D. V., *Tetrahedron Lett.* **38**, 973–976 (1997).

65. Mellor, S. L., McGuire, C. and Chan, W. C., *Tetrahedron Lett.* **38**, 3311–3314 (1997).

66. Hu, Y., Porco, J. A. Jr., Labadle, J. W., Gooding, O. W. and Trost, B. M., *J. Org. Chem.* **63**, 4518–4521 (1998).

67. Doi, T., Hijikuro, I. and Takahashi, T., *J. Am. Chem. Soc.* **121**, 6749–6750 (1999).

68. Estep, K. G., Neipp, C. E., Stramiello, L. M. S., Adam, M. D., Allen, M. P., Robinson, S. and Roskamp, E. J., *J. Org. Chem.* **63**, 5300–5301 (1998).

69. Lee, A., Huang, L. and Ellman, J. A., *J. Am. Chem. Soc.* **121**, 9907–9914 (1999).

70. Hall, D. G., Tailor, J. and Gravel, M., *Angew. Chem. Int. Ed.* **38**, 3064–3067 (1999).

71. Rabanal, F., Giralt, E. and Albericio, F., *Tetrahedron* **51**, 1449–1458 (1995).

72. Findeis, M. A. and Kaiser, E. T., *J. Org. Chem.* **54**, 3478–3482 (1989).

73. Bannwarth, W., Kueng, E. and Woherr, T., *Bioorg. Med. Chem. Lett.* **6**, 1525–1528 (1996).

74. Routledge, A., Stock, H. T., Flitsch, S. L. and Turner, L. J., *Tetrahedron Lett.* **38**, 8287–8290 (1997).

75. Parot, I., Werinuth, C.-G. and Hibert, M., *Tetrahedron Lett.* **40**, 7975–7978 (1999).

76. Rich, D. H. and Gurwara, S. K., *J. Am. Chem. Soc.* **97**, 1575–1579 (1975).

77. Wang, S.-S., *J. Org. Chem.* **41**, 3258–3261 (1976).

78. Hammer, R. P., Albericio, F., Gera, L. and Barany, G., *Int. J. Pept. Protein Res.* **36**, 31–45 (1990).

79. Ajayaghosh, A. and Pillai, V. N. R., *J. Org. Chem.* **55**, 2826–2829 (1990).

80. Holmes, C. P. and Jones, D. G., *J. Org. Chem.* **60**, 2318–2319 (1995).

81. Rodebaugh, R., Fraser-Reid, B. and Geysen, H. M., *Tetrahedron Lett.* **38**, 7653–7656 (1997).

82. Aurell, M. J., Boix, C., Ceita, M. L., Llopis, C., Tortajada, A. and Mestres, R., *J. Chem. Res., Synop.* 452–453 (1995).

83. Peukert, S. and Giese, B., *J. Org. Chem.* **63**, 9045–9051 (1998).

84. Backes, B. J. and Eliman, J. A., *J. Am. Chem. Soc.* **116**, 11171–11172 (1994).

85. Marshal, D. L. and Liener, I. E., *J. Org. Chem.* **35**, 867–868 (1970).

86. Kiso, Y., Fukui, T., Tanaka, S., Kimura, T. and Akaji, K., *Tetrahedron Lett.* **35**, 3571–3574 (1994).

87. Hoffman, S. and Frank, R., *Tetrahedron Lett.* **35**, 7763–7766 (1994).

88. Routledge, A., Abell, C. and Balasubramanian, S., *Tetrahedron Lett.* **38**, 1227–1230 (1997).

89. Gayo, L. M. and Suto, M. J., *Tetrahedron Lett.* **38**, 211–214 (1997).

90. Todd, M. H., Oliver, S. F. and Abell, C., *Org. Lett.* **1**, 1149–1151 (1999).

91. Zheng, A., Shan, D., Shi, X. and Wang, B., *J. Org. Chem.* **64**, 7459–7466 (1999).

92. Reitz, A. B., *Curr. Opin. Drug Discovery Dev.* **2**, 358–364 (1999).

93. Plunkett, M. J. and Ellman, J. A., *J Org. Chem.* **60**, 6006–6007 (1995).

94. Plunkett, M. J. and Ellman, J. A., *J. Org. Chem.* **62**, 2885–2893 (1997).

95. Hughes, I., *Tetrahedron Lett.* **37**, 7595–7598 (1996).

96. Morphy, J. R., Rankovic, Z. and Rees, D. C., *Tetrahedron Lett.* **37**, 3209–3212 (1996).

97. Craig, D., Robson, M. J. and Shaw, S. J., *Synlett* 1381–1383 (1998).

98. Gibson, S. E., Hales, N. J. and Peplow, M. A., *Tetrahedron Lett.* **40**, 1417–1420 (1999).

99. Maiorana, S., Seneci, P., Rossi, T., Baldoli, C., Ciraco, M., De Magistris, E., Licandro, E., Papagni, A. and Provera, S., *Tetrahedron Lett.* **40**, 3635–3638 (1999).

100. Comely, A. C., Gibson, S. E. and Hales, N. J., *Chem. Commun.* 2075–2076 (1999).

101. Cobb, J. M., Fiorini, M. T., Goddard, C. R., Theoclitou, M.-E. and Abell, C., *Tetrahedron Lett.* **40**, 1045–1048 (1999).

102. Stieber, F., Grether, U. and Waldmann, H., *Angew. Chem. Int. Ed.* **38**, 1073–1077 (1999).

103. Fujita, K.-I., Watanabe, K., Oishi, A., Ikeda, Y. and Taguchi, Y., *Synlett*, 1760–1762 (1999).

104. Koot, W.-J., *J. Comb. Chem.* **1**, 467–473 (1999).

105. Buchstaller, H.-P., *Tetrahedron* **54**, 3465–3470 (1998).

106. Piscopio, A. D., Miller, J. F. and Koch, K., *Tetrahedron Lett.* **39**, 2667–2670 (1998).

107. Matthews, J. and Rivero, R. A., *J. Org, Chem.* **63**, 4808–4810 (1998).

108. Kulkami, B. A. and Ganesan, A., *Tetrahedron Lett.* **39**, 4369–4372 (1998).

109. Romoff, T. T., Ma, L., Wang, Y. and Campbell, D. A., *Synlett* 1341–1342 (1998).

110. Brown, R. C. D. and Fisher, M., *Chem. Commun.* 1547–1548 (1999).

111. Smith, J. M. and Krchnak, V., *Tetrahedron Lett.* **40**, 7633–7636 (1999).

112. Mazurov, A., *Tetrahedron Lett.* **41**, 7–10 (2000).

113. Nicolaou, K. C., Wissinger, N., Pastor, J. and Murphy, F., *Angew. Chem. Int. Ed.* **37**, 2534–2537 (1998).

114. Nicolaou, K. C., Pastor, J., Wissinger, N. and Murphy, F., *J. Am. Chem. Soc.* **120**, 5132–5133 (1998).

115. van Maarseveen, J. H., *Comb. Chem. High Throughput Screening* **1**, 185–214 (1998).

116. Keifer, P. A., *Drugs Future* **23**, 301–317 (1998).

117. Bunin, B. A., *The Combinatorial Index*. Academic Press, San Diego, CA, 1998, pp. 213–235.

118. Kaiser, E., Colescott, R. L., Bossinger, C. D. and Cook, P. I., *Anal. Biochem.* **34**, 595–598 (1970).

119. Flegel, M. and Sheppard, R. C., *J. Chem. Soc., Chem. Commun.* 536–538 (1990).

120. Chu, S. S. and Reich, S. H., *Bioorg. Med. Chem. Lett.* **5**, 1053–1058 (1995).

121. Gisin, B. F., *Anal. Chim. Acta* **58**, 248–249 (1972).

122. Shah, A., de Biasi, V., Camilleri, P. and Rahman, S. S., *Anal. Commun.* **34**, 325–328 (1997).

123. Virgilio, A. A. and Ellman, J. A., *J. Am. Chem. Soc.* **116**, 11580–11581 (1994).

124. Sarin, V. K., Kent, S. B. H., Tam, J. B. and Merrifield, R. B., *Anal. Biochem.* **117**, 147–157 (1981).

125. Madder, A., Farcy, N., Hosten, N. G. C., De Muynck, H., De Clerq, P. J., Barry, J. and Davis, A. P., *Eur. J. Org. Chem.* 2787–2791 (1999).

126. Kuisle, O., Lolo, M., Quinoa, E. and Riguera, R., *Tetrahedron* **55**, 14807–14812 (1999).

127. Giralt, E., Rizo, J. and Pedroso, E., *Tetrahedron* **40**, 4141–4152 (1984).

128. Svensson, A., Fex, T. and Kihlberg, J., *Tetrahedron Lett.* **37**, 7649–7652 (1996).

129. Svensson, A., Bergquist, K.-E., Fex, T. and Kihlberg, J., *Tetrahedron Lett.* **39**, 7193–7196 (1998).

130. Stones, D., Miller, D. J, Beaton, M. W., Rutherford, T. J. and Gani, *Tetrahedron Lett.* **39**, 4875–4878 (1998).

131. Johnson, C. R. and Zhang, B., *Tetrahedron Lett.* **36**, 9253–9256 (1995).

132. Swayze, E. E., *Tetrahedron Lett.* **38**, 8643–8646 (1997).

133. Look, G. C., Holmes, C. P., Chinn, J. P. and Gallop, M. A., *J. Org. Chem.* **59**, 7588–7590 (1994).

134. Holmes, C. P., *J. Org. Chem.* **62**, 2370–2380 (1997).

135. Bam, D. R., Morphy, J. R. and Rees, D. C., *Tetrahedron Lett.* **37**, 3213–3216 (1996).

136. Marchioro, C. and Missio A., *personal communication*, March 1998.

137. Muddiman, D. C., Bakhtiar, R., Hofstadler, S. A. and Smith, R. D., *J. Chem. Educ.* **74**, 1288–1292 (1997).

138. Zambias, R. A., Boulton, D. A. and Griffin, P. R., *Tetrahedron Lett.* **35**, 4283 (1994).

139. Egner, B. J., Langley, G. J. and Bradley, M., *J Org. Chem.* **60**, 2652–2653 (1995).

140. Haskins, N. J., Hunter, D. J., Organ, A. J., Rahman, S. S. and Thom, C., *Rapid Commun. Mass Spectrom.* **9**, 1437–1440 (1995).

141. Egner, B. J. and Bradley, M., *Tetrahedron* **41**, 14021–10430 (1997).

142. Egner, B. J., Cardno, M. and Bradley, M., *J. Chem. Soc., Chem. Commun.* 2163–2164 (1995).

143. Fitzgerald, M. C., Harris, K., Shevlln, C. G. and Siuzdak, G., *Bioorg. Med. Chem. Lett.* **6**, 979–982 (1996).

144. Benninghoven, A., Wemer, H. W. and Rudenauer, F. G., *Secondary Ion Mass Spectrometry, Basic Concepts, Instrumental, Aspects, Application and Trends*. John Wiley, New York, 1986.

145. Enjabal, C., *Tetrahedron Lett.* **40**, 6217–6220 (1999).

146. Geysen, H. M., Wagner, C. D., Bodnar, W. M., Markworth, C. J., Parke, G. J., Schoenen, F. J., Wagner, D. S. and Kinder, D. S., *Chem. Biol.* **3**, 679–688 (1996).

147. Carrasco, M. R., Fitzgerald, M. C., Oda, Y. and Kent, S. B. H., *Tetrahedron Lett.* **38**, 6331–6334 (1997).

148. McKeown, S. C., Watson, S. P., Caff, R. A. E. and Marshall, P., *Tetrahedron Lett.* **40**, 2407–2410 (1999).

149. Murray, P. J., Kay, C., Scicinski, J. J., McKeown, S. C., Watson, S. P. and Carr, R. A. E., *Tetrahedron Lett.* **40**, 5609–5612 (1999).

150. Chen, C., Randall, L. A. A., Miller, R. B., Jones, A. D. and Kurth, M. J., *J. Am. Chem. Soc.* **116**, 2661–2662 (1994).

151. Hauske, J. R. and Dorff, P., *Tetrahedron Lett.* **36**, 1589–1592 (1995).

152. Yan, B., Kumaravel, G., Anjaria, H., Wu, A., Petter, R. C., Jewell. C. F. Jr., Wareing, J. R., *J. Org. Chem.* **60**, 5736–5738 (1995).

153. Yan, B. and Kumaravel, G., *Tetrahedron* **52**, 843–848 (1996).

154. Yan, B. and Li, W., *J. Org. Chem.* **62**, 9354–9357 (1997).

155. Yan, B., Sun, Q., Wareing, J. R. and Jewell, C. F., *J. Org. Chem.* **61**, 8765–8770 (1996).

156. Yan, B., Liu, L., Astor, C. A. and Tang, Q., *Anal. Chem.* **71**, 4564–4571 (1999).

157. Pivonka, D. E., *J. Comb. Chem.* **1**, 294–296 (1999).

158. Pivonka, D. E., Palmer, D. L. and Gero, T. W., *Appl. Spectrosc.* **53**, 1027–1032 (1999).

159. Haap, W. J., Kaiser, D., Walk, T. B. and Jung, G., *Tetrahedron* **54**, 3705–3724 (1998).

160. Huber, W., Bubendorf, A., Grieder, A. and Obrecht, D., *Anal. Chim. Acta* **393**, 213–221 (1999).

161. Rosse', G., Ouertani, F. and Schroder, H., *J. Comb. Chem.* **1**, 397–401 (1999).

162. Gosselin, F., Di Renzo, M., Ellis, T. H. and Lubell, W. D., *J. Org. Chem.* **61**, 7980–7981 (1996).

163. Chan, T. Y., Chen, R., Sofia, M. J., Smith, B. C. and Glennon, D., *Tetrahedron Lett.* **38**, 2821–2824 (1997).

164. Deben, I., Goorden, J, Van Doomum, E., Ovaa, H. and Kellenbach, E., *Eur. J. Org. Chem.*, 697–700 (1998).

165. Hammond, J., Moffat, A. C., Jee, R. D. and Kellam, B., *Anal. Commun.* **36**, 127–129 (1999).

166. Fischer, M. and Tran, C. D., *Anal. Chem.* **71**, 2255–2261 (1999).

167. Yan, B., Gremlich, H.-U., Moss, S., Coppola, G. M., Sun, Q., and Liu, L., *J. Comb. Chem.* **1**, 46–54 (1999).

168. Rahman, S. S., Busby, D. J. and Lee, D. C., *J. Org. Chem.* **63**, 6196–6199 (1998).

169. Yan, B. and Gremlich, H.-U., *J. Chromatogr., B: Biomed. Sci. Applic.* **725**, 91–102(1999).

170. Hochlowski, J., Whittem, D., Pan, J. and Swenson, R., *Drugs Future* **24**, 539–554 (1999).

171. Luo, Y., Ouyang, X., Armstrong, R. W. and Murphy, M. M., *J. Org. Chem.* **63**, 8719–8722 (1998).

172. Andrew, E. R., *Prog. Nucl. Magn. Reson. Spectrosc.* **8**, 1–39 (1971).

173. Fitch, W. L., Detre, G., Holmes, C. P., Shoolery, J. N. and Keifer, P. A., *J. Org. Chem.* **59**, 7955–7956 (1994).

174. Kempe, M., Keifer, P. A. and Barany, G., in *Proceedings from the Twenty-Fourth European Peptide Symposium*, R. Rampage and R. Epton (Eds.). Edinburgh, Scotland, September 8th–13th, 1996. Mayflower Scientific Ltd., West Midland, 1998, pp. 533–534.

175. Keifer, P. A., *J. Org. Chem.* **61**, 1558–1559 (1996).

176. Shapiro, M. J. and Wareing, J. R., *Curr. Opin. Chem. Biol.* **2**, 372–375 (1998).

177. Lippens, G., Bourdonneau, M., Dhalluin, C., Waffass, R., Richert, T., Seetharaman, C., Boutillon, C. and Piotto, M., *Curr. Org. Chem.* **3**, 147–169 (1999).

178. Riedl, R., Tappe, R. and Berkessel, A., *J. Am. Chem. Soc.* **120**, 8994–9000 (1998).

179. Warrass, R., Wieruszeski, J.-M. and Lippens, G., *J. Am. Chem. Soc.* **121**, 3787–3788 (1999).

180. Missio, A. and Marchioro, C., *personal communication*, March 1998.

181. Chin, J., Fell, B., Pochapsky, S., Shapiro, M. J. and Wareing, J. R., *J. Org. Chem.* **63**, 1309–1311 (1998).

182. Sarkar, S. K., Garigipati, R. S., Adams, J. L. and Keifer, P. A., *J. Am. Chem. Soc.* **118**, 2305–2306 (1996).

183. Pursch, M., Schlotterbeck, G., Tseng, L.-H., Albert, K. and Rapp, W., *Angew. Chem., Int. Ed. Engl.* **35**, 2867–2868 (1996).

2 Solid-Phase Synthesis: Oligomeric Molecules

The first preparation of a biologically relevant oligopeptide on a solid support by Merrifield (1) represented a landmark, not only for the field of SPS but also for organic chemistry as a whole. The use of this technique to prepare oligomeric molecules by repeated use of the same type of chemical reaction applied to similar building blocks rapidly gained popularity. The advantages of oligomeric SPS versus solution-phase synthesis of the same oligomers are self-evident: tedious and troublesome purifications after each step are avoided and the repetitive nature of the chemistry lends itself to automation of the process. Excellent yields are obtained by forcing the reactions to completion through the use of large excesses and/or high concentration of reagents in solution.

Peptides are the single most exploited oligomeric class of molecules in SPS, and the last 35 years have seen significant improvements in all aspects of SP peptide synthesis. The SPS of peptoids, peptidomimetics, peptide hybrids, and cyclic peptides has also gained in importance because of their improved properties when compared with natural peptides (e.g., stability and solubility). *Oligonucleotides* represent the other class of biologically important molecules in which considerable progress has been made in automated synthesis on SP. The field of combinatorial chemistry started and, for quite some time, remained strictly related to the SPS of peptides and to some extent oligonucleotides. *Oligosaccharides* have also been considered as targets for synthesis in the SP; however, their complex structures makes the chemistry of this class much more challenging due to problems of regio- and chemoselectivity as well as availability of suitable building blocks. Even today, this class remains underexploited.

In this chapter, a historical overview of the SPS of peptides, nucleotides, oligosaccharides, and related oligomeric molecules will be provided, summarizing the main efforts in this area. The strategy followed in designing and carrying out the SPS of an oligomeric biopolymer will be presented and discussed through the use of examples, and eventually, it will be compared with the strategy used for the SPS of small organic molecules (see Chapter 3).

2.1 PEPTIDES

2.1.1 Solid-Phase Synthesis of Peptides

The number of peptides prepared on SP since the pioneering work by Merrifield in the early 1960s is huge, and the relevance of peptide SPS is such that various books and reviews extensively covered this topic. The interested reader is referred to a recent book (2) that covers the main aspects of peptide SPS to expand some of the basic concepts provided in this chapter.

The scheme of classical peptide SPS is shown in Fig. 2.1. The first step is the attachment of the first amino acid onto the resin, usually via the carboxylic group with both the amino and the side-chain functions of the α-amino acid protected (step 1). The inverse strategy, where the amino group would be linked to the resin and the carboxyl would be protected, has seldom been used because extensive racemization occurs during repeated resin-bound carboxylate activations (3); this limits the accessibility of C-terminal modified peptides that are biologically relevant. However, several recent reports (4–7) describe the SP modification of the peptide C-terminus using the more reliable C-to-N direction for peptide synthesis.

The synthetic scheme consists of repeated deprotections of the amino terminus followed by coupling of an N-protected amino acid (with side chains suitably protected) in solution until the target peptide is obtained. A single cycle (step 2) is detailed in Fig. 2.1, while the following cycles are identical. After the assembly of the fully protected peptide, all the side chains are deprotected and the peptide is released into solution via cleavage of the linker (Fig. 2.1). The repetitive nature of the deprotection/coupling steps where an amino-protecting group is removed and an amide bond is formed does not require any assessment of the chemistry since it is so well-known. This allows the rapid synthesis of long oligopeptides with excellent yields and no detectable resin-bound side products. The robustness of the peptide amidic bond significantly reduces the number and the importance of side reactions.

The synthesis of natural oligopeptides (of nonbacterial origin) of various lengths requires only 20 building blocks, the natural L-amino acids. They are commercially available with a large variety of orthogonal protecting groups on the amino function and on the functionalized side chains, which can be removed selectively without affecting other protecting groups present. This allows many possible protecting group strategies to be followed, depending on the linker and the chemistry to be used.

Two major protecting group strategies (see Fig. 2.1) are used for P_1 in SP peptide synthesis. The *Boc protocol* (8) involves the use of *N*-Boc-protected amino acids and the iterative deprotections are performed with TFA; the *Fmoc protocol* (9) employs *N*-Fmoc-protected amino acids and the iterative deprotections are performed with 20% v/v piperidine in DMF.

The use of Boc as N-protecting group allows the anchoring of the first amino acid onto the Merrifield resin through a benzyl ester bond. Final cleavage with HF or hydrogenolysis gives the peptide acid or, alternatively, the C-terminal amide is obtained through ammonolysis. Groups in the side chains are protected with benzyl-based protecting groups (e.g., 2-chloroCBz for Tyr, 4-methylbenzyl for Cys), which are removed during the hydrogenolytic cleavage. Amine linkers, such as the 4-methylbenzhydrylamine (MBHA, benzyl amide bond), have also been used to release peptide primary amides.

The older Boc protocol has largely been superceded by the Fmoc protocol, which has several relevant advantages over it. The use of milder deprotection conditions for the Fmoc group and its complete orthogonality with a large number of side-chain

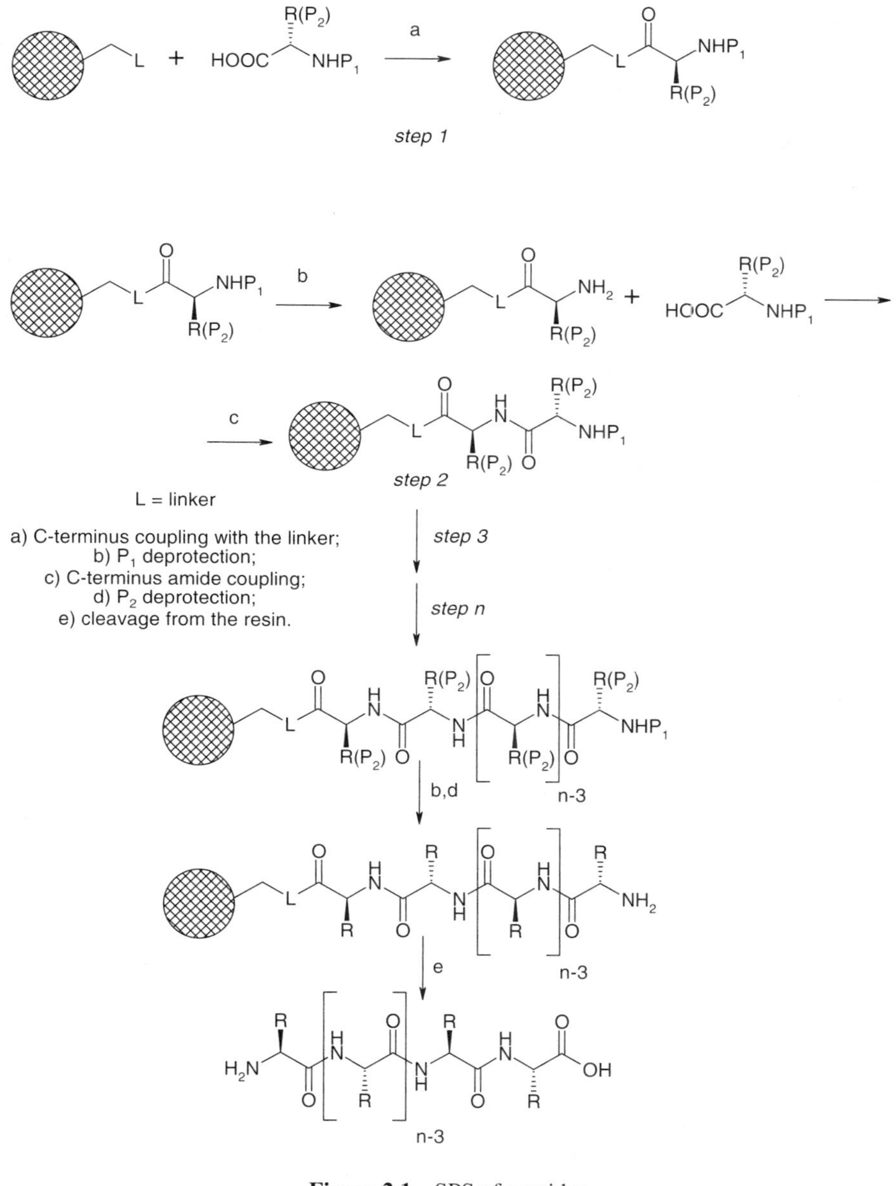

Figure 2.1 SPS of peptides.

protecting groups gives a better quality synthesis. Many different acid-sensitive SP linkers can be used, and the milder conditions allow the preparation of modified peptides containing acid-sensitive building blocks such as glycosylated or sulfated or phosphorylated peptides. Finally, the Fmoc reading of the cleavage solution as described in Section 1.4.2 allows for the easy quantitation of the coupling steps.

Many *solid supports* have been used in the SPS of peptides, and all of them are compatible with Fmoc-based chemistry. Cross-linked hydrophobic and hybrid PS-based resins are among the most used supports, but much SPS of peptides is performed by the so-called continuous-flow methodology (10) where sealed columns contain the resin and the peptide is assembled using flowing solutions of reagents for each step. This method is easily automated and is used by commercially available SP peptide synthesizers. The columns typically contain nonswelling resins such as the macroporous supports POLYHYPE and Kieselguhr described in Section 1.1.3 (11, 12). PS-based devices such as pins (13) have also been used as alternative supports together with paper (14), cotton (15), and sepharose (16).

Acid-labile *protecting groups* for side-chain functionalities allow the simultaneous final deprotection together with the cleavage of the acid-labile linker to produce the free peptide in solution in a single step. The most widely used protecting groups are *t*-butyl ethers (Ser, Thr, Tyr) or esters (Glu, Asp) and Boc carbamates (Lys). Acid-labile *linkers* with different acid stability are chosen depending on the stability of the target peptide and on the functional group to be released on the C-terminus (e.g., free acid, amide, hydrazide, or hydroxamate). The so-called Backbone Amide Linkage (BAL) (5) linker has recently found a wide application in SPS (17–19); a specific example will be described in details in Section 2.1.3. Examples of cyclative cleavage approaches to provide cyclic peptides or depsipeptides have also been reported (20).

The *coupling conditions* have been thoroughly studied and optimized. A lot of effort has gone into studying the influence of solvents and temperatures, the extent of racemization, and the best coupling reagents or carboxyl activating agents to use. An excellent review (21) summarizes the state-of-the-art and clearly shows the superior quality of peptide SPS. The synthesis of peptide 40mers with diverse structures using coupling cycles of just a few minutes and average coupling yields in excess of 99% can be considered a routine task. Further reduction of cycle times and improvement of their efficiency will probably allow the peptide SPS of even longer oligopeptides with lower quantities of impurities.

The quality of the SPS of a peptide can be determined by the analytical methods presented in Sections 1.3 and 1.4. Classical amino acid analyses following acid degradation (22), capillary electrophoresis (23), and MS-related protocols (24, 25) are also commonly employed. Qualitative monitoring of couplings can be carried out using either the ninhydrin test or bromophenol blue, which is particular for the secondary amino group of prolines. The yield of each iterative cycle of coupling and deprotection is given by the Fmoc reading. On-bead spectroscopic and spectrometric methods such as MALDI–TOF, FTIR, and MAS–HR NMR are also routinely used.

The sequencing of a peptide (26) uses the well-known Edman degradation (27) of peptides, as shown in Fig. 2.2. The reaction cycle consists of reaction with phenylisothiocyanate followed by treatment with anhydrous TFA to promote cyclization of the intermediate thiourea. Rearrangement induced by treatment with aqueous TFA produces the phenylthiohydantoin (PTH) of the N-terminal amino acid **2.2** and the truncated peptide **2.3**. The sequence is repeated through *n* cycles until the whole peptide is degraded. The PTHs produced are detected by HPLC-UV and their retention times are compared to those of 20 standard PTHs, one for each natural amino acid.

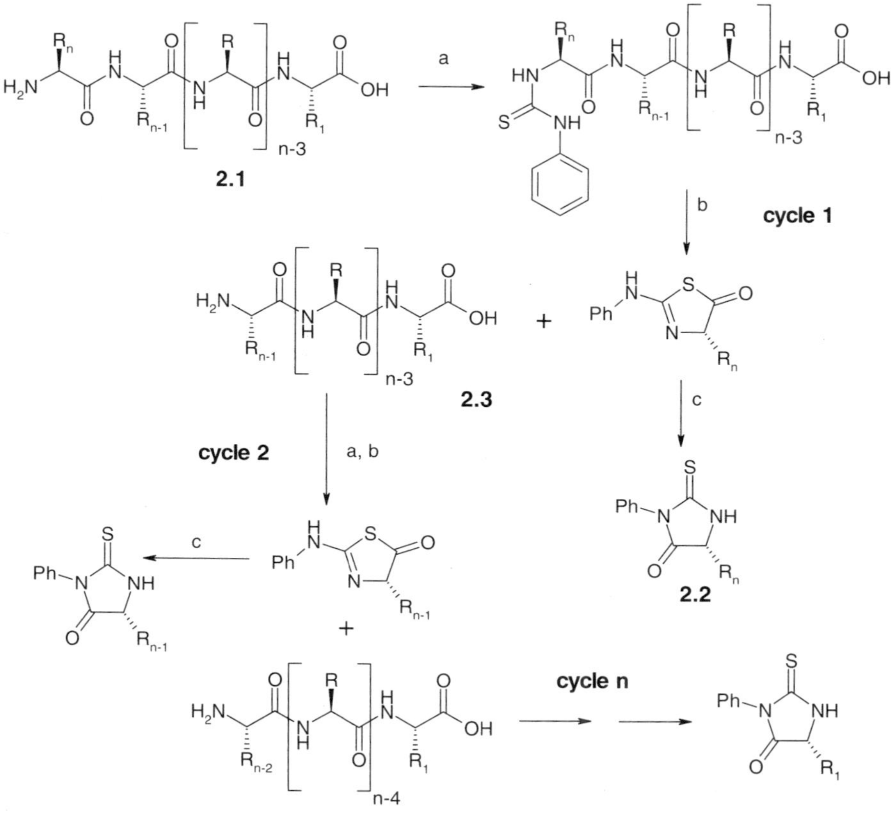

a) PhNCS, base, pH 9-10, 45°-50°C;
b) anhydrous TFA, 45°-55°C;
c) 25% aq. TFA, 60°-65°C.

Figure 2.2 Edman sequencing of peptides: iterative cycles.

Commercially available Edman sequencers perform a fully automated sequencing and HPLC separation and provide the HPLC traces for each reaction cycle. The presence of deletion peptides is easily spotted qualitatively by the appearance of secondary PTH peaks in each reaction cycle. The method is limited to α-amino acids, but polypeptides containing exotic α-amino acids can also be sequenced provided the appropriate standard PTH is available. The method is very sensitive, and usually a single bead is enough to allow an accurate determination of the sequence of a peptide, but the quantitation of each amino acid is difficult because of their different stabilities and properties under the conditions of the cleavage cycle. Edman sequencing can be used in conjunction with other techniques such as MS.

2.1.2 Solid-Phase Synthesis of Peptide-Related Oligomers

The wide use of peptides for biological applications has led to the SPS of many relevant sequences. The well-known drawbacks of natural peptides, such as their poor penetration through biological membranes and their poor stability in biological fluids, has stimulated the synthesis of modified peptides to overcome these limitations. There have been corresponding advances in SPS to accommodate altered structures, and some generic examples of peptide analogs prepared on SP are shown in Fig. 2.3.

The use of D-amino acids, or of *exotic* L- or D-amino acids such as **2.4**, does not significantly change the SPS procedure; however, the availability of suitable *N*-Fmoc-protected exotic α-amino acids may be limited; thus a method called solid-phase unnatural peptide synthesis (SP-UPS) has been reported. This method, which entails the SPS of unnatural α-amino acids by alkylation of Fmoc-Gly under mild conditions, has been integrated with traditional SP peptide chemistry (28).

The synthesis of *glycopeptides,* such as **2.5**, is compatible with the Fmoc strategy, and glycosylated Fmoc-protected amino acids such as Thr, Ser, Cys, and Tyr have been inserted into peptides prepared on SP (29). The sugar hydroxyls are normally protected with Fmoc-compatible acyl, benzyl, or silyl groups. Similarly, *phosphopeptides* **2.6** are prepared using Fmoc-protected phosphorylated Thr, Ser, and Tyr building blocks or by the so-called global phosphorylation method, which saturates the hydroxyl side chains of Thr, Ser, and Tyr in the resin-bound peptide (30). The phosphate residues may be protected as Fmoc-compatible *t*-butyl or benzyl esters.

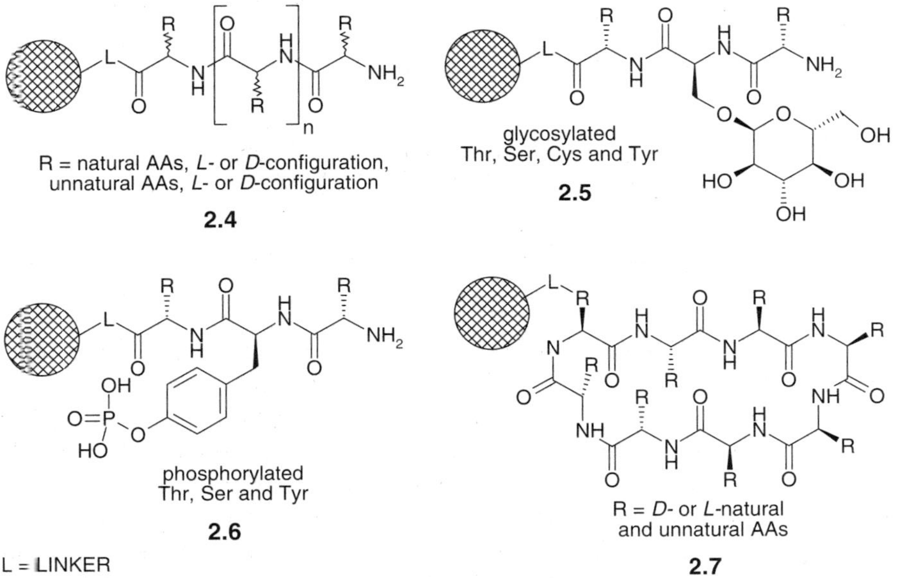

Figure 2.3 Structures of peptides containing non-natural amino acids (**2.4**), glycopeptides (**2.5**), phosphopeptides (**2.6**), and cyclic peptides (**2.7**).

Many biologically active peptides are cyclic in nature, and the SPS of this class of peptides, exemplified by **2.7**, has also received attention with several different strategies for the final cyclization. The phenomenon of pseudodilution on bead in which each resin site is essentially isolated from its neighbors favors the intramolecular cyclization reaction compared to the intermolecular dimerization, which occurs in solution even at high dilutions. The SPS of cyclic peptides has recently been covered in two excellent reviews (31, 32). The technique of cyclative cleavage via the N- or the C-terminus (see Section 1.2.7) has been used, as has anchoring through amino acid side chains with sequential cyclization and peptide release.

Solid-phase synthesis has also been used to make *peptoids*, some examples of which are shown in Fig. 2.4. Compounds of general structure **2.8**, where the amino acid side chain is on the nitrogen, have been prepared either by the corresponding Fmoc-protected N-alkylated glycines (33) or, in an improved method, via treatment of a resin-bound secondary amine with bromoacetic acid to produce the first peptoid building block, which is then elaborated via iteration of the procedure (34). Other modified β-peptoid structures such as **2.9** with repeating β-amino propionic units have been prepared by acylation of a resin-bound amine with acriloyl chloride followed by Michael addition of a primary amine. The cycle is repeated to build up the polymer (35).

Peptidomimetics, a class of compound in which the peptidic bond is replaced by a grouping of similar geometry, have also been prepared on SP. Two examples are shown in Fig. 2.4. The peptidomimetic **2.10**, in which the N-terminal amide has been substituted by an olefin bond, was prepared by the Wittig reaction of a resin-bound α-amino aldehyde and sequential peptide SPS to give oligomers of various lengths (36). The amidic bond of peptides has also been replaced by a urea, as in the

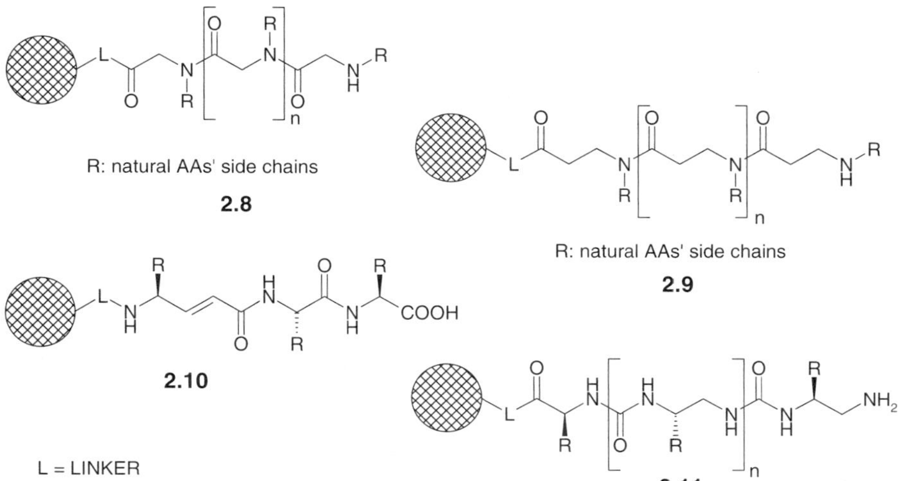

2.8 **2.9** **2.10** **2.11**

Figure 2.4 Structures of peptoids (**2.8, 2.9**), peptidomimetics (**2.10**), and oligoureas (**2.11**).

β-oligoureas **2.11** prepared from a resin-bound amino acid treated with a suitable *N*-phthaloyl amino isocyanate. Deprotection releases the terminal amine, which can be treated with another *N*-phthaloyl-protected amino isocyanate to repeat the cycle (37).

The importance of SPS for the rapid and reliable preparation of complex peptides, peptoids, and peptidomimetics cannot be underestimated. The exploitation of the SPS of this type of oligomer has allowed the birth of synthetic combinatorial chemistry and its subsequent exploitation in pharmaceutical research. Even now that the main accent of SPS and synthetic combinatorial chemistry has shifted toward the preparation of small nonoligomeric organic molecules, a steady flow of reports regarding the SPS of medium-long peptides or peptidelike molecules with significant biological properties is still appearing.

A brief survey of recently appeared papers in peptide SPS should highlight the trends related to this area. They include the synthesis of novel amino acidic building blocks such as sulfonylated aspartic and glutamic acids (38), ω-N-substituted (39) and thio (40) arginines, or phosphotyrosines (41); the introduction of novel orthogonal protecting groups such as the cyclohexyl ether for serine (42) or the 2-(4-nitrophenyl)sulfonylethoxycarbonyl (Nsc) carbamate for amines (43); the continuous improvements in experimental protocols, regarding the automated synthesis of small proteins (44), or the reduction of racemization (45, 46), or the introduction of the swelling volume coupling protocol (47); the introduction of novel analytical techniques to monitor and quantify peptide SPS (48); the introduction of novel resins (49) or linkers (50–52) for peptide SPs; the postcoupling modification of amino acid side chains on SP, as for the transformation of aspartyl or glutamyl carboxylates (53); the synthesis of biologically relevant peptide sequences (54, 55); the synthesis of novel cyclic peptides (56–58); and the synthesis of novel peptide analogues belonging to the classes of glycopeptides (59), of peptide and glycopeptide dendrimers (60, 61), of oligopyrrole–peptide conjugates (62), of (PNA)–peptide conjugates (63) of urea-based (64), trifluoromethyl ketone–containing (65) and constrained peptidomimetics (66). Clearly, everything related to peptide SPS is reliable and flexible, but improvements are always looked for in any specific section and will continue to be the topic of high-quality scientific research.

2.1.3 An Example: SPS of cyclo(Arg-D-Phe-Pro-Glu-Asp-Asn-Tyr-Glu-Ala-Ala)

The SPS of the cyclic homodetic peptide cyclo(Arg-D-Phe-Pro-Glu-Asp-Asn-Tyr-Glu-Ala-Ala) **2.12** was previously attempted by anchoring the starting amino acid to the resin through the side chain (via $\mathbf{L_{1-4}}$, Fig. 2.5) (67). The carboxylic groups of Glu^6, Glu^{10}, and Asp^7 and the amide of Asn^8 were anchored during several different attempts using the 4-hydroxymethylphenoxyacetic linker for the acids and the PAL linker for the amide bond. The best results were obtained using the Asn^8 handle and the PAL linker, with a 71% yield of the cyclized decapeptide (HPLC, purity calculated with an authentic standard sample).

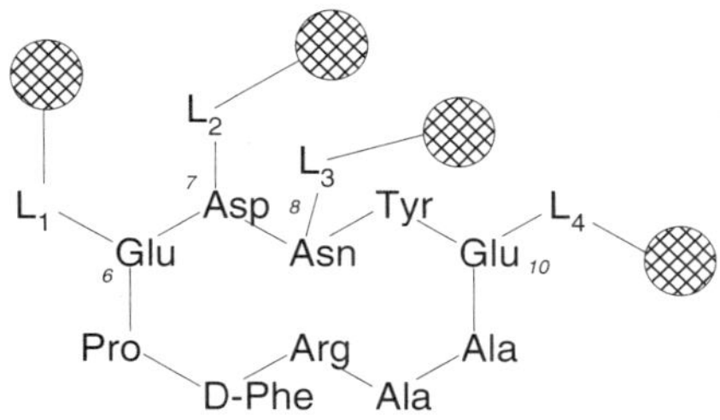

2.12

best results obtained with
L_3: PAL linker, 71% final yield

Figure 2.5 Structure of the cyclic peptide **2.12** and its possible side-chain attachments onto the support.

This synthesis has been recently revisited (5), and a new linker that allows the fastening of a backbone amide onto the support strategy has been applied. While this peptide represents a nice target that allows direct comparison of the quality of the SPS with previous attempts, its main advantage lies in the fact that use of the backbone amide linkage, or BAL (**2.13**, Fig. 2.6), allows anchoring of peptides other than Asp, Glu, Gln, or Asn, thus permitting the synthesis of analogues otherwise not accessible in the original work.

The linker was prepared by hooking the aldehyde **2.14** (68) onto the support and then performing a reductive amination with the first α-amino acid protected as an allyl

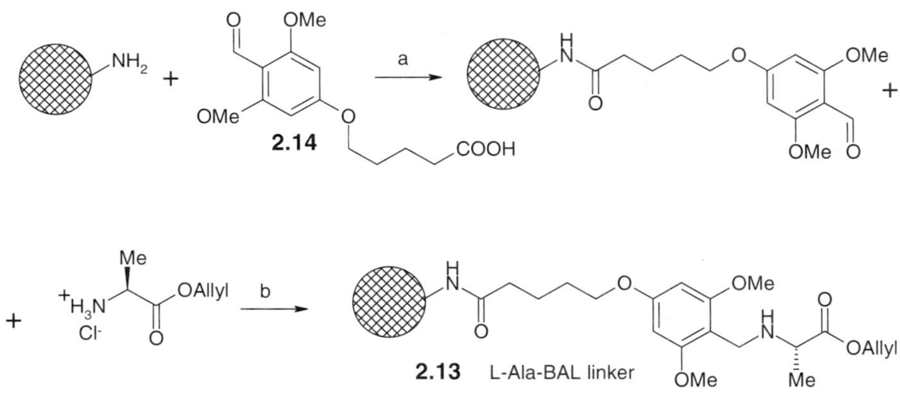

a) HATU/DIEA, DMF; b) $NaBH_3CN$, DMF.

Figure 2.6 Synthesis of the BAL linker **2.13**.

ester (Ala in this example). The peptide chain was grown in the C-to-N direction using Fmoc-protected amino acids. Acid-labile side-chain protection was used for Arg (Pmc), Glu and Asp (OtBu), Asn (Trt), and Tyr (tBu).

Attempts to incorporate the second amino acid were frustrated by the formation of the diketopiperazine **2.15** (Fig 2.7). Apparently, during the standard Fmoc protocol,

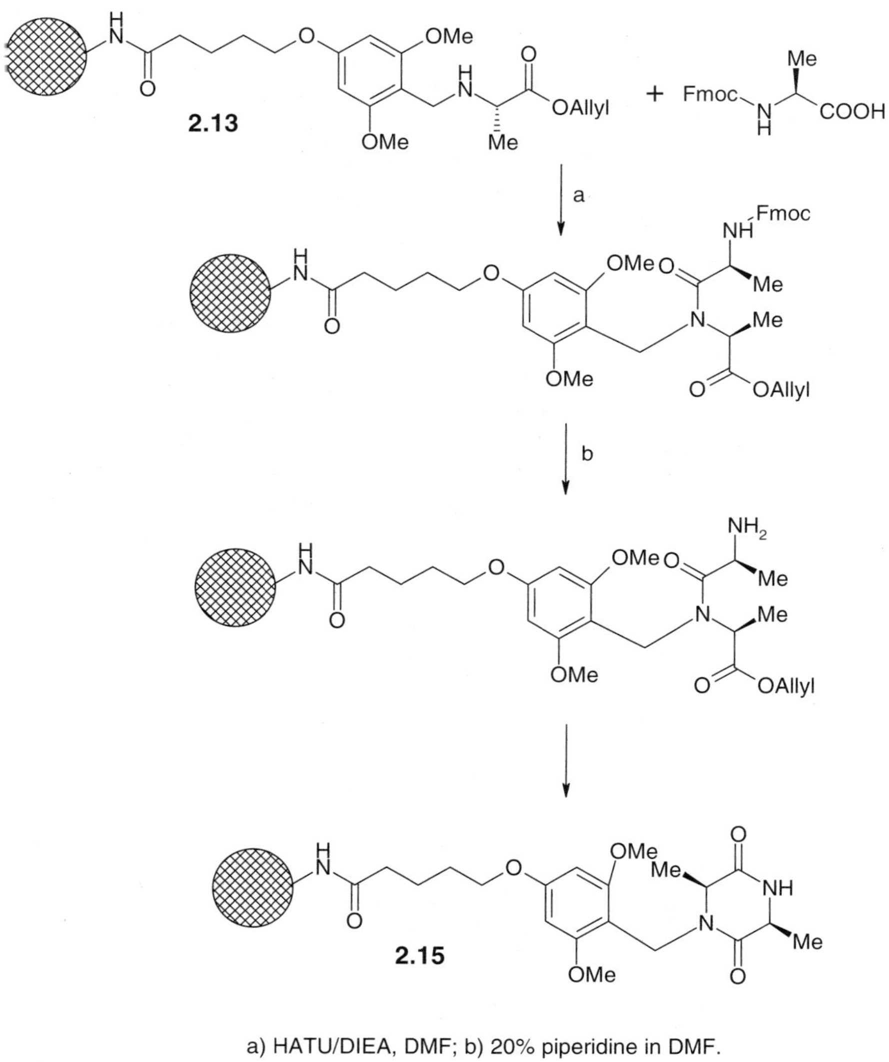

Figure 2.7 Side reaction of BAL-supported *N*-Fmoc amino acids during Fmoc deprotection: Diketopiperazines **2.15**.

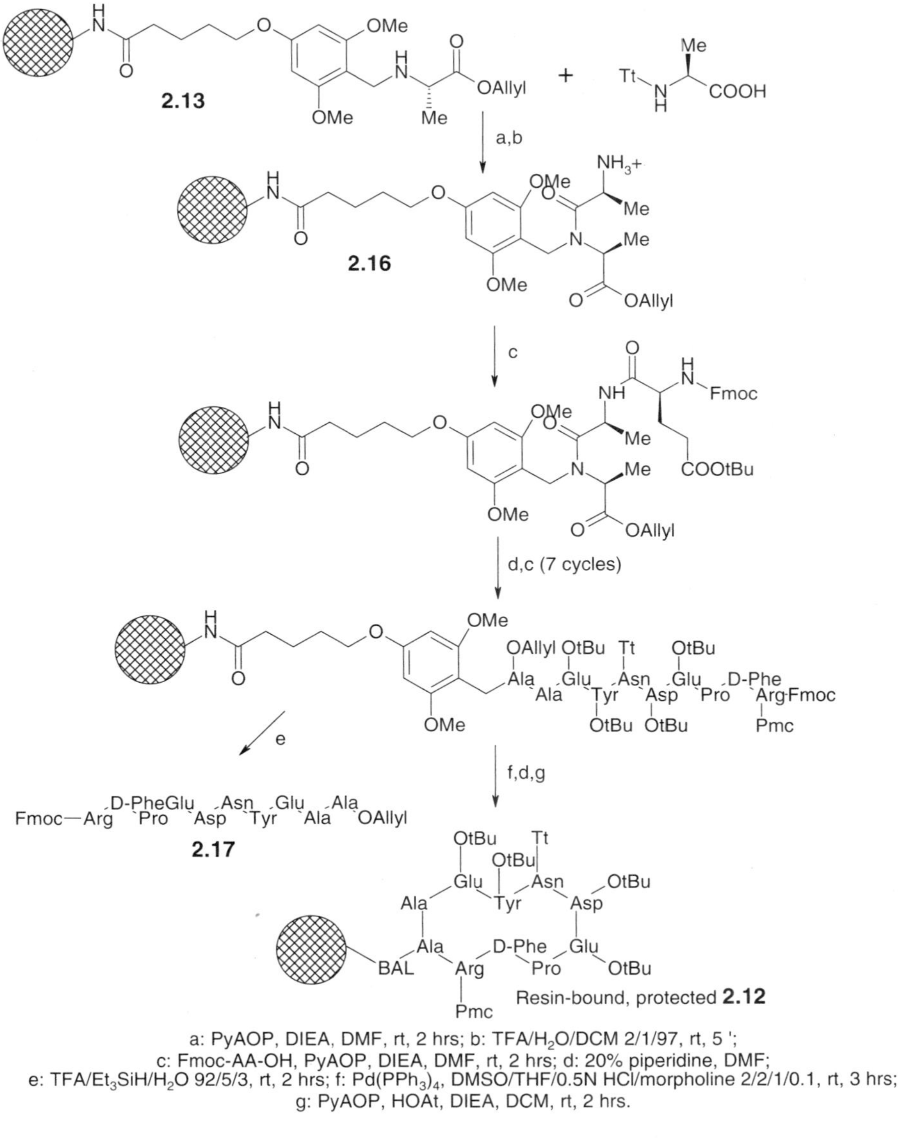

Figure 2.8 SPS of resin-bound fully protected cyclo(Arg-D-Phe-Pro-Glu-Asp-Asn-Tyr-Glu-Ala-Ala). **2.12** and release of deprotected **2.17**.

the deprotection gave a free nucleophilic amine that intramolecularly attacked the allyl ester to produce the unwanted diketopiperazine **2.15**. Although the use of a *t*-butyl ester prevented diketopiperazine formation, the need for a protection for the C-terminus that was removable under mild conditions and was orthogonal to the protected acid side chains did not suggest this selection.

It was found that the trityl (Tt) group was suitable for the amine in that it could be removed selectively after coupling to give a protonated amine **2.16** that was incapable of cyclization (Fig. 2.8). The SPS continued with standard Fmoc-based coupling/deprotection cycles using PyAOP [phosphonium tri-1-pyrrolidinyl (3H-1,2,3-triazolo[4,5-b] pyridin-3-yloxy)-hexafluoro phosphate] (69) [a phosphonium salt derived from HOAt (hydroxy benzo triazole)] as coupling agent together with DIEA to improve the reaction yields (8). A sample of the open decapeptide **2.17** was cleaved in greater than 85% and characterized off-bead as the *N*-Fmoc *C*-allyl ester derivative.

The cyclization on SP under a variety of experimental conditions was evaluated by monitoring the amounts of noncyclized residual decapeptide, the total yield of the desired product, and the degree of C-terminal epimerization due to the resin-bound activation of the carboxylic group. The results are summarized in Table 2.1. The best results were obtained using DIEA as base and DCM as solvent while other combinations of solvents and bases produced more extensive racemization (entries b, c, d, e) or significant amounts of open decapeptide (entries c, e).

The synthesis of the cyclic peptide **2.12** on SP was achieved in excellent yield and purity, demonstrating the flexibility of the SPS approach for a complex sequence containing difficult amino acids such as Glu, Asp, and Arg. A new linker strategy suitable for any peptidelike sequence with no problems of stability and able to permit the final modification of the C-terminus was developed. The iterative nature of peptide SPS and the robustness of these oligomeric structures were instrumental in reaching such good-quality results that are easily amenable to automated SPS procedures.

TABLE 2.1 Cyclization of the Linear Decapeptide on SP: Optimization Studies

Cyclization Conditions[a]	Base	Solvent	Noncyclized Peptide[b]	L–D ratio/epimerization
a	DIEA	DCM	ND	88/12
b	DIEA	DMF	ND	75/25
c	2,4,6-triMepyridine	DCM	26%	76/24
d	2,4,6-triMepyridine	DMF	ND	50/50
e	2,4,6-triMepyridine	$CHCl_3$/TFE 7/3	75	78/22

[a]PyAOP–HOAt as coupling agents.

[b]ND = not detectable.

2.2 OLIGONUCLEOTIDES

2.2.1 Solid-Phase Synthesis of Oligonucleotides

Oligonucleotides (ONs) are the other major class of important biopolymers that readily lends itself to automated synthesis on SP and as such has also received a great deal of attention, the first attempts (70) dating back to the mid 1960s. While the SPS of this class of natural oligomers has the same general advantages seen for peptides, the different properties of the two oligomeric backbones lead to different choices in terms of supports, linkers, protecting group strategy, and cleavage conditions.

There are three widely used chemistries for oligodeoxynucleotide synthesis, and all have been successfully transferred onto SP. The first is the so-called phosphotriester approach (71), which is shown in Fig. 2.9. In this method a 5′-protected nucleoside

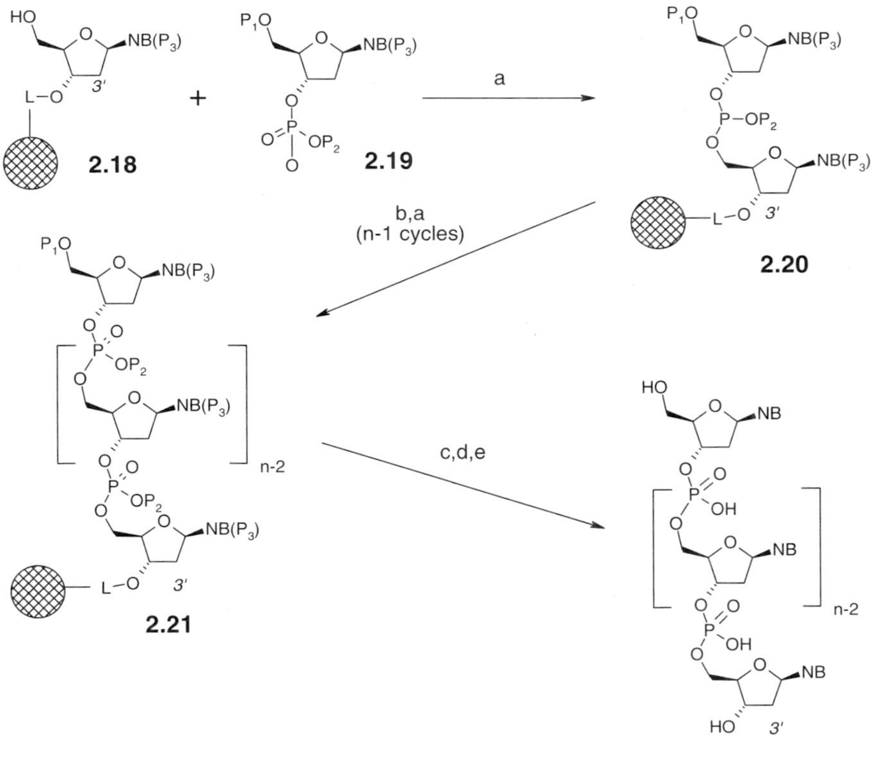

L = LINKER

a: sulfonyl chloride, 1-methylimidazole (cat); b: P_1 deprotection; c: P_2 deprotection; d: P_3 deprotection; e: cleavage from the support.

Figure 2.9 Phosphotriester protocol for oligonucleotide SPS.

phosphate diester **2.19** is reacted with a 3′-resin-bound nucleoside **2.18** in the presence of an appropriate sulfonyl chloride and 1-methyl imidazole as a catalyst to give the dinucleoside phosphate triester **2.20**. Deprotection of the 5′-protecting group gives the bound 5′-OH substrate ready for the next coupling reaction, and so on, until the fully protected target resin-bound oligonucleotide **2.21** is obtained. The phosphate ester protecting group (P_2, Fig. 2.9) is hydrolyzed in basic conditions, the protecting groups on the bases are removed, and finally, the nucleotide is cleaved from the resin. This method, which is still used sometimes for specific applications in SPS (72), has been superceded by the phosphonate and the phosphoramidite routes (see below).

The phosphonate route (73), which is shown in Fig. 2.10, involves the coupling of a 5′-protected nucleoside hydrogen phosphonate **2.22** with the 3′-resin-bound nucleoside **2.18** in the presence of adamantoyl chloride as activating agent to give the 5′-protected dinucleoside phosphonate **2.23**. The cycle continues with 5′-deprotection followed by coupling with another nucleoside ($n - 1$ cycles). The final phosphonate is oxidized with iodine in a basic environment to give the fully protected target ON **2.24**. The free oligonucleotide is obtained after deprotection of the nucleotide bases and cleavage from the solid support. This protocol has been successfully automated on SP (74), but the oxidation step is sometimes problematic and the total yields are generally lower than using the phosphoramidite routes, thus confining this method to short oligonucleotide sequences.

While the SPS phosphonate route is used for specific applications (75, 76), the phosphoramidite method (Fig. 2.11) is the most used method for the assembly of oligonucleotides both in solution (77) and on SP (78). The 5′-protected nucleoside phosphoramidite **2.25**, which is readily prepared from a 5′-protected 3′-hydroxy nucleoside and an amino chloromethoxyphosphine, is coupled with the resin-bound nucleoside **2.18** using tetrazole as an activating agent. The resulting phosphite is oxidized with *t*-butyl peroxide to produce the phosphite triester **2.26**. This compound is iteratively extended by deprotection/coupling/oxidation cycles to produce a fully protected ON **2.27** that is subjected to final deprotections before cleavage from the support. This method has been reviewed recently (79–81) and will now be considered in depth.

Many solid supports including macroporous resins (82), cellulose (83), and hydrophobic (84) and hybrid hydrophilic PS resins (72) have been used for the preparation of ONs. Early experiments (85) highlighted the incompatibility of hydrophobic supports with the chemistry used, and among the hydrophilic supports examined the fully inorganic silica-based CPG supports (78) rapidly became the most popular. They are nonswelling materials (see Section 1.1.3) and are ideally suited to continuous-flow SPS. A large variety of pre-packed CPG columns are now commercially available and are regularly employed for the automated synthesis of ONs with more than 100 residues. Over time, the optimization of the CPG structure and of the linker between the nucleoside and the silica core (86–88) has produced supports with pore sizes of 500–1000 Å and fully extended spacers of considerable lengths (>25 atoms). A typical example is the so-called long-chain alkylamine (LCAA) CPG support **2.28** (89) shown in Fig. 2.12. A recent contribution has introduced a novel polymer colloid supported

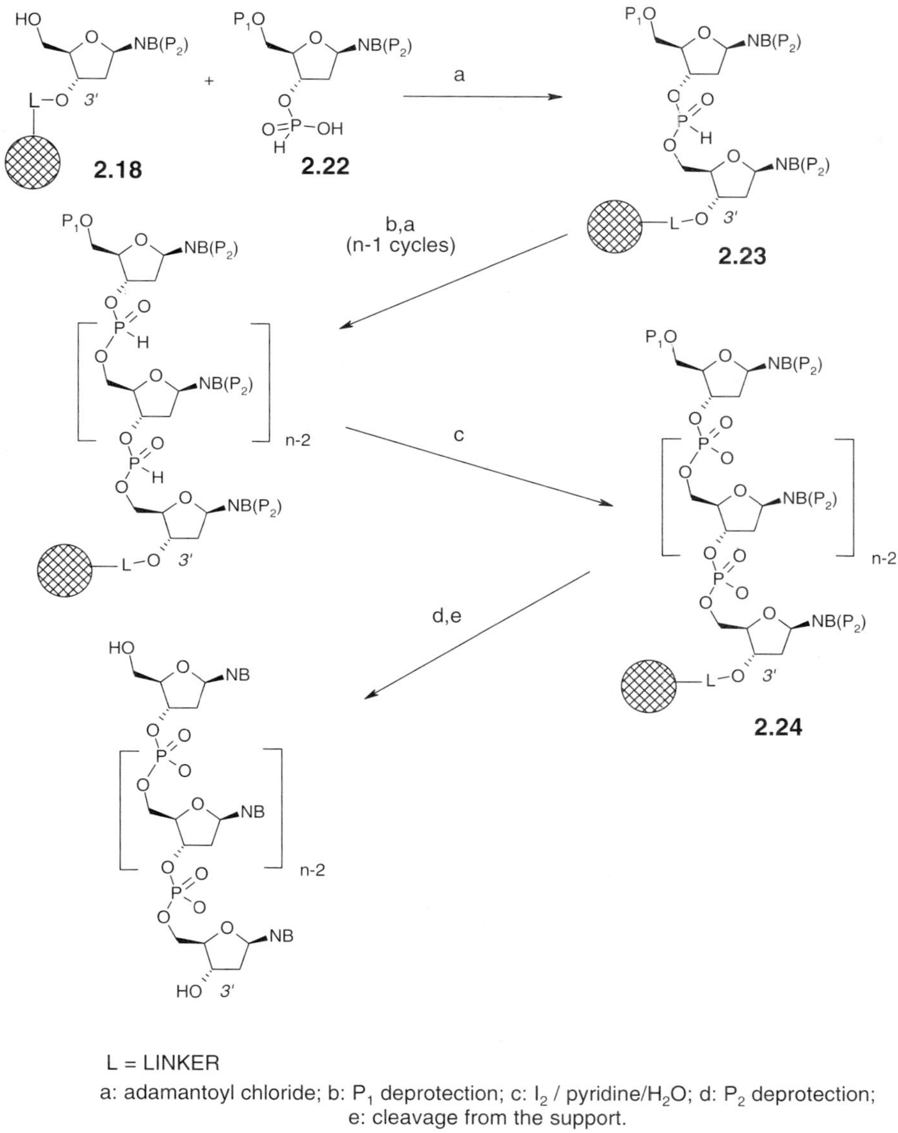

Figure 2.10 Phosphonate protocol for oligonucleotide SPS.

onto a polyethylene filter disk as an alternative support to CPG and resin beads in ON SPS with at least similar synthetic performances (90).

In contrast to SP peptide synthesis, the low acid stability of the ON backbone requires a different linker and chemistry strategy and the growing ON is attached to the support via base-labile bonds. Some examples of these linkers are shown in Fig.

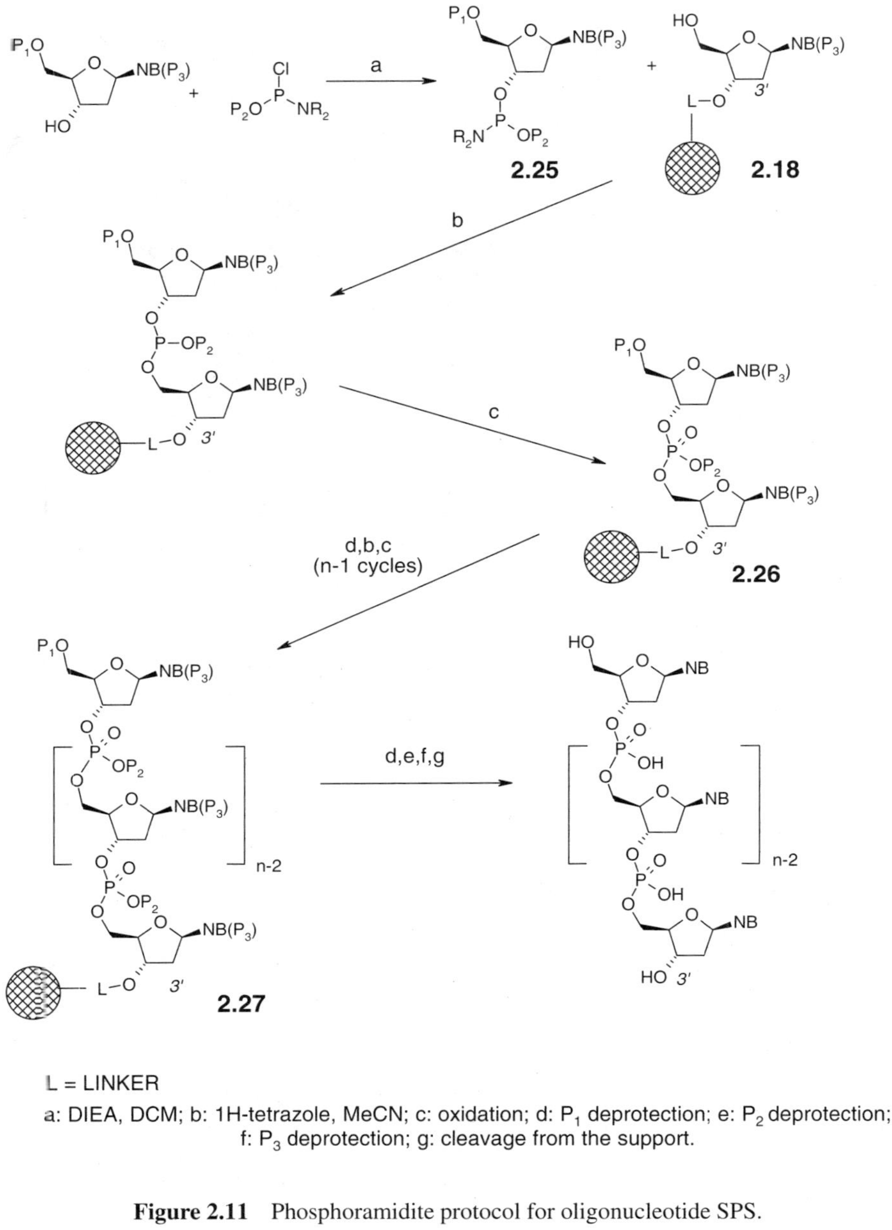

L = LINKER

a: DIEA, DCM; b: 1H-tetrazole, MeCN; c: oxidation; d: P_1 deprotection; e: P_2 deprotection; f: P_3 deprotection; g: cleavage from the support.

Figure 2.11 Phosphoramidite protocol for oligonucleotide SPS.

2.13 An ester bond is widely used, and the succinate **2.29** (91) is one of the most popular linkers. It is introduced onto the LCAA-CPG support as the 3′-monosuccinate ester of the first nucleoside, and it is cleaved under basic conditions in the final SPS step. The inclusion of a sarcosine spacer as in **2.30** (92) increases the linker stability

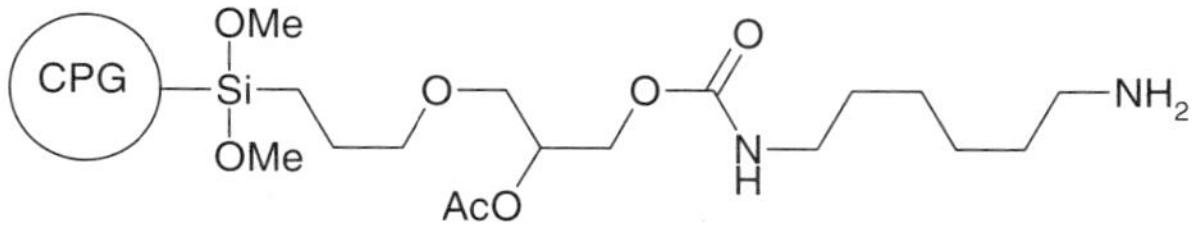

2.28 LCAA-CPG

Figure 2.12 LCAA CPG support **2.28** for oligonucleotide SPS.

toward base and allows the use of Fmoc-based chemistry at the 5′-position. A similar linker **2.31**, in which the succinyl moiety is replaced by the oxalyl group, is also available (84) and can be cleaved in the final step with 5% methanolic ammonia. The much more robust carbamate linker **2.32** (93) requires concentrated ammonium hydroxide at 56 °C for 2 days to achieve complete release of the target ON.

Four recent examples of universal linkers/supports, in which the first nucleoside is anchored onto the preformed linker–support construct, are shown in Fig. 2.14. The disulfide linker **2.33** has been used to prepare terminal 3′-phosphate ONs (94, 95) through cleavage with a solution of ammonia in dithiothreitol. The photolabile linker **2.34** (96) is used to prepare 3′-alkyl carboxylic acids. The allyl-based linker **2.35** (97) is used to prepare free 3′-OH ONs by cleavage with Pd(0) and treatment with an aqueous buffer at pH 10. The linker **2.36** (98) differs from those discussed so far in

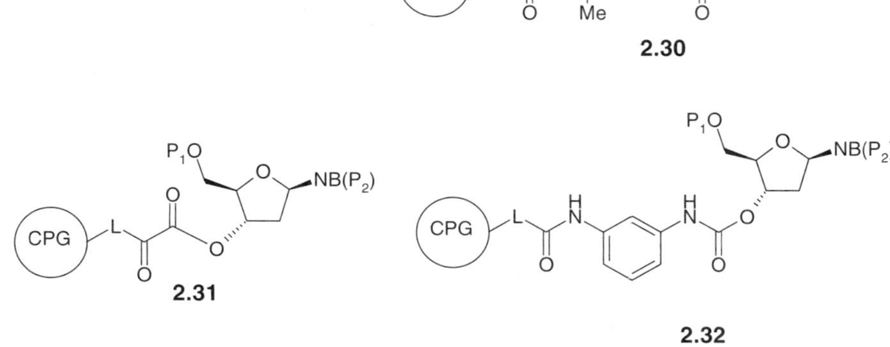

L = LINKER

Figure 2.13 Base-labile linkers **2.29–2.32** for oligonucleotide SPS.

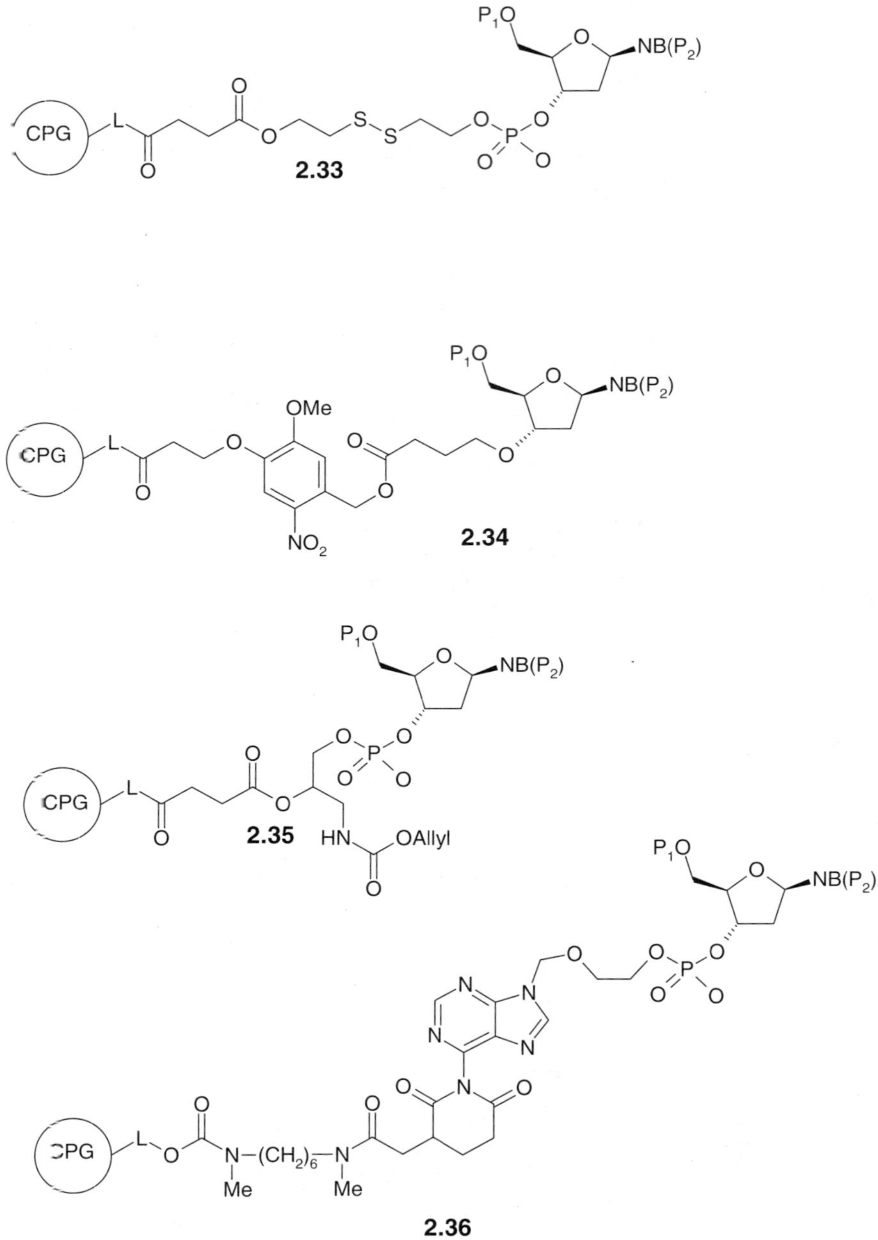

Figure 2.14 Universal linkers **2.33–2.36** for oligonucleotide SPS.

that it links the growing ON through the heterocyclic base and not through the sugar. This allows a flexible strategy for incorporation of modified sugars in the terminal position. Three general supports able to support any starting ON or ODN (oligodeoxyribonucleotides) were recently reported (99–101). The majority of the cleavage procedures for automated ON synthesis use hot aqueous condition bases such as ammonium hydroxide, potassium carbonate, or other bases. A recent paper (102) reported that cleavage could be effected with gaseous ammonia under pressure, thus allowing easier work-up and washing procedures and giving better purity of target ONs.

The most popular protecting group strategy uses trityl-based protecting groups **2.37–2.40**, which are reported in Fig. 2.15. The relative stability of ONs to acidic conditions means that mild acidic deprotection of the trityl-based protecting groups can be employed during the sequence, but stability during the coupling step is also required. The trityl **2.37** and trimethoxytrityl **2.40** groups are not suited to

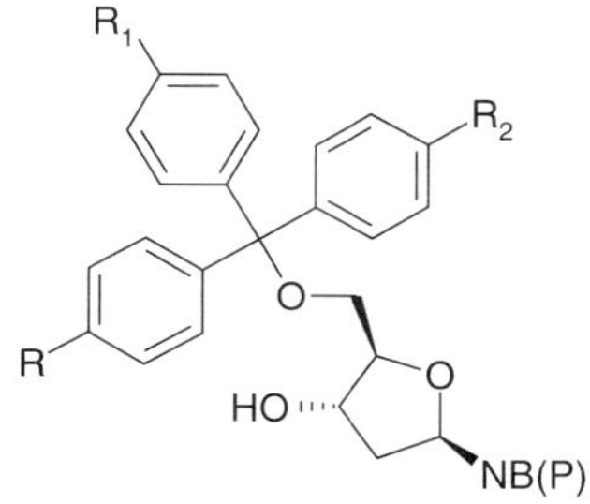

2.37 R=R$_1$=R$_2$=H Trityl (Tt)
2.38 R=OMe, R$_1$=R$_2$=H MeOTrityl (MMTt)
2.39 R=R$_1$=OMe, R$_2$=H DiMeOTrityl (DMTt)
2.40 R=R$_1$=R$_2$=OMe TriMeOTrityl (TMTt)

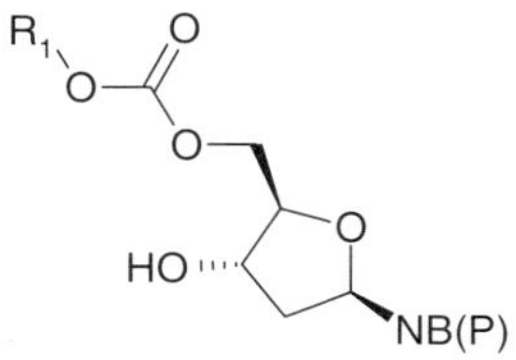

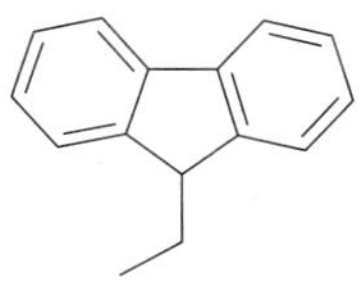

2.41 R$_1$=Fmoc

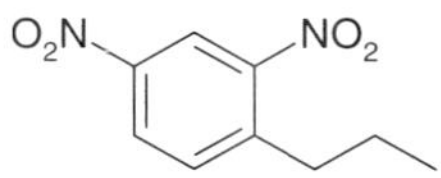

2.42 R$_1$=DNPEoc

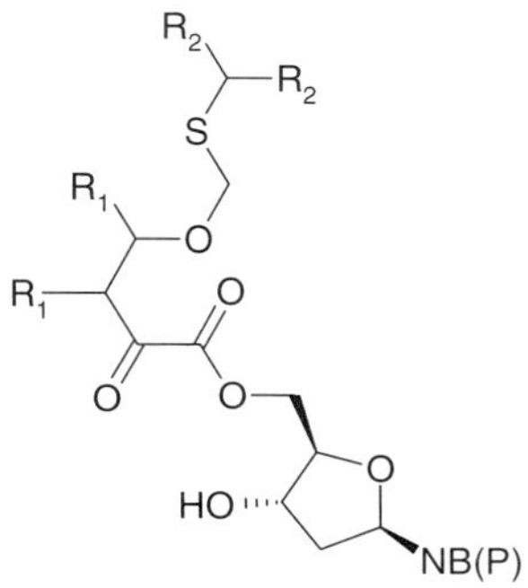

2.43 R$_1$=H, R$_2$=H
2.44 R$_1$=Ph(condensed), R$_2$=H
2.45 R$_1$=Ph(condensed), R$_2$=Me

Figure 2.15 5′-protecting groups **2.37–2.45** for oligonucleotide SPS.

SP ON synthesis, being too stable and unstable, respectively. However, the monomethoxytrityl (MMt) **2.38** and the dimethoxytrityl (DMTt) **2.39** protecting groups possess the desired stability characteristics (103) and are used as 5′-protecting groups. In particular, DMTt has rapidly become the 5′-protecting group of choice (104) as it is easily cleaved under a variety acidic conditions, including 80% aqueous AcOH, $BF_3\cdot Et_2O$, and ZnBr in various solvents or solvent mixtures (79). The base-labile 5′-protecting groups sometimes encountered are the carbonates, among them the Fmoc **2.41** (105, 106) and the DNPEoc (dinitro phenethyl carbonyl) **2.42** (107, 108) derivatives. The latter is cleaved by a TEA-promoted β-elimination. Finally, Hg-labile S-containing 5′-protecting groups such as **2.43–2.45** (Fig. 2.15) (109, 110) are more stable to basic hydrolysis but are readily cleaved with mercury salts. For example, **2.45** is cleaved by $Hg(ClO_4)_2$ in a few minutes.

A recent review (79) gives details of other 5′-protecting groups, 2′-protecting groups for RNA-based ONs, and the use of specific protecting groups for nucleobases. Among the more recent developments, *N*-pent-4-enoyl (NPT) amide has been advocated as a universal nucleobase protecting group (111) applied to the building blocks **2.46–2.51** (Fig. 2.16), which are suitable for ON SPS. Protected NPT A, C, and G nucleosides are stable and can be employed in both amidite and phosphonate chemistries for the preparation of standard ONs or modified oligonucleotide analogs. Cleavage is carried out under either oxidative (iodine-base) or basic aqueous/anhydrous conditions. Other recent reports highlight the use of phthaloyl protections for the amine

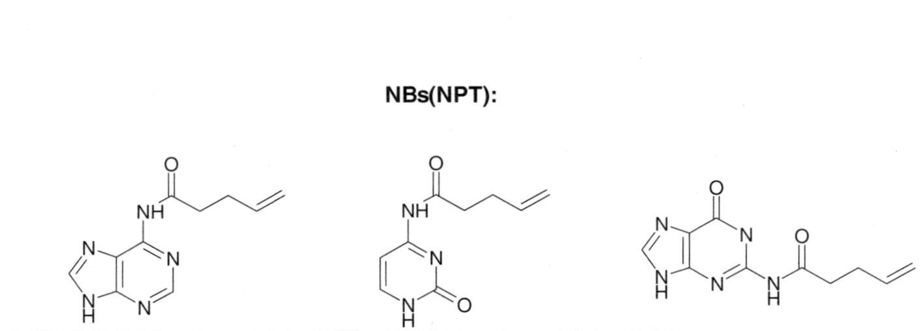

2.46 A-NPT (phosphoramidite) **2.47** C-NPT (phosphoramidite) **2.48** G-NPT (phosphoramidite)
2.49 A-NPT (phosphonate) **2.50** C-NPT (phosphonate) **2.51** G-NPT (phosphonate)

Figure 2.16 Nucleobase NPT-protected building blocks **2.46–2.51** and SP constructs for oligonucleotide SPS.

functions of the nucleobases (112) and of photolabile protections for both sugars and nucleobases (113, 114) in automated RNA and DNA synthesis.

The chemistry of the phosphoramidite coupling has been thoroughly studied, and the use of 1*H*-tetrazole (115) as an activating agent for the reaction (Fig. 2.17) with support-bound **2.52** allows the SPS of long ON sequences with high yields (116). The use of stable and easily handled *N,N*-diisopropylphosphoramidites such as **2.53** (117, 118) is important for the automation of the process (119). The base-labile

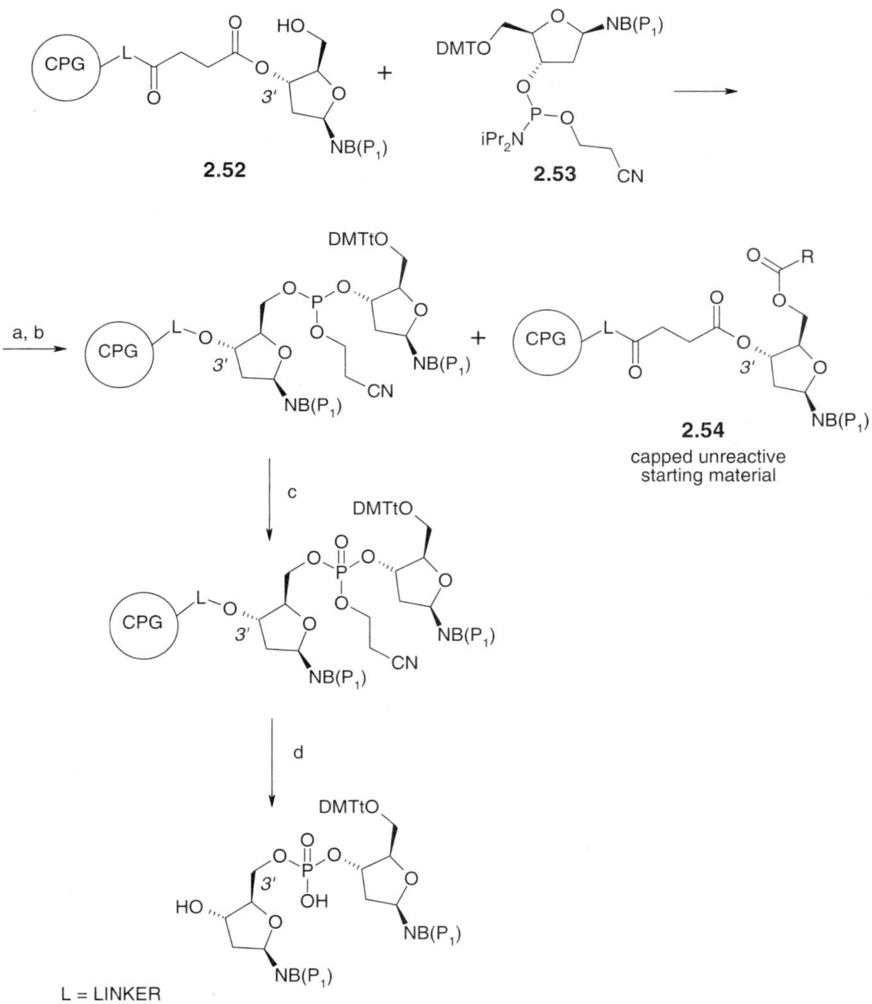

a: 1H-tetrazole; b: capping, $(RCO)_2O$ or RCOCl; c: I_2, pyridine, H_2O; d: aq. NH_4OH, heat.

Figure 2.17 Detailed phosphoramidite-based oligonucleotide SPS protocol.

β-cyanoethoxy protecting group is used for the phosphate function of **2.53** (120) and allows the simultaneous deprotection of this function and cleavage of the ON from the supports in the final step; other protected phosphoramidites were recently reported (121). Capping, that is, quenching of unreacted termini with either an acid anhydride or an acid chloride after each coupling cycle (step b, **2.54**, Fig. 2.17), terminates all the sequences that would produce shortened ONs (122). Even if the yield of the coupling is typically >98%, the introduction of a capping step significantly improves the quality of the SPS and is especially important for long ON sequences. The oxidation with aqueous iodine is performed after rather than before each capping step, reducing further the amount of impurities (112).

The monitoring of a reaction and the methods used in determination of the structures of the products are influenced by the nature of the nonswelling CPG support, which does not allow the acquisition of NMR spectra in the solid phase. Other on-bead or off-bead methods presented in Sections 1.3 and 1.4 are more suited to the characterization and monitoring of the synthesis. When DMTt is used as the 5′-protecting group, the reaction can be monitored by quantitating the release of the bright orange trityl carbocation in solution by spectrophotometry at 498 nm (104) in a similar manner to the Fmoc method used in peptide synthesis.

2.2.2 Solid-Phase Synthesis of Oligonucleotide-Related Oligomers

There are several drawbacks to ONs as potential drugs that seriously limit their potential therapeutic applications; however, their potential to be selective and potent agents acting on specific targets is very high. Major hurdles for the exploitation of ONs are their susceptibility to endo- and exonucleases, their poor solubility, and their poor pharmacokinetic properties.

Much effort has been expended toward the stabilization of ONs to render them more bioavailable while maintaining levels of selectivity and potency in binding to complementary nucleotide strands. A recent review (123) covers the preparation of analogues containing either modified nucleobases such as **2.55–2.57** or modified sugars **2.58–2.60** (Fig. 2.18). Newer approaches have included replacement of the sugars with a 4-hydroxy-*N*-acetylprolinols **2.61** (124) and 2′-*O*,3′-C-linked [3.3.0] or [3.2.0] bicycloarabino sugars such as **2.62** (125) and **2.63** (126). All of these analogues are accessible by SP and possess interesting biological and physicochemical properties. Recently several phosphoramidite building blocks suitable to introduce nucleobase- or sugar-modified ON building blocks were reported; among them 2′-*C*-β-methylcytidine (127), 4,8-dihydro-4-hydroxy-8-oxo-2′-deoxyguanosine (128), several 7-substituted 7,8-dideazaadenines (129) and (5′*S*,6*S*)-5′,6-cyclo-5,6-dihydro-5,6-dihydrothymidine (130).

Postsynthetic transformation of the nucleobases has also been reported using modified building blocks containing *o*-nitrobenzyl-protected acids or amines that after SP assembly and photolysis, could be selectively functionalized (131) or applying Pd-catalyzed cross-coupling reactions in carefully controlled reaction conditions on support-bound, preassembled ONs (132). Postsynthetic glycosidation of 5′-hydroxyls was also reported to give sugar-functionalized oligonucleotides (133).

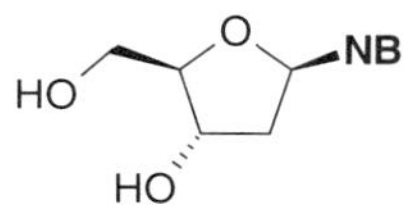

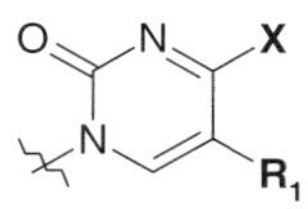

2.55

X=OH, NH_2
R_1=Me, Br, 2-pyridyl, 2-thiophenyl, 2-thiazolyl, 2-imidazolyl, 1-propinyl.

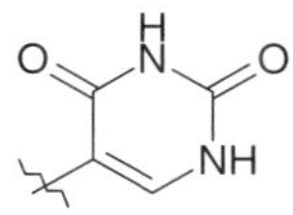

2.56

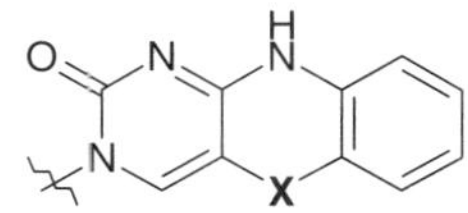

X=O, NH

2.57

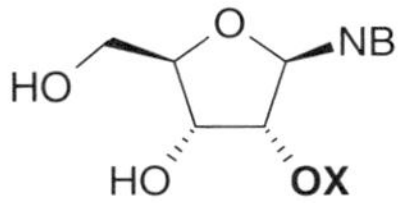

2.58

X=alkyl, subst. alkyl, aryl

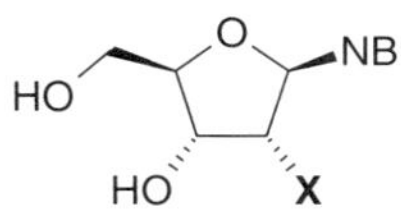

2.59

X=alkyl, subst. alkyl, aryl, alkenyl

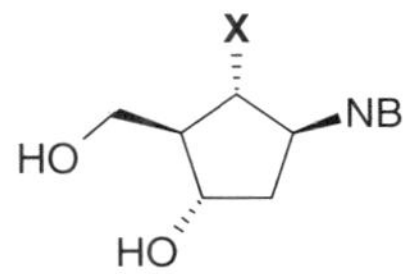

2.60

X=alkyl, OH, O-alkyl

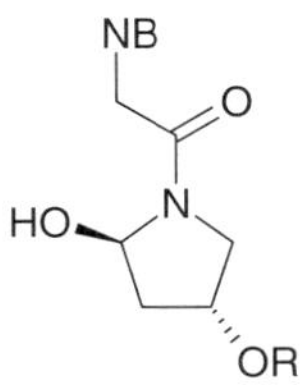

2.61

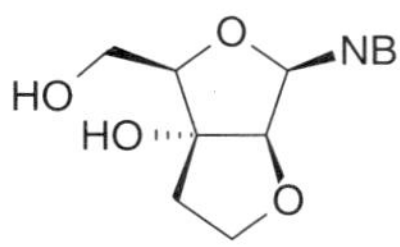

2.62

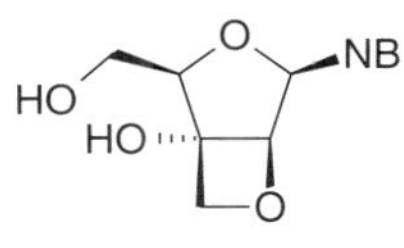

2.63

Figure 2.18 Modified nucleobases (**2.55–2.57**) and sugars (**2.58–2.63**) for the SPS of oligonucleotide analogues.

Modifications of the phosphodiester backbone have also been reviewed (123) and include phosphorus-containing replacements such as **2.64–2.66** (Fig. 2.19). Among these examples, the phosphorothioates **2.65** (134) and the phosphoramidates **2.66** (135) are particularly interesting as biologically active oligonucleotide analogues; several groups are working to improve the automated SPS of these and similar backbone-modified ONs (136, 137). The synthesis of a 15-mer phosphorothioate conjugate is reported in Section 2.2.3 as an example.

Cyclic ONs of small–medium size using the phosporamidite protocol to grow the ON chain and simple cyclization protocols to join the precursor ends were reported

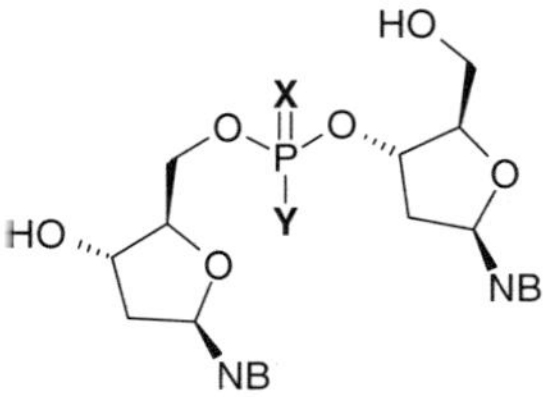

2.64

X=O, S Y=Me, NHR, OR, S

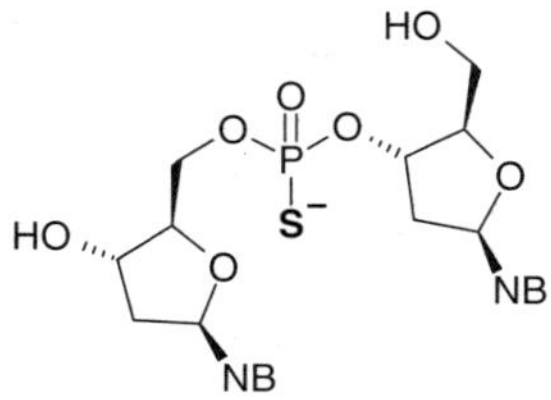

2.65

phosphorothioates

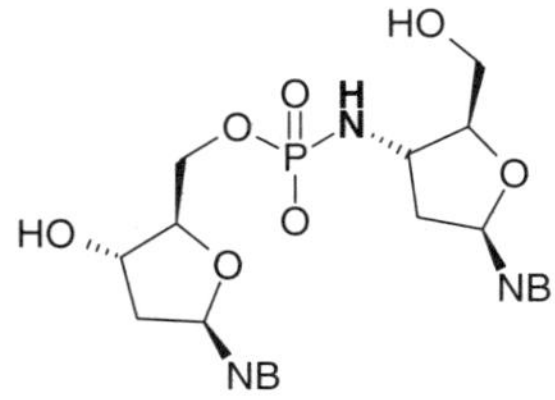

2.66

phosphoramidates

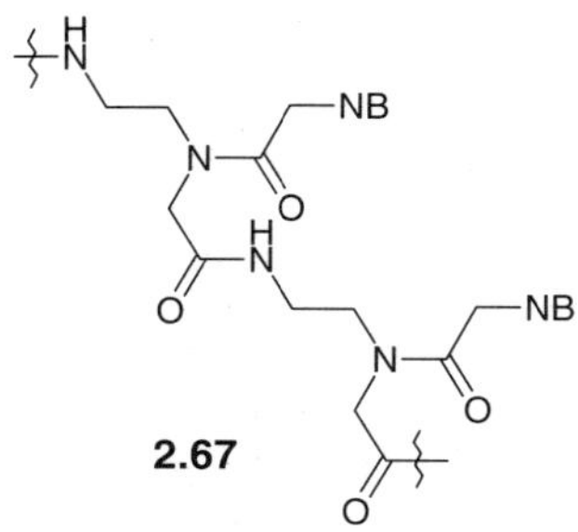

2.67

peptide nucleic acids
PNA

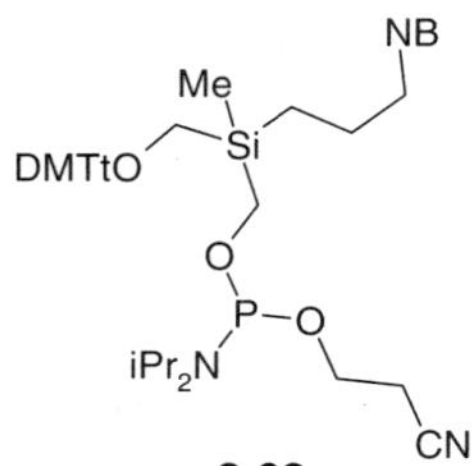

2.68

Figure 2.19 Phospodiester backbone replacements (**2.66–2.68**) commonly used for the SPS of oligonucleotide analogues.

(138, 139). The so-called peptide nucleic acids (PNAs) **2.67** (140–142) are examples of the replacement of the entire sugar–phosphate backbone with different functionalities that retain biological activity. The chemistry is similar to SP peptide synthesis employing Boc or Fmoc protocols (see Section 2.1). Peptide–ON hybrids, where either one (143) or two (144) of the ON termini are attached to peptide chains, have been reported. Acyclic silicon phosphoramidites **2.68** (145) have also been used for SPS of silicon-containing ON analogues.

A recent review (146) examines the features of several synthetic biopolymers mimicking the structures of both peptides and nucleotides and critically analyzes the future opportunities for each unnatural oligomeric family of compounds.

2.2.3 An Example: Synthesis of a Bioreversible ODN PS Conjugate

The synthesis of the target ON conjugate 5′-XTCTCACTACCTCTT (X = **2.72**, Fig. 2.20) was performed using the SP phosphoramidite protocol with PNT (N-pent-4-enoyl)-protected phosphoramidites **2.69** and **2.70**, the unprotected phosphoramidite **2.71**, and the conjugated phosphoramidite **2.72** (147). Their structure and synthesis (76) from natural nucleosides is reported in Fig. 2.20. The protection of the base was

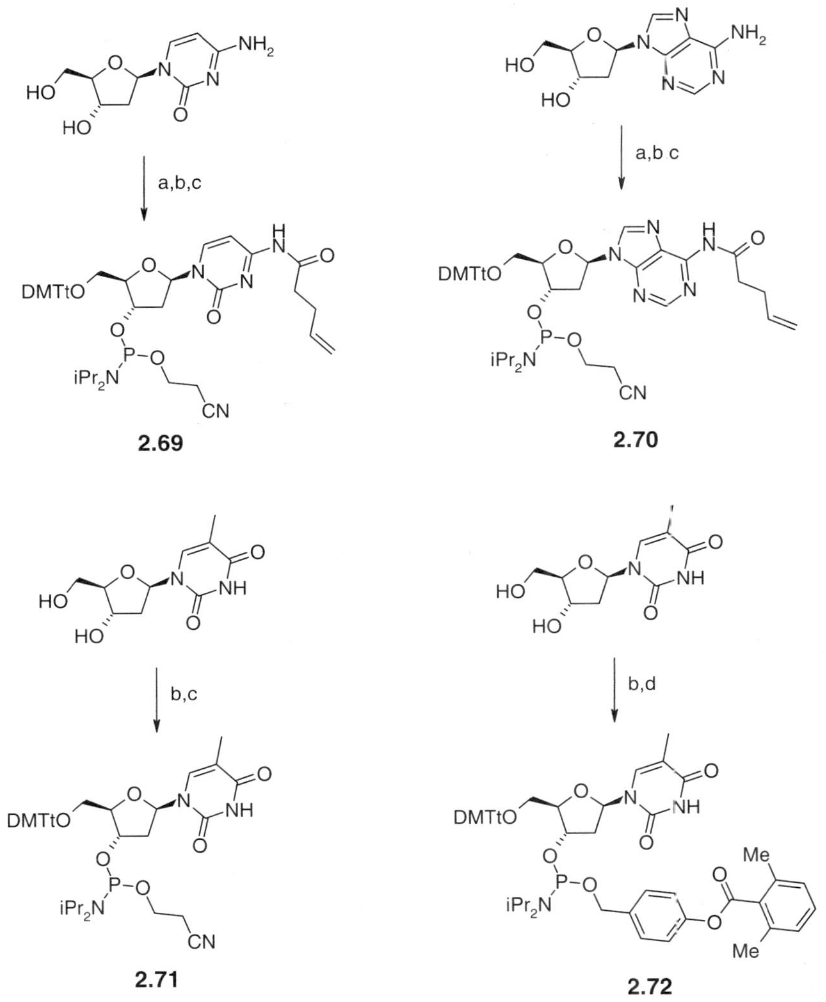

a: TMSCl, pyridine, pent-4-enoyl anhydride, 3 hrs, rt;
b: DMTt-Cl, pyridine, 1.5 hrs, rt; c: amidite, TEA, DCM, 3hrs, rt;
d: di-iPr phosphorus dichloride, 4-(2,6-dimethylbenzoyloxy)-benzyl alcohol, TEA, DCM, 40', 0°C to rt.

Figure 2.20 Structure and synthesis of nucleotidic building blocks **2.69–2.72**.

carried out through the corresponding anhydride (step a), followed by protection of the 5′-OH with DMTt-Cl (step b). The protected nucleoside was phosphitylated with a suitable chlorophosphoramidite (step c) to give the building blocks **2.69–2.70**. The thymidine-based building block **2.71** did not require nucleobase protection and was prepared by submitting thymidine to steps b and c. The conjugated phosphoramidite

2.72 was prepared from 5′-protected thymidine (step b) in a one-pot reaction with diisopropyl phosphorus dichloride and 4-(2,6-dimethyl)benzoyloxybenzyl alcohol (step d).

The SP assembly of the ON 15-mer was performed on a 1-μmol scale using an automated synthesizer. The succinate was coupled to the LCAA-CPG support (500 Å) to give **2.73** (Fig. 2.21), which was submitted to reaction cycles for each elongation. The first elongation cycle is shown in detail in Fig. 2.21 and begins with detritylation of the resin-bound intermediate (step a) followed by a washing step (step b). Coupling

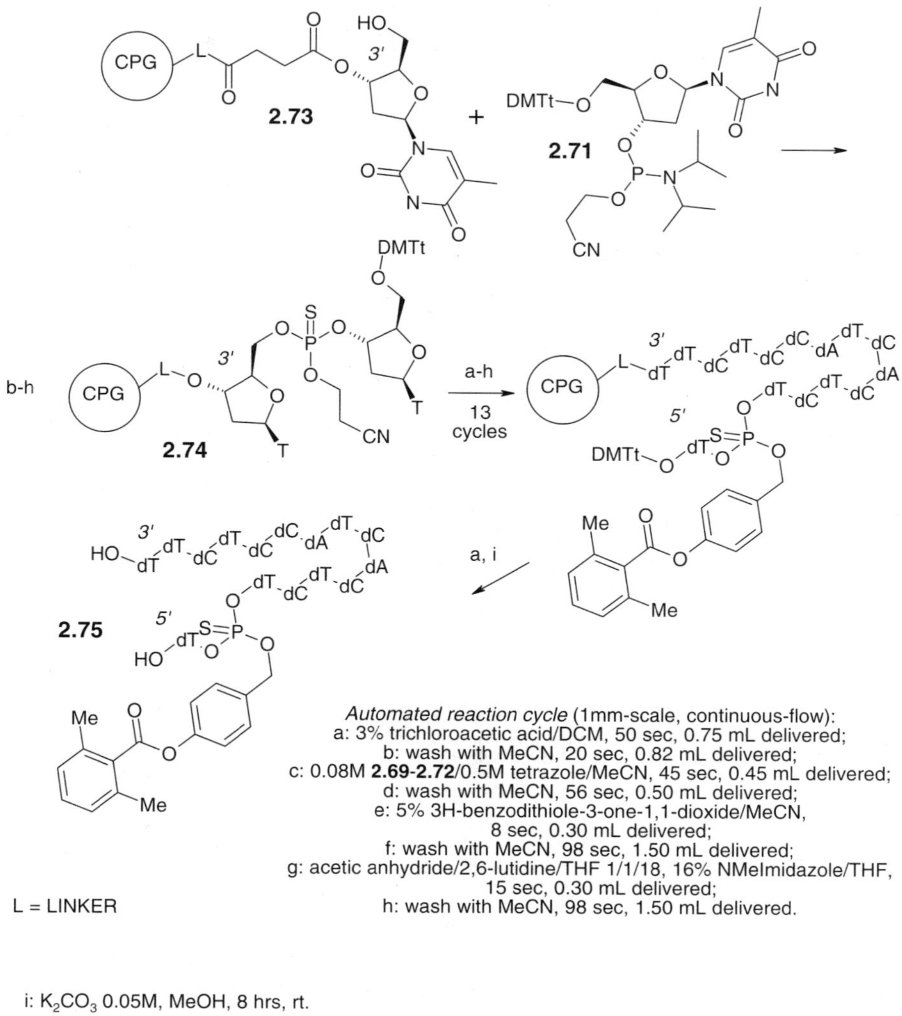

Figure 2.21 SPS of the ON phosphorothioate conjugate 5′-XTCTCACTACCTCTT. **2.75**.

with a suitable phosphoramidite catalyzed by 1*H*-tetrazole (step c), another washing (step d), and then oxidative sulfurization (step e) with 3*H*-benzodithiole-3-one-1,1-di-oxide (148), again followed by washing (step f), gave the elaborated oligonucleotide **2.74**. The unreacted sites on the support were capped (step g) and the resin was washed (step h) prior to iteration for 13 cycles. The conditions described were readily applied to automated continuous-flow SPS allowing for an extremely rapid and reliable preparation of the resin-bound target molecule (147). The simultaneous deprotection of the PNT group and the β-cyanoethyl group and cleavage from the solid support were finally carried out under mild conditions by treatment with methanolic potassium carbonate to give the target **2.75** (step i, Fig. 2.21).

The examples of SPS presented here and in Section 2.1.3 take advantage of multistep procedures that are routinely performed using an automated SP–peptide or SP–oligonucleotide synthesizer with reduced expenditure on optimization of conditions. Similar results with limited effort can be obtained for any oligomeric SPS (e.g., peptides, peptoids, oligonucleotides, oligoureas), providing that the repeating units form a backbone that is stable and robust during all of the synthetic steps. If the target molecule is not made up of repeating units but must be prepared by multistep SPS using diverse chemical reactions, the complexity of the SPS increases and the optimization of the chemistry on the SP becomes both crucial and time consuming, as we will see in the following chapter.

2.3 OLIGOSACCHARIDES

2.3.1 Solid-Phase Synthesis of Oligosaccharides

The SPS of oligosaccharides (OSs), like that of peptides and oligonucleotides, started in the early 1970s (149, 150) with the demonstration that the coupling between monosaccharide units was feasible on a solid support using different protocols (151–153). The higher complexity of the sugar scaffold, though, prevented the development of a reliable SP procedure for the assembly of carbohydrate building blocks in a regio- and stereospecific manner. In general, the multiple reactive sites and the stereochemical complexity have often hampered OS synthesis and, as of this writing, have prevented the growth of an automated technology for the SPS of OSs.

These limitations notwithstanding, the past 20 years have witnessed major achievements in SP OS synthesis, which now allows the reliable assembly of small oligomers (up to four to five saccharide units) in most of the cases with good yields and purity. A few glycosylation technologies for oligomer growth on SP that use different functionalities are available and among them the chemist may choose according to the needs of the specific synthesis; for example, the glycal (154), the sulfoxide (155), the trichloroacetylimidate (156), and the enzyme-based technology (Fig. 2.22) (157) have all been successfully employed. We will briefly describe the salient points of each method and show their usefulness through a critical review of their features with regard to the formation of α- or β-anomers, coupling with hydroxyls having different

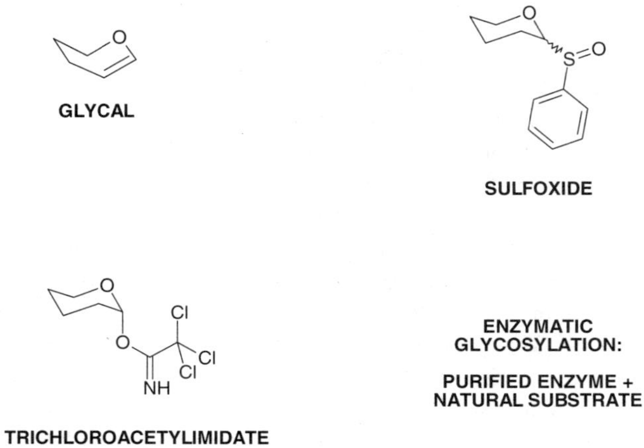

Figure 2.22 Common SP glycosylation reagents and methods.

reactivities and protecting groups that are compatible with their use (Fig. 2.22, bottom); a list of major issues to consider for OS SPS is reported in Fig. 2.23.

The glycal technology was first reported on SP by Danishefsky et al. (154) following a typical procedure reported in Fig. 2.24. The supported glycal **2.76**, where C_6–OH is used to link the glycal to the resin, was prepared in three steps from PS resin and was then epoxidized with DMDO (dimethyl dioxirane) to give the resin-bound glycosyl donor **2.77**. Treatment of **2.77** with a solution of a suitably protected glycosyl acceptor **A** in the presence of anhydrous zinc chloride produced the β-(1→6) galactosyl disaccharide **2.78**, which was further elaborated to the tetrasaccharide **2.79** using the same chemistry with different glycosyl acceptors (**A** and **B**, Fig. 2.24). Each free C_2–OH that originates from any glycosylation could be kept unprotected and eventually be used as a further reaction site (*vide infra*). The galactosyl anhydrosugar proved to be a suitable donor even for hindered glycosyl acceptors, but the glucosyl analogue proved to be unsuitable for all of the desired transformations due to side reactions, especially when coupled to less reactive acceptors. The elaboration of the resin-bound glycal **2.80** to a protected thioethyl donor **2.83** (158) via the anhydroglucose **2.81**

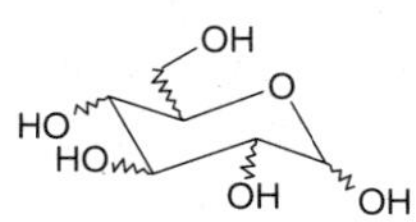

- anomeric carbon / stereochemistry?
- OH reactivities / regiochemistry?
- orthogonal protecting groups?
- support/linkers/cleavage compatible with OS structure?
- common, automation-friendly protocols?

Figure 2.23 Major issues in SP OS synthesis.

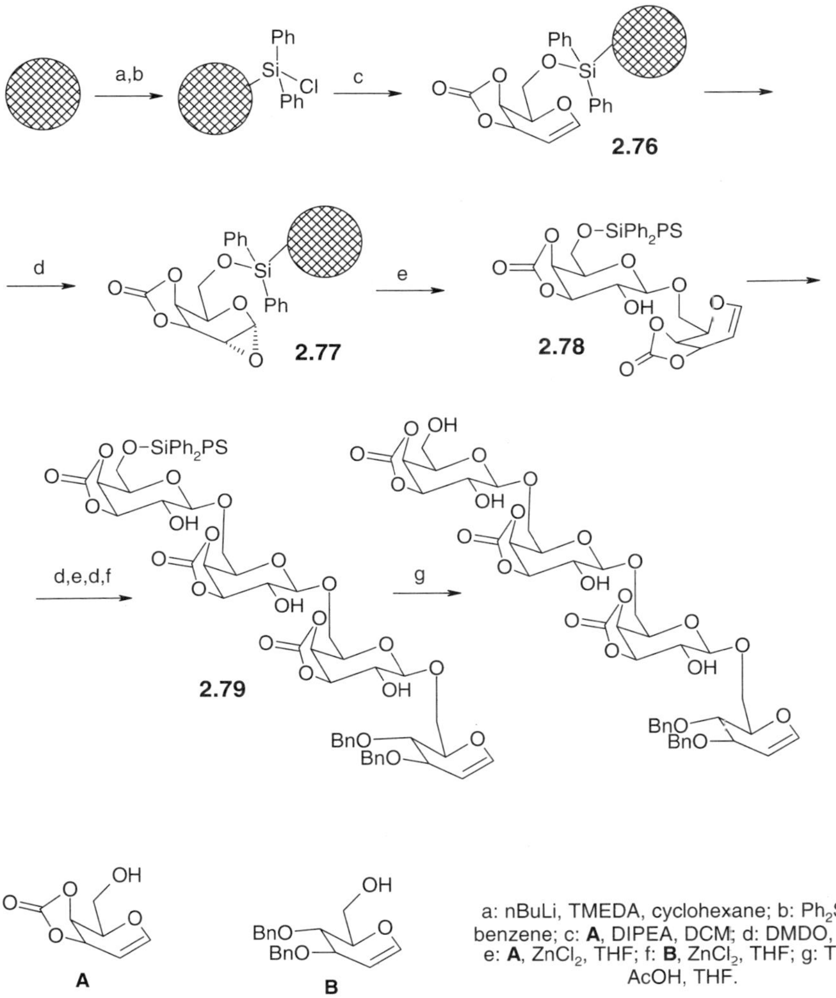

a: nBuLi, TMEDA, cyclohexane; b: Ph_2SiCl_2, benzene; c: **A**, DIPEA, DCM; d: DMDO, DCM; e: **A**, $ZnCl_2$, THF; f: **B**, $ZnCl_2$, THF; g: TBAF, AcOH, THF.

Figure 2.24 SP glycal protocol: SP synthesis of **2.79**.

allowed the formation of the α-glucosidic linkage **2.84** (Fig. 2.25). The functionalization of the less reactive C2–OH by using the thioethyl donor in solution after removal of the pivaloyl protecting group on SP was also used to give the branched trisaccharide **2.86** (Fig. 2.25). Removal of the protecting groups used during the synthesis and cleavage from the support produced the free C6–OH saccharide (*vide infra*). The glycal technology thus allows the construction of diversified oligosaccharides with various possible connections (159), including *N*-linked glycopeptides (160) and even branched oligomeric chains (161, 162). Unfortunately the α-stereochemistry at the anomeric center cannot be obtained as major glycosylation products with these protocols (163).

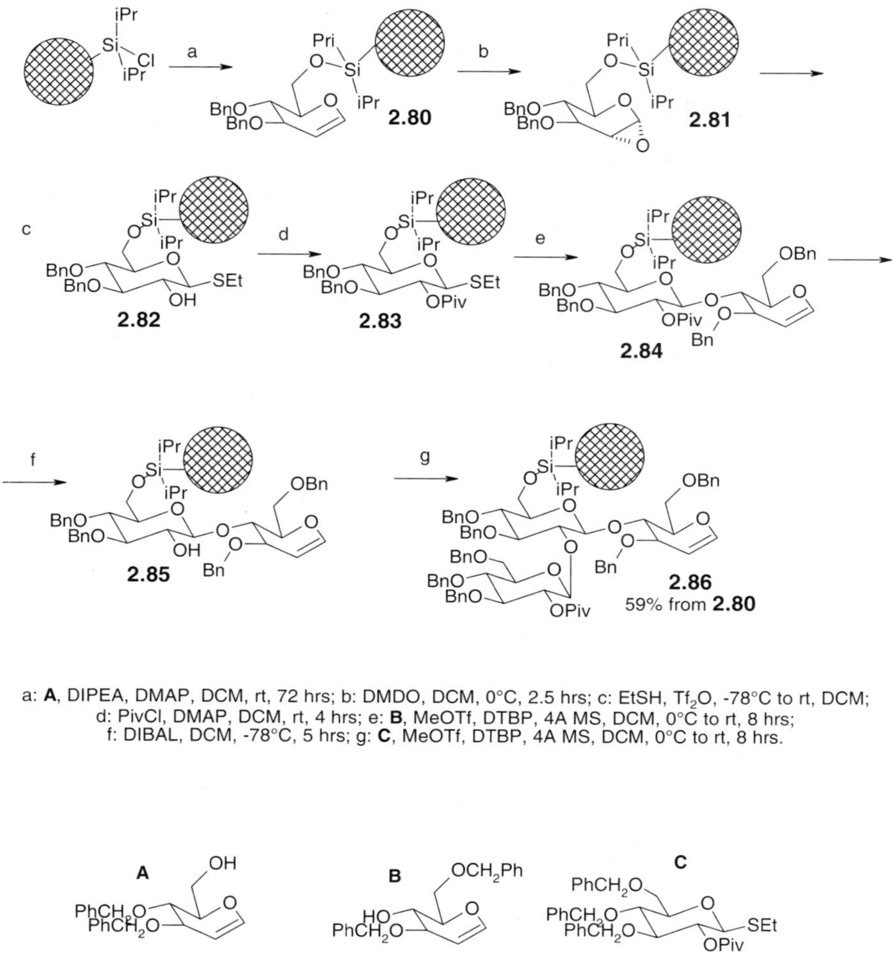

Figure 2.25 SP glycal protocol: thioethyl donors for the synthesis of **2.86**.

The sulfoxide technology was first reported by Yan et al. (155); its main features are shown schematically in Fig. 2.26. The resin-bound glycosyl acceptor **2.87**, easily prepared from Merrifield PS resin and the *p*-hydroxythiophenyl glycoside (164), was coupled with two different donors **A** and **B** in the presence of triflic anhydride and a hindered base. The two couplings produced respectively the α- and the β-linked disaccharides **2.88** and **2.89** in good yield. The presence of a C2-neighboring group able to participate in the glycosylation, such as the pivaloyl in **B**, assured the β-stereochemistry while the absence of such a group combined with low reaction temperature (–60 °C) invariably gave α-selectivity with secondary alcohols. Removal

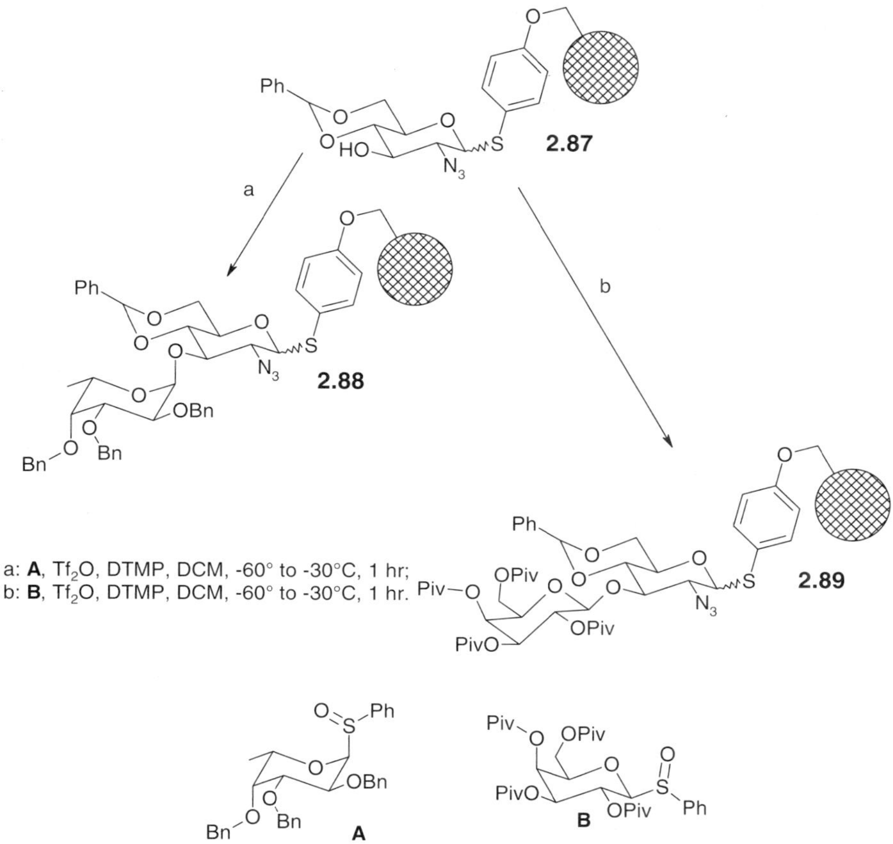

Figure 2.26 SP sulfoxide protocol: synthesis of **2.88** and **2.89**.

of the protecting groups and cleavage from the support produced the free reducing saccharide as an anomeric mixture (*vide infra*). Other SP sulfoxide-based syntheses have been reported (165), including an SP library preparation (166), and the method appears to be both flexible (e.g., both α-, β-, and diversely structured oligosaccharides can be prepared) and reliable.

The trichloroacetimidate route was first reported on soluble supports by Douglas et al. (167) and then on PS supports by Rademann et al. (156, 168) with the general scheme shown in Fig. 2.27. The polymer-bound protected acceptor **2.90** was prepared from the corresponding trichloroacetimidate **A** and an SH-functionalized PS resin. After deacetylation it was coupled with **A** in the presence of trimethylsilyl trifluoromethanesulfonate as catalyst to give the 1,2-*trans* α-dimannoside **2.92**, which was extended to the pentamannoside **2.93** by repeating three glycosylation/deacetylation cycles (Fig. 2.27). The free reducing saccharide or its methyl derivative (*vide infra*) was produced after deprotection and cleavage. Several other reports of trichlo-

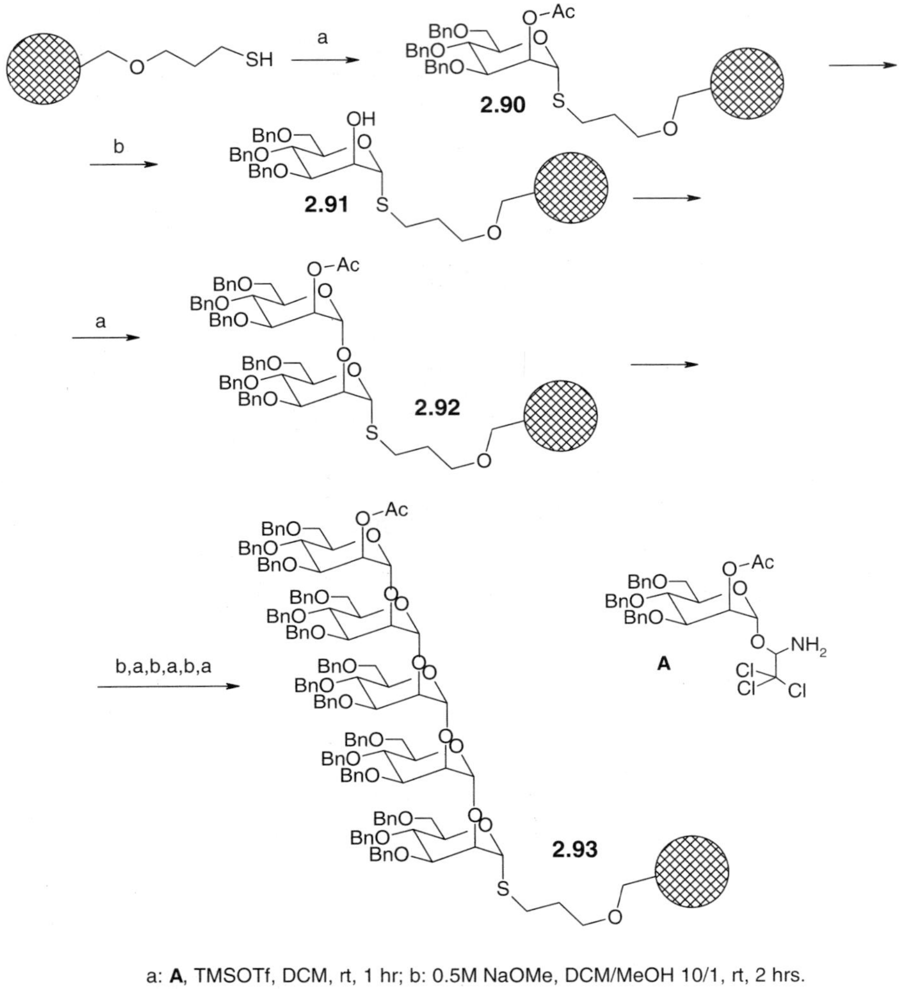

Figure 2.27 SP trichloroacetimidate protocol: synthesis of **2.93**.

roacetimidate SP glycosylation followed (169, 170) where both α- and β-stereochemistries were introduced in the same complex oligosaccharide structure, thus proving this method to be versatile enough for various SP synthetic applications.

Enzymatic glycosylation on SP was first reported by Schuster et al. (157) and since then has been used to perform multiple glycosylations. Two examples are reported in Fig. 2.28. The synthesis of oligosaccharides **2.94** and **2.95** related to the sialyl Lewis X antigen was carried out via glycosylations mediated by β-1,4-galactosyltransferase and α-2,3-sialyltransferase (171) and by β-1,4-galactosyltransferase, α-2,3-sialyltransferase, and fucosyltransferase (172), respectively. The use of unprotected saccha-

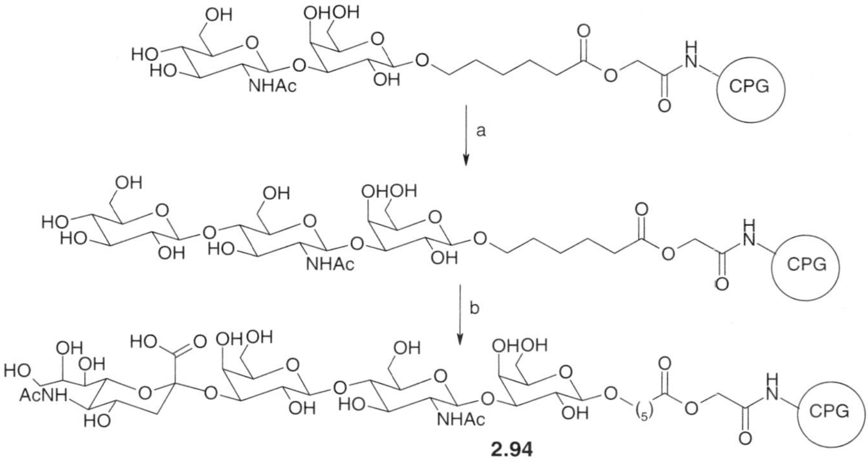

2.94

a: UDP-Galactose, β-(1-4)galactosyltransferase; b: CMP-Neu5Ac, α-(2-3)sialyltransferase.

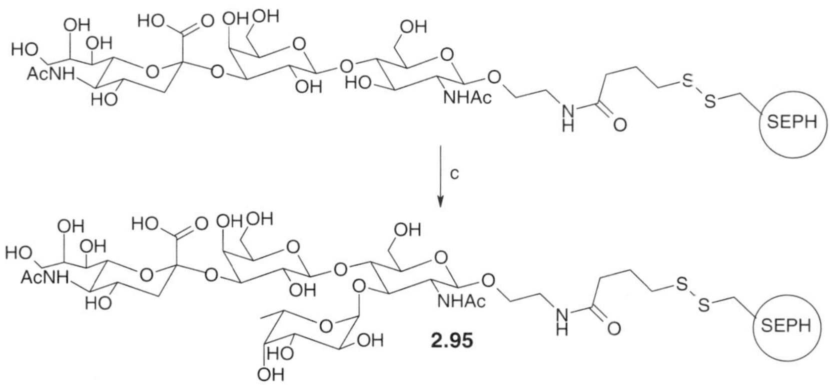

2.95

c: GDP-fucose, α-(1-3/4)fucosyltransferase

Figure 2.28 SP enzymatic glycosylation protocol: synthesis of **2.94** and **2.95**.

ride substrates, the specificity of the enzymes, and the availability of complex structures, which are currently too challenging for organic chemistry, are the main advantages of this method. However, the limited availability of the necessary enzymes, the constraints of using natural substrates for each purified enzyme, and the slow kinetics of SP enzymatic glycosylations are the drawbacks that confine this approach to specific applications where even kilograms of complex saccharide derivatives have been obtained (173).

A variety of solid supports have been used for OS SPS, including PS resins (155, 161, 170, 174), PS crowns (175), soluble PEG supports (167, 176, 177), CPG (169,

171), and sepharose (172). Each OS SP assembly method has preferentially adopted a specific support, most probably because of the expertise of the leading research group in the field. A variety of linkers have also been used and some of them are shown in Fig. 2.29 (**2.96–2.100**) and 2.30 (**2.101**, **2.102** and **2.105**). The Si-based linker **2.96** (156, 158–162) was prepared in three steps from PS resin and the glycal derivative **A**, then finally cleaved by fluoridolysis to give the pure α or β anomeric saccharide. Another similar, commercially available Si linker **2.97** used in SP (see Chapter 1, Fig. 1.8, **1.10**) was used in OS SPS (178) and cleaved with AcOH–THF–H_2O cocktails at 65 °C recovering pure disaccharides in good yields. The Si-based linker **2.98** (179) was prepared from silyl alcohol **B** (180) and trichloroacetimidate **C** in five steps and finally cleaved with acetic anhydride–Lewis acid to give the corresponding α,β anomeric acetate. The S-based linker **2.99** (155, 164) was prepared from Merrifield resin and a suitable *p*-hydroxythiophenyl glycoside **D** (164) and finally cleaved by mercury salts to give the α,β anomeric free oligosaccharide. The photolinker **2.100** (181) was prepared in two steps from Merrifield PS resin and the β-phenolic ester **E** and finally cleaved by irradiation at rt to give the pure α or β anomeric phenoxy ester.

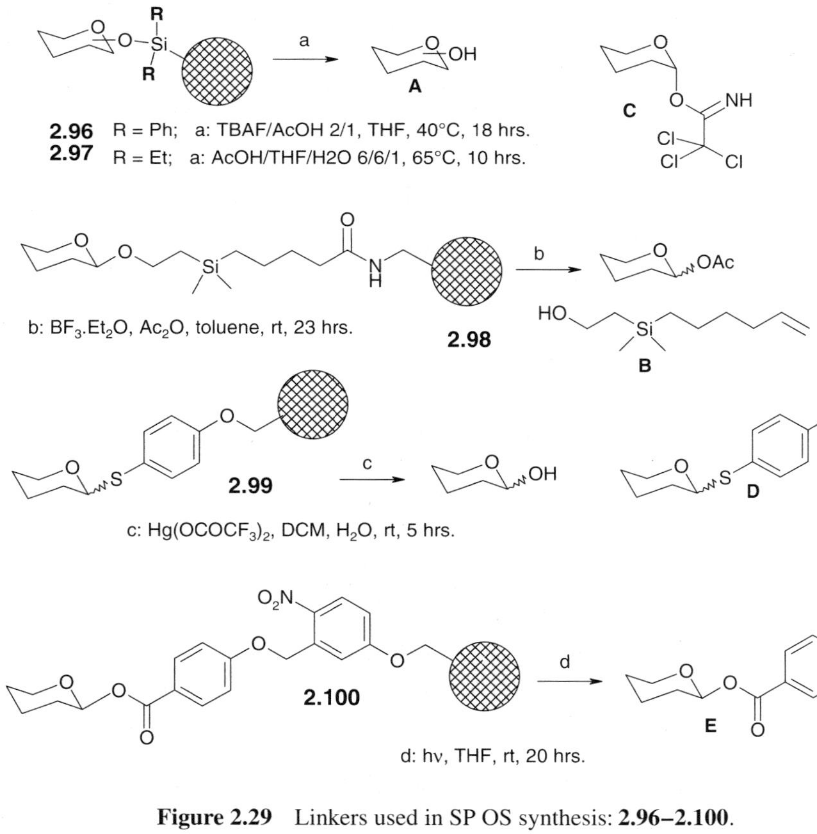

Figure 2.29 Linkers used in SP OS synthesis: **2.96–2.100**.

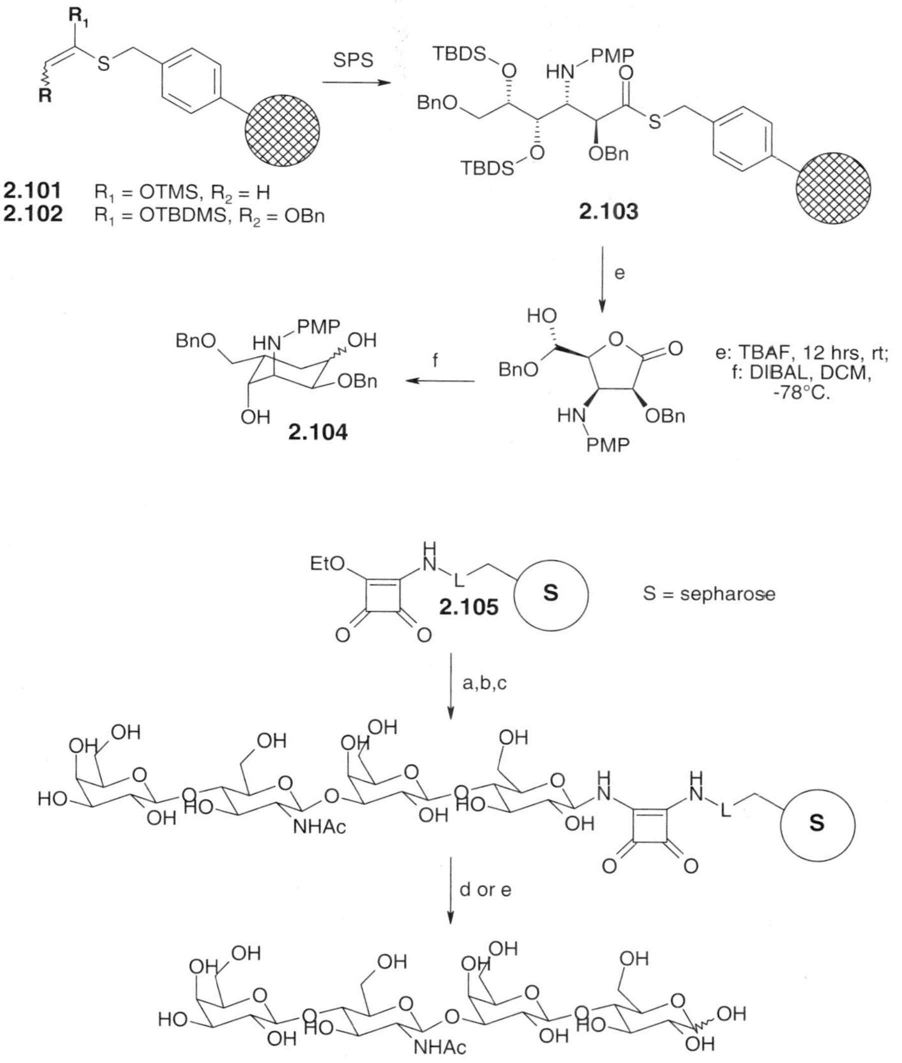

a: SP glycosylation (SP disaccharide); b,c: enzymatic SP glycosylations (to SP tetrasaccharide); d: Br_2, H_2O, 10', rt, then $NaBO_3$, ph 8, 1 hr, rt; e: $NaBO_3$, NH_4OH, ovn., rt.

Figure 2.30 Linker used in SP OS synthesis: **2.101**, **2.102** and **2.105**.

An intriguing report dealing with various monosaccharides made on SP and cleaved using a cyclative cleavage (CC) strategy was based on the silyl enol ether linkers **2.101** and **2.102**, easily obtained from thiol PS resin via acylation and O-alkylation; their SP elaboration to structures such as **2.103** was followed by cleavage using TBAF via TBDS (tert-butyl dimethyl silyl) deprotection and lactone formation to give, after reduction, the sugar **2.104** (182). The squarate linker **2.105** (183), prepared on Sepharose beads, was both chemically and enzymatically glycosylated to give a tetrasaccharide that was cleaved in good yield and purity either with aqueous bromine or with ammonia/ammonium borate.

A variety of protecting groups have also been used for the saccharide hydroxyls, including esters (acetate, benzoate, pivaloate), ethers (benzyl), silyl groups (TBDPS (tert-butyl diphenyl silyl)), carbonates, and benzylidene acetals; protecting groups tailored for the specific needs of OS SP have also been reported (184). Their careful selection as orthogonal protections for different OHs and their general compatibility with the above-mentioned and other SP linkers have allowed the successful SP synthesis of many complex oligosaccharides.

Reaction monitoring and structure determination have often used the more reliable off-bead analytical characterization. Isomeric oligosaccharides have the same MW, thus hampering MS techniques, and their NMR spectra are quite complex. Nevertheless, examples of significant ^{1}H MAS NMR spectra of resin-bound saccharides have been reported (185).

Several other glycosylation technologies have recently been reported, among them the methods by Nicolaou et al. (174, 181), Ito et al. (177), and Rodebaugh et al. (175) are worth mentioning and will probably become more widely used in the future. As a general comment, the extreme complexity of OS synthesis has required the development of a number of choices for the chemistry that are complementary rather than exclusive. The assessment of one, or more likely a few, universal SP routes for OS SP synthesis and their subsequent automation appears today to be an extremely ambitious, challenging, and eventually obtainable target, considering the increasing interest of key academic and industrial groups in SP oligosaccharide synthesis.

Several recent reviews (186–189) can provide the reader with more detail regarding all the previously examined aspects of oligosaccharide SPS. These reviews also cover the major aspects of oligosaccharide-related oligomers, which have either already been introduced, as for glycopeptides in Section 2.1.2 or for oligonucleotides in Section 2.2, or which will be described in other chapters, as for sugar-containing natural products described in Chapter 4 or for sugar scaffold analogues described as small organic molecules in Chapters 6–9.

2.3.2 Example: Synthesis of 3,4,6-Tri-*O*-benzyl-α-D-mannopyranosyl 3,4,6-tri-*O*-benzyl-α-D-mannopyranoside and 3,4-Di-*O*-acetyl-2-*O*-benzyl-α-L-fucopyranosyl 1-*O*-methyl-3,4,6-tri-*O*-benzyl-α-D-mannopyranoside

The synthesis of two 1,2 disaccharides **2.106** and **2.107**, with α-mannose and α-fucose linkages, respectively, was reported by Rademann and Schmidt (168) using the trichloroacetimidate technology via the coupling of two different glycosyl donors, **2.109** and **2.110** (Fig. 2.31), to the PS-supported linker **2.108**. The former was chosen because of the α-driving 2-OAc group (neighboring group participation in the glycosylation); the latter was chosen as a more challenging example for the generation of a 1,2-*cis* disaccharide. The synthesis of **2.108** is shown in Fig. 2.32. The two-step

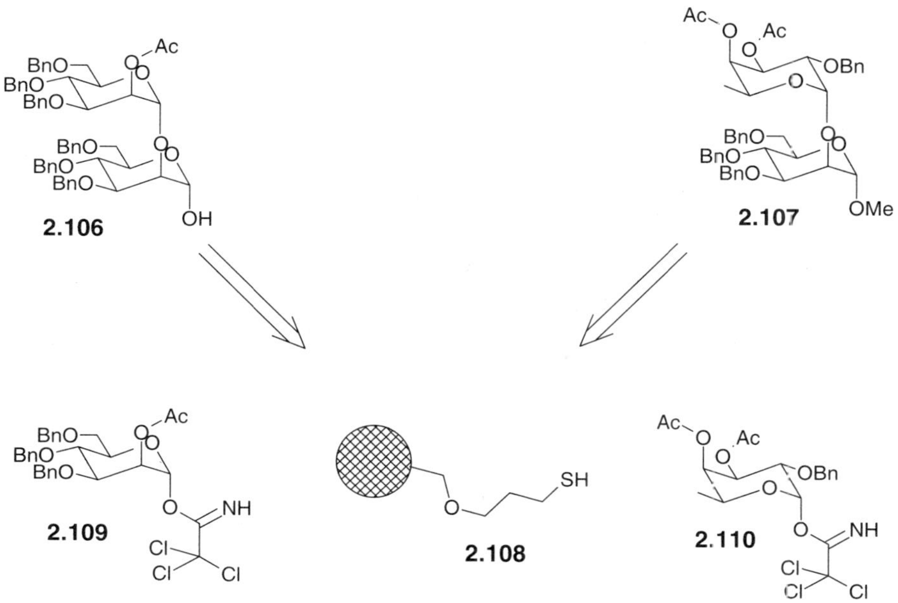

Figure 2.31 Retrosynthetic SP scheme for compounds **2.106** and **2.107**.

preparation of the S-protected linker **A** was followed by its coupling to Merrifield resin and final S-deprotection. The trichloroacetimidate glycosyl donors **2.109** and **2.110** were prepared from commercially available precursors using reported procedures (190, 191).

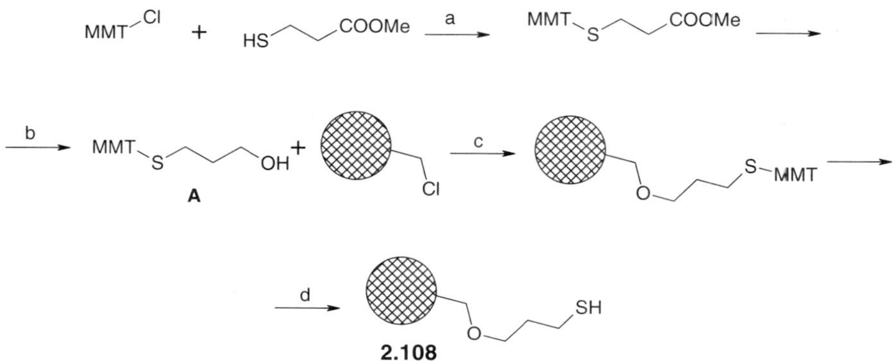

a: pyridine, 14 hrs, rt; b: $LiAlH_4$, 0°C, 2 hrs; c: NaH, 15-crown-5, THF, 60°C, 24 hrs;
d: washings with TFA/DCM 5/95, rt.

Figure 2.32 Synthesis of linker **2.108**.

The synthesis of **2.106**, which was partially described in Fig. 2.27 as a resin-bound intermediate to the pentamannoside **2.93**, is reported in Fig. 2.33. In this scheme, the resin-bound linker **2.108** was placed in a sealed glass tube (Teflon stopper) presaturated with dry nitrogen and the donor **2.109** (3 eq) was injected as a dry DCM solution. The reaction was shaken at rt for 15 min, then trimethylsilyl trifluoromethane sulfonate (TMSOTf, 0.5 *M* in DCM, 0.3 eq) was added and the vessel was stirred at rt for 1 h. The solution was then removed under nitrogen and several washing cycles with dry DCM–dry THF removed the soluble impurities. After drying the resin in vacuo, a small aliquot of the resin-bound protected saccharide **2.90** (1–2 mg) was cleaved (dry silver triflate, DCM–MeOH 10/1, ultrasound mixing, 10 min), mixed with a matrix solution and directly submitted to MALDI–TOF MS in the positive-ion mode to check the purity of the intermediate, which resulted to be excellent. Compound **2.90** was deacetylated with 0.5 *M* NaOMe–MeOH (6 eq) in DCM–MeOH (10/1) as a solvent with stirring at rt for 2 h. The solution was then removed and multiple washings with crown ether–AcOH in THF removed any trace of the deacetylation reagent. The usual DCM–THF washing cycles and drying in vacuo produced the resin-bound, deacetylated **2.91**. The same coupling–purity check cycle was repeated and the resin-bound, acetylated disaccharide **2.92** was finally dissolved into acetone–water 9/1 and oxidatively cleaved (NBS, DTBP, 4 eq) for 90 min at rt. The solution was recovered together with resin washings (THF–DCM), then triethylsilane was added to destroy the unreacted NBS. The solution was azeotroped with toluene and the residue was chromatographed (silicagel flash, then RP_{18} (reverse phase) flash) to give pure **2.106** with a 75% yield calculated on the resin loading of linker **2.108** (Fig. 2.33).

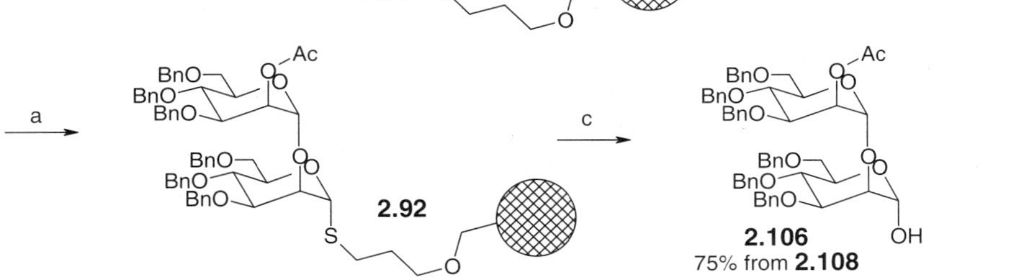

a: **2.109**, TMSOTf, DCM, rt, 1 hr; b: 0.5M NaOMe, DCM/MeOH 10/1, rt, 2 hrs;
c: NBS, DTBP, acetone/water 9/1, rt, 90'.

Figure 2.33 SP synthesis of compound **2.106**.

The synthesis of **2.107** is reported in Fig. 2.34. The first coupling–deacetylation cycle to give resin-bound **2.91** was carried out as above; then this intermediate was coupled with the donor **2.110** (3 eq) and TMSOTf (trimethyl silyl trifluorometan sulfonate) (0.3 eq) at –25 °C using a solvent mixture of dry DCM–dry 1,4-dioxane 1/1 to exploit the 1,2-*cis* driving anomeric effect of ethereal solvents. After stirring at –25 °C for 1 h the suspension was warmed to rt and the resin washed and dried as described previously to give the resin-bound, protected disaccharide **2.111**. This was cleaved using the same reagents and conditions (NBS, DTBP, 4 eq, rt, 90 min) with THF–MeOH as solvent mixture. The pure 1-*O*-methyl disaccharide **2.107** was prepared from **2.108** in 54% yield after chromatography (Fig. 2.34).

This example should have clearly shown the level of complexity involved in the oligomeric SP synthesis of OSs. If we compare it with the optimized, automated SP protocols commonly used to prepare dipeptides (Section 2.1) or dinucleotides (Section 2.2), it is clear that the protected reagents are more precious and must be prepared, rather than bought; the experimental conditions must be carefully controlled to ensure the desired regio- and stereochemical outcome of the reaction; and the cleavage/work-up/purification procedures require more attention to preserve the integrity of intermediates and final compounds. Despite all of this, the clear advantages provided by SPS, especially related to simplified work-up/purification of resin-bound compounds compared to solution procedures, make OS SP extremely appealing to obtain large quantities of complex and sensitive oligosaccharides with high yields and purities. We may reasonably expect significant technological improvements to provide the chemist

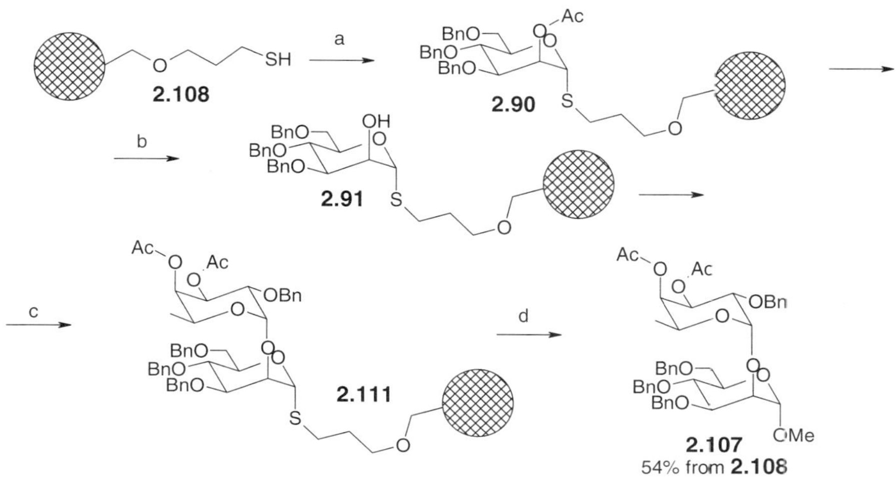

a: **2.109**, TMSOTf, DCM, rt, 1 hr; b: 0.5M NaOMe, DCM/MeOH 10/1, rt, 2 hrs;
c: **2.110**, TMSOTf, DCM, -25°C to rt, 2 hrs; d: NBS, DTBP, acetone/water 9/1, rt, 90'.

Figure 2.34 SP synthesis of compound **2.107**.

with better tools able to provide a large variety of oligosaccharides for the most diverse applications in the future.

REFERENCES

1. Merrifield, R. B., *J. Am. Chem. Soc.* **85**, 2149–2154 (1963).
2. Fields, G. B., *Methods Enzymol.* **289** (Solid-Phase Peptide Synthesis) 1997.
3. Henkel, B., Zhang, L. and Bayer, E., *Liebigs Ann./Recl.,* 2161–2168 (1997).
4. Davies, M. and Bradley, M., *Angew. Chem., Int. Ed. Engl.* **36**, 1097–1099 (1997).
5. Jensen, K. J., Alsina, J., Songster, M. F., Vagner, J., Albericio, F. and Barany, G., *J. Am. Chem. Soc.* **120**, 5441–5452 (1998).
6. Zheng, A., Shan, D. and Wang, B., *J. Org. Chem.* **64**, 156–161 (1999).
7. Davies, M. and Bradley, M., *Tetrahedron* **55**, 4733–4746 (1999).
8. Alewood, P., Alewood, D., Miranda, L., Love, S., Meutermans, W. and Wilson, D., *Methods Enzymol.* **289**, 14–29 (1997).
9. Wellings, D. A. and Atherton, E., *Methods Enzymol.* **289**, 44–67 (1997).
10. Sheppard, R. C. *J. Chem. Br.* **19**, 402–414 (1983).
11. Small, P. W. and Sherrington, D. C. *J. Chem. Soc. Chem. Commun.* 1589–1591 (1989).
12. Atherton, E., Brown, E., Sheppard, R. C. and Rosevear, A., *J. Chem. Soc. Chem. Commun.* 1151–1152 (1981).
13. Bray, A. M., Maeji, N. J. and Geysen, H. M., *Tetrahedron Lett.* **31**, 5811–5814 (1990).
14. Frank, R. and Doering, R., *Tetrahedron* **44**, 6031–6040 (1988).
15. Eichler, J., Bienert, M., Stierandova, A. and Lebl, M., *Pept. Res.* **4**, 296–307 (1991).
16. Orlowska, A., Holodowicz, E. and Drabarek, S., *Pol. J. Chem.* **55**, 2349–2354 (1981).
17. Royo, M., del Fresno, M., Frieden, A., Van Den Nest, W., Sanseverino, M., Alsina, J., Kates, S. A., Barany, G. and Albericio, F., *React. Funct. Polym.* **41**, 103–110 (1999).
18. Alsina, J., Jensen, K. J., Albericio, F. and Barany, G., *Chem. -Eur. J.* **5**, 2787–2795 (1999).
19. Alsina, J., Scott Yokum, T., Albericio, F. and Barany, G., *J. Org. Chem.* **64**, 8761–8769 (1999).
20. Suguro, T. and Yanai, M., *J. Antibiot.* **52**, 835–838 (1999).
21. Albericio, F. and Carpino, L. A., *Methods Enzymol.* **289**, 104–126 (1997).
22. Smith, A. J., *Methods Enzymol.* **289**, 419–426 (1997).
23. Sanchez, A. and Smith, A. J., *Methods Enzymol.* **289**, 469–478 (1997).
24. Beranova-Giorgianni, S. and Desiderio, D. M., *Methods Enzymol.* **289**, 478–499 (1997).
25. Burdick, D. J. and Stults, J. T., *Methods Enzymol.* **289**, 499–519 (1997).
26. Grant, G. A., Crankshaw, M. W. and Gorka, J., *Methods Enzymol.* **289**, 395–419 (1997).
27. Edman, P., *Acta Chem. Scand.* **4**, 283–286 (1950).
28. O'Donnell, M. J., Zhou, C. and Scott, W. L., *J. Am. Chem. Soc.* **118**, 6070–6071 (1996).
29. Kihlberg, J., *Methods Enzymol.* **289**, 221–244 (1997).
30. Perich, J. W., *Methods Enzymol.* **289**, 245–266 (1997).

31. Kates, S. A., Sole, N. A., Albericio, F. and Barany, G., in *Peptides: Design, Synthesis and Biological Activity*, C. Basava and G. M. Anantharamaiah (Eds.). Birkhauser, Boston, MA 1994, pp. 39–58.

32. Blackburn, C. and Kates, S. A., *Methods Enzymol.* **289**, 175–198 (1997).

33. Simon, R. J., Kania, R. S., Zuckermann, R. N., Huebner, V. D., Jewell, D. A., Banville, S., Ng, S., Wang, L., Rosenberg, S. et al., *Proc. Natl. Acad. Sci. USA* **89**, 9367–9371 (1992).

34. Zuckermann, R. N., Kerr, J. M., Kent, S. B. H. and Moos, W. H., *J. Am. Chem. Soc.* **114**, 10646–10647 (1992).

35. Hamper, B. C., Kolodziej, S. A., Scates, A. M., Smith, R. G. and Cortez, E., *J. Org. Chem.* **63**, 708–718 (1998).

36. Rotella, D. P., *J. Am. Chem. Soc.* **118**, 12246–12247 (1996).

37. Burgess, K., Ibarzo, J., Linthicum, D. S., Shin, H., Shitangkoon, A., Totani, R , Zhang, A. J., *J. Am. Chem. Soc.* **119**, 1556–1564 (1997).

38. Bischoff, L., David, C., Roques, B. P. and Fournie-Zaluski, M. C., *J. Org. Chem.* **64**, 1420–1423 (1999).

39. Szekely, Z., Zakhariev, S., Guarnaccia, C., Antcheva, N. and Pongor, S., *Tetrahedron Lett.* **40**, 4429–4432 (1999).

40. Buchinska, T. V., *J. Pept. Sci.* **53**, 314–321 (1999).

41. Krog-Jensen, C., Christensen, M. K. and Meldal, M., *Lett. Pept. Sci.* **6**, 193–197 (1999).

42. Nishiyama, Y. and Kurita, K., *Tetrahedron Lett.* **40**, 927–930 (1999).

43. Ramage, R., Jiang, L., Kim, Y.-D., Shaw, K., Park, J.-L. and Kim, H.-J., *J. Pept. Sci.* **5**, 195–200 (1999).

44. Miranda, L. P. and Alewood, P. F., *Proc. Natl. Acad. Sci. USA* **96**, 1181–1186 (1999).

45. Robertson, N., Jiang, L. and Ramage, R., *Tetrahedron* **55**, 2713–2720 (1999).

46. Angell, Y. My Han, Y., Albericio, F. and Barany, G., *Fifteenth Pept. Proc. Am. Pept. Symp.*, J. P. Tam and P. T. P. Kaumayai (Eds.). Kluwer, Dordrecht, Netherlands, 1999, pp. 339–340.

47. Rinnova, M., Lebl, M. and Soucek, M., *Lett. Pept. Sci.* **6**, 15–22 (1999).

48. Fischer, M. and Tran, C. D., *Anal. Chem.* **71**, 2255–2261 (1999).

49. Rademann, J., Grotli, M., Meldal, M. and Bock, K., *J. Am. Chem. Soc.* **121**, 5459–5466 (1999).

50. Rosenthal, K., Erlandsson, M. and Unden, A., *Tetrahedron Lett.* **40**, 377–380 (1999).

51. Bourne, G. T., Meutermans, W. D. F., Alewood, P. F., McGeary, R. P., Scanlon, M., Watson, A. A., Smythe, M. L., *J. Org. Chem.* **64**, 3095–3101 (1999).

52. Hargittai, B., Han, Y., Kates, S. A. and Barany, G., *Fifteenth Pept. Proc. Am. Pept. Symp.*, J. P. Tam and P. T. P. Kaumayai (Eds.). Kluwer, Dordrecht, Netherlands, 1999, pp. 273–274.

53. Paris, M., Douat, C., Heitz, A., Gibbons, W., Martinez, Y. and Fehrentz, Y.-A., *Tetrahedron Lett.* **40**, 5179–5182 (1999).

54. Milton, S. C. F., De Lisle Milton, R. C., Kates, S. A. and Glabe, C., *Lett. Pept. Sci.* **6**, 151–156 (1999).

55. Kaiser, T., Luppa, P. and Voelter, W., *J. Chromatogr., A* **852**, 189–195 (1999).

56. Sabatino, G., Chelli, M., Mazzucco, S., Ginanneschi, M. and Papini, A. M., *Tetrahedron Lett.* **40**, 809–812 (1999).

57. Cabrele, C., Langer, M. and Beck-Sickinger, A. G., J. *Org. Chem.* **64**, 4353–4361 (1999).

58. Moore, M. L., Newlander, K. A., Bryan, H., Bryan, W. M. and Huffman, W. F., *Fifteenth Pept. Proc. Am. Pept. Symp.*, J. P. Tam and P. T. P. Kaumayai (Eds.). Kluwer, Dordrecht, Netherlands, 1999, pp. 42–44.

59. Kutterer, K. M. K., Barnes, M. L. and Arya, P., *J. Comb. Chem.* **1**, 28–31 (1999).

60. Jezek, J., Velek, J., Veprek, P., Velkova, V., Trnka, T., Pecka, J., Ledvina, M., Vondrasek, J. and Pisacka, M., *J. Pept. Sci.* **5**, 46–55 (1999).

61. Veprek, P. and Jezek, J., *J. Pept. Sci.* **5**, 203–220 (1999).

62. Vazquez, E., Caamano, A. M., Castedo, L, Gramberg, D. and Mascarenas, J. L., *Tetrahedron Lett.* **40**, 3625–3628 (1999).

63. Mayfield, L., Katipally, R., Simmons, C. and Corey, D. R., *Fifteenth Pept. Proc. Am. Pept. Symp*, J. P. Tam and P. T. P. Kaumayai (Eds.). Kluwer, Dordrecht, Netherlands, 1999, pp. 299–300.

64. Hamuro, Y., Marshall, W. J. and Scialdone, M. A., *J. Comb. Chem.* **1**, 163–172 (1999).

65. Poupart, M.-A., Fazal, G., Goulet, S. and Mar, L. T., *J. Org. Chem.* **64**, 1356–1361 (1999).

66. Gennari, C., Mielgo, A., Potenza, D., Scolastico, C., Piarulli, U. and Manzoni, L., *Eur. J. Org. Chem.* 379–388 (1999).

67. Kates, S. A., Sole, N. A., Johnson, C. R., Hudson, D., Barany, G. and Albericio, F., *Tetrahedron Lett.* **34**, 1549–1552 (1993).

68. Albericio, F., Kneib-Cordonier, N., Biancalana, S., Gera, L., Masada, R. I., Hudson, D. and Barany, G., *J. Org. Chem.* **55**, 3730–3743 (1990).

69. Albericio, F., Cases, M., Alsina, J., Triolo, S. A., Carpino, L. A. and Kates, S. A., *Tetrahedron Lett.* **38**, 4853–4856 (1997).

70. Letsinger, R. L. and Mahadevan, V., *J. Am. Chem. Soc.* **88**, 5319–5324 (1966).

71. Letsinger, R. L. and Ogilvie, K. K., *J. Am. Chem. Soc.* **91**, 3350–3355 (1969).

72. Kehler, J., Pueschl, A. and Dahl, O., *Tetrahedron Lett.* **37**, 8041–8044 (1996).

73. Garegg, P. J., Regberg, T., Stawinski, J. and Stroemberg, R., *Chem. Scr.* **25**, 280–282 (1985).

74. Agrawal, S. and Tang, J.-Y., *Tetrahedron Lett.* **31**, 7541–7544 (1990).

75. Kung, P. P. and Jones, R. A., *Tetrahedron Lett.* **33**, 5869–5872 (1992).

76. Brill, W. K.-D., *Tetrahedron Lett.* **36**, 703–706 (1995).

77. Letsinger, R. L. and Lunsford, W. B., *J. Am. Chem. Soc.* **98**, 3655–3661 (1976).

78. Matteucci, M. D. and Caruthers, M. H., *J. Am. Chem. Soc.* **103**, 3185–3191 (1981).

79. Beaucage, S. L. and Iyer, R. P., *Tetrahedron* **48**, 2223–2311 (1992).

80. Beaucage, S. L. and Iyer, R. P., *Tetrahedron* **49**, 1925–1963 (1993).

81. Beaucage, S. L. and lyer, R. P., *Tetrahedron* **49**, 10441–10448 (1993).

82. Kaplan, B. E. and Itakura, K., in *Synthesis and Applications of DNA and RNA,* S. A. Narang (Ed.). Academic Press, Orlando, FL, 1997, pp. 9–23.

83. Ott, J. and Eckstein, F., *Nucl. Acids Res.* **12**, 9137–9142 (1984).

84. Alul, R. H., Singman, C. N., Zhang, G. and Letsinger, R. L., *Nucl. Acids Res.* **19**, 1527–1532 (1991).

85. Koster, H., *Tetrahedron Lett.* 1527–1530 (1972).

86. Katzhendler, J., Cohen, S., Rahamim, E., Weisz, M., Ringel, I. and Deutsch, J., *Tetrahedron* **45**, 2777–2792 (1989).

87. Van Aerschot, A., Herdewijn, P. and Vanderhaeghe, H., *Nucleosides Nucleotides* **7**, 75–90 (1988).

88. Efcavitch, J. W., *Biophosphates and their Analogues—Synthesis, Structure, Metabolism and Activity,* K. S. Bruzik and W. J. Stec (Eds.). Elsevier, Amsterdam, Holland, 1987, pp. 85–104.

89. Gough, G. R., Brunden, M. J. and Gilliam, P. T., *Tetrahedron Lett.* **22**, 4177–4180 (1981).

90. Devivar, R. V., Koontz, S. L., Peltier, W. J., Pearson, J. E., Guillory, T. A. and Fabricant, J. D., *Bioorg. Med. Chem. Lett.* **9**, 1239–1242 (1999).

91. Balgobin, N. and Chattopadhyaya, J., *Nucleosides Nucleotides* **6**, 461–463 (1987).

92. Brown, T., Pritchard, C. E., Turner, G. and Salisbury, S. A., *J. Chem. Soc. Chem. Commun.* 891–893 (1989).

93. Sproat, B. S. and Brown, D. M., *Nucl. Acids Res.* **13**, 2979–2987 (1985).

94. Asseline, U. and Thuong, N. T., *Tetrahedron Lett.* **30**, 2521–2524 (1989).

95. Asseline, U. and Thuong, N. T., *Tetrahedron Lett.* **31**, 81–84 (1990).

96. Yoo, D. J. and Greenberg, M. M., *J. Org. Chem.* **60**, 3358–3364 (1995).

97. Lyttle, M. H., Hudson, D. and Cook, R. M., *Nucl. Acids Res.* **24**, 2793–2798 (1996).

98. Waldvogel, S. R. and Pfleiderer, W., *Helv. Chim. Acta* **81**, 46–58 (1998).

99. Azhayev, A. V., *Tetrahedron* **55**, 787–800 (1999).

100. Kumar, P. and Gupta, K. C., *React. Funct. Polym.* **41**, 197–204 (1999).

101. Lyttle, M. H., Dick, D. J., Hudson, D. and Cook, R. M., *Nucleosides Nucleotides* **18**, 1809–1824 (1999).

102. Iyer, R. P., Yu, D., Xie, J., Zhou, W. and Agrawal, S., *Bioorg. Med. Chem. Lett.* **7**, 1443–1448 (1997).

103. Smith, M., Rammler, D. H., Goldberg, J. H. and Khorana, G., *J. Am. Chem. Soc.* **84**, 430–440 (1962).

104. Caruthers, M. H., Barone, A. D., Beaucage, S. L., Dodds, D. R., Fisher, E. F., McBride, L. J., Matteucci, M., Stabinsky, Z. and Tang, J. Y., *Methods Enzymol.* **154**, 287–313 (1987).

105. Kwiatkowski, M., Heikkilae, J., Welch, C. J. and Chattopadhyaya, J., *Stud. Org. Chem.* **20**, 259–274 (1985).

106. Ma, Y.-X. and Sonveaux, E., *Biopolymers* **28**, 965–973 (1989).

107. Pfleiderer, W., Schwarz, M. and Schirmeister, H., *Chem. Scr.* **26**, 147–154 (1986).

108. Pfleiderer, W., *Biophosphates and their Analogues—Synthesis, Structure, Metabolism and Activity,* K. S. Bruzik and W. J. Stec (Eds.). Elsevier, Amsterdam, Holland, 1987, pp. 133–146.

109. Brown, J. M., Christodoulou, C., Jones, S. S., Modak, A. S., Reese, C. B., Sibanda, S. and Ubasawa, A., *J. Chem. Soc., Perkin Trans. I,* 1735–1750 (1989).

110. Brown, J. M., Christodolou, C., Modak, A. S., Reese, C. B. and Serafinowska, H. T., *J. Chem. Soc., Perkin Trans. I,* 1751–1767 (1989).

111. Iyer, R. P., Yu, D., Habus, L, Ho, N.-H., Johnson, S., Devlin, T., Jiang, Z., Zhou, W., Xie, J. and Agrawal, S., *Tetrahedron* **53**, 2731–2750 (1997).

112. Beier, M. and Pfleiderer, W., *Helv. Chim. Acta* **82**, 633–644 (1999).

113. Stutz, A. and Pitsch, S., *Synlett*, 930–934 (1999).

114. Alvarez, K., Vasseur, J.-J., Beltran, T. and Imbach, J.-L., *J. Org. Chem.* **64**, 6319–6328 (1999).

115. Beaucage, S. L. and Caruthers, M. H., *Tetrahedron Lett.* **22**, 1859–1862 (1981).

116. Josephson, S., Lagerholm, E. and Palm, G., *Acta Chem. Scand. B* **38**, 539–545 (1984).

117. McBride, L. J. and Caruthers, M. H., *Tetrahedron Lett.* **24**, 245–248 (1983).

118. Adams, S. P., Kavka, K. S., Wykes, E. J., Holder, S. B. and Galluppi, G. R., *J. Am. Chem. Soc.* **105**, 661–663 (1983).

119. Caruthers, M. H., *Science* **230**, 281–285 (1985).

120. Sinha, N. D., Biernat, J., McManus, J. and Koester, H., *Nucl. Acids Res.* **12**, 4539–4557 (1984).

121. Cramer, H. and Pfleiderer, W., *Helv. Chim. Acta* **82**, 614–632 (1999).

122. Eadie, J. S. and Davidson, D. S., *Nucl. Acids Res.* **15**, 8333–8349 (1987).

123. De Mesmaeker, A., Haener, R., Martin, P. and Moser, H. E., *Acc. Chem. Res.* **28**, 366–374 (1995).

124. Ceulemans, G., Van Aerschot, A., Wroblowski, B., Rozenski, J., Hendrix, C. and Herdewijn, P., *Chem. - Eur. J.* **3**, 1997–2010 (1997).

125. Nielsen, P., Pfundheller, H. M., Olsen, C. E. and Wengel, J., *J. Chem. Soc., Perkin Trans. I* 3423–3434 (1997).

126. Christensen, N. K., Petersen, M., Nielsen, P., Jacobsen, J. P., Olsen, C. E. and Wengel, J., *J. Am. Chem. Soc.* **120**, 5458–5463 (1998).

127. Tang, X.-Q., Liao, X. and Piccirilli, J. A., *J. Org. Chem.* **64**, 747–754 (1999).

128. Romieu, A., Gasparutto, D., Molko, D., Ravanat, J.-L. and Cadet, J., *Eur. J. Org. Chem.*, 49–56 (1999).

129. Seela, F. and Zulauf, M., *J. Chem. Soc., Perkin Trans. I*, 479–488 (1999).

130. Romieu, A., Gasparutto, D. and Cadet, J., *J. Chem. Soc., Perkin Trans. I*, 1257–1264 (1999).

131. Kahl, J. D. and Greenberg, M. M., *J. Am. Chem. Soc.* **121**, 597–604 (1999).

132. Khan, S. I. and Grinstaff, M. W., *J. Am. Chem. Soc.* **121**, 4704–4705 (1999).

133. Adinolfi, M., Barone, G., De Napoli, L., Guariniello, L., Iadonisi, A. and Piccialli, G., *Tetrahedron Lett.* **40**, 2607–2610 (1999).

134. Stec, W. J., *J. Am. Chem. Soc.* **106**, 6077–6079 (1984).

135. Bannwarth, W., *Helv. Chim. Acta* **71**, 1517–1527 (1988).

136. Zhang, Z., Nichols, A., Tang, J. X., Han, Y. and Tang, J. Y., *Tetrahedron Lett.* **40**, 2095–2098 (1999).

137. Seeberger, P. H. and Caruthers, M. H., *Tetrahedron* **55**, 5759–5772 (1999).

138. Micura, R., *Chem. - Eur. J.* **5**, 2077–2082 (1999).

139. Frieden, M., Grandas, A. and Pedroso, E., *Chem. Commun.*, 1593–1594 (1999).

140. Nielsen, P. E., Egholm, M., Berg, R. H. and Buchardt, O., *Science* **254**, 1497–1500 (1991).

141. Thomson, S. A., Josey, J. A., Cadilla, R., Gaul, M. D., Hassman, C. F., Luzzio, M. J., Pipe, A. J., Reed, K. L., Ricca, D. J. et al., *Tetrahedron* **51**, 6179–6194 (1995).

142. De Napoli, L., Messere, A., Montesarchio, D., Piccialli, G., Benedetti, E., Bucci, E. and Rossi, F., *Bioorg. Med. Chem.* **7**, 395–400 (1999).

143. Schwope, I., Bleczinski, C. F. and Richert, C., *J. Org. Chem.* **64**, 4749–4761 (1999).

144. Corey, D. R., *Trends Biotechnol.* **15**, 224–229 (1997).

145. Thibon, J., Latxague, L. and Deleris, G., *J. Org. Chem.* **62**, 4635–4642 (1997).

146. Kirshenbaum, K., Zuckerman, R. N. and Dill, K. A., *Curr. Opin. Struct. Biol.* **9**, 530–535 (1999).

147. Iyer, R. P., Ho, N.-H., Yu, D. and Agrawal, S., *Bioorg. Med. Chem. Lett.* **7**, 871–876 (1997).

148. Iyer, R. P., Egan, W., Regan, J. B. and Beaucage, S. L., *J. Am. Chem. Soc.* **112**, 1253–1254 (1990).

149. Frechet, J. M. and Schuerch, C., *J. Am. Chem. Soc.* **93**, 492–496 (1971).

150. Frechet, J. M. and Schuerch, C., *J. Am. Chem. Soc.* **94**, 604–609 (1972).

151. Zehavi, U. and Patchornik, A., *J. Am. Chem. Soc.* **95**, 5673–5677 (1973).

152. Guthrie, R. D., Jenkins, A. D. and Roberts, G. A. F., *J. Chem. Soc., Perkin Trans. I*, 2414–2417 (1973).

153. Excoffier, G., Gagnaire, D., Utille, J. P. and Vignon, M., *Tetrahedron Lett.* 5065–5068 (1972).

154. Danishefsky, S. J., McClure, K. F., Randolph, J. T. and Ruggeri, R. B., *Science* **260**, 1307–1309 (1993).

155. Yan, L., Taylor, C. M., Goodnow, R. Jr. and Kahne, D., *J. Am. Chem. Soc.* **116**, 6953–6954 (1994).

156. Rademann, J. and Schmidt, R. R., *Tetrahedron Lett.* **37**, 3989–3990 (1996).

157. Schuster, M., Wang, P., Paulson, J. C. and Wong, C.-H., *J. Am. Chem. Soc.* **116**, 1135–1136 (1994).

158. Zheng, C., Seeberger, P. H. and Danishefsky, S. J., *J. Org. Chem.* **63**, 1126–1130 (1998).

159. Randolph, J. T. and Danishefsky, S. J., *Angew. Chem. Int. Ed. Engl.* **33**, 1470–1473 (1994).

160. Roberge, J. Y., Beebe, X. and Danishefsky, S. J., *Science* **269**, 202–204 (1995).

161. Seeberger, P. H., Eckhardt, M., Gutteridge, C. E. and Danishefsky, S. J., *J. Am. Chem. Soc.* **119**, 10064–10072 (1997).

162. Zheng, C., Seeberger, P. H. and Danishefsky, S. J., *Angew. Chem. Int. Ed.* **37**, 786–789 (1998).

163. Timmers, C. M., van der Marel, G. A. and van Boom, J. A., *Recl. Trav. Chim. Pays-Bas* **112**, 609–610 (1993).

164. Excoffier, G., Gagnaire, D., Utille, J. P. and Vignon, M., *Tetrahedron* **31**, 549–553 (1975).

165. Wang, Y., Zhang, H. and Voelter, W., *Chem. Lett.* 273–274 (1995).

166. Liang, R., Yan, L, Loebach, J., Ge, M., Uozumi, Y., Sekanina, K., Horan, N., Gildersleeve, J. and Thompson, C. et al., *Science* **274**, 1520–1522 (1996).

167. Douglas, S. P., Whitfield, D. M. and Krepinsky, J. J., *J. Am. Chem. Soc.* **113**, 5095–5097 (1991).

168. Rademann, J. and Schmidt, R. R., *J. Org. Chem.* **62**, 3650–3653 (1997).

169. Heckel, A., Mross, E., Jung, K.-H., Rademann, J. and Schmidt, R. R., *Synlett*, 171–173 (1998).

170. Rademann, J., Geyer, A. and Schmidt, R. R., *Angew. Chem. Int. Ed.* **37**, 1241–1245 (1998).

171. Halcomb, R. L., Huang, H. and Wong, C.-H., *J. Am. Chem. Soc.* **116**, 11315–11322 (1994).
172. Blixt, O. and Norberg, T., *J. Org. Chem.* **63**, 2705–2710 (1998).
173. Bayer, R., presented at the *Second CHI GlycoTechnology Conference,* Cambridge Healthtech Institute, La Jolla, CA, 1994.
174. Nicolaou, K. C., Wissinger, N., Pastor, J. and DeRoose, F., *J. Am. Chem. Soc.* **119**, 449–450 (1997).
175. Rodebaugh, R., Joshi, S., Fraser-Reid, B. and Geysen, H. M., *J. Org. Chem.* **62**, 5660–5661 (1997).
176. Ito, Y., Kanie, O. and Ogawa, T., *Angew. Chem. Int. Ed. Engl.* **35**, 2510–2512 (1996).
177. Ito, Y. and Ogawa, T., *J. Am. Chem. Soc.* **119**, 5562–5566 (1997).
178. Doi, T., Sugiki, M., Yamada, H., Takahashi, T. and Porco, J. A., Jr., *Tetrahedron Lett.* **40**, 2141–2144 (1999).
179. Weigelt, D. and Magnusson, G., *Tetrahedron Lett.* **39**, 2839–2842 (1998).
180. Wallberg, A, Weigelt, D., Falk, N. and Magnusson, G., *Tetrahedron Lett.* **38**, 4285–4286 (1997).
181. Nicolaou, K. C., Watanabe, N., Li, J., Pastor, J. and Winssinger, N., *Angew. Chem. Int. Ed.* **37**, 1559–1561 (1998).
182. Kobayashi, S., Wakabayashi, T. and Yasuda, M., *J. Org. Chem.* **63**, 4868–4869 (1998).
183. Blixt, O. and Norberg, T., *Carbohydr. Res.* **319**, 80–91 (1999).
184. Fukase, K., Egusa, K., Nakai, Y. and Kusumoto, S., *Mol. Diversity* **2**, 182–188 (1997).
185. Seeberger, P. H., Beebe, X., Sukenick, G. D., Pochapsky, S. and Danishefsky, S. D., *Angew. Chem. Int. Ed.* **36**, 491–493 (1997).
186. Sofia, M. J., in *Combinatorial Chemistry and Molecular Diversity in Drug Discovery,* E. M. Gordon and J. F. Kerwin, Jr. (Eds.). Wiley-Liss, New York, 1998, pp. 243–269.
187. Seeberger, P.H. and Danishefsky, S. J., *Acc. Chem. Res.* **31**, 685–695 (1998).
188. Osborn, H. M. I. and Khan, T. H., *Tetrahedron* **55**, 1807–1850 (1999).
189. Davis, B. G., *J. Chem. Soc., Perkin Trans. I*, 3215–3237 (1999).
190. Paulsen, H. and Helpap, B., *Carbohydr. Res.* **216**, 289–313 (1991).
191. Windmuller, R. and Schmidt, R. R., *Tetrahedron Lett.* **35**, 7927–7930 (1994).

3 Solid-Phase Synthesis: Small Organic Molecules

The high reliability of automated oligomer SPS derives from the capillary optimization of the SP reaction conditions used for the iterative cycles. All the parameters have been thoroughly studied, and improvements such as new coupling agents, new protected building blocks, or new linkers continuously increase the options for the synthetic chemist. The advantages of peptide and ON SPS versus solution-phase synthesis for the same sequence are evident: no need of intermediate purification, complete automation of reaction cycles, fast SP reactions, and negligible loss of material. Examples where significant amounts of pure 10- to 100-mers were prepared by SPS are abundant in the literature. The automation of the SP protocols is possible because of the relatively restricted variety of SP reaction conditions used: couplings, deprotections, protections, and cleavage. As an example, the Fmoc protocol for the synthesis of a decapeptide will require from 22 to 23 SP steps: the support of the starting material on SP, 10 Fmoc deprotections, 10 couplings, the final cleavage and deprotection of side chains plus, when necessary, the attachment of a linker onto the support. The couplings and the Fmoc deprotections are identical for each of the 20 steps, so that the classical conditions used in an Fmoc- protocol only sporadically require minor adjustments for specific oligomeric sequences. The automation of these steps requires only a few optimized experimental protocols to be repeated in iterative cycles.

The synthesis of nonoligomeric small organic molecules is quite different, and its characteristics do not help an automated transfer onto SP. The whole arsenal of organic reactions can be used to build a molecule, and even a short synthesis may require a few steps with completely different reaction conditions. While each reaction in solution is independent, allowing every intermediate to be purified and submitted to the following reaction conditions without any interference from the previous step, an SPS route requires the support and the linker to be perfectly stable during all the reaction conditions used. A wide variety of different supports and linkers is thus necessary to accomplish the SPS of small organic molecules. For these reasons one or a few common protocols for organic SPS cannot be defined. Instead, a specific study to define an SP-compatible synthetic route and to assess its feasibility must be performed every time. In this chapter the entire *process* from the selection of a target to its successful SP synthesis, including all of the intermediate steps, will be detailed (Sections 3.1 and 3.2). A scheme showing the important steps in this process is reported in Fig. 3.1. The *exploitation* of the SP route for a specific target, encompassing the design, assessment, and realization of the SPS of one or more combinatorial libraries (Section 3.3), will also be described. This will represent the introduction to the

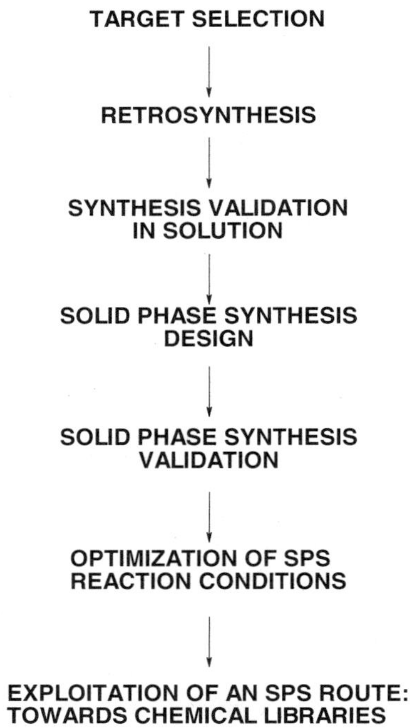

Figure 3.1 Logical scheme to a successful small organic molecule SPS and to related synthetic combinatorial libraries.

following chapters where combinatorial chemistry and combinatorial technologies will be extensively covered.

All of the above-mentioned aspects will also be illustrated through the detailed description of a specific *example* (1), which spans every phase of a successful SPS and shows its potential to generate synthetic organic libraries (Section 3.4). The same will happen for several other recent examples (Section 3.5) using the analysis model built in Sections 3.1–3.3.

3.1 SMALL ORGANIC MOLECULES ON SOLID PHASE: TARGET SELECTION AND SOLUTION STUDIES

3.1.1 Target Selection and Retrosynthesis

The *selection* of a small organic molecule as the target for a chemical synthetic route may arise from any potential application for this molecule in any specific field. The selection depends on the requirements of the individual chemical project and cannot be generalized in any way. Once the target is selected, the so-called *retrosynthetic process* identifies a plausible synthetic route to the target starting from commercially available precursors. At present the adopted process in SPS is identical to that for classical organic synthesis in solution. The vast majority of organic reactions have been developed as homogeneous reactions in solution, and it would be preposterous to design an SPS scheme using reactions that have not been described in contemporary organic synthesis. The same will happen for several other recent examples using the analysis model built in Sections 3.1–3.3 (Section 3.5).

An extensive knowledge of organic synthesis, both theoretical and practical, is instrumental for the chemist to select a target, to design a viable synthetic route, and to individuate, from the beginning, the issues for each chemical step employed. The reader may address a number of excellent books and reviews that deal extensively with retrosynthesis and strategies in synthetic organic chemistry (2–4) to expand these relevant topics.

3.1.2 Validation of the Synthetic Route in Solution

The *validation* of a planned synthetic scheme *in solution* is the necessary starting point to design, refine, assess and carry out successfully the corresponding SPS. A synthetic scheme in solution must provide all the intermediates and the target molecule with good to excellent yields before being transferred onto SP. Sensitivity to changes of selected parameters (e.g., temperature, concentration, and solvent) should be known for the relevant reactions, and several experimental options for each step are desirable. Side reactions and side products formed at each step should be known and understood, and their dependence on reaction parameters should be determined.

The chemist must use the notions of synthetic organic chemistry to initially select the most reasonable experimental conditions, to subsequently perform the whole synthetic route, and to eventually modify the reaction parameters, if this proves necessary, to obtain a reliable, robust, clean, and high-yielding synthesis of the target molecule in solution. Even if the required synthesis is reported in the literature, it would be advisable to check carefully the experimental conditions in order to solve possible issues arising from any reaction before moving to the SPS.

3.2 SMALL ORGANIC MOLECULES ON SOLID PHASE: SOLID-PHASE SYNTHESIS

3.2.1 Design of a Solid-Phase Synthesis

The *design* of a successful SPS requires the proper selection of many SPS-related entities that are not present in the corresponding solution synthesis but have to be compatible with the whole SPS sequence. A brief survey of these choices is reported here and summarized in Fig. 3.2.

The starting synthon for the SPS may or may not contain suitable functionalities to be anchored directly, or via a linker, onto the solid support. Most chemical functionalities can be coupled to an existing linker due to the large number of available linkers

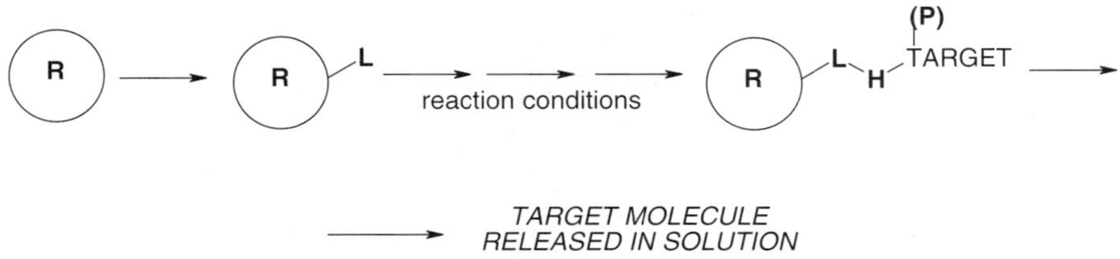

R = choice of the solid support

L = choice of the SP linker (when necessary)

H = choice of the handle on the target molecule

P = choice of the protecting groups

reaction conditions: adaptation to SPS features

Figure 3.2 Crucial factors for the design of small organic molecule SPS.

(see Section 1.2). If a molecule does not contain suitable functionalities (e.g., aromatics and biphenyls), a C–H bond of the target molecule can be replaced by a traceless linker. As a general rule, the most obvious choice of handle should be favored, rather than looking for more challenging but more complex solutions. If, even after considering all the available linking strategies, no obvious choice appears, the transfer of the synthesis onto SP should be avoided.

The same principle applies to the selection of a linker. If the molecule can be safely supported directly on the grafted group of a PS resin and eventually released after the SPS, there is no need for a linker. If an assessed and/or commercially available linker is suitable for the planned SPS, this should be preferred. Linkers that can be cleaved with clean cleavage reagents (e.g., light, TFA vapors, and ammonia vapors) should be chosen when possible. Approaches such as cyclative cleavages, which also allow the clean release of the final cyclic compound into solution, should be considered. More complex linker strategies, or even studies to create a new linker tailored to the SPS needs, are justified only if the validation of a specific linker for the SPS of the target and/or the creation of a new linker moiety are the scientific rationale of the project.

The choice of protecting groups, when they are necessary for a successful SPS, is strongly influenced both by the commercial availability of protected reagents and by the linker or the bond directly linking the target molecule to the support. In fact, it is preferable to have the simultaneous release of the target molecule from the support and deprotection of all the protecting groups as the final SPS step. When a protecting group is to be removed during the SPS, it must be orthogonal to the linker used and to the SPS intermediates. A typical example is the base-labile N-protecting Fmoc carbamate in peptide SPS using acid-labile linkers and Fmoc-compatible coupling conditions.

The solid support is normally chosen following the personal preferences of each scientist. Many supports have similar performances and often their use gives similar results. The use of PEG–PS resins is suggested when detailed on-bead NMR studies of the reaction outcome are planned, providing that the reaction conditions do not affect the stability of the PEG chains. Hydrophobic PS resins are normally used for the preparation of large batches of compounds due to their low price and when hydrophilic solvents are not used for the SPS. Macroporous supports are used for automated continuous-flow syntheses and when aqueous and/or hydrophilic conditions are necessary. The easy elimination of solvent residues from the macropores is an advantage if further water-sensitive SP steps must be performed. Macrosupports such as pins and crowns are used to simplify the handling of the support and to obtain large quantities of target molecule after cleavage. As a general rule, an SPS performed on a specific support cannot be transferred directly onto another support even when they are very similar, as the switch to another support often requires some optimization of the experimental conditions.

The reaction conditions used in solution synthesis are generally adapted to the SPS; the reaction times are usually increased, the reagents in solution are added in excess, and their concentration is increased to drive the SPS steps to completion. Typical experimental conditions are three- to fivefold excess of solution reagents and 0.2–1 *M* concentrations, while the optimal reaction time is evaluated via on-bead (FTIR) or off-bead reaction monitoring. If the reagents in solution are precious, a study to determine the minimum excess required to drive the reaction to completion will be necessary.

3.2.2 Validation of Solid-Phase Synthesis and Optimization of Reaction Conditions

Once the SPS has been designed, a first attempt to *validate* the solution-phase reaction conditions is made incorporating the adjustments required by the introduction of the support and the linker, including the selection of appropriate washing cycles and drying procedures after each step in order to remove all the reagents in solution and evaporate the residual solvents. The presence of the resin-bound intermediates is checked by means of the most appropriate on-bead or off-bead analytical methods, and, if the expected compound is absent, appropriate modifications of the reaction parameters are made. If a specific reaction step cannot be transferred to SPS after having tried all of the sensible options, an alternative SP synthetic route may be designed by changing the parameter(s) that are deleterious to the reaction outcome. Typically, new reagents with different protecting groups are used, different linkers or supports are employed or different handles on the starting material are exploited to support it on SP. If, even after these further trials, the expected reaction product is absent, the SPS strategy has to be either radically changed, with the new synthetic route being assessed first in solution, or completely avoided.

If the SPS gives high yields of pure target molecules by simply using the conditions employed in solution, optimization efforts are not necessary (see the example in Section 3.4). Usually, however, the expected intermediates and the final target molecule are detected, but the overall synthesis requires an *optimization* to increase the yields and to reduce/eliminate the on-bead side products formed. First of all, the reaction kinetics are evaluated by following the reaction course either on-bead (see

Section 1.3) or off-bead (parallel set of experiments quenched and cleaved at different times). Further experiments should also vary the amount and the concentration of the reagents in solution, the temperature, the solvent, and any other reaction parameter. The best compromise between high yields of the expected compound and the minimum amount of on-bead side products for each reaction parameter should be determined, and the relevant conditions introduced in the final experimental procedure, to provide a successful SPS scheme. This optimized SPS scheme is scaled up, when necessary, to prepare significant amounts of the target molecules for the application selected by the SP chemist.

3.3 SMALL ORGANIC MOLECULES ON SOLID PHASE: FROM SOLID-PHASE SYNTHESIS TO SYNTHETIC ORGANIC LIBRARIES

3.3.1 General Considerations

The advantages of SPS versus solution-phase synthesis for oligomers have been detailed in Chapter 2. The noniterative synthesis of small organic molecules generally requires a careful choice of linkers, reagents, and protecting groups as well as significant efforts to optimize the selected experimental conditions for each reaction step on SP. Usually SPS is not competitive with classical techniques in solution to prepare one or a few small organic molecules, a fact that is evident from the small number of SPS of nonoligomeric molecules reported until the early 1990s.

The advent of *combinatorial chemistry* has dramatically changed this scenario. The simultaneous synthesis of many compounds requires automated procedures both for the preparation and the purification of products, making even major efforts to optimize the SPS reaction conditions a minor factor when compared to the advantages provided by the automated synthesis and purification of large libraries of compounds. A thorough presentation of combinatorial technologies will follow in the coming chapters, and the close relationship between SPS and combinatorial technologies will be illustrated in detail and critically reviewed therein. For now, only the main steps for the conversion of an SPS strategy aimed toward a single target into the design and the validation of an SP synthetic route for a combinatorial library will be presented.

3.3.2 Decoration of a Target Molecule

When a multistep SPS has been set up successfully, the number of building blocks/reagents used during the whole process and the number of chemical functionalities present in the final target molecule are both known definitively. As outlined in Fig. 3.3 for a specific example (5), some building blocks can be *replaced* during the SPS by alternative compounds of the same chemical class, for example, primary amines (**R1**), *o*-hydroxyacetophenones (**R2**), and cyclic aminoketones (**R3**). The selection of these replacements for the original building block is driven by their commercial availability and by the rational design of the planned synthetic organic library (see Chapter 4 for more details). The final resin-bound molecule may then contain some functional groups that can be *decorated* either by modifying the chemical functionality (as for

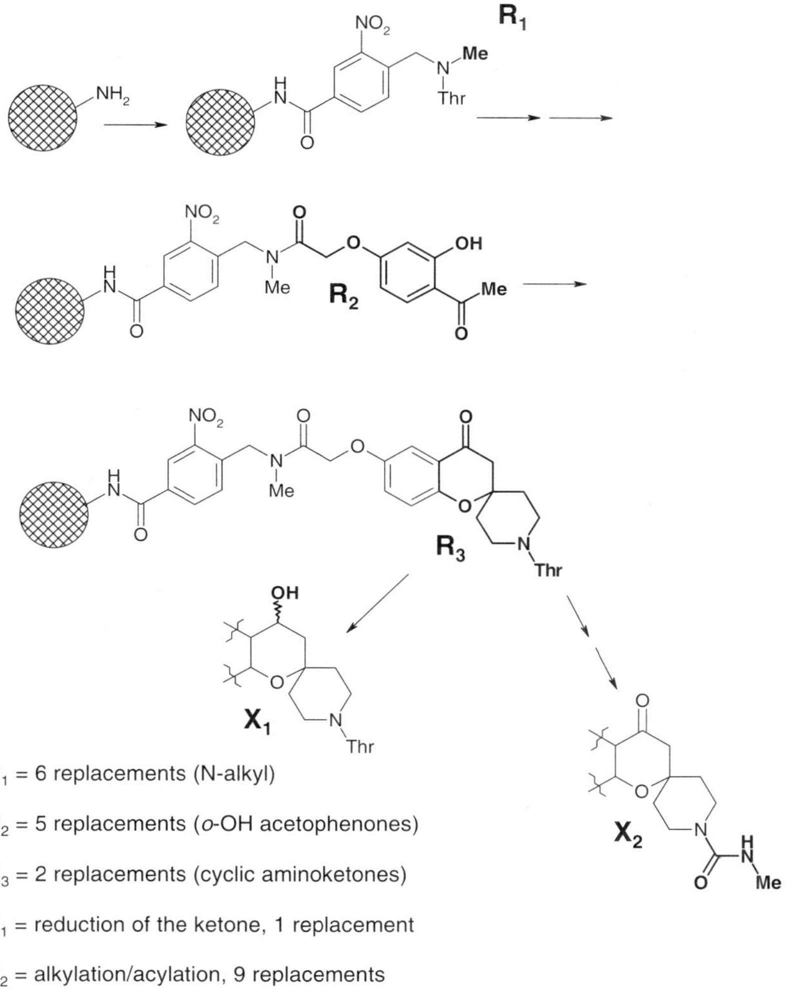

Figure 3.3 Monomer sets R_1–R_3 and decoration functions X_1, X_2 in small organic molecule SPS.

the ketone $\mathbf{X_1}$ in Fig. 3.3) or by reacting them with specific classes of reagents (as for the secondary amine $\mathbf{X_2}$ in Fig. 3.3). After an extensive chemical assessment, the exploitation of the available building blocks $\mathbf{R_1}$–$\mathbf{R_3}$ and the decoration of functions $\mathbf{X_1}$–$\mathbf{X_2}$, as in Fig. 3.3, produced a synthetic organic library composed of $7 \times 6 \times 3 \times 2 \times 10 = 2520$ compounds ($\mathbf{R_1} \times \mathbf{R_2} \times \mathbf{R_3} \times \mathbf{X_1} \times \mathbf{X_2}$ permutations).

The replacement of a building block with a similar reagent and the decoration of a chemical function in the final molecule may eventually produce a huge number of analogous structures. Most of the SPS strategies currently reported in the literature have been designed to produce a chemical library, rather than a specific target molecule. From the beginning, thus, the chemist also selects a specific SP route on the basis of its *versatility* with respect to the availability of similar building blocks to be used in specific SP steps, and the presence of functionalities to be decorated on the core structure of the molecules. While it may be difficult to illustrate the theory of this exploitation process, its potential should be clear through the previous example (5) and a more detailed description of another example (1), which will be reported in the next section.

3.4 AN EXAMPLE: SOLID-PHASE SYNTHESIS OF 1*H*-[2]PYRINDINONES

3.4.1 Target Selection, Retrosynthesis, and Validation in Solution

An excellent report from Parke-Davis (1) presented the SPS of constrained scaffolds based on the 1*H*-[2]pyrindinone ring shown in Fig. 3.4. Such a rigid structure, when substituted with versatile functionalities, can be used as a scaffold for the generation of constrained molecules with potential biological activities, through classical organic synthesis or via decoration to give combinatorial libraries.

Their *retrosynthetic study* was based around the Pauson-Khand cyclization (6), which couples an alkene, an alkyne, and a carbon monoxide source (typically dicobalt octacarbonyl) to give a cyclopentenone ring (Fig. 3.5, top). This reaction has been widely used for synthetic purposes, and some excellent reviews (7, 8) have covered its principal features and the recent improvements to its experimental conditions. This reaction, in its intramolecular version, is ideal for the assembly of the 1*H*-[2]pyrindinone scaffold in two distinct versions, differing in the stereochemistry of the ring junction (Fig. 3.5, bottom). Hence, the readily available unsaturated amino acid derivatives **3.1a,b** undergo intramolecular Pauson–Khand reaction to produce the two unsaturated scaffolds **3.2a,b**.

Once the strategy was selected, the *validation* of the relevant cyclization in solution and the determination of its stereochemical outcome and yield were carried out. The synthetic scheme is reported in Fig. 3.6. The commercially available allyl (**3.3**) and propargylglycines (**3.7**) were sequentially tosylated and alkylated with propargyl and allyl bromide, respectively, to give **3.5** and **3.9**. The intramolecular Pauson–Khand cyclization produced the two isomers **3.6** and **3.10,** with different stereochemistries, in a stereospecific reaction (the chiral allylglycine produced **3.6** as a single enantiomer,

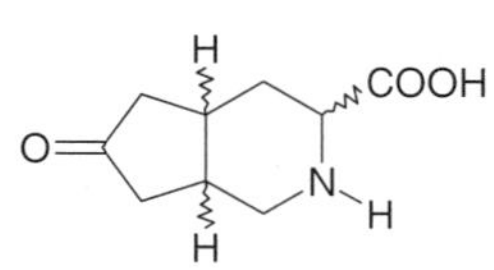

Figure 3.4 Structure of a functionalized 1*H*-[2]pyrindinone.

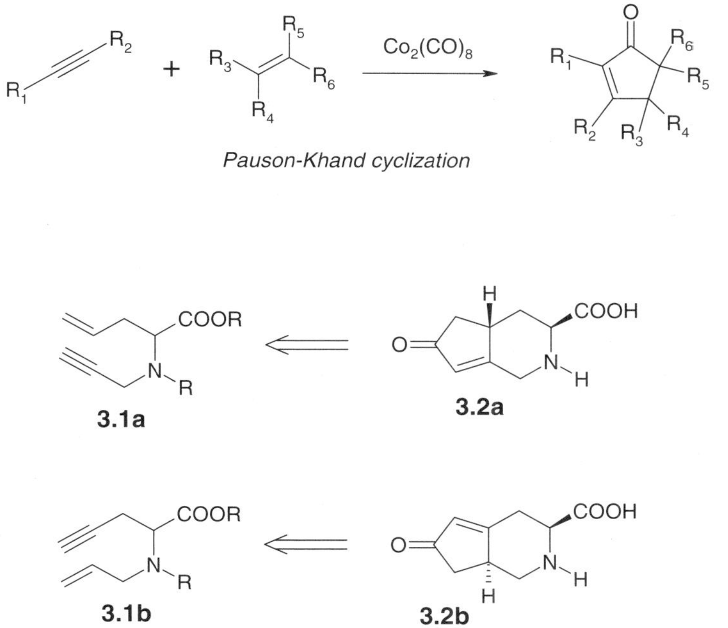

Figure 3.5 Retrosynthetic study: Pauson–Khand cyclization to 1*H*-[2]pyrindinones **3.2a,b**.

while racemic propargylglycine gave the trans-enantiomeric pair **3.10a,b**). The yield of each reaction step was satisfying following a single, nonoptimized experiment, and more importantly, no side products were detected after any of the reaction steps. While further investigations could have provided more information regarding the sensitivity to changes in the reaction conditions, Bolton et al. (1) switched immediately to the SPS applying a somewhat gradual approach to transfer the synthetic scheme onto SP.

3.4.2 Design and Validation of the Solid-Phase Synthesis

Some considerations regarding the design of an SPS of 1*H*-[2]pyrindinones, which can be made on the basis of the target structures and the solution synthetic route, are illustrated in Fig. 3.7 using the structure of **3.10a**.

Structures **3.3–3.10** possess two potential handles for their support on SP. The obvious choice is the carboxylic function, which could be linked either to a chloromethyl or to a hydroxymethyl PS resin through an ester bond. The insertion of a commercially available acid-labile linker, possibly already supported onto the resin, would allow the release of the target into solution under mild conditions. Different functionalities could be released by cleaving the acid-labile linker with, for example, TFA (free acid) and amines (amides). Another possible handle is the secondary amine, which could be anchored to resin-bound carboxylates or halides and finally released as an *N*-acyl or *N*-alkyl moiety. For both handles the protection of the other functional

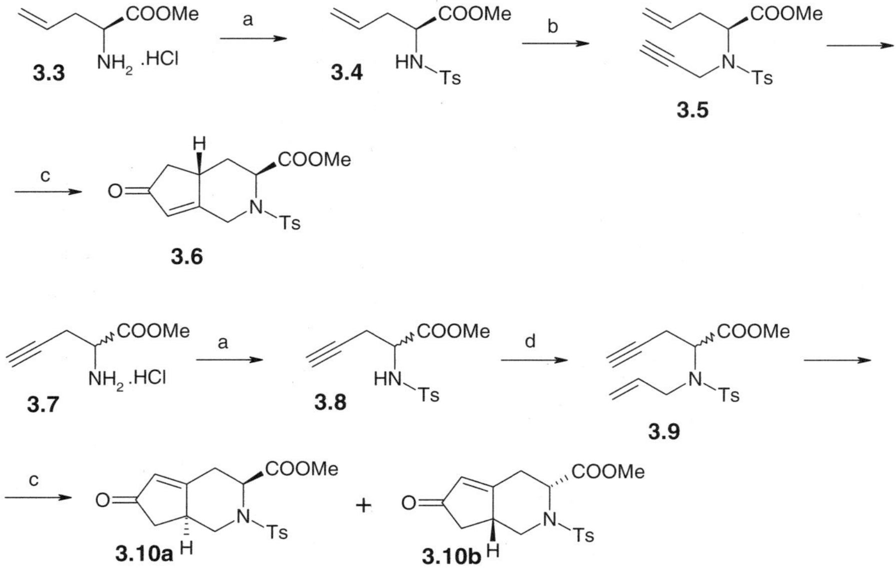

a) TsCl, TEA, DCM, 0°C, 3 hrs, then rt, 2 hrs, 74% (**3.4**) and 84% (**3.8**);
b) propargyl bromide, Cs_2CO_3, DMF, rt, 2 hrs, 97%;
c) $Co_2(CO)_8$, DCM, rt, 2 hrs, then N-methylmorpholine-N-oxide, DCM, rt, 2 hrs, 77% (**3.6**) and 71% (**3.10**);
d) allyl bromide, Cs_2CO_3, DMF, rt, 2 hrs, 93%.

Figure 3.6 Validation of the selected synthetic scheme in solution: synthesis of **3.6** and **3.10a,b**.

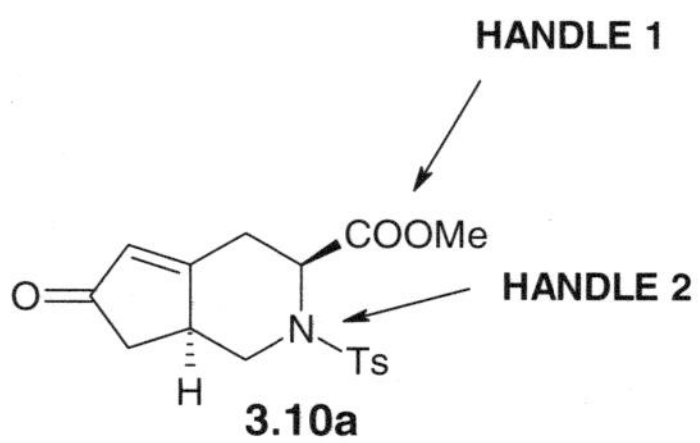

HANDLE 1: ester bond with the support, linker allowed but not necessary, release as free COOH or as ester/amide

HANDLE 2: amide bond with the support, linker necessary to cleave, release as N-alkyl or N-acyl

SUPPORT: no constraints related to reaction conditions

LINKER: no major constraints related to reaction conditions

REACTION MONITORING: possible, on-bead (FT-IR among others) or off-bead (stable intermediates)

Figure 3.7 Crucial factors for the design of a successful SPS of 1*H*-[2]pyrindinones.

group is necessary. Fmoc or Boc carbamates can be considered as protecting groups for the amine, while a *t*-butyl ester could be used on the carboxylic group if an acid-labile linker is chosen to cleave and deprotect simultaneously the final compound. As for the support, the reaction conditions used in solution do not favor any of the most popular choices in respect to the others. Similarly, the conditions appear compatible with the presence of many linkers, including the widely used acid-labile linkers, and they should only require tuning on the basis of the different reaction kinetics in SPS. The key cyclization looks the most challenging step, even if successful Pauson–Khand reactions on SP have been reported previously (9) and subsequently (10) to the work by Bolton et al. (1). Reaction monitoring and on-bead structure determination may make use of some groups such as the α,β-unsaturated ketone, formed during the Pauson–Khand step, but the cleavage of all the stable intermediates from the resin is also a reasonable option for an off-bead yield and purity estimation.

The authors chose, sensibly, to begin by checking the Pauson–Khand outcome on SP using an advanced intermediate, **3.11** (Fig. 3.8), prepared in solution by simple hydrolysis of the methyl ester **3.9**. The free carboxylic group was hooked onto the commercially available Wang–PS resin (acid-labile linker) utilizing the mixed anhy-

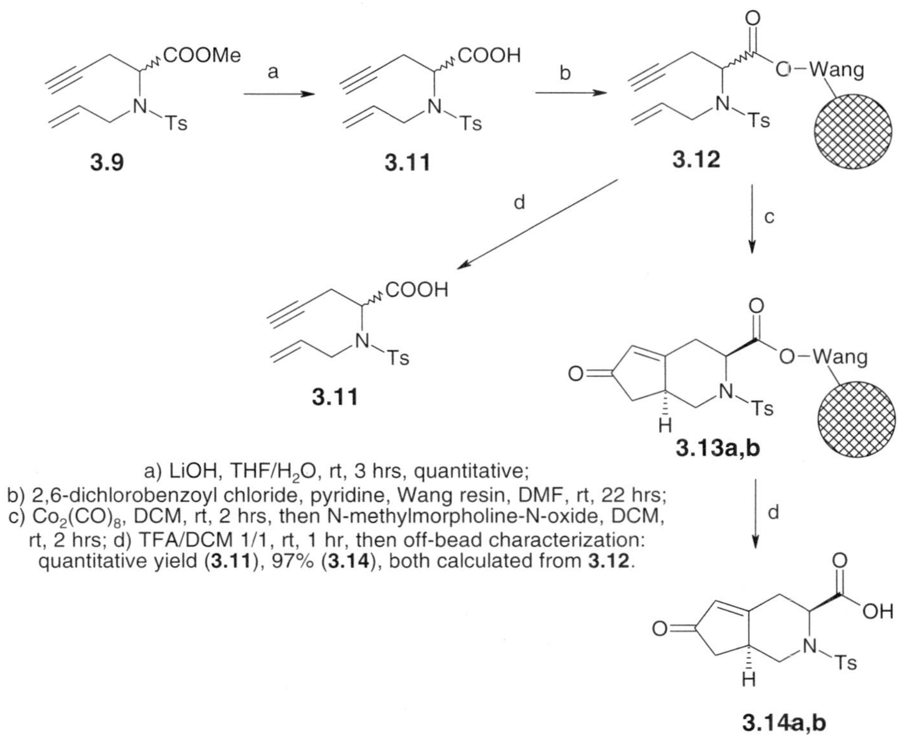

Figure 3.8 Intramolecular Pauson–Khand reaction on SP: chemical assessment and synthesis of **3.11** and **3.14a,b**.

dride method (11), and the yield, calculated after cleavage of a small aliquot of **3.12** with TFA (off-bead yield), was found to be almost quantitative. This resin was treated with dicobalt octacarbonyl in DCM and subsequently with NMMO, exactly as for the cyclization performed in solution. The off-bead yield obtained for **3.14a,b** (from now on only the enantiomer **a** will be shown in the figures) was around 97%. The other diastereomeric scaffold was not prepared, a similarly high yield being assumed for the analogous sequence.

Having accomplished the key cyclization on SP, the whole synthesis was attempted following the scheme reported in Fig. 3.9. The suitable *N*-Fmoc amino acids were coupled to the Wang resin, deprotected, tosylated, and N-alkylated to give **3.12** and **3.21**, respectively. These intermediates were cyclized, and the six-step SPS produced the two scaffolds **3.14a,b** and **3.23**, with calculated off-bead yields of 84 and 74%, respectively, from **3.12** and **3.21.** As a comparison, the three-step synthesis in solution gave 56 and 55% yields, which are comparable to the total yield for the six-step SPS. Besides the introduction of an N-protecting group in the early steps, the synthesis was performed exactly as in solution with the one exception of the N-alkylation to give **3.12** and **3.21**, which required a larger excess of reagents and an extended time (16 h rather than 2 h as in solution). Both the high yields and the simplicity of the work-up procedures for the SPS process (a single final chromatography was performed, while each solution step required extractions and chromatographies) make it more suitable than the corresponding solution-phase approach for the synthesis of 1*H*-[2]pyrindinones.

The high yields obtained by using similar reaction conditions to the solution-phase synthesis did not require major efforts to optimize the reaction conditions. The minor adjustments have already been described above.

3.4.3 Exploitation of the Solid-Phase Route: Toward Synthetic Organic Libraries

The SPS route to 1*H*-[2]pyrindinones, shown in Fig. 3.10, can be modified not only by introducing similar building blocks (Fig. 3.10, points 1 and 2) or new reaction steps (Fig. 3.10, 3), but also by reducing (Fig. 3.10, point 4) or decorating the target scaffold on the keto group (Fig. 3.10, 5) or on the carboxylic function (Fig. 3.10, point 6). The exploitation of these options, as presented in the original work (1), will be briefly discussed here. For simplicity, only the modifications performed on the scaffold leading to **3.13a,b** will be shown even if the scaffold leading to **3.22** was similarly exploited.

The use of different sulfonyl chlorides, reacted with the amine **3.16** after Fmoc deprotection (Fig. 3.10, point 1), was attempted with moderate success using a chloro-substituted aryl sulfonyl chloride (Fig. 3.11). A nitro derivative was also considered as a transient protecting group, allowing the eventual functionalization of the ring nitrogen with various alkylating reagents after cyclization, but reduction of the nitro group during the Pauson–Khand cyclization erased this option. The use of arenesulfonyl chlorides as analogues of the original *p*-toluenesulfonyl chloride could

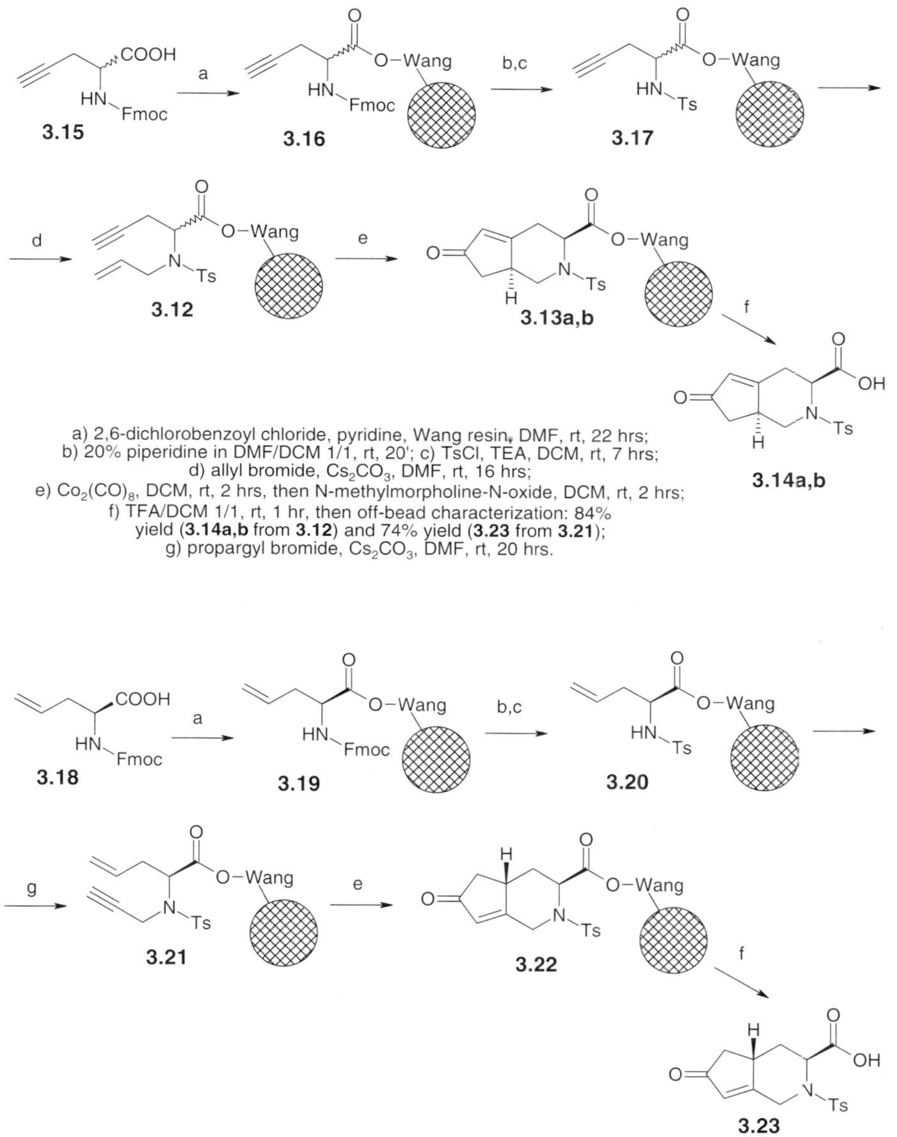

Figure 3.9 Solid-phase synthesis of two functionalized 1*H*-[2]pyrindinones **3.14a,b** and **3.23**.

increase the number of compounds available from this SPS, but significant efforts to improve the reaction yields would be necessary.

The substitution of allyl bromide as the alkylating reagent in the functionalization of **3.17** (Fig. 3.10, point 2) was investigated using 2- and 3-substituted allyl bromides (Fig. 3.12). The former gave modest results in the SP cyclization, while the latter

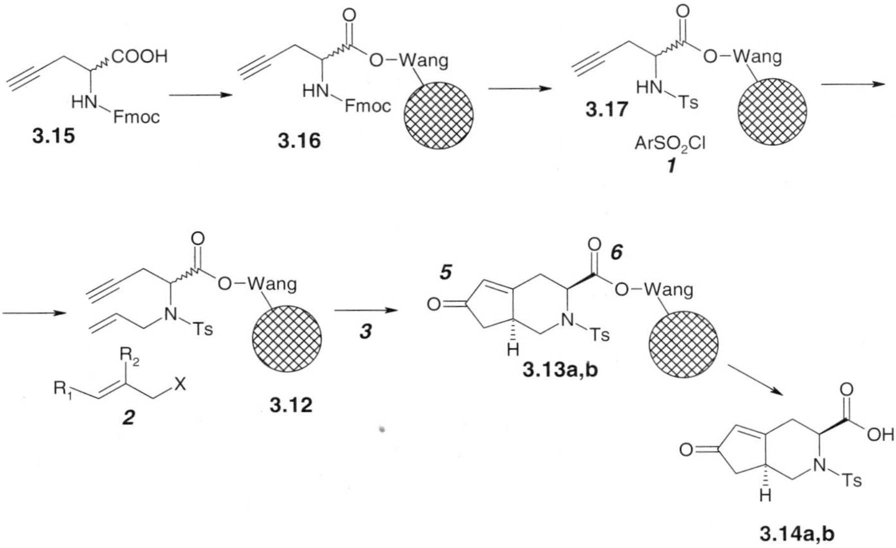

Figure 3.10 Combinatorial exploitation of successful SPS strategy: options 1–6.

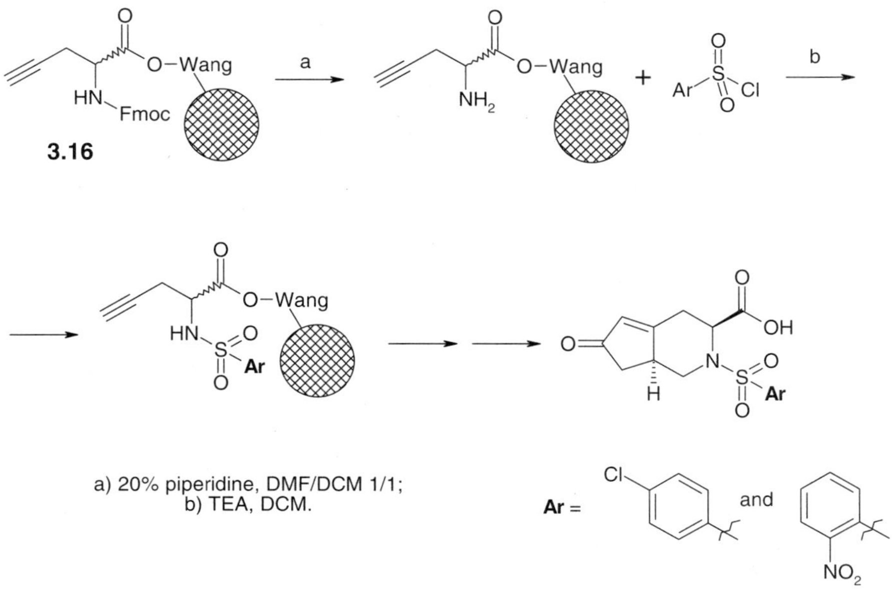

Figure 3.11 Diversification of sulfonyl chlorides (option 1): arenesulfonyl N-substituted 1*H*-[2]pyrindinones.

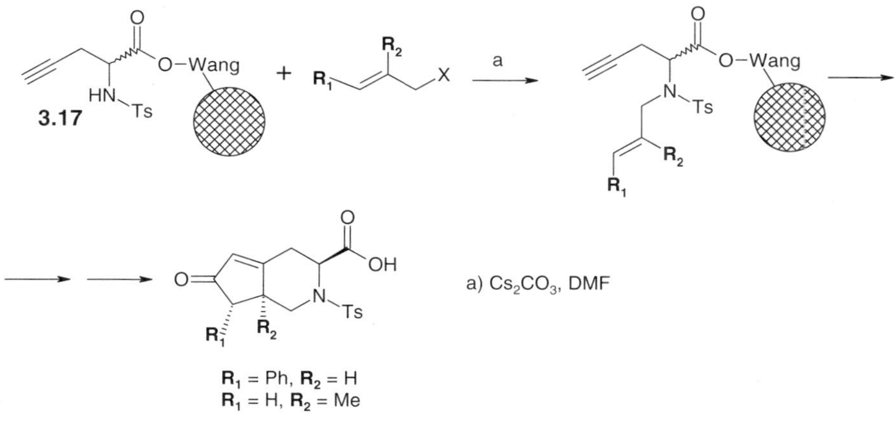

Figure 3.12 Diversification of allyl bromides (option 2): alkyl- or aryl-*C*-substituted 1*H*-[2]pyrindinones.

compound produced only traces of the expected cyclic scaffold. Although the reaction conditions could potentially be optimized to give better yields, this may require significant efforts and does not seem a suitable modification for the exploitation of the SPS strategy with respect to library generation.

When the advanced resin-bound intermediate **3.12** was reacted with an aryl iodide prior to cyclization (point 3, Fig. 3.10), an aryl group was inserted in the original scaffold. A number of examples showed reasonable to good yields of cleaved 1*H*-[2]pyrindinones could be obtained by inserting various aryl groups (Fig. 3.13).

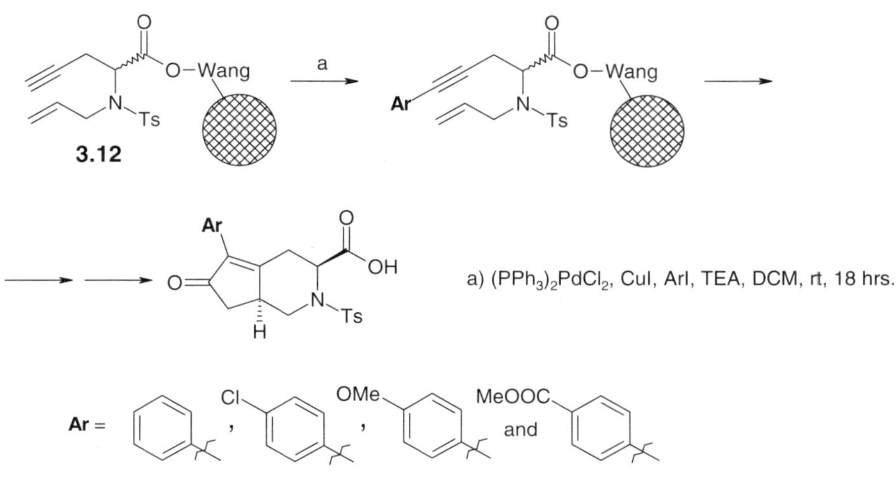

Figure 3.13 Addition of aryl iodides (option 3): aryl substituted 1*H*-[2]pyrindinones.

Therefore, the introduction of selected representatives from the many commercially available aryl iodides should make it possible to generate a population of analogues with good yields. This modification is thus suitable for combinatorial library generation (see the next chapters for the detailed description of these processes).

The final resin-bound scaffold contains two functional groups, ketone and carboxylic acid, amenable to chemical decoration. The ketone (Fig. 3.10, point 5), after reduction of the double bond to give the saturated scaffold **3.24a,b** (Fig. 3.10, point 4), was reductively aminated and capped with an acyl chloride or with an isocyanate (Fig. 3.14). Final cleavage from the resin produced diastereomeric decorated scaffolds with good yields opening a route to prepare a set of ketone-modified analogues of the original scaffold. Finally, an amino acid could be inserted between the Wang linker and the carboxylic function (Fig. 3.10, point 6) to produce, after cleavage, a high yield of a diastereomeric carboxyamide (Fig. 3.15) as a representative of another decorated family of analogues.

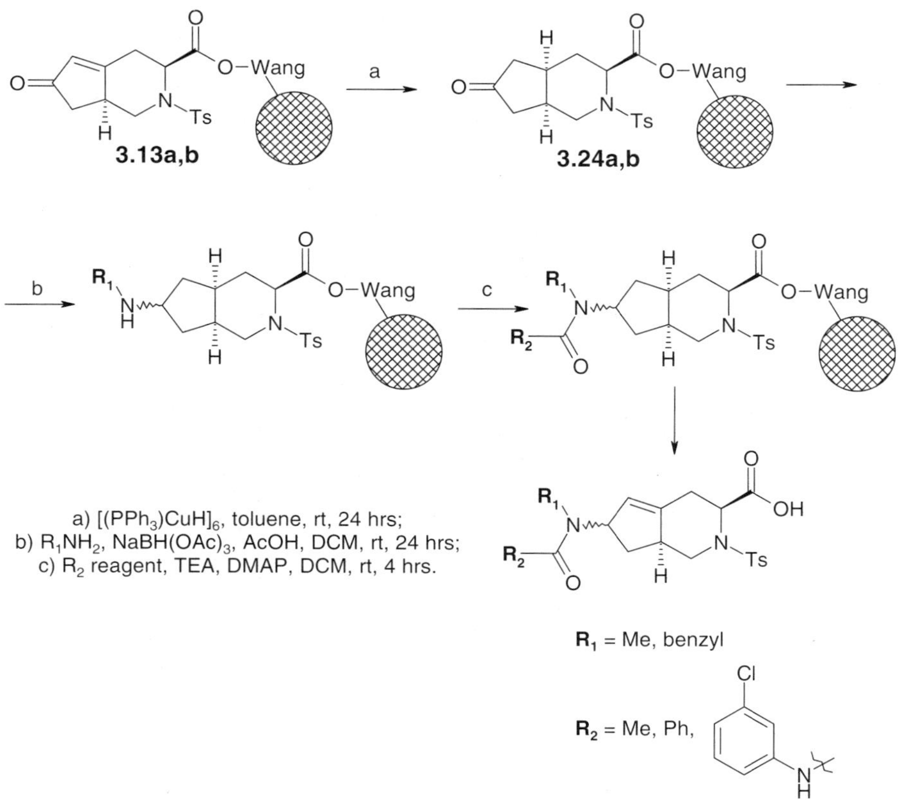

Figure 3.14 Reduction of endocyclic double bond and decoration of 1*H*-[2]pyrindinones on keto group (options 4 and 5).

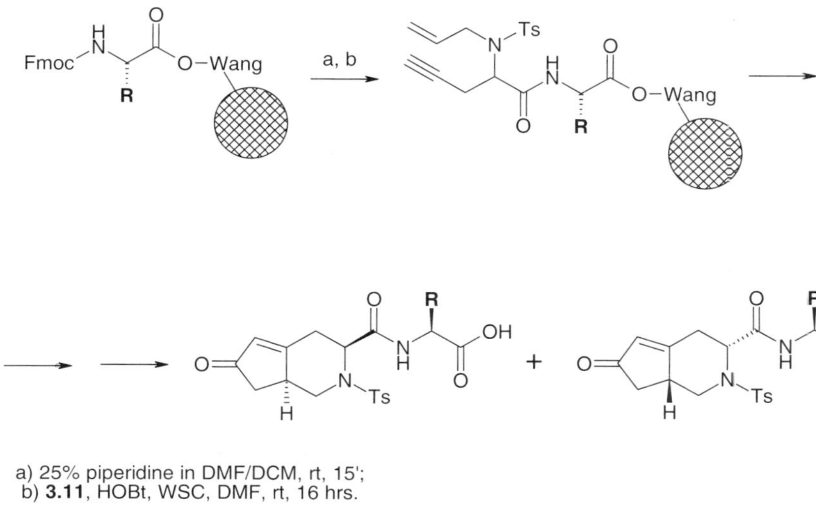

Figure 3.15 Decoration of 1*H*-[2]pyrindinones on the carboxylic group (option 6).

3.5 SOLID-PHASE SYNTHETIC STRATEGIES: SELECTED EXAMPLES

3.5.1 General Considerations

A full coverage and an adequate referencing of the most significant and recent SP efforts by leading academic and industrial groups would require a whole book to be accurate; the reader is referred to several updated reviews (12–18) and to a book (19) to navigate through this fascinating synthetic world at the interphase.

In this Section four examples are covered following the arrangement of an ideal process for the development of a successful solid-phase synthetic strategy as described in Sections 3.1–3.3. The rationale of the project and the structure of the synthetic target are presented and discussed first, followed by the design of a synthetic route in solution and by its validation; the design and assessment of a related SP route complete the process. Both assessments are critically examined and alternative options are presented; the choice of supports, linkers, reagents, solvents, concentration, temperature and reaction time is analyzed as much as possible; key experimental observations from the authors are discussed, while additional details can be found in the original paper. Each example shares the thoroughness used in the design and the assessment of a SP synthetic scheme, and also the sound project rationale to which the SPS is related; each project is inspired by completely different purposes and this is reflected by the diverse strategies used, hopefully providing the reader with a small but significant sampling of high quality, complex and challenging SPS. For each example the exploitation of the successful SPS scheme for combinatorial library synthesis is covered at the end, but is also an inspiring criterion throughout the whole example.

3.5.2 SPS of 2,2-Dimethylbicyclo[3.3.1]nonan-9-ones (20)

Rationale of the project:

- The 2,2-dimethylbicyclo[3.3.1]nonan-9-one scaffold exemplified by structure **3.25** (Fig. 3.16) is the core structural motif of several natural products (e.g. **3.26** and **3.27**, Fig. 3.16) with, among others, strong neurotrophic properties;
- Synthetic analogues containing the 2,2-dimethylbicyclo[3.3.1]nonan-9-one scaffold should likely possess relevant biological activities;
- The assembly of an efficient and versatile synthesis of a tricyclic nucleus with two quaternary bridgehead carbon atoms is a great challenge, although previous reports could provide assistance (21, 22);
- The structure of **3.25** looks suited for the introduction of decorating functions in various positions to create a combinatorial library based on a natural rigid scaffold;
- The full combinatorial exploitation of **3.25** requires an assessed SP synthetic method to build and decorate the scaffold for SP library generation;
- This approach may represent a novel application for recently reported multipurpose selenium-based SP linkers (21).

Target selection and synthetic strategy in solution: Compound **3.25** contains a condensed lactone and a selenium-based substituent; appropriate transformations of the two groups should give access to diverse tricyclic compounds. A more radical decoration of the nucleus (**3.28**, Fig. 3.17) could reasonably be conceived to fully exploit the bicyclo[3.3.1]nonan-9-one nucleus. Literature search found a related approach (22) where a similar, less congested tricyclic nucleus **3.29** was prepared by selenium-

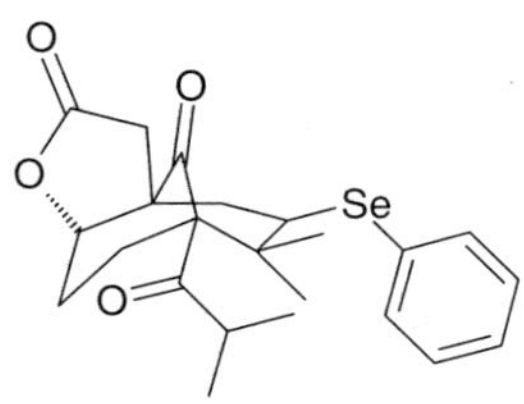

3.25

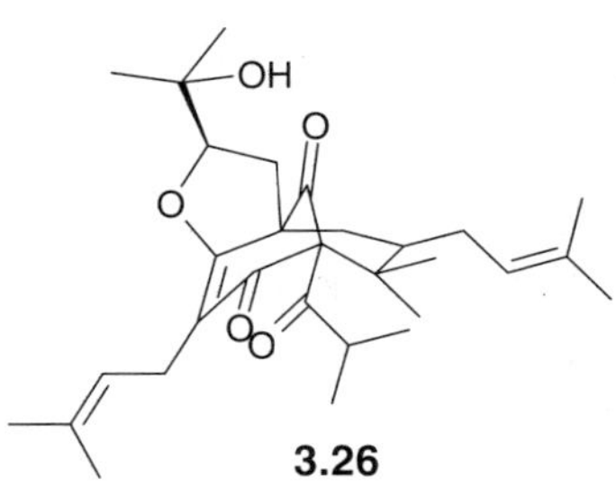

3.26

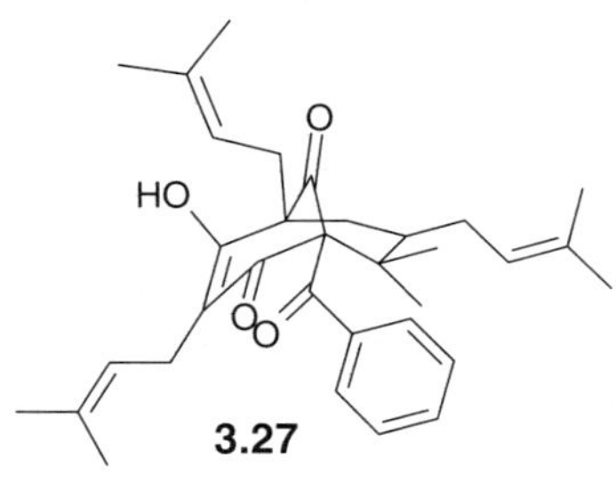

3.27

Figure 3.16.

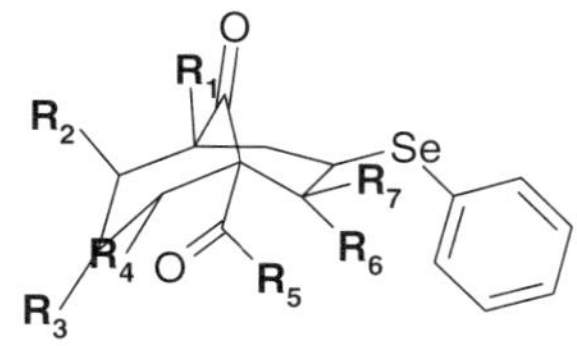

3.25 R_1, R_2 = $-CH_2COO-$; R_3, R_4 = H; R_5 = *i*-Pr; R_6, R_7 = Me
3.28 R_1-R_7 = alkyls, alkoxy, aryls, condensed cycles

Figure 3.17.

promoted, acid-catalyzed intramolecular C-endocyclization of an olefin onto a P-dicarbonyl system (Fig. 3.18).

The authors tried to validate this route for congested tricycles using the easily accessible β-ketoester **3.30** (Fig. 3.19), but the conversion of kinetically favored O-cyclized **3.31** into the desired **3.32** did not happen even forcing the reaction conditions. An elegant solution was found by masking the β-ketoester function of **3.30** as the acetate **3.33**, thus preventing O-cyclization; C-cyclization took place in extremely mild conditions, simultaneously releasing the acetate function and giving **3.32** in high yields and purities (Fig. 3.19).

The same strategy was applied for the synthesis in solution of **3.25** (Fig. 3.20). The suitable protected β-ketoester **3.35** was easily prepared according to published procedures (23, 24) with moderate yields. Selenium-mediated endocyclization to **3.25** proceeded rather cleanly at higher temperatures but only up to 40% conversion of **3.35**; any attempt to increase the conversion by further increasing the temperature or the reaction time only decomposed the starting material. The reported assessment in solution was considered satisfactory, and the authors moved to SP studies.

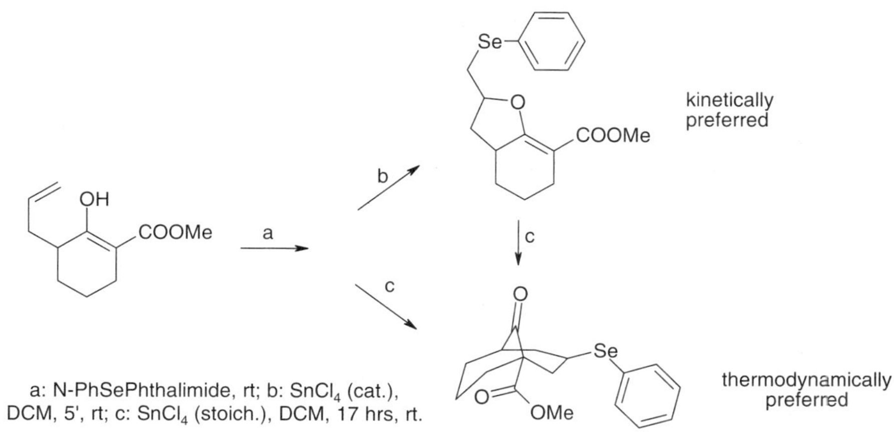

3.29

Figure 3.18.

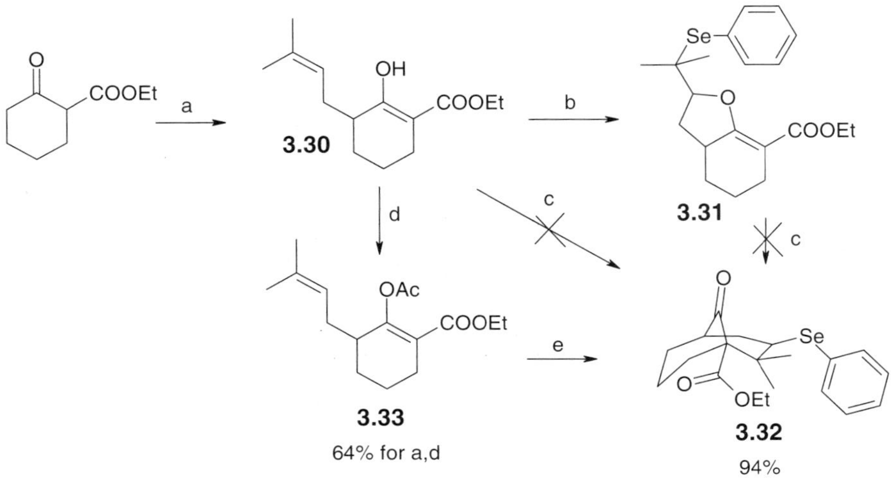

a: LDA, prenyl bromide, 1 hr, -78°C to 0°C; rt; b: N-PhSePhthalimide, $SnCl_4$ (cat.), DCM, 0°C;
c: $SnCl_4$ (stoich.), DCM, 17 hrs, rt, or harsher conditions; d: Ac_2O, DMAP, 30', 80°C;
e: N-PhSePhthalimide, $SnCl_4$ (stoich.), DCM, 5', -78°C.

Figure 3.19.

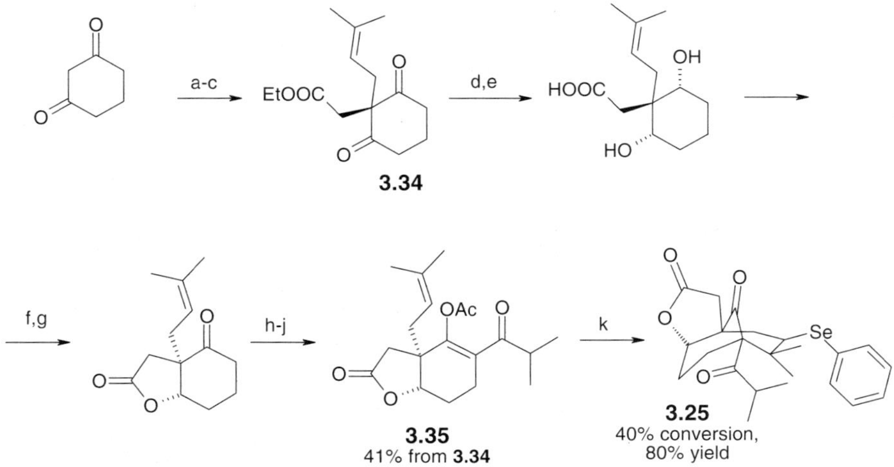

a: 1-Br-3-MeButene, Cu, KOH, H_2O, 2 hrs, rt; b: 6N HCl, pH 4; c: $BrCH_2COOEt$, DBU, LiI, THF, 24 hrs, 65°C; d: $LiAlH(Ot\text{-}Bu)_3$, THF, 2.5 hrs, 0°C; e: LiOH, THF/H_2O 4/1, 30', rt; f: DCC, DMAP, DCM, 30', rt; g: PDC, celite, DCM, 6 hrs, rt; h: LHDMS, HMPA, THF, 30', -78°C; i: *i*-PrCOCN, 15', -78°C, then 15', 0°C; j: Ac_2O, DMAP, 30', 80°C; k: N-PhSePhthalimide, $SnCl_4$ (stoich.), DCM, 15', -23°C.

Figure 3.20.

SP synthetic strategy to analogues of 3.25 and chemical assessment: Nicolaou et al. (21) recently reported the use of PS-supported selenium reagents **3.36** and **3.37** (Fig. 3.2 1). A simple reasoning prompted to substitute N-phenyl selenium phthalimide with **3.36** in the scheme to **3.25** (Fig. 3.21); unfortunately even an extensive scan of experimental conditions could not support any trace of tricyclic **3.38**. The reaction with supported bromide **3.37**, though, worked extremely well and the final, assessed protocol (Fig. 3.21) gave **3.38** with an 81% loading using an excess of dissolved reagents at –23 °C for 20 min. The higher conversion (81% versus 40% in solution) should have derived from the excess of **3.35** which probably drove the reaction to completion even with the decomposition of some starting material.

Cleavage of **3.38** was attempted with an oxidation-elimination double protocol (steps c,d, Fig. 3.21) which proved to be extremely efficient in releasing pure **3.39** with 91% yield. The use of supported Se reagents ensured higher conversions and cleaner workup procedures than their counterparts in solution; additional advantages provided by SPS for this project are reported below.

Exploitation of the synthetic route in solution to 3.25: The diversification of the 2,2-dimethylbicyclo[3.3.1]nonan-9-one scaffold using the same synthetic route, as intended by general structure **3.28** (Fig. 3.17), was done by preparing 16 tricyclic analogues. Their general structures (**3.44–3.59**) and their synthesis from the appropriate β-ketoesters (**3.40a–d**, **3.41a–d**, **3.42a,b**, **3.43a–f**) are reported in Fig. 3.22; yields

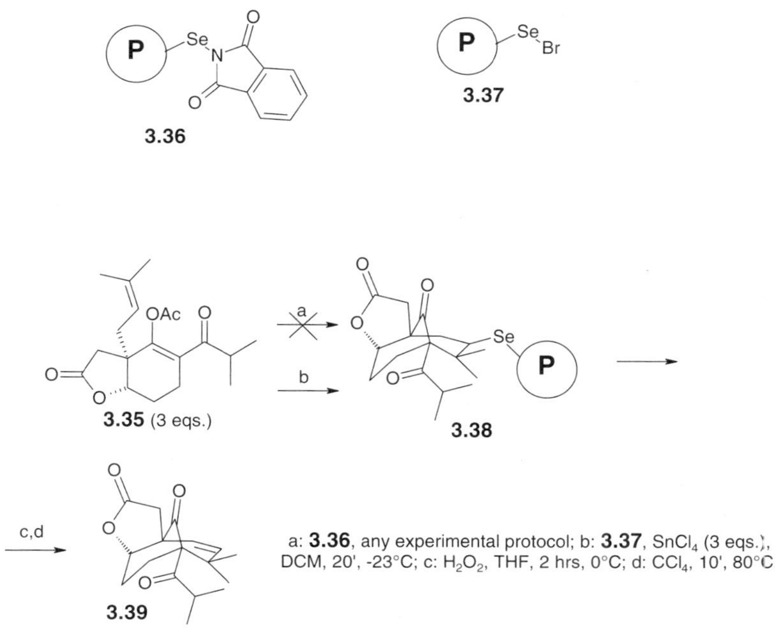

Figure 3.21.

and reaction conditions are also provided. Most of the introduced substituents gave the expected products with good yields and purities; the only exception was represented by substituted allyl chains (**3.43a–f**) which seem very stringent for the prenyl radical to prevent O-cyclization and orient towards C-cyclization (Fig. 3.22).

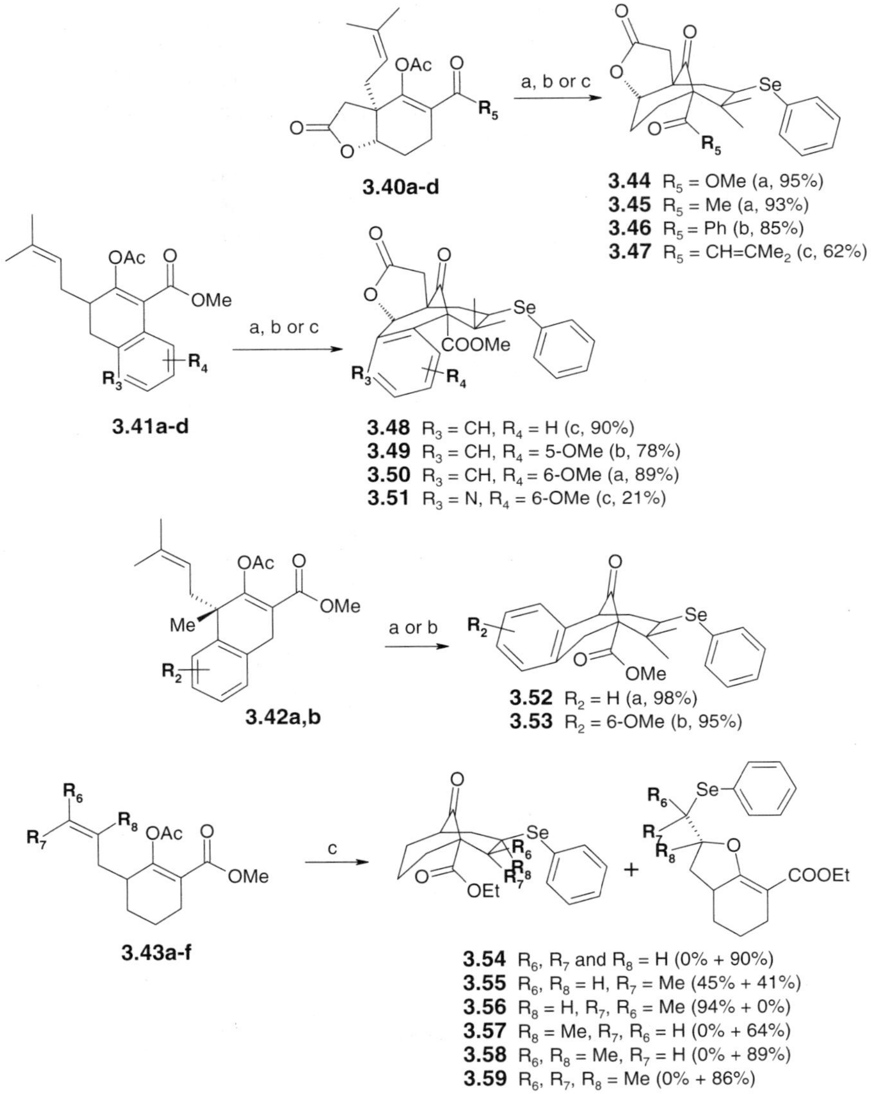

a: N-PhSephthalimide, $SnCl_4$, DCM, -23°C, 5 min; b: as a, 10 min; c: as a, 15 min.

Figure 3.22.

Exploitation of the SP route: The SP endocyclization of β-ketoesters **3.43a–f** was abandoned due to results in solution. Six condensation products **3.60–3.65** were obtained from the reaction of β-ketoesters **3.30**, **3.40a**, **3.41a**, **3.41c**, **3.42a** and **3.42b** with bromide **3.37** with similar yields and purities than their solution counterparts (Fig. 3.23); the same experimental protocol seen for **3.38** (Fig. 3.21) was applied.

Selenium linkers provide multiple cleavage options: oxidation/elimination released the alkene **3.39** (Fig. 3.21), but radical reactions could either lead to a traceless cleavage (hydride transfer) or to the allylation of the released molecule. These cleavage protocols were validated using **3.60** and **3.64** as substrates (Fig. 3.24). Both the multiple cleavage options and the loading of decorated β-ketoester on SP could lead to diverse sets of bicyclo[3.3.1]nonan-9-ones. Post-loading transformations of the tricycle prior to the cleavage should significantly extend the scope of this SP approach.

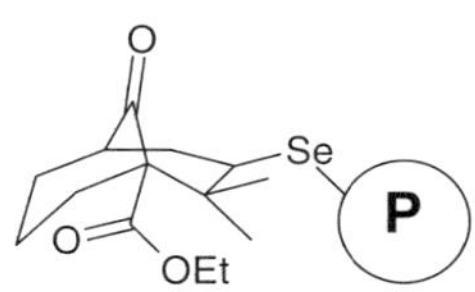

3.60
83% yield, 98% purity

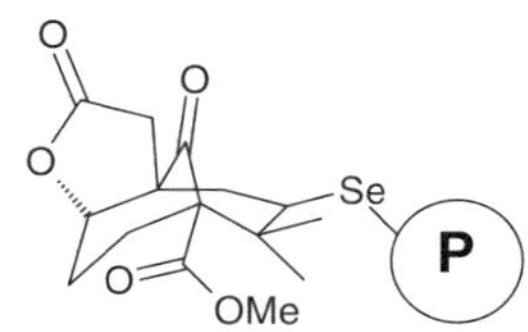

3.61
91% yield, 98% purity

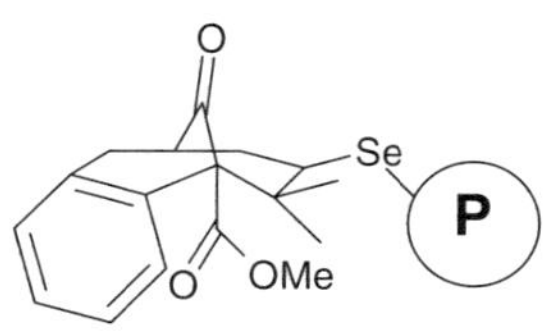

3.62
85% yield, 98% purity

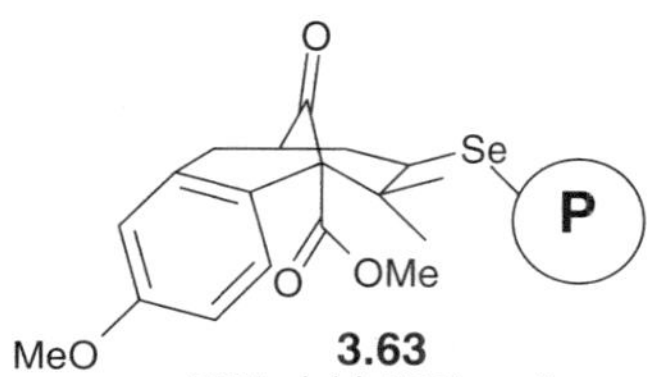

3.63
55% yield, 80% purity

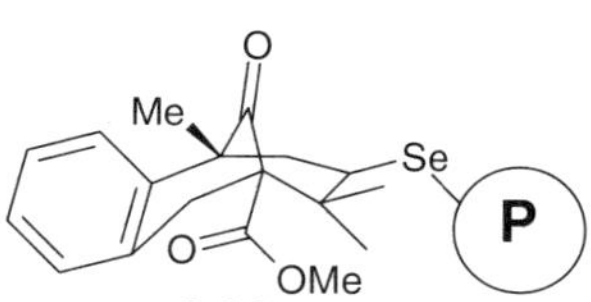

3.64
87% yield, 98% purity

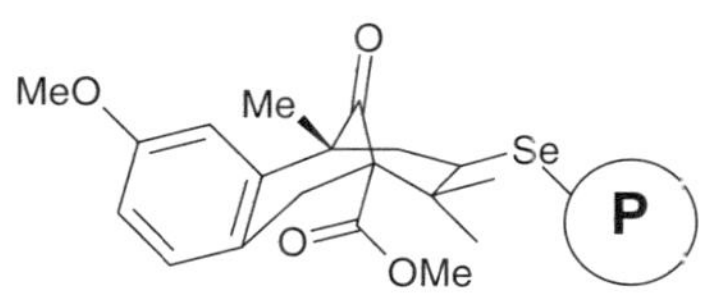

3.65
90% yield, 98% purity

Figure 3.23.

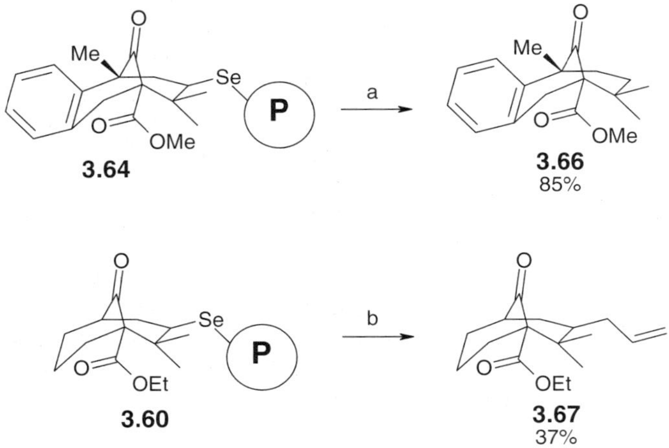

a: n-Bu_3SnH, AIBN, benzene, 2 hrs, 80°C; b: n-$Bu_3SnCH_2CH{=}CH_2$, AIBN, benzene, 2 hrs, 80°C

Figure 3.24.

3.5.3 SPS of Indolactam Derivatives (25).

Rationale of the project:

- Members of the protein kinase C (PKC) family are targets for the cure of several diseases and consequently compounds interacting with PKC members can become useful drugs;
- (–)-Indolactam V (**3.68**, Fig. 3.25) contains the key structural features of known natural PKC activators and is *per se* a PKC activator;
- Total synthesis of **3.68** in solution has been reported (26);
- The rigidity of **3.68** orients its substituents and most likely is essential to develop key interactions with biological receptors;

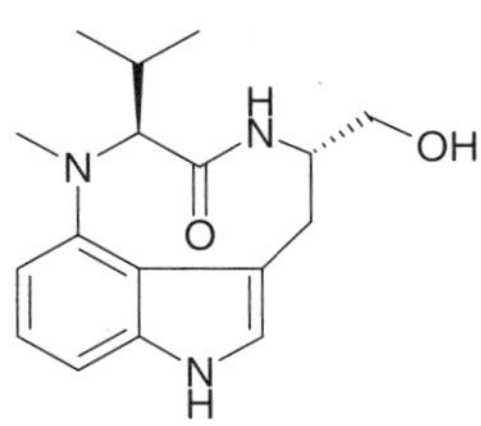

indolactam V
3.68

Figure 3.25.

- The structure of **3.68** looks suited for the introduction of decorating functions in various positions to create a combinatorial library based on a natural rigid scaffold;
- The full combinatorial exploitation of **3.68** requires an assessed SP synthetic method to build the indolactam scaffold and to generate an SP library.

Target selection and synthetic strategy in solution: Several scaffold substitutions and a free primary hydroxyl as in **3.68** have potentially positive effects on biological activity. The target structure for the project was thus drawn as in Fig. 3.26 (**3.69**, compare with **3.68**); the key synthons to **3.68** and to **3.69** are also reported (**3.70–3.73**, Fig. 3.26). As the solution synthesis of **3.68** was already known, its validation was not considered necessary.

SP synthetic strategy to 3.68: The advanced chiral intermediate **3.74** (Fig. 3.27), obtained with good yields and purities in solution, was selected to be anchored via a suitable linker onto SP. The primary OH function was ideal for SP anchorage, using the support as a protecting group during SP synthesis, and for the final release of the free primary alcohol **3.68**. The planned SP decoration route (Fig. 3.27) was compatible with PS supports that were thus chosen. The risk of racemization for the chiral scaffold

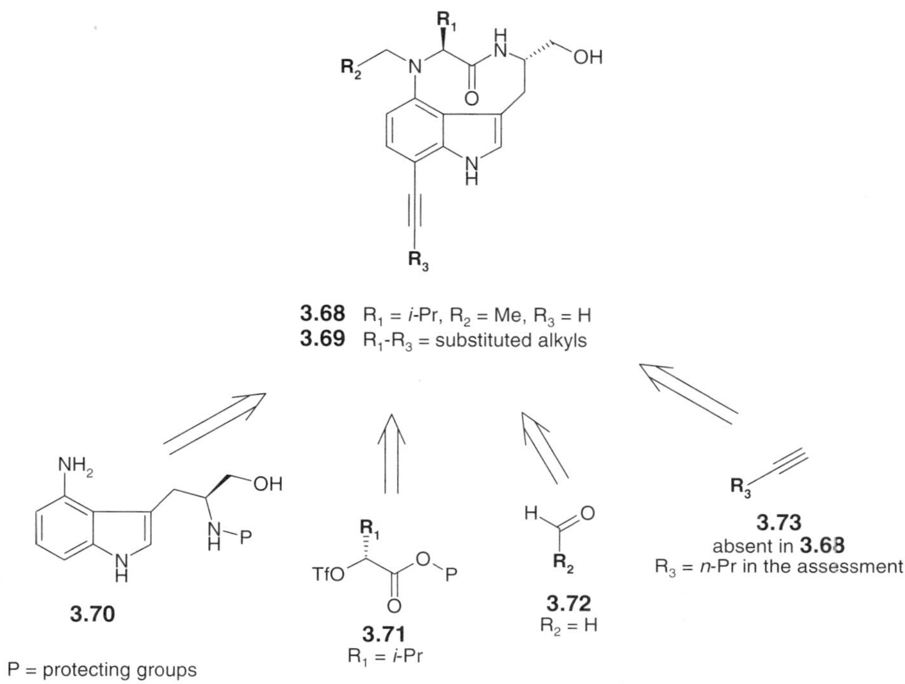

Figure 3.26.

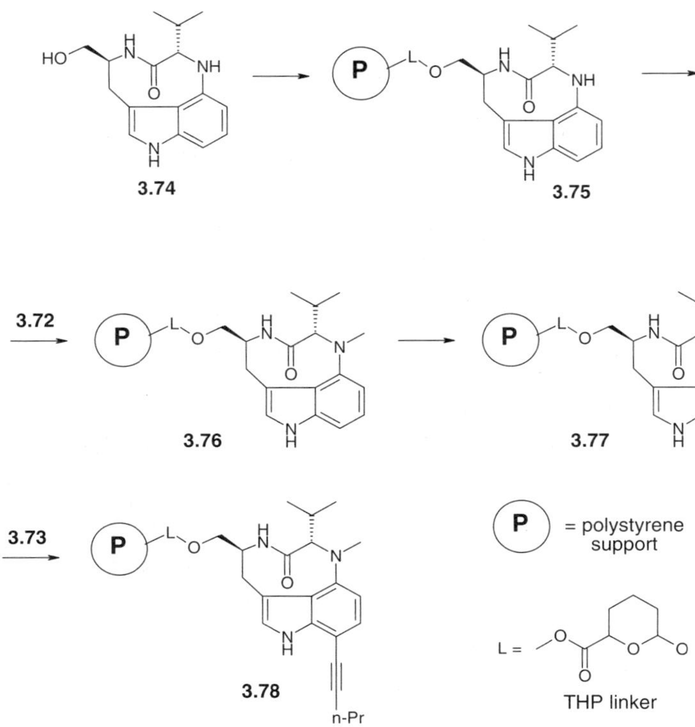

Figure 3.27.

in basic conditions and the planned SP protocols selected the acid-labile tetrahydropyranyl (THP) linker (27), whose general stability and sensitivity to mild acidic conditions appeared appropriate. The use of protecting groups was not required by the designed strategy. Reductive amination of the secondary aromatic amine in **3.75** with formaldehyde/borohydride was expected to give **3.76** (resin-bound **3.68**). 7-functionalization of supported indolactam **3.76** with alkynyl substituents was conceived by electrophilic iodination (**3.77**) and Sonogashira coupling with a terminal alkyne to give **3.78** (Fig. 3.26); this must be considered an SP representative example extendable to other metal-catalyzed C-C couplings (e.g. Suzuki, Heck, Stille).

SP chemistry assessment: The chiral intermediate **3.74** was prepared in solution with a 13-step revised strategy from protected amine **3.79** (25) (Fig. 3.28); its coupling with PS-supported THP linker gave poor loading results for unknown reasons. The problem was solved by coupling **3.74** in solution with the protected DHP derivative **3.80**, then supporting **3.81** onto Merrifield resin to give **3.75** (73% yield of recovered alcohol after cleavage). Both reductive amination to **3.76** and iodination/coupling to **3.78**

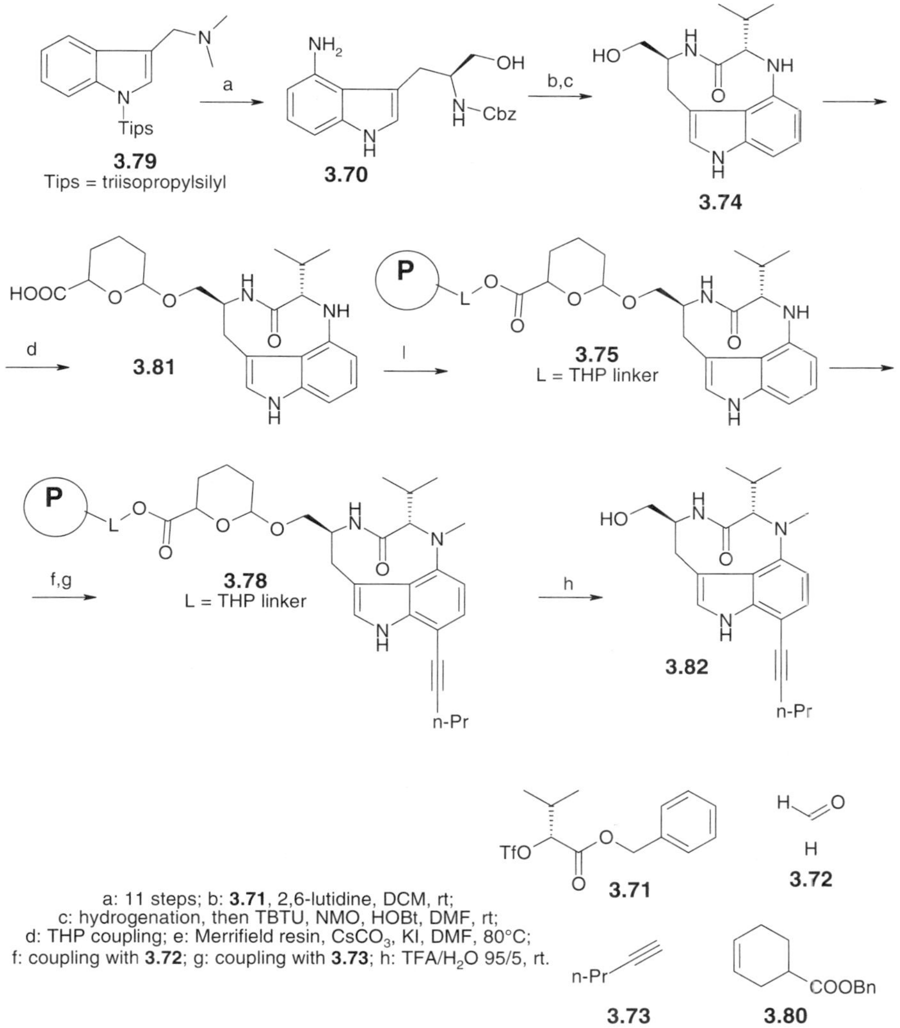

Figure 3.28.

proceeded smoothly using standard reaction conditions. Mild acidic cleavage released pure 7-*n*-propyl indolactam V **3.82** in moderate yield (Fig. 3.28).

Exploitation of the SP route: Three steps were identified to build a combinatorial SP route to indolactam-inspired libraries (Fig. 3.29):

- Coupling of **3.70** with various triflate benzyl esters $\mathbf{M_1}$;
- Reductive amination of **3.75**-like with aldehydes $\mathbf{M_2}$;

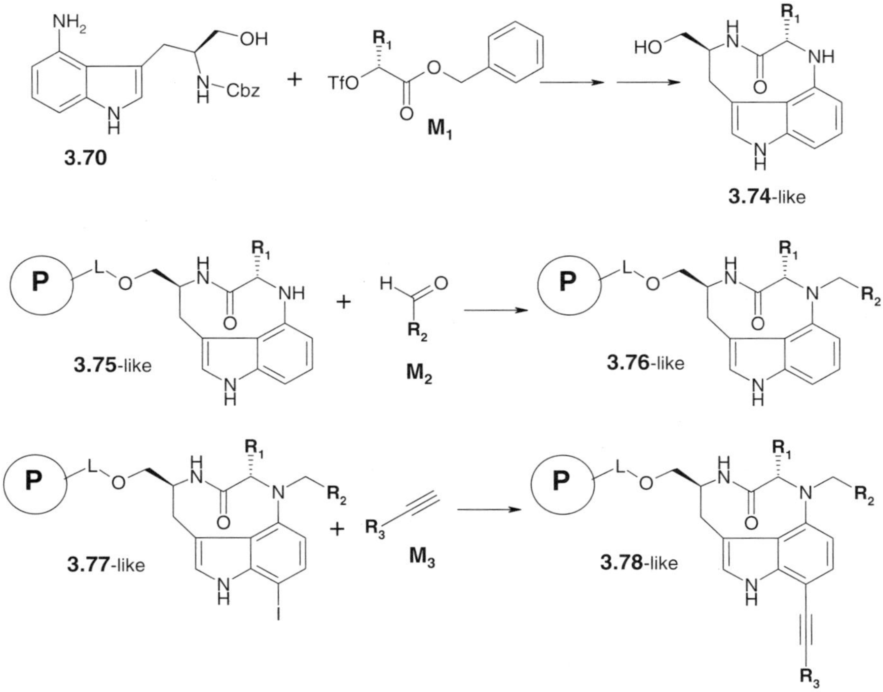

Figure 3.29.

- Sonogashira coupling of **3.77**-like with alkynes **M3**.

A limited number of commercial **M1–M3** representatives were reacted with the appropriate precursors and produced moderate to good yields of analogues after cleavage (Fig. 3.30). Several general comments are worth mentioning:

1. The SP route is of general use, although only similar **R1–R3** groups were used in the exploitation; consequently, the use of more diverse monomers (e.g., orthogonally protected bifunctional reagents) should be recommended;
2. A large number of indolactam analogues should be achievable using commercially available **M1–M3** monomers;
3. Additional derivatizations tolerated by known SAR could be conceived (e.g., N-indole derivatization), and an appropriate SP scheme designed.

3.5.4 SPS of Hexahydro-2,3a,7-Triazacyclopenta[c]Pentalene-1,3-Diones (28).

Rationale of the project:

- Pharmaceutical research craves for novel rigid polyfunctionalized scaffolds with drug-like properties exploitable via library synthesis;

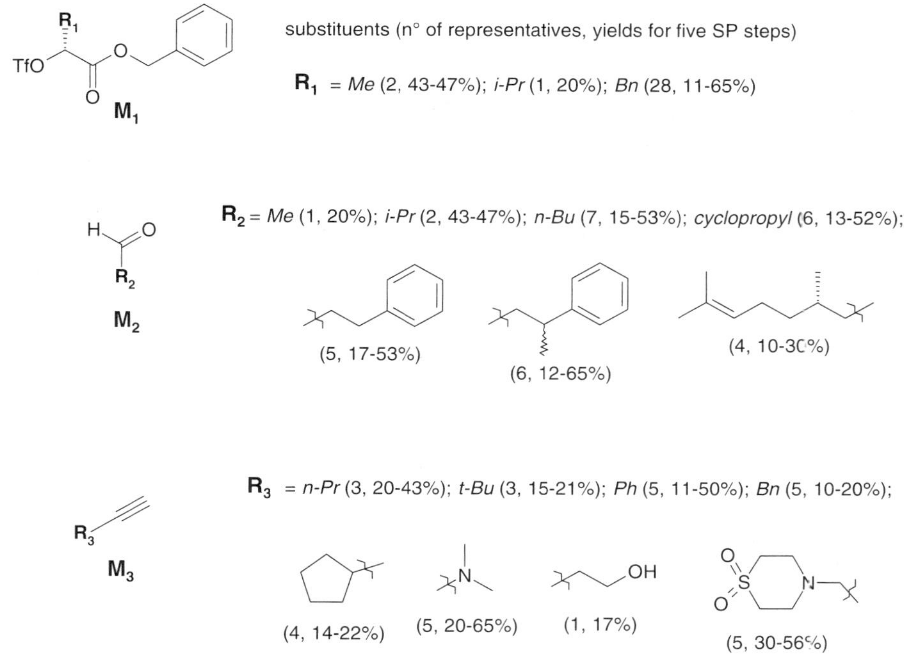

Figure 3.30.

- Among these, polyazaheterocycles are particularly appreciated, as they often recur in pharmacologically active synthetic drugs and natural products;
- Hexahydro-2,3a,7-triazacyclopenta[c]pentalene-1,3-diones such as **3.83** (Fig. 3.31) satisfy the above requirements and a reasonable synthetic scheme can be derived from previous reports (29);
- The full combinatorial exploitation of **3.83**-like scaffolds requires an assessed SP synthetic method to build the tricyclic scaffold and to generate an SP library;

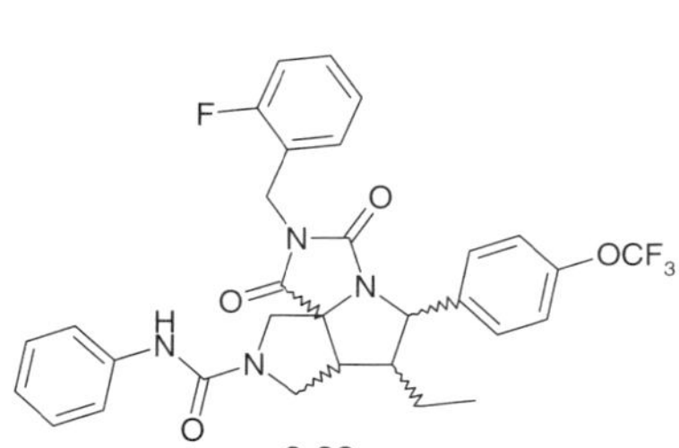

3.83

Figure 3.31.

- The cyclative cleavage approach can be applied building the hydantoin ring and simultaneously releasing in solution the pure tricycle at the end of the SPS.

Target selection and synthetic strategy in solution: The retrosynthetic study to **3.83** (Fig. 3.32) identified compounds **3.84–3.88** as the precursors to the desired nucleus; the aim of this project being the full combinatorial exploitation of a novel scaffold, an assessed SP protocol to general structures **3.89** was necessary. The synthesis in solution of related compounds (**3.90a,b**, Fig. 3.33; compare with **3.89**, Fig. 3.32) being known (29), its adjustment to lead to **3.83** was deemed straightforward; for this reason the authors moved directly to SPS.

SP synthetic strategy to 3.83: The designed synthesis included as a final step the formation of an urea on **3.89** with simultaneous intramolecular cyclization on the carboxylic ester (Fig. 3.34); the use of an ester function to support the **3.89**-like intermediate would have allowed the cyclative cleavage of the desired hexahydro-2,3a,7-triazacyclopenta[c]pentalene-1,3-dione. The use of classical PS resins was not prevented by any of the transformations needed to give **3.83**.

The cyclative cleavage determined the SP synthetic scheme (Fig. 3.35) which supported commercially available **3.84** on hydroxymethyl PS resin, built the appropriate resin-bound cycloaddition substrate **3.93** via orthogonal N-deprotection and functionalization and led to the bicycle **3.94**; this was to be coupled with **3.87**, deprotected and decorated with **3.88**, and cyclatively cleaved to **3.83** (Fig. 3.35). The choice of the two N-protecting groups is crucial for the success of the SP scheme: the α-N-Boc on **3.84** was considered acceptable, being early removed in the SPS, but the β-N-Fmoc

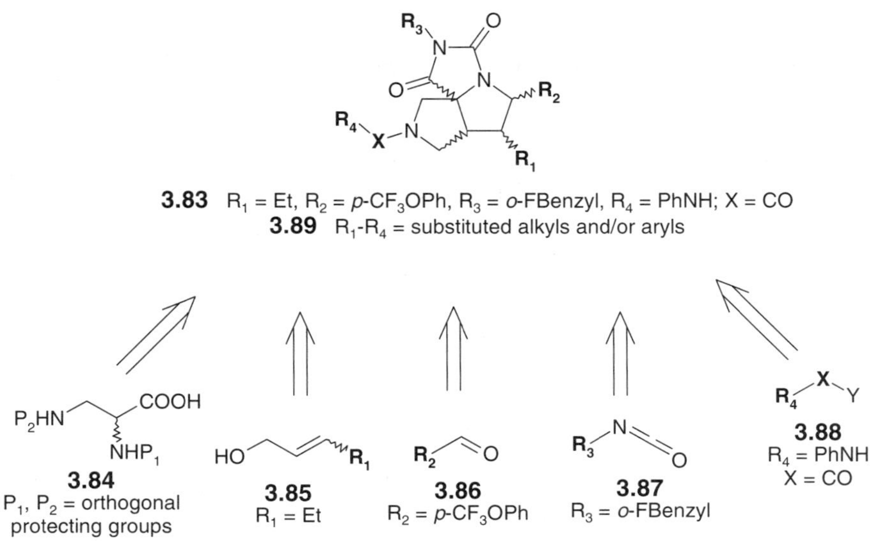

Figure 3.32.

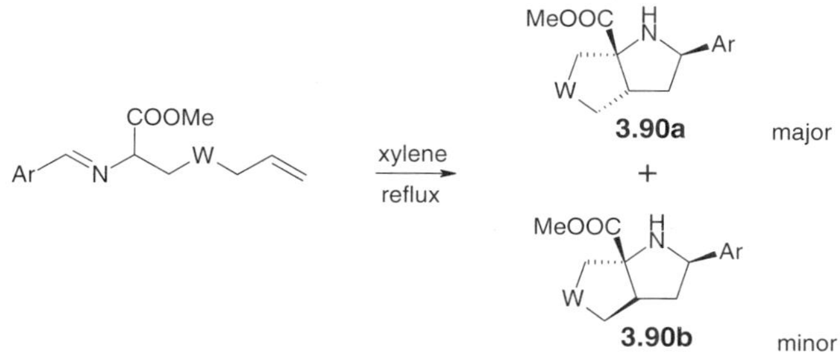

Figure 3.33.

protection was likely to be sensitive during the SP scheme and its replacement with an acid-stable sulfonamide group was planned (Fig. 3.35). The use of a cyclative cleavage approach should have provided pure final compounds, with all the side products still anchored onto SP.

SP chemistry assessment: The designed SP route was successfully validated, using standard reaction conditions, until the formation of the cycloaddition substrate **3.93**; the experimental conditions reported in Fig. 3.36 do not deserve further comments. The cycloaddition to **3.94** was thoroughly studied using a variety of Lewis acids/tertiary bases, and the best results were obtained using 10 eqs. of zinc acetate/DBU in dry acetonitrile. The yield and purity of the **3.84–3.94** conversion was unequivocably determined by cleavage of the ester and release in solution of **3.97** which was isolated in an excellent 56% yield (seven steps plus the cleavage). An unexpected advantage of the SP cycloaddition was represented by its complete regioselectivity in contrast with the results obtained with similar reactions in solution (see **3.90a,b**, Fig. 3.33).

The formation of **3.95** proved to be challenging, and even an extensive assessment with various experimental protocols could not optimize the coupling of isocyanate **3.87** to an acceptable level; a clever solution was found by first coupling the hindered nitrogen of **3.94** with highly reactive and small phosgene, then coupling the imidoyl chloride **3.98** with the primary amine **3.99** (Fig. 3.37). The obtained urea **3.95** reentered the original SP scheme that led with no major problems to the final SP intermediate **3.96** (Fig. 3.37).

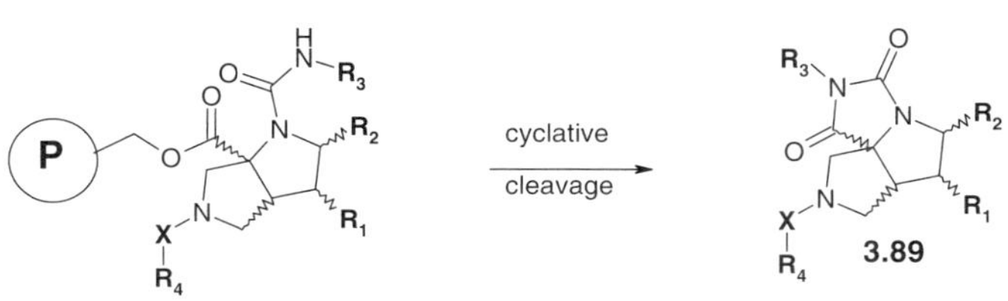

Figure 3.34.

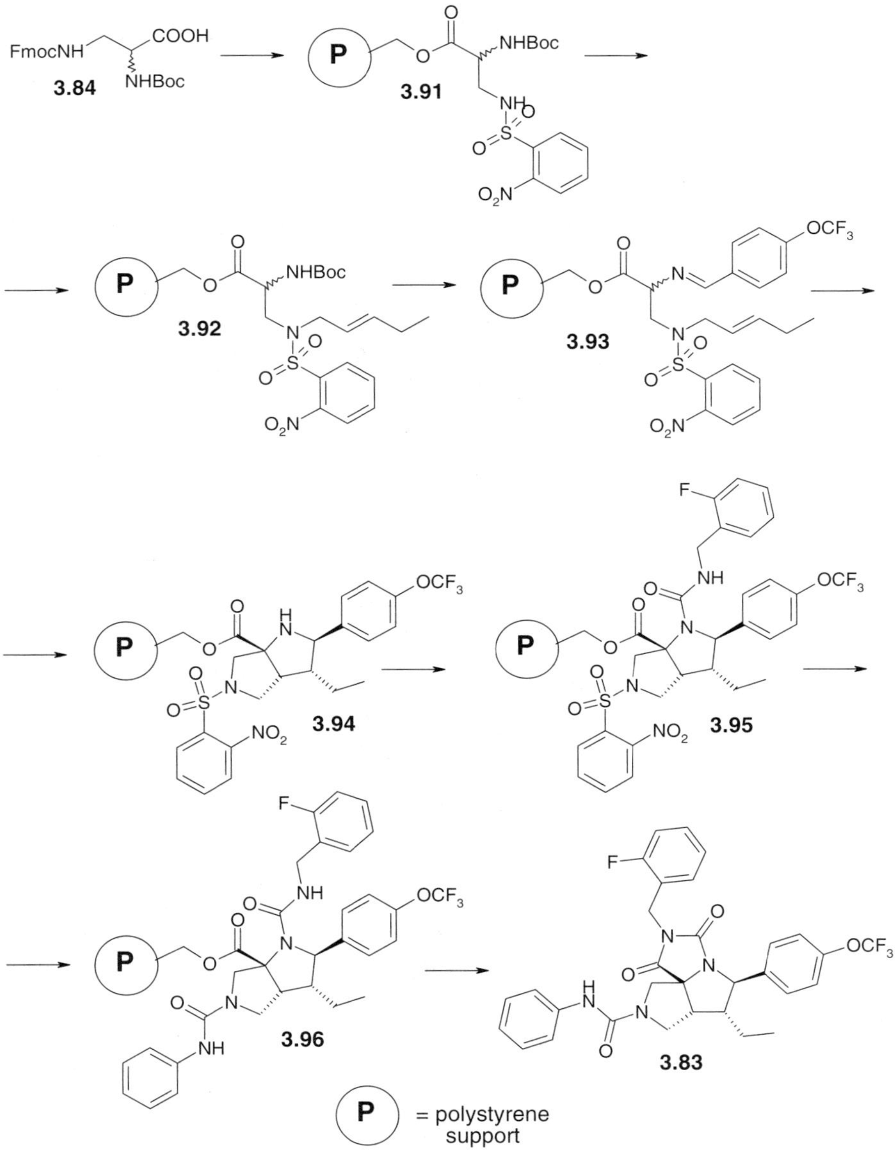
FmocNH
COOH
NHBoc
3.84
P
NHBoc
3.91
NH
O
S
O
O
O
O₂N
3.92
N
3.93
OCF₃
N
3.94
NO₂
H
F
NH
3.95
3.96
3.83
= polystyrene support

Figure 3.35.

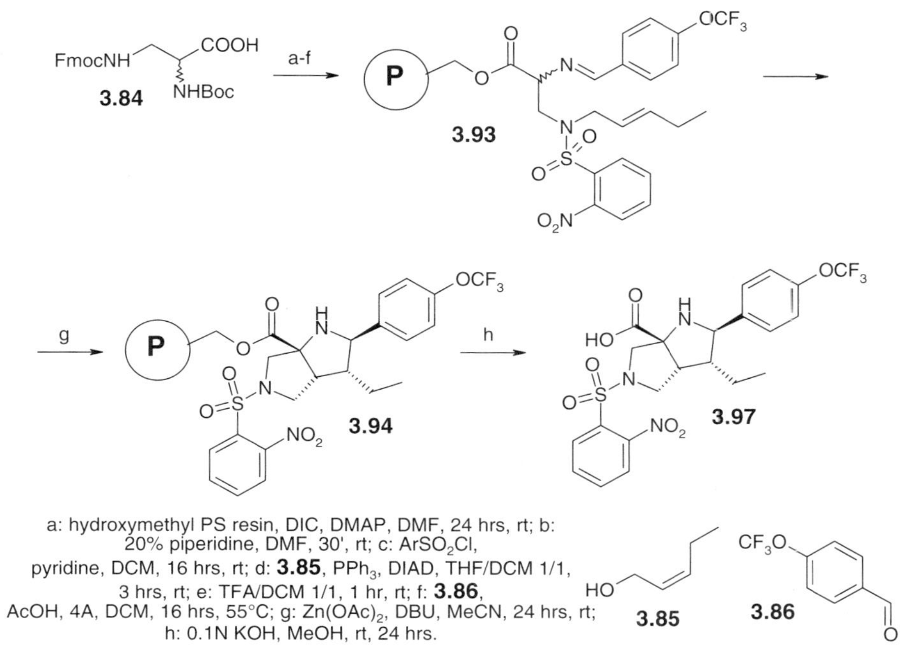

Figure 3.36.

The formation of **3.83** was highly sensitive to experimental conditions, and the use of previously reported protocols for SP intramolecular cyclizations to hydantoins failed. The most abundant isolated side-product was the acid urea **3.100**, derived from basic hydrolysis of the ester function; this finding led to the successful validation of an anhydrous cyclative cleavage protocol using five equivalents of potassium *t*-butoxide in anhydrous THF at RT under an Ar atmosphere for 1 hr (Fig. 3.38). Compound **3.83** was isolated in a satisfactory overall 20% yield (12 steps).

Exploitation of the SP route: Four steps were identified to build a combinatorial SP route to **3.83**-inspired libraries (Fig. 3.39):

- Coupling of **3.91** with various allyl alcohols $\mathbf{M_1}$;
- Reductive amination of deprotected **3.92**-like with aldehydes $\mathbf{M_2}$;
- Reaction of **3.98**-like with primary amines $\mathbf{M_3}$;
- Acylation or reductive amination of deprotected **3.95**-like with acylating agents $\mathbf{M_4}$ or aldehydes $\mathbf{M_5}$.

A set of fully characterized representative compounds was reported (see for example structures **3.101–3.106**, Fig. 3.40); the synthesis of >1000-member libraries was

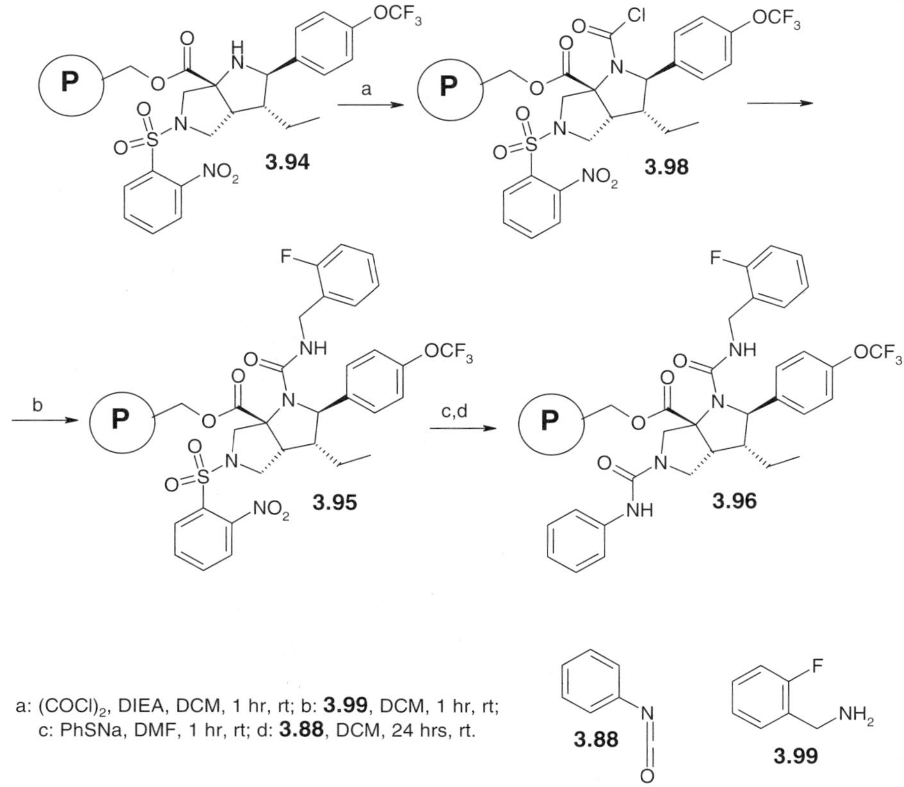

Figure 3.37.

claimed but not described by the authors. Several general comments are worth mentioning:

1. The SP route is of general use, and several monomer classes (**M2**, **M3** and **M5**) were extensively rehearsed. Monomer classes **M1** and **M4** should be better rehearsed in the SP scheme;
2. A large number of hexahydro-2,3a,7-triazacyclopenta[c]pentalene-1,3-diones analogues should be achievable using commercially available **M1**–**M5** monomers;
3. Successfully rehearsed **M2**: aromatic or heteroaromatic, apart from ortho-substituted, strong electron-withdrawing substituents, and non-enolizable aliphatic aldehydes;
4. Successfully rehearsed **M3**: all except poorly nucleophilic and sterically hindered amines;
5. Successfully rehearsed **M5**: no major restrictions.

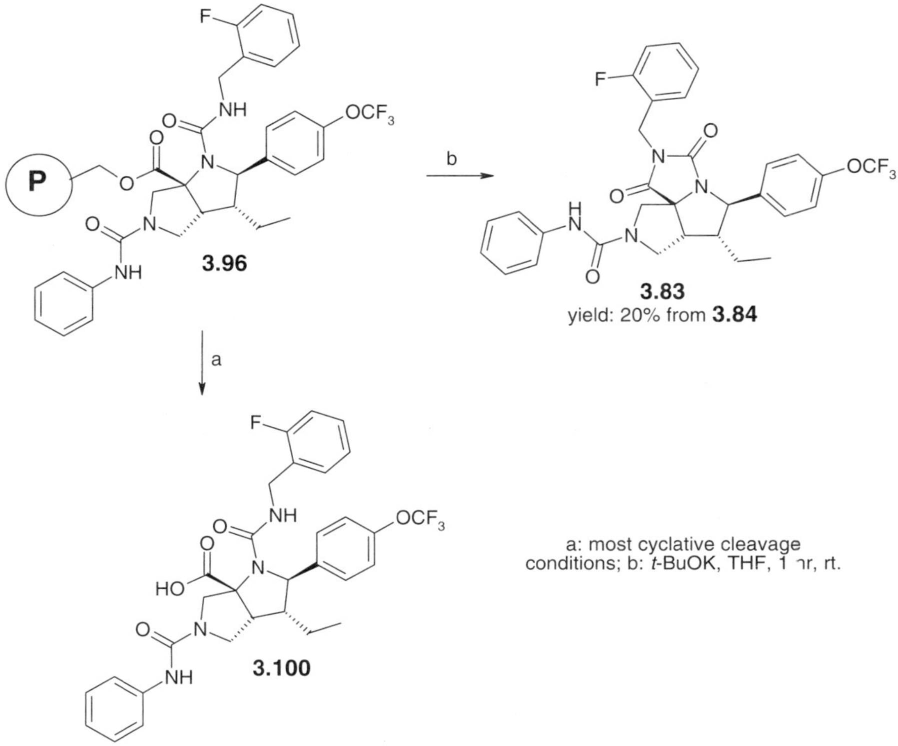

Figure 3.38.

3.5.5 SPS of Prostaglandin E$_1$ (PGE$_1$) analogues (30).

Rationale of the project:

- Prostaglandins are an under-exploited, well known class of biologically active compounds for many clinical applications;
- Rationally designed prostaglandin analogues can be tailored towards specific receptors; an example is **3.107** (Fig. 3.41), a potential binder to prostaglandin EP3 receptor inspired by the structures of known EP3 binders PGE$_1$ (**3.108**), PGE$_1$ methyl ester (**3.109**) and sulprostone (**3.110**);
- The rigid, substituent-orienting prostaglandin core is an ideal scaffold for combinatorial library generation;
- An assessed protocol amenable to high throughput synthesis of prostaglandins does not exist;
- The full combinatorial exploitation of **3.107** requires a solid, assessed SP synthetic method to build the prostaglanding scaffold and to generate an SP library;

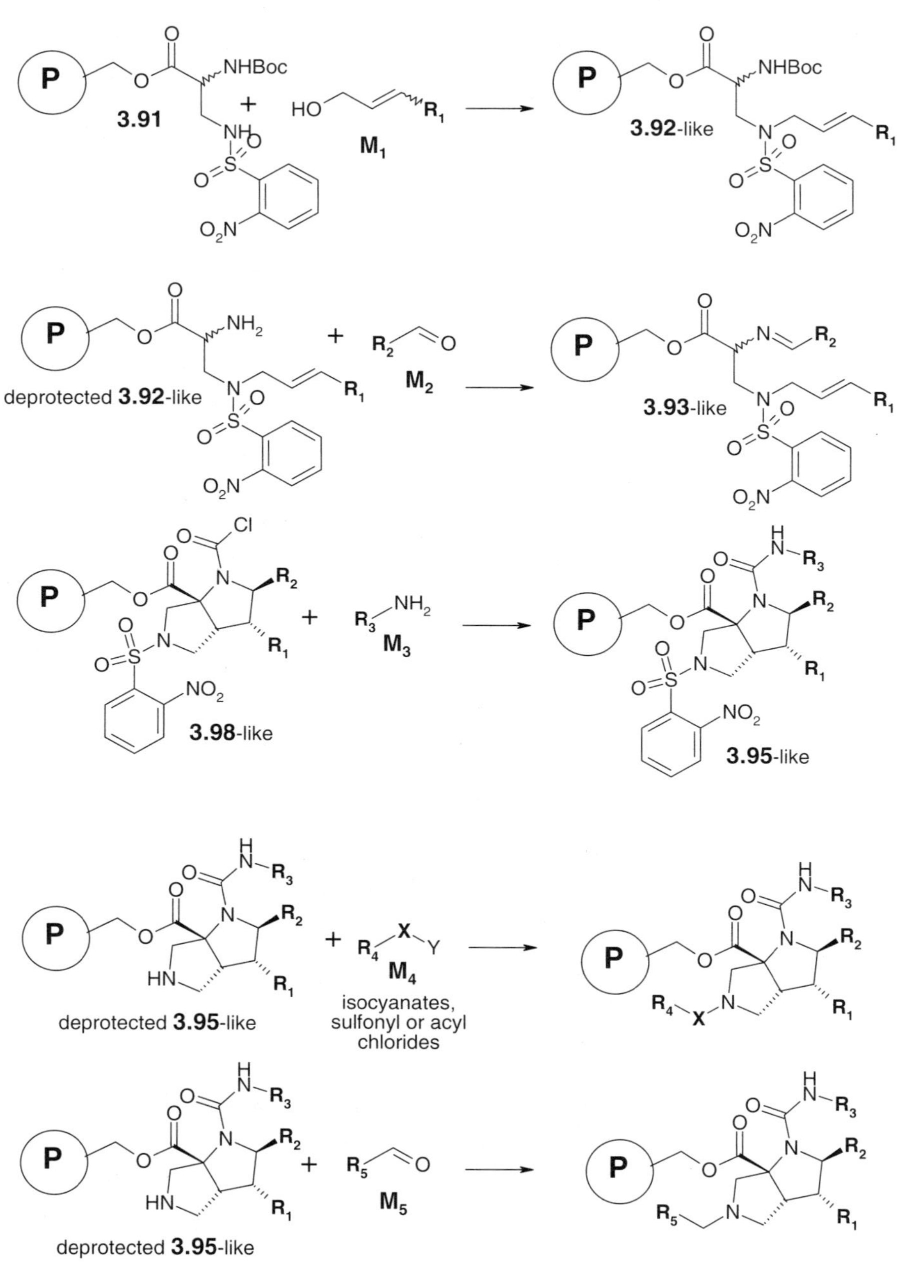
3.91
M1
3.92-like
deprotected 3.92-like
M2
3.93-like
3.98-like
M3
3.95-like
deprotected 3.95-like
M4
isocyanates, sulfonyl or acyl chlorides
deprotected 3.95-like
M5

Figure 3.39.

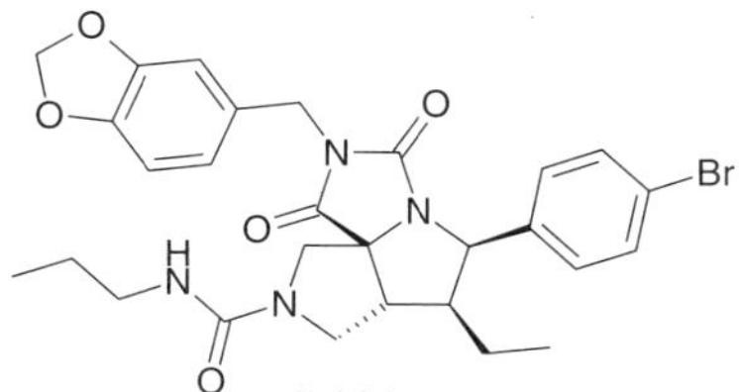

3.101
yield: 19% from **3.84**

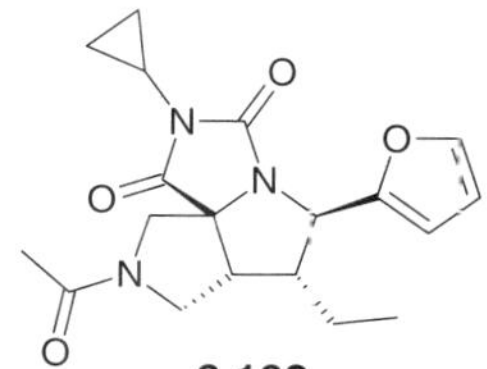

3.102
yield: 25% from **3.84**

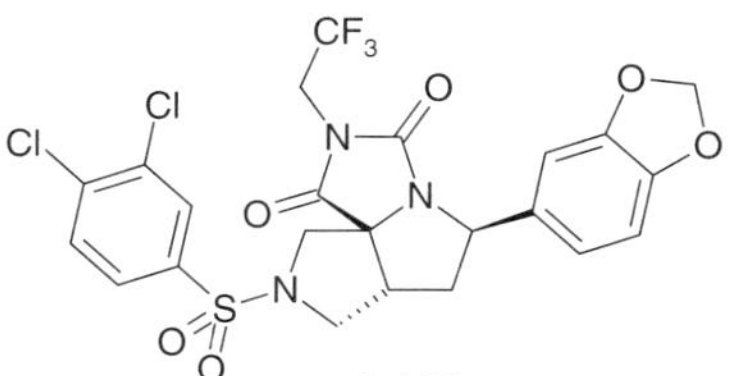

3.103
yield: 21% from **3.84**

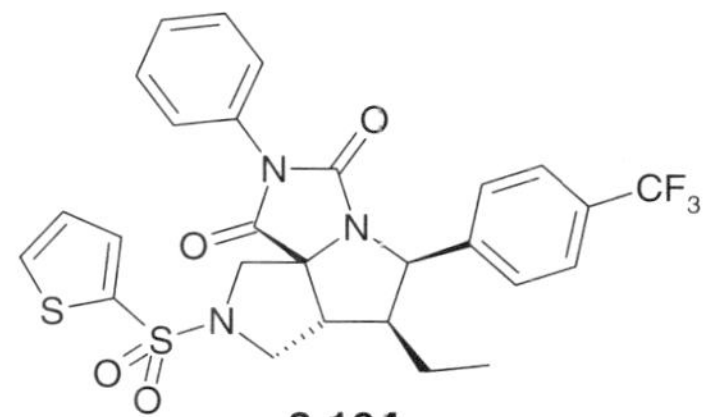

3.104
yield: 19% from **3.84**

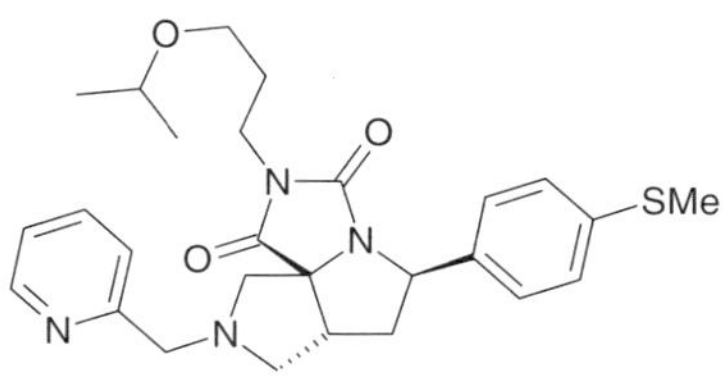

3.105
yield: 21% from **3.84**

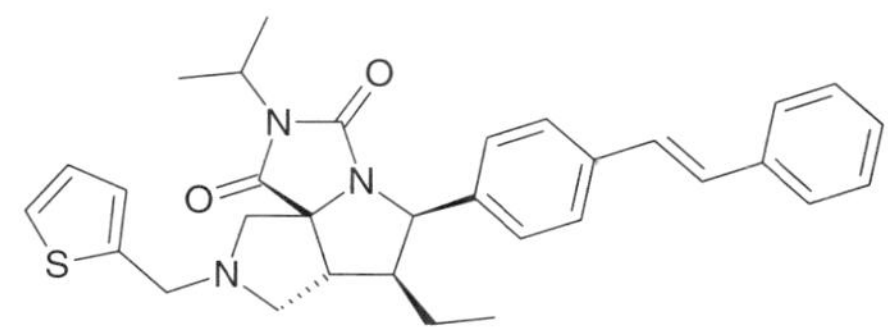

3.106
yield: 16% from **3.84**

Figure 3.40.

- This method should then be adjustable for SPS of other tailored prostaglandins.

Target selection and synthetic strategy in solution: Compound **3.107** represents an example of EP3-targeted prostaglandins to be tested on in vitro assays; other compounds were similarly designed based on **3.108** and **3.109** (*vide infra*). The main structural drivers of the project were the replacement of the carboxylic function of **3.108** to increase bioavailability and affinity compared to the ester **3.109**, and to use decorating side-chains either identical or inspired by EP3-active known compounds; thus, general structure **3.111** can be seen as the project target. Retrosynthesis from **3.107** selected the easily accessible key intermediates **3.112–3.116** (Fig. 3.42); the extensive knowledge about prostaglandin synthesis in solution and previous SP assessments for similar compounds (31) prompted the authors to move directly to SPS.

SP synthetic strategy to 3.107: The designed SP synthetic scheme (Fig. 3.43) supported the intermediate **3.112** via its free hydroxyl function, built the precursor of the

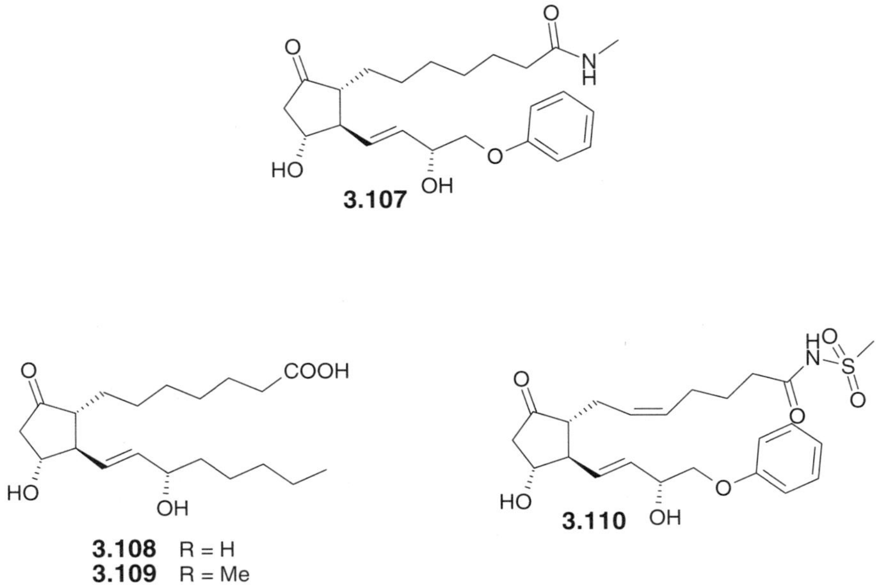

Figure 3.41.

first chain through Suzuki coupling with **3.114** (**3.118**) and then built the second chain through 1,4 conjugate addition of **3.116** on an α,β-unsaturated system obtained after hydroxyl deprotection and oxidation (**3.119**). Conversion of the N-acylsulfonamido group to the desired carboxamide as in **3.107** was planned through N-cyanomethylation (as for traceless linkers, Section 1.2.6) and substitution with amine **3.115** to give the resin-bound prostaglandin **3.121**.

Supporting the hydroxyl function of **3.112** solved the issue of protecting this group during the synthesis; protecting groups other than TMTt (*vide infra*) were not necessary. This scheme did not prevent the use of PS resins, while a careful evaluation of available linkers was requested: stability to basic, nucleophilic and mild acidic conditions was requested, as was a clean cleavage protocol to recover pure materials with no contamination from cleavage reagents. Commercially available, fluoride-labile polystyrene diethyl silyl resin (PS-DES) (32) theoretically satisfies the above requisites, and was selected by the authors. The mild acidic cleavage requested by TMTt should have caused only a negligible loss of loading, while the most common dimethoxytrityl (DMTt) protection would have significantly affected the SPS performance.

SP chemistry assessment: Compound **3.112** was supported onto PS-DES using standard conditions which proceed through chlorination of the resin and *in situ* reaction with **3.112**; moisture was excluded from the reaction mixture and a good loading was obtained (>75%, Fig. 3.44). Suzuki coupling to give **3.118** employed standard conditions; terminal alkene **3.114** was converted *in situ* to the alkyl 9-BBN reagent (step c).

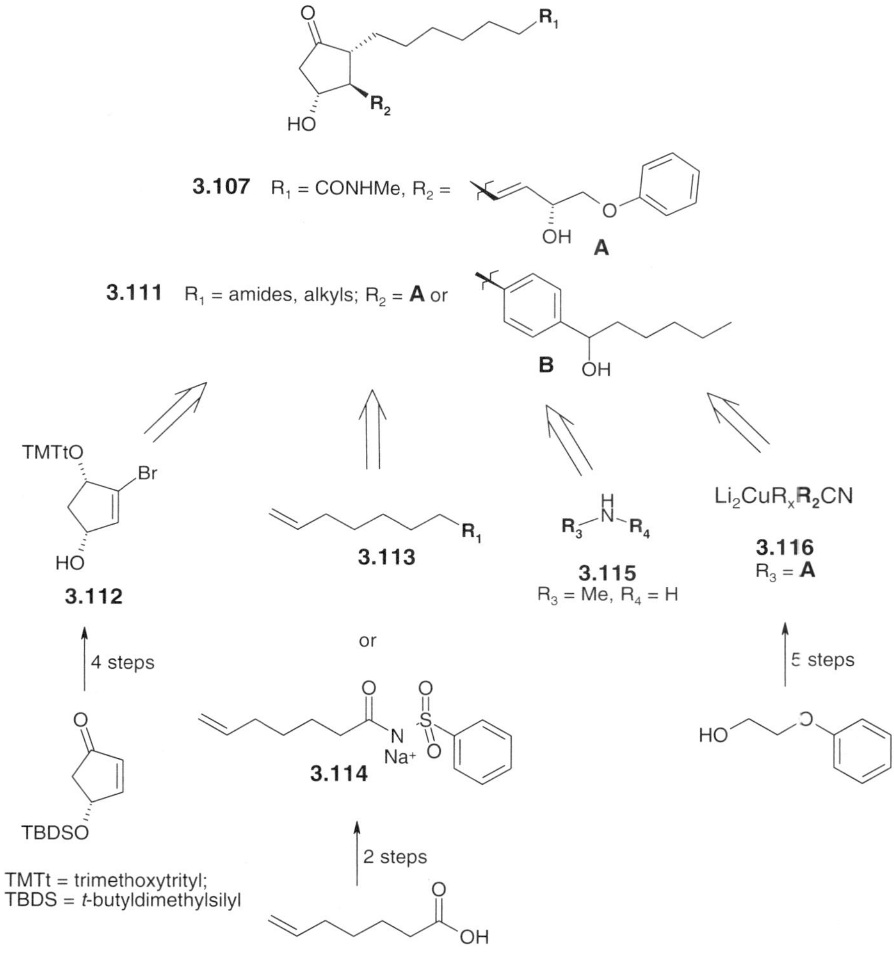

Figure 3.42.

TMTt deprotection was performed in extremely careful conditions (1 min stirring at rt, five cycles, extensive washing) to avoid the cleavage of the DES linker; oxidation with Dess-Martin reagent was performed successfully to give **3.123** (Fig. 3.44).

The 1,4 conjugate addition resulted extremely sensitive to experimental conditions. The detailed procedure reported in Fig. 3.44 gave good overall yields, and represented the best compromise among low temperatures (lower amounts of released material from the support, higher stereocontrol, but negligible resin swelling and reaction rates) and high temperatures and extended reaction times (good swelling and reaction rates, but loss of material and formation of other stereoisomers). Thiophene as an auxiliary

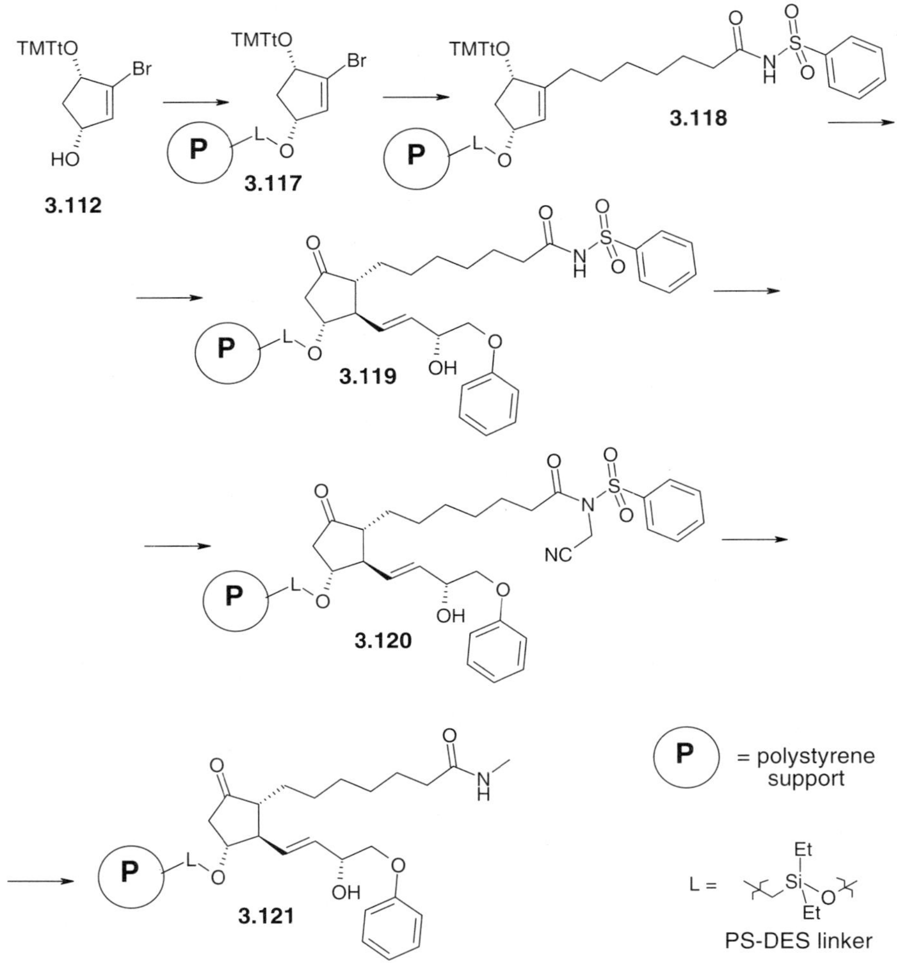

Figure 3.43.

ligand $\mathbf{R_X}$ for the cuprate ensured both a good alkenyl $\mathbf{R_2}$ transfer and absence of competition for the conjugate addition.

Elaboration of **3.119** (prepared as in Fig. 3.45) to **3.121** (steps a,b, Fig. 3.46) was performed in standard experimental conditions. Final cleavage of **3.121** to **3.107** (steps c,d) was better performed using a recently developed "traceless residue" protocol (32) which converts fluoride anions to volatile silyl fluorides which are evaporated from the solution together with pyridine. The product **3.107** was obtained in a good 37% overall yield, calculated for the entire 14-steps SP protocol starting with the loading of **3.112** onto PS-DES resin.

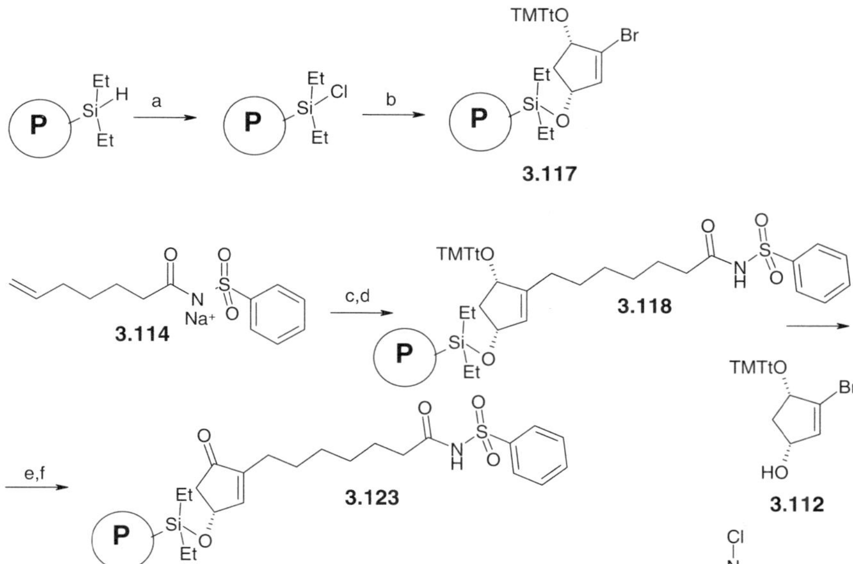

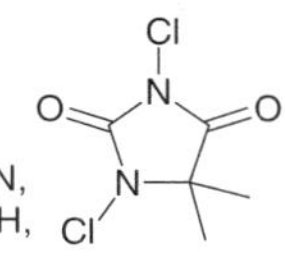

a: **3.122**, anhydrous DCM, Ar, 1 hr, rt; b: **3.112**, imidazole, 4 hrs, rt; c: 9-BBN, THF, 6 hrs, rt; d: **3.117**, $Pd(PPh_3)_4$, Na_2CO_3, THF,12 hrs, 70°C; e: 1M HCOOH, DCM, five cycles, 1' each, rt; f: Dess Martin reagent, DCM, 2 hrs, reflux.

3.122

Figure 3.44.

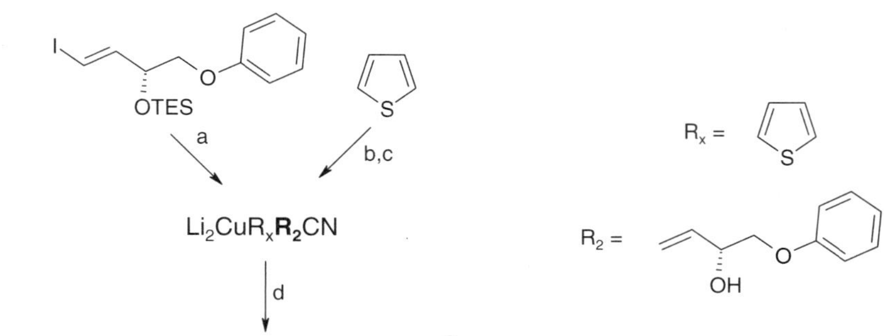

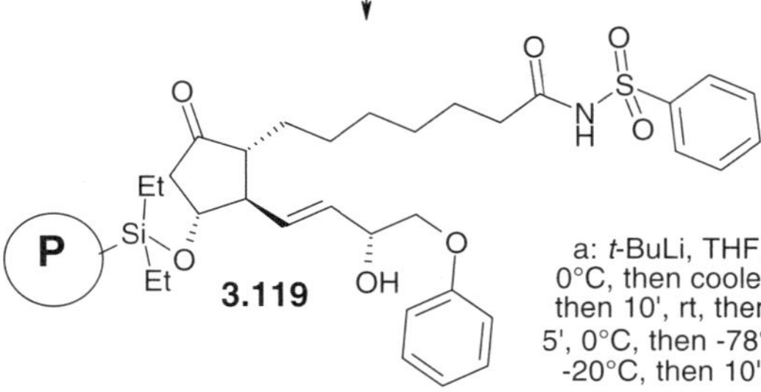

a: *t*-BuLi, THF, 3 hrs, -78°C; b: *n*-BuLi, THF, 1 hr, 0°C, then cooled to -78°C; c: CuCN, THF, 5', -78°C, then 10', rt, then 10', -78°C; d: warming to 0°C, then 5', 0°C, then -78°C, then **3.123**, 20', -78°C, then 1 hr, -20°C, then 10', - 78°C, then quenched with AcOH.

Figure 3.45.

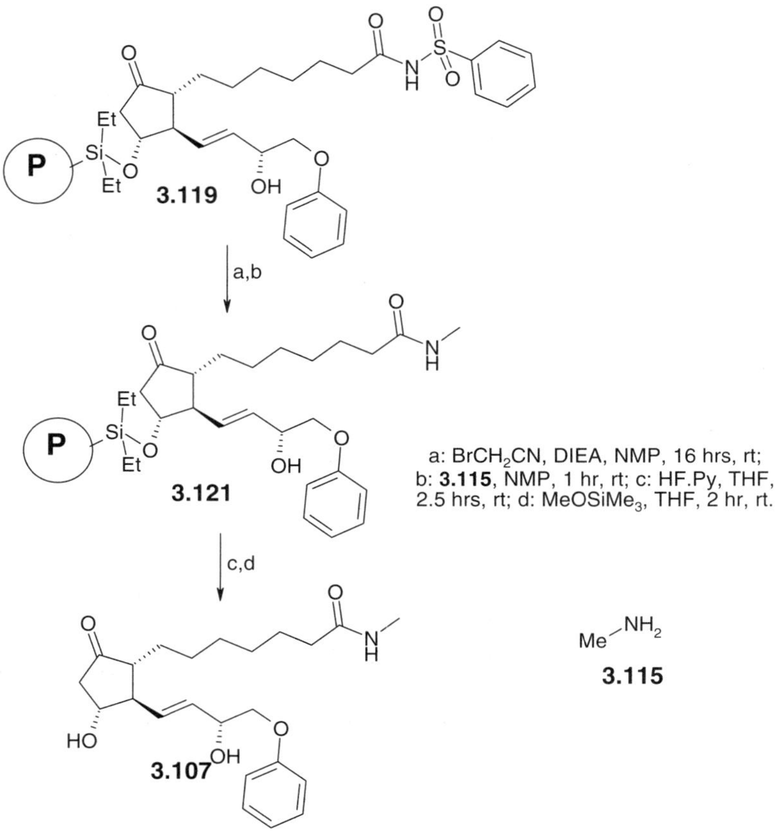

Figure 3.46.

Exploitation of the SP route: The focused nature of this library restricted the modifications to the assessed SP route to two functionalities of **3.107** (Fig. 3.47):

- R2 was imposed as **A** (as for **3.107**) or **B** by addition of the respective cuprates (see the project rationale);
- Compound **3.117** was reacted with **3.114**, then transformed on SP as seen for **3.107** and diversified with amines R3R4NH to give **3.125** (twenty amides, route a);
- Alternatively, compound **3.117** was reacted with terminal alkenes **3.113** to provide compounds **3.127** (six representatives, route b).

The 26 fully characterized representatives were obtained in moderate to good yields. Several general comments are worth mentioning:

1. The focused set of substituents was perfectly suited for the selected and assessed SP route;

2. Depending on biological results on the prostaglandin EP3 receptor a larger set of compounds could easily be obtained using other amines or N-nucleophiles (route a) or terminal alkenes (route b) or cuprates (both routes);

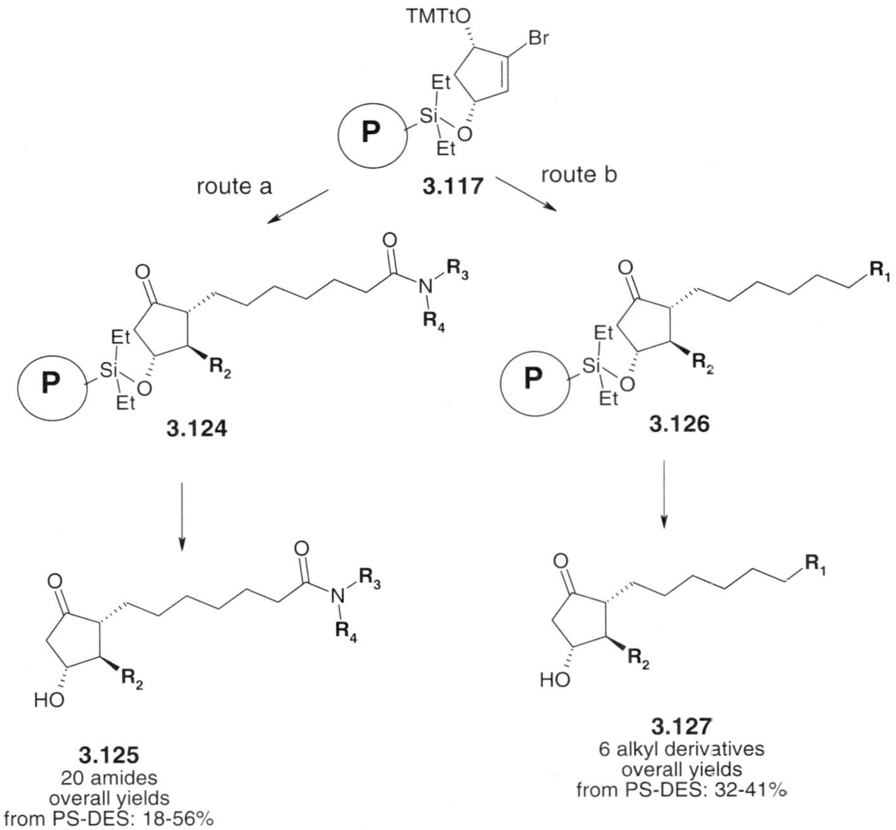

R_1 = H, Me, Et; R_2 = **A** or **B**; HNR_3R_4 = 8 primary and 2 secondary alkyl amines

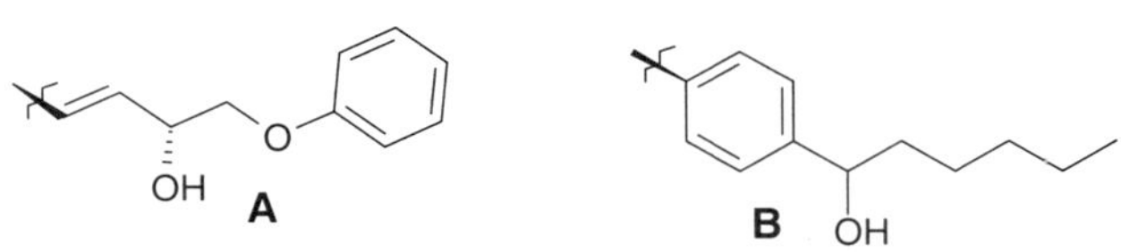

Figure 3.47.

3. The SP route should be of general use, and different classes of monomer sets could focus either towards other prostaglandin receptors or even generate large libraries of diverse prostaglandins.

REFERENCES

1. Bolton, G. L., Hodges, J. C. and Rubin, J. R., *Tetrahedron* **53**, 6611–6634 (1997).
2. Corey, E. J. and Cheng, X. M., *The Logic of Chemical Synthesis*. Wiley-Interscience, New York, 1989.
3. March, J., *Advanced Organic Chemistry: Reactions, Mechanisms, and Structure*, fourth edition. John Wiley & Sons, New York, 1992.
4. Bruice, P. Y., *Organic Chemistry*, second edition. Prentice Hall Press, Englewood Cliffs, NJ, 1998.
5. Baldwin, J. J., *Mol. Diversity* **2**, 81–88 (1996).
6. Khand, I. U., Knox, G. R., Pauson, P. L. and Watts, W. E., *J. Chem. Soc., Perkin Trans.* 977–981 (1973).
7. Ingate, S. T. and Marco-Contelles, J., *Org. Prep. Proced. Int.* **30**, 121–143 (1998).
8. Geis, O. and Schmalz, H.-G., *Angew. Chem., Int. Ed. Engl.* **37**, 911–914 (1998).
9. Schore, N. E. and Najdi, S. D., *J Am. Chem. Soc.* **112**, 441–442 (1990).
10. Spitzer, J. L., Kurth, M. J., Schore, N. D. and Najdi, S. D., *Tetrahedron* **53**, 6791–6808 (1997).
11. Sieber, P., *Tetrahedron Lett.* **28**, 6147–6150 (1987).
12. Fruechtel, J. S. and Jung, G., *Angew. Chem., Int. Ed. Engl.* **35**, 17–42 (1996).
13. Hermkens, P. H. H., Ottenheijm, H. C. J. and Rees, D. C., *Tetrahedron* **52**, 4527–4554 (1996).
14. Hermkens, P. H. H., Ottenheijm, H. C. J. and Rees, D. C., *Tetrahedron* **53**, 5643–5678 (1997).
15. Booth, S., Hermkens, P. H. H., Ottenheijm, H. C. J. and Rees, D. C., *Tetrahedron* **54**, 15385–15443 (1998).
16. Hall, S. E., *Annu. Rep. Combi. Chem. Mol. Div.* **2**, 15–26 (1999).
17. Andres, C. J., Denhart, D. J., Desphande, M. S. and Gillman, *Comb. Chem. High Throughput Screen.* **2**, 191–210 (1999).
18. Kingsbury, C. L., Mehrman, S. J. and Takacs, J. M., *Curr. Org. Chem.* **3**, 497–555 (1999).
19. Burgess, K., *Solid-Phase Organic Synthesis*. Wiley Interscience, New York, 2000.
20. Nicolaou, K. C., Pfefferkom, J. A., Cao, G.-Q., Kim, S. and Kessabi, J., *Org. Lett.* **1**, 807–810 (1999).
21. Nicolaou, K. C., Pastor, J., Barluenga, S. and Wissinger, N., *Chem. Commun.* 1947–1948 (1999).
22. Jackson, W. P., Ley, S. V. and Morton, J. A., *Tetrahedron Lett.* **22**, 2601–2604 (1981).
23. Verhe', R., Schamp, N. and De Buyck, L., *Synthesis* 392–393 (1975).
24. Nicolaou, K. C., Pfefferkom, J. A., Kim, S. and Wei, H. X., *J. Am. Chem. Soc.* **121**, 4724–4725 (1999).

25. Meseguer, B., Alonso.Diaz, D., Griebenow, N., Herget, T. and Waldmann, H., *Angew. Chem. Int. Ed.* **38**, 2902–2906 (1999).
26. Kogan, T. P., Somers, T. C. and Venuti, M. C., *Tetrahedron* **46**, 6623–6632 (1990).
27. Thompson, L. A. and Eliman, J. A., *Tetrahedron Lett.* **35**, 6333–6336 (1994).
28. Peng, G., Sohn, A. and Gallop, M. A., *J. Org. Chem.* **64**, 8342–8349 (1999).
29. Armstrong, P., Grigg, R., Jordan, M. W. and Malone, J. F., *Tetrahedron* **41**, 3547–3558 (1985).
30. Dragoli, D. R., Thompson, L. A., O'Brien, J. and Ellman, J. A., *J. Comb. Chem.* **1**, 534–539 (1999).
31. Thompson, L. A., Moore, F. L., Moon, Y. C. and Ellman, J. A., *J. Org. Chem.* **63**, 2066–2067 (1998).
32. Hu, Y. H., Porco, J. A., Labadie, J. W., Gooding, O. W. and Trost, B. M., *J. Org. Chem.* **63**, 4518–4521 (1998).

4 Combinatorial Technologies: Basic Principles

This and the following seven chapters will cover the main topic of this book, combinatorial technologies. The field of combinatorial technology was born only very recently; however, it has rapidly captured the imagination of an enormous number of research workers generating a vast literature in the process. A discipline that was primarily confined to applications in the pharmaceutical industry is now being widely applied in other fields and has attracted the attention of world renowned scientists such as Nicolaou, Ley, Boger, Weinberg, Burgess, Schultz, Curran, and Geysen, to name but a few of those who have contributed substantially to the expansion of combinatorial technologies.

This introductory chapter will be devoted to three main topics, presenting the reader with the most useful basic principles of the most relevant aspects of combinatorial technologies. To begin with, a *glossary* will be provided, in which the most relevant concepts of combinatorial technologies will be briefly defined to familiarize the reader with the terms and with their meaning used in the area. This will be followed by *a historical background*, which will describe the main motives leading to the birth of combinatorial chemistry and its attendant technologies through the milestone contributions presented in chronological order. The different *classes* of combinatorial libraries will then be briefly presented and discussed, highlighting their relative merits and usefulness for various applications.

4.1 COMBINATORIAL TECHNOLOGIES

4.1.1 Glossary

The purpose of this section is to present the reader with basic definitions of some of the most commonly encountered terms in the area of combinatorial technologies rather than attempting to provide an exhaustive list. Many concepts introduced here will appear often in the following chapters.

Combinatorial chemistry refers to the synthetic chemical process that generates a set or sets (combinatorial libraries) of compounds in a simultaneous rather than a sequential manner. The broader concept of *combinatorial technologies* includes all of the disciplines involved in the synthesis and the applications of synthetic chemical libraries, such as their attendant analytical and computational chemistry methods and high-throughput screening (HTS, *vide infra*), as well as the disciplines related to other libraries that have found applications in the biological disciplines and also in materials science.

A *combinatorial library* is a single entity composed of many individuals (typically hundreds to millions) that can be prepared in a variety of formats through the use of different techniques. A *synthetic organic library* is prepared using standard organic chemistry, either in solution or on SP. A *biosynthetic library* is usually composed of natural oligomers and is prepared by natural means (e.g., phage display libraries) or through the use of biological or biochemical reagents (e.g., enzymes and oligonucleotide amplification systems). A *materials science library* is made from inorganic compounds and is prepared using techniques peculiar to this field, such as sputtering deposition of thin films, electron beam evaporation, and moving-masks techniques, among others.

A *primary*, or unbiased, *library* is a large set of compounds (typically thousands to millions) based on diversity and aimed at the discovery of samples of interest for targets for which little, if any, information is available. *Diversity* is a concept unrelated to the library size that attempts to evaluate the representation of chemical space by a chemical library using computational methods: If this space is sampled evenly by the components of a library, then this library is considered to be diversity based (Fig. 4.1, left). A *focused*, or biased, *library* is a similarity-based set of compounds (typically hundreds to thousands) aimed at the discovery and optimization of lead structures for a target for which a structural model on which to design the library is available. *Similarity* is a concept unrelated to the library size that is opposite to diversity: if the library components are clustered around the model structure **A**, the library is similarity based (see Fig. 4.1, right).

A *discrete library* is a set of compounds that are obtained as individuals at the end of the library synthesis. *Parallel synthesis* leads to a discrete library by simultaneous addition of reactants in different reaction vessels and parallel handling of each library sample (see the example in Fig. 4.2, where 15 discretes are prepared from the common intermediate **A** with two parallel reaction steps). Conversely, a *pool library* is a set of compounds that are obtained as mixtures, or library pools, at the end of the synthesis. *Mix and split* (or divide and recombine) is the process leading to an SP pool library

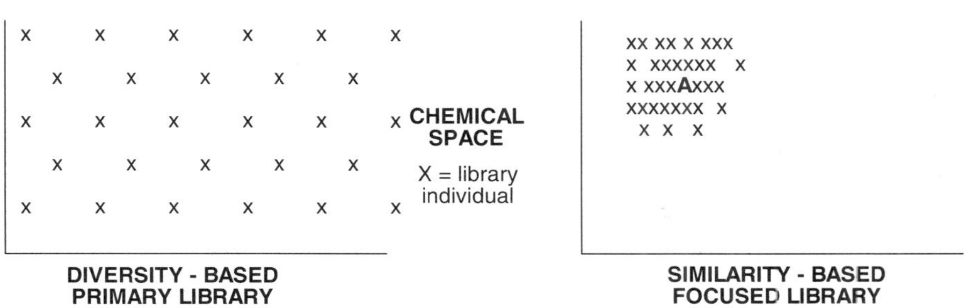

Figure 4.1 Chemical diversity and similarity: primary and focused libraries in the chemical space.

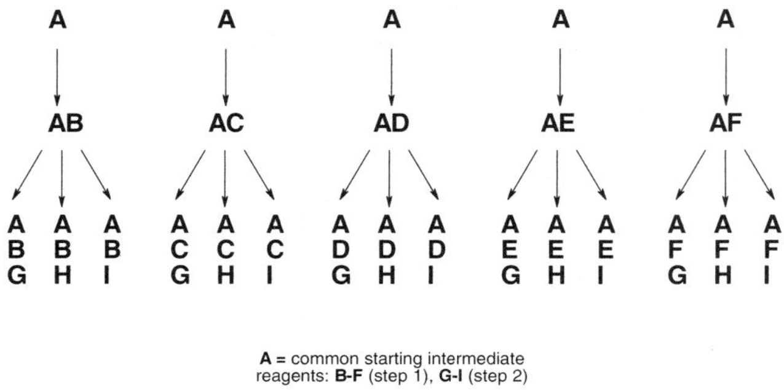

Figure 4.2 Parallel synthesis of a 15-member discrete library ABG-AFI.

by simultaneous addition of reactants in different reaction vessels followed by mixing and splitting the resin portions to eventually produce the final library pools in which each resin bead carries a single library component (see Section 7.1.1).

A *scaffold* is the common structural element contained in all of the individuals of the library: It can be an iterated chemical bond (e.g., in oligomeric libraries such as peptides it is a repeating amide backbone), a functional group (e.g., in libraries of substituted guanidines), or a ring motif (e.g., in libraries of substituted benzodiazepines as in Fig. 4.3). A *building block* is a reagent used during the synthesis of a library (e.g., the lithium salt of acetanilide; Fig. 4.3). A *monomer set* is a class of reagents with a common functional group that is used either to produce substituted scaffolds or to decorate preexisting scaffolds (e.g., substituted anthranilic acids, alkyl halides, and α-amino acids; Fig. 4.3). A *monomer* is a reagent that is part of a monomer set (e.g., 5-chloroanthranilic acid, 2-methoxybenzyl bromide, and 2-thienylalanine methyl ester; Fig. 4.3). A *randomization point* ($\mathbf{R}_1$, $\mathbf{R}_2$, and $\mathbf{R}_3$ in Fig. 4.3) is a position where a monomer can be inserted into a library during the construction of the scaffold or where it can be coupled to a preexisting scaffold to produce library components containing all of the possible combinations of the selected monomers.

The term *chemical assessment* describes the process through which the reaction scheme to arrive at a target molecule is combinatorialized. This process may include the transfer of a reaction from solution onto SP and/or the adaptation of the reaction conditions to the use of many monomers with different reactivities and stabilities for library synthesis. *Monomer rehearsal* is an accurate check of the reactivity of a monomer set in the synthetic scheme for the buildup of the library so that the unreactive/difficult monomers are removed from the set. A *model library* is a small set of discretes, or a small pool, that is prepared using the planned synthetic route for the library and is fully characterized by the appropriate analytical tools: Only if the results are satisfactory is the library synthesis carried out. *Quality control* determines the analytical profile of a library as a single entity, but data from each library individual, or a significant percentage of library individuals, are acquired. A library with 80% confirmed pure compounds is a good-quality library, but the 20% of samples that are

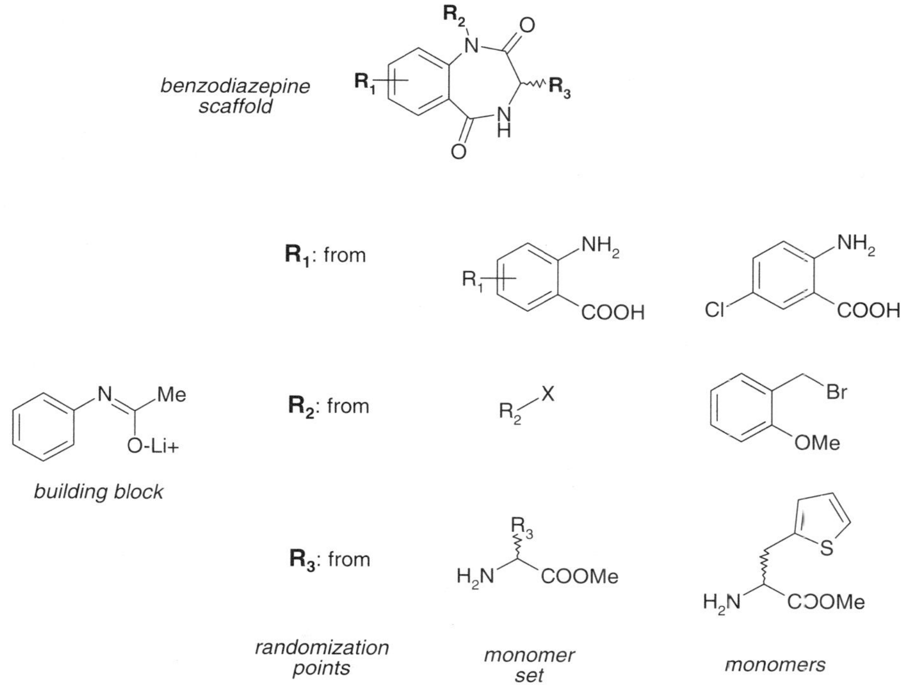

Figure 4.3 Structural elements of a library: scaffolds, building blocks, monomers, monomer sets, and randomization points.

not confirmed, or are impure, are known and discarded before the library screening (*vide infra*).

Structure determination links a desired activity observed during the screening of the library with the structure of the active library component and is relatively straightforward for discrete libraries, while several techniques can be applied for the structure determination of compounds in pooled libraries. *Deconvolution* is the process whereby the complexity of an active SP pool is unraveled through iterative synthesis of smaller pools while tracking the new location of the desired activity. The final iteration is made of individual compounds, and the structure of the active component can be determined. *Chemical encoding* is a process in which the structure of every library component is unambiguously encoded by another chemical structure, called a *tag* or *code*, that is easily identified by automated procedures. This allows the indirect structure determination of the active library component after the library screening. The principle of *nonchemical encoding* is identical, but the tag is not a chemical entity; examples include radiofrequency tags or optical tags.

Library screening is the process where the activity of each library individual is measured and reliably determined; it may consist of a biological assay (for pharmaceutical purposes) or an analytical measurement (for materials science, for catalysis, or for polymer applications among others). *High-throughput screening* (HTS) is the fast, automated version of classical screening protocols and is the ideal counterpart of combinatorial chemistry: Large libraries made by combinatorial chemistry are screened by HTS methods and the whole process is fast and reliable. A *positive* is a library component that shows a significant activity during the library screening and whose structure is unequivocally determined via the above-mentioned structure determination methods.

4.1.2 Historical Background

Historically, the use of compounds derived from natural sources satisfied many of the medical and technological needs of society, but the advent of rational scientific disciplines and of chemistry in particular has culminated in the development of synthetic chemical methods that can supply compounds designed to possess specific attributes suited to a wide range of applications. The classical synthetic process, which is still used in many areas of chemistry, implies the synthesis of a compound, its purification to determine structure and quality, and its characterization in the test system for which it was prepared to determine its properties. This latter step is performed through reliable, low-throughput assays that match the low throughput of classical synthetic methods. An example related to pharmaceutical research would be the synthesis and thorough evaluation of around 100 druglike molecules per year by a small group of chemists that could be considered successful using classical synthetic and pharmacological methods.

The gradual shift to the so-called HTS methods, driven in part by the needs of pharmaceutical research, has revolutionized the concept of the drug discovery process. Implementation of assays aimed at molecular targets, requiring minimal quantities of compound, and that are easy to automate became common in the late 1980s. The consequence of this was to significantly increase the demand for chemical entities to be tested in the high-throughput biological assays. Every major pharmaceutical company possesses a large compound collection containing many proprietary compounds (typically hundreds of thousands) that were previously only tested for a specific mechanism of action in a specific assay. With the advent of this new screening philosophy, these collections were tested randomly on many different assays in the hope of producing novel and relevant biologically activity hits. An interesting experimental study has recently analyzed the usefulness of compound collections in drug discovery in providing novel, unpredictable active structures on specific targets (1).

While useful, these collections were generally biased by the previous activities of each company, which largely depend on the historical involvement in certain markets to the exclusion of others. The repeated use of these collections for HTS eventually brought about the consumption of many of their components. Methods to produce large sets of structurally related compounds in a high-throughput manner therefore became extremely important because of the need to replenish and extend available compound collections for testing for potential pharmaceutical applications. These methods, which aimed at providing a compound flow that could match the expectations of HTS, produced libraries of compounds with defined structures either to be used for generic applications or that were designed to include specific structural features related to selected biological targets. Both industrial and academic groups started to become active in this field around the mid-1980s.

Initially attention was focused on natural oligomeric molecules such as peptides and oligonucleotides because the preparation of libraries based on these structures, either by chemical synthesis or by biosynthetic methods, was easily carried out while large sets of other classes of compounds required more effort and/or more complex procedures.

The first attempts to create peptide synthetic libraries were reported by Furka (2) in an obscure communication in 1982, but not having been published in a major journal, their importance was completely underestimated. In 1984 Geysen et al. (3) reported the synthesis and the biological characterization of a peptide library made up of several hundred discrete compounds using a rack of plastic polystyrene pins as supports (Section 1.1.4). A year later, Houghten (4) disclosed the use of the so-called tea bags, each containing 10–20 mg of resin beads, to prepare and biologically characterize as discretes a few hundred peptide 13-mers. After several other reports by Furka et al. (5), Frank et al. (6), and Houghten et al. (7), an important breakthrough was achieved by Lam et al. (8) in 1991, who described the synthesis of around 2,500,000 peptides using the so-called mix-and-split method and introduced the "one bead, one peptide" concept. In the same year, Fodor et al. (9) reported the synthesis of thousands of discrete peptides in 10 steps using the so-called spatially addressable parallel synthesis on glass support. In the following years the synthesis of large peptide libraries containing up to several million compounds has become a somewhat routine task using automated SPS.

Large sets of peptides or oligonucleotides have also been obtained from biological sources. The so-called phage display technique was introduced by Smith in 1985 (10), who then described its potential application to library generation in 1988 (11), which was eventually exploited in 1990 (12, 13). The expression of large, random assortments of peptides on the surface of gram-negative bacteria using fusion protein strategies and their application as peptide libraries, vaccines, or antibodies was reported. There have been many accounts of the creation of phage display libraries in the literature since these first reports (Section 10.1).

While complementary DNA (cDNA) and genomic libraries have long been used in molecular biology, the breakthrough for oligonucleotide libraries came when Tuerk and Gold (14) introduced SELEX (systematic evolution of ligands by exponential) enrichment to prepare large, amplifiable oligonucleotide libraries and to select from their populations sequences of nucleic acids capable of binding to specific targets (aptamers). This methodology, which was inspired by previous reports in the late 1980s (15–17), has become very popular. A similar selection–amplification process was reported by Pan and Uhlenbeck (18) and Bartel and Szostak (19) to find catalytic nucleic acid sequences called ribozymes, which are able to catalyze the cleavage or formation of various chemical bonds. Many reports of aptamer or ribozyme libraries appeared in the literature in the last 10 years (Section 10.2).

The use of oligonucleotide and especially peptide libraries was instrumental in uncovering to find new sequences able to interact with molecular targets and also in finding oligomeric structures suitable for use as a variety of biological tools. These molecules, though, suffer from many serious drawbacks when considered as potential drugs and the need for high-throughput synthesis of more "druglike" molecules for pharmaceutical applications became more pressing. The so-called small organic molecule (SOM) combinatorial libraries first appeared in 1992 (20) in a paper by Bunin and Ellman in which the preparation of a few 1,4-benzodiazepines on SP was described. This SPS was extended in 1994 (21) to give a discrete library of 192 1,4-benzodiazepines on plastic pins. The last few years have witnessed an enormous growth in interest for SOM libraries with the production of many molecules of interest for a wide range of pharmaceutical applications. The SOM libraries offer a more diverse range of structures when compared to the oligomeric libraries and can be tailored more precisely to specific needs. These libraries have gradually become the main target for combinatorial chemists for pharmaceutical applications, and many applications in other fields have also been reported. The accurate description of these libraries and of the related combinatorial methodologies will span five full chapters (5–9) of this book.

Finally, the use of libraries in the field of materials science libraries is also growing steadily after the first report by Xiang et al. in 1995 (22), which described the parallel synthesis of spatially addressable libraries of solid-state materials and their screening to produce novel materials with superconducting properties. This and other groups have reported the discovery of magnetoresistive, photoluminescent, and dielectric materials among others (Section 11.1 and 11.2). A significant number of polymer libraries were also reported in the last five years; they will be extensively covered in Section 11.3.

4.2 COMBINATORIAL LIBRARIES

4.2.1 Synthetic Organic Libraries: Natural and Unnatural Oligomers

The synthesis of oligomeric libraries, and especially peptide libraries, has been described in a large number of reports. We will discuss the main properties and applications of such libraries using the four examples reported in Figs. 4.4–4.8.

A linear hexapeptide amide library (**L1**, Fig. 4.4) was prepared on SP by Dooley et al. (23) by randomizing the 20 natural L-amino acids in positions X_1 and X_2 and 19 of them in positions X_3–X_6 (Cys omitted). The library consisted of $20 \times 20 \times 19 \times 19 \times 19 \times 19 = 52{,}128{,}400$ different hexapeptides and was prepared as a pool library made up of 400 subsets, or library pools, of $19 \times 19 \times 19 \times 19 = 130{,}321$ individuals. Among these pools, $YPX_3X_4X_5X_6$ (**L2**, Fig. 4.4) was the most active as inhibitor of the binding of an enkephalin-related peptide to rat brain homogenates and was further deconvoluted to find a few active sequences (Fig. 4.4) among which YPFGFR–$CONH_2$ was the most active inhibitory concentration ($IC_{50} = 13$ n*M*). The huge number of possible library components that can be made from the 20 commercially available precursors, together with the robust and readily available automated SPS of peptides (see Section 2.1), make synthetic peptide libraries with up to eight randomized positions a useful reservoir for "ligand fishing," that is, to identify peptidic sequences that are able to interact with a selected target. Biosynthetic peptide libraries (Section 11.1) are more

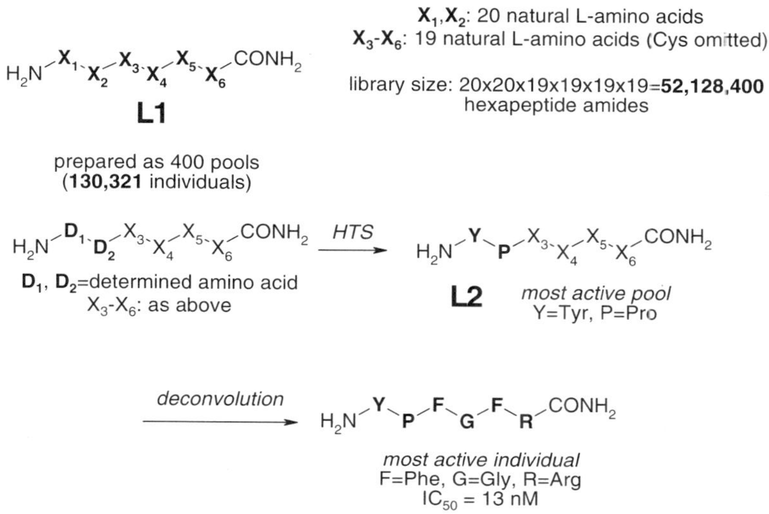

Figure 4.4 Structure and deconvolution of the linear SP hexapeptide amide library **L1**.

appropriate for the preparation of large libraries of completely randomized peptides, antibodies, vaccines, or small proteins.

Many reports in which a structure (or some structural motif) has been identified from large synthetic pool libraries of peptides have appeared in the literature in the last 10–15 years, and some excellent reviews covering this field have appeared recently (24–26). These structures can be used to refine the binding mode hypothesis for an unknown receptor through biostructural studies (e.g., X-ray co-crystallization studies and NMR studies), or they can be used as assay tools to set up a HTS campaign for the specific receptor. Unfortunately, peptide ligands are generally not suitable as drug candidates because of their poor stability and pharmacokinetic profiles, among other reasons. Their use as starting points for a chemical modification program to give druglike analogues devoid of these drawbacks has not given generally good results as of today. Moreover, even a huge peptide library does not sample diversity in an appropriate way because the α-amino acidic–based oligomeric backbone makes each library component essentially similar to the others. Significantly smaller small organic molecule libraries can be much more diverse and are potentially more promising sources of valuable positives in a primary screen.

A cyclic pentapeptide SP library (**L3**, Fig. 4.5, top) was prepared by Spatola and Crozet (27) by randomizing 12 D- and L-α-amino acids in four positions and keeping D-Asp as a common fifth residue. The total number of individuals was $12 \times 12 \times 12 \times 12 = 20{,}736$, and they were prepared using the positional scanning technique (28), which produced four library copies in 48 pools containing $12 \times 12 \times 12 = 1728$ individuals (Fig. 4.5, top). The structure of an endothelin antagonist obtained from this

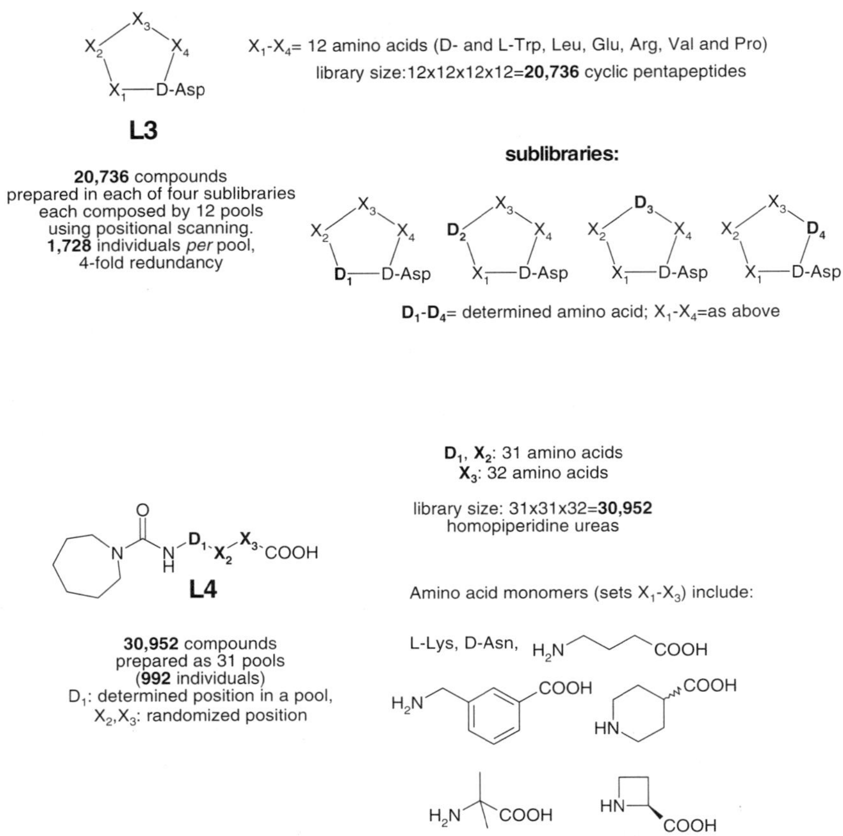

Figure 4.5 Structure of a cyclic SP pentapeptide library **L3** and of a trimeric SP peptide-related library **L4**.

library was identical to a known optimized structure, thus confirming the potential of pool screening to find active compounds.

A trimeric SP peptide-related library **L4** was reported by Terrett et al. (29), where $32 \times 31 \times 31 = 30{,}752$ compounds were produced using various amino acids as building blocks with final N-capping to produce the homopiperidine ureas (Fig. 4.5, bottom). The library was prepared as 31 library subsets (X_1 fixed) of 992 individuals, and deconvolution produced a few peptidomimetic endothelin antagonists with nanomolar activity. Both these libraries show that the introduction of D-α-amino acids, exotic amino acidic building blocks, ureido bonds, and/or cyclization do not reduce the synthetic opportunities for library preparation (the availability of exotic building blocks may be the only limitation) while the diversity of the library components is enhanced and the poor profile of natural peptides as druglike compounds can be improved.

Oligonucleotide libraries are commonly used either to encode biosynthetic display peptide libraries (Section 11.1) or to screen for novel nucleotide ligands (aptamer) or catalytic activities (ribozymes). They are produced using high-quality automated ON SP protocols seen in Section 2.2, but their amplification and selection are performed via biological protocols (see Sections 4.2.4 and 11.2).

The SP phosphoramidate library **L5** (Fig. 4.6) reported by Fathi et al. (30) is an example of a oligonucleotide–peptide hybrid library consisting of $20 \times 20 \times 2 \times 11 = 8800$ products, including stereoisomers at each phosphoramidate P atom, prepared as 11 pools of 800 compounds. This ON-modified library was used as a generic source of lead compounds to be used on many biological targets. In this case the chemical assessment was more difficult because of differences between oligonucleotide and peptide synthesis and because of the stabilities of the products. Monomer rehearsal required the elimination of some building blocks that were unsuitable for incorporation into the final synthetic scheme. However, the final library was more diverse than a peptide, peptidelike, or oligonucleotide library. Such approaches have become popular and allow a more efficient coverage of diversity through the use of various oligomer-related libraries.

Oligomeric libraries as sources of biological tools (ligands, catalytic antibodies, and metal coordination sequences, among others) were, and still are, very popular even if the development of SP and solution-phase techniques to generate SOM libraries has shifted the attention of combinatorial chemists toward these latter targets. A steady flow of reports concerning oligomeric libraries from various sources (Section 4.2.4) is nevertheless to be expected in the future, and major areas of interest will probably be the synthesis of new building blocks and the optimization of reaction conditions to produce new homogeneous or hybrid oligomeric structures.

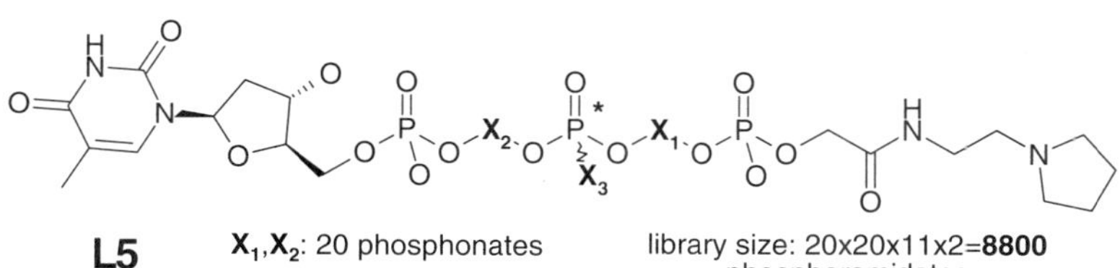

L5 X_1,X_2: 20 phosphonates X_3: 11 natural L-α-amino acids

library size: 20x20x11x2=**8800** phosphoramidates

prepared as 11 pools of **800** individuals (20x20=400 monomers, two phosphoramidate isomers)

*: chiral P atom

Monomeric phosphonates include:

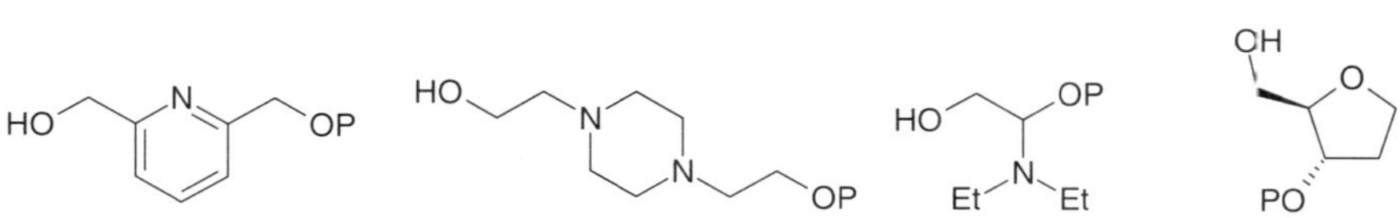

Figure 4.6 Structure of an SP phosphoramidate library **L5**.

We have already examined the inherent differences and the higher degree of complexity embedded into oligosaccharides when compared to peptides and oligonucleotides (Section 2.3). Oligosaccharides are peculiar in the sense that their SP exploitation has still to happen, and major efforts leading to significant results are to be expected in the near future. Nevertheless, the extreme biological relevance of carbohydrate structures, or of sugar-decorated synthetic or natural products, has produced several significant combinatorial oligosaccharide libraries and will stimulate further efforts from leading groups.

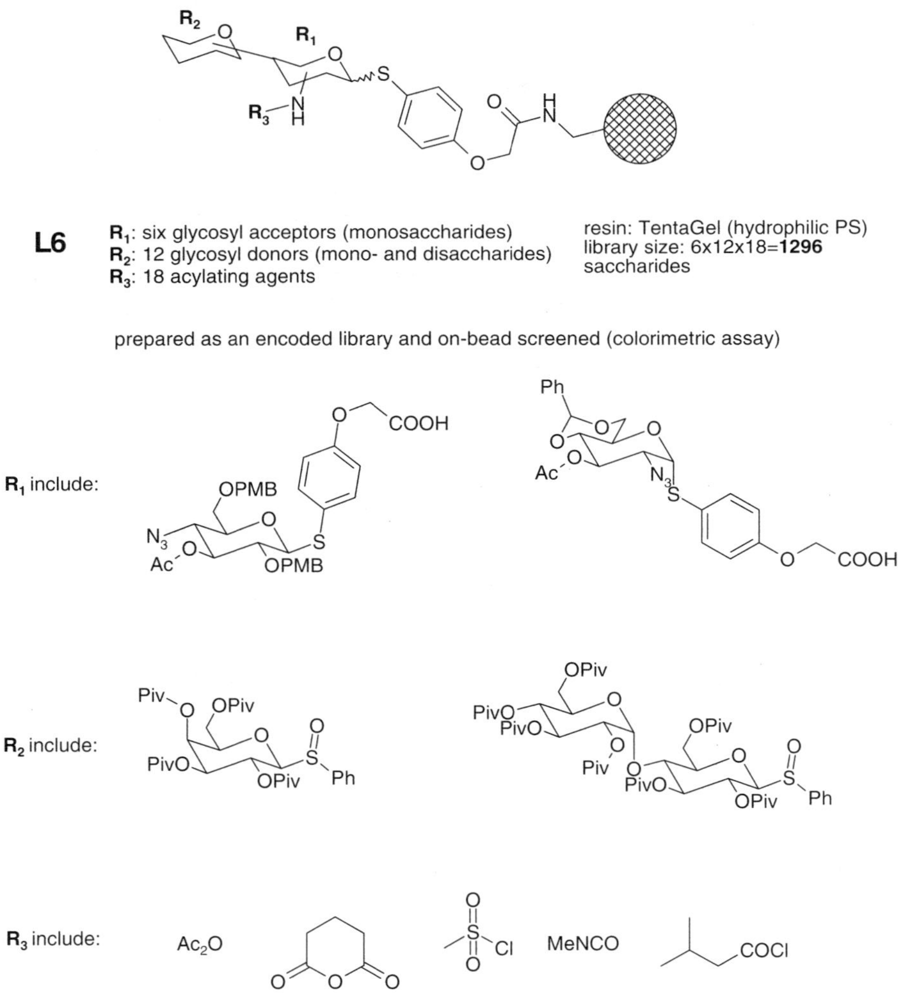

Figure 4.7 Structure of an SP oligosaccharide encoded library **L6**.

A list of recent papers (31, 32) and reviews (33–38) dealing with combinatorial libraries of oligosaccharides should give a flavor of the present and the future to the interested reader.

The first oligosaccharide SP-encoded (39) library **L6** was reported by Liang et al. (40) and consisted of 1296 carbohydrates that were tested on-bead for their lectin binding activity using hydrophilic PS resin as a support. The structure of the library and selected building blocks are shown in Fig. 4.7. A high-quality diverse library of

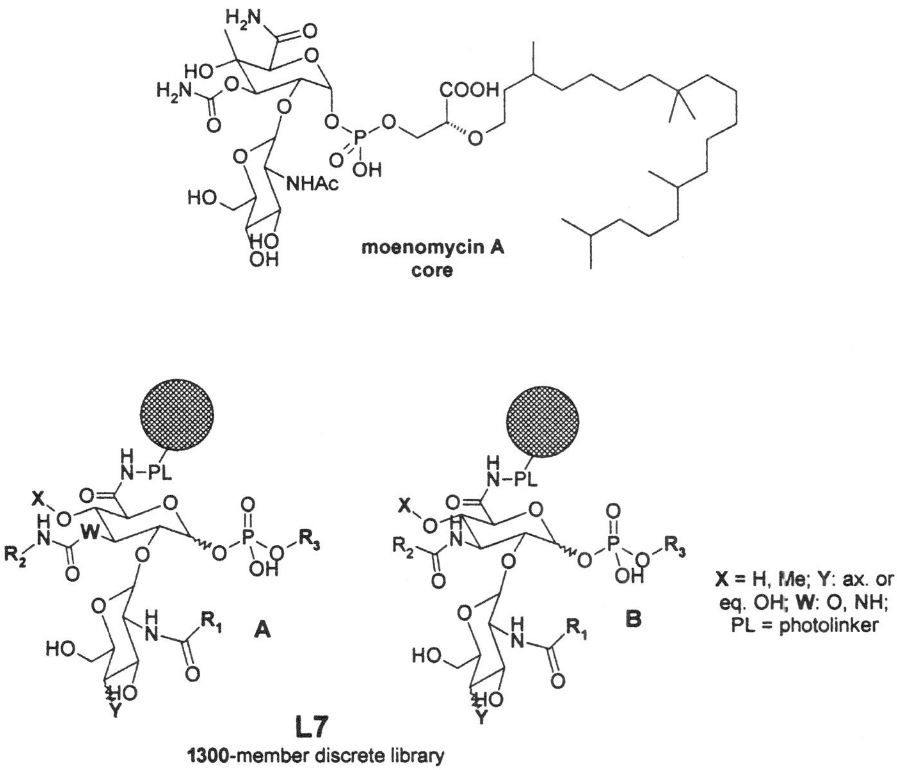

R_1: 13 carboxylic acids (aliphatic, aromatic, heterocyclic)

R_2A: 24 isocyanates (aliphatic, aromatic)

R_2B: 4 carboxylic acids (bifunctional)

R_3: 8 lipid phosphoramidites (aliphatic, aromatic)

Figure 4.8 Structure of a moenomycin-inspired SP discrete library **L7**.

di- and trisaccharides was assembled through the reaction of glycosyl acceptors, including the handle function, with glycosyl sulfoxide donors and acylating agents. It is clear that the degree of complexity of the chemistry involved is much greater when compared to either ON or peptide libraries and that the protected glycosidic building blocks must be built through complex syntheses. Another important point is that the reaction conditions must take into account the relative sensitivity of all the intermediates and of the final, resin-bound library components.

Sofia et al. (41) reported a SP library of disaccharides **L7** made by 1300 individuals and inspired by the disaccharide core of moenomycin A, a bacterial cell wall inhibitor (Fig. 4.8). Both the structure of the library and its main features are shown in Fig. 4.8. The four major disaccharide scaffolds (**X**, **Y**, and **W** variations) were either built on resin via glycosylation or attached onto the photolabile linker, and subsequently decorated by introduction of $\mathbf{R_1}$, $\mathbf{R_2}$, and $\mathbf{R_3}$; radiofrequency encoding (42) was used to obtain a large number of discretes with an affordable number of reactions. The

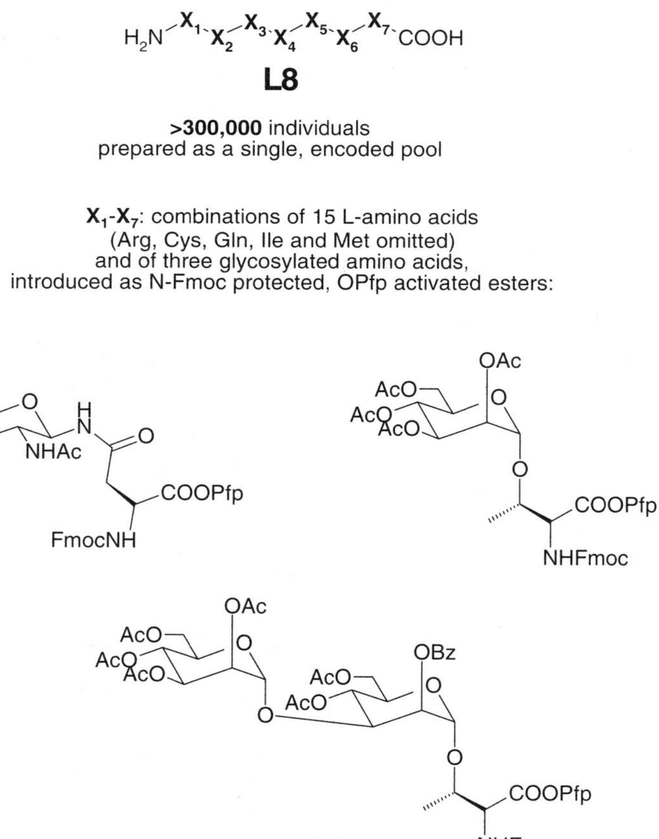

Figure 4.9 Structure of an SP glycopeptide encoded library **L8**.

library screening discovered a novel class of simplified moenomycin A analogs as antibacterial agents; their detailed characterization and structural optimization by the same group is to be expected soon.

A 300,000-member, encoded library of heptaglycopeptides **L8** was reported by St. Hilaire et al. (43). The structure of **L8** and of the monomers used to build the oligomeric library are reported in Fig. 4.9. The easier peptide SP protocols to assemble the library allowed the preparation of a large set of glycopeptides, from which several active molecules were identified (43). The preparation of several other glycosylated amino acids could easily increase the diversity of such a library and would add new possibilities of recognition of library members by membrane receptors. The considerable technical challenges posed by oligosaccharides notwithstanding, the field of oligosaccharide and oligosaccharide-related SP combinatorial chemistry has the potential to produce valuable libraries as sources of relevant drugs or for a variety of other applications in many diverse fields and is currently receiving a great deal of attention.

4.2.2 Synthetic Organic Libraries: Small Organic Molecules

The synthesis of SOM libraries is now by far the most important component of combinatorial chemistry. The potential application of any reaction from the field of classical organic chemistry for the preparation of such libraries confers a widespread appeal to this class. We will describe the structures of a few SOM libraries, such as those shown in Figs. 4.10–4.13.

The very first primary SOM library was prepared by Bunn and Ellman in 1992 (20), and its further exploitation by Boojamra et al. (44) produced an SP library **L9** of discrete 1,4-benzodiazepine-2,5-diones containing $12 \times 11 \times 10 = 1320$ individuals (2508 compounds taking into account the 9 racemates among the 10 α-amino acids used; Fig. 4.10). The building blocks used were commercially available α-amino acids **R1**, anthranilic acids **R2**, and alkylating agents **R3**, which could easily be expanded to give larger libraries. A few representative structures of some of the library components are reported in Fig. 4.10. The decoration produced by the monomers **R1**–**R3** around the benzodiazepine core increased the diversity of the library when compared to the oligomeric libraries discussed previously. The benzodiazepine class is a well-known source of pharmaceutically relevant compounds, and a positive from a specific HTS assay from this type of library could potentially require significantly less effort in subsequent elaboration to arrive at a drug candidate than a positive coming from an oligomeric library. In this case, more effort was initially required to transfer the chemistry onto SP and then into a combinatorial library format for the monomer rehearsal stage. The balance between the effort required and the potential applications for a designed SOM library may sometimes encourage the synthesis, while in other instances the library must be rethought to make use of more combinatorial-friendly chemistries. All of these considerations are common for any SOM library.

A focused SP library **L10** of pools of 1,4-dihydropyridines (DHP) was reported by Gordeev et al. (45) and tested as a source of calcium channel blockers. The library consisted of $10 \times 3 \times 10 = 300$ members prepared as 30 pools of 10 compounds (Fig. 4.11) whose deconvolution produced several new compounds of interest. The mono-

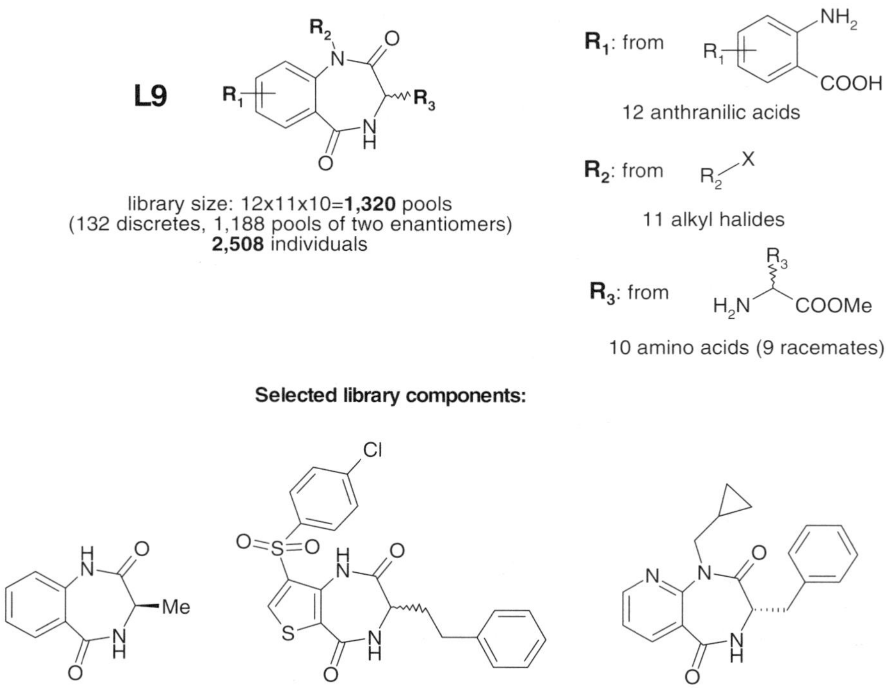

Figure 4.10 Structure of an SP 1,4-benzodiazepine-2,5-dione library **L9**.

mer sets used were commercially available and easily expandable α-keto esters (**R1**), α-dicarbonyl compounds (**R2**), and aromatic aldehydes (**R3**). Some examples from the library components, shown in Fig. 4.11, highlight their similarity with the structure of the known active compound nifedipine that inspired their synthesis. This and the previous library built the core structure (benzodiazepine or DHP) during the synthesis, thus allowing a large degree of flexibility in substituting a monomer class with similar analogues to produce diverse scaffolds, even though a significant effort to reoptimize the SP reaction conditions was necessary.

The decoration of a central scaffold is illustrated by the triazine library **L11** reported by Gustafson et al. (46) where trichlorotriazine was sequentially substituted with anilines (**R1**), primary aliphatic amines (**R2**), and secondary aliphatic amines (**R3**). The library, made up of $20 \times 16 \times 20 = 12{,}800$ compounds, was prepared as discretes (Fig. 4.12) and used as a primary library to be screened on various assays. The introduction of sugars, dipeptides, and α-ketoamides among other amine substituents in **L11** created diversity in the components even though each of them shares a common 2,4,6-triaminotriazine scaffold. Examples of individuals from **L11** are provided in Fig. 4.12. Such a decoration library can be made when a suitable scaffold is available in large amounts, either commercially or through a simple synthetic route. Constrained

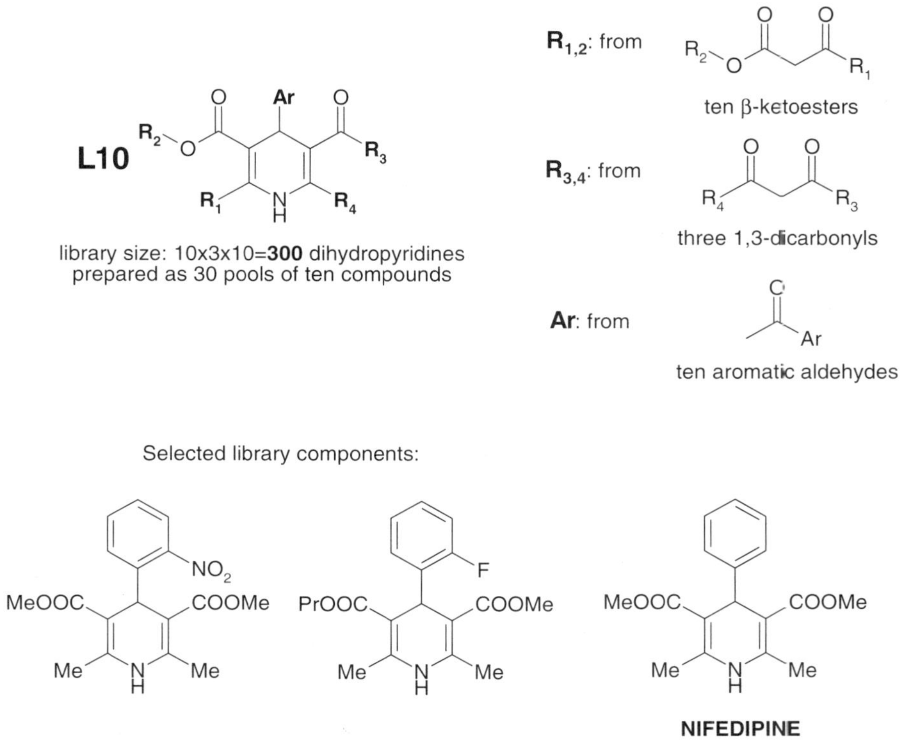

Figure 4.11 Structure of a nifedipine-inspired SP 1,4-dihydropyridine library **L10**.

scaffolds are often preferred because their decoration with meaningful monomers allows the creation of rigid structures that are able to explore the pharmacophoric space of various molecular targets. Many examples have been reported and among them scaffolds derived from natural products deserve a special mention (see next section).

A primary *modular* library was reported by Powers et al. (47), and its structure is shown in Fig. 4.13. The key chalcone intermediates were prepared as a discrete library **L12** made up of $32 \times 40 = 1280$ individuals using acetophenones (**R1**) and aromatic aldehydes (**R2**) as monomer sets. This library was used to generate nine further libraries by addition of different reagents or monomer sets onto the α,β-unsaturated system (libraries **L13**–**L15**) or small monomer sets (**L16**–**L21**). The assembly of subsets **L12**–**L21**, composed of 1280 + 1280 + 1280 + 1280 + 7680 + 7680 + 12,800 + 7680 + 7680 + 25,600 = 74,240 discretes, can be considered as a modular library where the diversity of the chemical space is spanned by 10 different decorated scaffolds produced from the same intermediate library. This approach, which generates daughter libraries from a parent library, has an enormous potential to generate diverse primary libraries that can be focused on a specific scaffold during a second iteration of library synthesis after finding active structures from an HTS campaign.

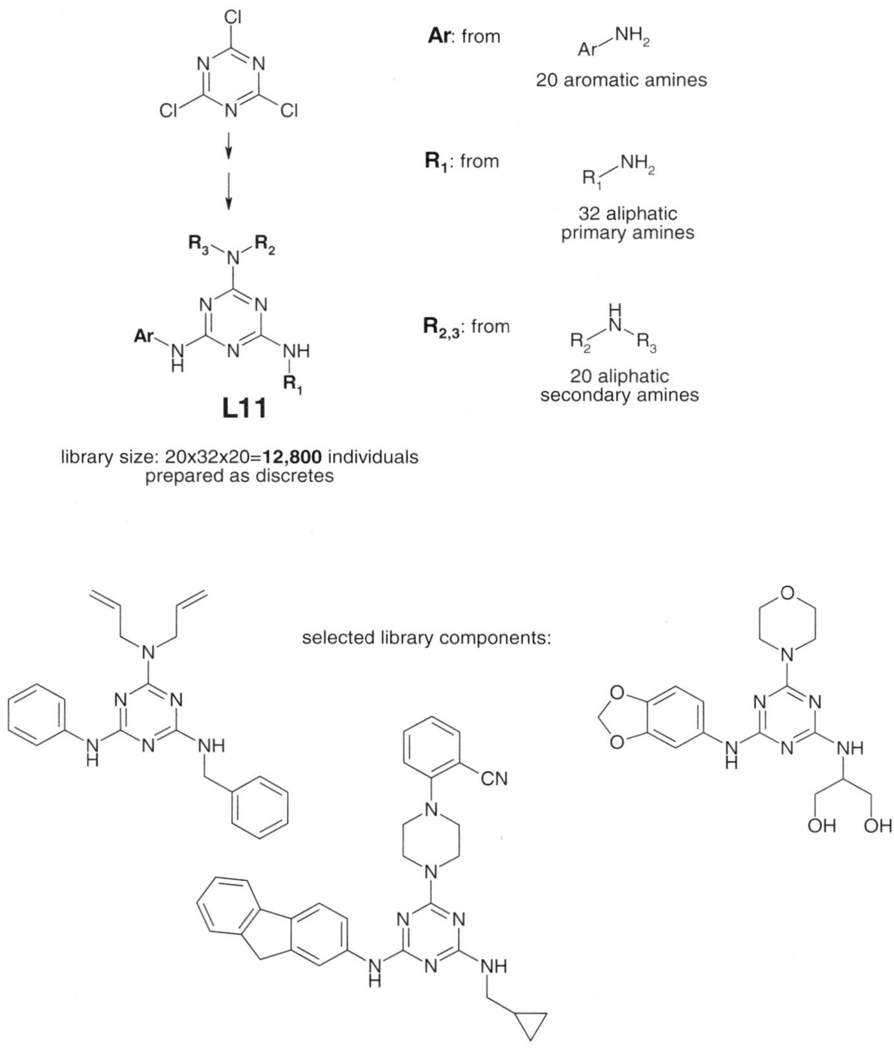

Figure 4.12 Structure of a solution-phase triazine library **L11**.

4.2.3 Synthetic Organic Libraries: Natural Products

A specific class of libraries in which biologically active natural products are either built during the synthesis or used as scaffolds to be decorated has recently emerged and has gained considerable attention for pharmaceutical purposes. The biological information contained in the natural scaffold increases the chances of discovering novel active structures. Some examples depicted in Figs. 4.14–4.18 demonstrate the potential of natural product libraries.

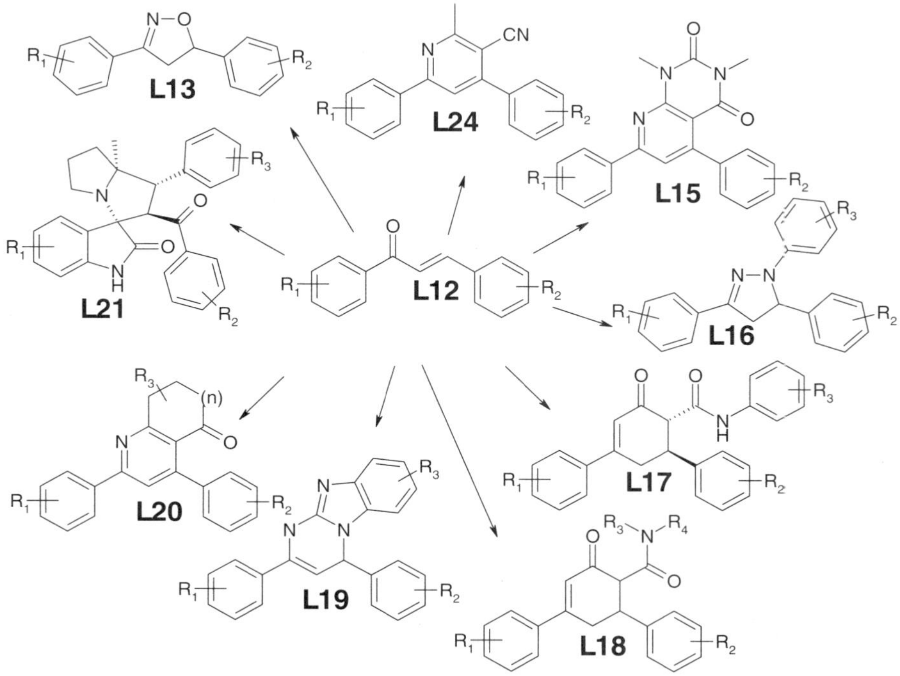

L12 R_1: 32 acetophenones, R_2: 40 aromatic aldehydes library size: 32x40=**1,280** discrete chalcones

L13 L12 reacted with NH_2OH library size: 1,280x1=**1,280** discrete isoxazolines

L14 L12 reacted with 3-aminocrotononitrile library size: 1,280x1=**1,280** discrete pyridines

L15 L12 reacted with 6-NH_2-1,3-diMeuracyl library size: 1,280x1=**1,280** discrete byciclic compounds

L16 L12 reacted with 6 phenylhydrazines library size: 1,280x6=**7,680** discrete pyrazolines

L17 L12 reacted with 6 acetoacetanilides library size: 1,280x6=**7,680** discrete cyclohexenones

L18 L12 (320 cpds) with 40 acetoacetamides library size: 320x40=**12,800** discrete cyclohexenones

L19 L12 with 6 aminobenzimidazoles library size: 1,280x6=**7,680** discrete tricyclic compounds

L20 L12 with 6 cyclic ketones and NH_3 library size: 1,280x6=**7,680** discrete byciclic compounds

L21 L12 (80 cpds) reacted with 16 isatins and 20 amino acids library size: 80x16x20=**1,280** discrete spiro compounds

Figure 4.13 Structure of a solution-phase primary modular library **L12** based on a chalcone core and of several derived libraries **L13–L21**.

The combinatorial modification of yohimbinic acid, a natural alkaloid, was reported by Atuegbu et al. (48), and the structure of the derived SP pool library **L22** is shown in Fig. 4.14. The library was made by 36 × 22 = 792 compounds as 22 pools of 36 compounds, and the monomer sets used were L-α-amino acids (**R1**) and carboxylic

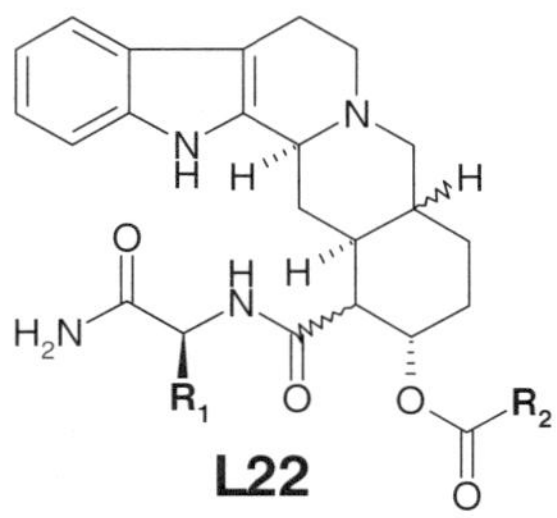

Figure 4.14 Structure of an SP yohimbinic acid decoration library **L22**.

acids (**R2**). The biological evaluation of the library was not reported. The commercially available natural scaffold is known to possess a wide range of biological activities, and the possibility to further expand the two monomer sets make such a primary library a likely source for new biologically active analogues. There is a real possibility of building a preliminary structure activity relationship (SAR) and of finding substituents that favor different biological activities providing that the decoration of the natural product diversifies the library components enough.

Another decoration pool library **L23** was reported by Nestler (49), who presented the appendage of a peptidic chain to the two hydroxylic functions of a steroid scaffold (Fig. 4.15). The library was made up of $10 \times 10 \times 10 \times 10 = 10{,}000$ individuals using a chemical encoding method (39) and L-α-amino acids as monomer sets (**R1**–**R4**). The assay of the library as a source of artificial two-armed receptors for enkephalin-related peptides produced positives with micromolar affinity. Much larger libraries could be obtained by simply increasing the monomer sets and the length of the two arms; this could lead to a primary library of peptide-binding artificial receptors; similar scaffolds have been repeatedly exploited for combinatorial purposes (50–52).

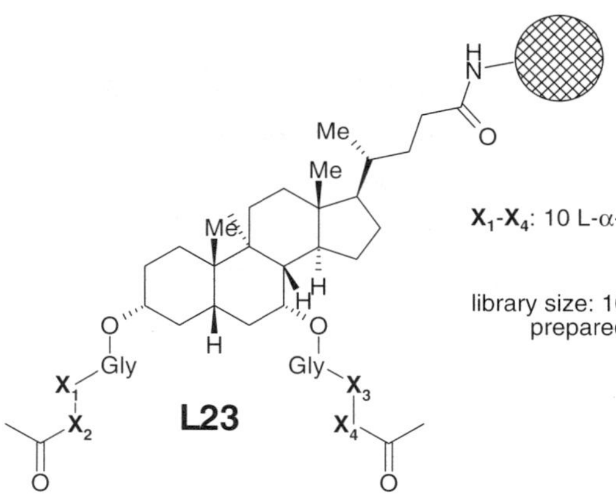

Figure 4.15 Structure of an SP steroid decoration library **L23**.

Quinic acid was decorated on the C_1-carboxylic acid and on the C_3, C_4, C_5-hydroxyls by Phoon and Abell (53) as in Fig. 4.16, producing two small arrays **L24** (ten monoester-monoamide compounds) and **L25** (six triester-monoamide compounds). The reliability and the robustness of both the scaffold and of the acylation/esterifications ensure an expansion of quinate-based libraries, which has already been reported for shikimate-based libraries (54).

The natural scaffold, or a scaffold resembling a natural product, can also be built during the synthesis of the library. An example presented by Nielsen and Lyngsoe (55) involved the synthesis of a library **L26** inspired by the structure of balanol (Fig. 4.17). A disconnection study revealed α-amino alcohols $\mathbf{R_1}$, aromatic dicarboxylic acids $\mathbf{R_2}$, and aromatic carboxylic acids $\mathbf{R_3}$ as monomer sets to mimic the natural constituents of balanol, and the small library **L26** was prepared. It was made up of $2 \times 4 \times 4 = 32$ compounds prepared as four pools of eight compounds that could be expanded by using enlarged monomer sets to produce primary balanol-biased combinatorial libraries. Introduction of modified building blocks (aliphatic diacids or monoacids, acyclic or aromatic amino alcohols, etc.) could lead, after substantial assessment of the chemistry, to a modular library of different balanol-inspired compounds with a high level of diversity.

A more complex combinatorial strategy based on a natural scaffold was reported by Nicolaou et al. (56) with the synthesis of an epothilone-based SP library **L27** (Fig. 4.18). This biased library was prepared as a $3 \times 3 \times 5 = 45$-member collection, but the final reaction vessels could, theoretically contain 4 different isomeric desoxyepothilones (Fig. 4.18), thus leading to a total number of $45 \times 4 = 180$ components. The library was prepared using a radiofrequency encoding technique (42), and the final

$\mathbf{R_1}$: 3 primary amines (*n*-propyl, benzyl and cyclohexyl)

$\mathbf{R_2}$: 5 carboxylic acids (acetic, benzoic, 4-pyridyl, phenylacetic and acrylic)

$\mathbf{R_3}$: 2 carboxylic acids (acetic and benzoic)

Figure 4.16 Structure of two quinicacid-inspired SP small arrays **L24** and **L25**.

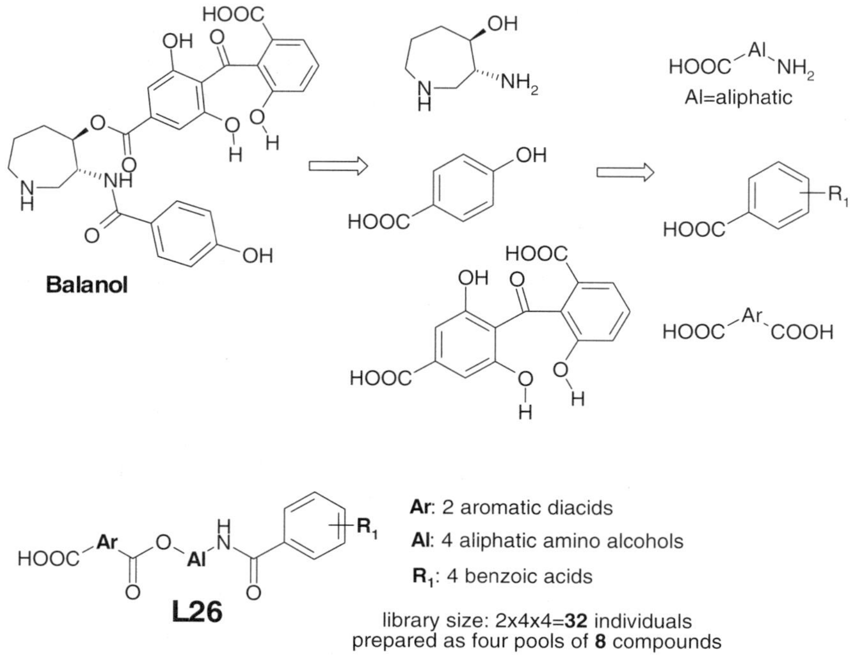

Figure 4.17 Structure of an SP balanol-inspired library **L26**.

cleavage solutions from each vessel were purified to give pure single isomers that were assayed to obtain a detailed SAR (56). The monomer sets consisted of three protected hydroxy aldehydes (**R1**), three protected hydroxy keto acids (**R2**), and five heterocyclic secondary alcohols (**R3**) prepared using published multistep routes, and the expansion of these sets would require significant synthetic effort. Once such an ambitious synthetic scheme has been set up, it would be desirable to move away from the initially used biased monomer sets and to introduce more diverse commercially available monomers to build larger, less epothilone-like libraries with the same synthetic strategy that could be used as primary libraries to search for other biological activities. This example shows that even complex natural scaffolds can be dissected and reconstructed using appropriate reactions and monomer sets. The balance between the appeal of the designed library and the resources/effort needed to set up the library synthesis should drive the strategic decision to go for the synthesis or not.

4.2.4 Biological Libraries

Libraries produced by microorganisms or through biochemical techniques have been frequently used to find either peptidic or oligonucleotidic sequences that bind certain receptors, or enzymes, and/or possess catalytic properties (catalytic antibodies and ribozymes, among others). Biological libraries can also be produced by combinatorial

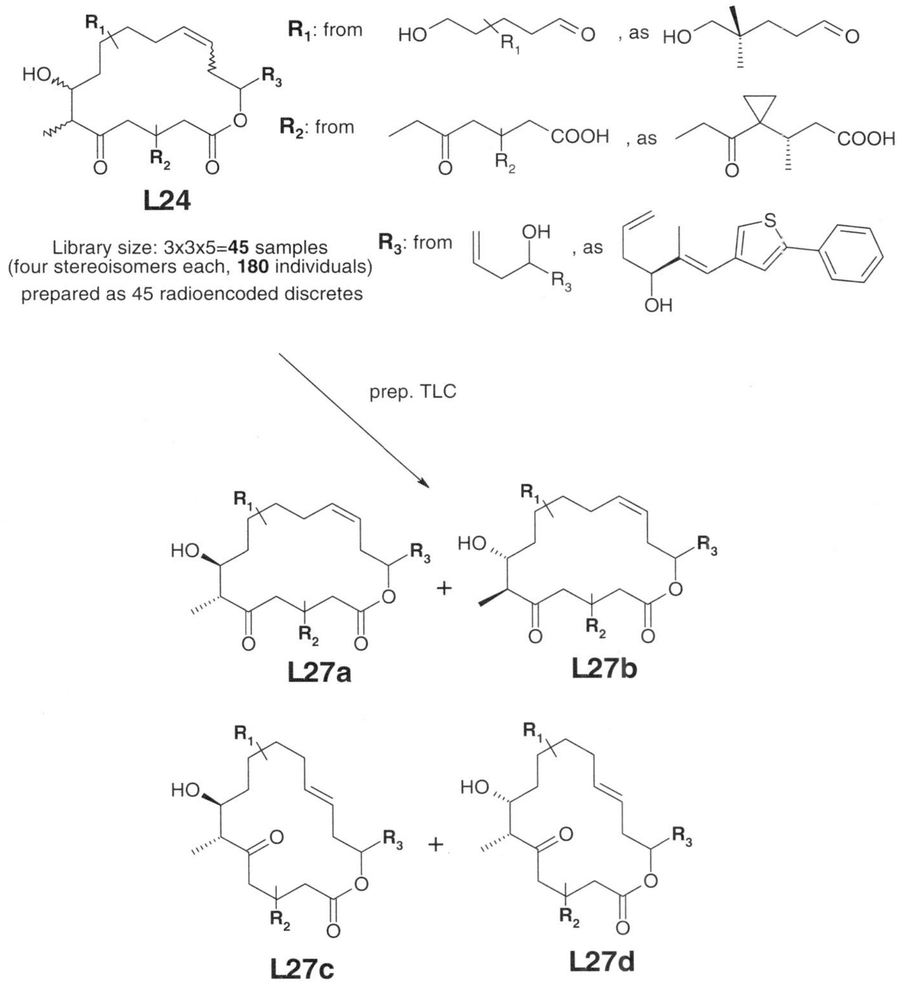

Figure 4.18 Structure of an SP epothilone-inspired library **L27**.

modifications of chemical scaffolds performed by isolated enzymes, whole microorganisms, and even by the so-called combinatorial biosynthesis, which alters the enzymatic machinery responsible for the synthesis of natural products, thus allowing the preparation of combinatorial biosynthetic analogues. Some biological libraries are reported in Figs. 4.19–4.22.

An application of phage display libraries to the identification of peptidic sequences that selectively target tumor blood vessels was reported by Arap et al. (57). Three cyclic peptide libraries were produced by inserting the corresponding degenerate oligonucleotide sequences into a vector (fUSE 5), then transforming MC1061 cells by

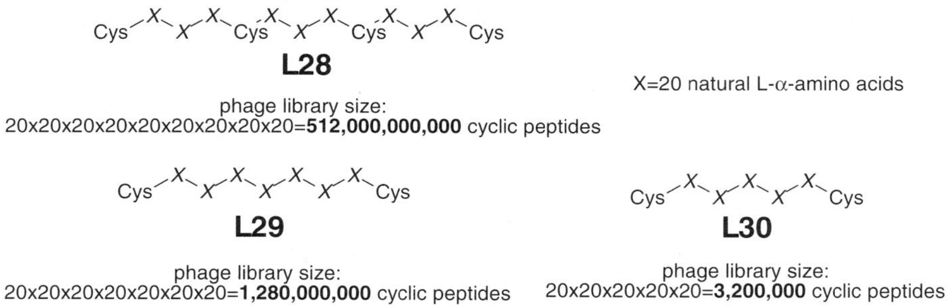

Figure 4.19 Structure of phage display S–S cyclic peptide libraries **L28–L30**.

electroporation. The libraries produced contained two terminal cysteines to provide S–S cyclization of the peptides: the $CX_3CX_3CX_3C$ library (**L28**), the CX_7C library (**L29**), and the CX_5C library (**L30**) (Fig. 4.19). Three distinct structural motifs were recovered after in vivo selection in tumor-bearing mice; NGR (in C**NGR**CVSGCAGRC from **L28**), RGD (in CDC**RGD**CFC from **L29**), and GSL (in C**GSL**VRC from **L30**). All three of them delivered peptidic sequences preferentially to the tumor and seemed to bind to different tumor receptors. Their use as drug delivery systems for the development of targeted chemotherapy looks extremely promising. The usefulness of phage display libraries in producing enormous numbers of peptidic sequences of various lengths combined with the possibility of tailoring the structures of the displayed population and the commercial availability of several phage systems makes this biological library source extremely appealing. Production of biochemical reagents (e.g., antibodies, diagnostics, and vaccines) or tools (e.g., to validate targets via their inhibition and to produce peptidic ligands) and gathering structural information about receptors by finding peptide agonists or antagonists have all been reported. Undoubtedly, the exploitation of phage libraries will increase and significant breakthroughs are to be expected in the future.

An example of deoxyribozyme (deoxyoligonucleotidic enzymatic activity) selection applied to the isolation of a broad specificity RNA-cleaving activity was reported by Santoro and Joyce (58). A specific construct was built up by assembling a 5′-biotin moiety (**A**), attached by a short ODN spacer (**B**), to a target 12-mer ribonucleotide containing the target cleavage sequence (**C**) and a 50-mer random ODN (**E**). This latter sequence was surrounded by fixed-sequence ODN spacers (**D**) on the 3′ and 5′ ends (Fig. 4.20). The corresponding library **L31** (10^{14} ON–ODN components) was applied to a streptavidin-coated support and bound tightly to it via the biotin–streptavidin recognition. The few RNA-cleaving DNA sequences released only the cleaved constructs, which, after 10 rounds of polymerase chain reaction (PCR) amplification, eventually produced two catalytic motifs (**DNA 1** and **2**, Fig. 4.20) which were thoroughly studied and refined. Many other reports of ribozymes and deoxyribozymes have been reported, as the use of PCR amplification to select ON–ODN aptamers that bind to receptors/small molecules.

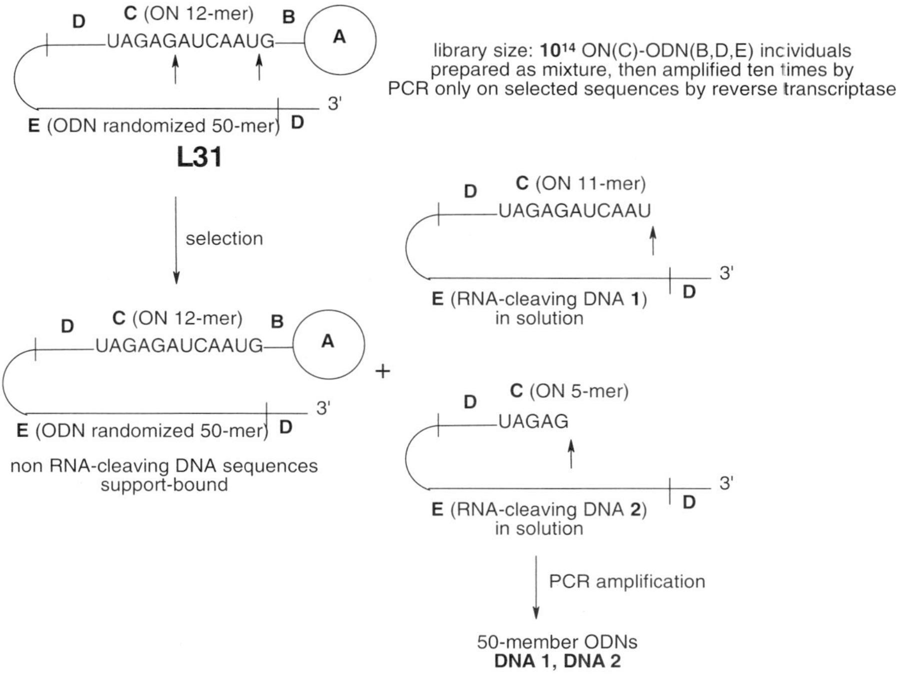

Figure 4.20 Structure of an SP ON/ODN library **L31**.

The generation of biased libraries from synthetic or natural precursors using either purified enzymes or whole microorganisms was reported by Khmelnitsky et al. (59) adopting the so-called combinatorial biocatalysis approach. The library **L32**, based on (±)-(2-endo,3-exo)-bicyclo[2.2.2]oct-5-ene-2,3-dimethanol (BOD), was obtained by submitting BOD to a panel of multistep enzymatic transformations. A total of 1222 compounds were obtained and some of the representative structures are shown in Fig. 4.21. This approach allows the generation of very diverse and complex compounds from a common scaffold employing mild enzymatic transformations that can also accommodate sensitive substrates. These transformations are often stereoselective and regioselective, so that complex derivatives may be obtained without protection/deprotection steps. A lot of effort is required to process the reaction samples and to isolate, purify, and structurally characterize the final library components. However, this strategy is particularly useful when a lead compound with a complex structure, which has already proven its value for a specific application, must be optimized for progression as a drug candidate.

Combinatorial biosynthesis is an approach aimed at the modification of the cellular machinery involved in the biosynthetic pathways that produce natural products. There are several groups active in the field, and an example reported by McDaniel et al. (60)

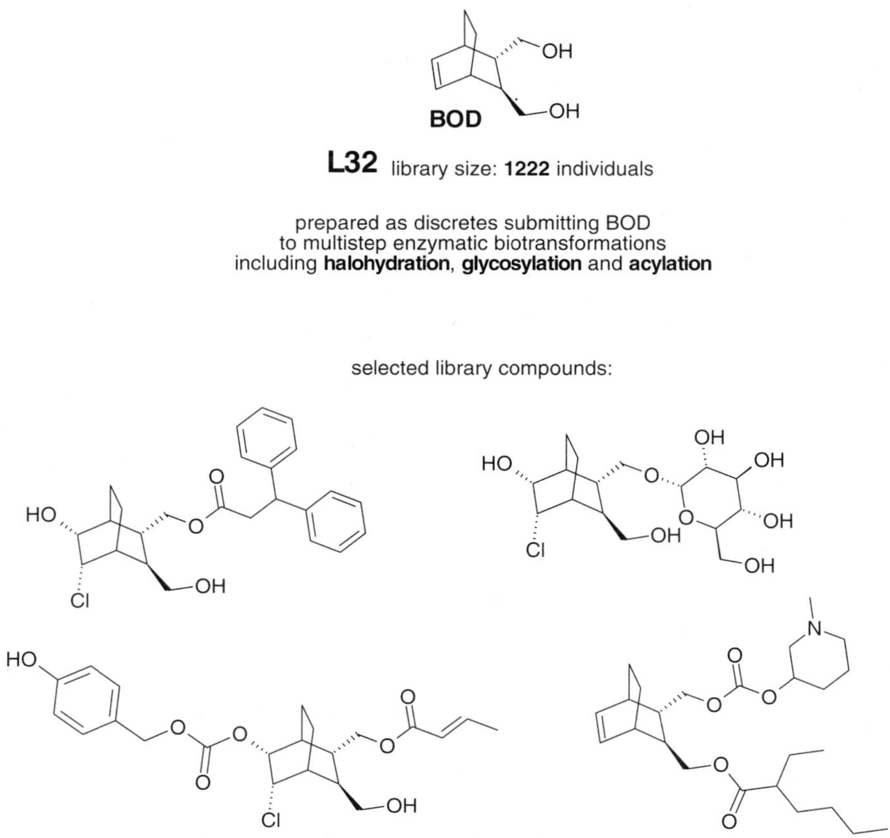

Figure 4.21 Structure of a combinatorial biocatalysis BOD-derived library **L32**.

demonstrates the generation of rationally engineered biosynthesis of polyketides. A scheme highlighting the current possibilities offered by biosynthetic manipulation including the biosynthetic modifications involved and the intermediates and the final compound produced is shown in Fig. 4.22. This field has enormous potential and will benefit from efforts aimed at making biosynthetic pathways other than the polyketide available to the armamentarium of combinatorial chemistry.

4.2.5 Materials Science Libraries

Libraries composed of inorganic components, normally referred to as materials science libraries, surely represent one of the most intriguing and interesting class of combinatorial libraries. Since their recent appearance in combinatorial chemistry, they have been applied to many problems in the field of materials science, and an example is

Engineered polyketide biosynthesis

Manipulation of the following steps currently feasible:

CHAIN LENGTH of the product

KETOREDUCTION (yes or no)

1st RING STEREOSPECIFICITY

1st RING AROMATIZATION

2nd RING CYCLIZATION (yes or no)

↓ combinations of the above

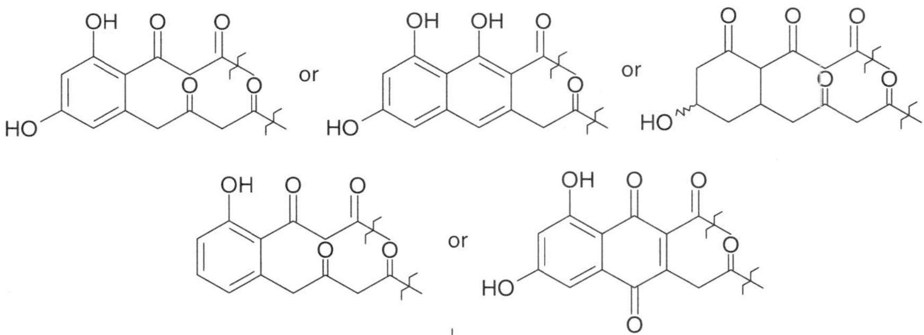

↓ uncatalyzed end reaction

RATIONALLY ENGINEERED POLYKETIDES

Figure 4.22 Combinatorial biosynthesis: manipulation of the aromatic polyketide pathway.

reported in Fig. 4.23, which will be used to briefly define the properties and opportunities provided by these libraries.

A library **L33** made up of different compositions and stoichiometries of Ln_xM_y-CoO_z (Ln = La or Y, M = Ba, Sr, Ca, or Pb) was reported by Briceno et al. (61) as a source of new magnetoresistant materials. It was composed of 128 discretes (Fig. 4.23) and was prepared in duplicate using a combination of thin-film deposition and physical masking to vary the individual components. Each library copy was analyzed for magnetoresistance using a different procedure. A significant number of individuals showed a meaningful (>5%) magnetoresistant effect, and a few of them were characterized in detail. Automated synthesis of such discrete material libraries, their analysis for different applications in material sciences, and the automated and reliable readout of the results make this class of libraries extremely useful, and in fact similar approaches are appearing more and more frequently in the literature.

$Ln_xM_yCoO_z$

L33

prepared in duplicate as discrete individuals using materials science techniques

Ln= 2 lanthanides (La, Y)

M= 4 metals (Ba, Sr, Ca, Pb)

x/y= 16 different ratios

library size: 2x4x16=**128** materials

composition of the 128 materials:

Ln= 4 different compositions: 0.62, 0.45, 0.33 and 0.27

M= 4 different compositions for each Ln and Ln value:

Ln=La, 0.62 M=1.38, 1.14, 0.89, 0.64
Ln=La, 0.45 M=1.00, 0.83, 0.65, 0.47
Ln=La, 0.33 M=0.73, 0.60, 0.47, 0.34
Ln=La, 0.27 M=0.61, 0.50, 0.39, 0.28

Ln=Y, 0.62 M=1.24, 1.00, 0.74, 0.50
Ln=Y, 0.45 M=0.90, 0.72, 0.54, 0.36
Ln=Y, 0.33 M=0.66, 0.52, 0.40, 0.26
Ln=Y, 0.27 M=0.55, 0.45, 0.33, 0.22

Figure 4.23 Structure of a materials science library **L33**.

REFERENCES

1. Spencer, R. W., *Biotechnol. Bioeng.* **61**, 61–67 (1998).
2. Furka, A., *Notarized document file number 36237/1982*, Dr. Judit Bokai, state notary public, June 15th, 1982, Budapest.
3. Geysen, H. M., Meloen, R. H. and Barteling, S. J., *Proc. Natl. Acad. Sci. USA* **81**, 3998–4002 (1984).
4. Houghten, R. A., *Proc. Natl. Acad. Sci. USA* **82**, 5131–5135 (1985).
5. Furka, A., Sebestyen, F., Asgedom, M. and Dibo, G., *Int. J. Pept. Prot. Res.* **37**, 487–493 (1991).
6. Frank, R., Gueler, S., Krause, S. and Lindenmaier, W., *Peptides 1990: Proceedings of Twenty First European Peptide Symposium*, E. Giralt and D. Andrew (Eds.). ESCOM, Leiden, Netherlands, 1991, pp. 151–152.
7. Houghten, R., Pinilla, C., Blondelle, S. E., Appel, J. R., Dooley, C. T. and Cuervo, J. H., *Nature* **354**, 84–86 (1991).
8. Lam, K. S., Salmon, S. E., Hersh, E. M., Hruby, V. J., Kazmierski, W. M. and Knapp, R. J., *Nature* **354**, 82–84 (1991).
9. Fodor, S. P. A., Read, J. L., Pirrung, M. C., Stryer, L., Lu, A. T. and Solas, D., *Science* **251**, 767–773 (1991).
10. Smith, G. P., *Science* **228**, 1315–1317 (1985).
11. Parmley, S. F. and Smith, G. P., *Gene* **73**, 305–318 (1988).
12. Devlin, J. J., Panganiban, L. C. and Devlin, P. E., *Science* **249**, 404–406 (1990).

13. Cwirla, S. E., Peters, E. A., Barrett, R. W. and Dower, W. J., *Proc. Natl. Acad. Sci. USA* **87**, 6378–6382 (1990).
14. Tuerk, C. and Gold, L., *Science* **249**, 505–510 (1990).
15. Oliphant, A. R., Nussbaum, A. L. and Struhl, K., *Gene* **44**, 177–183 (1986).
16. Oliphant, A. R., Brandl, C. J. and Struhl, K., *Mol. Cell. Biol.* **9**, 2944–2949 (1989).
17. Kinzler, K. W. and Vogelstein, B., *Nucl. Acid Res.* **17**, 3645–3653 (1989).
18. Pan, T. and Uhlenbeck, O. C., *Biochemistry* **31**, 3887–3895 (1992).
19. Bartel, D. P. and Szostak, J. W., *Science* **261**, 1411–1418 (1993).
20. Bunin, B. A. and Ellman, J., *J. Am. Chem. Soc.* **114**, 10997–10998 (1992).
21. Bunin, B. A. and Ellman, J., *Proc. Natl. Acad. Sci. USA* **91**, 4708–4712 (1994).
22. Xiang, X.-D., Sun, X., Briceno, G., Lou, Y., Wang, K.-A., Chang, H., Wallace-Freedman, W. G., Chen, S.-W. and Schultz, P. G., *Science* **268**, 1738–1740 (1995).
23. Dooley, C. T., Kaplan, R. A., Chung, N. N., Schiller, P. W., Bidlack, J. M. and Houghten, R. A., *Peptide Res.* **8**, 124–137 (1995).
24. Rodda, S., Tribbick, G. and Geysen, M., in *Combinatorial Peptidic and Nonpeptidic Libraries*, G. Jung (Ed.). VCH, Weinheim, Germany, 1996, pp. 303–326.
25. Lebl, M. and Khrchnak, V., *Methods Enzymol.* **287**, 336–392 (1997).
26. Zhao, Z.-G. and Lam, K. S., *Annu. Rep. Combi. Chem. Mol. Div.* **1**, 192–209 (1997).
27. Spatola, A. F. and Crozet, Y., *J. Med. Chem.* **39**, 3842–3846 (1996).
28. Dooley, C.T. and Houghten, R. A., *Life Sci.* **52**, 1509–1517 (1993).
29. Terrett, N. K., Bojanic, D., Brown, D., Bungay, P. J., Gardner, M., Gordon, D. W., Mayers, C. and Steele, J., *Bioorg. Med. Chem. Lett.* **5**, 917–922 (1995).
30. Fathi, R., Patel, R. and Cook, A. F., *Mol. Diversity* **2**, 125–134 (1997).
31. Minke, W. E., Hong, F., Verlinde, C. L. M. J., Hol, W. G. J. and Fan, E., *J. Biol. Chem.* **274**, 33469–33473 (1999).
32. Lubineau, A. and Bonnaffe', D., *Eur. J. Org. Chem.*, 2523–2532 (1999).
33. Kahne, D., *Curr. Opin. Chem. Biol.* **1**, 130–135 (1997).
34. Sofia, M. J., *Mol. Diversity* **3**, 75–94 (1998).
35. Schweizer, F. and Hindsgaul, O., *Curr. Opin. Chem. Biol.* **3**, 291–298 (1999).
36. Shinkai, S. and Takeuchi, M., *NATO ASI Sci. Ser., C*, **527**, 157–159 (1999).
37. Sofia, M. J., *Annu. Rep. Combi. Chem. Mol. Div.* **2**, 41–52 (1999).
38. Sofia, M. J. and Silva, D. J., *Curr. Opin. Drug Discovery Dev.* **2**, 365–376 (1999).
39. Ohlmeyer, M. H. J., Swanson, R. N., Dillard, L., Reader, J. C., Asouline, G., Kobayashi, R., Wigler, M. and Still, W. Clark., *Proc. Natl. Acad. Sci. USA* **90**, 10922–10926 (1993).
40. Liang, R., Yan, L., Loebach, J., Ge, M., Uozumi, Y., Sekanina, K., Horan, N., Gildersleeve, J., Thompson, C. et al., *Science* **274**, 1520–1522 (1996).
41. Sofia, M. J., Allanson, N., Hatzenbuhler, N. T., Jain, R., Kakarla, R., Kogan, N., Liang, R., Liu, D., Silva, D. J., Wang, H., Gange, D., Anderson, J., Chen, A., Chi, F., Dulina, R., Huang, B., Kamau, M., Wang, C., Baizman, E., Branstrom, A., Bristol, N., Goldman, R., Han, K., Longley, C., Midha, S. and Axelrod, H. R., *J. Med. Chem.* **42**, 3193–3198 (1999).
42. Nicolaou, K. C., Xiao, X. Y., Parandoosh, Z., Senyei, A. and Nova, M. P., *Angew. Chem., Int. Ed. Engl.* **34**, 2289–2291 (1995).

43. St. Hilaire, P. M., Lowary, T. L., Meldal, M. and Bock, K., *J. Am. Chem. Soc.* **120**, 13312–13320 (1998).
44. Boojamra, C. G., Burow, K. M., Thompson, L. A. and Ellman, J. A., *J. Org. Chem.* **62**, 1240–1256 (1997).
45. Gordeev, M. F., Patel, D. V., England, B. P., Jonnalagadda, S., Combs, J. D., and Gordon, E. M., *Bioorg. Med. Chem.* **6**, 883–889 (1998).
46. Gustafson, G. R., Baldino, C. M., O'Donnell, M.-M. E., Sheldon, A., Tarsa, R. J., Verni, C. J. and Coffen, D. L., *Tetrahedron* **54**, 4051–4065 (1998).
47. Powers, D. G., Casebier, D. S., Fokas, D., Ryan, W. J., Troth, J. R. and Coffen, D. L., *Tetrahedron* **54**, 4085–4096 (1998).
48. Atuegbu, A., Maclean, D., Nguyen, C., Gordon, E. M. and Jacobs, J. W., *Bioorg. Med. Chem.* **4**, 1097–1106 (1996).
49. Nestler, H. P., *Mol. Diversity* **2**, 35–40 (1996).
50. Wess, G., Bock, K., Kleine, H., Kurz, M., Guba, W., Hemmerle, H., Lopez-Calle, E., Baringhaus, K.-H., Glombik, H., Ehnsen, A. and Kramer, W., *Angew. Chem., Int. Ed. Engl.* **35**, 2222–2224 (1996).
51. Tremblay, M. R. and Poirier, D., *Tetrahedron Lett.* **40**, 1277–1280 (1999).
52. Rehman, A., Li, C., Budge, L. P., Street, S. E. and Savage, P. B., *Tetrahedron Lett.* **40**, 1865–1868 (1999).
53. Phoon, C. W. and Abell, C., *J. Comb. Chem.* **1**, 485–492 (1999).
54. Tan, D. S., Foley, M. A., Shair, M. D. and Schreiber, S. L., *J. Am. Chem. Soc.* **120**, 8565–8566 (1998).
55. Nielsen, J. and Lyngsoe, L. O., *Tetrahedron Lett.* **37**, 8439–8442 (1996).
56. Nicolaou, K. C., Vourloumis, D., Li, T., Pastor, J., Winssinger, N., He, Y., Ninkovic, S., Sarabia, F., Vallberg, H., Roschangar, F., King, N. P., Finlay, M. R. V., Giannakakou, P., Verdier-Pinard, P. and Hamel, E., *Angew. Chem., Int. Ed. Engl.* **36**, 2097–2103 (1997).
57. Arap, W., Pasqualini, R. and Ruoslahti, E., *Science* **279**, 377–380 (1998).
58. Santoro, S. W. and Joyce, G. F., *Proc. Natl. Acad. Sci. USA* **94**, 4262–4266 (1997).
59. Khmelnitsky, Y. L., Michels, P. C., Dordick, J. S. and Clark, D. S., in *Molecular Diversity and Combinatorial Chemistry: Libraries and Drug Discovery*, I. M. Chaiken and K. D. Janda (Eds.). ACS, Washington, DC 1996, pp. 144–157.
60. McDaniel, R., Ebert-Khosia, S., Hopwood, D. A. and Khosla, C., *Nature* **375**, 549–554 (1995).
61. Briceno, G., Chang, H., Sun, X., Schultz, P. G. and Xiang, X.-D., *Science* **270**, 273–275 (1995).

5 Synthetic Organic Libraries: Library Design and Properties

From now until Chapter 9 our attention will focus on *synthetic organic libraries*, which are by far the most relevant libraries produced by combinatorial technologies. Their basic properties and their design by means of computational techniques will be described in this chapter as an introduction to the different synthetic organic library formats, which will be covered in the next chapters.

So-called *primary*, or unbiased, *libraries* will be described first, then *focused*, or biased or targeted, *libraries* will be presented and thoroughly discussed. A somewhat intermediate class of synthetic organic libraries, which we call *biased-targeted*, will also be introduced, and their properties and their usefulness will be highlighted. For each of these three classes there will be a discussion of their *rationale*, their *synthesis*, and the *selection* of the scaffolds and the monomer sets used to build the libraries.

The fourth section of this chapter will address a fundamental topic, the use of *computational chemistry* methods to help and to drive the design of each library class. A section for each library class will briefly illustrate all the relevant methodologies and their applications through the discussion of specific examples.

5.1 PRIMARY LIBRARIES: SHOOTING IN THE DARK?

5.1.1 Properties

We call primary, or unbiased, or diversity based, the libraries that are not designed on the basis of structural information, but rather those intended as diverse collections of molecules as sources of active compounds for different targets. Examples of such libraries have been reported frequently in the literature (1–3), and their main properties are shown in Fig. 5.1.

These libraries contain a *large number of individuals* (typically thousands or more) and are often prepared as *pool libraries*, even though the recent progress of automated parallel synthesis allows the preparation of thousands of discretes in a timely manner. Primary libraries are predominantly prepared *on SP* because of the easier preparation of high-quality pools on SP and the negligible work-up required for each SP sample. The library components are prepared in small quantities, often significantly less than a milligram. They are *diversity-based* (see Section 5.4.1) and are usually tested on *many different targets*: if their design is successful, they represent a source of positives for most of the screenable targets. The positives obtained from these libraries, called

PRIMARY LIBRARIES

- LARGE (thousands to millions)
- FREQUENTLY ON SOLID PHASE
- FREQUENTLY IN POOLS
- <<1mg *PER* LIBRARY INDIVIDUAL
- DIVERSITY-BASED
- TESTED ON MANY TARGETS
- DESIRED OUTCOME: **HIT**

Figure 5.1 Primary libraries: main features.

hits, typically have a moderate or even weak activity and represent the starting points for further structure optimization.

5.1.2 Rationale

Whenever a library is designed in a particular format, some basic questions should be addressed to justify its synthesis. We will address them first in theory; then we will apply the same principles to a published example of a primary library.

- **Why**? A primary library can be planned for many purposes. Sometimes the project goal is purely academic, that is, simply to open a new combinatorial synthetic route and to report it. This respectable option allows complete freedom to prepare any primary library, with no constraints dictated by parameters such as application, cost, or resources. We will not comment further upon this approach in this section. More often, though, a primary library is prepared as a source of relevant active molecules on various biological targets. This obviously necessitates the availability of several robust and reliable HTS assays for these targets, as well as the synthetic resources, the analytical resources, and the instrumentation to ensure the successful synthesis and analytical characterization of the library.
- **When**? The combinatorialization of the planned synthesis is often long and arduous for primary libraries, so several applications must be available for the library to compensate for this major undertaking: A single target would not justify the efforts required, and hence another library format would be preferable.
- **How**? The large number of components of primary libraries is more suitable for SP pool libraries, but each library has to be designed keeping in mind the available equipment and the expertise of the chemical resource(s). The same is true for the analytical methods used to check every step of the library synthesis,

which must produce a reliable quality profile for all or most of the library components. It is important to have a certain flexibility, that is, to be able to use different library formats for different applications without negative consequences on the quality of the prepared primary libraries.

- **What**? The chemistry required for a primary library synthesis must be extremely robust and assessed to accommodate a wide range of monomers/scaffolds in order to produce the large numbers of diverse individuals needed. A compromise between "easy" reaction schemes, which readily produce a reliable library but may not lead to novel positives, and "hard" ones, where the library components are novel but the optimization of reaction conditions is difficult and requires major efforts, should be found for each specific project.
- **How Much**? Every synthesis has its costs, both in terms of resources and reagents: A primary library synthesis employs a large number of chemicals, and its total cost will always be significant. The introduction of expensive reagents, monomers, or scaffolds should be avoided as a general rule, using commercially available chemicals whenever possible and utilizing computational tools to improve selection among them to help prepare a diversity-based primary library (see Section 5.4.1). The same is true for monomers/scaffolds requiring a long synthesis, which are better suited for more focused libraries, as we will describe in the next section.

The primary library shown in Fig. 5.2, which was reported by Baldwin (1), was clearly designed as a source of biologically active molecules on several targets (why), thus compensating the efforts required for the chemical assessment and for a satisfactory characterization (when). The SP library was prepared in pools using chemical encoding (4, 5) to produce a population of around 62,000 individuals (how). The synthetic scheme was composed of both simple and more complex SP steps, and several monomer sets (**A–F**, Fig. 5.2) were used (what). These monomers were either commercially available or easily prepared from commercial precursors, while the library benzopyranic core was formed during the synthesis (how much).

5.1.3 Synthesis and Characterization

After the design of a library in the most suitable format, the synthetic process starts with the *chemical assessment*. Usually the designed synthetic route is first tested in solution on one or a few representative structures, then its transfer onto SP is accomplished while optimizing the reaction conditions for model individuals. This process already has been described in detail in Chapter 3.

The preparation of a large primary library requires the use of some monomer sets containing a considerable number of monomers, and before embarking on the library synthesis, a so-called *monomer rehearsal* (6) is necessary. The monomer set is split into homogeneous subsets, and a representative of each subset is used in the relevant SP reaction step. A theoretical example is shown in Fig. 5.3, where eight representative aldehydes (M_1-CHO–M_8-CHO) are examined for a reductive amination with R–NH_2. If even after significant efforts to adapt the poorly performing monomers (e.g., change

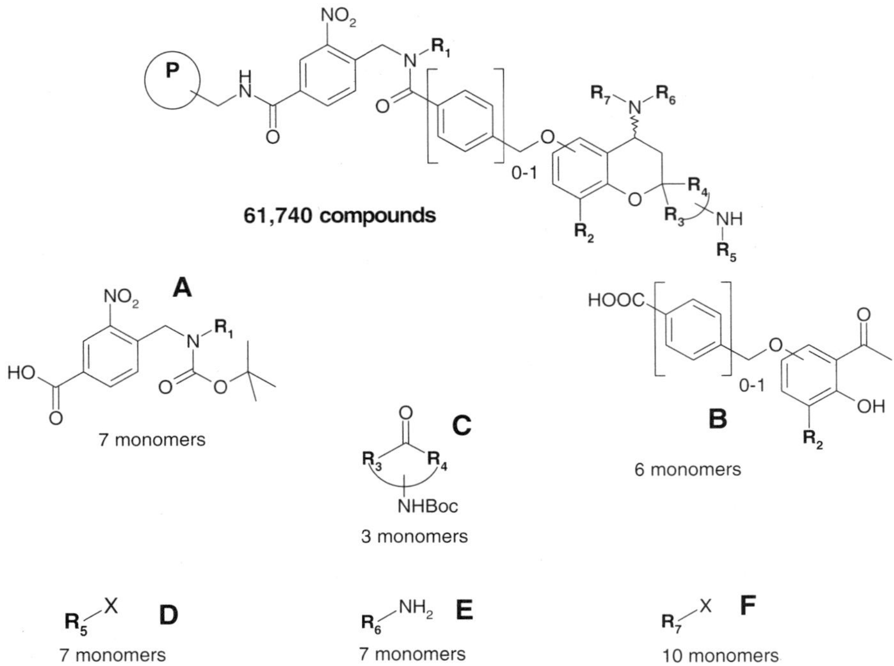

Figure 5.2 Example of a benzopyran primary library.

of protecting groups) their reactivity remains poor, they are removed from the whole set: In our example only six aldehydes pass the rehearsal, producing the expected reaction product with good yields and purities (Fig. 5.3). During the monomer rehearsal an extensive analytical characterization is essential, using on-bead and off-bead methods, to detect the side products formed, to assess the purity of the resin-bound intermediates, and to understand the influence of each modification of the reaction conditions on the outcome. Rather often this process is not extensively reported in published works, as for (1), but the unequivocal basis of a good-quality pool library lies in careful monomer rehearsal.

The next step in primary library synthesis is to prepare the so-called *model library*, which contains a small number of compounds (typically from 5 to 50) prepared using the very same instrumentation, techniques, and format planned for the library synthesis (6, 7). Simply moving from single compounds to simultaneous multiple synthesis requires adaptation of the technical operations (addition of solids or liquids, evaporation of solvents, stirring, reaction times, etc.), and the impact of these modifications is sometimes significant: A complete analytical characterization of the model library using the most appropriate techniques allows selection of the best operating conditions to be used for the library synthesis. The direct synthesis of a large library after monomer rehearsal may sometimes produce poor-quality results for the above-mentioned rea-

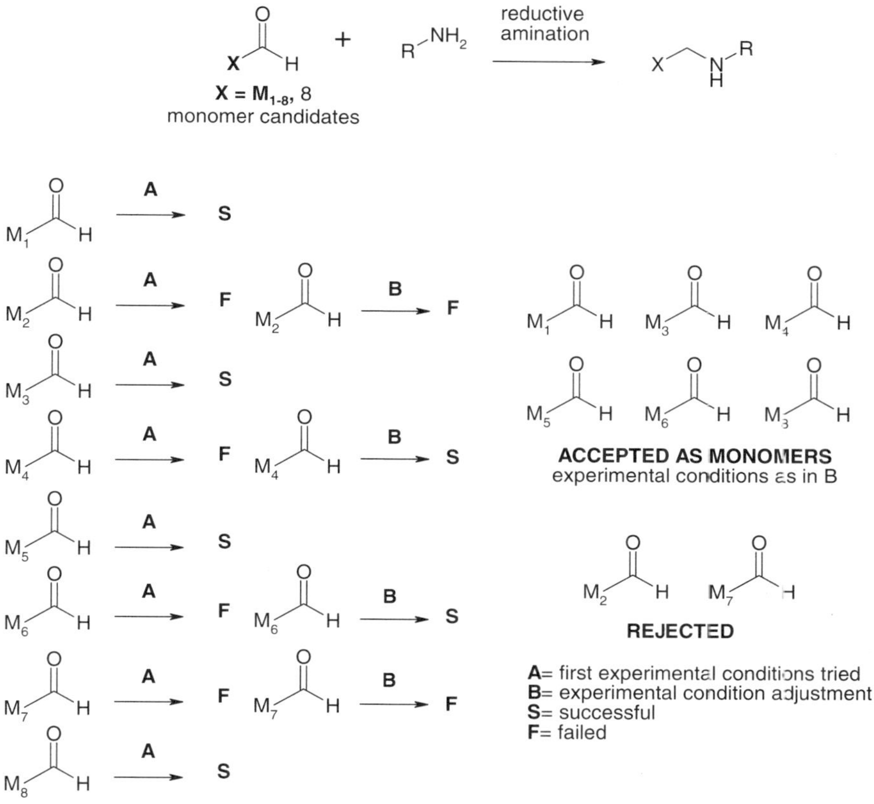

Figure 5.3 Monomer rehearsal: A hypothetical example for reductive amination.

sons, wasting substantial amounts of precious materials and the significant efforts of both synthetic and analytical chemists. It is highly recommendable to follow this gradual scale-up process via high-quality model libraries to minimize the risk of synthesis failures.

The primary library synthesis is then carried out using any of the techniques that will be illustrated in detail in the next three chapters for solution- or solid-phase, pool or discrete libraries. Many choices in terms of instrumentation or automation are available, and each of them may afford high-quality results. The same is true for analytical quality control, which determines the synthesis outcome and decides if the library is suitable for screening against desired targets.

5.1.4 Scaffolds and Monomers

A synthetic organic library is characterized by a *common core structure* for all its components: This motif may be used to start a library synthesis that consists of the decoration of this *scaffold* or it may be assembled from the various synthons to give the scaffold. For both options *monomer sets* will be used, either to decorate the scaffold through its randomization points (decoration libraries) or to build the scaffold and simultaneously introduce diversity in the library components (assembly libraries).

The assessment of a solid and reliable synthetic route for a primary library is essential, but it is not the only requirement to fulfill. A large library requires the availability of large monomer sets that should, if possible, be composed of commercially available chemicals. The costly and time-intensive synthesis of several monomers can be justified only when the compounds produced using them have a high probability of being useful, not when diversity is the only criteria driving the library synthesis. Examples of monomer classes that are suitable for large library synthesis include alcohols, amines, aldehydes, carboxylic acids, α-amino acids, and other common functionalities. The appetite for new proprietary monomers and for new monomer sets is increasing with the interest for combinatorial technologies; when a serious and continuous involvement in this new discipline is planned, even the synthesis of precious proprietary monomers (e.g., bifunctional monomers with orthogonal protection and poorly represented functionalities) in large amounts to be used repeatedly for high-value primary libraries becomes convenient and even desirable due to the novelty they ensure to library components. We will discuss the computational methods available for the selection of monomers for a primary library within Section 5.4.

A different scenario must be considered for a scaffold, especially when the library is made by its decoration. A highly functionalized proprietary scaffold is a very desirable asset, and a difficult synthesis may be accepted because even easily accessible commercial monomers will produce original library individuals by its decoration. Significant efforts to design hindered and constrained polyfunctional scaffolds, often taking advantage of the structures of natural products, are commonly accepted and may represent the key factor to prepare a high-quality meaningful primary library.

5.2 FOCUSED LIBRARIES: HIGH-THROUGHPUT STRUCTURE–ACTIVITY RELATIONSHIPS

5.2.1 Properties

We call focused, or biased, or similarity based the libraries designed using structural information (a compound active on the target, or on a similar target) or a rational hypothesis and allowing the library structure to focus on very specific features. The main properties of such libraries are shown in Fig. 5.4.

These libraries contain a relatively *small number of individuals* (typically tens to hundreds) and are almost always prepared as *discrete libraries* using parallel synthesis and automated or semiautomated devices. Focused libraries are predominantly prepared *in solution* because of the easier shift from classical organic synthesis to solution-phase combinatorial chemistry, while automated purification procedures for relatively small arrays of discrete compounds in solution are common nowadays. The

FOCUSED LIBRARIES

- MEDIUM/SMALL (tens to thousands)
- FREQUENTLY IN SOLUTION
- FREQUENTLY AS DISCRETES
- >1mg *PER* LIBRARY INDIVIDUAL
- SIMILARITY-BASED
- TESTED ON A SPECIFIC TARGET
- DESIRED OUTCOME: **LEAD**

Figure 5.4 Focused libraries: main features.

quantities of single library components are usually a few milligrams but may even reach a *few tens of milligrams*. They are *similarity based* (see Section 5.4.2), and they are usually tested only *on the target/assay for which the library was prepared*. If their design is successful, they provide a relatively accurate, even if preliminary, structure–activity relationship for the specific target allowing the selection of a compound, or a few compounds, called *lead(s)*, with significant activity on the target.

5.2.2 Rationale

- **Why**? A focused library is used as a source of lead compounds for a specific target, with a better activity profiling than for hits from primary libraries: smaller numbers allow testing of the library components in medium- or even low-throughput assays. Other properties besides the potency on the target are evaluated for active library components to determine their usefulness (solubility, toxicity, aspecificity, and physicochemical properties, among others). Measurement and evaluation of these parameters must be fast to allow rapid project progression. The synthetic and analytical instruments required to synthesize and analytically characterize the library are significantly less complex than for a primary library; even the normal equipment of an organic synthesis laboratory is often appropriate.
- **When**? The combinatorialization, usually in small solution- or solid-phase arrays, of the planned synthesis is not particularly demanding for focused libraries. Moreover, an active compound from a primary library is often the structural information around which the focused library has been designed so that a detailed chemical assessment has already been performed. The use of focused libraries should become more and more frequent given the simplicity of their synthesis and the significant impact they may have on the progression of many projects.

- **How**? Each library has to be designed according to the equipment and the expertise of the combinatorial chemist(s) but also, if possible, using the knowledge accumulated during the synthesis of related primary libraries. The same is true for the analytical methods used to check every step of the library synthesis; the components of a focused library must be thoroughly evaluated, their purity requirements are high, and their analytical characterization may be slower but more accurate than for a primary library. The flexibility of using different library formats for different applications is especially important for the automated purification procedures that are becoming popular and allow isolation of significant amounts of every library component with a high degree of purity.
- **What**? The chemistry required for a focused library synthesis may be extremely robust and assessed, but even less robust synthetic schemes can be adapted to accommodate the needs of a small focused library synthesis. All the library components are inspired by the structural information available, but nevertheless they must provide a detailed exploration of the structure–activity relationship for analogues of the parent structure.
- **How Much**? A focused library is by definition composed of a manageable number of entities, and its total cost will not be significant most of the time. The introduction of meaningful, expensive reagents, monomers, or scaffolds must be accepted and encouraged, even when they require significant synthetic efforts, providing that less expensive or commercial replacements are not available.

A small focused library, shown in Fig. 5.5, which was reported by Wipf and co-workers (8, 9), was designed to exploit the information provided by natural inhibitors of serine/threonine protein phosphatases (why) and to gain additional

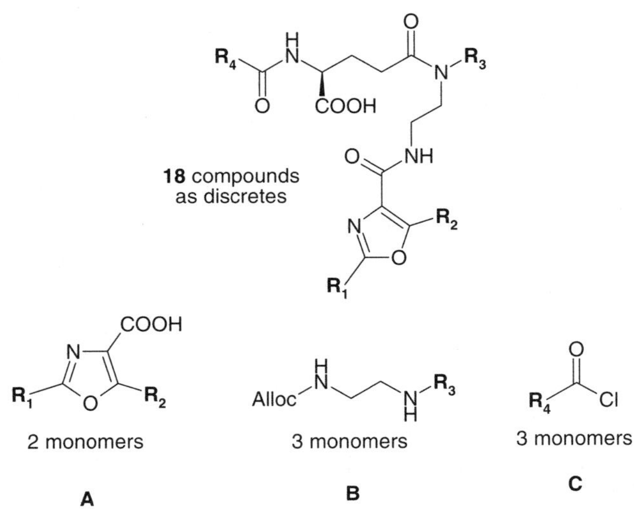

Figure 5.5 Example of an oxazole focused library.

information on the library components, compensating the synthetic efforts and the satisfactory analytical characterization (when). The library was prepared as discretes and was composed of 18 compounds (how). The synthetic scheme was composed of several SP steps with different complexities, and both noncommercial (**A**, **B**) and commercial (**C**) small monomer sets were used (what). Neither the time and effort required nor the chemicals used for the focused library synthesis represented a burdensome investment (how much).

5.2.3 Synthesis and Characterization

After having assessed the planned synthetic route on a single standard (when the focused library will be prepared in solution, this step is not necessary), the *combinatorialization* process takes place. This process is significantly less demanding than for primary libraries because of the reduced diversity of the monomers composing the sets, which typically show similar reactivity. It is therefore easier to find optimal reaction conditions, meaning the monomer rehearsal should be faster and produce only a few, if any, rejected monomers for the focused library synthesis. The high *purity* requirements for the focused library individuals will demand a meticulous characterization of any side reaction or side product arising in order to obtain pure final compounds in good yields. Particular attention will be devoted to automated or semiautomated purification techniques when solution-phase focused libraries are involved.

The synthesis of a model library is not strictly necessary because of the reduced complexity of a focused library, but it may be appropriate to prepare a few discretes with the very same equipment to be used for the focused library synthesis. The library is then prepared, purified when necessary, and fully characterized. The purity of all the compounds will be determined using the most appropriate analytical techniques and only after having satisfied all the selected criteria, the library individuals will be tested on the target, and the active molecules progressed as lead compounds for a specific application.

5.2.4 Scaffolds and Monomers

A focused library, even if much simpler in architecture when compared to a primary library, is a very valuable tool to speed up the discovery of a lead compound or its optimization. Its synthesis takes place when positive indications regarding the structure of its components are already known, so that the probability of finding relevant activities on a specific target is higher than for a primary library, where moderate activities can be expected at best. This justifies any effort for *rationally selected monomers/scaffolds* (see Section 5.4.2), which are useful to gather additional information on the target and to lead to the requisite activities. The selection of noncommercial monomers and/or novel scaffolds should actually be preferred in that the novelty embedded in the resulting structure provides an advantage over competitors interested in the same target.

5.3 BIASED-TARGETED LIBRARIES: INFORMATION-RICH PRIMARY LIBRARIES

5.3.1 Properties

It could be safely assumed that nowadays diversity-inspired, pure primary libraries are not synthesized any more: in fact, although they can produce a large number of compounds their representation of the huge "chemical diversity space" remains necessarily limited, and the need to focus large libraries on "meaningful diversity" is more and more urgent.

A compromise between large unbiased and small focused libraries has become popular, especially for pharmaceutical applications. These biased-targeted libraries are not inspired by a precise structural information, but rather by general information regarding similar classes of targets (e.g., kinases or 7-transmembrane receptors) or by the desired activity-unrelated profile that a drug must possess (e.g., the molecular weight, the partition coefficient, the water solubility, and other physicochemical properties). Their main properties are listed in Fig. 5.6.

Biased-targeted libraries are similar to primary libraries in terms of their architecture and format. In fact, the reported primary libraries have often been designed taking into account some of the above-mentioned filters (*information biased*) and thus should actually be called pseudofocused libraries. They are tested on as many targets as possible, as for primary libraries, but the information gathered in terms of activity (*valuable hits*) should be more easily transformed into relevant active leads to be developed further.

BIASED-TARGETED LIBRARIES

- MEDIUM/LARGE (thousands to hundreds of thousands)
- FREQUENTLY ON SOLID PHASE
- FREQUENTLY AS POOLS
- <<1mg *PER* LIBRARY INDIVIDUAL
- INFORMATION-BIASED
- DIVERSITY-BASED
- TESTED ON MANY TARGETS
- DESIRED OUTCOME: **VALUABLE HIT**

Figure 5.6 Biased-targeted libraries: main features.

5.3.2 Rationale

- **Why**? A biased-targeted library is used as a source of relevant activities on various targets, and the availability of several robust and reliable HTS assays and the instrumentation allowing the successful synthesis and analytical characterization of the library are necessary, as mentioned for primary libraries. Moreover, some target-related or some physicochemical filters are introduced to select more valuable compounds. Knowledge of the appropriate selection techniques and the necessary equipment must be available (e.g., commercial computational databases and proprietary or published information on target classes; see Section 5.4.3).
- **When**? **How**? **What**? **How Much**? The answers to all these questions are identical to those given above for primary libraries and will not be commented upon further. It is worth saying that any generic information that can be transformed into a structural filter to select valuable library components should be applied whenever possible. The cost of the library synthesis in terms of time and effort will not vary significantly when compared to a primary library.

An example of a biased library, shown in Fig. 5.7, which was reported by Boger et al. (10), was designed to be a source of tools to probe protein–protein interactions (why). The effort required for the chemical assessment and for a satisfactory charac-

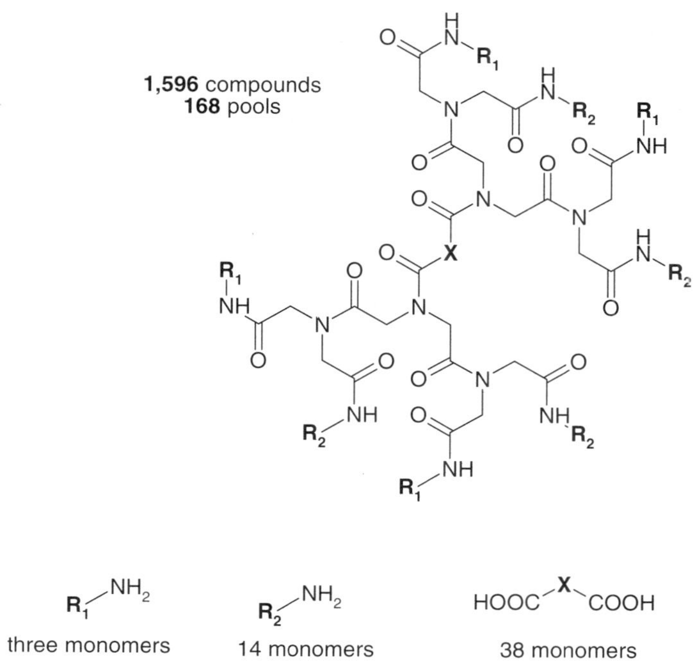

Figure 5.7 Example of a polyamide biased-targeted library.

terization was significant, but the successful preparation of a library aimed at a general class of biological targets more than accounted for this effort (when). The library was prepared as 168 pools in solution (1596 iminodiacetic acid diamide tetramers; how). The synthetic scheme was a mix of different reaction conditions in solution, and some biased monomer sets were used (what). These monomers were either commercially available or easily prepared from commercial precursors, while the library core was the commercially available iminodiacetic acid (how much).

5.3.3 Synthesis and Characterization

All the aspects, from the design of a biased-targeted library to its preparation, analytical characterization, and assay on various targets, are similar to the ones already described for primary libraries and will not be discussed further.

5.3.4 Scaffolds and Monomers

While the number of monomer sets employed for the synthesis of biased-targeted libraries is similar to that seen for primary libraries, some additional selection criteria are applied. If the library design is driven by structural information, the monomers will be chosen accordingly (e.g., hydrogen bond donors or acceptors and lipophilic pockets, to be placed in specific parts of the final library individuals). Furthermore, the scaffold, either as the starting point of the library synthesis or formed during the synthetic scheme, will need to satisfy the structural needs of the target biological class.

When the library individuals are filtered with physicochemical parameters, the nature of the scaffold becomes fundamental. As an example, if a maximum accepted value of log P (partition coefficient between n-octanol and water) of 4 is set as a limit, the use of a functionalized scaffold with log $P = 6$ will enormously limit the selection of monomers to highly hydrophilic structures, while the selection of a more appropriate scaffold would allow a higher degree of diversity while respecting the imposed filter. The same is true for monomers. For example, if an upper limit of 600 is imposed for the molecular weight (MW) of the final library components, the use of monomers with an MW higher than 250–300 will not be acceptable.

5.4 LIBRARY DESIGN VIA COMPUTATIONAL TOOLS

5.4.1 Primary Libraries

The use of computational tools to increase the quality of combinatorial libraries is becoming more and more popular. These methods can be applied to *any library format*, and their careful use may add significant value to the library. Their main application is the selection of individuals/monomers for a library from the many virtual monomers/final compounds theoretically available by using all the easily accessible monomers. We will examine in detail the usefulness of computational tools for each library format through the presentation of the main reported approaches by leading groups in the field. These tools have also been widely used to determine the diversity of compound collections (11–14), either proprietary or commercial, and to select individuals to add to these collections to improve their representativity of chemical space (15–17). While the methods are the same, some typical implications for database selection will be discussed when appropriate.

Primary libraries are not inspired by any structural information, and their purpose is to contain the *maximum of chemical diversity* so as to function as a potential source of active compounds for many different applications. To consider a library size and its diversity as directly proportional entities is totally wrong, as the simple example in Fig. 5.8 shows. The library on the left, composed of 25 molecules, is much more "diverse" than the 50-member library on the right because it spans more of the chemical space reported in the figure. While the concept of chemical diversity is intuitive, in order to measure the diversity of library components and to select from them the most representative, we must define some key features and methods.

A *virtual library* is defined as a computer-generated library containing all the compounds obtained by all the permutations of each monomer set composed of all the available monomers possessing the specific chemical functionalities. The available virtual monomers are gathered from commercial databases and sometimes from internal accessible collections. All the virtual library compounds are enumerated and the generated data are stored. This library undergoes a so-called *virtual screening*, that is, the characterization of each library individual and the selection of the components to be included in the actual library that will be synthesized. This computational screening requires the identification of the appropriate *molecular descriptors* that characterize a molecule and the determination of indices, or numerical values, that allow a comparison between the various molecules. Once the molecules have been characterized and compared in terms of the selected descriptors, *selection methods* are necessary to pick out of the virtual library members the actual library components that will be synthesized.

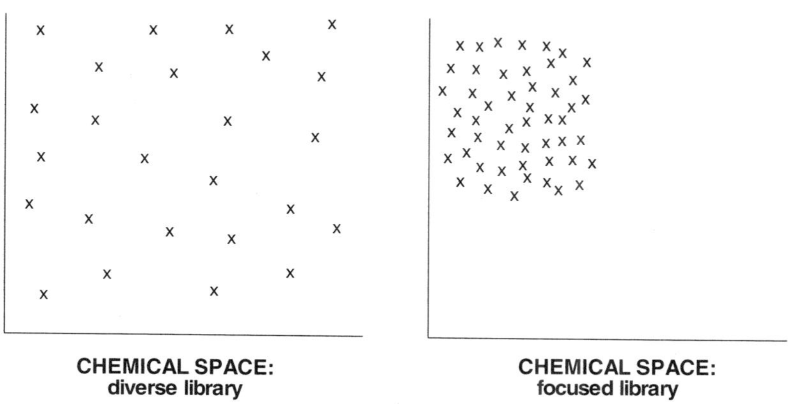

Figure 5.8 Chemical diversity in the chemical space.

The selection of virtual monomers is made through the electronically accessible databases of chemicals and, where possible, proprietary collections of compounds. The final list will contain compounds that possess the specific functionality, and additional filters may be introduced to consider only reactive potential monomers (e.g., aromatic aldehydes where aliphatic ones are known to be problematic). This filtering function is driven by the *chemical expertise* and aims to remove inadequate reagents from the very beginning. Once the virtual monomer sets are prepared, the library is *enumerated*, that is, all the virtual components are prepared "on screen" (18, 19). The storage space required for the complete enumeration of a large virtual primary library is so huge that two more compact representations are normally used: *Markush* representations, which cover a library with common structural features, through either line notations (20–22) or connection tables (23–25), and *reaction-based* representations, which are commercialized by various companies (26–28). Both methods use the set of precursors/reactants/monomers and the generic reactions used to prepare the library to characterize the whole virtual set. Examples of both representations are shown in Fig. 5.9; a more detailed description of these representations can be found in the references mentioned above (18–28).

These molecules are then characterized through various molecular descriptors. Such a process can be performed on two different compound sets, product- or reagent-based sets, as shown in the hypothetical example reported in Fig. 5.10. Considering a library where a common scaffold is decorated using three large monomer sets composed respectively of 2000, 1000, and 500 monomers just using commercial compounds, our virtual library would consist of $2000 \times 1000 \times 500 = 1 \times 10^9$ individuals. Thousands of candidates are commercially available for the most common monomers such as amines, carboxylic acids, alcohols, or aldehydes, and even bigger virtual libraries can be easily built. All of these virtual products should be enumerated and their properties/descriptors calculated using a *product-based* selection, requiring significant efforts for the management of such a large number of structures. An alternative is selection based on the *structures of the reagents* included in the monomer sets, which would involve only the evaluation of three sets totalling 3500 compounds. While it has been proven that the latter method is less accurate in generating diverse libraries (29), it can nevertheless produce a sound selection of the most diverse monomers in a specific class within a reasonable time and has been used mostly for large virtual diversity-based libraries (30). The accuracy lost in examining only the reactant space is more than compensated for by the reduced computing time required and by the effective reduction of each monomer set size.

If we consider the example of Fig. 5.10, when the reagents are selected to produce a $20 \times 10 \times 10 = 2000$-membered library we end up with combined monomer sets composed of 40, rather than 3500, individuals (Fig. 5.11, top). If the 2000 most diverse library components are selected from the product space (Fig. 5.11, bottom), it may be that a significant fraction, or even all, the virtual monomers from one set must be used to prepare them, so that the library cost, in terms of both reagents and human effort, is not significantly reduced. This selection problem was noticed (31) during the selection from a virtual library of the 1600 most diverse amides, which were found to contain 137 amine and 146 carboxylic acid building blocks (19,992 possible combi-

Markush Representation

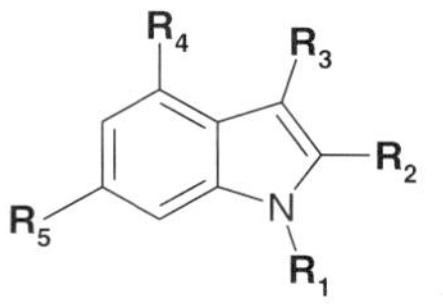

100,000-membered indole library

R$_1$ = H, Me, Et, n-Pr, cHex, Bn, Phet, Ph, 4-MeOPh, 4-MePh

R$_2$ = H, Cl, Br, Me, Ph, CHO, COOH, COOMe, $CONH_2$, $CONMe_2$

R$_3$ = H, Cl, Br, I, OMe, OBn, NMe_2, NBn_2, Me, Et

R$_4$ = H, Cl, Br, I, Me, CN, COOH, COOMe, $CONH_2$, $CONMe_2$

R$_5$ = H, Cl, Br, Me, Ph, CN, NO_2, COOMe, CF_3, COPh

Line Notations: scaffold identification: *1H-Indole12346*, then descriptors for 5 randomization points: *[10R$_1$]1.[10R$_2$]2.[10R$_3$]3.[10R$_4$]4.[10R$_5$]6*, then monomer definitions: *Me=C, Et=CC, CF$_3$=C(F)(F)F* and so on.

Reaction-based Representation

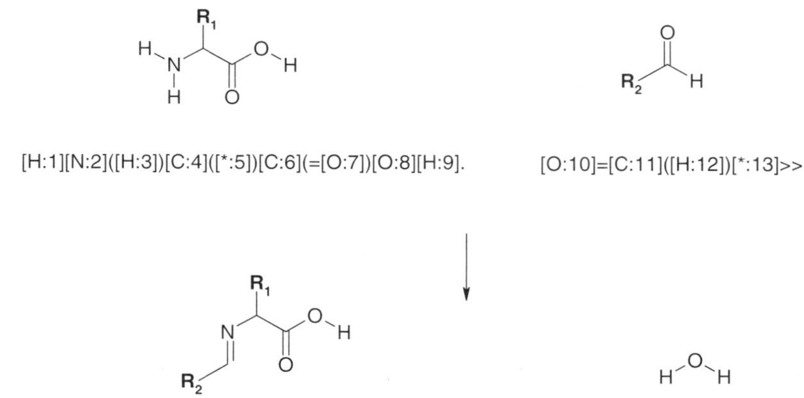

* = Randomization point; >> = ⟶

Figure 5.9 Representation of chemical libraries: Markush and reaction-based approaches.

nations). Such a problem disappears when considering selections among commercial/proprietary databases where the virtual compounds to be ordered/withdrawn for a specific assay are structurally unrelated and need not be synthesized.

Monomer selection may be significantly assisted by a different perspective, that is the computational evaluation of the reactivity of each potential monomer belonging to a virtual set. In fact, it may well happen that a suitable, diversity-adding selected monomer has a poor reactivity in the library reaction scheme that will eventually

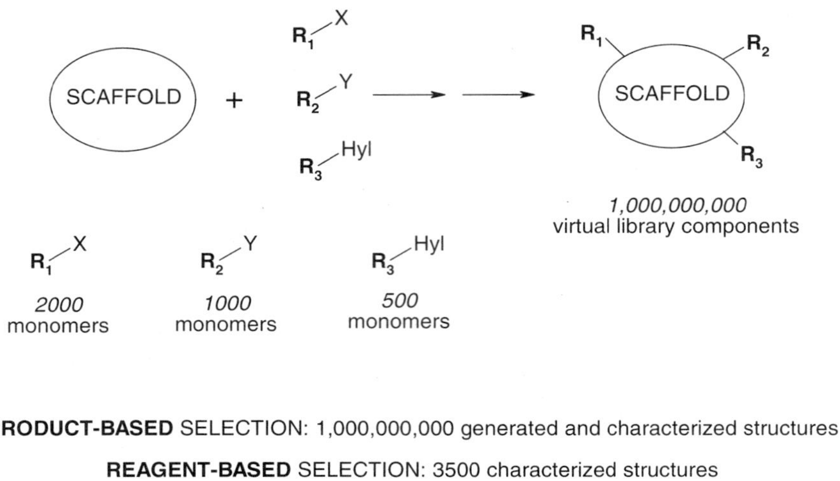

Figure 5.10 Virtual libraries: product-based and reagent-based selection methods.

surface during the library synthesis; the *a priori* selection of a similar, better reacting monomer would be highly beneficial and would save time, efforts and money. Horvath (32) has reported a reactivity prediction model for any chemical transformation, and has successfully applied it to the selection of carboxylic acids for the synthesis of an amide library; 100 out of the virtual 150 monomers were selected according to their reactivity, and the error of predicted *versus* experimental reactivity in the acylation for the acids was on average less than 10% and never more than 20%.

Examples of library selections based on virtual products, rather than monomers, have been reported, especially when the computational burden is reduced by fixing the scaffold orientation and optimizing only the randomization points (33). From now on we will not mention specifically if selections are performed on virtual sets of reagents or products, providing that the reader remembers throughout the rest of this section the relevance of this issue and its dependence on project-related factors (number, availability of hardware/software, and so on), rather than on dogmatic assumptions.

The selection among virtual monomers, or products, is made by considering their *properties* as determined by a set of molecular descriptors, and then selecting the most diverse representatives. Many descriptors have been reported (34), and they can be grouped into three main classes: one-, two-, and three-dimensional *(1D, 2D, and 3D) descriptors*. One-dimensional descriptors are represented by a single value integer, as for the so-called topological indices (35–37, 11), which characterize the bonding pattern of a molecule, or by a real value calculated for global molecular properties (38, 39, 30), such as *molecular weight*, *lipophilicity*, and *solvation energy*. Their calculation is easy and fast, thus reducing the time required for the virtual screening.

Two-dimensional descriptors are normally represented by linear bit strings indicating the presence or the absence of some properties in the molecule: Examples include structural fragments (*structural keys*; 40, 41, 36), specific atom paths with predefined

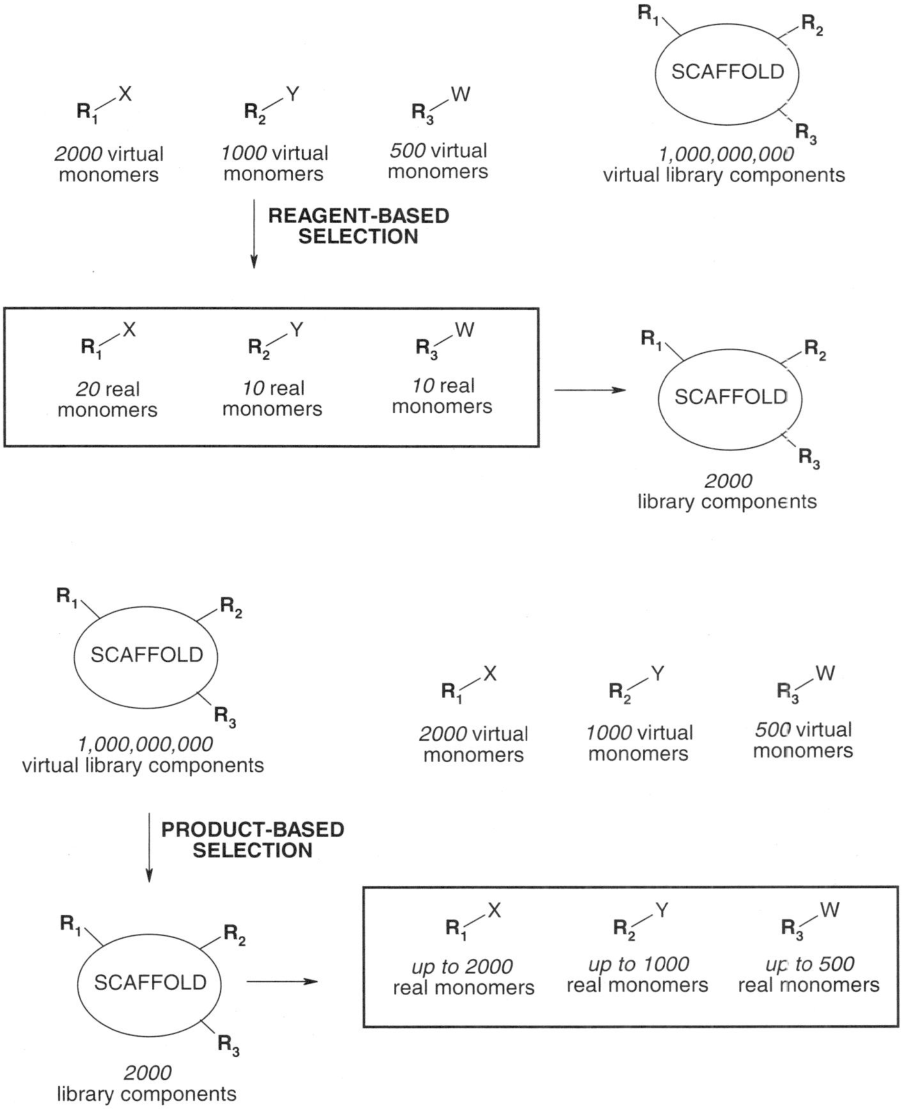

Figure 5.11 Product-based and reagent-based selection methods: impact on monomer selection.

lengths (*hashed fingerprints*; 42–45), and intermolecular interactions (*binding properties*; 46–48). Their calculation is relatively fast, and when chosen for a monomer/product selection, they do not represent a bottleneck for the project. An example of 2D descriptors is reported in Fig. 5.12.

Three-dimensional descriptors represent spatial relationships, such as *distances* and *angles*, between key functionalities in a structure, and they are encoded by linear bit

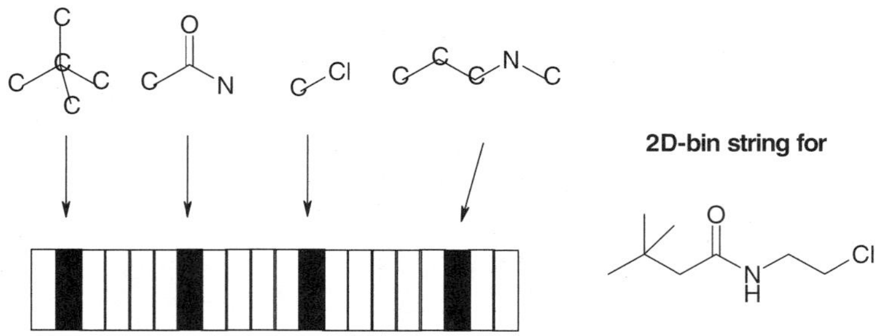

Figure 5.12 Two-dimensional molecular descriptors: an example.

strings, where a range is first defined (e.g., 2–10 Å for a distance or 30°–90° for an angle) and then a certain increment defines the bin width (e.g., 1 Å or 5°). These distances and angles are measured between atom pairs, or between functionalities, and define the so-called *pharmacophores* (three- or four-point pharmacophores), which are stored and used to compare molecules in terms of diversity or similarity. Examples of 3D descriptors are reported in Fig. 5.13. Many different research groups have defined different sets of 3D descriptors/pharmacophores (49–62), and all of them have been used to select diverse sets of compounds.

A recent method called 4D-QSAR (63) builds and optimizes a set of relevant 3D-pharmacophores for a specific target and effectively samples the available conformational space and identifies a sound QSAR in terms of binding mode of each compound examined. The method has been successfully validated in the design and virtual screening of libraries aimed towards the thromboxane A_2 receptor (64) and glycogen phosphorylase *b* (65, 66); both examples have produced virtual hits which, once prepared and tested on the target, were confirmed as novel bioactive compounds.

While a 3D descriptor should represent the 3D interaction between a molecule and a biological target more accurately, the time required to define the 3D descriptors for a large virtual set of structures is significantly higher than for 1D or 2D descriptors, and often this cannot be tolerated by the project. Three-dimensional descriptors can be obtained from a fixed low-energy conformation for a given structure (*rigid descriptors*) or by the same distances/angles obtained on a number of conformations for the same structure (*flexible descriptors*). The former are easily calculated in a timely manner but are significantly less information rich than the latter. An example (41) showed how 2D descriptors can perform significantly better than rigid 3D descriptors in separating biologically active molecules from inactives, thus hopefully leading to a good compound selection; other comparisons between descriptors and descriptor classes have also been recently reported (67–70).

The selection of the most appropriate descriptors out of this wide range of choices is often driven by project-related parameters (number of virtual library components or monomers) or by the available computational equipment (software and hardware). It is important to avoid the use of different descriptors that describe the same property or are biased by similar contributions because this would increase their weight in the

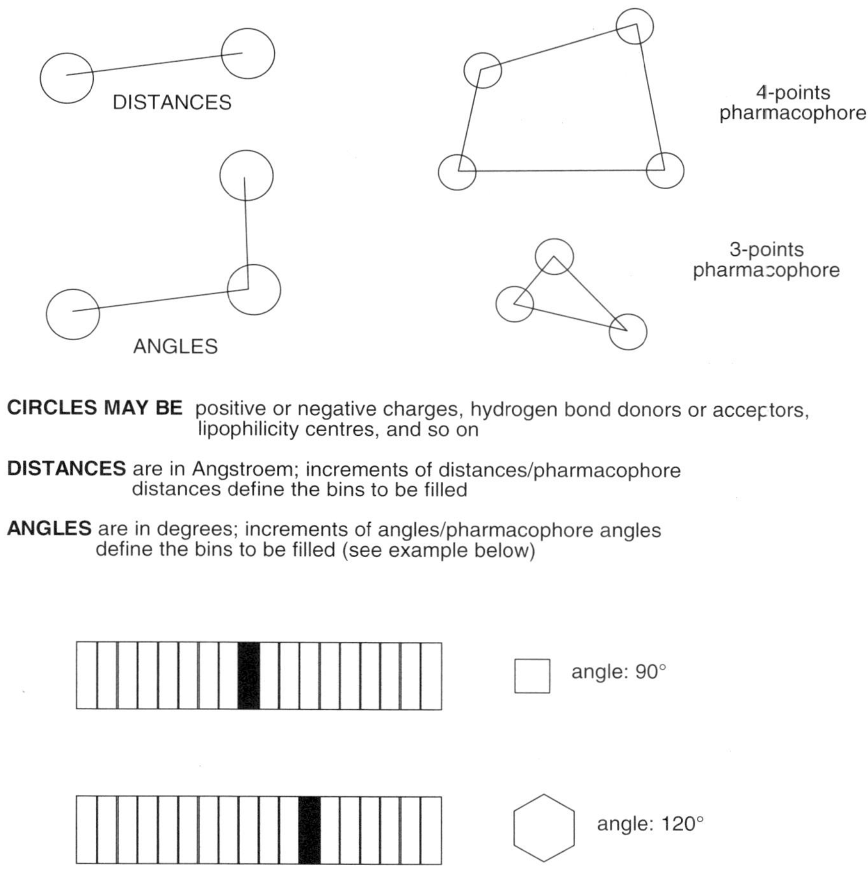

Figure 5.13 Three-dimensional molecular descriptors: several examples.

overall evaluation of the molecules and would overestimate the contribution of this property to the determination of global diversity for a virtual library. *All classes of descriptors have proven useful* to select diverse sets of compounds when compared to random selections (29), but the use of flexible 3D descriptors/pharmacophores for huge virtual libraries is, as of today, prevented by the computational time and memory required.

Any type of selected descriptor will provide a more or less complex characterization of each virtual library component. The use of similarity indices offers a straightforward method to evaluate similarities between virtual compounds. These indices use a bit-string representation for any descriptor (distances, fingerprints, pharmacophores, and so on) and, by simply counting the presence or the absence of specific bins and comparing the bit strings of virtual compounds, provide a numerical similarity index. The formula for the commonly used *Tanimoto* similarity index (71, 43), which can readily be transformed into the complementary diversity index, is reported in Fig. 5.14

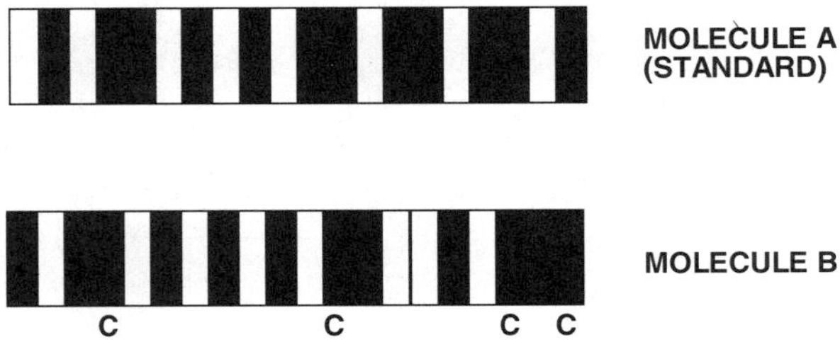

$$\text{Tanimoto similarity index:} \quad \frac{N_{common}}{N_{common} + N_{different}} = \frac{4}{20} = \mathbf{0.2}$$

$$\text{Tanimoto dissimilarity index:} \quad \frac{N_{different}}{N_{common} + N_{different}} = \frac{16}{20} = \mathbf{0.8}$$

Figure 5.14 Similarity/dissimilarity comparison of two molecules: the Tanimoto index.

(molecules **A** and **B** are highly dissimilar in this hypothetical example). Such an index, even if questionable when applied to dissimilarity evaluations (54), allows fast comparisons between structures or subsets of molecules.

Compound selection is a core process of library design, and three main methods can be mentioned. *Dissimilarity-based methods* select compounds in terms of similarity/distance between individuals in chemical space. *Clustering methods* first group compounds into clusters based on similarity/distance and then choose representative compounds from different clusters. *Partitioning methods* first create a uniform cell space that subdivides the "chemical space," then assign all virtual compounds to the relative cells according to their properties, and finally choose representative compounds from different cells.

Dissimilarity-based methods are the fastest selection techniques, as can be easily seen from the hypothetical example reported in Fig. 5.15. This method, based on the *maximum dissimilarity* approach (72–74), must first define a minimum acceptable dissimilarity threshold *R* and a maximum subset *M* of compounds to be selected. The paradigm starts randomly from any virtual component (**A**, Fig. 5.15) and then searches for the most dissimilar individual in the virtual population (**B**, Fig. 5.15). The following selection rounds choose the individual most dissimilar from all those already selected, and this process is repeated (**A–E**, Fig. 5.15) until either the last selection produces a candidate whose dissimilarity to the other selected components is less than *R* or until *M* number of virtual components (five in the example) have been selected. Such a method is easy and fast, does not require significant computing time, and, while intrinsically biased toward picking extreme compounds from a virtual library (73), allows reasonably diverse selections even in very large virtual libraries. Other dissimilarity-based selection methods, which can complement nicely the maximum dissimilarity approach, have been reported in the past few years (75–77); their performance in

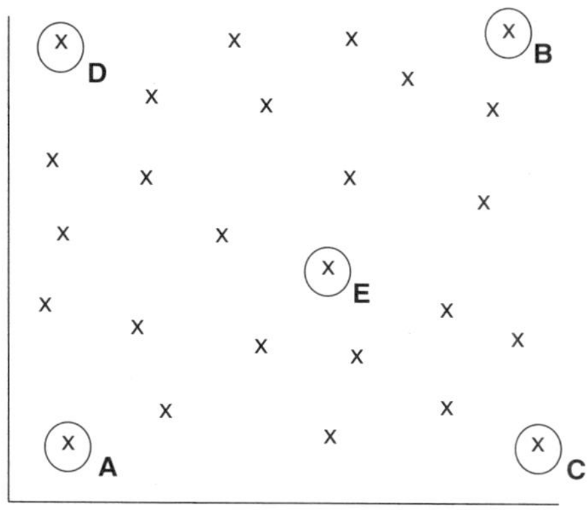

Figure 5.15 Dissimilarity-based selection methods: maximum dissimilarity approach.

selecting active compounds has been recently compared using available databases of drug molecules (78).

The clustering method (79–81) first groups the virtual library components in clusters, such that all the individuals belonging to the same cluster have a high similarity and the individuals in different clusters have a high degree of dissimilarity. Each cluster may contain many compounds or even just one compound, called a singleton. The purpose of clustering is to select the most representative set of diverse compounds, composed of a relatively small number of individuals: A method that reduces the number of singleton clusters is highly recommended. One class of clustering methods is called *nonhierarchical*, an example of which is depicted in Fig. 5.16. To begin with, a search of the nearest neighbors (the most similar compounds to the selected structure in the virtual set) is performed for two compounds (**A** and **B**, Fig. 5.16) and a previously defined number X of nearest neighbors (8, included in circles, for **A** and **B**). Both compounds must be in each others nearest-neighbor sets (**B** in **A**'s neighbors and **A** in **B**'s neighbors), and a predefined minimum number of virtual compounds must be present in both neighbors sets (5 out of 8, including **A** and **B**, in Fig. 5.16: compounds in bold characters). If these conditions are satisfied, as for **A** and **B** in the example, the two compounds are clustered together. If a compound is not among the nearest-neighbor set of another (as **C** for **B**, Fig. 5.16), or if the minimum number of shared nearest neighbors is not satisfied (as **C** for **A**, being only 4 out of 8 in common), the compounds are considered members of two different clusters. This strategy does not take into account the relationship between different clusters, limiting its usefulness, and usually gives high numbers of singletons, which makes the selection of small sets more complex, but it is fast and useful for very large sets of virtual libraries. The so-called *Jarvis–Patrick* (82, 83) clustering technique is the most commonly used and has been applied to several compound selections (84, 40). Another

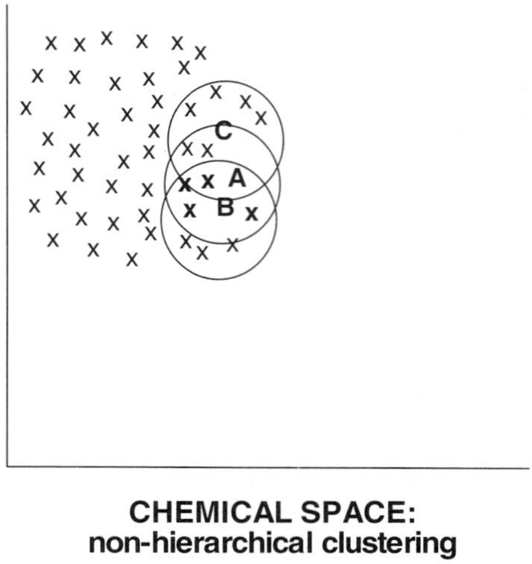

Figure 5.16 Clustering selection methods: nonhierarchical approaches.

class of clustering methods, called *hierarchical methods*, provides parent–daughter relationships between clusters and allows clustering according to the selection needs; an example of these methods is shown in Fig. 5.17. The treelike dendrogram can be cut at any level to give clusters of different number and size as necessary (cut 1, 10 clusters; cut 2, 32 clusters; Fig. 5.17). The dendrograms can be generated from the first huge cluster either by going to the highest fragmentation (*hierarchical*) or by working

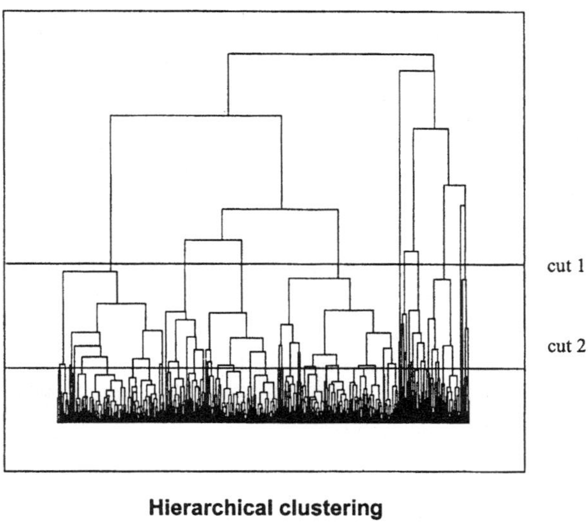

Figure 5.17 Clustering selection methods: hierarchical approaches.

from the bottom up (*agglomerative*). These methods are more information rich than the nonhierarchical techniques but are also significantly more computationally time consuming. A popular hierarchical technique, the *Ward* method (85), has been used frequently for compound selections (84, 86, 87). Clustering methods, while outperforming dissimilarity-based methods, are generally computationally *time consuming and arbitrary*, measuring diversity among a given virtual set without framing it in the chemical space generated by the selection descriptors; nevertheless successful applications of clustering protocols to medium–large databases are made by known druglike compounds and library individuals (88). It is difficult to add new virtual compounds designed to increase the set diversity using clustering because a total reclustering process should be performed after each compound addition.

Partitioning methods (89–91) first subdivide chemical space into predefined bins (Fig. 5.18) by simply defining *incremental values* of the descriptors used and then assign each virtual library component to its proper bin. The schematic examples in Fig. 5.18 consider two properties/descriptors **Y** (8 increments, *A–H*) and **X** (12 different increments, 1–12) that generate 96 different bins, or cells. The library partitioned in Fig. 5.18, top, is clearly diverse, spanning the large majority of the bins, while underneath we have a library containing a large number of compounds in a small area of chemical space producing a nondiverse library. These methods, compared to clustering, are *less computationally demanding* and allow a better evaluation of a library in terms of diversity. In fact, the voids in the space are evident (see Fig. 5.18), and additional virtual compounds can be designed to fit into the desired bin. As a drawback, very similar compounds can sometimes fall into two different cells: for example, if the molecular weight is one of the considered properties and two increments are $200 < MW < 300$ and $300 < MW < 400$, two compounds weighing respectively 299 and 301 daltons would occupy two different bins. Recent examples of large library diversity analysis by partitioning techniques have been reported (92, 93).

In general, dissimilarity-based selections are less time and computationally consuming but are also less accurate in predicting and/or selecting diverse sets; clustering methods are good when large agglomerations of similar compounds are present in the virtual set, as they discriminate well among them; partitioning methods are reasonably fast and effectively represent large varied sets of virtual candidates. It is very difficult to give absolute rules regarding the best choice in terms of descriptors, indices, and selection methods to select the most diverse libraries: This process is highly project specific, and the best solution readily changes from case to case. However, as mentioned earlier, it has been thoroughly proven that a selection using computational diversity/similarity analysis (see also the next Section) is more successful than a random selection in producing a more meaningful library.

5.4.2 Focused Libraries

We could say that it is enough to substitute the concept of dissimilarity with that of similarity in the previous section to determine the application of computational library design to focused libraries. In fact, the tools used to describe and to select among virtual

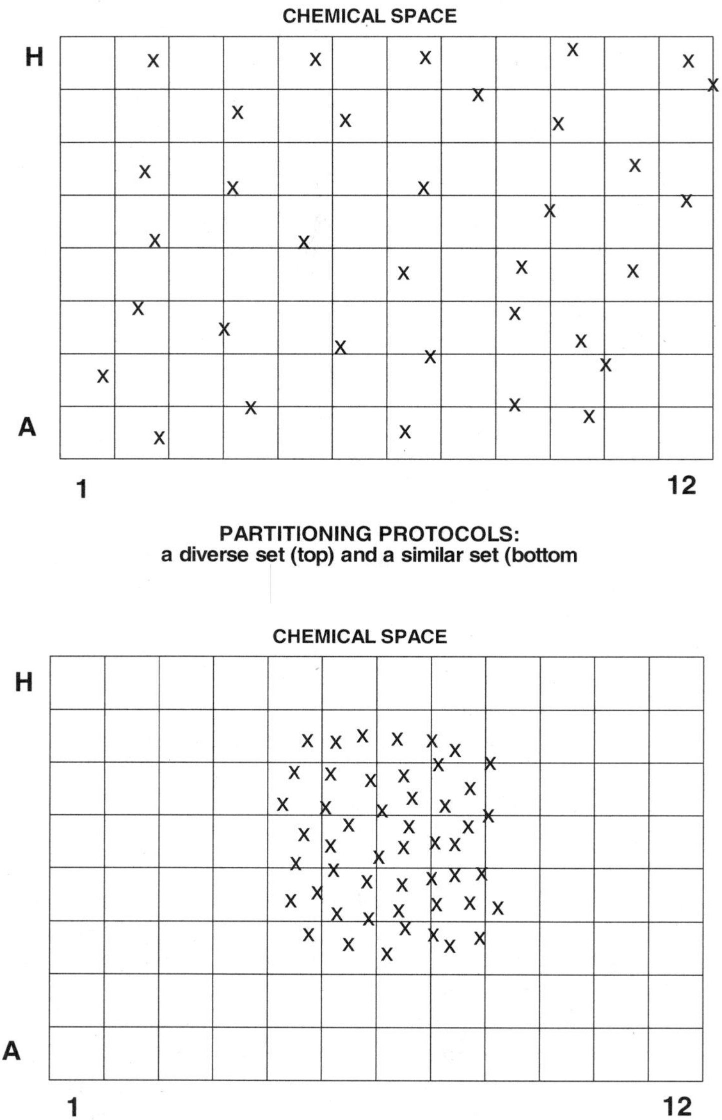

Figure 5.18 Partitioning selection methods: diverse (top) and similar (bottom) sets of compounds.

molecules may be the same as seen previously, while the goal is now to find the *most similar sets of reagents/products* to produce a library similar in respect to a known structure (compound **A**, Fig. 5.19), or to a structural hypothesis (e.g., fit into a pharmacophore). We should always remember, though, that chemical and biological diversity, or similarity, are not the same thing. Very similar structures may have a

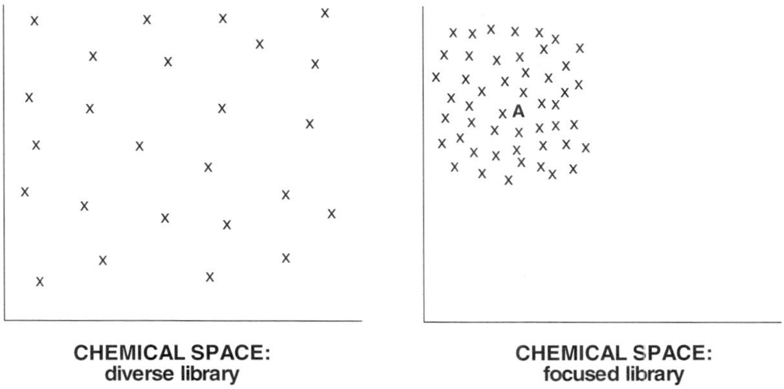

Figure 5.19 Chemical similarity in the chemical space.

Tanimoto index > 0.9 but may be very different in terms of activity (chemically similar, biologically diverse), while completely different structures are known to have the same biological activity (chemically diverse, biologically similar). This intrinsic drawback to the computational screening of virtual libraries should always be considered when interpreting screening results of a computationally designed library, and real data should be used to refine any virtual SAR information based on chemical similarity or dissimilarity.

The number of individuals in a focused library is usually small so that it is easy to rapidly design a small set, then prepare and test it. Once this preliminary quantitative structure activity relationship (QSAR) is known, the available information can be used to produce a more significant second focused library, and the iterative cycle can continue until an active product with the desired potency is obtained. A few other points are worth mentioning. The significantly smaller set of virtual library components usually allows the use of *3D descriptors*, especially as three- or four-point pharmacophores. Rather than using similarity/dissimilarity indices, such as Tanimoto's, the comparison of *fingerprints/pharmacophores* can be carried out and even stored in most cases. The virtual focused library will contain mostly clusters of very similar individuals, and *clustering* becomes the selection method of choice where the similarity parameters can be tailored so as to obtain highly diversified subsets.

Four recent examples of computational selection of the most similar compounds in a virtual library set will be now reviewed. Either a few different lead structures were available as starting points for the selection or a detailed knowledge of the target allowed to design and select a medium/small focused library.

The first two examples, which imply that similarity between compounds is related to similar activity profiles, use two popular selection methods: *genetic algorithms* (GA; 96–100) and *simulated annealing* (SA; 101, 102).

Pentapeptides were used (94) as a training set for the design of bradykinin-potentiating peptides. The two known active sequences **5.1** and **5.2** (103, 104) were used to design and compare virtual focused libraries, as shown in Fig. 5.20. The authors randomly selected 100 pentapeptides, calculated their similarity with the two lead peptides using either amino acid–based descriptors (reagent sets) or topological descriptors (product sets) developed by Hall and Kier (35), and used a GA to select the most similar compounds out of the 3,200,000 virtual pentapeptides obtained by using the 20 natural L-amino acids as building blocks. The GA-based techniques apply evolutionary concepts to library design and optimization, as shown schematically in Fig. 5.21. Two parent sequences (**27** and **74** in the example) are randomly selected from the first 100-membered random pool (**a**, Fig. 5.21), and submitted to *crossover* by mixing the two sequences (**b**, Fig. 5.21). One of them is then *mutated* by randomly replacing one building block with all the other 19 L-amino acids of the monomer set (**c**, Fig. 5.21). Each newly generated sequence is compared, in terms of similarity, with the two lead peptides, then with the two parent compounds, and the two highest scoring peptides are stored in the 100-membered pool in place of the two parent sequences: After this selection cycle either the two parent peptides are more similar to the leads and the initial set is unchanged (**d**, Fig. 5.21) or two of the newly generated sequences are better and replace their parents in the 100-membered set (**e**, Fig. 5.21). This extremely rapid iterative process resembles evolution in producing negative (**d**, Fig. 5.21) or positive (**e**, Fig. 5.21) mutations, and its iteration (**f**, Fig. 5.21) leads to a

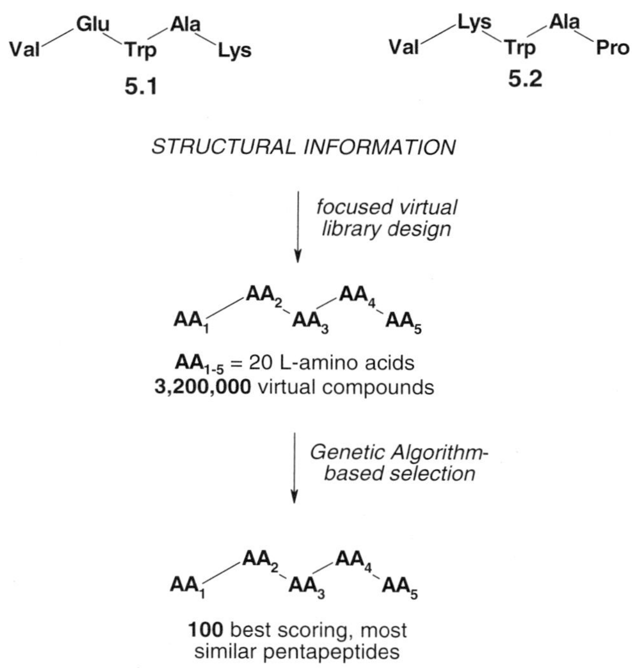

Figure 5.20 Virtual screening of a 3,200,000-membered pentapeptide virtual library inspired by structures **5.1** and **5.2** using GA-based selection.

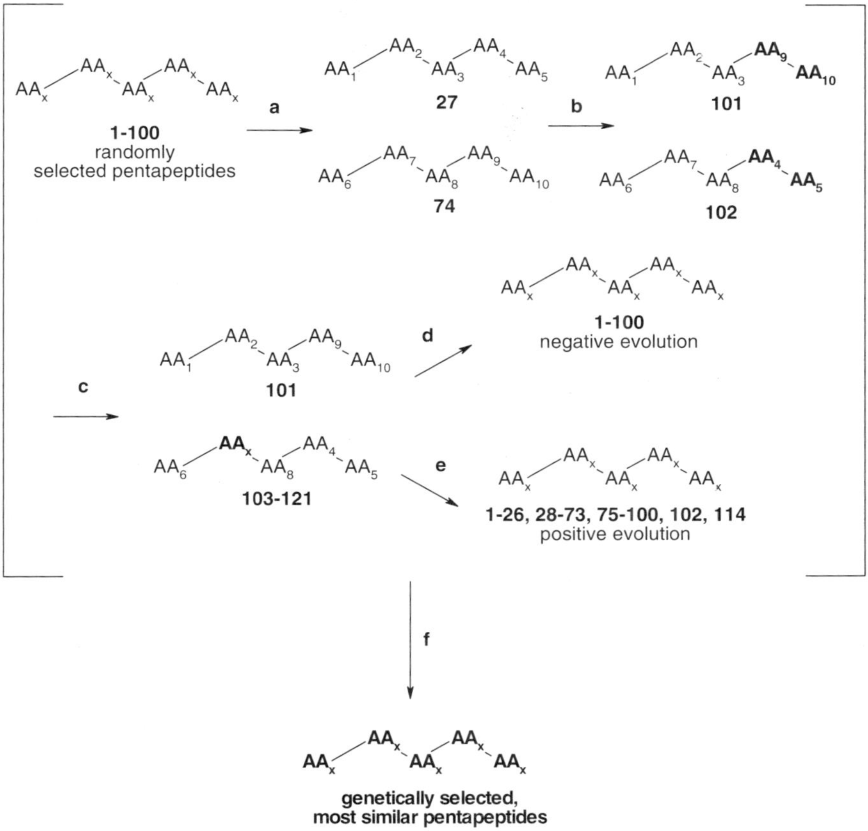

a: random selection; **b**: crossover; **c**: single point mutation; **d,e**: selection of the two best scoring sequences among **27**, **74** and **101-121**; **f**: repeat **a-e** (2000 times).

Figure 5.21 GA-based selection: a detailed protocol for a 3,200,000-membered pentapeptide virtual library.

100-membered population of similarity-based pentapeptides. In the reported example, the 100 most similar pentapeptides generated after 2000 iterations bore a striking resemblance to the 100-membered set generated after exhaustive comparison of all the 3,200,000 virtual pentapeptides. The amino acid composition for each position was *almost identical* for the final GA iteration and for the exhaustive search of the virtual set. This impressive result, which derived from the use of topological descriptors related to the properties of the whole pentapeptides for similarity comparisons (the reagent-based descriptors performance was significantly worse), was possible because only several thousands of structures were considered and evaluated during the 2000 iterative cycles, allowing a significant reduction of the necessary computational efforts for the 3,200,000 total virtual set. This fact, along with the possibility to simultaneously optimize all the properties of virtual library individuals, makes GA methods very useful. The same procedure, using dissimilarity comparisons, can be applied for the selection of diverse subsets of compounds from extremely large databases: The size of the actual library can also be tailored by the organic chemist that will eventually perform the synthesis. Recent examples of genetic algorithm-driven library design and screening have appeared (105, 106), and the application of this technique to combinatorial technologies has been reviewed (107, 108).

Another example from the same group (95) used an alternative stochastic method, simulated annealing (101, 102), to rapidly select the most similar virtual library individuals when a lead compound structure was available. A virtual library of tripeptoids was considered for the generation of compounds with α_1-adrenergic activity, in accordance with a previous report (109). The structure of the virtual library (15,625 individuals), the building blocks used, and the structure of one of the three original tripeptoid hits, **5.3**, are reported in Fig. 5.22. Structure **5.3** was used as a lead compound in a validation study whose strategy is reported in Fig. 5.23. A random starting point was chosen (**a**, Fig. 5.23), its similarity to the lead calculated (**b**, Fig. 5.23), and an iterative cycle of substitution of one or two of the substituents started (**c**, Fig. 5.23). If the similarity of the new compound to the lead was higher than that of the parent, the structure was carried forward (**d**, Fig. 5.23), while the parent was kept if the new compound was less similar to the original lead (**e**, Fig. 5.23). Four series of cycles were performed, from which three identified the tripeptoid lead compound after a few tens of iterative cycles, while the fourth required a longer path. The validation having been successful, the authors used both a natural peptide opiate, *met*-enkephalin (**5.4**, Fig. 5.24), and morphine (**5.5**, Fig. 5.24) as lead compounds for μ-opiate receptor

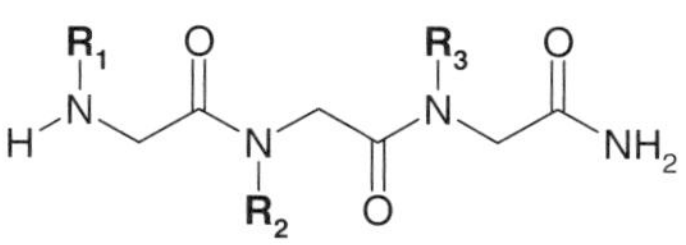

R_{1-3} = 25 different alkyls or aryls

15,625 virtual library components

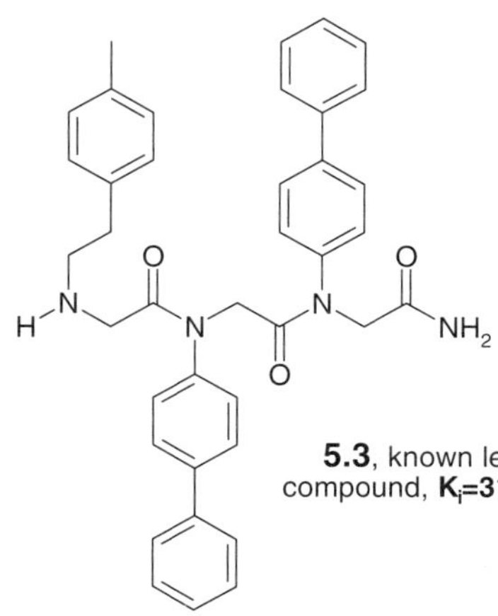

5.3, known lead compound, **K_i=310 nM**

Figure 5.22 Virtual library of 15,625 tripeptoids with potential μ-opiate activity containing the active lead compound **5.3**.

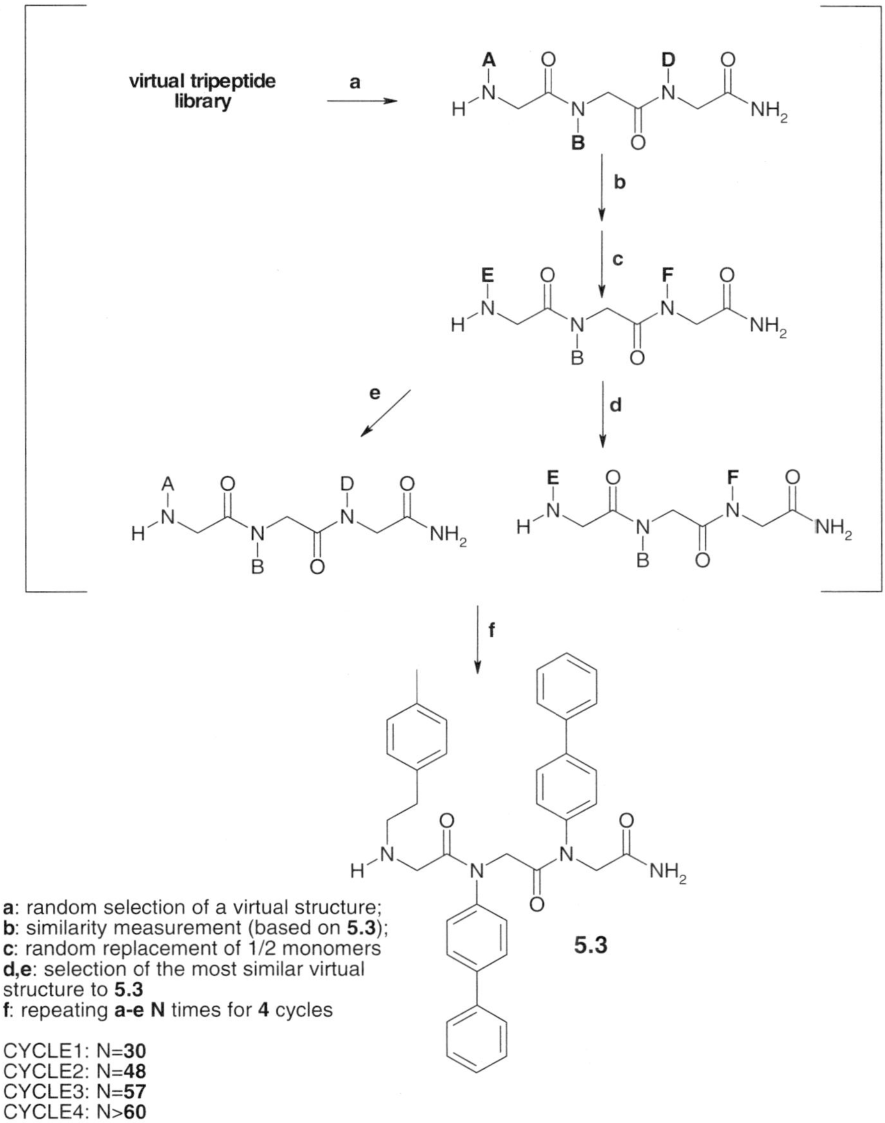

Figure 5.23 Virtual screening of a 15,625-membered tripeptoid virtual library using SA-based selection: **5.3** as a lead compound for computational selection.

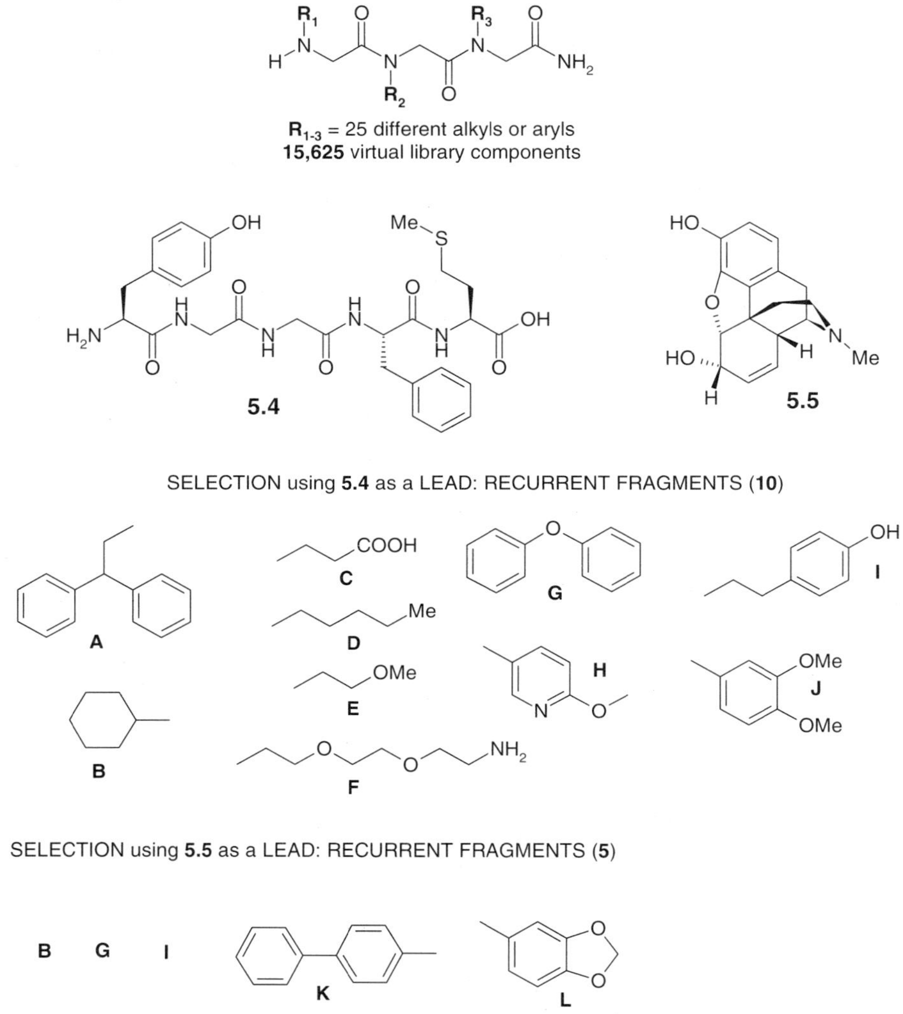

Figure 5.24 Virtual screening of a 15,625-membered tripeptoid virtual library using SA-based selection: *met*-enkephalin **5.4** and morphine **5.5** as lead compounds for computational selection.

binding activity, and for each performed four series of 40 SA similarity-based selection cycles on the same 15,625-membered virtual tripeptoid library. The results of these series, expressed as the occurrence of each building block in the 50 most similar structures found, are reported for both lead structures in Fig. 5.24. The two nonpeptoid leads produced complementary patterns, and the selection of the 12 building blocks **A–L** with a statistically significant occurrence would have found all the three active tripeptoids (Fig. 5.25) out of the original library tested for μ-opiate activity (109). A

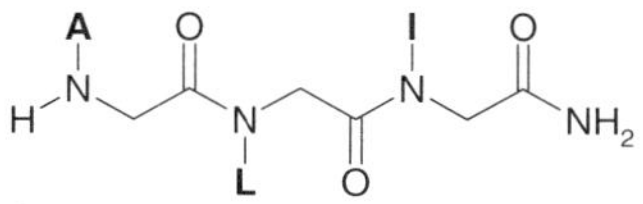

5.6
6 nM
μ-opiate

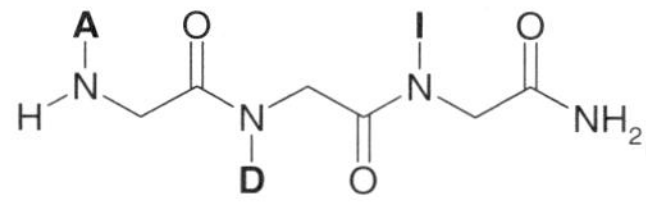

5.7
46 nM
μ-opiate

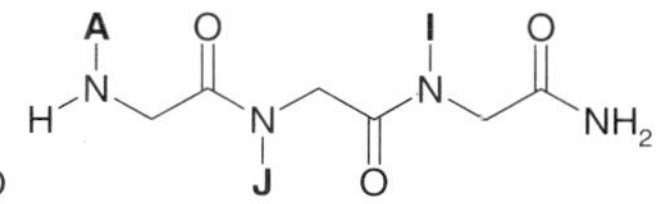

5.8
31 nM
μ-opiate

SELECTION using **5.4** (met-enkephalin) as a LEAD structure: **A, D, I, J** (4 out of 5) detected

SELECTION using **5.5** (morphine) as a LEAD structure: **I, L** (2 out of 5) detected

SELECTION using **5.4** and **5.5** as LEAD STRUCTURES: 12 substituents detected, 5 out of 5 from μ-opiate positives

Figure 5.25 Virtual screening of a 15,625-membered tripeptoid virtual library using SA-based selection: final outcome leading to nonpeptoidic leads.

library of 12^3 = 1728 compounds would have thus been sufficient to cover the significant activity/similarity profile embedded in the whole virtual library (25^3 = 15,625 virtual components). Another report highlighting the potential of simulated annealing guided evaluation (SAGE) in library design has recently been published (110).

Two other examples of rational design of focused libraries are related to the exploitation of target information, and use either Genetic Optimization for Ligand Docking (GOLD) (111–113) to screen virtual combinatorial libraries for their docking with a partially flexible target using a genetic algorithm; or the ligand design program LUDI (114, 115) to map the active site of a target and to extract the structural information needed to build a target-focused small library.

Jones et al. (113) selected as target a lipase from *Candida rugosa* with known active site coordinates (116) and built a 44,730-member virtual library of amides **L1** from commercially available acids ($\mathbf{M_1}$, 426 candidates) and amines ($\mathbf{M_2}$, 105 candidates) (Fig. 5.26). The selection was performed on reagents, as the library size would have prevented the docking of each reaction product; each monomer, though, was "adjusted" to resemble the final amides by converting acids $\mathbf{M_1}$ to N-methylamides $\mathbf{M_1R}$ and amines $\mathbf{M_2}$ to acetylamino compounds $\mathbf{M_2R}$ (**a**, Fig. 5.26). Each adjusted monomer was docked into the lipase active site using GA to select the ideal GOLD fit; three 50,000 operation-GA runs *per* monomer were performed to select three different fits, unless the first two results were already extremely similar (**b**, Fig. 5.26). A total of 1076 GOLD fits for the acids and 270 fits for amines were identified. The fits were then merged to build the more meaningful amides from the virtual library: merging was done deleting the overlapping parts of $\mathbf{M_1R}$ and $\mathbf{M_2R}$ and joining the remaining fragments, providing that they were in the correct orientation to be linked and that the resulting amides did not clash with the active site structure. Only 311 combinations

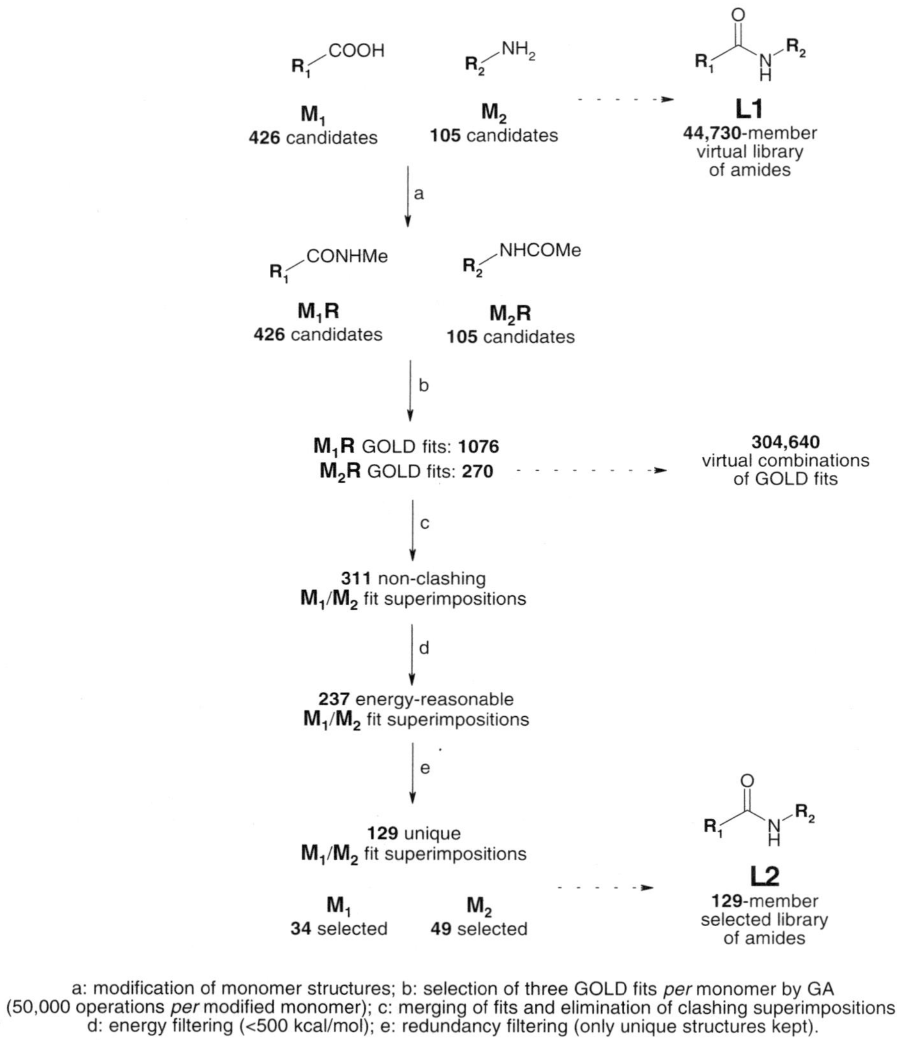

Figure 5.26 Virtual screening of a combinatorial amide library **L1** targeted towards *Candida rugosa* lipase using Genetic Optimization for Ligand Docking (GOLD).

out of the theoretical 304,640 possible superimpositions of **M1R** and **M2R** survived this step (**c**, Fig. 5.26). Among them only the ones with reasonable energies (cut-off of 500 kcal/mol, **d**, Fig. 5.26) and representing unique compounds (**e**, Fig. 5.26) were retained, eventually affording a 129-member library **L2** from 34 acids **M1** and 49 amines **M2**. The process required 159 hours of CPU time, and allowed both a >30 reduction in the actual *versus* virtual library size and a significant reduction in the number of actual *versus* virtual monomers (>10 for acids, >2 for amines); a confirmation of the virtual screening results via resynthesis and testing of the selected individuals would be crucial to validate the process. The authors cross-checked computationally the results by docking the 129 selected compounds into the lipase active site using a more rigorous GA-selection procedure to find the preferred docking conformations (10 GA runs *per* compound, 100,000 operations *per* run); the vast majority (>85%) of compounds gave a good correlation between reagent-based and product-based binding mode.

Lew and Chamberlin (115) selected the human T cell Kv1.3 potassium channel as a target for novel immunosuppressive agents (117). Using LUDI (114, 115) the target active site was screened by docking into it over 1,000 random structural fragments (e.g., amines, acids, phenyl rings); the essential active site residues for specific Kv1.3 blockade and the optimal location of hydrophobic, hydrophilic and hydrogen bond donor/acceptor groups in the site were determined (see ref. 115 for more details). The specificity of the built model was successfully validated by docking known, selective Kv1.3 blockers (complete match with the LUDI model) and aspecific blockers (partial mismatch). The identified fragments were then connected respecting their orientation and distances; this resulted in the selection of a phenyl stilbene scaffold (**5.9**, Fig. 5.27) as a compromise between ideal docking, restricted conformational freedom and reasonable binding energies. The general fit of the scaffold to the target active site was exploited by combinatorial modification of three randomization points (**R$_1$**–**R$_3$**), and a synthetic scheme for a scaffold **5.9**-derived library employing commercially available reagents and monomers was designed and assessed (Fig. 5.27). A 400-member discrete library **L3** was prepared (Fig. 5.28) and screened discovering several weakly active channel blockers (e.g., **5.10–5.13**, Fig. 5.28); a preliminary, but assessed SAR for the phenyl stilbene scaffold which could be used for further optimizations was also obtained.

Computational design/selection tools for focused libraries can significantly reduce the efforts required to optimize preliminary structural information while ensuring the exploitation of the same structural input. The use of these results to refine the structural requirements and to eventually generate better focused libraries will become a key tool for combinatorial scientists working with synthetic organic libraries for different applications.

5.4.3 Biased-Targeted Libraries

Concurrent to the increasingly popular use of diverse libraries to sample chemical space and to provide chemical leads for various applications, the search for areas inside chemical space that are more promising for specific applications has also gained relevance (118–120). We will review the example of pharmaceutical research, where the so-called *druglike properties* of a molecule are the subject of frequent reports by various groups.

It has been observed that most of the currently available drugs are small organic molecules, and their *molecular weight* is typically between 250 and 600 daltons. With this constraint introduced for library design, and knowing that at least two monomer

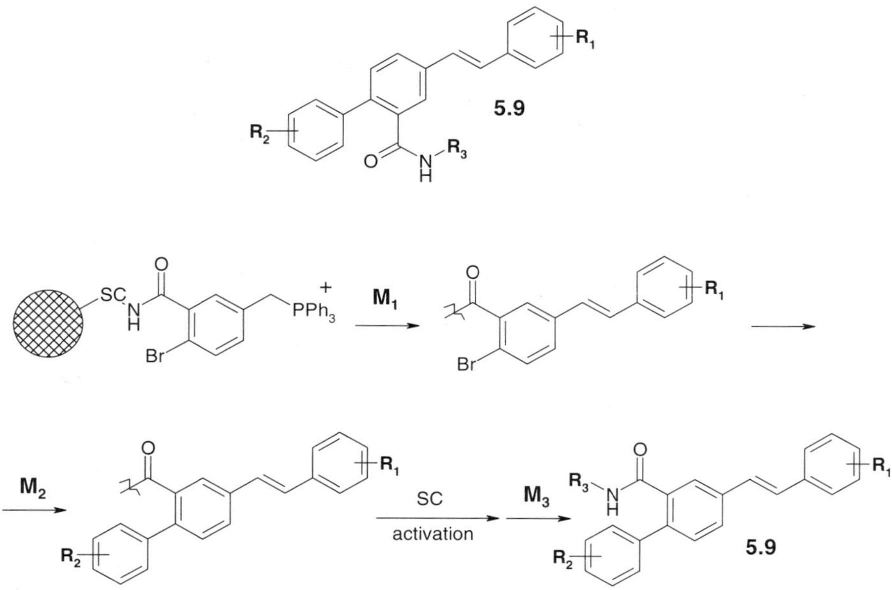

SC = safety-catch linker

M_1 = aromatic aldehydes

M_2 = aromatic boronic acids

M_3 = nucleophiles

Figure 5.27 Design of an SP route to a focused SP library of phenyl stilbenes selected with the program LUDI for the Kv1.3 potassium channel.

sets are to be used for a library synthesis, each monomer must weigh 250 daltons or even less, thus strongly limiting the number of candidate monomers for each chemical class. The *lipophilicity* of organic molecules is also extremely important to determine their usefulness as drugs. Typically, this property, expressed as the partition coefficient of a molecule between *n*-octanol and water (log *P*), is ideal for drug discovery when it is between the logarithmic values of 2 and 4–5, handicapping both extremely hydrophilic and lipophilic building blocks. Another popular filter is the *number of rotatable bonds* contained in a library individual: When compounds are extremely flexible, their binding capacity, related to a specific conformation, becomes weaker so that significant libraries should always contain a reasonable degree of rigidity (possibly embedded into the scaffold, but also present from the monomer) to act effectively as a source of relevant hits. Computational chemistry may help in both evaluating the degree of flexibility of molecules and subsets and selecting compounds having predefined profiles in terms of degrees of freedom. These and other properties can

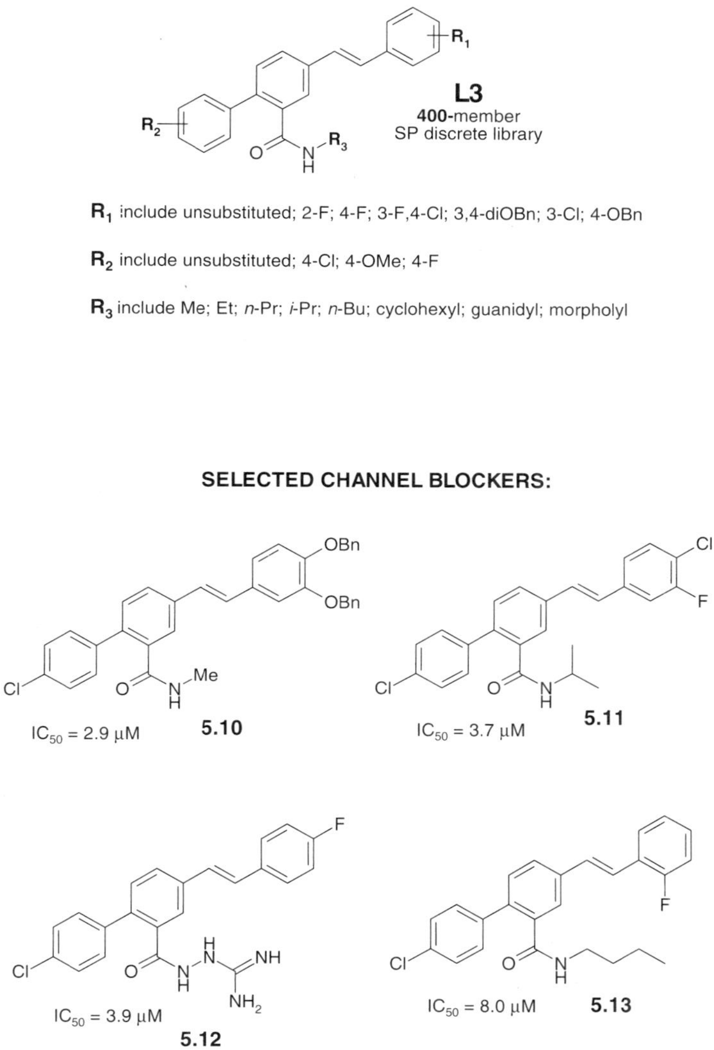

Figure 5.28 Structure of the phenyl stilbene focused library **L3** and of several positives from screening (**5.10**–**5.13**).

easily be predicted by software programs, as shown in several papers (75, 121–123), to aid selection among the virtual sets comprised in the druglike chemical space.

A recent report presented the so-called MultiLevel Chemical Compatibility (MLCC) to determine the similarity of any structure to known drug-compatible compounds from representative databases (124); although by definition incomplete, as all the "drug-like" space is far from being determined, this and similar tools can increase the confidence and the probability of success in progressing a class of drug-like compounds.

Another important class of targeted libraries is aimed at specific families of targets. Examples are kinases (125, 126), proteases (127–129), and G-protein coupled receptors (130, 131). These libraries are driven by generic structural information on the family of proteins, and the correct use of this information (creation of a loose pharmacophore, selection of monomers/products on the basis of the similarity fit into the pharmacophoric model) is similar to the process observed for focused library design.

We will describe here a specific example where Kick and co-workers (132, 33) designed and prepared a library of inhibitors of aspartyl proteases. The general structure of the library, as inspired by the transition state of the enzymatic reaction, and the three monomer sets used to prepare it (amines, **R1**, and carboxylic acids, **R2** and **R3**) are reported in Fig. 5.29. The benzyl substituent originated from the structure of pepstatin, a known inhibitor of cathepsin D, the specific target for this work. The virtual monomer sets were filtered by selecting commercially available amines and acids with a maximum MW of 275 daltons. This led to a final list of around 700 amines (**R1**) and 1900 acylating agents (**R2** and **R3**, sulfonyl chlorides and isocyanates were also included to give sulfonamides and ureas, respectively), and consequently to a virtual library exceeding 10^9 compounds. The library design started by modeling the (*S*)-hydroxyethylene scaffold in the enzyme active site, as for the enzyme–pepstatin complex (**a**, Fig. 5.30); then a number of low-energy conformations were generated for the scaffold and clustered into four families (**b**, Fig. 5.30). For each of these families a thorough conformational search was performed for each substituent independently (**c–e**, Fig. 5.30), and the conformations with **R1–R2** clashes were discarded (**f**, Fig. 5.30). The monomers exceeding \$35 per gram were removed and the 50 highest scoring components from all the conformational families were merged to create the corresponding virtual compounds (**g**, Fig. 5.30). The compounds were clustered (**h**, Fig. 5.30), and the 10 highest scoring monomers from different clusters were finally selected for each randomization point (**i**, Fig. 5.30) to give the biased-targeted 1000-member library **L4**. Another 1000-member diverse library **L5** was generated from the same monomer sets using clustering to select the most diverse monomers for each position (**j**, Fig. 5.30). This library was used to measure the effectiveness of targeted design in discovering actives on cathepsin D. The results for both libraries are reported in Table 5.1. Library **L4** was more successful both in terms of the number and the

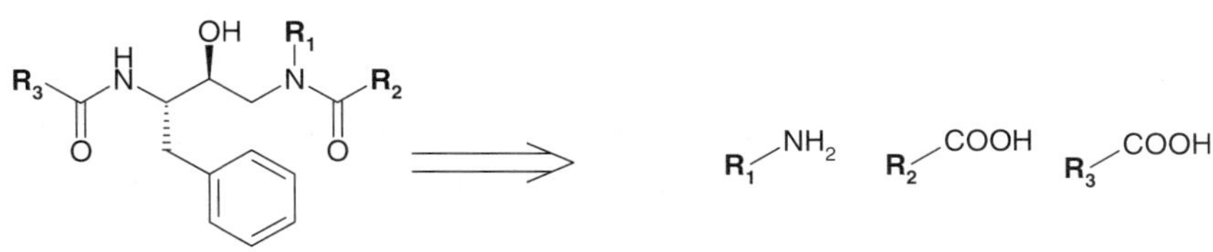

Figure 5.29 Retrosynthesis of a biased targeted library aimed at aspartyl protease inhibitors.

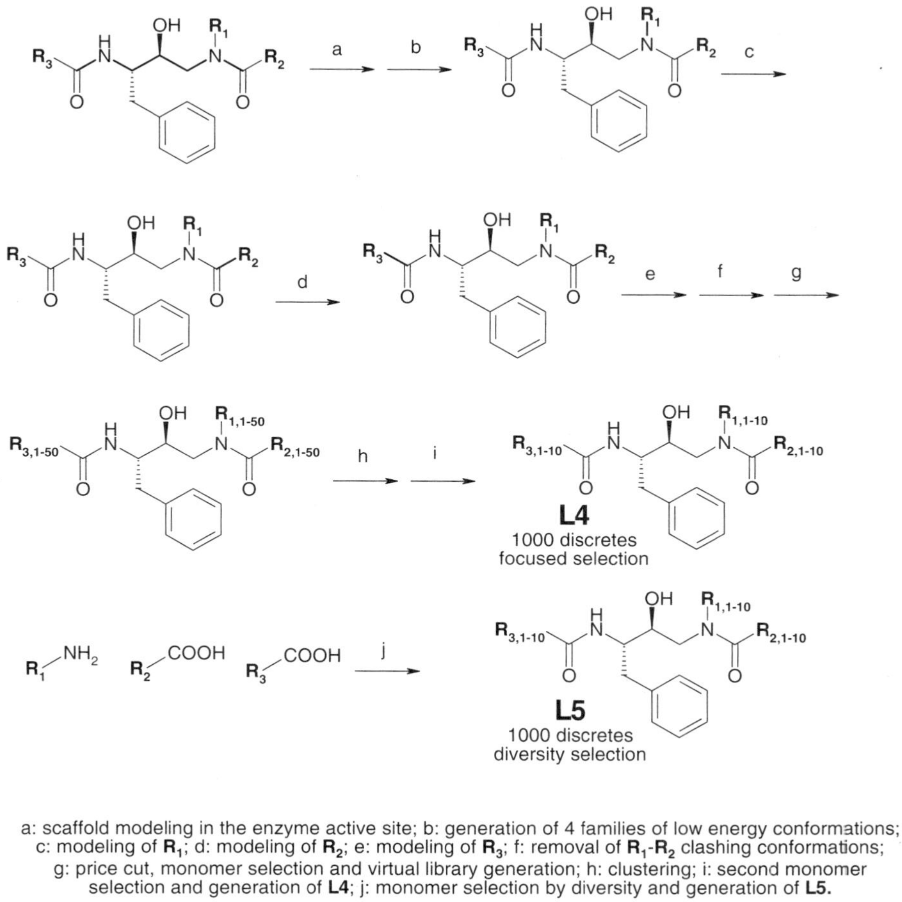

Figure 5.30 Rational design of a biased-targeted library **L4** and a diverse library **L5** using chemical filters and clustering selection methods.

TABLE 5.1 Inhibition of Cathepsin D by Discrete Libraries L4-L6: Number of Positives at Different Inhibitor Concentrations

Concentration[a]	**L4**[b]	**L5**[b]	**L6**[c]
100 n*M*	7	1	7
330 n*M*	23	3	NT
1 μ*M*	67	26	36

[a]Concentration of the inhibitor.

[b]1000 compounds tested.

[c]39 compounds tested.

potency of hits: Resynthesis of its hits as pure compounds produced **5.14** with a K_{i} = 73 ± 9 nM and several other inhibitors with $K_{\mathrm{i}}^{\$}$ around 100–200 nM. Library **L5** produced **5.15** as its best inhibitor with a K_{i} = 356 ± 31 nM and other micromolar activities (Fig. 5.31). These results were then optimized by making a focused library **L6** (39 discretes) using the monomers clustered together with the ones producing the most active compounds in **L4**. Several low nanomolar inhibitors such as **5.16** (K_{i} = 14 ± 2 nM) and **5.17** (K_{i} = 18 ± 2 nM) were obtained (Fig. 5.31), and a large percentage of compounds showed significant enzyme inhibition (Table 5.1).

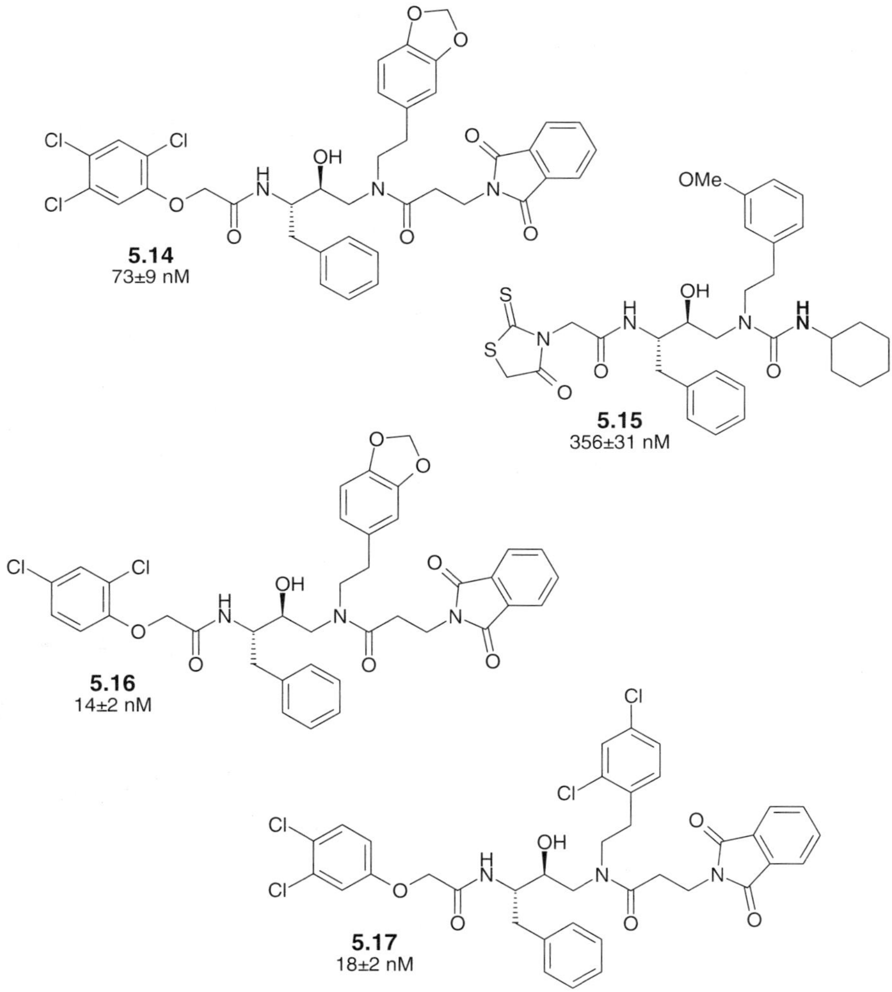

Figure 5.31 Structure of cathepsin D inhibitors **5.16–5.17** from libraries **L4**, **L5**, and **L6**.

The use of both druglike filters for monomers and structural information for target classes can easily produce high-quality libraries targeted to these receptors. In our example the replacement of the embedded pepstatinlike features could bias the obtained libraries toward other representatives of the aspartyl proteases. These first-round targeted libraries may then be further optimized, thus speeding enormously the discovery of novel potent, biologically active molecules.

5.4.4 Recent Advancements

The use of computational methods in combinatorial technologies is extremely popular, and many contributions have appeared recently. A brief list of them, together with a minimal description of their content, should be of use for all the readers to expand their knowledge about any specific topic.

Molecular descriptors are stimulating many groups of researchers, and novel descriptors are generated and validated at a fast pace. Pogliani (133) has used molecular connectivity terms to model properties such as aqueous solubility and the pH at the isoelectric point; Randic and Basak (134) have optimized the relative weight of weighted path numbers to be better suited for structure–property studies; Benigni et al. (135) have studied the potential of IR-embedded information in QSAR analysis, comparing the fingerprint region of IR spectra with other known QSAR descriptors; Filimonov et al. (136) have reported the use and the validation of multilevel neighborhoods of atoms as a set of descriptors and have successfully proven their ability to predict several physicochemical and biological properties; Eisfeld and Maurer (137) have established a good correlation between quantum-chemical calculated descriptors and the octanol/water partition coefficient of several diverse chemicals.

Descriptors have been applied to specific examples, successfully discriminating between sets of active and inactive molecules. Stanton (138) has used the burden, chemical abstract service, University of Texas (BCUT) descriptors (71) to study and rank a set of inhibitors of dihydrofolate reductase; Robert et al. (139) have used descriptors derived from steric and electrostatic quantum similarity measures to predict satisfactorily the corticosteroid-binding globular binding affinity of a family of steroids; the partial least-squares (PLS) projections to latent structures technique has been used to predict the blood–brain distribution of a set of structurally diverse drugs by Luco (140) and to predict the intestinal absorption in humans for a set of druglike compounds by Norinder et al. (141); Mestres et al. (142) have applied the a software program, based on molecular field-based similarity methods, to the creation of a pharmacophoric model for binding to HIV-1 reverse transcriptase from three non-nucleoside inhibitors.

The selection of compounds from molecular databases to assemble screening sets or to select sets to be acquired was reported by Sadowski (143), comparing different selection methods based on dissimilarity or clustering protocols; the selection of the most diverse subsets of reagents to create a diverse combinatorial library by using various algorithms was reported by Gardiner et al. (144); the same task was approached by Mount et al. (145) using a dissimilarity-based program which considers the conformational flexibility of reactants and of final compounds.

Klebe and Abraham (146) reported the evaluation and ranking of virtual combinatorial libraries as sources for thermolysin inhibition through their scoring for enzyme affinity; Muegge et al. (147) scored a database of several thousands of small molecules for FK506 binding protein (FKBP) binding and compared the results with NMR-calculated binding affinities using the acquired information to design meaningful focused libraries; de Julian-Ortiz et al. (148) screened two large virtual libraries of phenol esters and anilides for their anti-herpes activity and selected several individuals which confirmed a moderate, micromolar activity against HSV-1 and could be considered novel leads to be further optimized.

Several reviews dealing with different areas of library design have been published (149–159); their content and the references cited therein should further illustrate the future tendencies of computer-assisted combinatorial chemistry.

REFERENCES

1. Baldwin, J. J., *Molecular Diversity* **2**, 81–88 (1996).
2. Griffith, M. C., Dooley, C. T., Houghten, R. A. and Kiely, J. S., in *Molecular Diversity and Combinatorial Chemistry: Libraries and Drug Discovery*, I. W. Chaiken and K. D. Janda (Eds.). ACS, Washington, DC, 1996, pp. 50–57.
3. Powers, D. G., Casebier, D. S., Fokas, D., Ryan, W. J., Troth, J. R. and Coffen, D. L., *Tetrahedron* **54**, 4085–4096 (1998).
4. Ohlmeyer, M. H. J., Swanson, R. N., Dillard, L., Reader, J. C., Asouline, G., Kobayashi, R., Wigler, M. and Still, W. Clark, *Proc. Natl. Acad. Sci. USA* **90**, 10922–10926 (1993).
5. Nestler, H. P., Bartlett, P. A. and Still, W. Clark, *J. Org. Chem.* **59**, 4723–4724 (1994).
6. Fitch, W. L., Look, G. C. and Detre, G., in *Combinatorial Chemistry and Molecular Diversity in Drug Discovery*, E. M. Gordon and J. F. Kerwin, Jr. (Eds.). John Wiley and Sons, New York, 1998, pp. 349–368.
7. An, H., Haly, B. D., Fraser, A. S., Guinosso, C. J. and Cook, P. D., *J. Org. Chem.* **62**, 5156–5164 (1997).
8. Wipf, P., Cunningham, A., Rice, R. L. and Lazo, J. S., *Bioorg. Med. Chem.* **5**, 165–177 (1997).
9. Rice, R. L., Rusnak, J. M., Yokokawa, F., Yokokawa, S., Messner, D. J., Boynton, A. L., Wipf, P. and Lazo, J. S., *Biochemistry* **36**, 15965–15974 (1997).
10. Boger, D. L., Goldberg, J., Jiang, W., Chai, W., Ducray, P., Lee, J. K., Ozer, R. S., Andersson, C.-M., *Bioorg. Med. Chem.* **6**, 1347–1378 (1998).
11. Cummins, D. J., Andrews, C. W., Bentley, J. A. and Cory, M., *J. Chem. Inf. Comput. Sci.* **36**, 750–763 (1996).
12. Turner, D. B., Tyrrell, S. M. and Willett, P., *J. Chem. Inf. Comput. Sci.* **37**, 18–22 (1997).
13. Higgs, R. E., Bemis, K. G., Watson, I. A. and Wikel, J. H., *J. Chem. Inf. Comput. Sci.* **37**, 861–870 (1997).
14. Mason, J. S. and Hermsmeier, M. A., *Curr. Opin. Chem. Biol.* **3**, 342–349 (1999).
15. Shemetulskis, N. E., Dunbar, J. B., Jr., Dunbar, B. W., Moreland, D. W. and Humblet, C., *J. Comput.-Aided Mol. Design* **9**, 407–416 (1995).

16. Matter, H., *J. Med. Chem.* **40**, 1219–1229 (1997).

17. Bayada, D. M., Hamersma, H. and van Geerestein, V. J., *J. Chem. Inf. Comput. Sci.* **39**, 1–10 (1999).

18. Barnard, J. M. and Downs, G. M., *Perspect. Drug Discov. Design*, 13–30 (1997).

19. Leach, A. R., Bradshaw, J., Green, D. S. V., Hann, M. M. and Delany, J. J. III, *J. Chem. Inf. Comput. Sci.* **39**, 1161–1172 (1999).

20. Weininger, D., *J. Chem. Inf. Comput. Sci.* **28**, 31–36 (1988).

21. Siani, M. A., Weininger, D. and Blaney, J. M., *J. Chem. Inf. Comput. Sci.* **34**, 588–593 (1994).

22. Siani, M. A., Weininger, D., James, C. A. and Blaney, J. M., *J. Chem. Inf. Comput. Sci.* **35**, 1026–1033 (1995).

23. Barnard, J. M., Lynch, M. F. and Welford, S. M., *J. Chem. Inf. Comput. Sci.* **22**, 160–164 (1982).

24. Barnard, J. M., in *Chemical Structure Synthesis*, J. E. Ash, W. A. Warr and P. Willett (Eds.). Ellis Horwood, Chichester, UK, 1991, pp. 9–30.

25. Downs, G. M. and Barnard, J. M., *J. Chem. Inf. Comput. Sci.* **37**, 59–61 (1997).

26. Chem-X, Chemical Design Ltd., *http://www.awod.com/netsci/Companies/CDL/.*

27. Daylight, *http://www.daylight.com/dayhtml/doc/theory/theory.rxn.html.*

28. C^2 Diversity, Molecular Simulations Inc., *http://www.msi.com.*

29. Gillet, V. J., Willett, P. and Bradshaw, J., *J. Chem. Inf. Comput. Sci.* **37**, 731–740 (1997).

30. Martin, E. J., Blaney, J. M., Siani, M. A., Spellmeyer, D. C., Wong, A. K. and Moos, W. H., *J. Med. Chem.* **38**, 1431–1436 (1995).

31. Cribbs, C. M., Menius, J. A., Cummins, D. and Young, S. S., *Book of Abstract, 211st ACS Meeting*, New Orleans, *COMP-074*, March 26th–28th, ACS, Washington, DC, 1996.

32. Braban, M., Pop, I., Willard, X. and Horvath, D., *J. Chem. Inf. Comput. Sci.* **39**, 1119–1127 (1999).

33. Kick, E. K., Roe, D. C., Skillman, A. G., Liu, G., Ewing, T. J. A., Sun, Y., Kuntz, I. D. and Ellman, J. A., *Chem. Biol.* **4**, 297–307 (1997).

34. Brown, R. D., *Perspect. Drug Discov. Design*, 31–49 (1997).

35. Hall, L. H. and Kier, L. B., in *Reviews in Computational Chemistry*, K. B. Lipkowitz and D. B. Boyd (Eds.), Vol. 2. VCH, New York, 1991.

36. Cheng, C., Maggiora, G., Lajiness, M. and Johnson, M., *J. Chem. Inf. Comput. Sci.* **36**, 909–915 (1996).

37. Gupta, S., Singh, M. and Madan, A. K., *J. Chem. Inf. Comput. Sci.* **39**, 272–277 (1999).

38. Downs, G. M., Willett, P. and Fisanick, W., *J. Chem. Inf. Comput. Sci.* **34**, 1094–1102 (1994).

39. Hudson, B. D., Hyde, R. M., Rahr, E., Wood, J. and Osman, J., *Quant. Struct.-Act. Relat.* **15**, 285–289 (1996).

40. Brown, R. D. and Martin, Y. C., *J. Chem. Inf. Comput. Sci.* **36**, 572–584 (1996).

41. Downs, G. M. and Willett, P., in *Applied Multivariate Analysis in SAR and Environmental Studies*, J. Devilliers and W. Karcher (Eds.). ECSC, Brussels, Belgium, 1991, pp. 247–261.

42. Patterson, D. E., Cramer, R. D., Ferguson, A. M., Clark, R. D. and Weinberger, L. E., *J. Med. Chem.* **39**, 3049–3059 (1996).

43. Cramer, R. D., Clark, R. D., Patterson, D. E. and Ferguson, A. M., *J. Med. Chem.* **39**, 3060–3069 (1996).

44. Briem, H. and Kuntz, I. D., *J. Med. Chem.* **39**, 3401–3408 (1996).

45. Schneider, G., Neidhart, W., Giller, T. and Schmid, G., *Angew. Chem., Int. Ed. Engl.* **38**, 2894–2896 (1999).

46. Davies, K. and Briant, C., *Network Science*, July 1995, *http://www.awod.com/netsci.*

47. Kearsley, S. K., Sallamack, S., Fluder, E. M., Andose, J. D., Mosley, R. T. and Sheridan, R. P., *J. Chem. Inf. Comput. Sci.* **36**, 118–127 (1996).

48. Pickett, S. D., Mason, J. S. and McLay, I. M., *J. Chem. Inf. Comput. Sci.* **36**, 1214–1223 (1996).

49. Sheridan, R. P., Nilakantan, R., Rusinko, A. III, Bauman, N., Haraki, K. S. and Venkataraghavan, R., *J. Chem. Inf. Comput. Sci.* **29**, 251–255 (1989).

50. Martin, Y. C., Bures, M. G., Danaher, E. A., DeLazzer, J., Lico, I. and Pavlik, P. A., *J. Comput.-Aided Mol. Design* **7**, 83–102 (1993).

51. Hassan, M., Bielawski, J. P., Hempel, J. C. and Waldman, M., *Mol. Diversity* **2**, 64–74 (1996).

52. Sheridan, R. P., Miller, M. D., Underwood, D. J. and Kearsley, S. K., *J. Chem. Inf. Comput. Sci.* **36**, 128–136 (1998).

53. Chapman, D., *J. Comput.-Aided Mol. Design* **10**, 501–512 (1996).

54. Willett, P., *Perspect. Drug Discov. Design*, 1–11 (1997).

55. Turner, D. B., Willett, P., Ferguson, A. M. and Heritage, T. W., *J. Comput.-Aided Mol. Des.* **13**, 271–296 (1999).

56. Zaliani, A. and Gancia, E., *J. Chem. Inf. Comput. Sci.* **39**, 525–533 (1999).

57. Son, S. H., Han, C. K., Ahn, S. K., Yoon, J. H. and No, K. T., *J. Chem. Inf. Comput. Sci.* **39**, 601–609 (1999).

58. Tuppurainen, K., *SAR QSAR Environ. Res.* **10**, 39–46 (1999).

59. Blankley, C. J., *Pharm. Pharmacol. Commun.* **4**, 139–146 (1998).

60. Dixon, S. L. and Villar, H. O., *J. Chem. Inf. Comput. Sci.* **38**, 1192–1203 (1998).

61. McGregor, M. J. and Muskal, S. M., *J. Chem. Inf. Comput. Sci.* **39**, 569–574 (1999).

62. Chen, X., Rusinko, A. III, Tropsha, A. and Young, S. S., *J. Chem. Inf. Comput. Sci.* **39**, 887–896 (1999).

63. Hopfinger, A. J., Wang, S., Tokarski, J. S., Jin, B., Albuquerque, M. and Madhav, P. J., *J. Am. Chem. Soc.* **119**, 10509–10524 (1997).

64. Albuquerque, M. G., Hopfinger, A., Barreiro, E. J. and de Alencastro, R. B., *J. Chem. Inf. Comput. Sci.* **38**, 925–938 (1998).

65. Venkatarangan, P. and Hopfinger, A. J., *J. Chem. Inf. Comput. Sci.* **39**, 1141–1150 (1999).

66. Hopfinger, A. J., Reaka, A., Venkatarangan, P., Duca, J. S. and Wang, S., *J. Chem. Inf. Comput. Sci.* **39**, 1141–1150 (1999).

67. Xue, L., Godden, J., Gao, H. and Bajorath, J., *J. Chem. Inf. Comput. Sci.* **39**, 699–704 (1999).

68. Gao, H. and Bajorath, J., *Mol. Diversity* **4**, 115–130 (1999).

69. Dixon, S. L. and Koehler, R. T., *J. Med. Chem.* **42**, 2887–2900 (1999).
70. Matter, H. and Potter, T., *J. Chem. Inf. Comput. Sci.* **39**, 1211–1225 (1999).
71. Pearlman, R. S. and Smith, K. M., *Perspect. Drug Discov. Design*, 339–353 (1998).
72. Brannigan, L. H. and Duewer, D. L., *Pharmacochem. Libr.* **16**, 553–556 (1991).
73. Clark, R. D., *J. Chem. Inf. Comput. Sci.* **37**, 1181–1188 (1997).
74. Lajiness, M. S., *Perspect. Drug Discov. Design*, 65–84 (1997).
75. Barnard, J. M. and Downs, G. M., *J. Chem. Inf. Comput. Sci.* **32**, 644–649 (1992).
76. Holliday, J. D., Ranade, S. S. and Willett, P., *Quant. Struct.-Activ. Relat.* **14**, 501–506 (1995).
77. Holliday, J. D. and Willett, P., *J. Biomol. Screening* **1**, 145–151 (1996).
78. Snarey, M., Terrett, N. K., Willett, P. and Wilton, D. J., *J. Mol. Graphics Modell.* **15**, 372–385 (1998).
79. Sneath, P. H. A. and Sokal, R. R., *Numerical Taxonomy*. Freeman, San Francisco, CA, 1973.
80. Willett, P., *Similarity and Clustering in Chemical Information Systems*. Research Studies Press, Letchworth, UK, 1987.
81. Dunbar, J. B. Jr., *Perspect. Drug Discov. Design*, 51–63 (1997).
82. Jarvis, R. A. and Patrick, E. A., *IEEE Trans. Comput.* **C-22**, 1025–1031 (1973).
83. Menard, P. R., Lewis, R. A. and Mason, J. S., *J. Chem. Inf. Comput. Sci.* **38**, 497–505 (1998).
84. Brown, R. D. and Martin, Y. C., *SAR QSAR Environ. Res.* **8**, 23–39 (1998).
85. Ward, J. H., *J. Amer. Statistical Assoc.* **58**, 236–240 (1963).
86. Matter, H. and Larssen, D., *Chim. Oggi* **14**, 9–15 (1996).
87. Clark, R. D. and Langton, W. J., *J. Chem. Inf. Comput. Sci.* **38**, 1079–1086 (1998).
88. Gao, H., Godden, J. and Bajorath, J., *Book of Abstracts, 217th ACS National Meeting, COMP-103* Anaheim, March 21st–25th, 1999, ACS, Washington, DC.
89. Lewis, R. A., Mason, J. S. and McLay, I. M., *J. Chem. Inf. Comput. Sci.* **37**, 599–614 (1997).
90. Holliday, J. D. and Lynch, M. F., *J. Chem. Inf. Comput. Sci.* **35**, 659–662 (1995).
91. Mason, J. S. and Pickett, S. D., *Perspect. Drug Discov. Design*, 85–114 (1997).
92. Schnur, D., *J. Chem. Inf. Comput. Sci.* **39**, 36–45 (1999).
93. Rusinko, A. III, Farmen, M. W., Lambert, C. G., Brown, P. L. and Young, S. S., *J. Chem. Inf. Comput. Sci.* **39**, 1017–1026 (1999).
94. Cho, S. J., Zheng, W. and Tropsha, A., *J. Chem. Inf. Comput. Sci.* **38**, 259–268 (1998).
95. Zheng, W., Cho, S. J. and Tropsha, A., *J. Chem. Inf. Comput. Sci.* **38**, 251–258 (1998).
96. Forrest, S., *Science* **261**, 872–878 (1993).
97. Sheridan, R. P. and Kearsley, S. K., *J. Chem. Inf. Comput. Sci.* **35**, 310–320 (1995).
98. Brown, R. D. and Martin, Y. C., *J. Med. Chem.* **40**, 2304–2313 (1997).
99. Weber, L., *Curr. Opin. Chem. Biol.* **2**, 381–385 (1998).
100. Tropsha, A., Cho, S. J. and Zheng, W., *ACS Symp. Ser.* **719**, 198–211 (1999).
101. Bohachevsky, T., *Technometrics* **28**, 209–212 (1986).
102. Sun, L.-X., *Computers Chem.* **18**, 103–105 (1994).

103. Ufkes, J. G. R., Visser, B. J., Heuver, G. and Van der Meer, C., *Eur. J. Pharm.* **50**, 119–122 (1978).
104. Hellberg, S., Sjoestroem, M., Skagerberg, B. and Wold, S., *J. Med. Chem.* **30**, 1126–1135 (1987).
105. Gordon, D. J., Bellott, E. M. and Tenenbaum, B., *Book of Abstracts, 217th ACS National Meeting, COMP-128*, Anaheim, March 21st–25th, 1999, ACS, Washington, DC.
106. Reynolds, C. H., *J. Comb. Chem.* **1**, 297–306 (1999).
107. Brown, R. D. and Clark, D. E., *Expert Opin. Ther. Pat.* **8**, 1447–1460 (1998).
108. Clark, D. E., *ACS Symp. Ser.* **719**, 255–270 (1999).
109. Zuckermann, R. N., Martin, E. J., Spellmeyer, D. C., Stauber, G. B., Shoemaker, K. R., Kerr, J. M., Figliozzi, G. M., Goff, D. A., Siani, M. A. et al., *J. Med. Chem.* **37**, 2678–2685 (1994).
110. Zheng, W., Cho, S. J., Waller, C. L. and Tropsha, A., *J. Chem. Inf. Comput. Sci.* **39**, 738–746 (1999).
111. Jones, G., Willett, P. and Glen, R. C., *J. Mol. Biol.* **254**, 43–53.
112. Jones, G., Willett, P., Glen, R. C., Leach, A. R. and Taylor, R., *J. Mol. Biol.* **267**, 727–748 (1999).
113. Jones, G., Willett, P., Glen, R. C., Leach, A. R. and Taylor, R., *ACS Symp. Ser.* **719**, 271–291 (1999).
114. Bohm, H. J., *J. Comp. Aided Molec. Design* **6**, 61–78 (1992).
115. Lew, A. and Chamberlin, A. R., *Bioorg. Med. Chem. Lett.* **9**, 3267–3272 (1999).
116. Grochulsky, P., Li, Y., Schrag, J. D., Bouthillier, F., Smith, P., Harisson, D., Rubin, M. and Cygler, M., *J. Biol. Chem.* **268**, 12843–12847 (1993).
117. Leonard, R., Garcia, M. L., Slaughter, R. S. and Reuben, J. P., *Proc. Natl. Acad. Sci. USA* **89**, 10094–10098 (1992).
118. Ajay, Walters, W. P. and Murcko, M. A., *J. Med. Chem.* **41**, 3314–3324 (1998).
119. Lewell, X. Q., Judd, D., Watson, S. and Hann, M., *J. Chem. Inf. Comput. Sci.* **38**, 511–522 (1998).
120. Walters, W. P., Ajay and Murcko, M. A., *Curr. Opin. Chem. Biol.* **3**, 384–387 (1999).
121. Lewell, X. Q. and Smith, R., *J. Mol. Graphics Modell.* **15**, 43–48 (1997).
122. Martin, E. J. and Critchlow, R. E., *J. Comb. Chem.* **1**, 32–45 (1999).
123. Gillet, V. J., Willett, P., Bradshaw, J. and Green, D. V. S., *J. Chem. Inf. Comput. Sci.* **39**, 169–177 (1999).
124. Wang, J. and Ramnarayan, K., *J. Comb. Chem.* **1**, 524–533 (1999).
125. Norman, T. C., Gray, N. S., Koh, J. T. and Schultz, P. G., *J. Am. Chem. Soc.* **118**, 7430–7431 (1996).
126. Gray, N. S., Wodicka, L., Thunnissen, A.-M. W. H., Norman, T. C., Kwon, S., Espinoza, F. H., Morgan, D.O., Barnes, G., LeClerc, S., Meijer, L., Kim, S.-H., Lockhart, D. J. and Schultz, P. G., *Science* **281**, 533–538 (1998).
127. Lee, C. E., Kick, E. K. and Ellman, J. A., *J. Am. Chem. Soc.* **120**, 9735–9747 (1998).
128. Liebeschuetz, J. W., *Book of Abstracts, 217th ACS National Meeting, COMP-217*, Anaheim, March 21st–25th, 1999, ACS, Washington, DC.
129. Gehlhaar, D. K., Bouzida, D. and Rejto, P. A., *ACS Symp. Ser.* **719**, 292–311 (1999).

130. Bradley, E. K., Beroza, P., Eksterowicz, J., Genin, H., Grootenhuis, P., Miller, D., Miller, J., Penzotti, J., Shirley, W., Spellmeyer, D., Srinivasan, J. and Stanton, R., *Book of Abstracts, 217th ACS National Meeting COMP-201*, Anaheim, March 21st–25th, ACS, Washington, DC.

131. Mason, J. S., Morize, I., Menard, P. R., Cheney, D. L., Hulme, C. and Labaudiniere, R. F., *J. Med. Chem.* **42**, 3251–3264 (1999).

132. Kick, E. K. and Ellman, J. A., *J. Med. Chem.* **38**, 1427–1430 (1995).

133. Pogliani, L., *J. Chem. Inf. Comput. Sci.* **39**, 104–111 (1999).

134. Randic, M. and Basak, S. C., *J. Chem. Inf. Comput. Sci.* **39**, 261–266 (1999).

135. Benigni, R., Passerini, L., Livingstone, D. J., Johnson, M. A. and Giuliani, A., *J. Chem. Inf. Comput. Sci.* **39**, 558–562 (1999).

136. Filimonov, D., Poroikov, V., Borodina, Y. and Gloriozova, T., *J. Chem. Inf. Comput. Sci.* **39**, 666–670 (1999).

137. Eisfeld, W. and Maurer, G., *J. Phys. Chem. B* **103**, 5716–5729 (1999).

138. Stanton, D. T., *J. Chem. Inf. Comput. Sci.* **39**, 11–20 (1999).

139. Robert, D., Amat, L. and Carbo-Dorca, R., *J. Chem. Inf. Comput. Sci.* **39**, 333–344 (1999).

140. Luco, J. M., *J. Chem. Inf. Comput. Sci.* **39**, 396–404 (1999).

141. Norinder, U., Osterberg, T. and Artursson, P., *Eur. J. Pharm. Sci.* **8**, 49–56 (1999).

142. Mestres, J., Rohrer, D. C. and Maggiora, G. M., *J. Comput.-Aided Mol. Des.* **13**, 79–93 (1999).

143. Sadowski, J., *Pharm. Pharmacol. Commun.* **4**, 153–155 (1998).

144. Gardiner, E. J., Holliday, J. D., Willett, P., Wilton, D. J. and Artymiuk, P. J., *Quant. Struct-Act. Relat.* **17**, 232–236 (1998).

145. Mount, J., Ruppert, J., Welch, W. and Jain, A. N., *J. Med. Chem.* **42**, 60–66 (1999).

146. Klebe, G. and Abraham, U., *J. Comput.-Aided Mol. Des.* **13**, 1–10 (1999).

147. Muegge, I., Martin, Y. C., Hajduk, P. J. and Fesik, S. W., *J. Med. Chem.* **42**, 2498–2503 (1999).

148. de Julian-Ortiz, J. V., Galvez, J., Munoz-Collado, C., Garcia-Domenech, R. and Gimeno-Cardona, C., *J. Med. Chem.* **42**, 3308–3314 (1999).

149. Bures, M. G. and Martin, Y. C., *Curr. Opin. Chem. Biol.* **2**, 376–380 (1998).

150. Agrafiotis, D. K., Myslik, J. C. and Salemme, F. R., *Mol. Diversity* **4**, 1–22 (1999).

151. Ghose, A. K., Viswanadhan, V. N. and Wendoloski, J. J., *ACS Symp. Ser.* **719**, 226–238 (1999).

152. Drewry, D. H. and Young, S. S., *Chemom. Intell. Lab. Syst.* **48**, 1–20 (1999).

153. Ferguson, A. M. and Cramer, R. D., *Adv. Med. Chem.* **4**, 219–244 (1999).

154. Polinsky, A., *Curr. Opin. Drug Discovery Dev.* **2**, 197–203 (1999).

155. Antel, J., *Curr. Opin. Drug Discovery Dev.* **2**, 224–233 (1999).

156. Hann, M. and Green, R., *Curr. Opin. Chem. Biol.* **3**, 379–383 (1999).

157. Dixon, S. L. and Villar, H. O., *J. Comput. Aided-Mol. Des.* **13**, 533–545 (1999).

158. Kirkpatrick, D. L., Watson, S. and Ulhaq, S., *Comb. Chem. High Throughput Screening* **2**, 211–221 (1999).

159. Gillet, V. J., in *Combinatorial Chemistry and Technology: Principles, Methods, and Applications*, S. Miertus and G. Fassina (Eds.). Marcel Dekker, New York, 1999, pp. 251–274.

6 Synthetic Organic Libraries: Solid-Phase Discrete Libraries

This chapter will cover the subject of SP discrete libraries, which is the most popular library format currently in use in many industrial and academic laboratories. The most appealing features of this format are the significantly increased throughput and quality of individual compounds through the use of methods that allow better control of the reaction intermediates and the final library components.

The main aspects of the preparation of discrete libraries on SP will be covered in the first section, which will deal with the area of so-called parallel synthesis and discuss its application to different chemical problems. The choice of manual versus automated or semiautomated parallel synthesis will also be discussed through the detailed description of selected examples.

The analytical techniques and purification procedures used for quality control and structure determination of the intermediates and final library components will be treated in Section 6.2. A section will be devoted to five examples of SP discrete libraries, presenting in detail different approaches and solutions to obtain libraries of various size and complexity.

Finally, a section will be devoted to emerging trends in SP-supported discrete library synthesis such as the development of new supports and techniques aimed at increasing the throughput of parallel synthesis by taking advantage of miniaturization.

6.1 SYNTHESIS OF SOLID-PHASE DISCRETE LIBRARIES

6.1.1 Parallel Synthesis

The use of heterogeneous supports for organic synthesis and the SPS of small organic molecules were extensively discussed in Chapters 1 and 3, respectively. When a number of compounds are prepared simultaneously rather than sequentially, as in classical organic synthesis, the synthetic procedure can be described as being a *parallel synthesis* (1–5), a concept that is valid for both SP and solution-phase libraries of discretes, as we will see in Chapter 8. Thus, it is not the number of compounds produced that determines if we are synthesizing a discrete library, but rather the fact that the whole synthesis is run in parallel. This means that the preparation of reagents and solutions, the addition of monomer sets to separate reaction vessels, the reaction monitoring, and all the work-up and purification procedures are each carried out on all of the components simultaneously for any given step. For example, we could prepare a small array of 5 compounds in parallel as a discrete library, while the sequential synthesis of 100 compounds cannot be defined as a library but rather as a set of derivatives prepared via classical organic synthesis. The special features of parallel synthesis will be discussed in the following sections.

6.1.2 High-Throughput Organic Synthesis: Manual, SemiAutomated and Automated Devices

The main purpose of parallel high-throughput organic synthesis (HTOS) is, as already mentioned in Chapter 4, to prepare focused libraries of discrete compounds that can be used to assess a fast, preliminary structure–activity relationship for a specific target. The components of these libraries may vary in number from tens to thousands of compounds; therefore different instrumentation and expertise are required depending on the library size.

Many examples, among which a few are cited here (6–10), report SP discrete libraries prepared by *manual techniques* using nonautomated laboratory equipment. This may include glassware such as reaction flasks or, more commonly, glass vials in conjunction with the typical synthetic organic chemist's arsenal of tools such as plastic syringes with porous frits, Eppendorf tubes, pipettes, and pipette tips (all of which are cheap); some basic analytical instrumentation to monitor reactions off-bead after cleavage of the intermediates; and some equipment for work-up and purification such as manifolds to connect syringes to the vacuum and rotary evaporators. With just a simple mind switch and some inexpensive materials, the synthesis of 10- or even 100-member discrete libraries is quickly and effectively accomplished in any organic chemistry laboratory. The addition of commercially available multiple synthetic devices, often based on 96-well plate architectures (Fig. 6.1; 11–13), of smart handling protocols of alternate supports (14) and of multichannel pipettes for solution handling allows the parallel synthesis, work-up, and purification of hundreds- to thousands-member libraries of discretes with reasonable effort and moderate expense.

Repetitive operations are the rule, rather than the exception, in SP library synthesis, especially when the library size increases, and sometime they become tedious for the operator and can become a significant source of experimental error. While manual synthesis can be effective and inexpensive, the introduction of *partially or fully automated procedures* allows a more reliable experimental protocol and often a higher throughput. Moreover, the automated steps do not require the presence of the chemist, who thus can focus his or her attention on the more scientifically challenging problems of chemical assessment and library design. When automation is introduced in some of the steps of library preparation, we talk of *semiautomated techniques*. Normally the repetitive operations are automated while other steps (e.g., addition of solid reagents and evaporation of cleavage solutions) are still performed manually by the chemist. For example, a popular semiautomated synthetic device (15–17) has been designed to automate operations such as washing cycles, bubbling of gas, and stirring of resin aliquots, thus decreasing (or eliminating) the most tedious steps, which require the attention of the chemist. Other operations with a more significant variance are performed manually. Other similar devices (18–20) are suited not only to the produc-

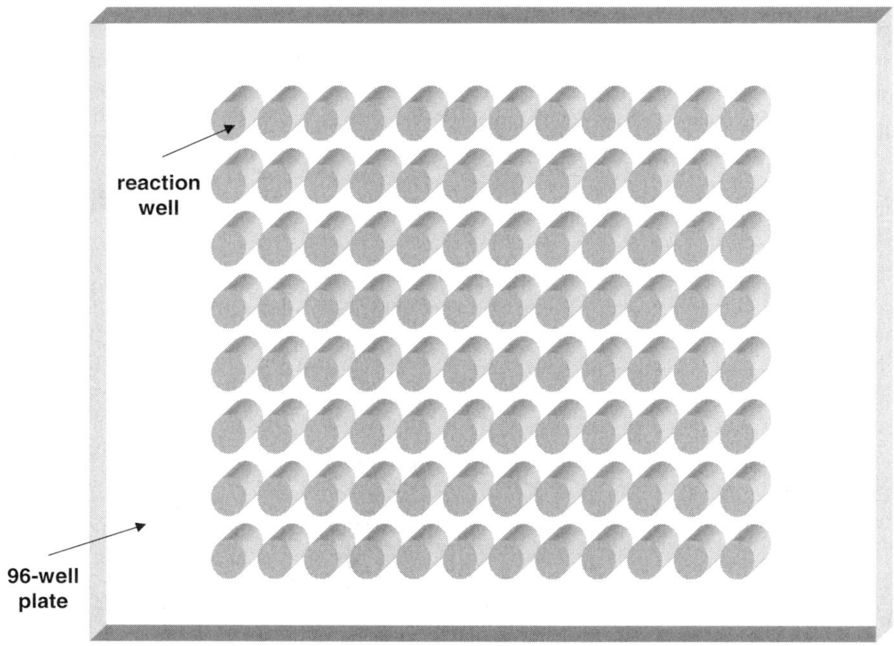

Figure 6.1 A 96-well architecture of a microtiter plate for SP parallel synthesis.

tion of small discrete libraries but also to semiautomated optimization studies to assess the best SP conditions for a synthetic strategy. They are flexible and can be easily located in a chemistry cupboard and, although their cost is significant, this is balanced by their usefulness in a laboratory that produces several focused libraries per year. A higher throughput, simple semi-automated robotic station based on surface suction rather than on filtration of liquids is suited to wash in parallel thousands of SP discretes but also, in particular conditions, to perform basic SP reaction protocols (21).

More complex instrumentation that allows the automation of every step of an SPS of discrete libraries is also available from several commercial sources. These fully automated machines can be grouped into instruments that are best suited to *chemical assessment* studies (22) in which the reaction conditions of each reaction vessel can be varied and instruments that are directed toward *library production* (23–27) in which an automated protocol is used to generate sets of compounds on a common reaction block, usually made up of 96 reaction wells. It is important to note that, while some operations are significantly accelerated using automation, others require the same or even more time than when performed manually. The advantage of a fully automated synthesizer is that the system controls and drives everything and the operator supposedly only collects the outcome of the synthetic procedure at the end of the operations. In terms of throughput, automated SP synthesizers can vary from low (chemical assessment) to medium (library production) while *high-throughput SP organic syn-*

thesis of primary libraries, as realized by SP pool techniques and made of hundreds of thousands or even millions of individuals (see Chapter 7), is currently not possible using parallel synthesis. Even some recently designed multiblock synthesizers (28, 29) cannot prepare more than several hundred discretes in a given synthetic run. Innovative approaches that are beginning to address this crucial issue will be reported in Section 6.4.

These instruments are engineered to a high degree of complexity and allow the reliable control of the multiple operations and reaction conditions encountered in the automated synthesis protocol. This topic has been extensively covered in some recent reviews (30–40), and we will limit our discussion of automated SP discrete synthesizers to the implications of their use for the synthesis of small organic molecules on SP.

The most demanding requirement of a SP synthesizer is the necessity to perform a large variety of organic reactions. This means being able to deal with a wide temperature range (usually from –40 °C to +150°C) and the use of inert atmosphere for moisture-sensitive reaction conditions (the reaction blocks can be kept under an argon blanket or sealed with various materials) and also the ability to accommodate all types of reagents including acids, bases, and corrosive substances. Typically, glass and Teflon are used as inert construction materials for the reaction blocks, frits, tubing, and valves.

Complex hardware components are needed to constitute automated SP synthesizers, but their complexity should be reduced as much as possible, without slowing the synthetic protocols, to minimize the maintenance and the failures of the system. For example, a single valve or robotic pipette that fails to deliver or to withdraw a reagent compromises the whole automated synthesis, and so the minimum acceptable number of such devices should be used and overautomation should be avoided. The number of automated operations such as delivering and withdrawing solutions, washing, stirring, heating, and cooling should also be reduced to the minimum; additional manual steps such as the addition of solid reagents and the filtration of suspensions are only introduced in an automated SP library synthesis protocol if they are necessary to perform a high-quality synthesis.

Another important feature of an automated system is the software that controls the synthesizer. It must be very user friendly since engineers and programmers may be needed for maintenance, but their presence should not be required to run an automated SP protocol! Even though the program may have a complex architecture, the interface with the chemist should be straightforward. Usually the system is designed to automatically record any anomaly that may occur during the automated protocol such that an electronic logbook allows immediate troubleshooting. These instruments are quite expensive, and their use in nonspecialized chemical laboratories could be problematic. On the contrary, the use of several complementary SP synthesizers in a dedicated combinatorial laboratory allows the fast and reliable production of libraries. The production of libraries is often coupled with automated SP assessment studies that speed up the preparation of focused libraries of discretes.

6.1.3 Analytical Methods for SP Discrete Library Synthesis

The synthesis of a library of discretes in SP makes extensive use of analytical tools to check the validity of the synthetic route and to monitor the reaction course for each library individual. Every reaction vessel contains a single entity with the possible presence of side products that can be fully characterized at any step in the synthesis. We will briefly review the most useful analytical techniques that can be employed for the SP discrete library synthesis. For a detailed description of each technique the reader is referred to Sections 1.3 and 1.4 and to the references cited therein.

The exact nature of the *monomer rehearsal* of the library strongly depends on the size of the library. For example, if the planned library is small, the preparation of a few target compounds may suffice, while the number of compounds prepared during rehearsal may be larger when the discrete library is made by thousands of components. In both cases these compounds are prepared with the same procedure that will be used for the library synthesis, and thus the same steps will be present in both the monomer rehearsal and the final library synthesis. This step is crucial to rule out monomers that perform poorly and to determine the optimized reaction conditions.

A strong analytical support is required in this phase, and, ideally, all of the off-bead and on-bead techniques described in Chapter 1 should be used to determine the quality of the synthesis. Each reaction during the rehearsal must be carefully monitored to determine the ideal conditions for each representative monomer and to eventually define the most general conditions. Simple colorimetric techniques are used when possible, but on-bead monitoring methods based on IR spectroscopy (using either normal instruments by grinding a few milligrams of bead into a KBr pellet or single-bead techniques) or MAS-NMR spectroscopy (when this expensive and sophisticated instrumentation is available) are ideal techniques that not only allow the reaction to be followed but also allow the monitoring of side products and selection of the best conditions to minimize side reactions. The quantity of resin employed for the rehearsal of the synthesis may be relatively large so that off-bead analytical methods can also be used. The cleavage and the analytical characterization of each intermediate step should be performed, and the results compared with on-bead analysis. The use of several on-bead and off-bead techniques allows the selection of suitable monomers and reagents for the library as well as the determination of the most useful synthetic conditions for library production and of the most suitable cleavage conditions. In general, the more analytical information gathered in this phase, the higher the chance of success for the planned library.

A recent report (41) has highlighted the usefulness of CE to determine the enantiomeric purity of individuals coming from parallel synthesis, either during chemical assessment to help revising the experimental protocols and as a library QC tool to determine the stereochemical quality and purity of chiral libraries. The authors studied the Pictet-Spengler reaction of chiral tryptophans with carbonyl compounds on SP to provide tetrahydro-β-carbolines, and clearly highlighted the reaction conditions-dependent racemization of the reaction products.

The same techniques can also be used during the library synthesis, and, depending on the size of the library, either all or only some of the intermediates may be analyzed at each step and any anomalous behavior recorded and double-checked during the quality control of the final library. The significant increase in the number of samples to be processed makes HPLC/MS, which is easily automated and fast, the technique of choice for off-bead characterization of the components of the final library after cleavage.

6.2 STRUCTURE DETERMINATION, QUALITY CONTROL, AND PURIFICATION OF SOLID-PHASE DISCRETE LIBRARIES

6.2.1 Structure Determination

The determination of the structure of a library member is not a problem for SP libraries of discretes as the structure of any given library component is *positionally encoded*, that is, by simply tracking the locations in which the different monomers are added, the structures of the *expected* final products are known. When dealing with automated syntheses and with large libraries, the software controlling the synthesizer determines the location of each monomer and delivers all the solutions of reagents accordingly. This information is provided to the chemist through spreadsheets or tables that represent the whole reaction block. When medium–large libraries are prepared manually, the tracking of monomer addition is made easier by segregating this monomer in specific areas of the plate or plates. A hypothetical example shown in Fig. 6.2 presents the synthesis of a 960-member SP discrete library in ten 96-well plates where the use of each monomer in the first set (10 compounds, A_1–A_{10}) is coupled to a different plate; then each monomer of the second set (8 compounds, B_1–B_8) is coupled with the same row of each plate; and finally the representatives of the third set (12 compounds, C_1–C_{12}) are coupled with the same column of each plate (Fig. 6.2).

6.2.2 Quality Control

After the final cleavage from the beads, the solutions containing the discrete library individuals are submitted to a work-up procedure and then the pure individuals are tested against one or a few selected targets. The results of the assays will, hopefully, create useful information that will allow further research to be focused on active structures. However, these results must be coupled with *quality control*, that is, the complete analytical characterization of the library. This allows the purity of each positive compound to be determined to check if the observed activity is due to the presence of impurities (*false positives*) and to locate the wells where the expected library individuals are absent (*false negatives*). Moreover, the analysis of the whole library will determine if a final purification of the compounds is required.

In theory, it would be necessary to provide a profile of analyses for each member of the library to have a reliable and exhaustive quality control. However, this is feasible only for small libraries (a few tens of components), and when the numbers of components increase to hundreds or thousands, this process would be too time consuming. The best compromise is provided by a *fully automated technique* that can

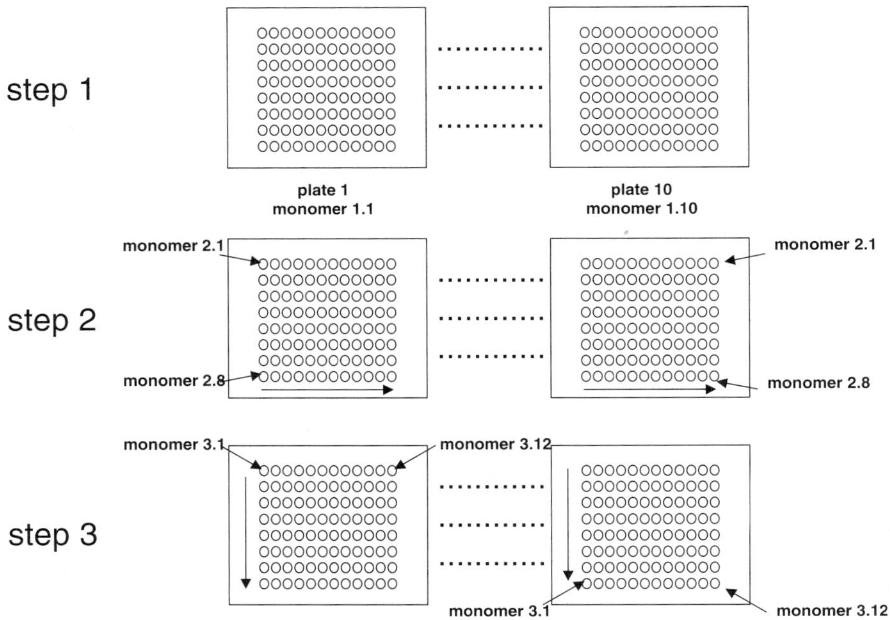

Figure 6.2 Monomer segregation in a 960-member SP discrete library synthesis.

be used for most, if not all, libraries of small organic molecules. This technique couples an efficient, fast, reliable separation method with one or more detection systems that allow the identification of library components with different physicochemical properties. Reverse-phase high-performance liquid chromatography (RP-HPLC) is generally accepted as being the most efficient chromatographic method for the separation of the expected compounds from the impurities, while coupling with MS permits the measurement of the molecular weight of each analyte. The combined use of these two techniques has proven very useful in processing chemical libraries (42–46) and in detecting/quantifying synthetic by-products (47). The main features of automated quality control, its limitations, and possible alternative solutions are discussed and presented below.

Classical HPLC analyses last around 20–30 min and are not rapid enough for fast high-throughput quality control. Extensive studies (48, 49) have demonstrated that the analysis time can be drastically reduced to around 1 min, including reequilibration time, without significant loss of the separation power by shortening the length of column (e.g., 3 cm), increasing the flow rate, appropriate choice of solvents and stationary phases, and modification of the solvent gradient. As an example, the separation obtained with a classical HPLC analysis (analysis time 30 min, column length 250 mm; Fig. 6.3, top) and with a faster, automation-friendly HPLC analysis protocol (analysis time 7.5 min, column length 50 mm; Fig. 6.3, bottom) is comparable for a complex mixture (50). Full automation of HPLC systems is common in many

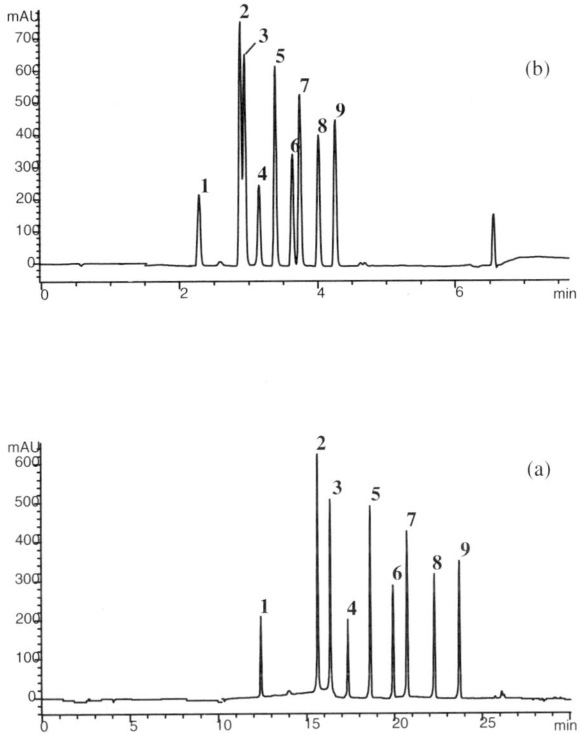

(b)

Column	Phenomenex Luna C18 (2)		
	50 x 4.6 mm, 3 micron		
Mobile phase	A: $NH_4H_2PO_4$ 10mM, pH=2.5		
	B: Acetonitrile		
Gradient:	0.0 - 0.5 min	A=100	B=0
	0.5 - 4.0 min	A=0	B=100
	4.0 - 5.5 min	A=0	B=100
	5.5 - 5.7 min	A=100	B=0
	5.7 - 7.5 min	A=100	B=0
UV	230 nm		
Flow rate	1.5 mL/min		
Injection	5 uL		
Analysis time	7.5 min		
	(5.5 + 2.0 min reequilibration time)		

(a)

Column	Supelcosil LC-18		
	250 x 4.6 mm, 5 micron		
Mobile phase	A: $NH_4H_2PO_4$ 10mM, pH=2.5		
	B: Acetonitrile		
Gradient:	0.0 - 5.0 min	A=100	B=0
	5.0 - 25 min	A=0	B=100
	25 - 30 min	A=0	B=100
UV	230 nm		
Flow rate	1.0 mL/min		
Injection	5 uL		
Analysis time	30 min		

Figure 6.3 HPLC chromatograms of a sample using classical (bottom) and accelerated (top) analytical conditions.

commercially available instruments, and no further comment will be made here; however, it is worth mentioning that some autosamplers for automated HPLCs use the same 96-well format as used for the synthesis of discrete libraries, thus reducing the number of operations and transfers necessary for the analytical characterization of the library. Several recent reports have highlighted the use of capillary LC, in conjunction with MS, to accelerate the analysis throughput using the so-called rapid back-flush microseparation protocol (51); the use of supercritical fluid chromatography/mass spectrometry (SFC/MS) to successfully analyze and characterize a discrete library of thiohydantoins in high-throughput mode (52, 53); and the use of electrospray ionization Fourier transform ion cyclotron resonance mass spectrometry (ESI-FT-ICR-MS), either alone or coupled with RP-HPLC, to analyze several hundreds of combinatorial discretes per *day* (54).

Flow injection analysis mass spectrometry (FIA-MS) has been reported to be a fast method for the characterization of combinatorial libraries (55, 56). The method verifies the presence of the molecular ions of the expected product and side products or impurities but does not provide information on the quality of the analyzed samples. Significant improvements related to the increased analytical throughput, obtained by reducing the time between each injection without increasing the intersample carry-over from each analysis, were recently reported (57, 58). When coupled with RP-HPLC, FIA-MS allows the separation and the determination of the molecular weight of the components of each sample. This is normally enough to unequivocally attribute the structure of the expected library component and of any side products from a library synthesis.

The methods of detection most commonly used are based on the UV and MS properties of the compounds. The acquisition of a UV chromatogram using a diode array detector with a wide wavelength window and of the total ion current (TIC) usually allows detection of all the components of a sample and allows an approximate but reliable quantitation. As an example, the UV and MS spectra of two compounds from an SP discrete library are reported in Fig. 6.4 (59); the different properties of the desired product and of an impurity are highlighted by their different UV and ionization behavior. Some classes of compounds, though, have poor or no chromophoric groups present in their structures (and hence no UV absorption) or ionize only poorly, thus giving no MS signal. For this reason alternative detection methods that can be coupled with HPLC/MS/UV are also used. *Evaporative light-scattering detection* (ELSD; 60–63) is based on the light-scattering properties of nonvolatile compounds. In this technique, the recorded signal is roughly proportional to the molecular weight of the molecule and does not significantly vary in intensity between members of the same chemical class; however, it has the major drawbacks of being relatively insensitive and being readily influenced by sudden changes in the gradient of the mobile phase. *Chemiluminescent nitrogen detection* (CLND; 64–66) is based on the total combustion of nitrogen and thus is able to detect organic compounds containing at least one nitrogen atom producing a reliable chromatogram with a high sensitivity and without any influence from other structural factors. ELSD- and CLND-based detectors are commercially available and can be easily coupled to automated RP-HPLC systems.

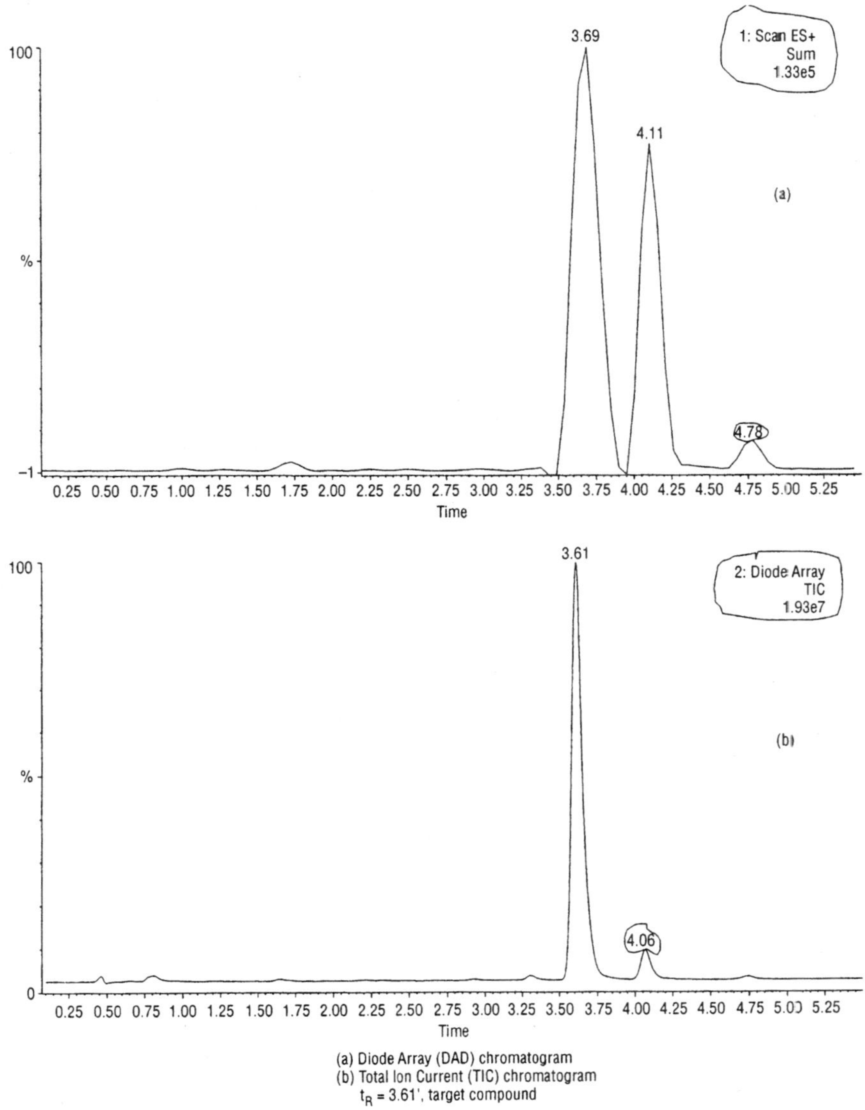

Figure 6.4 UV (top) and MS (bottom) detection for an analytical sample.

As these detection methods are complementary rather than mutually exclusive, multiple detection is recommended for laboratories that routinely perform the synthesis of discrete libraries and can be realized by splitting the eluate of the HPLC/UV system in two or three aliquots that are sent to the mass spectrometer and the CLND and/or the ELSD instrument. Examples of multiple UV/MS/CLND (67) and UV/MS/ELSD (68) show how the different methods of detection are useful in determining and quantifying the composition of an injected sample. Two examples of different sensitivities for the multiple system UV/MS/ELSD are reported in Figs. 6.5 and 6.6 (68). The former consists of a well from a discrete carbohydrate library containing two major components, the more hydrophilic of which ($R_t = 9.85$) does not contain a chromophore and is not detected by UV (Fig. 6.5b) but is clearly visible using ELSD or MS detection (respectively Figs. 6.5a and c). The latter consists of a well from a benzoic acid–inspired library containing two major components, the more lipophilic of which ($R_t = 2.20$) requires a detection method with high sensitivity and thus is not spotted by ELSD (Fig. 6.6a) but is clearly visible using UV or MS detection (respectively Figs. 6.6b and c).

The application of the powerful and information-rich combination of RP-HPLC and NMR, which has previously been used above all for the determination of drug metabolites (69, 70), to the problem of library analysis has been reviewed recently (71). NMR as a detection method is able to provide more detailed structural information than any other technique. Integrated HPLC-NMR systems can be created using commercially available flow NMR probes (72), and small libraries have been successfully analyzed with this method (73, 74). Quality control may be carried out either on-line, recording the NMR spectra during the elution, or by stopping the HPLC flow when a chromatographic peak elutes. Both methods are relatively insensitive and require significantly more substance and time than HPLC-MS, but the structural information provided can be unique, especially when isomeric compounds or compounds with the same molecular weight are present in the library. A recent example (74) reported the separation and structure determination of four isomers by analysis of their ^{1}H-NMR spectra (two of the peaks are partially overlapping, but determination of the structures was still possible), while a discrete 96-member library of methylene malonic acids has been fully characterized by HPLC-NMR (75). This method will become increasingly important for the quality control of libraries, and a multicomponent detection system that includes an NMR spectrometer is the method of choice for discrete library quality control in dedicated combinatorial laboratories.

The use of "ultrafast" HPLC gradients and of automated integrated instruments allows the processing of up to several hundred samples per day, which is satisfactory for most of the SP discrete libraries produced. If a large library of discretes made up of several thousand members is being considered, a significant percentage (typically 10–25% of the samples) is processed and the quality control obtained on this sample is assumed to reflect the quality of the library as a whole. If one or more positives from the screening of the library are among the nonprocessed samples, their characterization will follow immediately. Another useful procedure is to first run a fast FIA-MS analysis and discard the library wells/samples where the expected components are absent and only afterward analyze the confirmed library samples by HPLC with UV detection.

The processing of several hundreds of samples per day using a multidetection system means the acquisition of more than a thousand chromatograms, the evaluation of which by the chemist would require substantial time and effort. Software that is able to discriminate between "good" and "bad" library components without human intervention is desirable in order to speed up the process of quality control. Some software of this type has been developed and is commercially available (76, 77) or, alternatively,

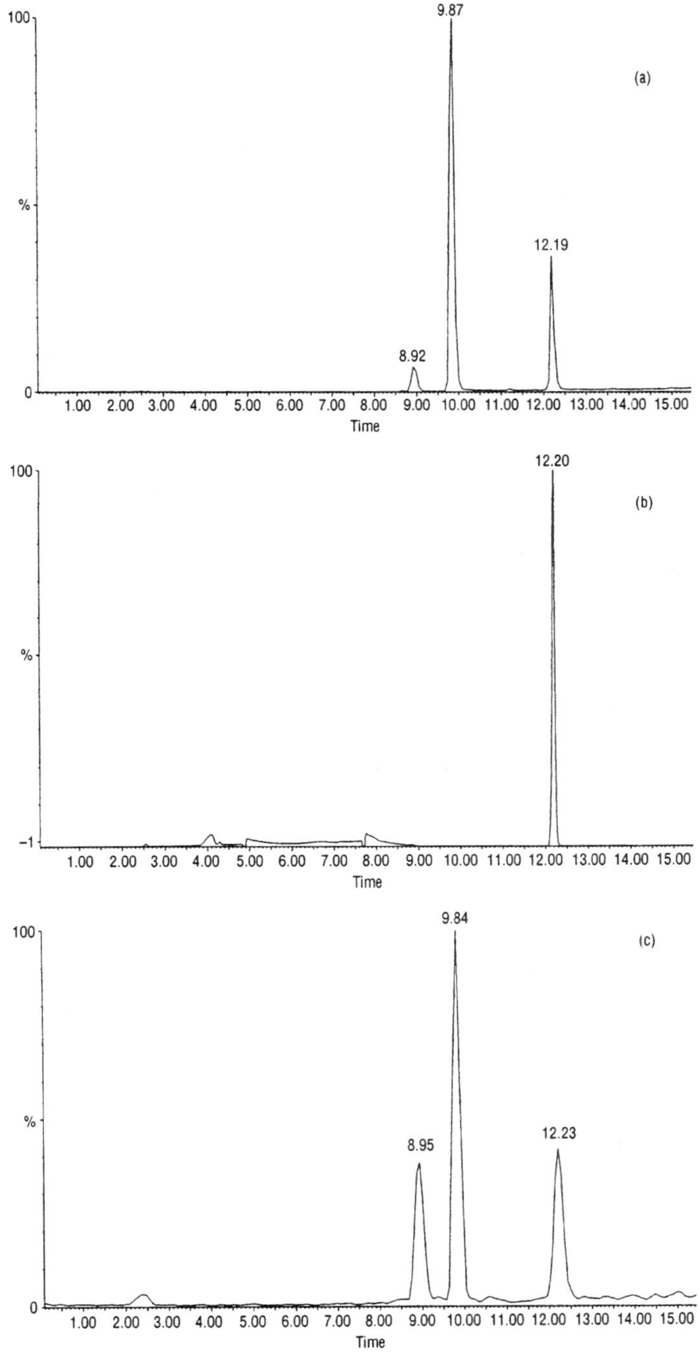

Figure 6.5 Multiple UV/MS/ELS detection: non-chromophoric carbohydrates.

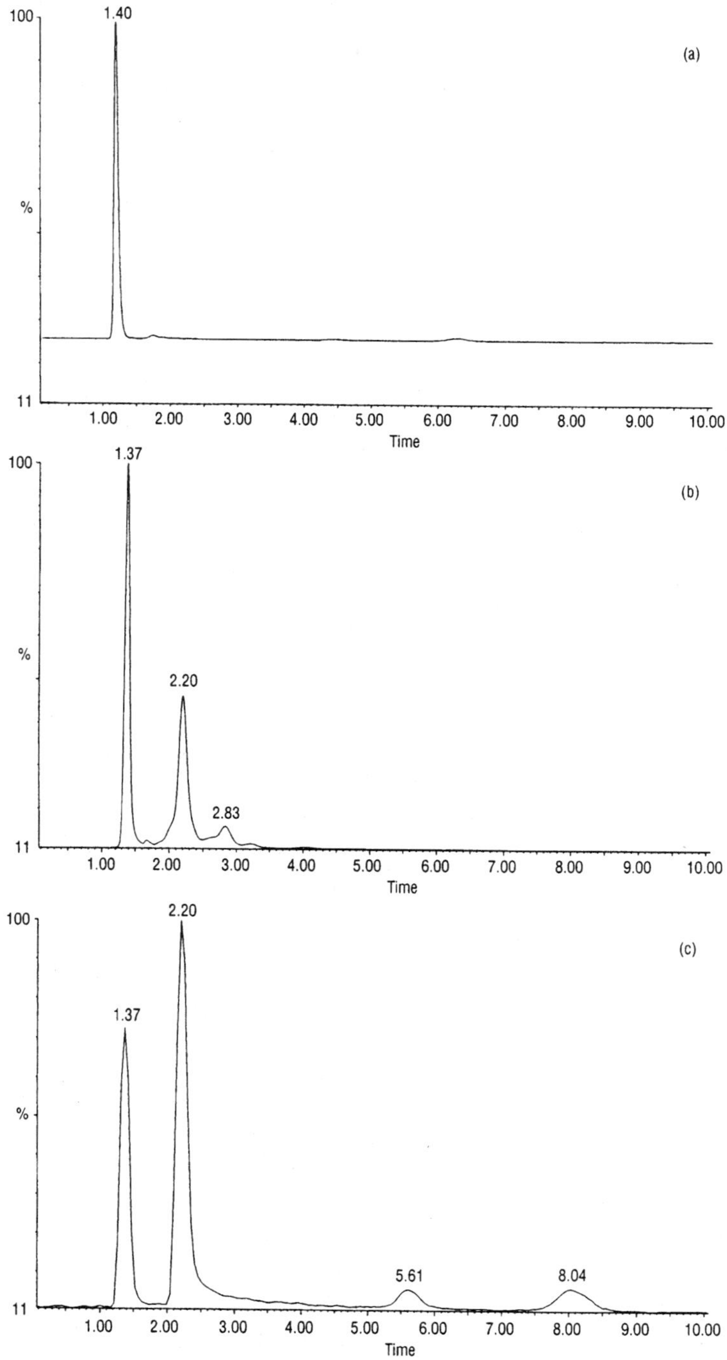

Figure 6.6 Multiple UV/MS/ELS detection: benzoic acids.

can be developed in-house together with high-throughput analytical techniques (78–82). Usually, these programs detect the presence of the predicted molecular weight for a library component in the sample and may also include minimum threshold purity criteria. The example shown in Fig. 6.7 (83) depicts the analysis of two reaction plates of 96 compounds where the plate on the left contains most of the expected reaction products (green spots, purity higher than the set standard as measured on the UV chromatogram) while the plate on the right is a poor-quality library where a large number of wells do not contain the expected library component or contain it as a mixture with impurities.

Several recent reviews (84–88) cover the characterization of discrete libraries from parallel synthesis with automated systems and using different analytical and/or detection methods; their perspective may help the reader to expand his or her knowledge of this area.

6.2.3 Purification

The SPS of libraries allows the automation of all of the work-up procedures and the elimination of tedious and time-consuming purification steps during the synthesis by simply filtering off the reaction solution and washing the beads repeatedly with fresh solvents. For all SP discrete libraries, the last synthetic step involves the cleavage of resin-bound individuals from the support and their release into solution. Careful selection of linkers and reagents usually allows the elimination of possible by-products from the cleavage, but often the recovered crude materials do not satisfy the purity requirements for screening. The purification of the cleaved compounds from an SP library of discretes is conceptually *identical to the purification* of the individuals *from a solution-phase library of discretes* (Chapter 8). However, this latter library format requires the automated parallel work-up/purification of intermediates during most of the synthetic steps. This important topic will be covered more thoroughly in Section 8.3.

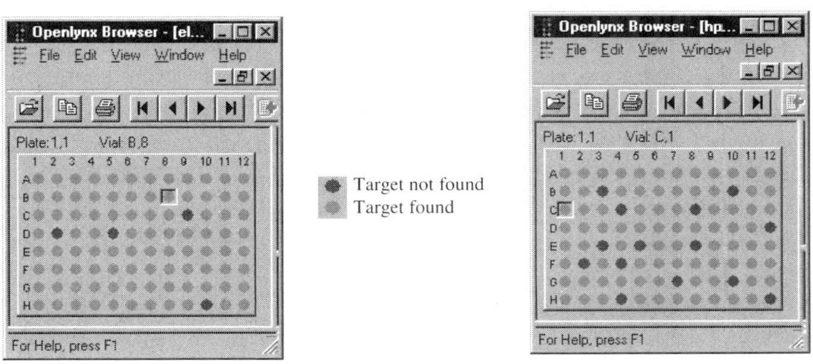

Figure 6.7 Visualization of analytical results for a 96-well reaction plate: good-quality (left) and poor-quality (right) libraries.

A peculiar topic is the handling of solid and liquid samples during the final cleavage to release the library in solution. Often semi- or fully automated SP synthesizers are not ideally suited for solution handling, even though they sometimes are provided with automated cleavage stations. Thus, cleavage protocols are better performed off-side with the help of commercially available 96-well plates equipped with a bottom frit; resin slurries are transferred either manually or with robotic pipettes to the sealed cleavage plates and the cleavage is performed, then the seal is removed and the liquid aliquots containing the library discretes are collected and processed (e.g., evaporation, extraction, see Section 8.3). Suitable cleavage reagents do not require complex steps for their removal from the released compounds; besides the ideal photolytic deprotection, where no cleavage reagents are necessary, the use of easily removable gaseous reagents (e.g., TFA, HF, ammonia) and the automation of the cleavage/washing operations to handle and to recover in parallel massive numbers of compounds is becoming widespread (89, 90).

6.3 EXAMPLES OF SOLID-PHASE DISCRETE LIBRARY SYNTHESIS

6.3.1 Manual Parallel Synthesis: Two Libraries of Tetrahydroquinolines

A recent paper by Kiselyov et al. (91) reported the synthesis of two small SP discrete libraries of *tetrahydroquinolines* **L1** and **L2** inspired by a three-component condensation reaction in solution (92–94) as depicted in Fig. 6.8. The scope of this condensation had been expanded by other groups (95–97), and the use of polymer-supported metal catalysts had also been reported (98, 99); however, an SP route with a broad tolerance for different amines, aldehydes, and olefins had still not been defined prior to this work.

The authors first considered the attachment of the aldehyde component to the SP using the AMEBA (acid-sensitive methoxy benzaldehyde) PS resin (100). 4-Carboxybenzaldehyde and two different amines were reacted to produce the resin-bound intermediates **6.1** and **6.2** on a 100-g scale according to the reaction scheme shown in Fig. 6.9, with good yields resulting from an optimization of the reaction conditions. Resin-bound **6.1** was reacted with the standard olefin (cyclopentadiene, **6.3**) and an aromatic amine (aniline, **6.9**) and, after optimization of the reaction conditions, the desired tetrahydroquinoline **6.14** was obtained in good yield and purity after cleavage.

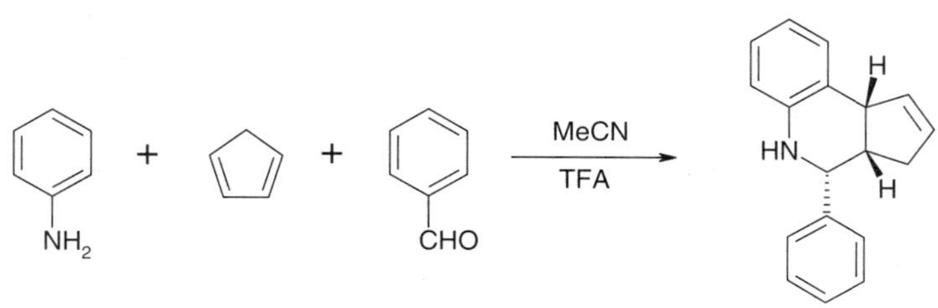

Figure 6.8 Synthesis of tetrahydroquinolines via a three-component condensation.

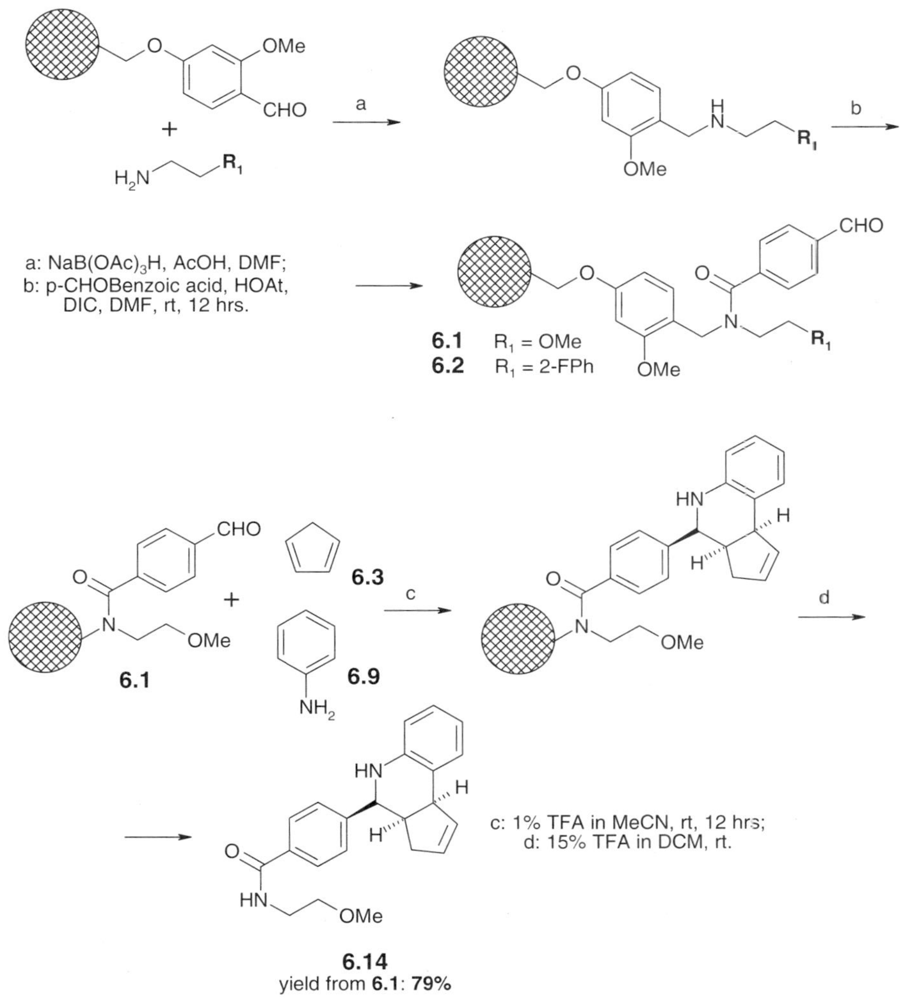

Figure 6.9 SP chemical assessment for the tetrahydroquinoline library **L1** from resin-bound aldehydes **6.1**, **6.2**.

The corresponding Schiff base was obtained as the only impurity in amounts ranging from 2 to 7% during the various chemical assessment runs. Compound **6.14** was formed as a single diastereomer with the relative orientation of the three chiral centers as shown.

These encouraging results prompted the authors to prepare an SP discrete library of 60 individuals, **L1**, using the two solid-supported aldehydes **6.1** and **6.2**, six olefins (**6.3–6.8**), and five substituted anilines (**6.9–6.13**), whose structures are reported in

TABLE 6.1 Yields of Selected Tetrahydroquinolines from the SP Discrete Library L1[a]

Anilines\Olefins	6.9	6.10	6.11	6.12	6.13
6.3	74/69[b]	72/73	81/82	88/79	86/81
6.4	71/68	65/61	79/82	82/83	86/75
6.5	60/62	64/62	72/77	79/82	77/79
6.6	69/68	71/66	82/84	80/76	87/82
6.7	65/68	69/71	82/81	76/84	79/81
6.8	—[c]/NT	–/NT	10/NT	69/NT	76/NT

[a]NT = not performed.
[b]Yield from reaction with aldehyde **6.1**/aldehyde **6.2**.
[c]Failed.

Fig. 6.10. The resin (15 g) was partitioned into 60 aliquots of 250 mg in 60 glass vials, and each was treated with an aniline and an olefin as 0.5 *M* solutions in acetonitrile in the presence of 2% TFA at room temperature for 24 h. The resin in each vial was filtered, washed, and dried and the product cleaved from the support with 15% TFA–DCM. The library individuals were obtained as solids (>80% by HPLC) after evaporation and trituration of the crude residue with Et_2O, with the yields reported in Table 6.1. The reactions proceeded well for alkenes **6.3–6.7** (60–84% yields), the only detectable impurities being the corresponding Schiff bases (5–10%). Alkene **6.8** gave poor yields of the desired product or failed to react completely in some cases (see Table 6.1). All of the tetrahydroquinolines were obtained as single diastereomers, as shown for compounds **6.14** and **6.15** (Fig. 6.10).

A similar library **L2** was prepared from the Wang-resin-bound alkene reagents **6.17** and **6.18** (105 g each, obtained via phosphoranes **6.16**) by reacting them with anilines and aldehydes. Compound **6.25**, which was formed as a single diastereomer, was prepared as a model in good yield (64%) and purity (Fig. 6.11). The use of four anilines (**6.9**, **6.10**, **6.19**, and **6.20**; Fig. 6.10) and four aldehydes (**6.21–6.24**; Fig. 6.12) produced a 32-member SP discrete library on a 250-mg resin scale for each individual using identical conditions to those used for **L1**. The final compounds were obtained in good yields (61–85%) and purities (>85% by HPLC), and as an improvement with respect to **L1,** the Schiff base side products remained in solution, even though small quantities of unreacted styrenes could sometimes be found in the final library components. The yields for library **L2** are reported in Table 6.2, and the diastereopure outcome of the reaction is shown by the structures of **6.25** and **6.26** (Fig. 6.12).

The parallel synthesis of several tens of compounds with two complementary SP routes employing a multicomponent reaction and producing diverse tetrahydroquinolines as decorated, biologically relevant scaffolds (101–103) was realized using simple laboratory equipment and commercially available reagents as monomers. The assessment work to produce compounds **6.14** and **6.25** with the optimized reaction conditions was deemed to be sufficiently robust to pass immediately to library production. This

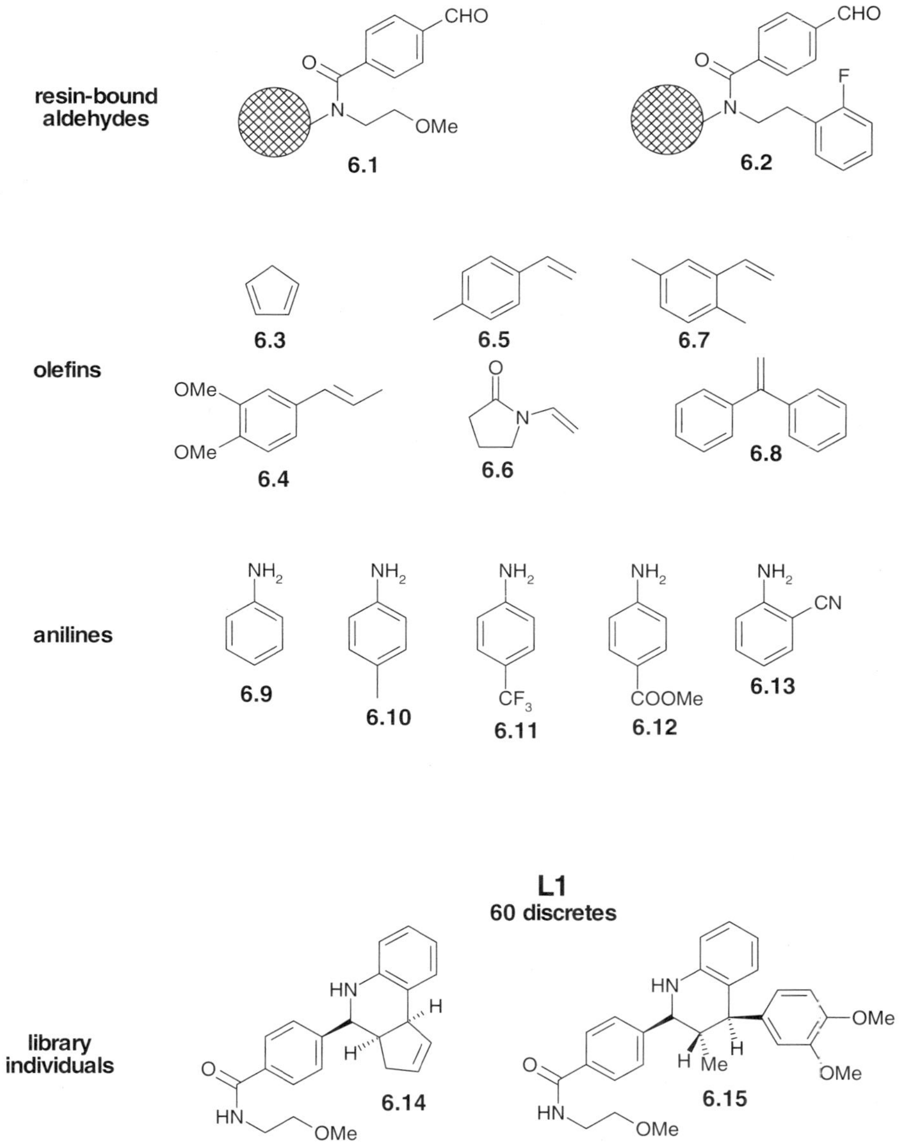

Figure 6.10 Monomers used for the synthesis of the tetrahydroquinoline library **L1** and selected library individuals.

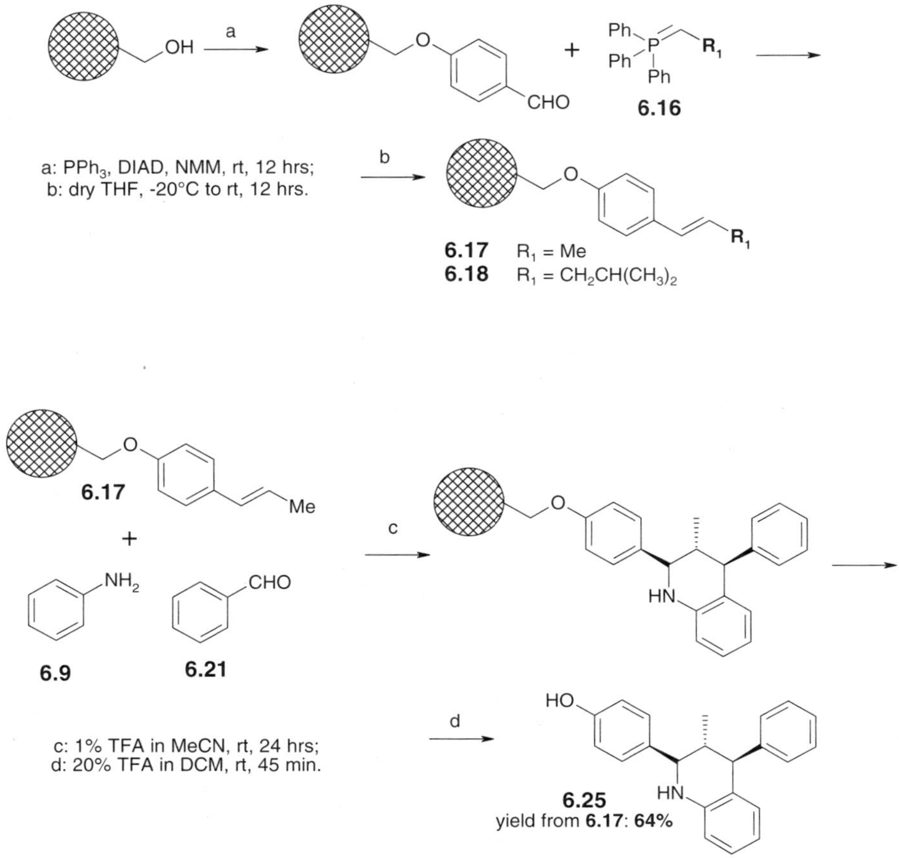

Figure 6.11 SP chemical assessment for the tetrahydroquinoline library **L2** from resin-bound olefins **6.17**, **6.18**.

was possible because analytical characterization of all the library individuals allowed the removal of compounds that did not meet the purity standards (see monomer **6.8**, Table 6.1) before screening.

6.3.2 Manual Parallel Synthesis: A Library of Benzopyrans

Breitenbucher and Hui (104) have recently reported the SP synthesis of a medium–large discrete library **L3** of 8448 *benzopyrans* using the reductive amination cocktail formed from titanium isopropoxide and sodium triacetoxyborohydride, known in solution but whose applications to SP were rare and limited to single experiments (105, 106). The benzopyran scaffold is present in a number of biologically active compounds (107–110), and this library was tested in several biological assays (111).

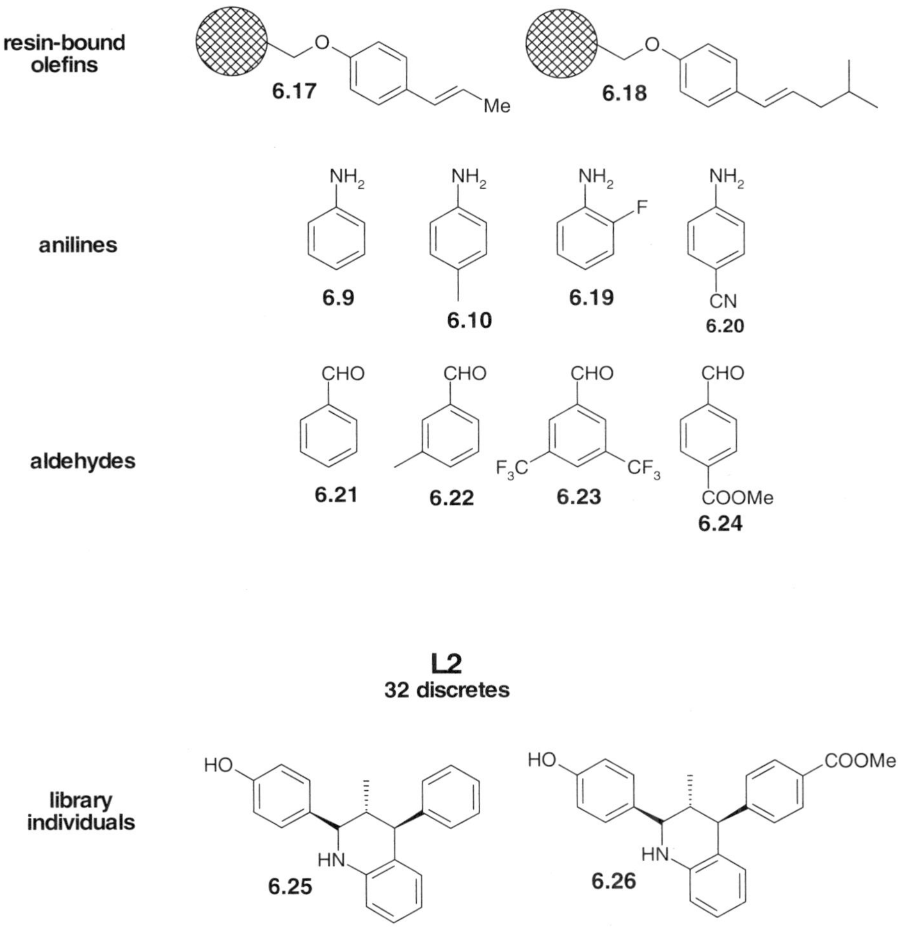

Figure 6.12 Monomers used for the synthesis of the tetrahydroquinoline library **L2** and selected library individuals.

TABLE 6.2 Yields of Selected Tetrahydroquinolines from the SP Discrete Library L2

Aldehydes\Anilines	**6.21**	**6.22**	**6.23**	**6.24**
6.9	64/61[a]	67/63	71/66	78/72
6.10	61/65	69/63	76/72	75/77
6.11	72/66	70/69	78/74	71/79
6.12	73/71	68/62	82/70	84/85

[a]Yield from reaction with olefin **6.17**/olefin **6.18**.

The generic structure of the library is reported in Fig. 6.13; a retrosynthetic analysis arrived at three acids (e.g., **6.27**; only the α-dimethyl was reported as a structure in the original article) as suitable precursors prepared from commercially available materials according to a reported procedure (Fig. 6.13) (112). The cleavage of an ester bond with amines was used to generate diversity and simultaneously release the compounds into solution, and a suitable linker was the 4-hydroxythiophenol linker **6.28** (Fig. 6.13) (113, 114).

The chemistry assessment is reported in Fig. 6.14. The linker **6.28** was coupled with the carboxybenzopyranone **6.27a** using standard SP esterification conditions, and the supported benzopyranone **6.29** was reductively aminated. At first, classical SP reductive amination conditions were used, but either incomplete reduction was obtained or extensive cleavage of the benzopyrans from the resin was observed through the use of forcing conditions to drive the reaction to completion. The authors then examined the $Ti(OiPr)_4/Na(OAc)_3H$ as reducing agent and obtained excellent yields of supported amine **6.30** using benzylamine and anhydrous toluene in an inert atmosphere after 2 h at room temperature with $Ti(OiPr)_4$ and subsequent addition of the borohydride with further stirring at room temperature. The outcome of the reaction was studied by single-bead IR spectroscopy, and the absence of cleaved intermediates from the solid support (<10%) was confirmed by HPLC analysis of the supernatant. An important finding was the negligible amount of insoluble titanium salts formed; the insoluble

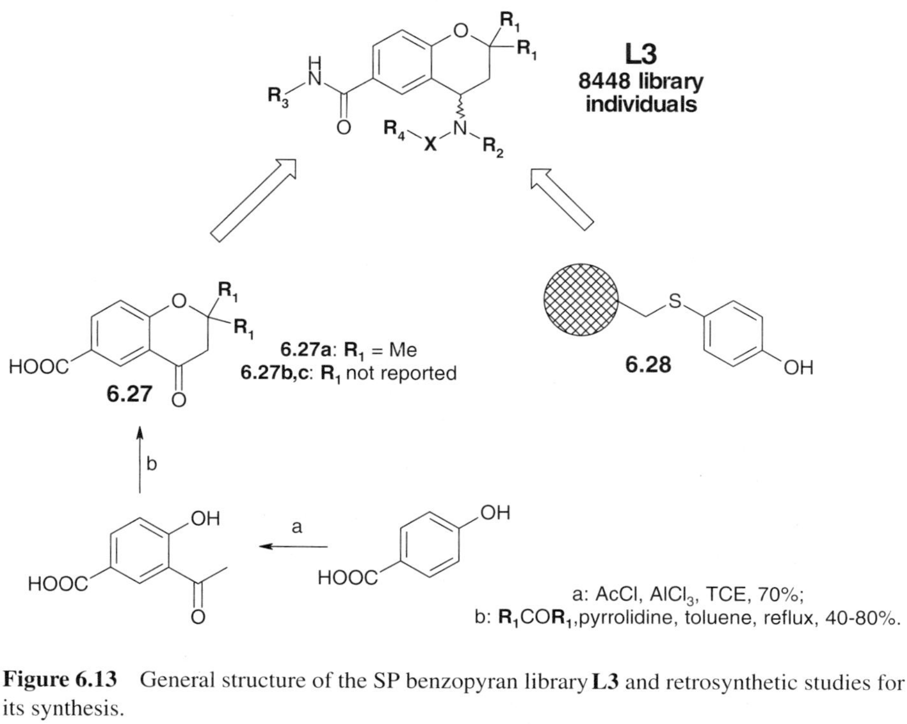

Figure 6.13 General structure of the SP benzopyran library **L3** and retrosynthetic studies for its synthesis.

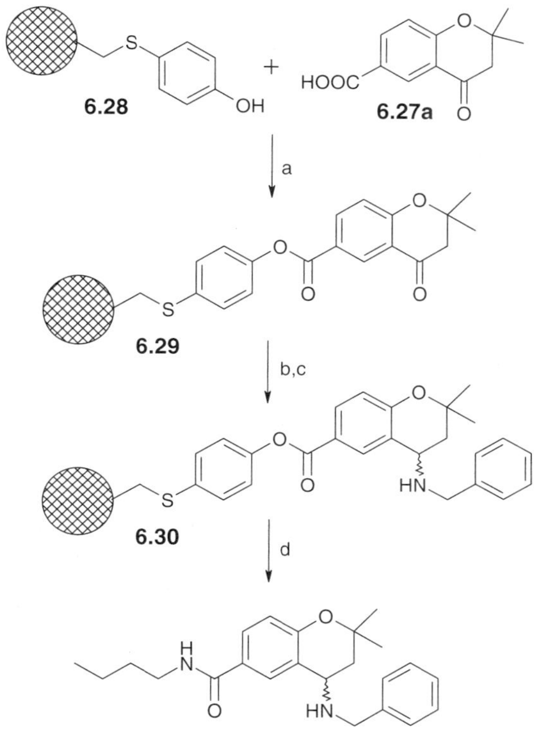

a: DIC, DMAP, NMM, rt; b: $Ti(OiPr)_4$, $BnNH_2$, dry toluene, rt, 2 hrs;
c: $Na(OAc)_3BH$, AcOH, dry toluene/dry THF, rt, 24 hrs; d: n-$BuNH_2$, Py, rt, 24 hrs.

Figure 6.14 SP chemical assessment for the benzopyran library **L3** from resin-bound benzopyranone 6.29.

salts could have significantly complicated the washing and filtration of the resin by using anhydrous reaction conditions. Resin-bound **6.30** was finally washed and cleaved with *n*-butylamine in pyridine at room temperature for 24 h to give compound **6.31** (Fig. 6.14).

The design of the library **L3** and the synthetic scheme are reported in Fig. 6.15; filtration of the solutions and washing of the resin aliquots were performed after each step. The three supported benzopyranones **6.32** (step a, from the monomer set $\mathbf{M_1}$, three ketones) were obtained in a similar manner to **6.29** on a 300-g scale. Each resin-supported benzopyranone was split into 8 separate peptide synthesis flasks containing 35 g of resin and all of the 24 samples were reductively aminated (steps b and c, monomer set $\mathbf{M_2}$, eight amines) to produce 24 resin-bound intermediates **6.33**. Each intermediate was then divided into 352 portions of 100 mg of resin that were distributed into four filter-bottom 96-well microtiter plates (2.0 mL capacity per well).

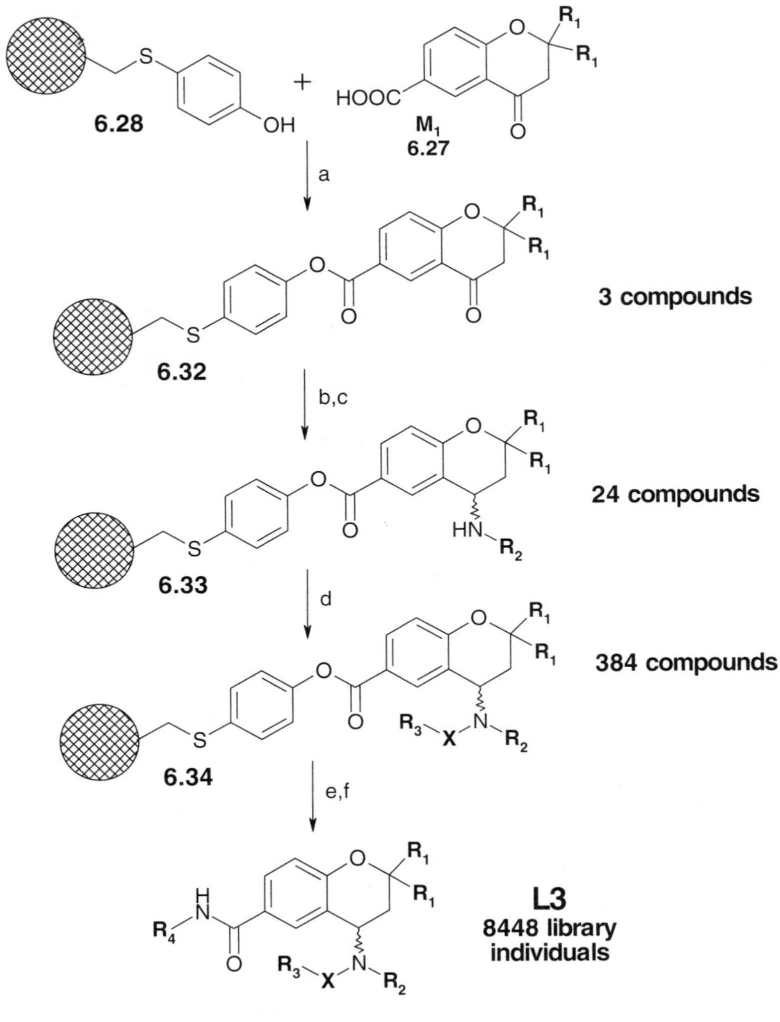

a: DIC, DMAP, NMM, rt; b: Ti(OiPr)$_4$, **M$_2$**, dry toluene, rt, 2 hrs;
c: Na(OAc)$_3$BH, AcOH, dry toluene/dry THF, rt, 24 hrs; d: **M$_3$**, DIPEA, rt;
e: **M$_4$**, Py, rt, 48 hrs; f: Supported Liquid Extraction (SLE).

Figure 6.15 SP synthesis of the benzopyran library **L3**.

These plates allowed the reaction of resin aliquots with reagents in solution by sealing the filter bottom and to discard the solution and wash the resin by filtering the solutions after removal of the seal. The samples **6.33** were then acylated (step d, monomer set **M3**, 16 acylating agents, including isocyanates, isothiocyanates, and acid chlorides) to produce 384 resin-bound intermediates **6.34**, each as a discrete distributed into 22 different wells. Final cleavage from the support was performed (step e, monomer set

M4, 22 amines), and after the reaction the cleavage solutions were recovered by filtration into 96 corresponding microtiter plates to give crude **L3** as 8448 spatially separated crude benzopyrans that were submitted to an automated purification procedure (step f) (115) and concentrated using commercially available vacuum centrifuges to give the final pure library (Fig. 6.15).

The dimensions of the library prevented its full analytical characterization, but enough data were acquired to judge its overall quality. Twelve percent of the individuals, that is, 1056 randomly selected compounds, were analyzed by electrospray mass spectrometry, and 83% of the analyzed samples showed the expected molecular ion as the principal MS peak. Eighty-five randomly selected compounds (corresponding to 1% of the total) were analyzed by HPLC-MS and the average purity was determined as 73% area/area. Finally, six discretes (**6.35–6.40**) were randomly picked and fully characterized. An authentic sample of each was prepared in solution, purified, and used as a standard for the quantitative determination of the six library components. The structures of the six compounds and the overall satisfactory results from their analytical characterization are reported in Fig. 6.16.

In this example, the use of reaction vessels of varying dimensions, 96-well microtiter plates, and simple equipment for purification and concentration of the solutions

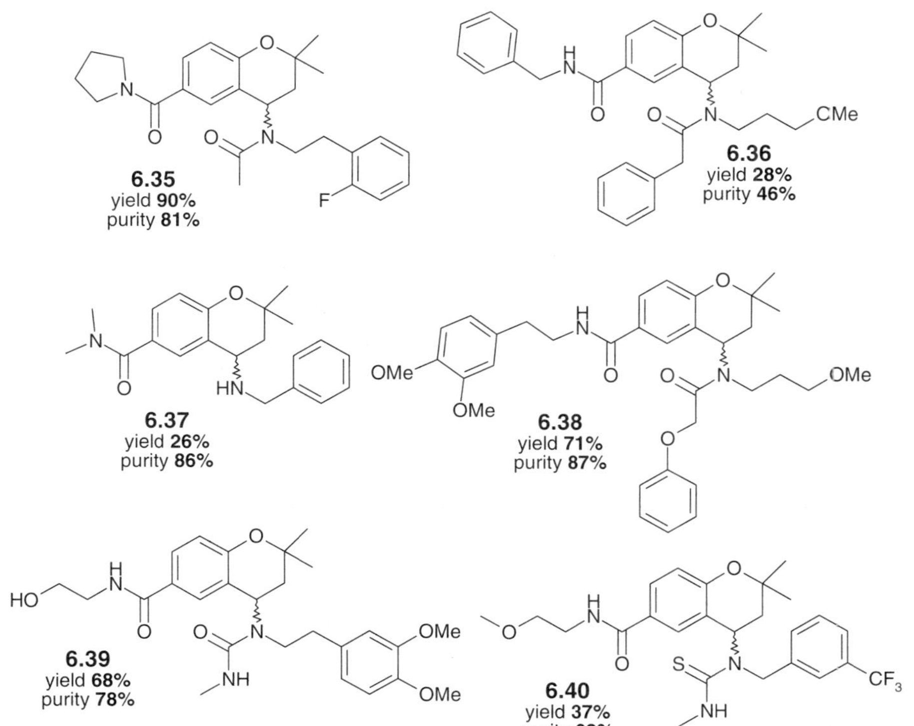

Figure 6.16 Structure of fully characterized discretes **6.35–6.40** from the benzopyran library **L3**.

allowed the preparation of a large SP discrete library that was tested "in a variety of high throughput biological screens" (104). All the monomers used were either commercially available or easily prepared from commercial sources. As often happens in published papers reporting the synthesis of medium–large library, only a few structures of the monomers used and even fewer structures of the library individuals are provided. This keeps the diversity generated in the library as proprietary knowledge and little or no data regarding the activity are usually provided for the libraries. This means that suitably active compounds can be tested further and eventually patented with a more defined property profile. We will cover this topic, which is related to the pharmaceutical applications of chemical libraries, in Section 9.1.

Many other reports of manual parallel SP synthesis leading to significant arrays of small organic molecules have appeared in the last five years, and most of them may be found by consulting the excellent reviews cited in references 1–5. Nevertheless, some recent papers in this area are worth mentioning. For example, Kiselyov et al. have reported the synthesis of two 12-membered library of 14-membered macrocycles, the first (116) and prepared from allyl bromide, bromoacetic acid, and aliphatic primary amines with 54–74% yields and >90% purity, the second (117) made by 30 representation and prepared from functional, primary amines and bromoacetic acid. A library of over 1000 thiazolidinones prepared from diamines, aldehydes, mercaptosuccinic acid, and amines with >65% average yields and high purity has been described by Munson et al. (118). Albert et al. (119) have published the synthesis of a 24-membered library of isoxazolylthioamides prepared from phenylacetic acid derivatives and *p*-substituted aryl isothiocyanates. A 96-membered hydantoin library, which was further expanded to a library of more than 10,000 members, was prepared from α-amino acids with >60% yields and >65% purity (120). Wong et al. (121) have described the synthesis of a 45-membered library of oligosaccharides from orthogonally protected monosaccharides with varying yields (5–90%) and purity. The Ugi four-component coupling (4CC) has been used by Li et al. for the preparation of an 108-membered library of α,α,α-difluoromethylene phosphonic acids from isonitriles, aldehydes, Rink amine resin, and a phosphonic acid with yields ranging from 10 to 95% and >70% purity (122). The same reaction has also been employed by Kim et al. (123), who used it to prepare a library of 96 peptidomimetics from aldehydes, carboxylic acids, Rink amine resin, and a single isonitrile with varying yields and >90% purity. Peptidomimetic-based libraries have been described by Souers et al. (124) and Ogbu et al. (125). The former highlighted the preparation of a 5500-member library of β-turn mimetics from primary amines, α-amino acids, and α-halo acids, whereas Ogbu et al. used the Diels–Alder cycloaddition of 1,2,4-triazolinediones and dienoic esters or amides to construct a 1500-member library of constrained β-strand mimetics. The venerable Pictet–Spengler cyclization of tryptophan, an aldehyde, and a primary amine was used by Fantauzzi and Yager to prepare a library of 345 tetrahydro-β-carbolines in high yield and purity (126). A 24-member aminopyrazoline library was prepared in unspecified yield from α,β-unsaturated nitriles and hydrazines with >80% purity (127). Mohan et al. (128) has shown that nucleophilic aromatic substitution can be useful for the synthesis of an amidinophenoxypyridine library. A large library of benzimidazoles was prepared from fluorobenzoic acid, aldehydes, and amines with >80% yields and >70% purity by Mayer et al. (129). Lin and Ganesan (130) have described the preparation of a 1000-member library of carbamoylguanidines with moderate to good yields and >85% purity. Van Loevezijn et al. (131) have reported the synthesis of a 42-member library of polycondensed heterocyclic analogues of indolyl diketopiperazine alkaloids with 50–100% yields and 70–99% NMR purity after recovery. A robust SP and solution-phase protocol for parallel synthesis of >100 member libraries of substituted ureas with high purity (>80%, MS) has been reported by Nieuwenhuijzen et al. (132). Romoff et al. (133) published an efficient protocol for SP parallel synthesis of polyfunctionalized 3-acyltetramic acids using a cyclative cleavage approach with average good yield and excellent HPLC/MS purity. Another cyclative cleavage approach was reported by Villalgordo et al. (134) to prepare SP discrete libraries of 3*H*-quinazolin-4-ones with average high yield (>70%, recovered material) and >95% HPLC purity. Nefzi (135) reported the synthesis of two 97-member libraries of diethylenetriamines using Houghten's tea-bag synthetic protocol (136) and obtaining good average yield and purity for the library individuals. The protocol to obtain parallel aldehyde libraries was reported, with good yield and purity, by Salvino et al. (137); the further use of these cleaved libraries as reagents for a "library from library" approach in a Horner–Emmons protocol using a supported phosphonate was also reported. Hennequin and Piva-Le Blanc (138) reported the procedure to prepare a discrete library of oxindole quinazolines, providing several examples with moderate to good yield and excellent purity. Roussel et al. (139) presented a 43-member discrete library of amidinonaphthols with an average purity of 60% (HPLC) that was used as a model library for mix-and-split synthesis of SP pool libraries. A small 30-member library of 2,5-dihydrofurans and 1,3-dihydroisobenzofurans was obtained with average 80% yield and >90% purity via iodine–magnesium exchange and further Grignard reaction by Rottlander and Knochel (140). Paio et al. (141) reported the SP protocol for the synthesis of polysubstituted amine libraries with moderate to good yield and good purity using supported benzotriazoles as key reaction intermediates.

Johnson et al. (142) reported the SP synthesis of 346 discrete benzothiophenes as potential thrombin inhibitors from the decoration of a core structure with alcohols and acylating agents with good yields and purities after parallel silica gel chromatography on 96-well plates. Tremblay et al. (143) presented the synthesis of a 20-member SP model library of 7-modified estradiols as potential estrogen receptor antagonists, using standard peptide SPS protocols and obtaining moderate yields (around 30%) and high purities (typically >85% by HPLC) for the five step protocol. Haskell-Luevano et al. (144) prepared a medium, >900-member SP library based on the known β-turn motif using a final cyclative cleavage and obtaining novel melanocortin-1 receptor activators with good yields and purities.

6.3.3 Semiautomated Parallel Synthesis: A Library of 1,4-Benzodiazepine-2,5-Diones

A recent paper by Boojamra et al. (145) reported the synthesis of a 2508-membered SP discrete library **L4** of *1,4-benzodiazepine-2,5-diones*, further exploiting the first reported SP synthesis of a library of small organic molecules (146, 147) based on a final cyclization step. The generic structure of the library is shown in Fig. 6.17, together with the three monomer classes used and their components. Nine of 10 chosen α-amino acid esters were coupled as racemic mixtures or were racemized during the synthesis, so that the library was prepared as 1320 samples containing 132 discretes and 1188 enantiomeric pairs.

An extensive assessment of the chemistry was performed, and several individual compounds were prepared to demonstrate the chemistry. The optimized synthetic route and reaction conditions are reported in Fig. 6.18 for a specific example. The acid-labile dimethoxybenzaldehyde **6.41**-based linker was chosen and linked to a Merrifield resin to give **6.42** (step a), which was reductively aminated to **6.43** under racemizing conditions to obtain the two enantiomers from L-leucine methyl ester (representative of monomer set $\mathbf{M_1}$, step b). The resin-bound secondary amine was then acylated with unprotected 4-chloro anthranilic acids (representative of monomer set $\mathbf{M_2}$, step c) to give **6.44** using EDC (1-(3-dimethylaminopropyl)-3-ethyl carbodiimide hydrochloride) as the only effective coupling agent. Cyclization occurred using lithium acetanilide as the base with an optimal pK_a to give the resin-bound cyclic anion **6.45** (step d). This was subsequently quenched with ethyl iodide (representative of monomer set $\mathbf{M_3}$, step e) to give the resin-bound benzodiazepinedione **6.46**, which was then released under acidic conditions (step f) to provide **6.47** in a 75% overall yield based on the amino acid ester loading (Fig. 6.18). The assessment allowed satisfactory reaction conditions for the whole library synthesis to be determined (*vide infra*) but also served to discard potential monomers on the basis of their reactivity. In this case, serine and valine were rejected from $\mathbf{M_1}$ due to α-elimination of the hydroxyl group and incomplete acylation of the secondary amine in the following step, respectively. *p*-Substituted anthranilic acids bearing electron-donating groups were rejected from $\mathbf{M_2}$ because they led to significant acylation of the unprotected aromatic amine function and formation of higher order anthranilic oligomers. The racemizing conditions used for the reductive amination step (preequilibration of the amino ester hydrochloride, DIEA (diisopropyl ethyl amine), and the resin for 6 h in DMF, then addition of the reducing agent and acetic acid) were easily changed for nonracemizing conditions (sequential addition of acetic acid, reducing agent, and amino ester to a stirred slurry of resin in DMF), which allowed the synthesis of enantiomerically pure benzodiazepinediones. This allowed the "chiral deconvolution" of each enantiomeric mixture showing through the preparation of the two single enantiomers. A total of 23 compounds were prepared as discretes in yields varying from 40 to 92% for the purified material; some of them are shown in Fig. 6.18. These results encouraged the authors to pursue synthesis of a medium–large library **L4**. These compounds were fully characterized by means of HPLC, MS, and NMR and served as standards for the quality control of the library as a whole (*vide infra*).

The first steps in the synthesis of the library were performed manually, and the usual filtration/washing cycles were carried out after each step. The aldehyde linker was coupled to 50 g of Merrifield resin in a 2 L-flask; then ten 3.5-g portions of resin were placed into ten separate 100-mL flasks and four batches reductively aminated under nonracemizing conditions using racemic mixtures of Ala, 2-naphthylAla and

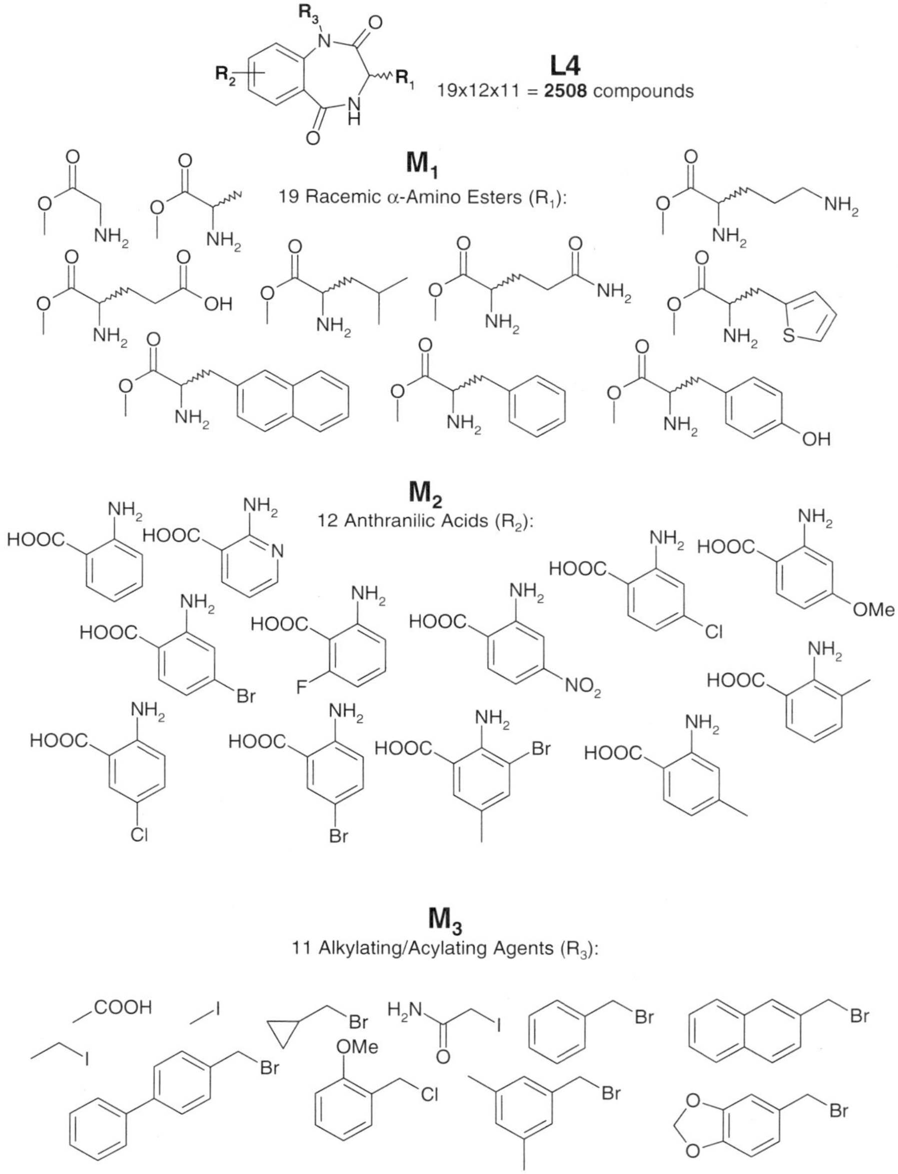

Figure 6.17 General structure of the benzodiazepine library **L4** and of the used monomer sets **M1–M3.**

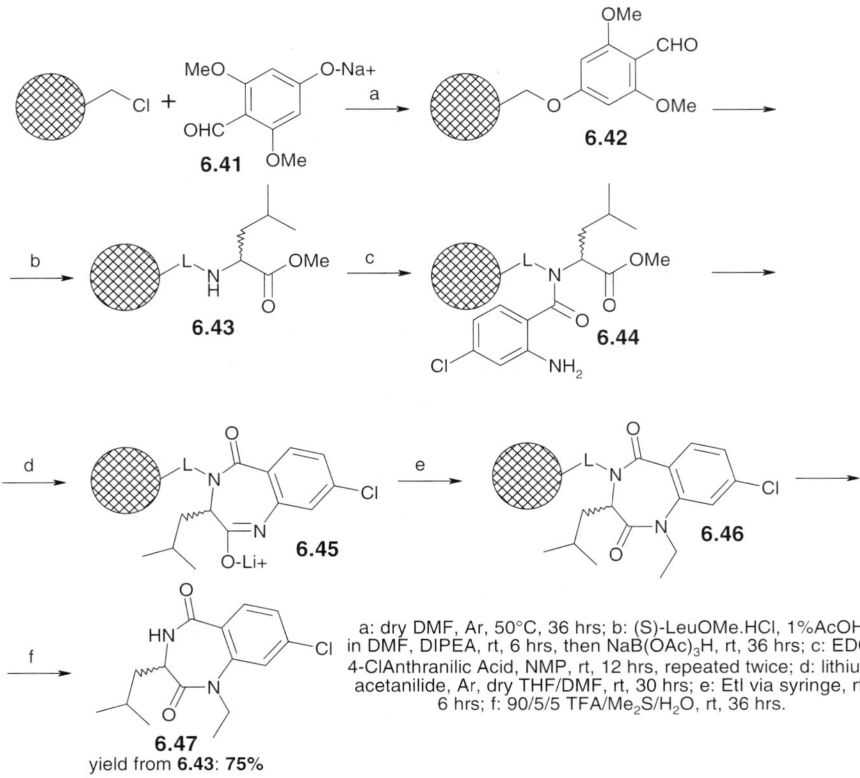

Other discretes from chemical assessment (α-Amino Ester, Anthranilic Acid (AA), Alkylating Agent, Yield of purified material):

Ala, AA, AcOH, 77%
Phe, 5-ClAA, AcOH, 89%
Leu, 4-ClAA, AcOH, 89%
Leu, 5-BrAA, EtI, 71%
Leu, 4-MeOAA, EtI, 81%
Glu, 5-ClAA, BnBr, 52%
Gln, 5-ClAA, EtI, 71%
Leu, $4\text{-}NO_2AA$, EtI, 92%
Leu, 3-Br,5-MeAA, MeI, 71%
2-ThieAla, $4\text{-}NO_2AA$, cxpropylBr, 81%
Gly, 4-MeOAA, o-MeOBnBr, 40%
Phe, 6-FAA, $IAcNH_2$, 89%
Ala, 5-FAA, 3,5-diMeBnBr, 62%
Phe, 5-ClAA, AllylBr, 89%
Leu, 4-MeOAA, cxpropylBr, 79%
Lys, 5-ClAA, AllylBr, 63%
Leu, 4-ClAA, EtI, 75%
Phe, PyrAA, cxpropylBr, 69%

Figure 6.18 SP chemical assessment for the benzodiazepine library **L4** from resin-bound α-amino ester **6.43**.

2-thienylAla, and Gly. The remaining six batches were reductively aminated under racemizing conditions. Each portion of resin was further divided into 12 aliquots and transferred to 120 filter cartridges. According to the recovered mass, after the reductive amination the resin aliquots varied in weight between 266 mg (3.55-g portion, Leu) and 316 mg (4.21-g portion, Gln). Each chosen acylating agent was reacted with the cartridges containing the 10 different resin-bound amino esters to give 120 intermediates for cyclization and alkylation. A small volume of a solution of DCE–DMF 3/2 (same density in respect to the resin beads) was then added to all of the cartridges to create an isopycnic suspension for further equimolar division into the final reaction vessels (11 wells per resin aliquot, 1320 total wells).

The following steps were performed on 1320 aliquots of resin, corresponding to the number of samples/racemic mixtures in the library. The authors expressly built a 96-well plate-based device in order to speed the handling of the relatively large number of resin portions while ensuring a good quality for the synthesis. A schematic representation of this device, called the *multitube apparatus*, is shown in Fig. 6.19. A thin polyethylene plate with 96 holes, obtained by sawing off the top of a 96-well, 1-mL microtiter plate, was used to hold 96 × 7-mm glass tubes; the bottoms of the tubes had been cut and replaced by a hydrophobic polyethylene frit. The tubes were cut to a length of around 6.5 cm, and the frits were sealed at their end by heating; then each tube was soaked in THF and inserted into the bracket, leaving only one-quarter of the tube above the plate. The authors prepared 17 of these low-cost devices rapidly, allowing the handling of >1500 SP discrete reactions.

Each of the 120 isopycnic suspensions was aliquoted in 240-μL portions that were delivered into each tube of the apparatus, filling 80 of the 96 tubes for each multiplate, leaving the first and last columns empty and thus using 16.5 multiplates in total. The aliquots were divided between 11 plates, each subsequently coupled with a single alkylating agent, and the remaining ones were used to add different alkylating reagents in the same plate. The isopycnic suspensions were filtered through the frits, and the multiplates were dipped into deep polyethylene trays (formed by the covers of the original 96-well plates) containing 100 mL of a solution of lithium acetanilide that freely penetrated into all tubes to induce cyclization of the bound substrate (corresponding to step d, Fig. 6.17). After filtering and washing, 11 multiplates were treated with solutions of 11 different alkylating agents, again using the polyethylene covers as reaction vessels, while the tubes of the remaining six multiplates were soaked into six 96-well 2-mL microtiter plates containing the desired alkylating solution in each

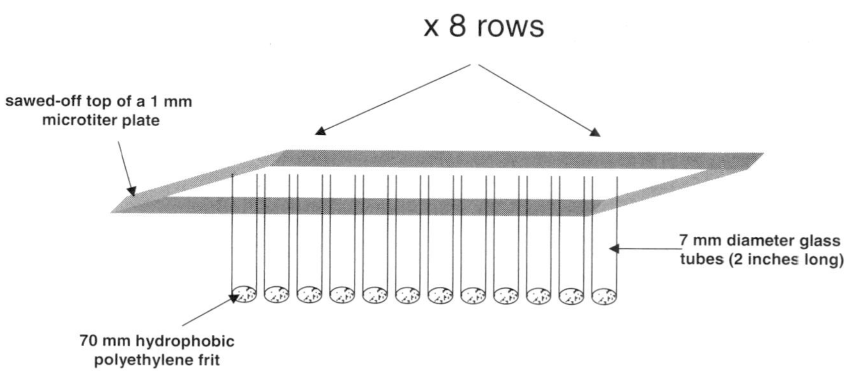

Figure 6.19 Multitube SP library parallel synthesizer.

well. After 8 h, the resin aliquots were rapidly washed, and the final products were cleaved by soaking the 17 multiplates into seventeen 96-well 2-mL microtiter plates containing the cleavage cocktail (1 mL/well) for 40 h. The multitubes were removed slowly to allow all of the liquid to drain into the corresponding well below and then submerged into a second set of 17 plates containing a rinse cocktail (DCE–DMF 1/1, 1 mL/well). The solutions were concentrated in a plate concentrator, the rinse cocktails were also added by using semiautomated multipipetters, and the residues were finally obtained and statistically characterized for yields and purity by HPLC in the presence of an external standard (multiple randomly selected wells) or NMR (36 randomly selected wells).

The use of such a synthetic strategy in which a semiautomated device allowed a significant improvement in the speed of the synthesis and ease of handling of the 1320 library samples, while using common glassware or laboratory equipment for the early steps, proved to be a good choice for the synthesis of this library.

Other groups have recently reported the preparation of discrete libraries in which part of the synthesis has involved the use of semiautomated or fully automated devices. Examples include the work of MacDonald et al. (148), who described the preparation of an eight-member quinolone library using the so-called Diversomer kit (15) and that of Shankar et al. (149), who reported an eight-member isoxazoline library using the same technology. An 80-member *N*-(alkoxyacyl)amino alcohol library using a 96-deep-well microtiter plate, a multichannel automated pipetter to deliver reagents and resin slurries, and a multichannel automated washing station has been reported by Krchnak et al. (150), while Wilson et al. (151) have demonstrated the use of a sophisticated liquid-handling robot modified to handle and deliver corrosive solutions and reagents for the synthesis of several 3-thio-1,2,4-triazole libraries. Boeijen et al. (152) reported the preparation of a 42-member hydantoin library using a semiautomated multiblock device (153). Krchnak (154) has recently reported the use of the Domino reaction block (20) to prepare several hundred amino acid–derived library individuals. The semiautomated synthesis of around 700 diketopiperazines from multicomponent reaction protocols using in-house developed instrumentation was reported by Szardenings et al. (155). Finally, Souers et al. (156) presented an 176-member β-turn mimetic library prepared using a semiautomated apparatus very similar to the multitube device described in the above-mentioned example. A 400-member phenyl stilbene library (157) was prepared using a semi-automated commercially available synthesizer (158) employing Wittig and Suzuki couplings. Han et al. (159) reported a small library of highly substituted thiophenes prepared via the prototype of a popular SP synthesizer (160).

6.3.4 Automated Parallel Synthesis: A Library of Tricyclic Compounds from the Tsuge Reaction

Bicknell et al. (161) has recently reported the successful transfer onto SP of a modified version of the Tsuge reaction (162) to produce a 96-member discrete library **L5** of tricyclic compounds by means of a commercially available automated SP multiple organic synthesizer (163). In this synthetic adaptation, the Tsuge reaction, that is, cycloaddition of pyridinium methilides with olefinic dipolarophiles, as shown in

Fig. 6.20, is followed in situ by reaction with a nitrile oxide to give tricyclic compounds containing a fused isoxazole ring.

The assessment of the chemistry was performed manually, as outlined in Fig. 6.21, and all of the reactions except the final cleavage from the resin were performed under an inert atmosphere. The methyl ester was replaced with an ester bond to Wang resin by first reacting the resin with chloroacetyl chloride (step a), then with pyridine (step b), to give the resin-bound pyridinium chloride salt **6.48**. Resin-bound **6.48** reacted readily with N-substituted maleimides and TEA (step c) to produce intermediates of general structure **6.49**, which were reacted in situ with an imidoyl chloride and TEA (step d) to give the expected resin-bound tetracycles **6.50**. Cleavage of the Wang linker produced the tricyclic compounds **6.51** in good yields (>70%) and purities (>80%) via fragmentation of the isoxazoline ring, aromatization, and hydrolysis of the oxime to the ketone (step e). A list of the characterized discretes is given in Fig. 6.21.

In this case, the SP protocol was suitable for complete automation as the reaction conditions throughout the synthesis were relatively mild, with the exception of the formation of the pyridinium methide and the nitrile oxide by addition of TEA. In a normal preparation, the base could be added to the reaction vessel dropwise; however, most automated SP synthesizers cannot simultaneously perform the stirring and addition of solutions. The automated protocol developed included the sequential addition of the N-substituted maleimides in dry THF, stirring for 1 min, addition of half of the required TEA in dry THF, stirring for 5 min, followed by addition of the second aliquot of base, and finally stirring for 1 h. The same procedure was also used for the addition of the imidoyl chloride and TEA (step d, Fig. 6.21). The purity and yields of the maleimide-fused indolizinium carboxylates obtained were generally good (average yields between 70 and 90%, HPLC purity between 65 and 97%).

A batch of resin-bound **6.48** was prepared in a reaction flask and divided into 50-mg portions in each well of the 96-well reaction block of the synthesizer. The synthesis of the library on the 96-well reaction block required around 8 h of instrument time and could have easily been repeated with different monomers in order to enrich the chemical diversity. Fully automated protocols provide access to 24-hr/day, seven-days/week operation and can thus maximize the use of an automated SP synthesizer; the same protocol has been subsequently adapted to a higher throughput SP synthesizer (164) and has provided a 3072-member discrete SP library of tricyclic compounds (165).

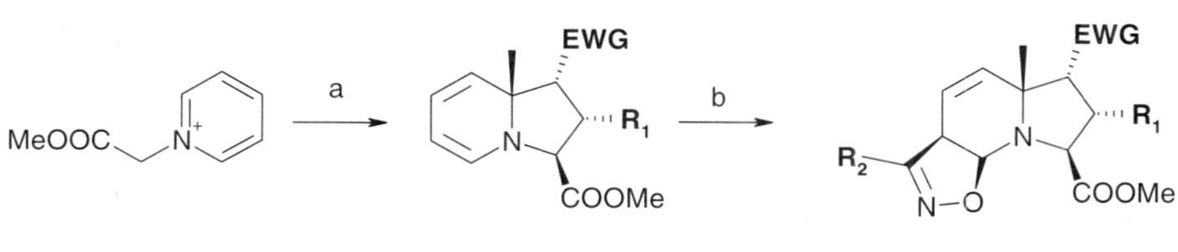

a: TEA, **EWG**CH=CH$\mathbf{R_1}$; b: TEA, Cl($\mathbf{R_2}$)C=NOH.

Figure 6.20 Tsuge reaction.

Other recent reports of the automated synthesis of SP libraries of discretes include those of Wilson et al. (166), who prepared a library of more than 1000 aminohydantoins from α-hydrazino amino acids, amines, and aldehydes; Perumattam et al. (167), who reported a 200-member library based on an anhydride template using anhydrides, primary amines, and α-amino acids; Smith et al. (168), who described the synthesis of a library of more than 1000 piperazinediones from α-amino acids; Crawshaw et al. (169), who presented a >200-member library of cyclohexanones from maleimides, nitrostyrenes, and aminobutadienes; Shao et al. (170), who described a 96-member library of quinazolinediones from anhydrides and amines; Lebl et al. (21), presenting a 30,816-member

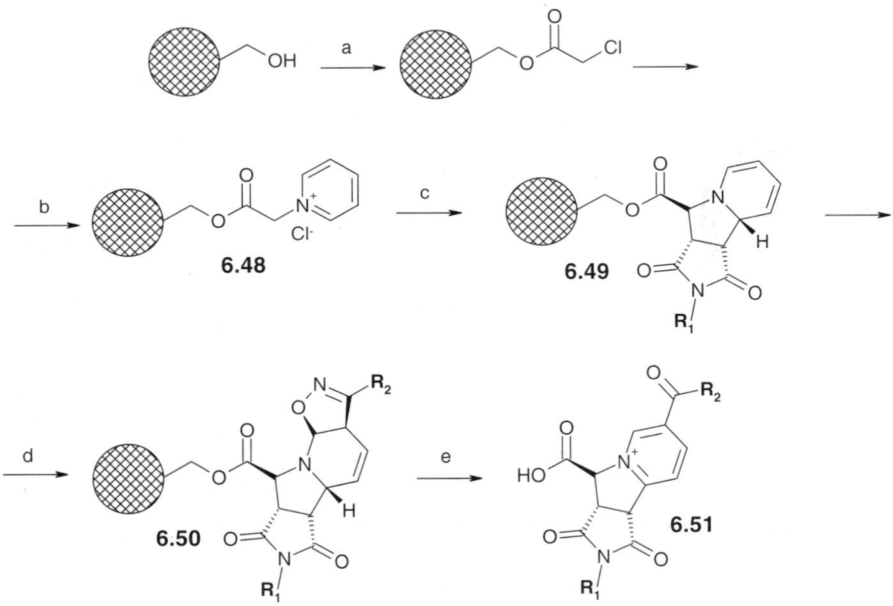

a: $ClCH_2COCl$, DCM, 0°C, 2 hrs; b: Py, DCM, rt, 24 hrs; c: N-subst. maleimide, dry THF, rt, 1', then TEA, rt, 1 hr; d: $R_2CClNOH$, dry THF, rt, 1', then TEA, rt, 2 hrs; e: TFA/DCM 1/1, rt.

Discretes prepared during the chemical assessment (R_1, R_2, yield, HPLC purity):

Me, Ph, 76%, 63%
Me, 3-PhOPh, 80%, 90%
Me, 4-tBuPh, 81%, 96%
Ph, 4-tBuPh, 83%, 79%
4-AcPh, 4-tBuPh, 92%, 83%
$PhCH_2$, 4-tBuPh, 88%, 76%

Figure 6.21 SP chemistry assessment for the tricyclic library **L5** from resin-bound pyridinium methilide **6.48**.

discrete library of tetrahydroisoquinolinones prepared both by surface suction and by tilted centrifugation (29) using an in-house developed automated synthesizer.

6.3.5 Automated Parallel Synthesis: High-Throughput Parallel Optimization of SP Reaction Conditions

A recent review by Porco et al. (171) highlighted the use of an automated SP chemistry development workstation (172) for the rapid optimization of SP reaction conditions in an automated parallel manner. Two examples will be described here: first, the influence of temperature on the generation of ureas from an oxime resin and, second, optimization of the length of time of reaction of the Suzuki biaryl coupling on an Argogel–Rink resin.

The SP reaction scheme for the generation of a *urea* library is shown in Fig. 6.22. One percent of cross-linked PS resin was reacted with *p*-nitrobenzoyl chloride, and

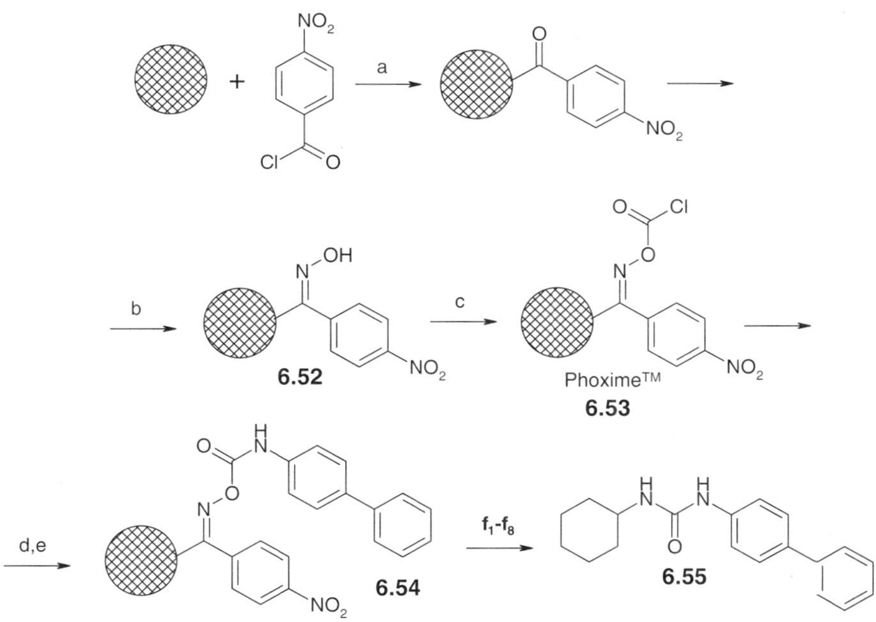

a: $AlCl_3$, DCM; b: $HONH_2$.HCl, Py, EtOH, D; c: triphosgene, DCM, rt; d: resin split in eight portions; e: 4-biphenylNH_2, dry DCM, rt, 16 hrs; **f_1-f_8**: cyclohexylNH_2, dry toluene, heating (b_1=50°C, then 10 increments up to b_8=120°C)

Recovered **6.53**: 35% (e_1), 55% (e_2), 72% (e_3), **80% (e_4)**, 75% (e_5), 82% (e_6), 84% (e_7), 86% (e_8).

Selected temperature for library synthesis: **80°C**

Figure 6.22 SP automated chemistry assessment for the urea library **L6** from Phoxime resin **6.53**.

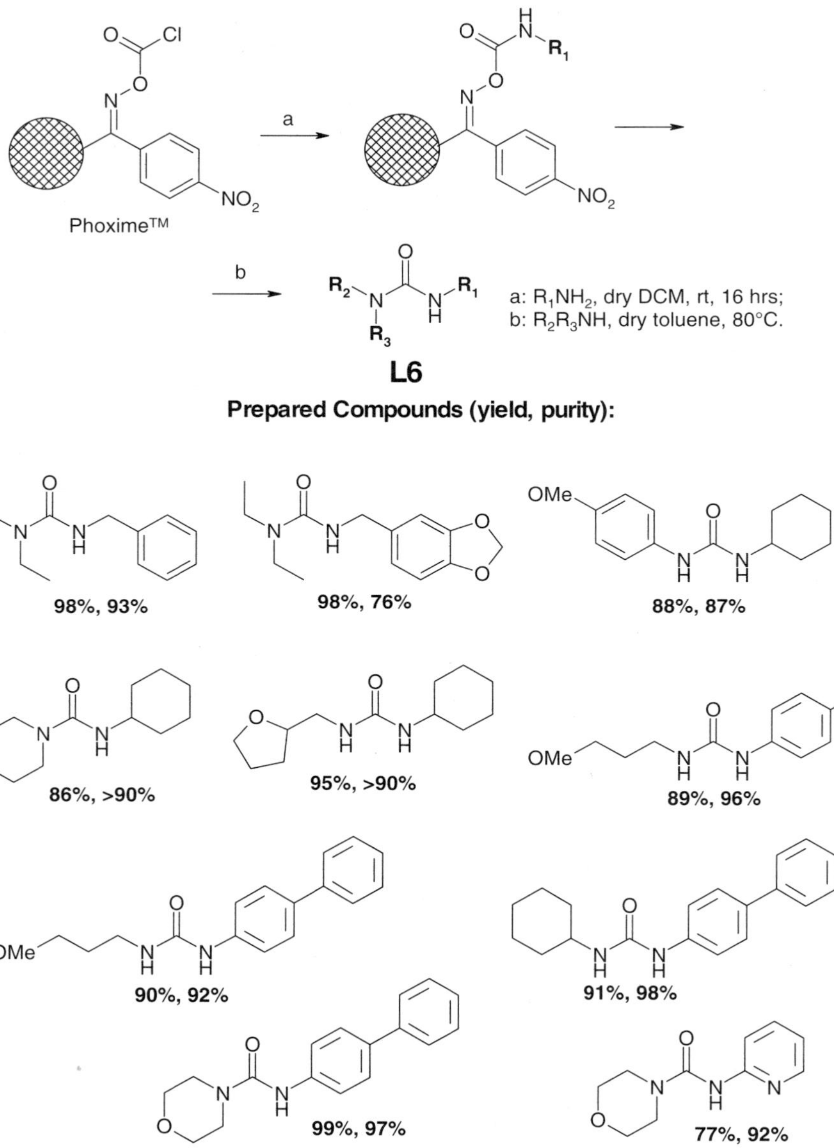

Figure 6.23 General structure of the SP urea library **L6** and selected library individuals.

the corresponding oxime resin was prepared in a 100-g scale according to a published procedure (steps a and b) (173). This resin was converted to Phoxime resin (step c) (174), which was then to be reacted with primary amines to provide the first class of resin-bound monomer. In this scheme, the formation of the urea with simultaneous cleavage from the support was the final step of the synthesis, which required optimization of the reaction temperature among other parameters. Eight 200-mg aliquots of Phoxime resin were placed in eight of the 24 independent reactors of the synthesizer and then coupled for 2 h at room temperature with 4-biphenylamine as a single standard (steps d and e, Fig. 6.22). The intermediate resin-bound carbamate **6.54** was successfully characterized. After the usual washing cycles, cyclohexylamine was added and the eight reaction mixtures were heated at different temperatures ranging from 50 to 120 °C (steps f_1–f_8). This experiment gave higher yields of recovered urea **6.55** with a temperature of 80 °C or above (Fig. 6.22). A small 10-member array **L6** was prepared at 80 °C, and the optimized conditions proved to be quite general for the production of ureas **6.53**; the structure and the quality of each library individual are reported in Fig. 6.23.

In the second example, optimization of a *Suzuki coupling* on SP was carried out. Two boronic acids were coupled with resin-bound iodide **6.56** (Fig. 6.24) at 90 °C for increasing reaction times of 1, 3, 6, and 10 h and the yields of the products **6.58** and **6.59** were calculated after cleavage of the product from the resin. The results are shown in Table 6.3. The less hindered thiophene boronic acid reacted completely after 1 h, while the more hindered tolyl derivative required 3 h to drive the reaction to completion. In this case it was found that while 3 h was an optimal length of time for both substrates, longer reaction times tended to decrease the reaction yields.

This type of SP synthesizer is ideal for more complex optimization studies in which multivariate optimizations, where more than one variable is changed at a time, could rapidly lead to optimized general SP reaction conditions. The high cost of this and other fully automated SP synthesizers may well be balanced by their regular application to library production and multivariate optimization in a dedicated combinatorial

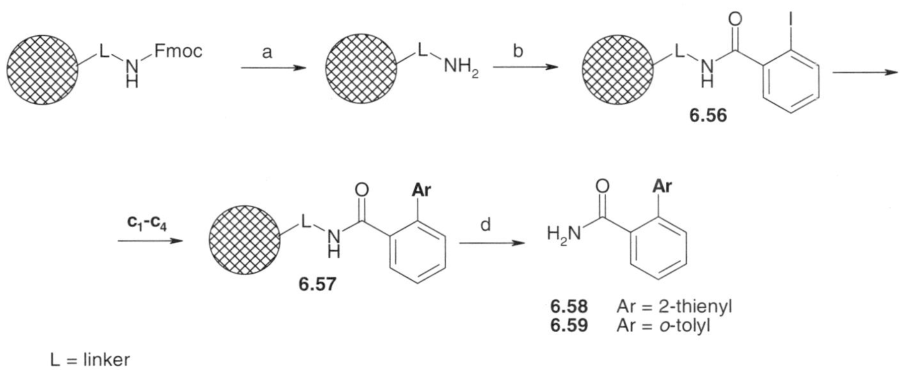

a: 20% piperidine in DMF, rt; b: 2-IBzCOOH, DIC, HOBt, rt; **c_1-c_4**: $ArB(OH)_2$, $PdCl_2(PPh_3)_2$, Na_2CO_3, DME, NMP, 90°C, 1 hr (**c_1**), 3 hrs (**c_2**), 6 hrs (**c_3**), 10 hrs (**c_4**); d: TFA/H_2O 95/5, rt, 30'.

Figure 6.24 SP automated chemistry assessment for the Suzuki coupling of supported aryl iodine **6.56** with boronic acids.

TABLE 6.3 Time Dependence of SP Suzuki Coupling to Give Biaryl Amides 6.58 and 6.59

Final Compound	Reaction Time (hr)	Yield[a]	Purity[b]
6.58	1	90	100
6.59	1	99	90
6.58	3	95	100
6.59	3	100	98
6.58	6	91	99
6.59	6	96	98
6.58	10	67	100
6.59	10	62	99

[a]Yield from **6.56**.
[b]Calculated by HPLC.

synthesis laboratory. A recent report (175) has presented the SP chemical assessment and the synthesis of a small array of 1,2-diamines in large quantities (>100 mg) using the same synthesizer.

6.4 NEW TRENDS IN SOLID-PHASE DISCRETE LIBRARY SYNTHESIS

6.4.1 Solid Supports for High-Throughput Organic Synthesis: Two-Dimensional SP Synthesis

Recently, the need for high-quality medium–large SP libraries of discretes has significantly increased, mostly because of problems encountered when trying to derive useful and reliable information from large SP libraries of pools, thus giving rise to general skepticism in many chemists. While we will address the issue of preparing quality pool libraries on SP in the next chapter, two emerging approaches for the synthesis of discrete libraries on dimensional supports will be presented here and in the next section. Resin beads, in fact, correspond to a zero-dimensional support in that the reactor is completely symmetrical and its portions undistinguishable; planar solid supports, covered in the section, have a bi-dimensional architecture which confers specific features to the SP as do mono-dimensional, thread-like supports covered in the next section.

Several years ago Fodor et al. (176) introduced the so-called VLSIPS (very large scale immobilized peptide synthesis) technique using a combination of α-amino acid–derivatized glass surface bearing amino groups protected with photocleavable protecting groups and photolithographic masks to direct the peptide synthesis. The principle of this technique is depicted in Fig. 6.25, where only the protecting groups X exposed to the light are removed and functionalized with the first α-amino acid. The production of a library using this method is shown in Fig. 6.26, where a spatially encoded library of 16 tetrapeptides is prepared using eight building blocks A–H (two

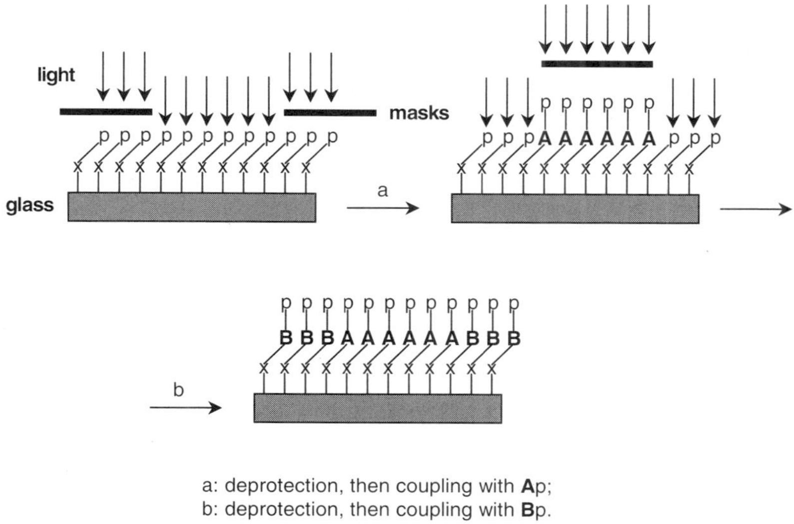

Figure 6.25 Basic principles of the SP very large scale immobilized peptide synthesis VLSIPS technique.

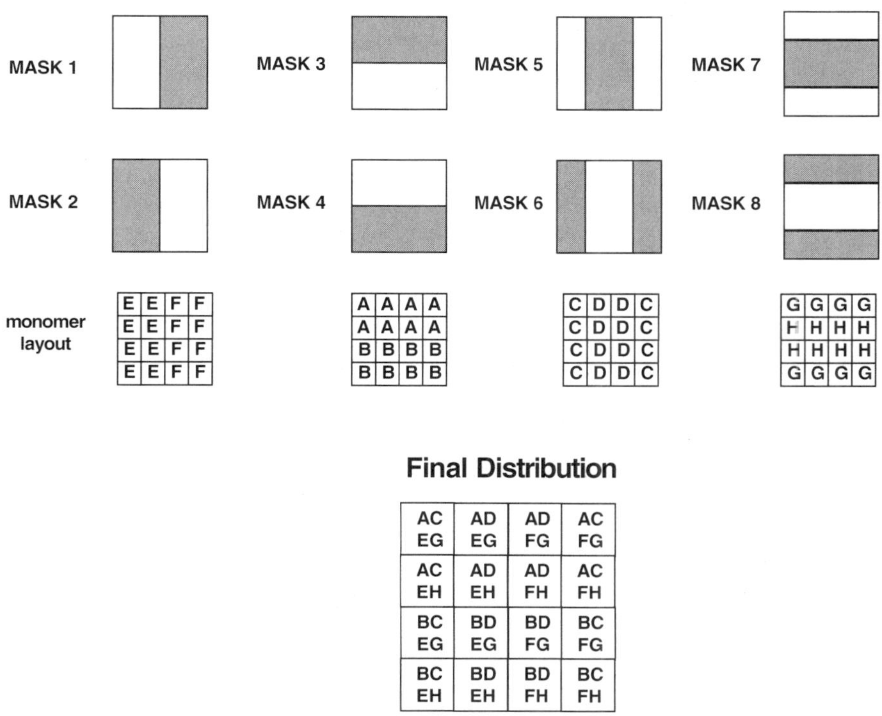

Figure 6.26 SP synthesis of a 16-member library using the VLSIPS masking method.

for each coupling) and eight masking/coupling steps 1–8. VLSIPS has received a lot of attention for the synthesis of oligomers, especially for nucleotides, and has recently been reviewed (177, 178). Advantages of this method are the high throughput and the miniaturization of the synthetic device [e.g., a recent example reported the preparation of more than 10,000 oligonucleotides on chips just slightly larger than 1 cm^2 (179)]; its application to materials science library will be thoroughly illustrated in Chapter 11.

A novel Maskless Array Synthesizer (MAS) based on an individually addressable digital micromirror array used in computer display projection systems has been recently reported for the synthesis of tens of thousands DNA sequences, each segregated on a 16 μm^2 area (180). Its usefulness for small organic molecule SPS, though, has largely been prevented by the impossibility of cleaving the final products to test them in solution without the concomitant loss of spatial encoding and by the limited range of chemical conditions that are applicable to this method. Alternative cheaper planar supports such as *paper* (181, 182) or *polyvinylidene fluoride* (183) have also been reported for peptide synthesis. The latter showed good stability and greater flexibility under different reaction conditions when compared to glass.

Recently two groups (184–186) have reported a modification to the above methods in which portions of resin beads are sandwiched between two inert woven polypropylene sheets that are then fused together to give a so-called *laminar solid phase*. The use of this construct as a support for SP chemistry is shown in Fig. 6.27 in a hypothetical example using five sheets. Each sheet is divided into 50 small squares (typically 2 cm^2) that are unequivocally marked with an inert ink (a–e for the sheets, 1–50 for the positions, step 1). Each sheet is then reacted with a different monomer of the first set (A_1–A_5, step 2), the columns of each sheet are cut and collated according to their position on each sheet, and each collated pool of columns is reacted with a different monomer of the second set (B_1–B_{10}, step 3). Finally, the columns are cut into squares, and those bearing the same initial numbers are pooled together, threaded, and reacted with the third monomer set (1–10, 11–20, 21–30, 31–40, 41–50, C_1–C_5, step 4, Fig. 6.27). By increasing the number of sheets and/or the squares in a sheet, it is possible to prepare tens of thousands of discretes using cheap materials. The other main advantage to this technique is the use of resin beads embedded in a fully permeable reaction support, which allows to transfer directly to laminar library production the chemical assessment performed on single resin portions in a conventional protocol.

A more sophisticated apparatus, called the *chemical inkjet printer*, has been reported by Lemmo et al. (187) and consists of a sheet of polypropylene (ChemSheet) divided into 2304 (48 × 48) shallow wells the bottoms of which are grafted with chemical functionalities. The principle with which SP parallel synthesis was carried out using this technique is identical to that of the laminar SP shown in Fig. 6.27 except that by using 48 ChemSheets and three sets of 48 monomers, more than 110,000 compounds could be prepared in parallel and then released into the wells. Each well, with a capacity of 8 μL, could produce around 5 nmol of final compound, and if we consider a classical solution dispensing system, with a dispensing time of 2 s per well, around 75 min would be necessary to fill all the wells with a single solution. Moreover,

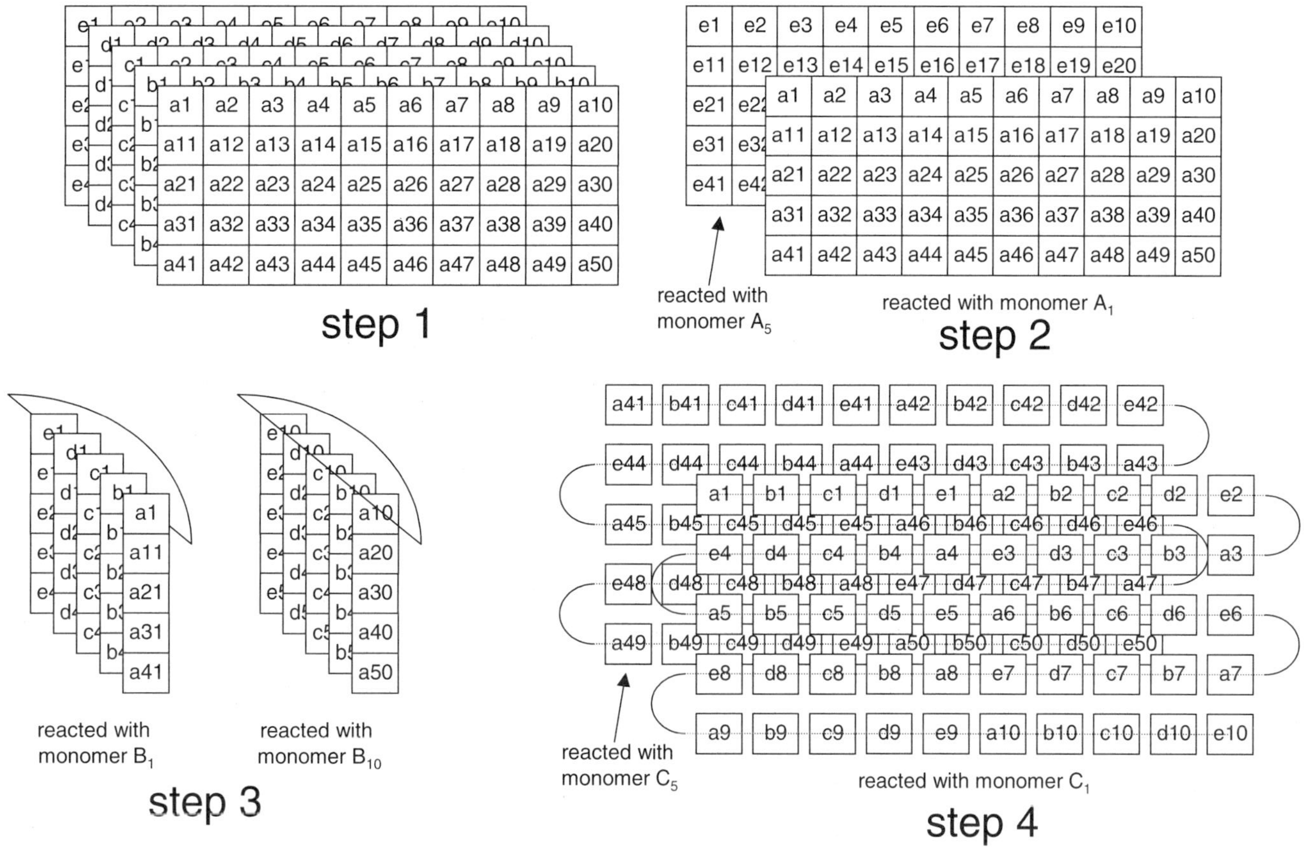

Figure 6.27 SP synthesis of a 250-member library on two-dimensional solid support.

the careful control of such a small volume with no cross-contamination of other cells would be problematic.

To get around these limitations, the authors used an inkjet microdispenser, first developed for printing applications but that has been widely used for scientific purposes (188–194). A pressurized solvent/reagent source is connected to a solenoid inkjet valve via a ribbon cable, and the valve is connected in the same way to a restricted orifice. When the valve is opened for a few milliseconds, as the pressurized solution flows into the cable and finally reaches the orifice bottleneck, its speed increases significantly, causing a jet of liquid. This is a noncontact technique; therefore, the orifice does not require rinsing prior to sampling again; the dispensing time is extremely short (around 10–30 msec); the system is extremely reliable in terms of reproducibility of aliquots even in the submicromolar range, and the cost of the valve system is not unreasonable. This technique shows considerable promise for combinatorial applications.

The authors realized and built an apparatus for parallel synthesis, as shown in Fig. 6.28, where two blocks of 48 valves were grouped into a dispensing head and each valve was connected to an orifice: The resulting 96-port microdispenser head was directed by a computer that controlled the parallel dispensing of chemicals from the reagent delivery system. Each ChemSheet plate was placed below the dispenser head. The microdispenser was made up of two manifolds per block, each comprising 24 reagent chambers that were connected to the valves. A single reagent reservoir is represented and schematically inserted into the whole channel structure in Fig. 6.29. The delivery system could be connected to the pressurized gas line to deliver reagents or to a vacuum line to recover them, and a common washing line was used to clean the whole ChemSheet at the end of the synthesis.

While the synthesis of an SP library has not yet been reported (187), the system has been carefully tested and the compatibility and reproducibility of dispensing various

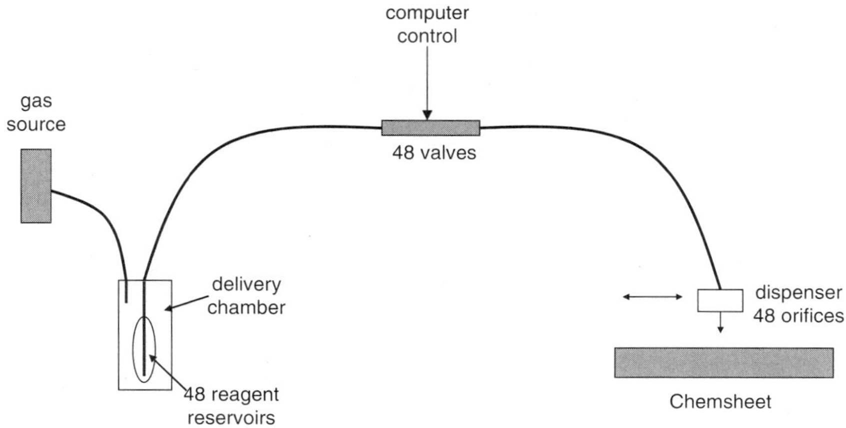

Figure 6.28 Schematic representation of the channel of an inkjet SP library synthesizer.

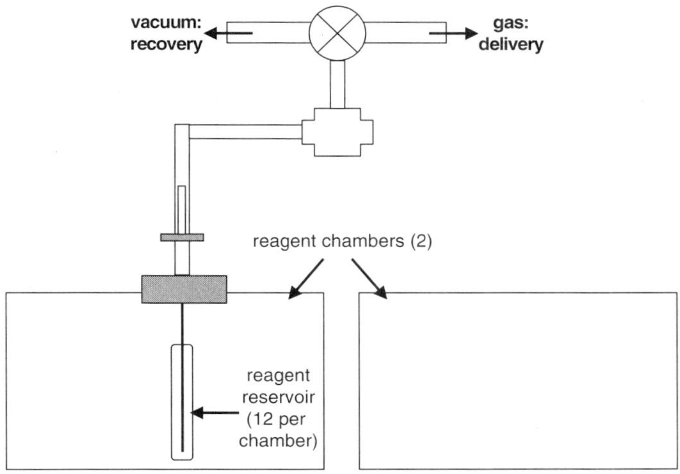

Figure 6.29 Details of a reagent delivery manifold in an inkjet microdispenser.

common solvents for organic reactions have been demonstrated. The complete filling of a ChemSheet plate by the microdispenser was successfully performed in less than 10 sec, thus theoretically allowing the processing of hundreds of thousands samples in parallel in very short amounts of time. It is likely that this technique will be more widely used in the future, together with some other miniaturized systems described below, once they have been adapted to accept the normal range of reaction conditions used in parallel organic synthesis. A comprehensive review dealing with microdispensing instrumentation and protocols in drug discovery has recently been published (195).

6.4.2 Planar Solid Supports for High Throughput Organic Synthesis: Mono-Dimensional SP Synthesis

A mono-dimensional support for SPS has been presented recently by Schwabacher et al. (196). A flexible thread grafted with appropriate chemical functionalities is wrapped around a cylinder in a spiral layer (cylinder 1, step 1, Fig. 6.30), placing each thread turn at the same distance. Four parallel lines are drawn lengthwise along the cylinder, and a wax barrier is deposed with a hot glue gun to segregate each of the four cylinder regions. The first monomer set (four representatives) is coupled to each cylinder region, and cross-contamination is prevented by the wax barrier (Fig. 6.30). The thread is then unwrapped and prepared for the second combinatorial step, which his performed on a cylinder with a different diameter (cylinder 2, step 2, Fig. 6.30). The second monomer set (six representatives) is coupled as seen before, and each cylinder region has the same width as before, but the different cylinder dimensions and number of monomers cause the formation of different adducts in the thread regions (Fig. 6.30). The authors used a cotton thread grafted with an amine function, and performed the SP synthesis of a 35-member hepta/octapeptide library Ac-**X**$_2$-**X**$_1$-Pro-Gln-Phe-Ala-Ala-Ala (191) with two randomized positions (**X**$_1$ = His, Ser, Asp, Ala or Phe; **X**$_2$ = Leu, Phe, His, Glu, Gly, benzoyl or acetyl), using for **X**$_1$ a 5 cm-circumference cylinder and for **X**$_2$ a 7-cm-circumference cylinder and dividing them in 1 cm-wide regions. Multiple couplings, both on the whole thread for the common residues and per thread region for **X**$_1$ and **X**$_2$, were performed to increase the coupling efficiency, and capping with acetic anhydride was done to abort deletion sequences.

The use of several cylinders allows the performance of multi-step SP syntheses and the production of small to large continuous libraries which are positionally encoded, in that each individual is determined on a specific thread segment simply by "unwrapping" the cylinders used in each synthetic step. Each monomer is repeated at constant distances on the thread, and the direct read-out of *on-thread assays* (equivalent to on-bead screening, see section 7.2.3) can be immediately related to SAR via Fourier Transform of the detection signal. Positive and negative effects of monomers, or of modifications in specific areas of the library individuals are thus immediately highlighted and further optimization efforts can be planned. The authors screened the peptide library for its affinity to fluorescent-labeled streptavidin, having inserted the known His-Pro-Gln binding motif in the library structure; on-thread screen confirmed the presence of positives in correspondence of **X**$_1$ = His and was able to discriminate among secondary effects of **X**$_2$ substituents (196).

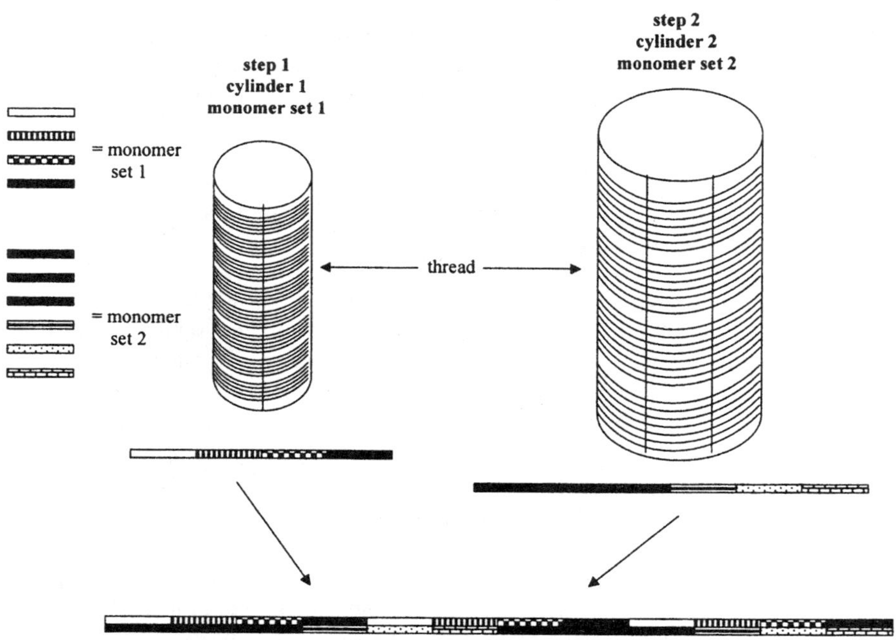

Figure 6.30 SP synthesis of a library on monodimensional solid supports.

6.4.3 Single-Bead Parallel Synthesis: Miniaturization

Apart from the chemical inkjet printer, the syntheses of SP discrete libraries reported in the literature typically require quantities of between a few milligrams and hundred of milligrams of resin for each library individual corresponding to at least several thousand beads functionalized with the same structure. Considering each resin bead as a single entity (i.e., a *microreactor,* the microreaction vessel of solid-phase organic chemistry), *miniaturization* of SP chemistry down to the scale of the individual bead would have two major effects. First, the repeated synthesis of medium–large libraries would not consume significant amounts of precious monomers, resins, and reagents due to the reduced amounts of materials necessary. Second, providing that a suitable automated single-bead synthesizer can be built and used, the throughput of the SPS could be theoretically increased to allow the preparation of very large discrete libraries. For example, a single reaction block containing hundreds of beads could have the dimensions of a chip and a normal hood could host the parallel synthesis of hundreds of thousands or even millions of individual compounds.

Rapp recently reported (197, 198) the use of one to four 800-μm Tentagel *macrobeads in a capillary tube microreactor* containing 25 μL of liquid with a sinter glass filter at the bottom for washing and filtration of reaction solutions. A small array of hydantoins was prepared, monitoring the synthesis by gel-phase ^{1}H-NMR and finally cleaved to give pure compounds in 50–100-nmol amounts. While the reduction in the quantities of building block needed for the preparation of a library of discretes on single macrobeads is significant, the necessary equipment to handle reactions in capillary tubes restricts the throughput of the synthesis to relatively small numbers of compounds.

This latter aspect is being addressed using microchips as the SP microreactors. Several leading companies are developing this technology for library synthesis and other applications. In a recent presentation (199) the *lab-in-a-chip concept* was illustrated, with examples of *six-well chips*, which are schematically depicted in Fig. 6.31. Every microwell has a multilayer structure with a total volume of 300 nL, can host up to 10 average SP beads, or up to 3 macrobeads, and can produce up to several nanomoles of compound per well. It has a drain for waste removal and inlets for fluid delivery (Fig. 6.31). The dispensing of nanoliter aliquots of reagent solutions is controlled by *electrohydrodynamic pumps* (200–202), whereas *electroosmotic valves* control the delivery of solvents and reagents. The apparatus is compatible with most organic solvents, and the closed system prevents evaporation of the solvents. More complex versions of the microchip reactors containing 144 wells have also been presented together with an automated system for the massive parallel synthesis of SP discrete libraries of more than 10,000 members using one to five beads per microwell. The synthetic device consists of a liquid-delivery plate on top, formed by a network of channels, reservoirs, and electric contacts, and a liquid-recipient plate underneath that carries the 144 well-chips on a standard microplate footprint. A multichip reaction block can be controlled together with the other components of the automated system (e.g., bead handling, filtration, and liquid handling) and can perform even the cleavage

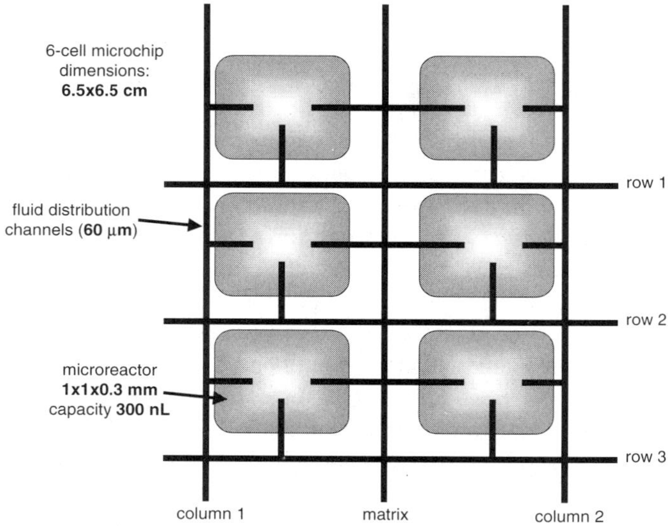

Figure 6.31 Architecture of a six-well microchip for SP synthesis.

and the repeated test of a library, including archiving functions for copies of the library individuals to be stored before the assay.

The synthesis of a model library of 100 amides from anhydrides on microchips has been described with quality control indicating excellent yields of pure amides and no cross-contamination of the wells (199). This test validated the chip structure, the seals of the system, and the electrohydrodynamic pumping system in the presence of solvents such as DMF and methanol and reagents such as the anhydrides, piperidine, and DIPEA. More challenging chemistries and larger libraries have also been presented.

Many issues still remain to be resolved in the field of microchip-based SP synthesis. While the reactors can be heated or cooled, a reliable and, possibly, independent temperature control over the wide range of temperature often required in organic synthesis is still problematic. The creation of parallel analytical techniques able to monitor the reactions and determine the structures without consuming most of the library individuals is awaited, and poor solubility of reagents in the reaction mixtures can compromise the quality of the whole synthesis. It is also clear that the information management of a massive parallel system (>100,000 individuals) where any compound or monomer addition is automatically tracked represents a major task. Nevertheless, this represents an important avenue of research for the future (203), and technological improvements will probably allow the eventual construction of reliable and flexible microchip-based library synthesis equipment.

The area of microchips/microreactors is very dynamic, and many research groups (including nonchemical groups) are active in this field. The interested reader is referred to details supplied in references 204–213, but several fields are being actively exploited: devices which allow handling, delivering and mixing of microvolumes (214, 215); analytical methods in a microenvironment (216); micro-assays for library testing (217, 218); micro-contact printing for chemical library synthesis (219). A recent review (220) summarizes briefly but exhaustively the state of the art for chemical miniaturization.

REFERENCES

1. Southern, E. M. and Maskos, U., *J. Biotechnol.* **35**, 217–227 (1994).
2. Meyers, H. V., Dilley, G. J., Powers, T. S., Winssinger, N. A. and Pavia, M. R., *Methods Mol. Cell. Biol.* **6**, 1–7 (1996).
3. Sarshar, S. and Mjalli, A. M. M., *Annu. Rep. Combi. Chem. Mol. Div.* **1**, 19–29 (1997).
4. Labaudiniere, R. F., *Drug Disc. Today* **3**, 511–515 (1998).
5. Hall, S. E., in *Combinatorial Chemistry and Molecular Diversity in Drug Discovery*, E. M. Gordon and J. Kerwin (Eds.). Wiley-Liss, New York, 1998, pp. 291–306.
6. Chucholowski, A., Masquelin, T., Obrecht, D., Stadlwieser, J., and Villalgordo, J. M., *Chimia* **50**, 525–530 (1996).
7. Zhang, H. -C. and Maryanoff, B. E., *J. Org. Chem.* **62**, 1804–1809 (1997).
8. Brown, S. D. and Armstrong, R. W., *J. Org. Chem.* **62**, 7076–7077 (1997).
9. Rice, R. L., Rusnak, J. M., Yokokawa, F., Yokokawa, S., Messner, D. J., Boynton, A. L., Wipf, P. and Lazo, J. S., *Biochemistry* **36**, 15965–15974 (1997).
10. Lampe, T. F. J., Weitz-Schmidt, G. and Wong, C.-H., *Angew. Chem. Int. Ed.* **37**, 1707–1711 (1998).
11. Williard, R., Jammalamadaka, V., Zava, D., Benz, C. C., Hunt, C. A., Kushner, P. J. and Scanlan, T. S., *Chem. Biol.* **2**, 45–51 (1995).
12. Tempest, P. A., Brown, S. D. and Armstrong, R. W., *Angew. Chem., Int. Ed. Engl.* **35**, 640–642 (1996).
13. Collins, J. L., Blanchard, S. G., Boswell, G. E., Charifson, P. S., Cobb, J. E., Henke, B. R., Hull-Ryde, E. A., Kazmierski, W. M., Lake, D. H., Leesnitzer, L. M., Lehmann, J., Lenhard, J. M., Orband-Miller, L. A., Gray-Nunez, Y., Parks, D. J., Plunkett, K. D. and Tong, W.-Q., *J. Med. Chem.* **41**, 5037–5054 (1998).
14. Smith, J. M., Gard, J., Cummings, W., Kaniszai, A. and Krchnak, V., *J. Comb. Chem.* **1**, 368–370 (1999).
15. DeWitt, S. H., Kiely, J. S., Stankovic, C. J., Schroeder, M. C., Cody, D. M. R., Pavia, M. R., *Proc. Natl. Acad. Sci. USA* **90**, 6909–6913 (1993).
16. DeWitt, S. H., Schroeder, M. C. and Stankovic, C. J., *Am. Lab.*, 26AA–26DD (1995).
17. Ghosh, S., Hogan, E. and DeWitt, S., *Drug News Perspect.* **9**, 82–86 (1996).
18. DeWitt, S. H. and Czarnik, A. W., *Acc. Chem. Res.* **29**, 114–122 (1996).
19. Gani, D., Akhtar, M., Kroll, F. E. K., Smith, C. F. M. and Stones, D., *Tetrahedron Lett.* **38**, 8577–8580 (1997).
20. Krchnak, V. and Padera, V., *Bioorg. Med. Chem. Lett.* **8**, 3261–3264 (1998).
21. Lebl, M., Krchnak, V., Ibrahim, G., Pires, J., Burger, C., Ni, Y., Chen, Y., Podue, D., Mudra, P., Pokorny, V., Poncar, P. and Zenisek, K., *Synthesis*, 1971–1978 (1999).

22. Porco, J. A. Jr., Deegan, T., Devonport, W., Gooding, O. W., Heisler, K., Labadie, J. W., Newcomb, B., Nguyen, C., van Eikeren, P., Wong, J. and Wright, P., *Mol. Diversity* **2**, 197–206 (1997).
23. Metivier, P., Josses, P., Bulliot, H., Corbet, J. P. and Joux, B., *Chemometr. Intell. Lab. Syst.* **17**, 137–143 (1992).
24. Gausepohl, H., Boulin, C., Kraft, M. and Frank, R. W., *Pept. Res.* **5**, 315–320 (1992).
25. Zuckermann, R. N., Siani, M. A. and Banville, S. C., *Lab. Rob. Autom.* **4**, 183–192 (1992).
26. Bartak, Z., Bolf, J., Kalousek, J., Mudra, P., Pavlik, M., Pokorny, V., Rinnova, M., Voburka, Z., Zenisek, K. et al., *Methods: Meth. Enzymol.* **6**, 432–437 (1994).
27. Brennan, T., Biddison, G., Frauendorf, A., Schwarcz, L., Keen, B., Ecker, D. J., Davis, P. W., Tinder, R. and Swayze, E. E., *Biotechnol. Bioeng.* **61**, 33–45 (1998).
28. Sophas M Solid-Phase Synthesizer, Zinsser Analytic. *http://www.zinsser-analytic.com/*
29. Lebl, M., *Bioorg. Med. Chem. Lett.* **9**, 1305–1310 (1999).
30. DeWitt, S. H. and Czarnik, A. W., *Curr. Opin. Biotechnol.* **6**, 640–645 (1995).
31. Hardin, J. H. and Smietana, F. R., *Mol. Diversity* **1**, 270–274 (1996).
32. Harness, J. R., in *Molecular Diversity and Combinatorial Chemistry*, Chaiken, I. M. and Janda, K. D. (Eds.). ACS, Washington, DC, 1996, pp. 188–198.
33. DeWitt, S. H., *Annu. Rep. Combi. Chem. Mol. Div.* **1**, 69–77 (1997).
34. Bondy, S. S., *Curr. Opin. Drug Disc. Dev.* **1**, 116–119 (1998).
35. Merritt, A. T., *Drug Disc. Today* **3**, 505–510 (1998).
36. Rivero, R. A., in *A Practical Guide to Combinatorial Chemistry*, A. W. Czarnik and S. H. DeWitt (Eds.). ACS, Washington, DC, 1997, pp. 281–307.
37. Mjalli, A. M. M., in *A Practical Guide to Combinatorial Chemistry*, A. W. Czarnik and S. H. DeWitt (Eds.). ACS, Washington, DC, 1997, pp. 327–354.
38. Calvert, S., Stewart, S. P., Swarna, K. and Wiseman, J. S., *Curr. Opin. Drug Discovery Dev.* **2**, 234–238 (1999).
39. Hird, N. W., *Drug Discovery Today* **4**, 265–274 (1999).
40. Antonenko, V. V., in *Combinatorial Chemistry and Technology: Principles, Methods, and Applications*, S. Miertus and G. Fassina (Eds.). Marcel Dekker, New York, 1999, pp. 205–232.
41. Cheng, C. C. and Chu, Y.-H., *J. Comb. Chem.* **1**, 461–466 (1999).
42. Zeng, L., Burton, L., Yung, K., Shushan, B. and Kassel, D. B., *J. Chromatogr. A* **749**, 3–13 (1998).
43. Zeng, L., Wang, X., Wang, T. and Kassel, D. B., *Comb. Chem. High Throughput Scr.* **1**, 101–111 (1998).
44. Zeng, L. and Kassel, D. B., *Anal. Chem.* **70**, 4380–4388 (1998).
45. Burg, J., *GIT Spez. Sep.* **19**, 47–48 (1999).
46. de Biasi, V., Haskins, N., Organ, A., Bateman, R., Giles, K. and Jarvis, S., *Rapid Commun. Mass Spectrometry* **13**, 1165–1168 (1999).
47. Aubagnac, J.-L., Amblard, M., Enjalbal, C., Subra, G., Martinez, J., Durand, P. and Renault, P., *Comb. Chem. High Throughput Screening* **2**, 289–296 (1999).
48. Rahn, P. C. and Bickler, J. R., *Book of Abstracts, Pittcon '97*, Atlanta, GA, USA, 1997.
49. Goetzinger, W. K. and Kyranos, J. N., *Am. Lab.*, 27–37 (1998).

50. Zaramella, A., Gehanne, S., Dal Cin, M. and Conti, N., *personal communication*, April 1998.

51. Marshall, P. S., *Rapid Commun. Mass Spectrometry* **13**, 778–781 (1999).

52. Ventura, M. C., Farrell, W. P., Aurigemma, C. M. and Greig, M. J., *Anal. Chem.* **71**, 2410–2416 (1999).

53. Ventura, M. C., Farrell, W. P., Aurigemma, C. M. and Greig, M. J., *Anal. Chem.* **71**, 4223–4231 (1999).

54. Walk, T. B., Trautwein, A. W., Richter, H. and Jung, G., *Angew. Chem. Int. Ed.* **38**, 1763–1765 (1999).

55. Issakova, O. L., communication at *SRI Symposium*, Dallas, TX, March 1998.

56. Wang, T., Zeng, L., Strader, T., Burton, L. and Kassel, D. B., *Rapid Commun. Mass Spectrom.* **12**, 1123–1129 (1998).

57. Richmond, R. and Gorlach, E., *Anal. Chim. Acta* **390**, 175–183 (1999).

58. Richmond, R. and Gorlach, E., *Anal. Chim. Acta* **394**, 33–42 (1999).

59. Dal Cin, M., Zaramella, A., Conti, N. and Gehanne, S., *personal communication*, February 1998.

60. Asmus, P. A. and Landis, J. B., *J. Chromatog.* **316**, 461–472 (1984).

61. Kibbey, C. E., *Mol. Diversity* **1**, 247–258 (1996).

62. Kohler, M., Haerdi, W., Christen, P. and Veuthey, J.-L., *Trends Anal. Chem.* **16**, 475–484 (1997).

63. Hsu, B. H., Orton, E., Tang, S.-Y. and Carlton, R. A., *J. Chromatogr., B: Biomed. Sci. Appl.* **725**, 103–112 (1999).

64. Fujinari, E. M. and Courthaudon, L. O., *J. Chromatog. A.* **592**, 209–214 (1992).

65. Bizanek, R., Damon, J., Fujinari, M. and Fujinari, E. M., *Pept. Res.* **9**, 40–44 (1996).

66. Fitch, W. L., Szardenings, A. K. and Fujinari, E. M., *Tetrahedron Lett.* **38**, 1689–1692 (1997).

67. Taylor, E. W., Qian, M. G. and Dollinger, G. D., *Anal Chem.* **70**, 3339–3347 (1998).

68. Gehanne, S., Conti, N., Zaramella, A. and Dal Cin, M., *personal communication*, March 1998.

69. Bayer, E., Albert, K., Nieder, M., Grom, E. and Keller, T., *J. Chromatog.* **186**, 497–507 (1979).

70. Nicholson, J. K., Holmes, E., Sidelmann, U., Lindon, J. C. and Wilson, I. D., *Pharm. Sci.* **2**, 127–130 (1996).

71. Albert, K., *J. Chromatogr. A* **703**, 123–147 (1995).

72. Bruker Analytik GmbH. *http://www.bruker.de/analytic/nmr-dep/nmr-dep.htm*

73. Lindon, J. C., Farrant, R. D., Sanderson, P. N., Doyle, P. M., Gough, S. L., Spraul, M., Hofmann, M. and Nicholson, J. K., *Magn. Reson. Chem.* **33**, 857–863 (1995).

74. Chin, J., *J. Org. Chem.* **63**, 386–390 (1998).

75. Hamper, B. C., Snyderman, D. M., Owen, T. J., Scates, A. M., Owsley, D. C., Kesselring, A. S. and Chott, R. C., *J. Comb. Chem.* **1**, 140–150 (1999).

76. OpenLynx Diversity Software, Micromass, Waters Corporation. *http://www.micromass.co.uk/*

77. HP1100 Series Combinatorial Chemistry Analysis System, Hewlett Packard. *http://chem.external.hp.com/cag/products/pharmasol.asp*

78. Gorlach, E., Richmond, R. and Lewis, I., *Anal. Chem.* **70**, 3227–3234 (1998).

79. Richmond, R., Gorlach, E. and Seifert, J.-M., *J. Chromatogr. A* **835**, 29–39 (1999).

80. Huang, N., Siegel, M. M., Kruppa, G. H. and Laukien, F. H., *J. Am. Soc. Mass Spectrom.* **10**, 1166–1173 (1999).

81. Tong, H., Bell, D., Tabei, K. and Siegel, M. M., F. H., *J. Am. Soc. Mass Spectrom.* **10**, 1174–1187 (1999).

82. Huang, N., Siegel, M. M., Muenster, H. and Weissenberg, K., *J. Am. Soc. Mass Spectrom.* **10**, 1212–1216 (1999).

83. Dal Cin, M. and Conti, N., *personal communication*, March 1998.

84. Kyranos, J. N. and Hogan, J. C., Jr., *Anal. Chem.* **70**, 389A–395A (1998).

85. Dulery, B. D., Veme-Mismer, J., Wolf, E., Kugel, C. and Van Hijfte, L., *J. Chromatogr., B: Biomed. Sci. Appl.* **725**, 39–47 (1999).

86. Swali, V., Langley, G. J. and Bradley, M., *Curr. Opin. Chem. Biol.* **3**, 337–341 (1999).

87. Sepetov, N. and Issakova, O., in *Combinatorial Chemistry and Technology: Principles, Methods, and Applications*, S. Miertus and G. Fassina (Eds.). Marcel Dekker, New York, 1999, pp. 169–203.

88. Iyer, R. P., Yu, D., Xie, J., Zhou, W. and Agrawal, S., *Bioorg. Med. Chem. Lett.* **7**, 1443–1448 (1997).

89. Lebl, M., Pires, J., Poncar, P. and Pokorny, V., *J. Comb. Chem.* **1**, 474–479 (1999).

90. Kerschen, A., Kaniszai, A., Botros, I. and Krchnak, V., *J. Comb. Chem.* **1**, 480–484 (1999).

91. Kiselyov, A. S., Smith, L. II, Virgilio, A. and Armstrong, R. W., *Tetrahedron* **54**, 7987–7996 (1998).

92. Grieco, P. and Bahsas, A., *Tetrahedron Lett.* **29**, 5855–5858 (1988).

93. Hayes, T. K., Kiely, J. S., Forood, B. and Brady T. P., *Book of Abstracts, 213th ACS National Meeting*, ACS, Washington, DC, ACS, San Francisco, April 13th-17th, *ORGN-275* 1997.

94. Campos, P. J., Lamaza, I., Rodriguez, M. A. and Canal, G., *Tetrahedron Lett.* **38**, 6741–6744 (1997).

95. Cheng, Y.-S., Ho, E., Mariano, P. S. and Ammon, H. L., *J. Org. Chem.* **50**, 5678–5686 (1985).

96. Lucchini, V., Prato, M., Scorrano, G. and Tecilla, P., *J. Org. Chem.* **53**, 2251–2258 (1988).

97. Annunziata, R., Cinquini, M., Cozzi, F., Molteni, V. and Schupp, O., *Tetrahedron* **53**, 9715–9726 (1997).

98. Kobayashi, S. and Nagayama, S., *J. Org. Chem.* **61**, 2256–2257 (1996).

99. Kobayashi, S. and Nagayama, S., *J. Am. Chem. Soc.* **118**, 8977–8978 (1996).

100. Fivush, A. M. and Willson, T. M., *Tetrahedron Lett.* **38**, 7151–7154 (1997).

101. Leeson, P. D., Carling, R. W., Moore, K. W., Moseley, A. M., Smith, J. D., Stevenson, G., Chan, T., Baker, R., Foster, A. C. et al., *J. Med. Chem.* **35**, 1954–1968 (1992).

102. Sarges, R., Gallagher, A., Chambers, T. J. and Yeh, L. A., *J. Med. Chem.* **36**, 2828–2830 (1993).

103. Anzini, M., Cappelli, A., Vomero, S., Giorgi, G., Langer, T., Hamon, M., Merahi, N., Emerit, B. M., Cagnotto, A. et al., *J. Med Chem.* **38**, 2692–2704 (1995).

104. Breitenbucher, J. G. and Hui, H. C., *Tetrahedron Lett.* **39**, 8207–8210 (1998)

105. Mattson, R. J., Pham, K. M., Leuck, D. J. and Cowen, K. A., *J. Org. Chem.* **55**, 2552–2554 (1990).

106. Khan, N. M., Arumugam, V. and Balasubramanian, S., *Tetrahedron Lett.* **37**, 4819–4822 (1996).

107. Ashwood, V. A., Buckingham, R. E., Cassidy, F., Evans, J. M., Faruk, E. A., Hamilton, T. C., Nash, D. J., Stemp, G. and Willcocks, K., *J. Med. Chem.* **29**, 2194–2201 (1986).

108. Lipinski, C. A., Aldinger, C. E., Beyer, T. A., Bordner, J., Burdi, D. F., Bussolotti, D. L., Inskeep, P. B. and Siegel, T. W., *J. Med. Chem.* **35**, 2169–2177 (1992).

109. Gadwood, R. C., Kamdar, B. V., Dubray, L. A. C., Wolfe, M. L., Smith, M. P., Watt, W., Mizsak, S. A. and Groppi, V. E., *J. Med. Chem.* **36**, 1480–1487 (1993).

110. Atwal, K. S., Grover, G. J., Ahmed, S. Z., Sleph, P. G., Dzwonczyk, S., Baird, A. J. and Normandin, D. E., *J. Med. Chem.* **38**, 3236–3245 (1995).

111. Baldwin, J. J., *Mol. Diversity* **2**, 81–88 (1996).

112. Cho, H., Katoh, S., Sayama, S., Murakami, K., Nakanishi, H., Kajimoto, Y., Ueno, H., Kawasaki, H., Aisaka, K. and Uchida, I., *J. Med. Chem.* **39**, 3797–3805 (1996).

113. Marshall, D. L. and Liener, I. E., *J. Org. Chem.* **35**, 867–868 (1970).

114. Breitenbucher, J. G., Johnson, C. R., Haight, M. and Phelan, J. C., *Tetrahedron Lett.* **39**, 1295–1298 (1998).

115. Johnson, C. R., Zhang, B., Fantauzzi, P., Hocker, M. and Yager, K. M., *Tetrahedron* **54**, 4097–4106 (1998).

116. Kiselyov, A. S., Eisenberg, S. and Luo, Y., *Tetrahedron* **54**, 10635–10640 (1998).

117. Kiselyov, A. S., Smith, L. II and Tempest, P., *Tetrahedron* **55**, 14813–14822 (1999).

118. Munson, M. C., Cook, A. W., Josey, J. A. and Rao, C., *Tetrahedron Lett.* **39**, 7223–7226 (1998).

119. Albert, R., Knecht, H., Andersen, E., Hungerford, V., Schreier, M. H. and Papageorgiou, C., *Bioorg. Med Chem. Lett.* **8**, 2203–2208 (1998).

120. Bhalay, G., Cowell, D., Hone, N. D., Scobie, M. and Baxter, A. D., *Mol. Diversity* **3**, 195–198 (1998).

121. Wong, C.-H., Ye, X.-S. and Zhang, Z., *J. Am. Chem. Soc.* **120**, 7137–7138 (1998).

122. Li, Z., Yeo, S. L., Pallen, C. J. and Ganesan, A., *Bioorg. Med. Chem. Lett.* **8**, 2443–2446 (1998).

123. Kim, S. W., Bauer, S. M. and Armstrong, R. W., *Tetrahedron Lett.* **39**, 6993–6996 (1998).

124. Souers, A. J., Virgilio, A. A., Schurer, S. S., Ellman, J. A., Kogan, T. P., West, H. E., Ankener, W. and Vanderslice, P., *Bioorg. Med. Chem. Lett.* **8**, 2297–2302 (1998).

125. Ogbu, C. O., Qabar, M. N., Boatman, P. D., Urban, J., Meara, J. P., Ferguson, M. D., Tulinsky, J., Lum, C., Babu, S., Blaskovich, M. A., Nakanishi, H., Ruan, F., Cao, B., Minarik, R., Little, T., Nelson, S., Nguyen, M., Gall, A. and Kahn, M., *Bioorg. Med. Chem. Lett.* **8**, 2321–2326 (1998).

126. Fantauzzi, P. P. and Yager, K. M., *Tetrahedron Lett.* **39**, 1291–1294 (1998).

127. Lyngso, L. O. and Nielsen, J., *Tetrahedron Lett.* **39**, 5845–5848 (1998).

128. Mohan, R., Yun, W, Buckman, B. O., Liang, A., Trinh, L. and Morrissey, M. M., *Bioorg. Med. Chem. Lett.* **8**, 1877–1882 (1998).
129. Mayer, J. P., Lewis, G. S., McGee, C. and Bankaitis-Davis, D., *Tetrahedron Lett.* **39**, 6655–6658 (1998).
130. Lin, P. and Ganesan, A., *Tetrahedron Lett.* **39**, 9789–9792 (1998).
131. van Loevezijn, A., Van Maarseveen, J. H., Stegman, K., Visser, G. M. and Koomen, G.-J., *Tetrahedron Lett.* **39**, 4737–4740 (1998).
132. Nieuwenhuijzen, J. W., Conti, P. G. M., Ottenheijm, H. C. J and Linders, J. T. M., *Tetrahedron Lett.* **39**, 7811–7814 (1998).
133. Romoff, T. T., Ma, L., Wang, Y, and Campbell, D. A., *Synlett*, 1341–1342 (1998).
134. Villalgordo, J. M., Obrecht, D. and Chucholowsky, A., *Synlett*, 1405–1407 (1998).
135. Nefzi, A., Ostresh, J. M. and Houghten, R. A., *Tetrahedron* **55**, 335–344 (1999).
136. Houghten, R. A., *Proc. Natl. Acad. Sci. USA* **82**, 5131–5135 (1985).
137. Salvino, J. M., Mervic, M., Mason, H. J., Kiesow, T., Teager, D., Airey, J. and Labaudiniere, R., *J. Org. Chem.* **64**, 1823–1830 (1999).
138. Hennequin, L. F. and Piva-Le Blanc, S., *Tetrahedron Lett.* **40**, 3881–3884 (1999).
139. Roussel, P., Bradley, M., Kane, P., Bailey, C., Arnold, R. and Cross, A., *Tetrahedron* **55**, 6219–6230 (1999).
140. Rottlander, M. and Knochel, P., *J. Comb. Chem.* **1**, 181–183 (1999).
141. Paio, A., Zaramella, A., Ferritto, R., Conti, N., Marchioro, C. and Seneci, P., *J. Comb. Chem.* **1**, 317–325 (1999).
142. Johnson, M. G., Bronson, D. D., Gillespie, J. E., Gifford-Moore, D. S., Kalter, K., Lynch, M. P., McCowan, J. R., Redick, C. C., Sall, D. J., Smith, G. F. and Foglesong, R. J., *Tetrahedron* **55**, 11641–11652 (1999).
143. Tremblay, M. R., Simard, J. and Poirier, D., *Bioorg. Med. Chem. Lett.* **9**, 2827–2832 (1999).
144. Haskell-Luevano, C., Rosenquist, A., Souers, A., Khong, K. C., Ellman, J. A. and Cone, R. D., *J. Med. Chem.* **42**, 4380–4387 (1999).
145. Boojamra, C. G., Burow, K. M., Thompson, L. A. and Ellman, J. A., *J. Org. Chem.* **62**, 1240–1256 (1997).
146. Bunin, B. A. and Ellman, J. A., *J. Am. Chem. Soc.* **114**, 10997–10998 (1992).
147. Bunin, B. A., Plunkett, M. J. and Ellman, J. A., *Proc. Natl. Acad. Sci. USA* **91**, 4708–4712 (1994).
148. MacDonald, A., DeWitt, S. H., Hogan, E. and Ramage, R., *Tetrahedron Lett.* **37**, 4815–4818 (1996).
149. Shankar, B. B., Yang, D. Y., Girton, S. and Ganguly, A. K., *Tetrahedron Lett.* **39**, 2447–2450 (1998).
150. Krchnak, V., Weichsel, A. S., Lebl, M. and Felder, S., *Bioorg. Med. Chem. Lett.* **7**, 1013–1016 (1997).
151. Wilson, M. W., Hernandez, A. S., Calvet, A. P. and Hodges, J. C., *Mol. Diversity* **3**, 95–112 (1998).
152. Boeijen, A., Kruijtzer, A. W. and Liskamp, R. M. J., *Bioorg. Med. Chem. Lett*, **8**, 2375–2380 (1998).
153. Coshisoft/Peptisearch, P.O. Box 68212, Tucson, AZ 85737, USA.

154. Krchnak, V., *Biotechnol. Bioeng.* **61**, 135–141 (1999).

155. Szardenings, A. K., Antonenko, V., Campbell, D. A., DeFrancisco, N., Ida, S., Shi, L., Sharkov, N., Tien, D., Wang, Y. and Navre, M., *J. Med. Chem.* **42**, 1348–1357 (1999).

156. Souers, A. J., Virgilio, A. A., Rosenquist, S. A., Fenuik, W. and Ellman, J. A., *J. Am. Chem. Soc.* **121**, 1817–1825 (1999).

157. Lew, A. and Chamberlin, A. R., *Bioorg. Med. Chem. Lett.* **9**, 3267–3272 (1999).

158. ACT ReacTech, Advanced ChemTech, 5609 Fem Valley Road, Louisville, KY 40228, US.

159. Han, Y., Giroux, A., Lepine, C., Lalliberte, F., Huang, Z., Perrier, H., Bayly, C. I. and Young, R. N., *Tetrahedron* **55**, 11669–11685 (1999).

160. Merck Temperature Controller Manual Synthesizer, then evolved into the Quest 210 by Argonaut Technologies, Inc., 887 Industrial Road, Suite G, San Carlos, CA 94070, US.

161. Bicknell, A. J., Hird, N. W. and Readshaw, S. A., *Tetrahedron Lett.* **39**, 5869–5872 (1998).

162. Tsuge, O., Kanemasa, S. and Takenaka, S., *Bull. Chem. Soc. Jpn.* **59**, 3631–3635 (1986).

163. ACT496 MOS, Advanced ChemTech, 5609 Fem Valley Road, Louisville, KY 40228, US.

164. Miriad Core System, The Technology Partnership.

165. Brooking, P., Crawshaw, M., Hird, N. W., Jones, C., MacLachlan, W. S., Readshaw, S. A. and Wilding, S., *Synthesis*, 1986–1992 (1999).

166. Wilson, L. J., Li, M. and Portlock, D. E., *Tetrahedron Lett.* **39**, 5135–5138 (1998).

167. Perumattam, J., Chakravarty, S., Mcenroe, G. A., Goehring, R. R., Mavunkel, B., Suravajjala, S., Smith, W. W. and Chen, B., *Mol. Diversity* **3**, 121–128 (1998).

168. Smith, R. A., Bobko, L. A. and Lee, W., *Bioorg. Med. Chem. Lett.* **8**, 2369–2374 (1998).

169. Crawshaw, M., Hird, N. W., Irie, K. and Nagai, K., *Tetrahedron Lett.* **38**, 7115–7118 (1997).

170. Shao, H., Colucci, M., Tong, S., Zhang, H. and Castelhano, A. L., *Tetrahedron Lett.* **39**, 7235–7238 (1998).

171. Porco, J. A. Jr., Deegan, T. L., Devonport, W., Gooding, O. W., Labadie, J. W., MacDonald, A. A., Newcomb, W. S. and Van Eikeren, P., *Drugs Future* **23**, 71–78 (1998).

172. Nautilus 2400, Argonaut Technologies, Inc., 887 Industrial Road, Suite G, San Carlos, CA 94070, US.

173. DeGrado, W. F. and Kaiser, E. T., *J. Org. Chem.* **47**, 3258–3261 (1982).

174. Scialdone, M. A., Shuey, S. W., Soper, P., Hamuro, Y. and Burns, D. M., *J. Org. Chem.* **63**, 4802–4807 (1998).

175. Doi, T., Hijikuro, I. and Takahashi, T., *Synlett*, 1751–1753 (1999).

176. Fodor, S. P. A., Read, J. L., Pirrung, M. C., Stryer, L., Lu, A. T. and Solas, D., *Science* **251**, 767–773 (1991).

177. Pirrung, M. C., *Chem. Rev.* **97**, 473–488 (1997).

178. Niemeyer, C. M. and Blohm, D., *Angew. Chem. Int. Ed.* **38**, 2865–2869 (1999).

179. Pease, A. C., Solas, D., Sullivan, E. J., Cronin, M. T., Holmes, C. P. and Fodor, S. P. A., *Proc. Natl. Acad. Sci. USA* **91**, 5022–5026 (1994).

180. Singh-Gasson, S., Green, R. D., Yue, Y., Nelson, C., Blattner, F., Sussman, M. R. and Cerrina, F., *Nat. Biotechnol.* **17**, 974–978 (1999).

181. Frank, R., *Tetrahedron* **48**, 9217–9232 (1992).

182. Frank, R., *Bioorg. Med. Chem. Lett.* **3**, 425–430 (1993).

183. Gao, B. and Esnouf, M. P., *J. Biol. Chem.* **271**, 24634–24638 (1996).

184. Terrett, N. K., Gardner, M., Gordon, D. W., Kobylecki, R. J. and Steele, J., *Chem. Eur. J.* **3**, 1917–1920 (1997).

185. Terrett, N., *Combinatorial Chemistry*. Oxford Press, London, UK, 1998, pp. 33–54.

186. Tapolczay, D. J., Kobylecki, R. J., Payne, L. J. and Hall, B., *Chem. Ind.*, 772–775 (1998).

187. Lemmo, A. V., Fisher, J. T., Geysen, H. M. and Rose, D. J., *Anal. Chem.* **69**, 543–551 (1997).

188. Bernardini, G. L., *J. Neurosci. Methods* **38**, 81–86 (1991).

189. Newman, J. D. and Turner, A. P. F., *Anal. Chim. Acta* **262**, 13–17 (1992).

190. Nilsson, J., Szecsi, P. and Schafer-Nielsen, C., *J. Biochem. Biophys. Methods* **27**, 181–190 (1993).

191. Sziele, D., Brueggeman, O., Doering, R., Freitag, R. and Schuegerl, K., *J. Chromatog. A* **669**, 254–258 (1994).

192. Hager, D. B., Dovichi, N. J., Classen, J. and Kebarle, P., *Anal. Chem.* **66**, 3944–3949 (1994).

193. Perein, G. and Khuri-Yakub, B. T., *Proc.-Electrochem. Soc.* **98**, 87–93 (1998).

194. Chang, S.-C., Bharathan, J., Yang, Y., Helgeson, R., Wudl, F., Ramey, M. B., Reynolds, J. R., *Appl. Phys. Lett.* **73**, 2561–2563 (1998).

195. Rose, D., *Drug Discovery Today* **4**, 411–419 (1999).

196. Schwabacher, A. W., Shen, Y. and Johnson, C. W., *J. Am. Chem. Soc.* **121**, 8669–8670 (1999).

197. Rapp, W.E., in *Combinatorial Peptide and Nonpeptide Libraries*, G. Jung (Ed.). VCH, Weinheim, Germany, 1996, pp. 425–464.

198. Rapp, W. E., in *Combinatorial chemistry: synthesis and application*, S. R. Wilson and A. W. Czarnik (Eds.). Wiley, New York, 1997, pp. 65–93.

199. DeWitt, S. H., communication presented at *Strategies and Techniques for Identification of Novel Bioactive Compounds*, CHI, October 7th–8th 1998, Cambridge Healthtech Institute, Zurich, Switzerland.

200. Fuhr, G., Schnelle, T. and Wagner, B., *J. Micromech. Microeng.* **4**, 217–226 (1994).

201. Baba, Y., *J. Flow Injection Anal.* **14**, 131–140 (1997).

202. Zanzucchi, P. J., McBride, S. E., Burton, C. A. and Cherukuri, S. C., *US Patent 5632876*, May 27th, 1997.

203. Leach, M., *Drug Disc. Today* **2**, 253–254 (1997).

204. Schnur, J. M., *Science* **262**, 1669–1676 (1993).

205. Stevens, A. M. and Richards, C. J., *Tetrahedron Lett.* **38**, 7805–7808 (1997).

206. Delamarche, E., Bernard, A., Schmid, H., Bietsch, A., Michel, B. and Biebuyck, H., *J. Am. Chem. Soc.* **120**, 500–508 (1998).

207. Christensen, P. D., Johnson, S. W. P., McCreedy, T., Skelton, V. and Wilson, N. G., *Anal. Commun.* **35**, 341–343 (1998).

208. Jackman, R. J., Duffy, D. C., Ostuni, E., Willmore, N. D. and Whitesides, G. M. *Anal. Chem.* **70**, 2280–2287 (1998).

209. Duffy, D. C., McDonald, J. C., Schueller, O. J. A. and Whitesides, G. M., *Anal. Chem.* **70**, 4974–4984 (1998).

210. Kricka, L. J., *Clin. Chem.* **44**, 2008–2014 (1998).

211. Campbell, D. P., Moore, J. L., Cobb, J. M., Hartman, N. F., Schneider, B. H. and Venugopal, M. G., *Proc. SPIE-Int. Soc. Opt. Eng.* **3540**, 153–161 (1999).

212. Warren, M. E., Anex, D. S., Rakestraw, D. and Goudey, P. L., *SANDIA Report UC-706, SAND98-0509*. Sandia National Laboratories, Albuquerque, NM, and Livermore, CA, 1998.

213. Wilding, P. and Kricka, L. J., *Trends Biotechnol.* **17**, 465–468 (1999).

214. Jacobson, S. C., McKnight, T. E. and Ramsey, J. M., *Anal. Chem.* **71**, 4455–4459 (1999).

215. Hosokawa, K., Fujii, T. and Endo, I., *Anal. Chem.* **71**, 4781–4785 (1999).

216. Siebert, P., Petzold, G. and Muller, J., *Proc. SPIE-Int. Soc. Opt. Eng.* **3680**, 562–571 (1999).

217. Hadd, A. G., Jacobson, S. C. and Ramsey, J. M., *Anal. Chem.* **71**, 5206–5212 (1999).

218. MacBeath, G., Koehler, A. N. and Schreiber, S. L., *J. Am. Chem. Soc.* **121**, 7967–7968 (1999).

219. Ermantraut, E., Wolfl, S. and Saulz, H. P., *Biomethods* **10**, 255–264 (1999).

220. Fletcher, P. and Haswell, S., *Chem. Brit.*, 38–41 (1999).

7 Synthetic Organic Libraries: Solid-Phase Pool Libraries

This chapter will deal with the most complex format of synthetic libraries. The difficulties associated with such libraries are, however, tempered by their enormous potential to discover or to contribute to the discovery of active molecules for any application. Solid-phase pool libraries of biopolymers were reported first (1–3), followed some years later by small organic molecule SP pool libraries, as their potential was recognized in the early days of combinatorial technologies (4–6). This chapter will present and review the properties of an SP pool library, will highlight the main issues related to such libraries, and will also present examples of SOM SP pool libraries to help describe these concepts.

The first section will be devoted to the *synthesis* of these libraries using the so-called *mix-and-split* or divide-and-recombine approach (2, 3) and to their *analytical characterization*. The following sections will focus on different methods to determine the *structure* of an active component from an SP pool library: *direct structure determination* (Section 7.2) and *indirect structure determination*, via *deconvolutive methods* (Section 7.3) or *encoding methods* (Section 7.4), will be covered. Finally, a section will be devoted to *new trends* in SP pool libraries, paying particular attention to innovative methods for the fast and reliable discovery of new active structures through miniaturization (bead-based techniques).

7.1 SYNTHESIS OF SOLID-PHASE POOL LIBRARIES

7.1.1 Mix-and-Split SP Synthesis

The use of heterogeneous supports in organic synthesis has been extensively covered, and its relevant implications on the reaction outcome were discussed in Chapter 1. A major advantage of heterogeneous versus homogeneous reactions for the generation of pool libraries is related to the handling and the purification of intermediate or final pools.

A homogeneous reaction in solution is carried out in a reaction vessel, and the distribution of aliquots from this solution would lead to several starting points for more functionalized products following a second reaction step. Handling of liquid aliquots and the need for intermediate purifications make the splitting of reactions in solution to give large numbers of individuals a complex process. Despite this drawback, such techniques are gaining popularity among chemists for discrete libraries synthesis

(7–9). The synthesis of high-quality solution-phase pool libraries is more difficult, and, while some approaches have been reported (Chapter 8) (10–12), this library format is not yet fully exploitable.

If we examine the same reaction carried out in an appropriate reaction vessel in the presence of a heterogeneous support, we can assume that a single bead is the microreactor, or the single entity for the reaction. Having a synthetic method able to deliver a single compound on each bead, we could prepare even million-member pool libraries from a few grams of resin with a single, or a few, beads carrying each library individual! The use of resin beads would also accelerate the intermediate purification procedures and facilitate the handling/aliquoting of compounds into different reaction vessels. If such large SP pool libraries proved reliable sources of new active structures with high purity, their advantage over large discrete libraries in terms of reduced operations and smaller scales/quantities of reagents would be extremely significant.

The three main SPS strategies to produce combinatorial libraries will be described and critically reviewed in this section to clarify the role of each SP library format. We will use a hypothetical example, represented in Fig. 7.1, where 100 carboxylic acids and 100 amines are used to produce a 100 × 100 = 10,000-member SP amide library. The amines are first coupled with a resin bearing an aldehyde linker (50 g, step a); then the imines are reduced (step b) and the acylating agents are coupled to the resin-bound amines (step c, Fig. 7.1). The first approach, named *one compound per well* (Fig. 7.2), requires the initial coupling of each amine with the aldehyde resin in 100 separate reaction vessels (500 mg of resin in each vessel, steps a and b) and, then, after their reduction (step c), the splitting of each vessel into 100 (step a) and the reaction of each of these vessels with one of the 100 acylating agents, such that each of the 10,000 final vessels contains a different product (step d, 5 mg of resin in each vessel). This *parallel synthesis* has been described in detail in the previous chapter. Its application to large library synthesis requires a high degree of automation and, as pointed out earlier,

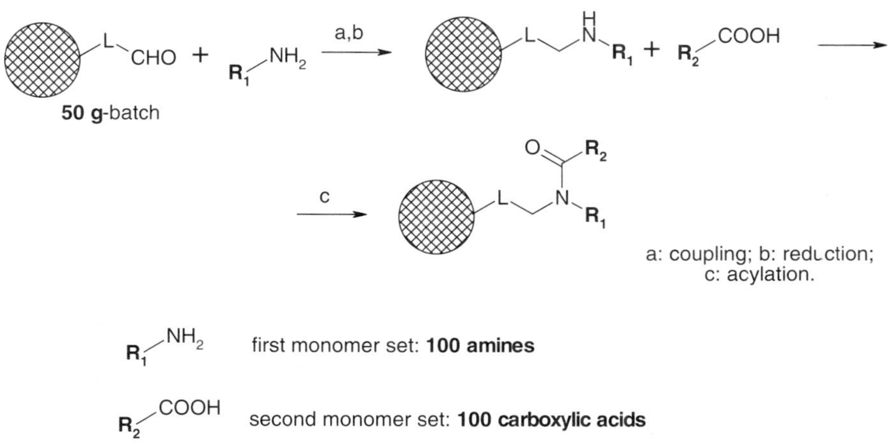

Figure 7.1 Synthetic pathway to a hypothetical SP 10,000-member amide pool library.

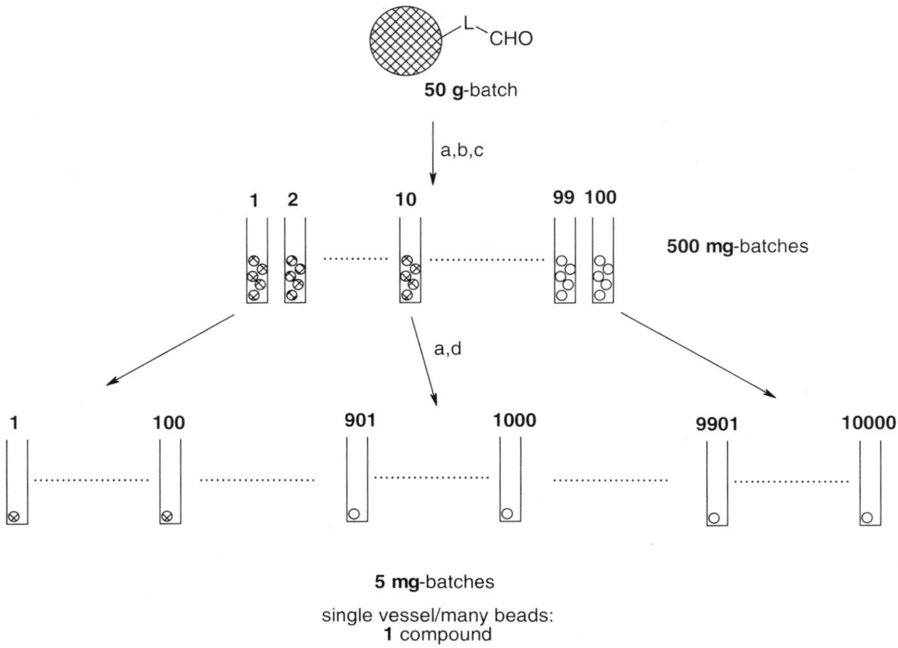

a: resin portioning (1 to 100); b: coupling with amines; c: reduction; d: coupling with acids.

Figure 7.2 SP parallel synthesis: one compound per well.

extremely large discrete libraries (hundreds of thousands or even millions of components) cannot be made in a timely manner as of today.

The second approach, named *many compounds per bead* (Fig. 7.3), starts by coupling the solid support in a single reaction vessel with an equimolar mixture of the 100 amines (step a); then the mixture is reduced (step b) and the resin-bound amines are reacted with an equimolar mixture of the 100 acylating agents (step c). The 10,000-member library is obtained as a single 50-g pool of resin, and each bead contains similar quantities of each library individual. A bead has typically 10^{14}–10^{15} reaction sites, so that each bead will contain an average of 10^{10}–10^{11} copies of each library individual. The library synthesis could technically be considered successful if all the monomers react properly and the 10,000 compounds are actually present, but the identification of positives from this library for any specific application is not feasible. In fact, the cleavage of resin-bound materials produces an equimolar mixture of all the components, whose activity, if any, is the activity of a 10,000-member unresolved mixture. As a consequence of this major limitation, this SPS approach is not used for library synthesis.

The third approach can be considered as a compromise between parallel synthesis and mixtures on SP. It is named *one compound per bead* (13) and allows the synthesis

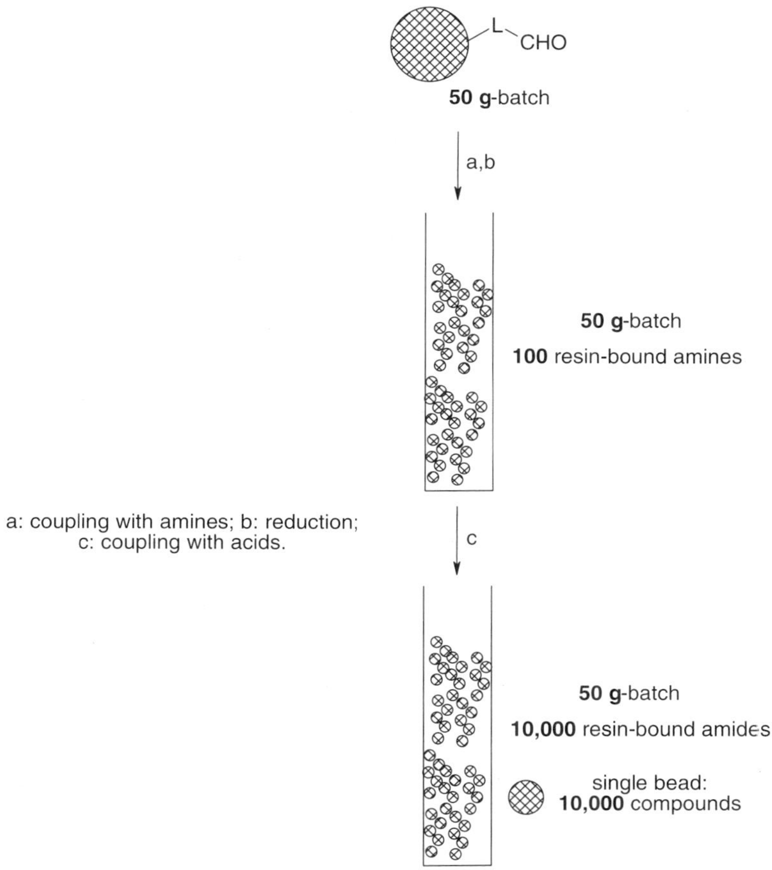

Figure 7.3 SP pool library synthesis: many compounds per bead.

of high-quality SP pool libraries from which structure determination of positives can be timely and effectively obtained. This method, usually called *mix and split* or divide and recombine, was reported at the very beginning of combinatorial technologies (2, 3) and has become extremely popular for the synthesis of oligomeric libraries. Even if the current trend for small organic molecule libraries is shifting toward discrete libraries, the synthesis of SP SOM pool libraries is still common and important, especially for encoded libraries (see Section 7.4), and has an ample margin left for quality improvements.

The synthesis of our hypothetical SP pool library of 10,000 amides by mix and split is reported in Fig. 7.4. Preparation of the 100 resin-bound amines follows an identical course to the one seen for parallel synthesis (steps a, b, and c), but then the 100 resin aliquots are mixed to give a single 50 g portion (step d). This portion is then split into 100 aliquots each containing all the 100 amines (step a), but each bead is loaded with

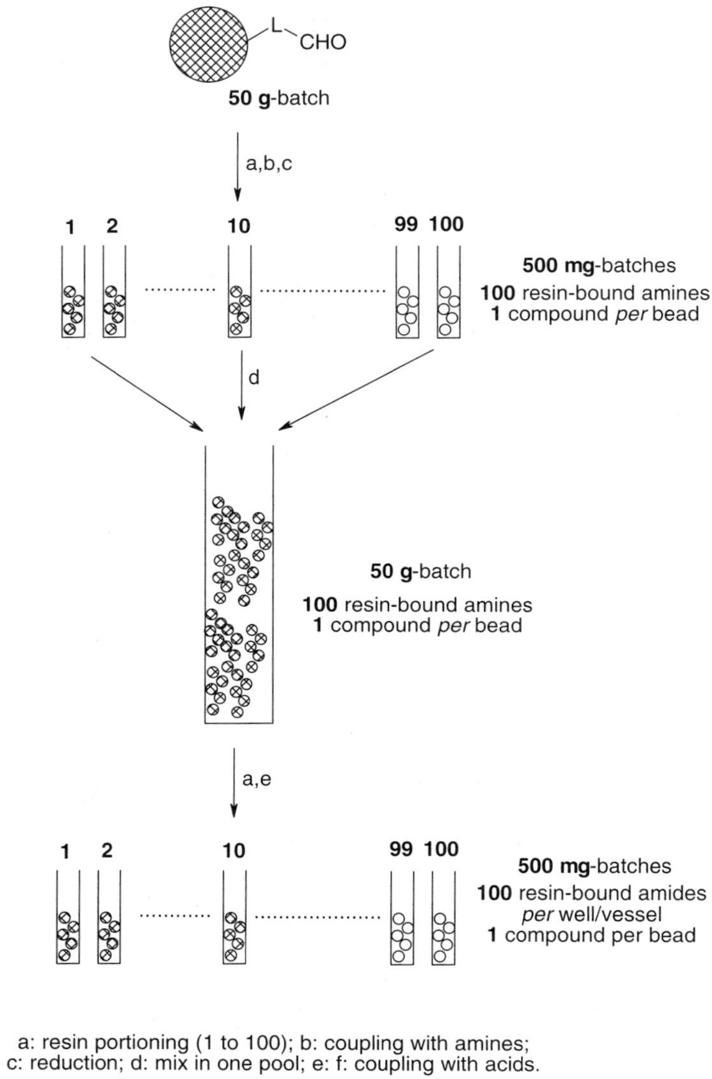

Figure 7.4 SP pool library synthesis: one compound per bead.

a single compound because the coupling of the amines to the solid support was performed in separate vessels. Each new aliquot is placed in a different vessel and the 100 acylating agents are added separately to each reactor (step e) to produce 100 pools of 100 compounds. Each pool contains only one acid monomer and the whole amine set, but each bead is loaded with a single individual.

Another hypothetical example of mix-and-split synthesis, involving a four-step process with five monomers for each of the four monomer sets, leads to a $5 \times 5 \times 5 \times$

5 = 625-member library made from 500 mg of resin (steps a–f, Fig. 7.5). The library is produced as five pools containing 125 compounds; the last monomer is determined and unique for each pool while the others are fully randomized. This feature of SP pool libraries is instrumental in the structure determination of positives, which will be described in the following sections. An SP pool library built by many mix-and-split steps and/or by large monomer sets can readily be obtained, providing that each monomer reacts properly and each library individual is represented. The more the library complexity is increased, the more the chemistry assessment and the monomer rehearsal before the library synthesis are essential for the production of a high-quality SP pool library.

An important property of SP pool libraries is called *redundancy*, measured as the ratio between the number of beads used in the library synthesis and the number of planned library individuals. Considering a 10,000-member library, using 10,000 resin beads we could theoretically have a single bead loaded with each library individual. However, as the mix-and-split process does not count exactly the number of beads in

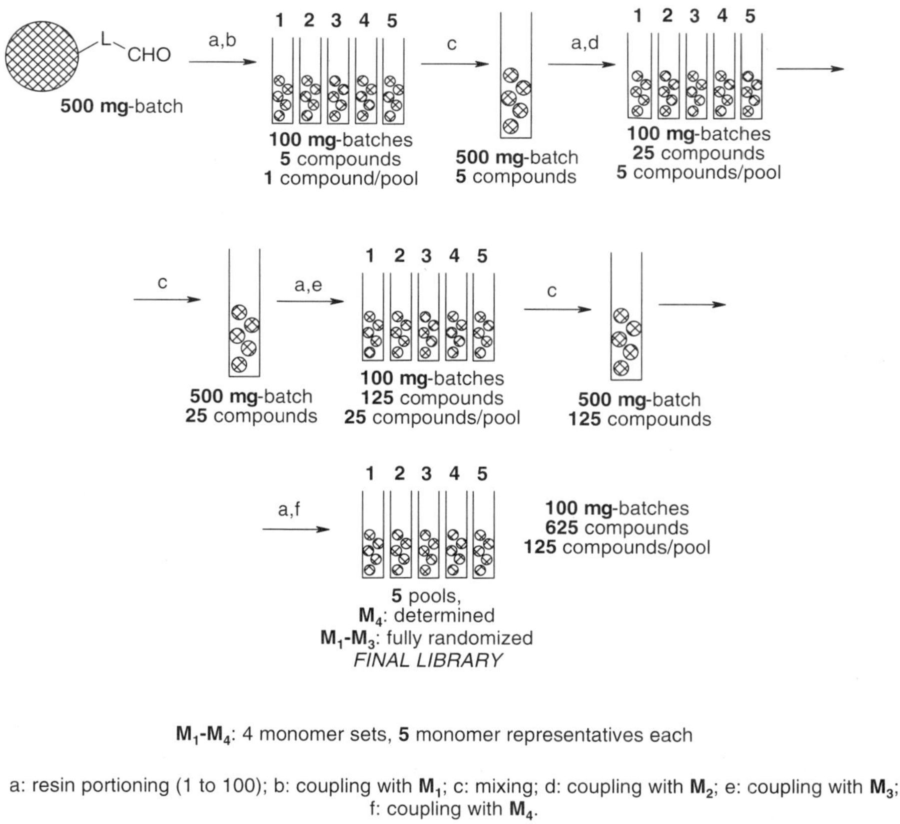

Figure 7.5 Mix and split: synthesis of a hypothetical 625-member SP pool library.

each aliquot but rather aliquots similar volumes of resin slurries (see next section), the library produced in such a case would contain several individuals that are represented more than once and others would be missing. Redundancy obviates this inconvenience: If 50 g of resin, that is, around 50,000,000 beads, is used for a 10,000-member library, a redundancy factor of 5000 ensures that the population of each library individual will vary roughly from double (10,000 beads) to half (2500 beads) of its theoretical amount at most but will probably be close to the theoretical 5000-bead representation. A redundancy of 5000 is generally too much, and typically lower redundancies (from 50 to 100) are suggested as the best compromise between fully represented libraries and cost savings. Redundancy can be safely decreased to 3–5 for a bead-based screening (*vide infra*) to ensure the presence of at least one bead loaded with each library component. If, as usual, the library will be screened on many assays, a redundancy of $5 \times N$, where N is the number of planned screenings, is needed.

7.1.2 Technical Aspects of SP Pool Libraries: Manual Versus Automated Synthesis

The general aspects of SPS, which have been examined in Chapters 1–3, are common for all the SPS procedures leading to pool libraries. The aspects of automated SPS, which were covered during the description of SP discrete libraries in Section 6.1.2 (e.g., architecture of SP synthesizers, inertness of the system, complexity of hardware and software, and automated versus semiautomated devices), are also important for SP pool library synthesis. The most distinctive aspect of SP pool synthesis is represented by the mixing and splitting of resin aliquots, which can be performed by handling small aliquots of beads, or even single beads, either manually or by automated devices. We will briefly describe the principles and the techniques of this crucial operation.

Resin aliquots can be handled as solids after elimination of solvents and soluble reagents by filtration and repeated washings and drying of the resin. In this case, if a mixing step of ten 100-mg portions is involved, the aliquots are combined and mixed as dry solids; then ten new portions are created by weighing ten 100-mg aliquots of dry resin and partitioning them in different reaction vessels. It is more usual, though, to handle the resin as a slurry, which is easily mixed by pouring all the separate slurries in a common vessel and collecting all the residual beads from each vessel by repeated washings, possibly with heavy solvents such as DCM, in which the residual beads float. The splitting step is then performed by sampling ten equivalent volumes of the slurry, either stirring the slurry to have an equal density of beads in the liquid volume or, better, using an isopycnic solvent mixture with the same density as the beads. This allows the sampling of ten equivalent aliquots with a pipette and their partitioning into the reaction vessels.

This *wet* procedure is preferable because of the easier and faster splitting of slurries by pipetting volumes rather than weighing dry aliquots of resin. Moreover, its partial or even total automation is significantly easier. The use of *robotic pipetting arms*, which withdraw slurried aliquots and distribute them in different reaction vessels, is widespread. They may have one or more pipetting arms to speed the process, which may be totally automated, as in an SP synthesizer for pool libraries, or partially automated using one of the commercial automated dispensing systems. Further details of such automatic synthesizers can be found in several reviews (14, 15). Another, more appealing solution for pool libraries consists of a closed system made of *valves* and *vessels/chambers*, where isopycnic beads–solvent mixtures can be easily carried from many reaction vessels to a mixing chamber and then partitioned again into the reaction vessels. This system allows an easier and faster handling of mix-and-split steps and is used in existing automated synthesizers (16). Other fully automated systems, including hybrids of the two techniques, have also been reported (17–19).

It is important to underline the ease of synthesis of even large SP pool libraries using only manual techniques and common laboratory glassware. Take, for example, the 12,000-member library of triazines **L1** reported by Stankova and Lebl (20) and shown in Fig. 7.6. First, a 12-member pool model library was prepared and fully characterized. Then, the first monomer set (20 α-amino acids) was coupled to the resin in 20 glass vials (steps a and b); then the 20 aliquots were mixed, the Fmoc group deprotected, and the cyanuric chloride coupled (steps c, d, and e) in a single reaction vessel. The resin-bound triazines were split into 30 portions as a slurry with a simple pipette; then the second monomer set was added to react at rt (30 amines, steps f and b) in 30 glass vials. The resin was then mixed as a slurry to give two portions (vials 1–15 were mixed together, as were vials 16–30 separately; step g), and each major

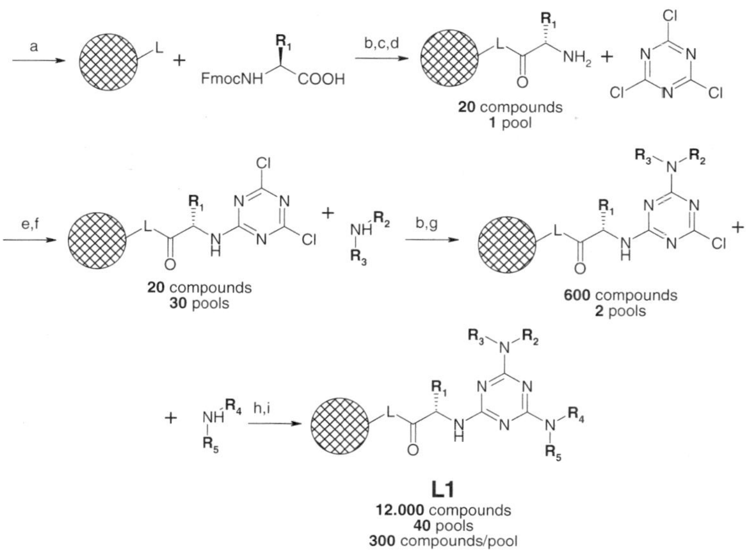

a: resin portioning (1 to 20); b: coupling, rt; c: Fmoc deprotection; d: mix in one pool;
e: coupling with cyanuric chloride; f: resin portioning (1 to 30); g: mix in two pools (1-15 and 16-30);
h: resin portioning (2 to 40); i: coupling, heating.

Figure 7.6 SP synthesis of a 12,000-member triazine SP pool library **L1**.

portion was split in 20 aliquots with a pipette (step h). The third monomer set was then coupled at 90 °C (20 amines and hydrazines, step i, Fig. 7.6) in 40 glass vials. No automation was used at any level, and the high-quality SP library **L1**, which was obtained as 40 pools of 300 compounds each, was used as a source of positives in many biological assays. When comparing this pool library to an equivalent 12,000-member SP discrete library, the ease of synthesis of the former should be underlined, especially when limited resources and budgets are available.

Another important technical problem for SP pool libraries is related to the handling of single beads. Often with encoded pool libraries or bead-based screening (*vide infra*), it is necessary at some stage to handle and deliver a certain number of single beads to single vessels/wells. This may well be done manually using a capillary tube and a microscope to check the pick of a single bead, but the process becomes tedious and time consuming when tens or hundreds of single beads must be processed. Automated instruments, called *bead pickers*, have been reported (21): they use microcapillaries mounted on a manifold (arrays of eight or more) that is switchable from vacuum (aspiration of a bead from a pool) to pressure (release of the single bead to a plate well) to pick and deliver up to 96 single beads in 5 min. The capillary ends are built so that the bead is retained by the vacuum but cannot enter because its diameter is bigger than the capillary end. The capillaries are rinsed in fresh solvents after delivering a bead and can then be used for another picking cycle. The percentage of errors, due either to the loss of bead during movement of the capillary to its plate destination or to multiple bead picking caused by bead lump formation, is reasonably low (<10% of total picking operations). These instruments can be built in a high-throughput version starting from commercially available 96-channel dispensing stations (22) and assembling around them a software/hardware system able to monitor the bead-picking operation, detecting the successful operations (one bead on one pipette tip), and discarding the unsuccessful ones (multiple beads or no beads on a tip).

7.1.3 Analytical Methods for the Synthesis, Quality Control, and Purification of Solid-Phase Pool Libraries

The synthesis of an SP pool library is more analytically challenging than the same library prepared as a discrete collection. The presence of many compounds in the same pool disturbs the qualitative and quantitative determination of purity and yields in all the synthetic steps, during the final quality control, and when necessary, during the off-bead purification of the pool. Several reviews (23–25) have dealt with the analytical chemistry aspects related to pool SP libraries; here we will review the analytical steps required for the synthesis, characterization, and purification of an SP pool library, while the structure determination of active components from a pool will be dealt with in the next three sections.

Monomer rehearsal is identical for SP discrete and pool libraries, but the generally larger dimensions of the latter require the use of more monomers and thus their qualification for the planned reaction scheme. The full analytical characterization of this step will be crucial. Discarding poorly performing monomers and identifying reliable monomers and optimized reaction conditions will guarantee the quality of the final pool library. Considering a hypothetical example where three monomer sets, each composed of 50 representatives, are used to build a $50 \times 50 \times 50 = 125{,}000$-member pool library on a common scaffold, a typical rehearsal procedure is shown in Fig. 7.7. Each monomer set representative (**$M_{1,1\text{–}50}$, $M_{2,1\text{–}50}$,** and **$M_{3,1\text{–}50}$**) is reacted according to the SPS scheme with the scaffold (for **M_1**) or with standard representatives of the other two sets (for **M_2** and **M_3**, path a, 150 reactions), and the monomers that give the

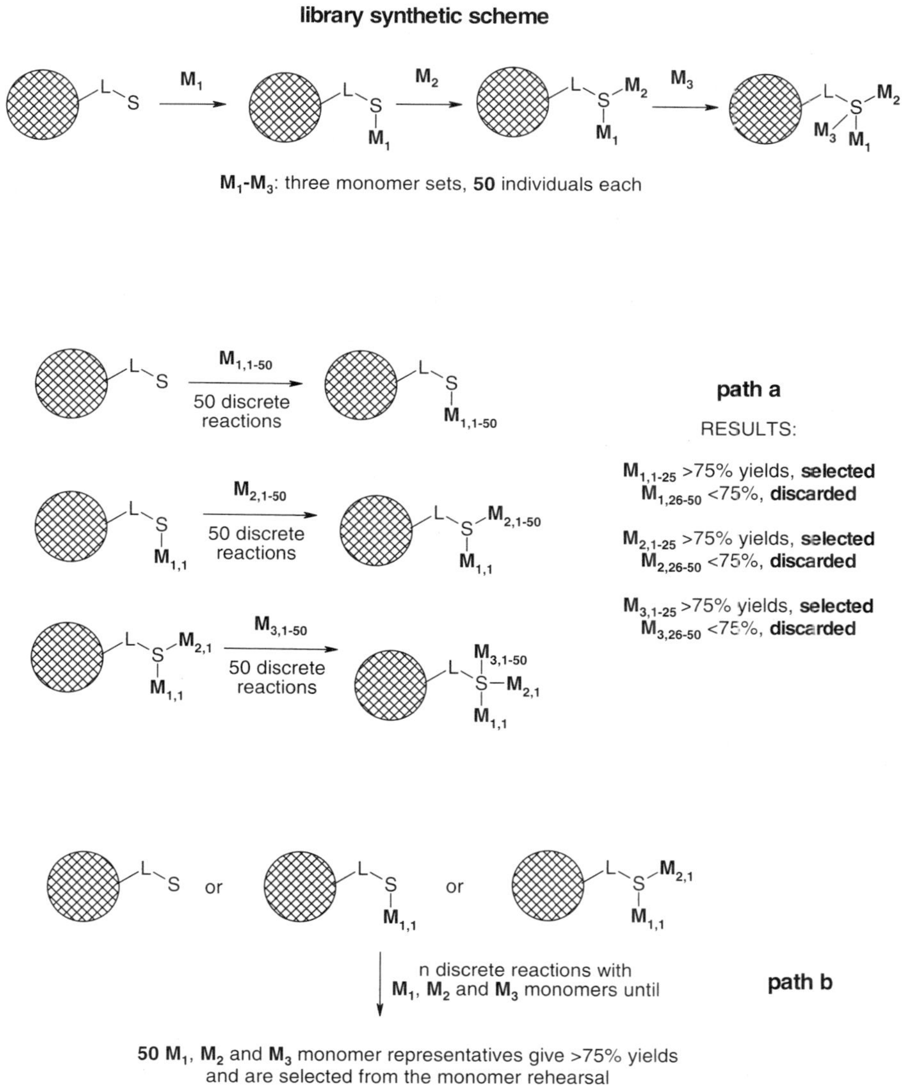

Figure 7.7 Monomer rehearsal for a hypothetical 125,000-member SP pool library.

expected reaction product with good yields and purity (typically >75%, 25 monomers for each set **M1–M3** in Fig. 7.7) are selected for library production. The rejected monomers are replaced with other candidates in each monomer set (path b, *n* reactions with the scaffold or standard representatives) until the three 50-member sets are selected. If a single monomer is misjudged, that is, is included in the selected sets but does not produce the expected library components, a total of $1 \times 50 \times 50 = 2500$ compounds will be lost and the library value will be diminished. In the example, a few hundred discrete reactions and their careful analytical characterization are the gateway to a high-quality 125,000-member SP pool library. Both off-bead and on-bead methods should be used, and depending on the nature of the library components, some techniques and/or detection methods will be more suitable, including sophisticated on-bead analytical methods (see Sections 1.3 and 1.4).

The following step is represented by the synthesis of a *model SP pool library*, which is prepared using the same equipment and strategies as planned for the real library but on a smaller scale (typically from 10 to 50 library individuals in small pools). This allows evaluation of the quality of the synthesis as it passes from discretes to mixtures while being able to fully characterize the small pools from an analytical standpoint. These pools are usually analyzed off-bead, and both the presence and the relative amounts of each pool component are determined. The influence of monomers with different reactivities on the library outcome can be checked by the careful selection of monomers for the model library, that is, if the general optimized reaction conditions are suitable both for reactive and less reactive monomers. Examples have been reported in the literature where HPLC (26, 27), MS (28, 29), HPLC/MS (30–33), GC/MS (34), and HPLC/NMR (35) were used to characterize small SP pool libraries of small organic molecules after cleavage from the beads. Most of these papers actually reported only the synthesis of a model SP pool library. It is probable that larger libraries were prepared in the same laboratories using the same synthetic scheme, but their full disclosure via publication was avoided to protect their proprietary chemical diversity.

When both the monomer rehearsal and the model library studies produce good results, the SPS of the planned pool library takes place. Reaction monitoring is now difficult because a mixture of many compounds is present in each well. Transformations related to specific functional groups can be monitored either on-bead (colorimetric methods such as the Kaiser test—appearance or disappearance of an amino group) or off-bead (UV reading, Fmoc deprotection of amines). Information may be obtained from off-bead methods after the cleavage of a small resin aliquot, but while the total or partial transformation of the starting materials can be determined (disappearance or appearance of diagnostic IR bands, or NMR signals), the absolute quantification of each pool component and of each remaining starting material is at best arduous. The rigorous rehearsal of monomers and a successful model library synthesis are the best guarantees for uneventful, high-quality SPS using the mix-and-split approach.

Structure determination is an extremely complex issue in SP pool libraries. Assuming as a hypothetical example a large library of 100,000 individuals in 1000 pools, each containing 100 compounds, we will know the location of each pool but we will not automatically have precise information on the composition of each pool. An accurate determination of each pool's content would require the cleavage of aliquots from each pool and their analytical characterization via multiple methods, but above all a strenuous effort to assign the structure of each individual to a specific set of signals, bands, and so on. Even if today's trend is to have pool libraries containing a small number of components per pool (typically 10 to 30–50), structure determination of all the library components is not possible in a timely and reasonable manner. Thus, accurate structure determination is performed after, rather than before, the library assay and only the positive compounds are structurally identified and confirmed. Several methods exist to perform this crucial operation, and they will be the subject of the next three sections.

This simplified and more time- and cost-effective strategy necessitates an accurate quality control (QC) for the library. While examples of MS characterization of several large peptide pools have been reported (36, 37), an easier and more significant method for SP pool library QC is needed. An interesting and emerging option is *bead-based QC*. Since mix and split produces beads loaded with a single compound, and a single bead analysis will determine the structure of the loaded compound and its purity, providing that the method is sensitive enough, this technique effectively allows the parallel analysis of discrete beads, with no interference from mixtures. The only difference from the characterization of SP discrete libraries would be the unknown structure of the supported individual. Off-bead MS analysis of a few hundred beads randomly selected from different library pools should allow the identification of library representatives through their molecular weight, and other on-bead techniques already available (FTIR), or to be expected in the near future (NMR, others), would allow the refinement of the bead-related qualitative (structure, presence of impurities) and quantitative (yields, loading of target and impurities) information. A sound proposal expressed by a leading combinatorial group (24) stipulates the lower limit of confirmed library individuals as at least 80% of the examined beads to define a good-quality SP pool library.

The separation and isolation of individual compounds is not in the spirit of a pool library, which is tested as a mixture using various techniques. A final purification by chromatography or extraction, though, may be useful to remove cleavage reagents or impurities that have extremely different physicochemical properties with respect to the library components (salts, greasy reagents, etc.). More details on final purification procedures for released SP pool libraries are reported in Section 8.3.

7.1.4 An Example: Synthesis of an SP Pool Library of Hexahydroisoindoles

A recent communication by Heerding et al. (38) reported the synthesis of a 3200-member SP pool library of isoindolines **L2** by means of a complex reaction scheme, including an intramolecular metathesis reaction and a Diels–Alder cycloaddition. The target bicyclic nucleus and the selected precursors to obtain the library are depicted in Fig. 7.8. The authors began by assessing a possible reaction scheme in solution using 4-benzyloxybenzoic acid **7.1** as a linker-mimicking synthon (Fig. 7.9). Reaction with allyl amine **7.2a** (step a), N-alkylation with the propargyl mesylate **7.4** (steps b,c), intramolecular metathesis (step d), and cycloaddition with maleimide (step e) produced the target cycloadduct **7.7** as a single diastereomer in reasonable yield and purity.

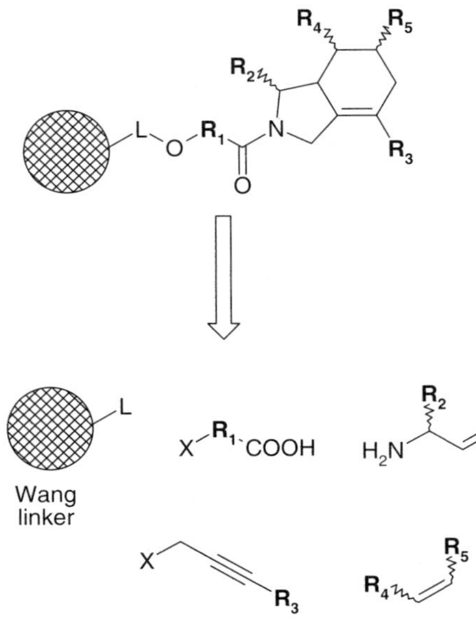

Figure 7.8 Retrosynthetic analysis for the isoindoline SP pool library **L2**.

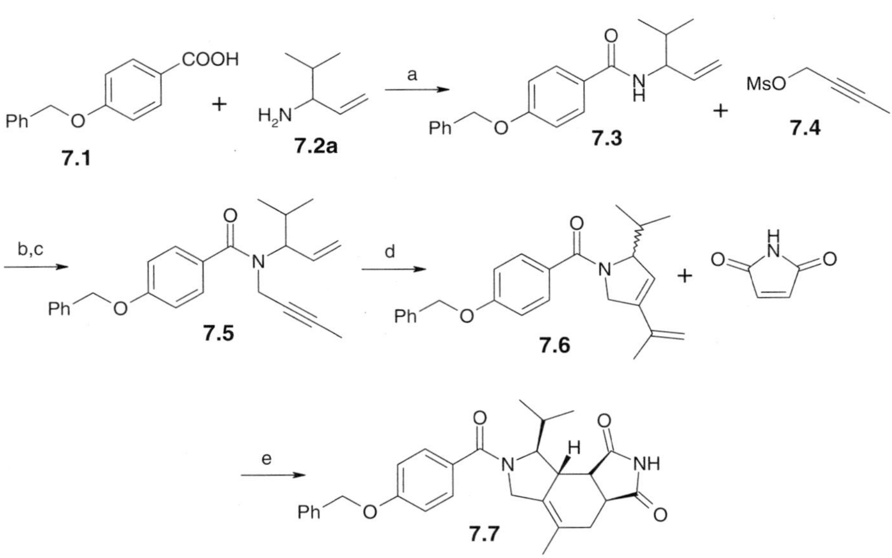

a: EDC, HOBt, TEA, DMF, rt, 18 hrs; b: NaH, DMF, rt, 30'; c: **7.4**, rt, 5 hrs;
d: $RuCl_2(=CHPh)(PCy_3)_2$ cat., benzene, reflux, 18 hrs; e: toluene, reflux, 18 hrs.

Figure 7.9 Validation in solution for the synthetic scheme to the isoindoline SP pool library **L2**: synthesis of **7.7**.

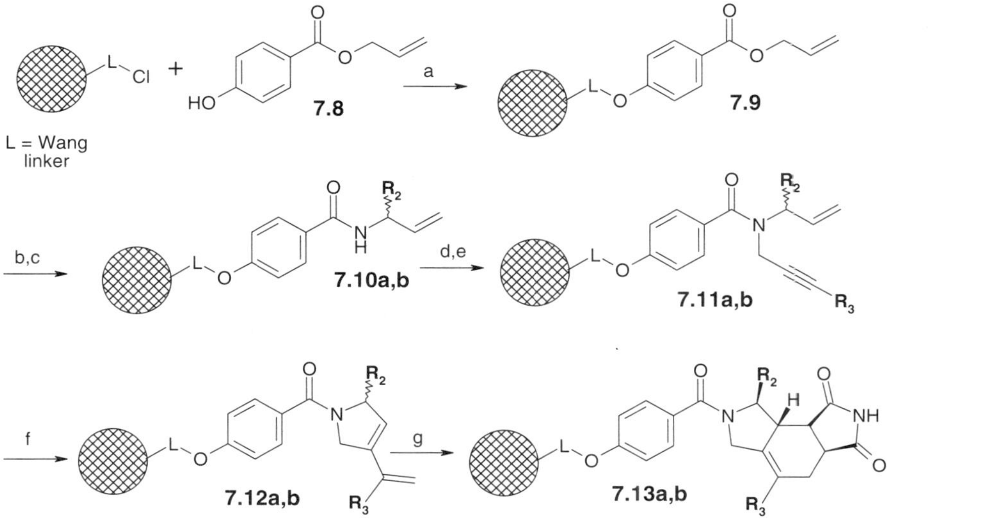

a: NaH, DMF; b: $(Ph_3P)_4Pd$, MeNHPh, DMSO/DMF 1/1, 55°C, 18 hrs; c: **7.2a,b**, EDC, HOBt, TEA, DMF, rt, 18 hrs; d: tBuOLi, THF, rt, 1.5 hrs; e: **7.4**, DMSO, rt, 5 hrs; f: $RuCl_2(=CHPh)(PCy_3)_2$ cat., benzene, reflux, 18 hrs; g: maleimide, toluene, reflux, 18 hrs.

Figure 7.10 SP chemistry assessment leading to the isoindoline pool library **L2**.

Some reactions did not go to completion, but the use of excess reagents on SP and the application of multiple reaction cycles were forecasted to drive the conversions to completion.

The same reaction scheme (Fig. 7.10) was followed on SP starting from the chloro derivative of Wang resin and coupling it with the sodium salt of protected phenolcarboxylate **7.8** to give the supported ester **7.9** (classical Mitsunobu reaction of Wang resin with the protected phenol was less successful). After deprotection with Pd (allyl performed better than simple methyl as a protecting group), the carboxylic function was coupled with two allyl amines (**7.2a,b**) to give the resin-bound secondary amides **7.10a,b**. These were N-alkylated by the mesylate **7.4** using two reaction cycles to convert all the starting material; lithium *t*-butoxide was used as a base, as other bases were either not soluble in the reaction medium or caused instability and degradation of **7.11a,b**. These intermediates were cyclized via ruthenium-catalyzed intramolecular metathesis without affecting the linker stability, and maleimide was condensed to give the final resin-bound adducts **7.13a,b** as single diastereomers with good yields and purity (Fig. 7.10). All the intermediates of the SP assessment were characterized either on-bead (^{1}H MAS NMR) or off-bead (MS, HPLC, NMR) after cleavage (TFA–water 95/5).

The next step was the synthesis of the pool library **L2**, made from 10 carboxylic acids (**R1**), 4 allyl amines (**R2**), 5 propargyl mesylates (**R3**), and 16 dienophiles (**R4**, **R5**), as 16 pools of 200 individuals where the dienophile position is defined (Fig. 7.11).

L = Wang linker

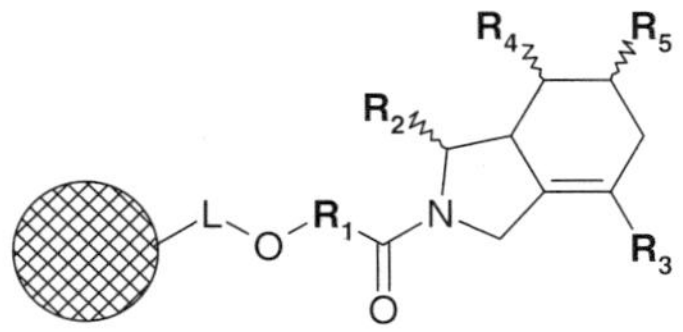

L2
3200 member pool library
16 pools
200 compounds/well
$\mathbf{R_4}$,$\mathbf{R_5}$ determined, $\mathbf{R_1}$-$\mathbf{R_3}$ randomized

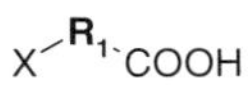

: **10 $\mathbf{R_1}$** substituents including aryl or arylalkyl groups optionally substituted on the aryl ring.

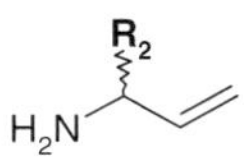

: **4 $\mathbf{R_2}$** substituents including H, alkyl or arylalkyl groups.

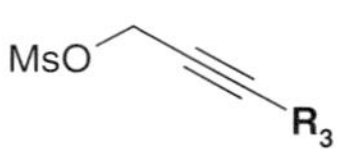

: **5 $\mathbf{R_3}$** substituents including H or alkyl groups.

: **16** dienophiles including cyclic, *cis* and *trans* disubstituted olefins containing at least one electron-withdrawing group, or alkynes containing at least 1 EWG.

Figure 7.11 General structure and composition of the 3200-member isoindoline SP pool library **L2**.

The jump from two discretes to a 3200-member pool library looks rather ambitious, but it is possible that more discretes and maybe even a small model library were prepared but not reported in the communication. The authors were, as usual, protecting the chemical diversity of the prepared library. In fact, the structures of the 10 + 4 + 5 + 16 = 35 monomers selected were not given in the paper. The library was QCed by randomly sampling 24 beads from each pool, cleaving them individually, and submitting the 384 solutions to LC/MS analysis; 14 out of the 16 pools produced >70% beads with >70% HPLC/MS purity, while the other 2 pools did not give any identifiable product and were discarded. It is safe to assume that 14 × 200 = 2800 individuals possessed a sufficient quality to be tested in various assays, and surely the major pharmaceutical company responsible for this communication had this SP pool library (and maybe expansions by using other monomers) tested on biologically relevant targets.

More examples of SP pool libraries will be presented in the next sections in relation to structure identification of positives from a library. Other references can be found in recent reviews (39–42), and we will mention here some very recent communications of the synthesis of SP pool libraries. Pei et al. (43) reported a 4140-member dihydroquinoline library prepared as 30 pools of 138 compounds from α-amino acids, *o*-nitrobenzaldehydes, and acyl chlorides; Krchnak and Weichsel (44) reported a 2720-member diazepine library from diamines, amino alcohols, aldehydes, and primary amines; Cao et al. (45) reported a 4608-member library of cyclopentanes prepared as 768 pools of six compounds from diamines, styrene derivatives, and amines using ring-opening cross-metathesis; Nefzi (46) reported a 38,800-member library of hydantoins and thiohydantoins from α-amino acids and alkylating agents; Neustadt et al. (47) reported a 6859-member biphenyl library prepared as 19 pools of 361 compounds from α-amino acids, including exotic amino acids; a 40-member library of 1,3,5-trisubstituted pyridinium salts prepared as pools of compounds from bromonicotinic acid performing a Suzuki coupling and an N-alkylation of pyridines was reported by Amparo Lago et al. (32); Zhu and Boons (33) reported a two-directional strategy to prepare a small library of trisaccharides as pools of compounds using thioglycoside building blocks; Gennari et al. (48) reported the synthesis of three SP pool libraries of N-alkylated vinylogous sulfonamido peptides as pools of compounds using mild N-alkylation conditions.

7.2 DIRECT STRUCTURE DETERMINATION OF POSITIVES FROM SOLID-PHASE POOL LIBRARIES

7.2.1 Structure Determination of Positives

A medium–large SP pool library is typically prepared in large quantities (up to 500 library equivalents), allowing it to be tested on many assays, and novel, relevant, identifiable structures should originate from each of these screening campaigns. While the library QC is performed only once after its synthesis and before screening, as seen in the previous section, fast and reliable methods to identify the active compound(s) from a pool and match activities with structures are needed after every screening.

Direct structure determination methods, where positives are characterized directly via off-bead or on-bead identification of their chemical structure, will be described in detail in this section. Indirect methods that determine the structure of positives from the library architecture will be covered later: they use either *deconvolutive methods* (Section 7.3), where the iterative synthesis of library pools with decreasing complexity via sequential determination of the best monomers leads to the identification of a positive structure, or *encoding methods* (Section 7.4), where, during the library synthesis, the structure of each component is coupled to a tag that can be read from a single bead after the library screening.

We will describe these methods illustrating their complementarity with respect to the skills and the equipment available in the laboratory where the SP pool library must be prepared, but most of all with respect to the specific synthetic scheme, format, and size that are planned for each library.

7.2.2 Off-Bead Direct Structure Determination of Positives

A high-quality SP pool library is usually partitioned into a few aliquots, each containing several library equivalents, after its synthesis and successful QC; then one or more of these portions are cleaved and immediately submitted for screening. Often the screening campaign measures the affinity of the library for the target, which is added to the assay medium as a purified soluble reagent. An off-bead, target-assisted structure determination of library positives is feasible if the assay functions as a separation method dividing the active compounds from the others and if a suitable analytical method can determine the structure of the positives in situ. Alternatively, an analytical technique may be able in determined experimental protocols to detect only the signal related to compounds binding to the target suppressing the signal of any unbound library individual, thus functioning as a screening technique in itself. We will describe the main features of both these methods and will comment on their usefulness in this section, considering that this strategy can be applied both to SP pool libraries, after their cleavage, and to solution-phase pool libraries.

The target may assist in structure determination via the formation of a noncovalent complex with the active library components. These high-MW complexes are then separated from the rest of the low-MW library individuals, and finally the complex is destroyed and the active library components are structurally characterized via a suitable analytical technique. The target acts as an affinity reagent, and care must be taken both in preventing nonspecific associations between the target and some library components and in finding a robust separation protocol to isolate the target-positive complexes, even when their binding constant and their affinity is relatively weak. The analytical technique must be extremely sensitive and soft, as it must be able to detect minimal quantities of the selected compounds without creating artifacts (fragmentation, degradation, etc.).

The majority of reports have used electrospray ionization mass spectroscopy (ESI-MS) as an analytical detection method because of its sensitivity and the soft nature of its ionization procedure, which generally only leads to the detection of the molecular ions of the positive library members. Many separation techniques have been coupled to ESI-MS, including affinity chromatography (49), size exclusion chromatography (50, 51), gel filtration (52), affinity capillary electrophoresis (53–58), capillary isoelectric focusing (59), immunoaffinity ultrafiltration (60), and immunoaffinity extraction (61). ESI-MS has also been used alone (62) to screen a small carbohydrate library. Other examples reported alternative analytical techniques such as MALDI MS, either alone (63, 64) or in conjunction with size exclusion methods (65), or HPLC coupled with immunoaffinity deletion (66).

As a case example we will consider the approach of van Breemen and co-workers (67–69), who applied pulsed ultrafiltration ESI-MS (70) to the screening for inhibitors of adenosine deaminase (67, 68) and dehydrofolate reductase (69). A schematic representation of the whole process is reported in Fig. 7.12. The authors used an HPLC/MS apparatus where the column was substituted with an ultrafiltration chamber (around 100 μL volume) equipped with a 10,000-MW cutoff membrane. Aqueous buffer solutions of the library components and the receptor were incubated, then loaded

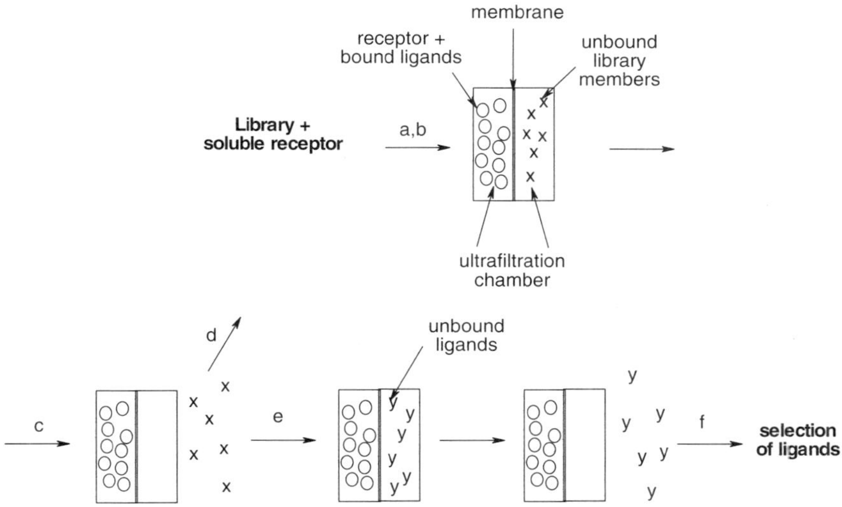

a: incubation; b: injection into the ultrafiltration chamber; c: elution of unbound library individuals with aqueous buffer; d: to waste; e: disruption of target-ligand complexes by elution with MeOH/H_2O/AcOH; f: detection of positives by ESI-MS spectrometry.

Figure 7.12 Off-bead structure determination: pulsed ultrafiltration/ESI mass spectrometry.

into the ultrafiltration chamber, which was flushed with water for several minutes to elute all the unbound, low-MW library components. The eluting mixture was then switched to MeOH–water–AcOH to disrupt the receptor–ligand complexes and the free ligands were identified on-line by ESI-MS (Fig. 7.12).

Identification of the noncovalent bound complex between the target and the ligands from the library, without the need to free the active library individual, would be an improvement for off-bead direct structure determination. The experimental protocol would be shortened, but more importantly the method would provide direct information on the strength of each ligand–target interaction and would thus be able to rank the library positives according to their potency. This principle was successfully applied by Bruce and co-workers (71–73), who used Fourier transform ion cyclotron resonance MS (FT-ICR-MS) to identify intact noncovalent complex ions between carbonic anhydrase II and derivatized peptide libraries and to rank them in terms of affinity; Hofstadler et al. (74) screened an artificial mixture of five aminoglycosides against several mass-tagged RNA targets using ESI-FTICR-MS, detecting several interactions between the ligands and prokaryotic RNA targets.

NMR has gained a lot of importance as a screening technique in off-bead target-assisted structure determination lately; several NMR-assisted screening methods providing either a qualitative or a quantitative estimation of the ligand-target binding strength using simple ligand mixtures will be briefly described here.

Fesik and co-workers reported the so-called SAR by NMR method (75, 76) which monitors the chemical shift perturbation of ^{15}N-^{1}H heteronuclear single quantum correlation (HSQC) data for small proteins in presence of small ligands. The method requires the use of uniformly labeled ^{15}N receptors, but can screen via sophisticated NMR CryoProbes (77) even 10,000 library components per day. Successful applications to identify inhibitors of Human Papillomavirus E2 protein (78), of stromelysin (79), of (80) and of Erm methyltransferases (81) have been reported by the same group. While the method is extremely reliable and accurate, often leading to relevant lead compounds, several bottlenecks must be highlighted: the need of significant quantities of uniformly ^{15}N labelled, purified receptors/proteins/enzymes; the restriction to low MW targets (only up to 30 kDa); the need of complete structure determination and NMR signal assignment for the target; the need of complex and expensive instrumentation which as of today confines this method to a few specialized laboratories. Its usefulness, thus, may be higher for later rather than early discovery phases via HTS.

A method named pulsed field gradient NMR (PFG-NMR) was recently reported by Shapiro and co-workers (82–86) as being able to discriminate between bound and unbound library individuals, which had different diffusion coefficients in solution when complexed with macromolecules. This property allowed the editing of the NMR spectrum to see only the signals of bound molecules. The authors applied this technique to detect interactions between small molecules (82–84), small molecules and vancomycin (85), and small molecules and medium-length oligonucleotides (86). A somewhat different application of PFG-NMR by Hajduk et al. (87) obtained the spectra of bound library components by subtraction of the spectra of the mixtures in the presence and absence of two different proteins, FKBP and stromelysin, and confirmed the detection of two known ligands in each NMR diffusion screening, thus validating this target-assisted screening protocol. The validation studies up to now reported have only hinted towards the usefulness of PFG-NMR, and the publication of a "real" library screening would reinforce the confidence in this method; as of today its low throughput (several hours per sample) and the risk of experimental error (up to 100%) limit its scope.

Meyer et al. recently reported transfer NOE (nuclear overhauser effect) (trNOE) (88), which detects the strong negative trNOE effect of receptor-bound molecules compared to the weak, positive trNOE of unbound compounds without the need of receptor labeling; the method was used to identify an E-selectin antagonist from an artificially assembled 10-member library of saccharides (89).

The same approach has been cleverly exploited by Fejzo et al. (90) by using the so-called SHAPES strategy. The authors selected among the CDC database (91) the meaningful frameworks which are most frequent in the structures of known drugs, i.e., the cyclic arrays of atoms constituting several rings and the connecting atoms working as linkers. A careful computational search and selection (92) limited to 41 the number of frameworks representatives of a large part of CDC. These core scaffolds were used to screen the ACD database (93) and to select commercially available decorated/modified scaffolds without toxic or unstable moieties; with side-chains conferring aqueous solubility; with at least one N or O atom. Several novel compounds obtainable with simple routes were finally added, leading to a SHAPES library of 132 compounds with MWs from 68 to 341 Da and with logPs from −2.2 to 5.5 (see refs. 90 and 92 for more details). This basic set was screened on several targets in a qualitative mode, that is without trying to quantify the trNOE phase switch of the ligand cross peaks but considering it a signal of ligand-target interactions; a reliable "yes-or-no" answer was unequivocably obtained for each SHAPES library individual. Several weak binding frameworks (micro- to low millimolar constants), which would have been missed by conventional biological screening protocols, were used to focus later larger screening efforts on biased collections or libraries containing these or similar frameworks; the frequency positives from SHAPES-derived screening sets was significantly increased if compared with random screening (90), proving the validity of the approach. The method does not require labeled receptors and is ideally suited for large macromolecules (>60 kDa); the amount of macromolecular target needed for a SHAPES screen is relatively limited (tens of mgs of large proteins); a complete NMR assignation for the macromolecular target is not required; the library size allows to use small mixtures of ligands (one to four ligands per sample), thus increasing reliability and simplicity; the SHAPES NMR screen takes only from several hours to a few days; a visual inspection of the spectra is enough to discriminate among binders and non-interacting compounds. Even though the information acquired is not quantitative (but can be quantified by progressing the positives through PFG-NMR experiments, as was done in ref. 90) and the binding specificity must be determined, this method will surely become a major asset for the early, information-poor phases of ligand fishing related to difficult targets for which no ligands have been found using traditional HTS campaigns. The method could in future be tailored also for non-pharmaceutical applications.

NOE pumping (94) was recently reported by Chen and Shapiro. After the application of a diffusion filter suppressing all the ligand signals, the NOE experiment started to "pump" the magnetization from the target to the binders in the ligand mixture. By increasing the mixing time t_m of the target and the ligand mixture the magnetization transfer gradually decreased the target signals (horse serum albumin in ref. 94), and gradually increased the ligand signals (salicylic acid) without interferences from nonbinders (glucose and ascorbic acid). Similar characteristics in respect with trNOE (no target labelling, no MW cut-offs for targets, no need of a complete NMR assignment for the target) make also this technique particularly appealing for NMR-assisted screening of mixtures.

Saturation transfer difference NMR (STD-NMR) was recently reported by Mayer and Meyer (95) as a fast and reliable screening method and was used to spot the binding of saccharides to wheat germ agglutinin (WGA); the same assay was repeated successfully anchoring WGA on SP and recording spectra from heterogeneous systems using magic angle spinning STD-NMR (96).

Several reviews dealing with analytical techniques and off-bead structure determination methods are available to the reader. The application of MS techniques to target-assisted, off-bead structure determination of library actives has been recently reviewed (97, 98), as happened also for NMR screening methods for positive identification (99–102). Two more general, screening-focused reviews dealing with target-assisted methods using immobilized (supported) or confined targets (segregated in a compartment) targets for identification of positives from mixtures have recently appeared (103, 104).

The majority of the above-mentioned examples dealt with small, assembled mixtures of commercial compounds, containing known ligands and inactive compounds, which were used to validate the ability of the target-assisted screening to detect known interactions. Other examples have tested small peptide or peptidomimetic libraries, and in general, all the reported studies have proved the applicability of off-bead target-assisted screening of model pool libraries. It is easy to predict, though, that the coming years will see an expansion of the efforts in this area. Eventually target-assisted screening and identification of positives for specific applications, especially in the pharmaceutical area, will become an important screening option.

7.2.3 On-Bead Direct Structure Determination of Positives

The identification of positives from an SP pool library usually requires cleavage from the beads, a screening campaign, and finally the crucial step of linking the observed activity with the structure of one, or more, of the library components. A powerful approach, generally named *bead based*, consists of submitting the libraries to assays that allow detection of the activities of single beads, and thus eventually to determine the structure of the individual that was loaded on the specific, active bead. Some formats for bead-based structure determination will be examined in detail in Sections 7.4 and 7.5, while here we will present the so-called *on-bead approach*.

Screening of a library is usually intended as a solution process, where both the library individuals and the target/receptor are dissolved in the same assay medium. An obvious drawback is that when the library aliquot (usually a few library equivalents, to ensure coverage all the chemical diversity) is cleaved and tested, only a few active compounds (if any) are found, but the whole library aliquot is then lost and a second screening campaign on a different target will require a new resin-bound library aliquot. If, however, the resin-bound library is on-bead screened, two major advantages are gained. First, the complex between the target and the library individual is formed at the solid–solution interface and, by using denaturating conditions, the target can be washed away and the same library aliquot can be tested on other targets. Second, positive beads are easily spotted, for example, by building a colorimetric or fluorescent detection method able to spot the target–ligand complex; after removal from the other library members of the spotted beads and complex denaturation, the structures of the ligand library individuals are directly determined by sensitive analytical methods.

On-bead screening and structure determination of positives from pool libraries was first reported by Lam et al. (2) using an enzyme-linked colorimetric assay. This detection method was subsequently used by the same group (105–110) and others (111–115) to probe the specificity of random peptide or peptidomimetic libraries. Other reports have used fluorescence-labeled (110, 116, 117) or radionucleotide-labeled (118, 119) targets. Aside from the above-mentioned binding assays, functional assays have also been performed on-bead (120–123), and hybrid off-bead/on-bead assays have been reported (124). The use of chromogenic test substrates to screen on-bead for Zr^{4+}-binding peptides as ligands to promote phosphate hydrolysis was also reported (125). All these examples were able to detect one or more peptide sequences, which were reconfirmed after their identification as true binders/inhibitors interacting with the desired target. Recently an example of large oligocarbamate libraries (one cyclic trimer library, 19,863 individuals, and two tetramer libraries, 531,441 individuals) screened to search for human thrombin ligands (126) and another of glycopeptide libraries (around 300,000 individuals) screened for lectin binding (127) somewhat enlarged the applicability of on-bead screening to library structures other than peptides.

Several considerations need to be made regarding the usefulness of on-bead screening and structure determination. The interested reader will find these treated in detail in some recent reviews (14, 105, 128). To date, on-bead screening has been used to detect and identify biological interactions using isolated receptors, enzymes, and antibodies as targets and aqueous environments. The use of hydrophilic supports is necessary to create the suitable aqueous environment for the target–ligand interaction while allowing proper swelling of the beads. Relevant studies (115, 119, 129–131) have been performed to determine the *accessibility of the resin sites* to macromolecular targets; enzymes such as trypsin (23.5 kDa), papain (23 kDa) and chymotrypsin (22 kDa) were used to cleave resin-bound peptide sequences (119, 129, 130) and to bind to resin-bound inhibitors (131).

The results of these studies showed that hydrophilic PS swelling resins, such as Tentagel or Argogel, are usually accessible to the enzymes only on their surface, while access to inner sites is possible (115, 131) but only through slow, complex exchange mechanisms with the surface sites and when the ligand has a potent binding activity. Other polyacrylamide-based PEGA hydrophilic resins (132) are completely accessible to the above-mentioned macromolecules, even at the inner sites and, at the time of writing, represent the resin of choice for on-bead screening and structure determination. Any other hydrophilic resin, though, can be used for colorimetric assays where binding to the surface stains the positive beads.

The technical details of on-bead screening are also extremely important, because *false positives* (beads that turned up positive but failed to reconfirm activity as resynthesized pure individuals) due to a specific binding of library individuals may be a major problem when testing millions of beads. Many possible causes have been reported for such a specific binding: extremely hydrophilic molecules or highly charged peptides, artificial dimeric or trimeric resin-bound individuals, which were not active as monomers, and even interactions with parts of the bead structure could produce a significant amount of false positives. Some general precautionary procedures [use of high-ionic-strength buffers or addition of proteins such as bovine serum albumin (BSA) to prevent aspecific lipophilic binding] have been suggested, and several protocols have also been developed where a double-staining procedure is performed to minimize aspecific interactions. These involve repeating the assay twice on positive beads after complex denaturation or after incubating them in the presence of a known ligand to show the competition for the receptor (107, 110). A dual-color detection scheme has also been designed and successfully applied (106).

The main issue is *support anchoring* the library individuals onto the SP. For peptides, adjustments can be made to support the molecules on a side-chain function that does not interfere with the binding and/or a spacer can be inserted between the resin and the peptide sequence. Small organic molecules, though, generally contain just a few functional groups, which are often essential for the activity on the target. Their use to support the library components will prevent binding and consequently the detection of positives. The same may also be true if the linking function is adjacent to a relevant portion of the active library compound, thus preventing the approach of the target and the correct docking between the two partners. It would be desirable to perform on-bead screening and structure determination on cleavable SP SOM pool libraries and then to cleave the library and test it in solution. A comparison of the results would show if on-bead screening is successful (similar screening results) or not (no on-bead activity, actives in solution). An even better solution to save precious library equivalents would be, when possible, to support a known active compound, belonging to the chemical class that will constitute the library, in exactly the same manner as planned in the SP scheme. If the active standard results on-bead active the library can be tested on-bead; otherwise an off-bead screening must be performed, followed by a more classical structure determination approach. Unfortunately, when active standards are available, the need of a large primary library is reduced, and a small–medium discrete focused library is usually preferred.

A comparison between different library screening formats was carried out using three identical, 576-member peptide polyamine conjugate libraries (133); one of them was screened in solution, one anchored onto Tentagel beads and the last as a PEGA-supported pool library. The libraries were screened for inhibition of trypanothione reductase used as a soluble target weighing around 55 kDa; in accordance with previous experiences (131), the PEGA-supported resin provided several medium–low nanomolar inhibitors, as did the soluble library, thus showing the good properties of this support for on-bead screening. In this specific example, the constraint imposed by the support and by the linkage to the bead did not prevent the biological interaction between the library individuals and the enzyme. The library supported onto Tentagel, on the contrary, did not show any result from screening, even if theoretically the identification of surface sites on the beads should have allowed to spot the positive beads. After this successful experiment, the same screening protocol applied to small organic molecule libraries supported onto PEGA-like resins is crucial to judge the future potential of on-bead direct screening. In the next section we will describe in detail an excellent example where peptide-derived libraries were screened on-bead and valuable active structures were rapidly determined.

7.2.4 An Example: On-Bead Screening and Structure Determination of Positives from SH3 Domain-Directed Libraries

Schreiber and co-workers recently reported (134, 135) the synthesis of two SP pool libraries **L3** and **L4** based on the structures of a phage display-derived dodecapeptide ligand of the SH3 domain of the tyrosine kinase Src (**7.14**, Fig. 7.13) (136), of its truncated version (**7.15**), and of a nonpeptide ligand derived from an encoded primary library (**7.16**, Fig. 7.13) (137). The authors kept the common motif proline-leucine-proline (PLP) in the biased library structures and explored the left (library **L3**, 2500 members, Fig. 7.13) and the right part of the ligand sequences (library **L4**, 125,000

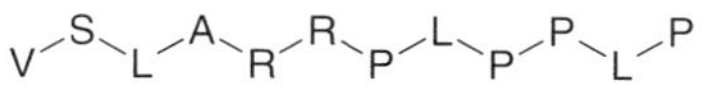

7.14

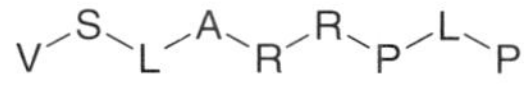

7.15

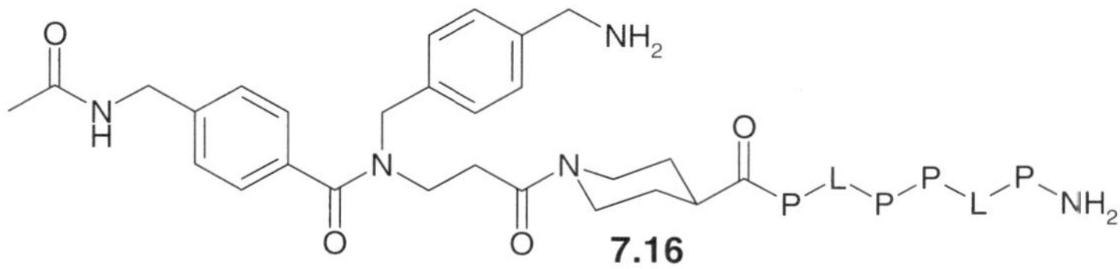

7.16

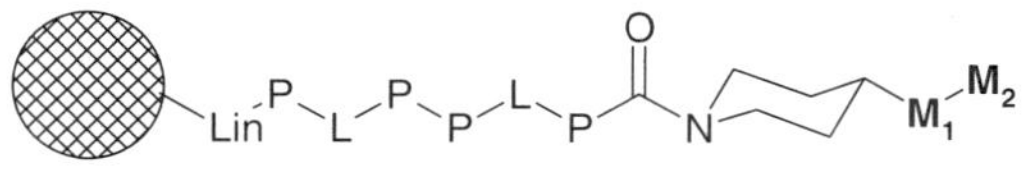

L3

2.500 individuals, encoded pool library
M_1: 50 monomers (amino acids)
M_2: 50 monomers (isocyanates, isothiocyanates, acyl chlorides, acids, anhydrides, amino acids)

Lin = linker

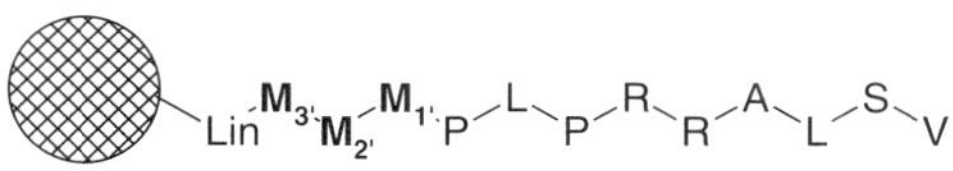

L4

125.000 individuals, encoded pool library
M_1-M_3: 50 monomers (amino acids)

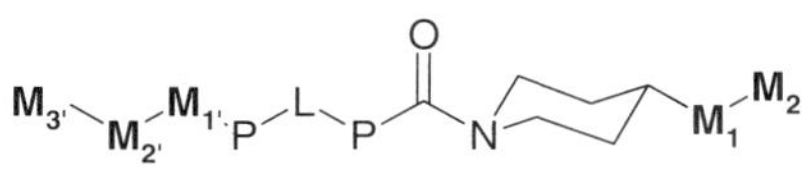

7.17

Figure 7.13 On-bead structure determination: structures of known SH3 peptidic and nonpeptidic ligands (**7.14–7.16**) of designed SP peptidomimetic libraries (**L3**, **L4**) and of the planned, optimized ligand **7.17**.

members, Fig. 7.13). Subsequent combination of the best findings from the two libraries should have produced a new, hybrid generic structure **7.17** (Fig. 7.13).

Both libraries were prepared using automated standard peptide coupling conditions for monomer additions using either commercially available or easily prepared monomers. The biasing library elements were either initially coupled to the resin (iNpe-PLPPLP, **L3**) and then elongated or used as a final capping reagent (VSLARRPLP, **L4**) for the trimeric randomized sequence. Randomizations were performed using mix-and-split techniques (2, 3), and each library component was encoded (111) to accelerate the structure determination of library individuals loaded onto positive beads.

Library **L3** was screened using a soluble N-biotinylated N-terminal residue of the Src SH3 domain (134), which had been previously incubated with the commercially available alkaline phosphatase conjugate with streptavidin (SAAP). The strong interaction between biotin and streptavidin formed a tight receptor complex, which was subsequently incubated with two batches of around 12,500 beads (five library equivalents), which had been previously incubated with the screening buffer to reduce nonspecific binding. After 12 h of incubation, the receptor solution was filtered off, the beads were thoroughly washed with the screening buffer and then treated with the alkaline phosphatase substrate BCIP (5-bromo, 6-chloro, 3-indolyl phosphate), and the staining agent nitro blue tetrazolium (NBT). The complexed alkaline phosphatase on positive beads dephosphorylated BCIP, and this eventually reduced NBT to the insoluble, deep blue diformazan, which mostly precipitated in proximity to the phosphatase, that is, on the active beads. These colored beads were pipetted out of the assay medium, resuspended in a new assay medium and then washed with a destaining, denaturating 6 *N* guanidinium hydrochloride solution to recondition them for new assay cycles. Treatment with SAAP alone allowed removal of beads interacting only with the alkaline phosphatase, while the others were submitted to a second, identical SH3 screening to reconfirm their activity.

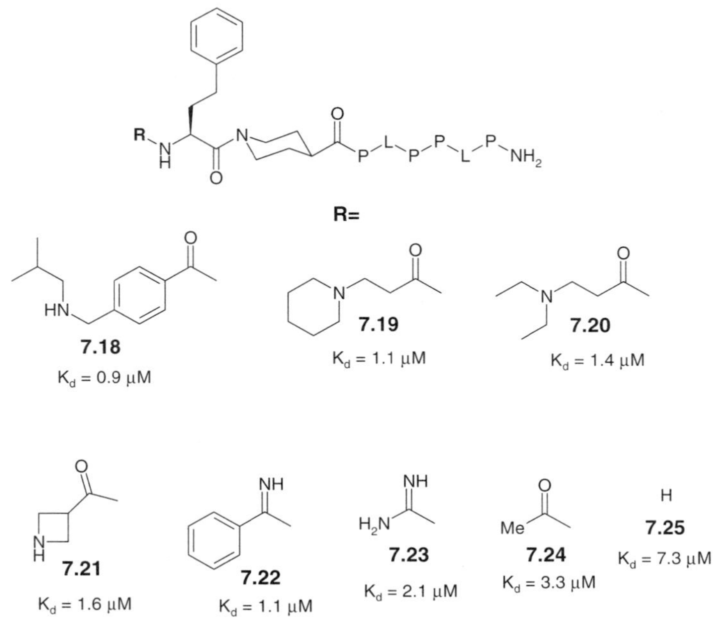

Figure 7.14 On-bead structure determination: resynthesized positives **7.18–7.25** from the SP peptidomimetic library **L3**.

On-bead screening of **L3** produced 32 positive beads, which were rapidly decoded (111). Thirty out of 32 beads contained a specific monomer in position **M1**, while position **M2** showed a higher tolerance for different monomers. The structures of resynthetized positives **7.18–7.25** with their binding constants are reported in Fig. 7.14. The larger library **L4** was tested and decoded using the very same protocol, and, while some sequences were discarded because they were unrelated to any other structure, a few related sequences (**7.26–7.28**, Fig. 7.15) were determined. The

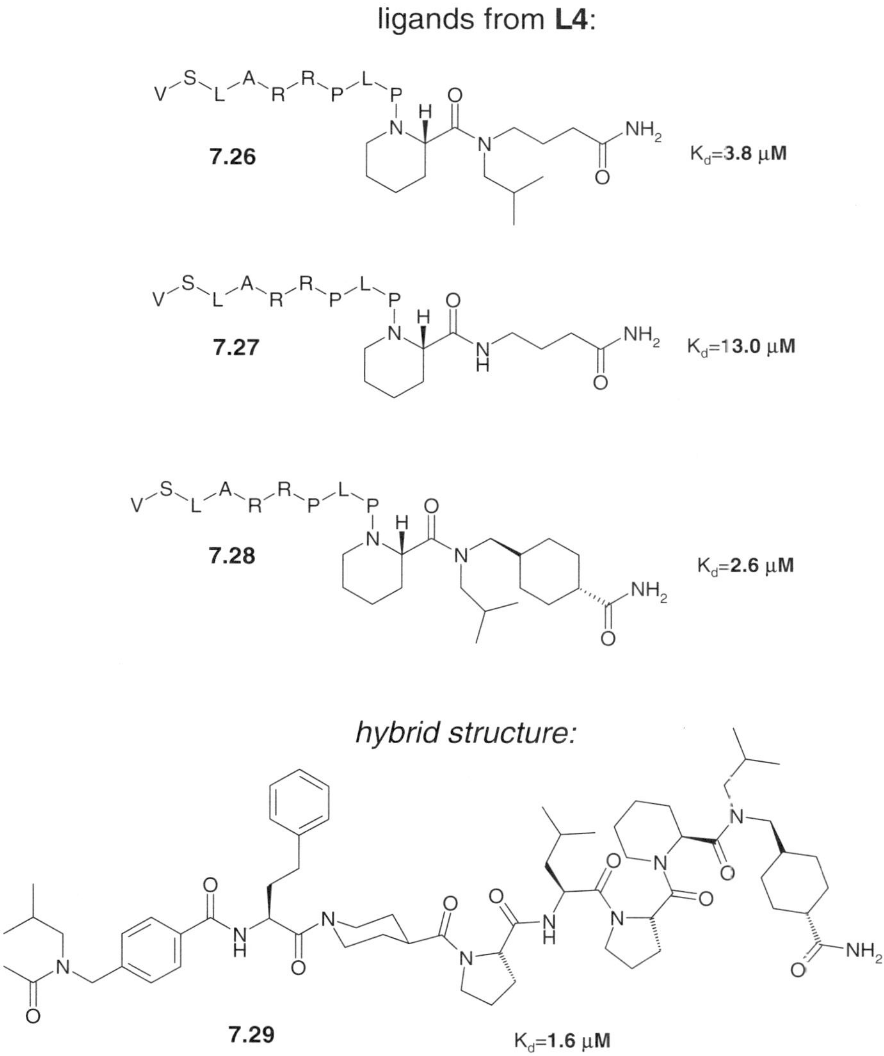

Figure 7.15 On-bead structure determination: resynthesized positives **7.26–7.28** from the SP peptidomimetic library **L4** and hybrid structure of an optimized **L3/L4** derived ligand **7.29**.

combination of structural information derived from **L3** (**7.18**) and **L4** (**7.28**) produced the "hybrid" ligand **7.29** (Fig. 7.15), which possessed a comparable binding activity to the original dodecapeptide **7.14** having replaced 9 of the 12 α-amino acids with nonnatural building blocks.

7.3 DECONVOLUTION METHODS FOR SOLID-PHASE POOL LIBRARIES

7.3.1 Iterative Deconvolution

While the structure determination methods seen in the previous section are based on the analytical identification of the positive library individuals, other methods exist to perform the so-called deconvolution of the library complexity and allow the eventual identification of one or more positives from the library without determining their structure by analytical methods. The most common approach is based on iterative synthesis cycles of less complex pools deriving from the original active pool(s) until single compounds are prepared and tested (*iterative deconvolution*) (3).

A detailed scheme for the iterative deconvolution of a 625-member hypothetical library, where four monomer sets, each composed of five representatives, are added to form the library members in four consecutive steps (Fig. 7.16), is reported in Fig. 7.17. Monomers **P–T**, **K–O**, and **F–J** are added respectively in the first, second, and third randomization steps using the mix-and-split technique; then monomers **A–E** are added in the fourth step and are kept divided. The library is thus prepared as five pools **A–E**, each containing 125 individuals, and its screening (step a, Fig. 7.17) produces four inactive pools and one active pool **D**, which is selected (step b). The four pools **A–C** and **E** are now discarded, and a first synthesis iteration (step c) produces an 125-member library where monomer **D** is always present in the fourth position. The five prepared pools **F–J**, each containing 25 individuals, have a determined monomer in the third position. Their screening (step a) produces two active pools, **G** (less active) and **J** (more active), and three inactive pools, from which the most active pool **J** is selected (step b). The second iterative synthesis (step d) produces a 25-member, **DJ**-biased library with five pools **K–O**. Among them pools **L** (more active) and **N** (less active) are active (step a) and the former is selected (step b). Finally the third synthesis iteration (step e) prepares 5 discretes **DJLP–DJLT**, and three actives **DJLP**, **DJLS** (most active), and **DJLT** are found (step b, Fig. 7.17). An attractive modification, named *recursive deconvolution* (138), archives an aliquot of each intermediate pool during the mix-and-split synthesis. This allows rapid iterative deconvolution cycles using archived samples rather than repreparing deconvolution libraries from the beginning. This modification has been successfully applied to a small carbohydrate library (139).

Iterative deconvolution selects the pools solely according to the screening results, and no analytical characterizations are performed at any deconvolution stage. This is particularly useful when sophisticated analytical instrumentation is not available. The measured activity is always the sum of each individual contribution. Assuming that the SP assessment studies, the monomer rehearsal, and the model library synthesis

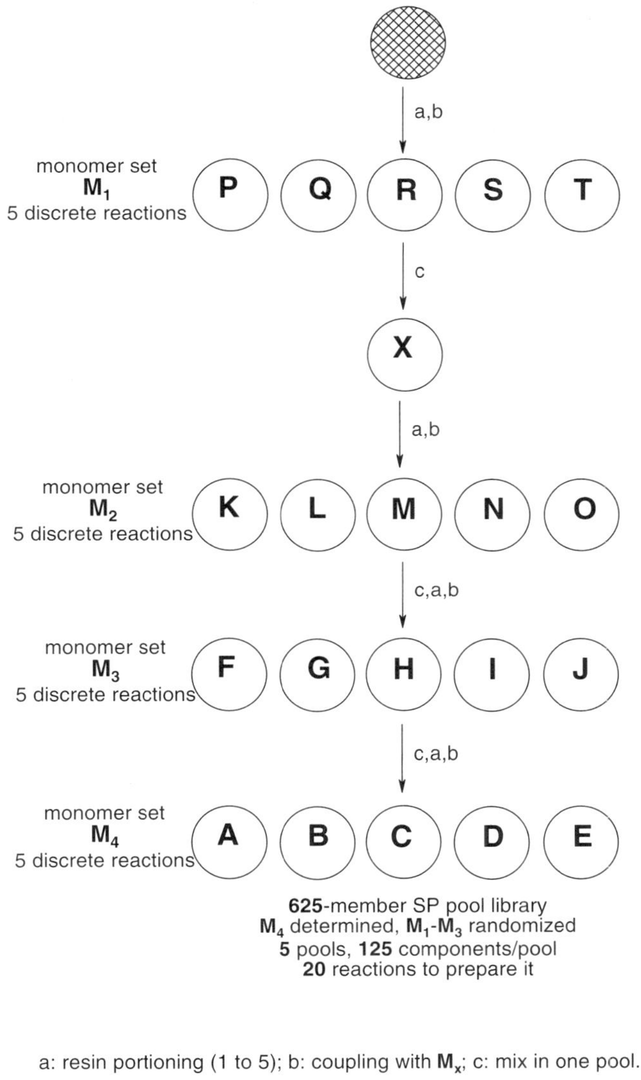

Figure 7.16 SP synthesis of a 625-member hypothetical pool library.

have produced satisfactory results, the population of each pool will be composed of the expected library members in roughly equivalent amounts. Nevertheless, if large pools are produced (>50 individuals), it is easy to lose even a very active individual that is contained in a pool of otherwise inactive compounds. The number of components in a pool should be determined on the basis of the screening method's sensitivity and of the final concentration of each pool component (highly sensitive methods may spot low concentrations of active compounds).

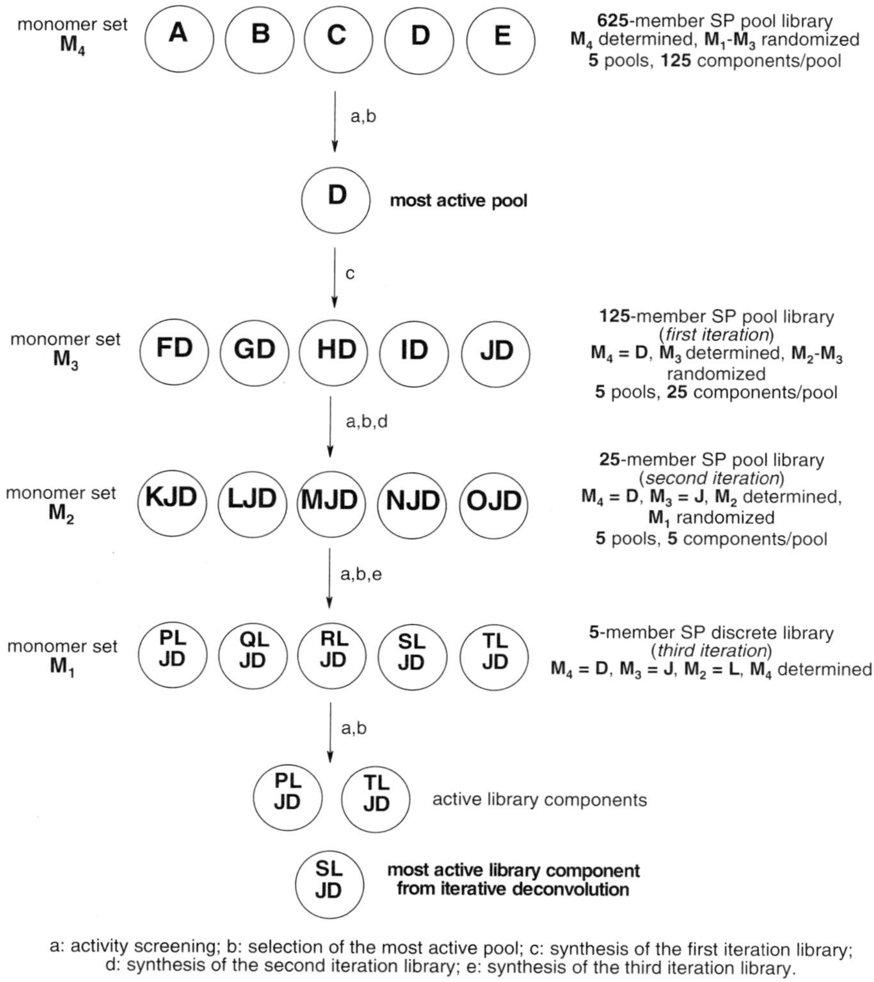

Figure 7.17 Iterative deconvolution of a 625-member hypothetical SP pool library.

Small organic molecule pools of >50 compounds are likely to cause problems in detecting and further deconvoluting activities. In fact, large pools often produce false positives as a result of interactions of pool components and are, in general, less reliable than small pools. Obviously the rationale of SP pool libraries is to increase the synthesis throughput, and small five-member pools would not justify a structure determination effort in comparison with a simpler, more reliable discrete library containing only five times more samples to be screened. Somewhat larger pools of 10–30 individuals represent a reasonable compromise between productivity and reliability of results. Often several pools, rather than a single one, show significant activity (Fig. 7.17, pools **G** and **J**, **L** and **N**). The most active one is progressed as a rule, but according to the complexity of the SP synthesis and the project needs, more than one pool per iteration may be deconvoluted. Sometimes a more active individual may result from a less active pool, and different families/structures of positives may be found. The deconvolution of many pools, though, would again lead to significant efforts that would probably make another library format (discretes) or structure determination method (bead-based, encoding) more effective.

7.3.2 An Example: Iterative Deconvolution of a 1,4-Dihydropyridine Library

Recently Gordeev et al. reported (140, 141) the synthesis and iterative deconvolution of a 300-member focused library of 1,4-dihydropyridines **L5** as potential calcium channel blockers inspired by the structure of nifedipine and other known bioactive compounds. The library was built on SP as 30 pools of 10 individuals, following the synthetic scheme shown in Fig. 7.18. Rink amine resin was deprotected, split into 10 portions (steps a and b), and treated with β-ketoesters (step c, $\mathbf{M_1}$, 10 representatives **A–J**, Fig. 7.19) to give 10 discrete N-supported β-enaminoesters **7.30**. The resin aliquots were then mixed (step d) and split again into 30 identical, 10-member pools (step e). Each resin portion was treated with a different combination of 1,3-dicarbonyls ($\mathbf{M_2}$, three representatives **K–M**, Fig. 7.19) and aldehydes ($\mathbf{M_3}$, 10 representatives **N–W**, Fig. 7.19) to give 30 pools of 10 open-chain precursors **7.31** (step f), which were cyclatively cleaved to give the final library **L3** (step g, Fig. 7.18). Library QC was performed by off-bead MS, identifying all the expected molecular ions in several pools.

The library was tested using a known competition assay (142), and binding activities for the 30 pools were acquired. While the pool complexity was low, as was therefore the possibility of false positives/artifacts, the extreme similarity of all the library components with known calcium channel blockers (compare the monomers in Fig. 7.19 leading to nifedipine, $\mathbf{M_1}$ = **A**, $\mathbf{M_2}$ = **K**, $\mathbf{M_3}$ = **T**, with all the others) meant a constant level of activity was to be expected for all pools. For such a small focused library, parallel synthesis would probably have been more suitable to acquire a refined SAR, but we will see how iterative deconvolution succeeded anyway in both identifying active individuals and showing significant activity differences for different pools. The screening results are reported in Table 7.1. Five pools showed activity > 1 μ*M*, 12 pools had an activity between 100 n*M* and 1 μ*M*, and 11 pools were active between 10 and 100 n*M*. Two pools showed an activity around 7–8 n*M*: They both contained methyl acetoacetate ($\mathbf{M_2}$, **K**) as well as 2-fluorobenzaldehyde ($\mathbf{M_3}$, **P**) and 2-nitrobenzaldehyde ($\mathbf{M_3}$, **T**), respectively.

Deconvolution of these pools consisted simply of a single iteration, preparing the 20 discrete components of pools **KP** and **KT**, which produced the results shown in Fig. 7.20. Nifedipine **7.32** was included in the library as a standard and was obtained by deconvolution of the pools together with other active compounds (**7.33–7.36**, Fig. 7.20). These results point toward an even more focused library where only *o*-substituted aromatic and heteroaromatic aldehydes would be included as aldehyde monomers.

Many cases of the iterative deconvolution of SP libraries have been reported, and the reader should refer to some recent reviews (143–145) to access further examples. We will mention only several very recent papers in which iterative deconvolution has

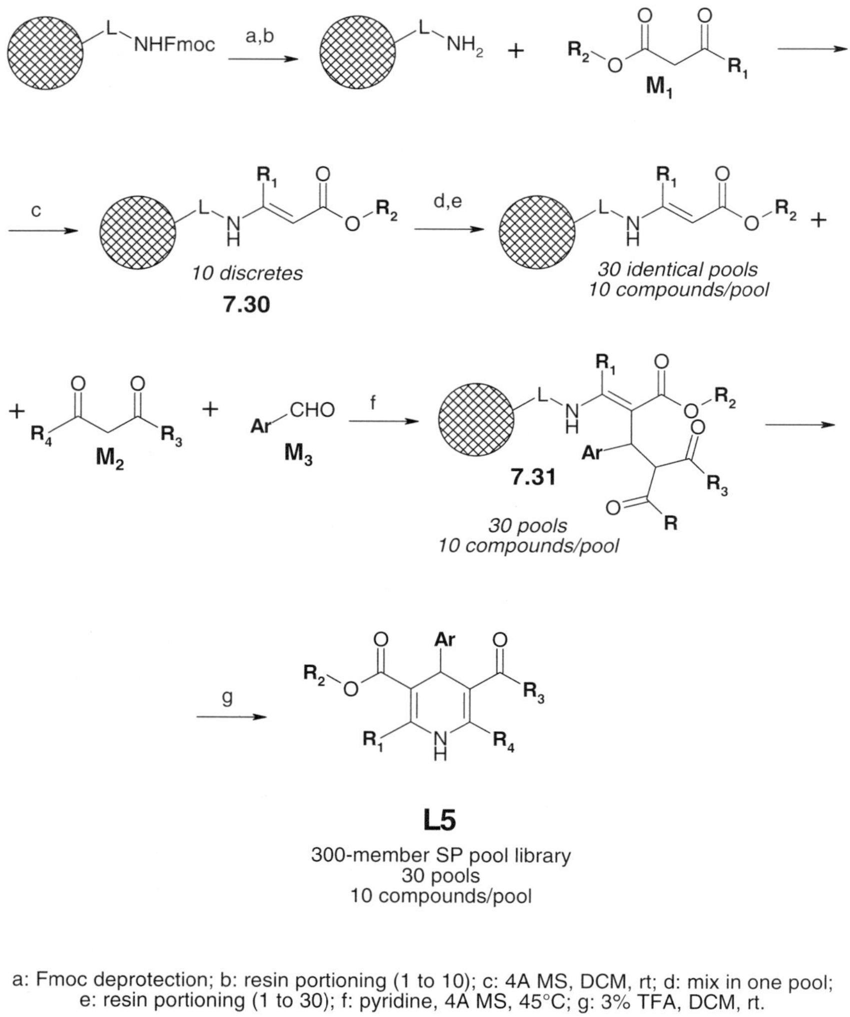

Figure 7.18 SP synthesis of the focused dihydropyridine pool library **L5**.

been successfully applied. Rohrer and co-workers (146–148) reported the iterative deconvolution of two libraries containing respectively 131,760 compounds as 79 pools of 1330 or 2660 individuals and >200,000 compounds as 147 pools. The deconvolution process identified several selective agonists of the somatostatin receptor which were further progressed by more focused efforts. Heizmann et al. (149) reported the synthesis and deconvolution of a 328,509-member SP pool tripeptoid library, made as 69 pools of 4761 compounds and screened for α-melanotropin (α-MSH) and for gastrin-releasing peptide (GRP)/bombesin affinity; the deconvolution process gave two structurally unrelated, low-micromolar ligands for the two targets. An 125-

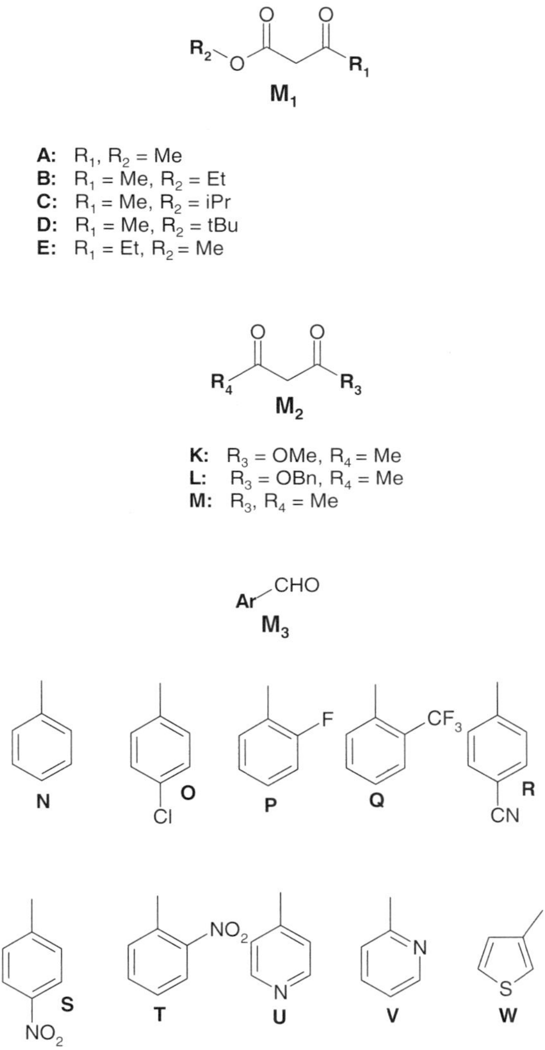

Figure 7.19 Monomer sets $\mathbf{M_1}$–$\mathbf{M_3}$ used for the synthesis of the focused dihydropyridine pool library **L5**.

member SP pool library of azasugar peptide conjugates was made by Lohse et al. (150, 151) as five pools of 25 compounds and screened for β-glucosidase inhibition; the simple deconvolution process yielded a medium–low micromolar enzyme inhibitor. An 144-member library of acylated amidinonaphthols was made by Roussel et al. (152) as 18 pools of eight compounds and tested for the inhibition of tissue factor (TF)/factor VIIa complex, providing a high-nanomolar inhibitor. Szardenings et al.

TABLE 7.1 Biological Activity of the SP Peptidomimetic Pool Library L3

IC_{50}	**K**	**K**	**K**	**K**	**K**	**K**	**K**	**K**	**K**	**K**	**L**	**L**	**L**	**L**	**L**	**L**	**L**	**L**	**L**	**L**	**M**	**M**	**M**	**M**	**M**	**M**	**M**	**M**	**M**	**M**
	N	**O**	**P**	**Q**	**R**	**S**	**T**	U	**V**	W	**N**	**O**	**P**	**Q**	**R**	**S**	**T**	U	**V**	**W**	**N**	**O**	**P**	**Q**	R	**S**	**T**	U	**V**	W
A–D[a]	C	A	**D**	C	B	B	**D**	C	C	C	B	B	B	C	B	B	C	C	C	B	A	A	C	B	A	A	B	B	C	B

[a] *A*: >1000 n*M*; *B*: 100–1000 n*M*; *C*: 10–100 n*M*; *D*: 1–10 n*M*.

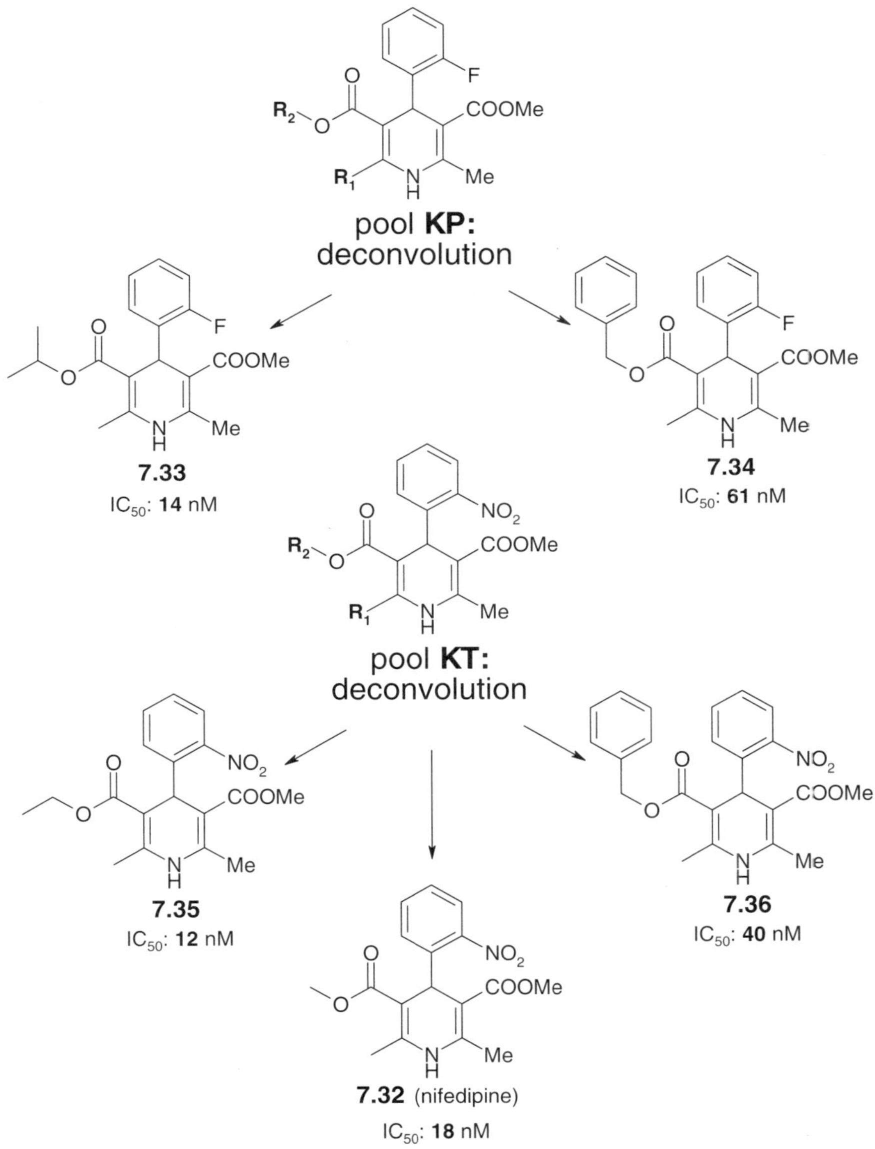

Figure 7.20 Iterative deconvolution of the focused dihydropyridine pool library **L5**: deconvoluted active structures **7.32–7.36**.

(153) reported the synthesis and the deconvolution of a 1225-member diketopiperazine library as 35 pools of 35 compounds; the pools were screened and deconvoluted for their activity on two matrix metalloproteinases (MMPs), gelatinase-B and collagenase-1. The process yielded several target-selective, low-nanomolar inhibitor.

Ng et al. (154) reported the identification of several antimicrobial peptoids from an 845-member SP pool library organized as 65 pools of 13 compounds inspired by previous, larger combinatorial efforts (155). A 400-member SP library based on a 1,3-disubstituted cyclohexanone scaffold, prepared as 20 pools of 20 compounds, was reported by Abato et al. (156); a selective inhibitor of a serine protease, plasmin, was identified from iterative deconvolution of the library.

Rather often, examples of the iterative deconvolution of large libraries do not contain all the details about the library structure, the monomer sets, and the positives obtained in order to protect the proprietary information extracted from the library. This makes it much more difficult to judge the real potential of iterative deconvolution in the structure determination of positives, and it remains only to accept the method as it is, with the practical considerations made in the previous section.

7.3.3 Other Deconvolutive Methods

Another popular deconvolution method is based on the synthesis and screening of multiple library copies, where different positions are determined to identify the best monomer for each library randomization point (*positional scanning*) (157). We can examine positional scanning with the same hypothetical 625-member library used to illustrate iterative deconvolution (monomers **A–T**, see Figs. 7.16 and 7.17). Rather than performing iterative cycles on one original pool library, positional scanning implies the up-front synthesis of four complete copies of the library (step a, Fig. 7.21). The corresponding 20 pools **P–T** (sublibrary **1**), **K–O** (sublibrary **2**), **F–J** (sublibrary **3**), and **A–E** (sublibrary **4**) are assayed (step b). While some pools are weakly active (**M**, **R**, **T**), the four most active pools of each library copy are respectively **D**, **J**, **L**, and **S**. Simply by combining the information derived from the screening and selecting the most active monomer representatives (step c), an active library individual, **DJLS**, is obtained (Fig. 7.21).

The up-front synthesis of multiple library copies, and the deconvolution of large pools to derive directly the active structure are the main features of positional scanning. The synthetic scheme is also more demanding with respect to typical mix-and-split methods: for example, sublibrary **1**, or **P–T**, requires 5 + 25 + 25 + 25 = 80 reactions (Fig. 7.22), while iterative deconvolution that corresponds to sublibrary **4**, or **A–E** (Fig. 7.16), requires only 5 + 5 + 5 + 5 = 20 reactions. Multiple-pool deconvolution is, however, much easier using positional scanning: Referring to the example (pools **M**, **R**, and **T** as less active than the selected most active pools), compounds **DJMS**, **DJLR**, **DJLT**, **DJMR**, and **DJMT** can be prepared as discretes together with **DJLS** with no additional deconvolution efforts. The observations made for iterative deconvolution on pool size (<50) and concentration of library individuals (depending on the screening sensitivity) are also valid for positional scanning.

This approach has often been applied to peptide libraries (158–161), but its use with small organic molecules has been, to date, restricted to two examples of 72- (162) and 54- (163) member solution libraries, which proved the validity of the technique with simple test cases; to a 1600-member SP amide library (164); to a set of large peptide, peptoid, and polyamine libraries (more than 70,000,000 total individuals) (165); and

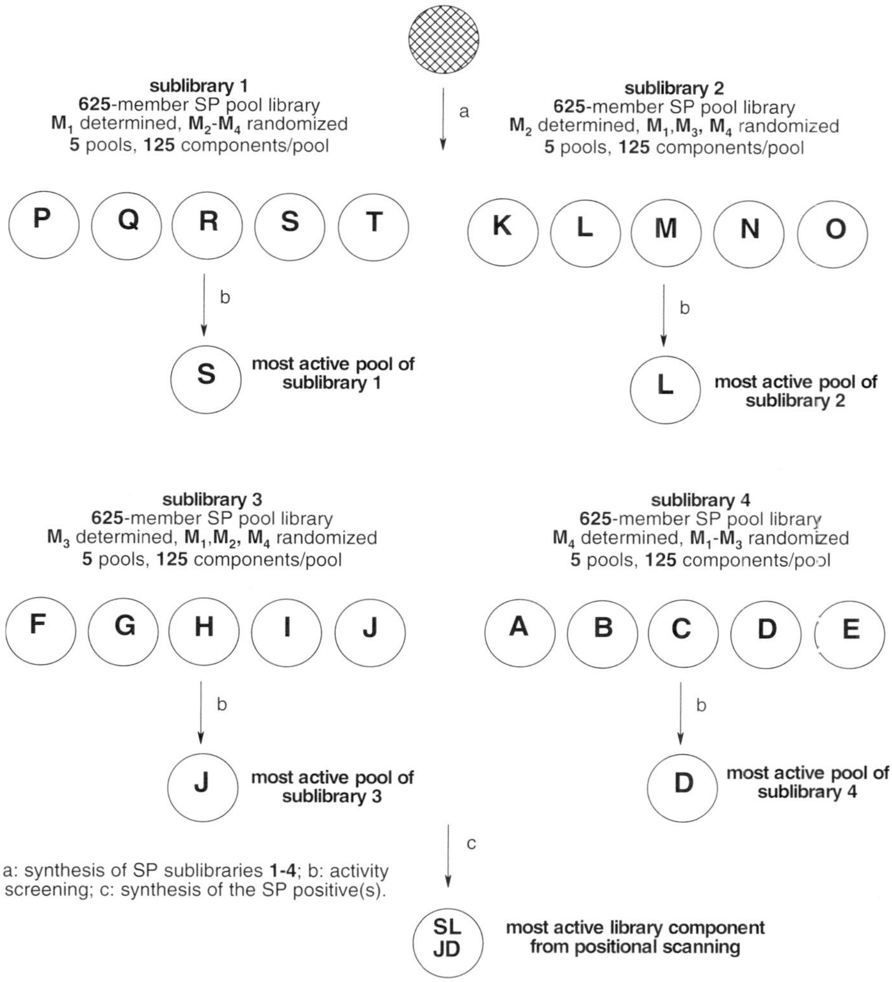

Figure 7.21 Positional scanning of a 625-member hypothetical SP pool library.

to an intriguing example of a > 100,000-member library of bicyclic guanidine-containing heterocycles (166). An interesting modification was reported (167) where a 9216-member peptidomimetic library was deconvoluted by positional scanning, preparing two library copies where two out of the four randomization points were fixed in each library copy. This accounted more accurately for the mutual influence of vicinal positions and gave more reliable results but required significant efforts to prepare the two library copies.

Several other deconvolution methods have been reported. Orthogonal libraries (168), subtractive deconvolution (169), omission libraries (170), bogus coin deconvolution (171), deletion synthesis deconvolution (172), and mutational SURF (Synthetic

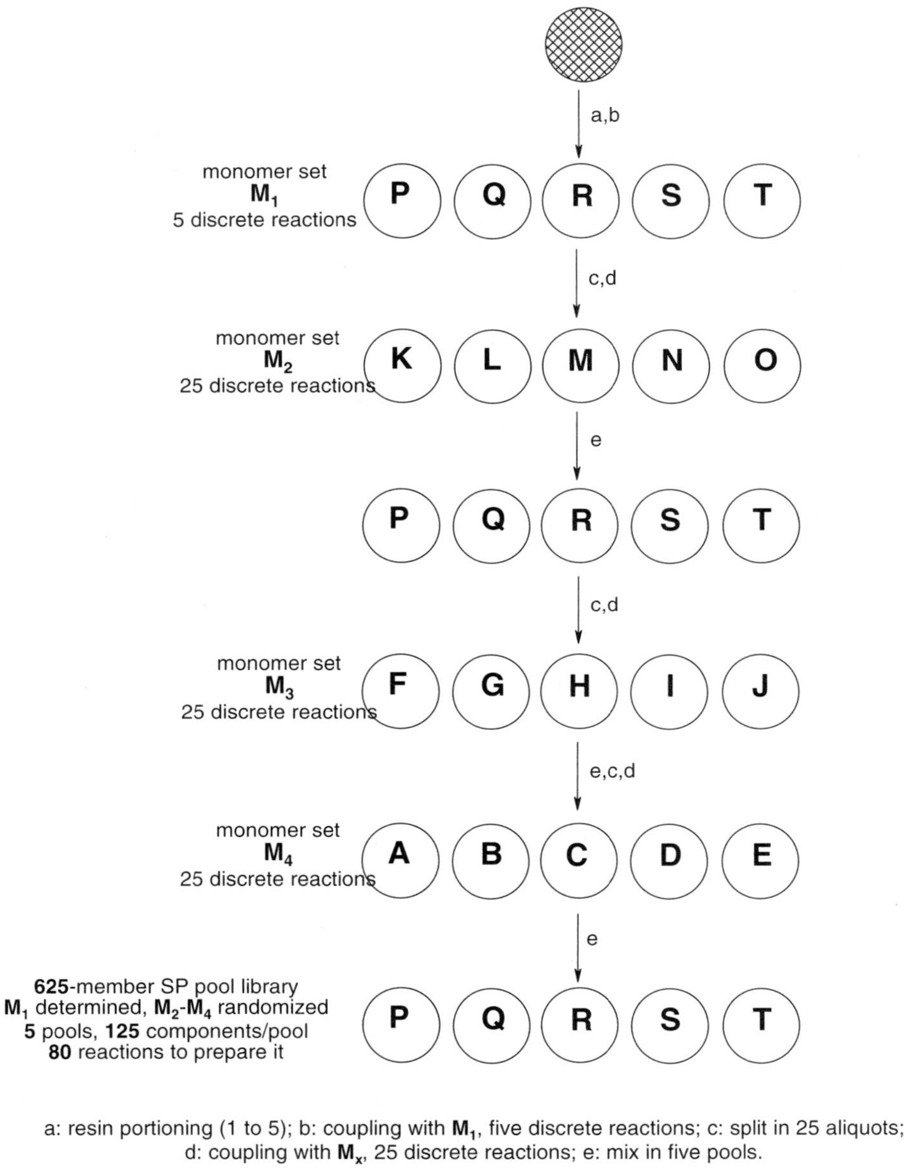

Figure 7.22 Positional scanning: synthesis of the four hypothetical sublibraries **1–4**.

Unrandomization of Random Oligomer Fragments) (173) have been applied either on SP or in solution to peptide or oligonucleotide library deconvolution, but none have found relevant applications in small organic molecule libraries. Several reviews dealing with deconvolution methods for SP libraries were mentioned in the previous section, and the interested reader may refer to them to expand this subject. A few papers (174–176) have tried to compare the efficacy of various deconvolution methods with other structure determination methods for pool libraries. Iterative deconvolution resulted as the method of choice, and positional scanning was similarly, if somewhat less, successful. All the other methods mentioned in this section were significantly less successful in finding the most active structures when the activity of library individuals (oligonucleotides) was known and virtual pools were constructed to make the identification of positives difficult.

Deconvolution was very popular in the early 1990s in combinatorial technologies. It allowed the screening of a significantly lower number of samples with a consequent reduction in costs when compared with the same library of discrete individuals, which were also more demanding in terms of synthesis, while obtaining the activity information on the whole library. The advent of medium–high throughput parallel synthesis and of bead-based SP pool libraries, however, has largely oriented the major active groups toward these libraries and has made the use of deconvolution methods to identify positives from medium–large primary libraries less appealing. These new library formats can be easily tested due to the general diffusion of high-throughput screening (HTS) methods, in pharmaceutical research and in other fields, which allow the rapid testing of large numbers of discrete or bead-based samples.

False positives have always hampered the deconvolution approaches, and often good positives were lost in the iterative/scanning process. These are, in fact, the only structure determination methods that do not identify all but only one or a few positives (sometimes not the best ones). Nevertheless, groups that enter this field and do not have adequate analytical instrumentation to prepare bead-based pool libraries, or HTS throughput screening methods to test them, but do have the need to prepare large SP pool libraries, and testing them for a specific application may still benefit largely from these low-cost methods.

7.4 ENCODING METHODS FOR SOLID-PHASE POOL LIBRARIES

7.4.1 Chemical Encoding

Structure determination of positives from SP pool libraries can also be made through the coupling of a tag, or code, that will unequivocally encode the structure of the library component to each library individual. These tags may be divided in two classes: *chemical tags*, which are compounds with a completely different chemical nature from the library compounds, and *nonchemical tags*, which use nonchemical entities to code for all the library structures.

Encoding methods are typically used for bead-based libraries, with a single exception (*vide infra*), and possess some attractive features. We will start with chemical encoding, which appeared early in the history of combinatorial chemistry and is still widely used by various research groups. The chemical tags are coupled to the resin beads in parallel with the library synthesis using the mix-and-split method, which ensures the presence of a single code and a single chemical structure on each resin bead. This variation of classical mix and split is shown in Fig. 7.23 for a hypothetical example where three monomer sets $\mathbf{M_1}$–$\mathbf{M_3}$ each composed of 10 representatives are used. The synthesis starts from an orthogonally protected solid support where most of the sites (*loading/library sites*) are protected by P_1 and a small portion (*tag sites*) are protected by P_2. The monomer sets $\mathbf{M_1}(P_1)$–$\mathbf{M_3}(P_1)$ are coupled (step c), after deprotection of loading sites (P_1, step a) and splitting into 10 aliquots (step b), while the tags $\mathbf{T_1}(P_2)$–$\mathbf{T_3}(P_2)$ are coupled after their respective monomer sets (step e), following deprotection of tag sites (P_2, step d). Each monomer set–tag coupling cycle is followed by mixing the resin aliquots into a single pool (step f, Fig. 7.23). The coupling of each monomer–tag pair may be inverted [$\mathbf{M_1}(P_1)$, then $\mathbf{T_1}(P_2)$ or $\mathbf{T_1}(P_2)$, then $\mathbf{M_1}(P_1)$] depending on the stability of resin-bound library intermediates. Sometimes the differentiation of loading and tag sites with orthogonal protecting groups may not be necessary (electrophoric tags, *vide infra*). After its synthesis and QC characterization, the library is tested using a bead-based assay, either off-bead or on-bead, and the positive beads are determined. Finally, the tags are cleaved, their structure is determined, and from this information the structure of the active library individual is obtained directly.

The tags are often used as binary-coded mixtures, as shown in Fig. 7.24 for a hypothetical example where six tags are available. Using tags to code individually for

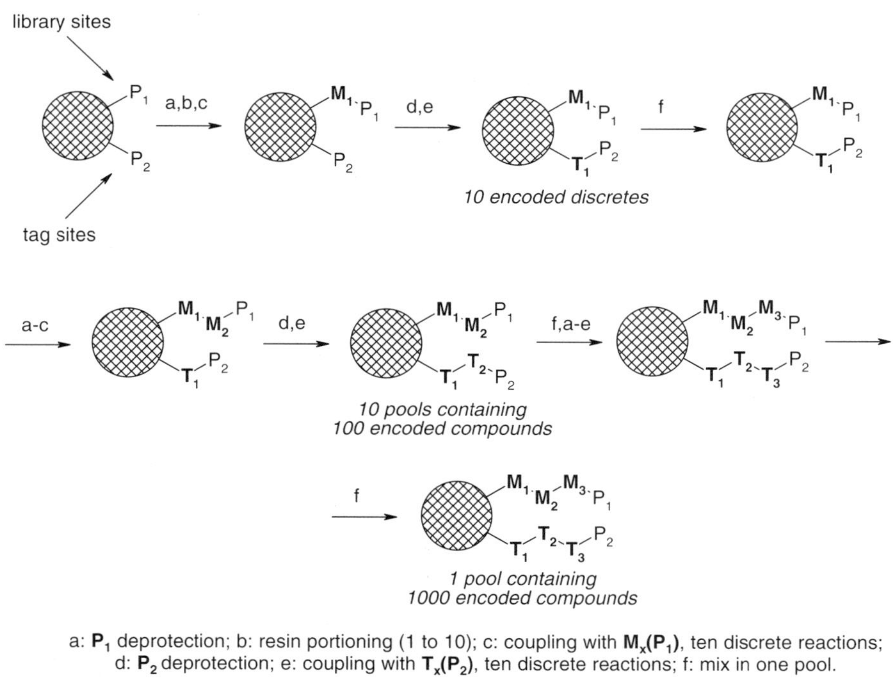

Figure 7.23 Hypothetical SP synthesis of a 1000-member pool library using chemical encoding.

INDIVIDUAL CODES:

$\mathbf{T_1}$ corresponds to $\mathbf{M_{1,1}}$; $\mathbf{T_2}$ to $\mathbf{M_{1,2}}$; $\mathbf{T_3}$ to $\mathbf{M_{1,3}}$;
$\mathbf{T_4}$ to $\mathbf{M_{1,4}}$; $\mathbf{T_5}$ to $\mathbf{M_{1,5}}$; $\mathbf{T_6}$ to $\mathbf{M_{1,6}}$

6 tags = 6 encoded monomers

BINARY CODES:

	T_1	T_2	T_3	T_4	T_5	T_6
M1,1	1	0	0	0	0	0
M1,2	0	1	0	0	0	0
M1,3	0	0	1	0	0	0
M1,4	0	0	0	1	0	0
M1,5	0	0	0	0	1	0
M1,6	0	0	0	0	0	1
M1,7	1	1	0	0	0	0
M1,8	1	0	1	0	0	0
M1,9	1	0	0	1	0	0
M1,10	1	0	0	0	1	0
M1,11	1	0	0	0	0	1
M1,12	0	1	1	0	0	0
M1,13	0	1	0	1	0	0
M1,14	0	1	0	0	1	0
M1,15	0	1	0	0	0	1
M1,16	0	0	1	1	0	0
M1,17	0	0	1	0	1	0
M1,18	0	0	1	0	0	1
M1,19	0	0	0	1	1	0
M1,20	0	0	0	1	0	1
M1,21	0	0	0	0	1	1

	T_1	T_2	T_3	T_4	T_5	T_6
M1,22	1	1	1	0	0	0
M1,23	1	1	0	1	0	0
M1,24	1	1	0	0	1	0
M1,25	1	1	0	0	0	1
M1,26	1	0	1	1	0	0
M1,27	1	0	1	0	1	0
M1,28	1	0	1	0	0	1
M1,29	1	0	0	1	1	0
M1,30	1	0	0	1	0	1
M1,31	1	0	0	0	1	1
M1,32	0	1	1	1	0	0
M1,33	0	1	1	0	1	0
M1,34	0	1	1	0	0	1
M1,35	0	1	0	1	1	0
M1,36	0	1	0	1	0	1
M1,37	0	1	0	0	1	1
M1,38	0	0	1	1	1	0
M1,39	0	0	1	1	0	1
M1,40	0	0	1	0	1	1
M1,41	0	0	0	1	1	1
M1,42	1	1	1	1	0	0

	T_1	T_2	T_3	T_4	T_5	T_6
M1,43	1	1	1	0	1	0
M1,44	1	1	1	0	0	1
M1,45	1	1	0	1	1	0
M1,46	1	1	0	1	0	1
M1,47	1	1	0	0	1	1
M1,48	1	0	1	1	1	0
M1,49	1	0	1	1	0	1
M1,50	1	0	1	0	1	1
M1,51	1	0	0	1	1	1
M1,52	0	1	1	1	1	0
M1,53	0	1	1	1	0	1
M1,54	0	1	1	0	1	1
M1,55	0	1	0	1	1	1
M1,56	0	0	1	1	1	1
M1,57	1	1	1	1	1	0
M1,58	1	1	1	1	0	1
M1,59	1	1	1	0	1	1
M1,60	1	1	0	1	1	1
M1,61	1	0	1	1	1	1
M1,62	0	1	1	1	1	1
M1,63	1	1	1	1	1	1

6 tags = 63 encoded monomers

Figure 7.24 Chemical encoding: individual and binary encoding systems.

monomers, we would encode just six monomers (Fig. 7.24, top), while the binary code would accommodate a much higher number of $2^n - 1$ monomers (63 from six code structures, removing the null code, Fig. 7.24, bottom).

A family of chemical tags should ideally possess a number of features. It should be stable to all, or to the large majority, of the reaction conditions that may be used in an SP library synthesis; its synthesis should be easy and almost quantitative using commercially available precursors; its presence should not interfere with the library synthesis scheme; it should be inert in library screening when on-bead methods are used; its structure should be determined with a fast and reliable method. We will now present the various chemical encoding strategies reported in the literature and will comment on their fulfillment of the above criteria.

The first encoding approaches used peptides (177) or oligonucleotides (178) as tags. We will consider two reported examples of resin-bound constructs able to support peptide-encoded (**7.37**, Fig. 7.25) (177) and oligonucleotide-encoded (**7.38**, Fig. 7.25) (179) peptide or peptidomimetic library synthesis. They both contain a core moiety, derived from L-lysine (**7.37**) or L-serine (**7.38**), which simultaneously multiplies the

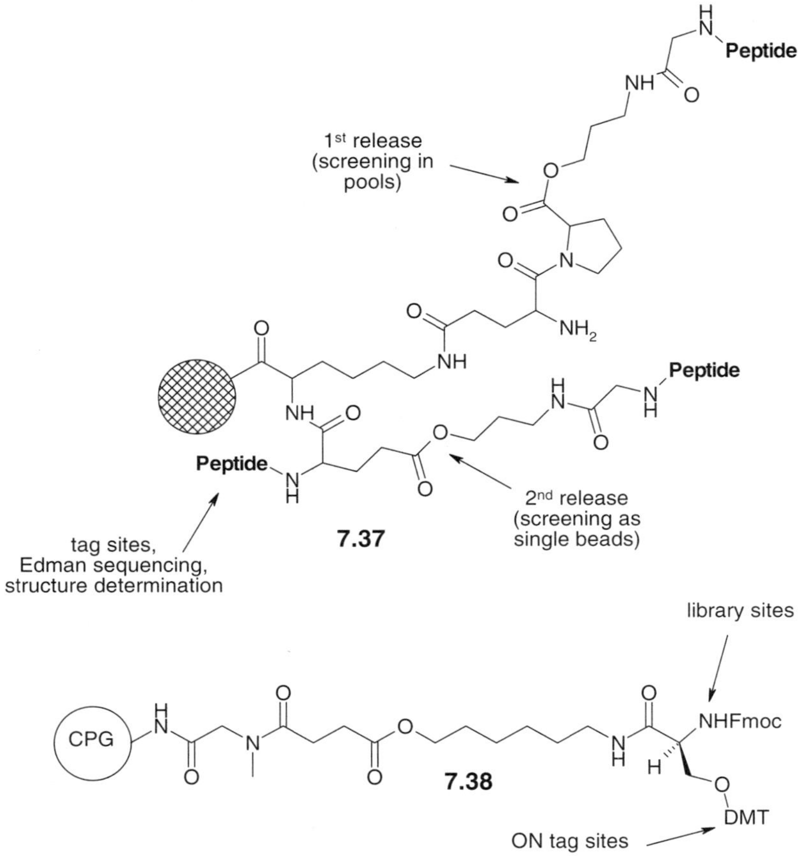

Figure 7.25 Peptide and oligonucleotide encoding: structures **7.37** and **7.38**.

loading of the resin bead and presents either orthogonally protected sites, which are used in turn to grow the library structure (NHfluorenylmethoxycarbonyl, NHFmoc) and to grow the tag structures (*O*-dimethoxytrityl, DMT), such as in **7.38**, or multiple cleavage sites such as in **7.37**, which allow the release and screening of library individuals with final sequencing of the positive beads and determination of positive structures via the remaining resin-bound peptide copies. Such tags are very convenient for the synthesis of on-bead screened oligomeric libraries. In fact, these libraries (usually made by peptides, peptoids, peptidomimetics, oligonucleotides) require standard reaction conditions, which allow the presence of peptide/ON codes without interfering in library, or tag, synthesis. The tags are made from commercial, cheap nucleotides or α-amino acids; the codes do not usually influence the biological activities of the library components, but when this does happen, this undesired effect can be easily spotted; the codes can be easily read with high sensitivity after the library screening by Edman degradation using an automated peptide sequencer for peptides or by polymerase chain reaction (PCR) for oligonucleotides. Several examples of the successful determination of active oligomeric sequences from SP-encoded pool libraries have been reported, with either oligonucleotide (180–184) or peptide (185–188) tags.

The use of such tags for small organic molecule libraries is prevented by the sensitivity of peptides and, even more so, of oligonucleotides to some of the reaction conditions that are common in organic synthesis: strong bases (proton abstractions, racemizations, alkylations), strong reducing agents (amide reductions), acidic conditions (sugar dehydrations), and so on. More stable tags were thus needed, and a major achievement was reported by Ohlmeyer et al. (111) and Nestler et al. (189) with the introduction of the so-called *electrophoric tags*. Their optimized structure is reported in Fig. 7.26 (**7.40**), together with their synthesis (path a), the mechanism of their

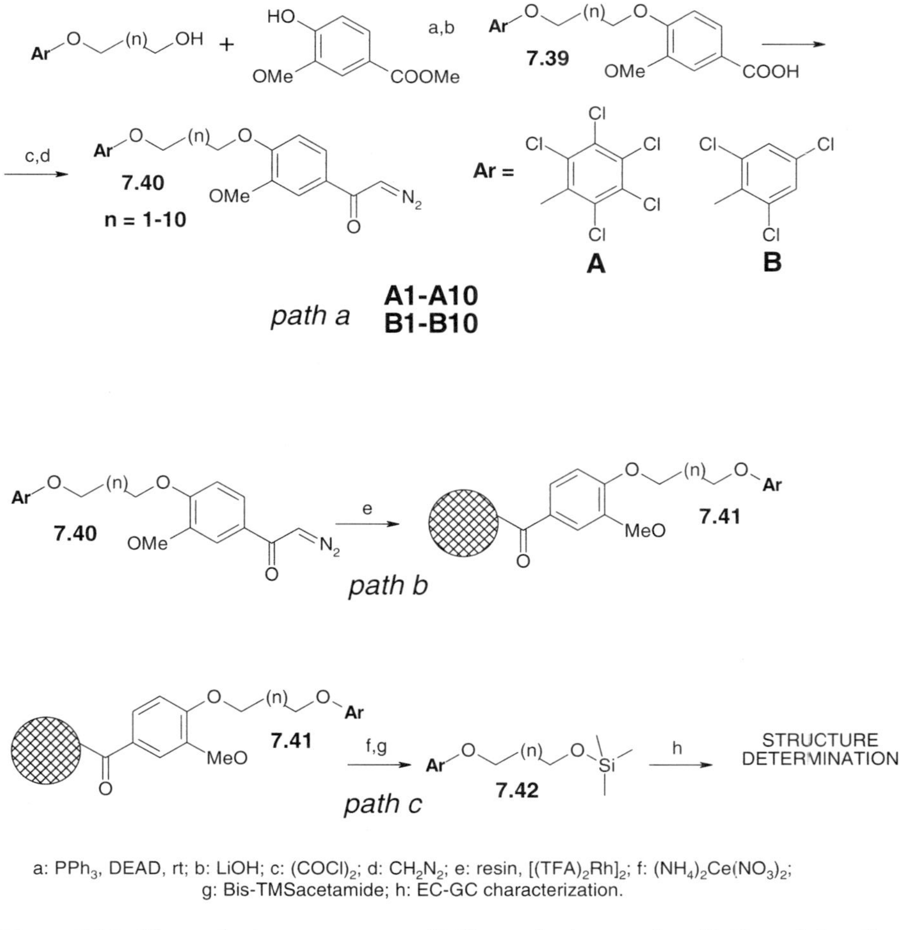

Figure 7.26 Electrophoric tags: structure (**7.40**), synthesis, encoding (**7.41**), and decoding (**7.42**) protocols.

support onto the solid phase (path b), and their cleavage (path c). The synthesis of the tags is straightforward, so that large quantities of $\mathbf{A}_{1-10}$ and $\mathbf{B}_{1-10}$ can be easily obtained, and they are stable and can be stored for months at room temperature. They are added to the support as acylcarbenes, generated by the treatment of diazoacetates **7.40** with rhodium trifluoroacetate, with no need for orthogonal sites on the resin beads, as the carbenes react aspecifically with the support (the aromatic rings) or with the library individuals. Fortunately, the large excess of the former versus the latter and the low amount of tags used (around 1% of the resin loading) will produce negligible amounts of modified library compounds while ensuring a correct encoding with tags **7.41**. Using the binary encoding system where tag mixtures encode for specific monomers, the acylcarbene addition is performed with equimolar amounts of different tags to follow the planned encoding scheme. Tag cleavage is performed, after library synthesis and screening, via oxidative cleavage with cerium ammonium nitrate (CAN) and silylation of the resulting alcohols to give **7.42**. The silyl derivatives **7.42** are then characterized via electron capture gas chromatography (EC-GC), and the retention times of each peak produce the structural information needed to identify the library positive(s).

This method employs tags that are stable to the large majority of the reaction conditions used in an SP library synthesis, their synthesis is easy using commercially available precursors, and their presence does not interfere with the library synthesis scheme; they are inert in library screening when on-bead methods are used; and their structure is determined with a fast and reliable method, even though EC-GC is not a very common analytical technique and the cleavage/derivatization procedure is somewhat time-consuming. This encoding system has been widely used by the group that discovered these tags (190–193) as well as by other researchers (194–196) and was instrumental in decoding the structure of positive library components for a wide panel of biological targets, employing both off-bead (194–196) and on-bead (190–194) screening methods.

Another important chemical encoding method, based on *secondary amine tags*, has been reported (197). The structure, synthesis (path a), and coupling to the solid support (path b) of these tags are reported in Fig. 7.27. The acid tags **7.44** are easily made from N-protected iminodiacetic acid anhydride **7.43** and require an amino function on the resin to be coupled. The use of orthogonally protected resin sites is also necessary; usually 90% of the sites are functionalized with a photolinker, then protected with Fmoc (P_1, Fig. 7.27), while the tag sites ($\approx$10%) are protected with an orthogonal group (either Boc or, more recently, Alloc; P_2, Fig. 7.27). The coupling of tag units or library units then takes advantage of orthogonal deprotections, as in path b, until the encoded library **7.45** is prepared and QCed. The binary coding system will require some monomers to be encoded by the coupling of equimolar tag mixtures.

After library cleavage, screening, and selection of positives (steps a–c, Fig. 7.28), the positive beads, such as **7.46**, are decoded simply by strong aqueous acid hydrolysis, neutralization, and dansylation of the residual amines to give the dansyl derivatives **7.47** (steps d–g), which are all distinct in an HPLC spectrum. Several recent reports have provided optimized HPLC/fluorescence decoding protocols (198), which shortened the average decoding procedure from 1 h to around 6 min, and also alternative

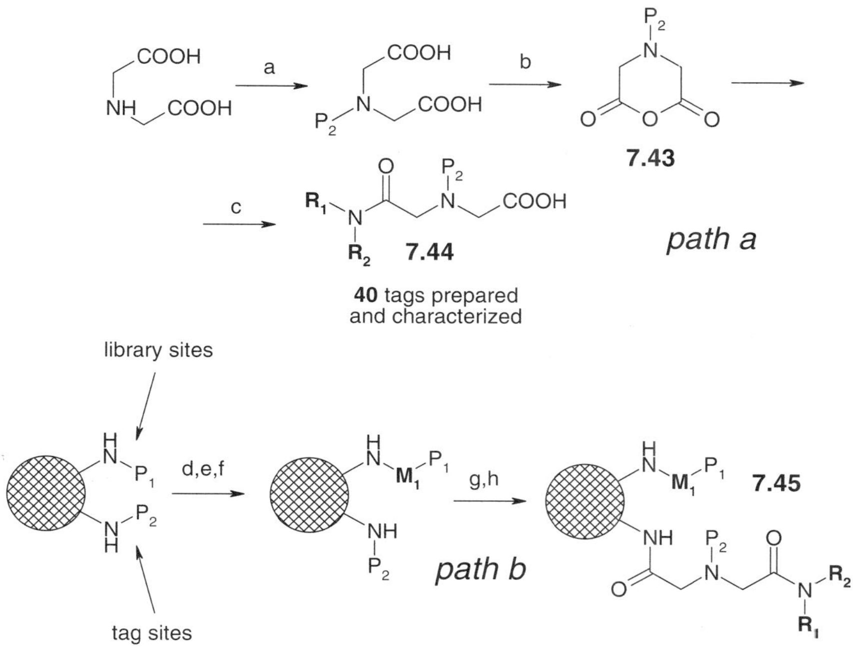

a: N-protection; b: $(COCl_2)_3$, TEA; c: coupling with $\mathbf{R_1R_2}$NH; d: P_1 deprotection; e: resin portioning (1 to n); f: coupling with $\mathbf{M_1}(P_1)$; g: P_2 deprotection; h: coupling with **7.44**.

Figure 7.27 Secondary amine tags: structure (**7.44**), synthesis, and encoding (**7.45**) protocols.

analytical decoding techniques involving capillary electrochromatography (CEC) (199, 200), so as to produce coding patterns that can be attributed to all the positive library individuals (step h, Fig. 7.28).

This method employs tags that are stable to many of the reaction conditions used in an SP library synthesis, even though they should be less stable in general than the aforementioned electrophoric tags; their synthesis is easy using commercially available precursors, and many tags can be prepared and characterized; their presence requires the use of ≈10% of the loading sites for tagging and a library synthesis scheme starting from a resin-bound amine function; and their structure is determined by fast and reliable methods. Researchers at Affymax have used this approach for several bead-based libraries (201–204) and have positively decoded the structures of active compounds.

Other approaches have been reported (127, 205–211), but their usefulness is, as of today, not comparable with the previously reported examples. The reader may refer to some recent reviews (212–217) to have more details on chemical encoding.

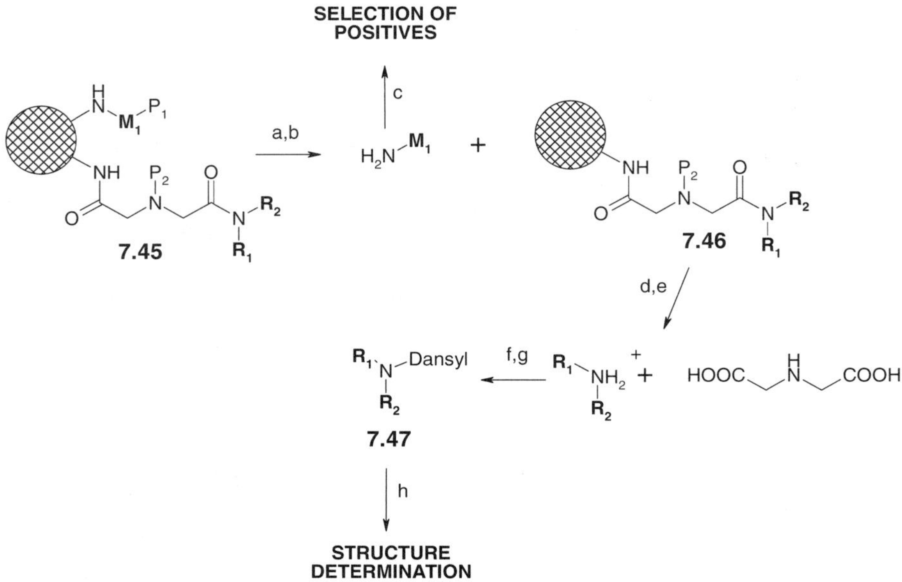

a: P_1 deprotection; b: cleavage of library components; c: screening; d: P_2 deprotection; e: HCl 6N, reflux, 15 hrs; f: Li_2CO_3;g: dansyl chloride; h: HPLC/CEC decoding.

Figure 7.28 Secondary amine tags: decoding (**7.46**) protocols.

7.4.2 An Example: Synthesis and Structure Determination of Positives from Encoded Dihydrobenzopyran Libraries

Baldwin (194) reported the synthesis of four encoded libraries of dihydropyrans **L6–L9**, whose synthetic scheme is shown in Figs. 7.29–7.32. The monomer sets used are $\mathbf{M_1}$–$\mathbf{M_6}$, shown in Figs. 7.33 and 7.34. The chemical diversity was built using amines (seven representatives, monomer set $\mathbf{M_1}$, step b, Fig. 7.29), carboxyacetophenones (six representatives, two three-member subsets, monomer sets $\mathbf{M_{2a,b}}$, step g, Fig. 7.29) and ketones (two subsets: seven ketones, monomer set $\mathbf{M_{3a}}$, step i, to give the intermediates **7.50**, and three cyclic N-protected aminoketones, monomer set $\mathbf{M_{3b}}$, step j, to give the intermediates **7.51**, Fig. 7.29). The intermediates **7.51** were treated with acylating/alkylating agents (seven representatives, monomer set $\mathbf{M_4}$, step b, Fig. 7.30) to give **7.52**. These intermediates were thoroughly mixed with resin **7.50** (step d), and the resin was split into 10 portions (step e, Fig. 7.30). Three of these portions were withdrawn (step a, Fig. 7.31): one was archived (**L6**, step b), one was reduced and archived (**L7**, steps c and b), and one was converted to a dithioketal and archived (**L6**, steps d and b, Fig. 7.31). The remaining seven portions were further diversified with amines (seven representatives, monomer set $\mathbf{M_5}$, steps a and b, Fig. 7.32) and, after encoding and mixing/splitting of the resin, with acylating agents (10 representatives, monomer set $\mathbf{M_6}$, to give **L9**, step f, Fig. 7.32).

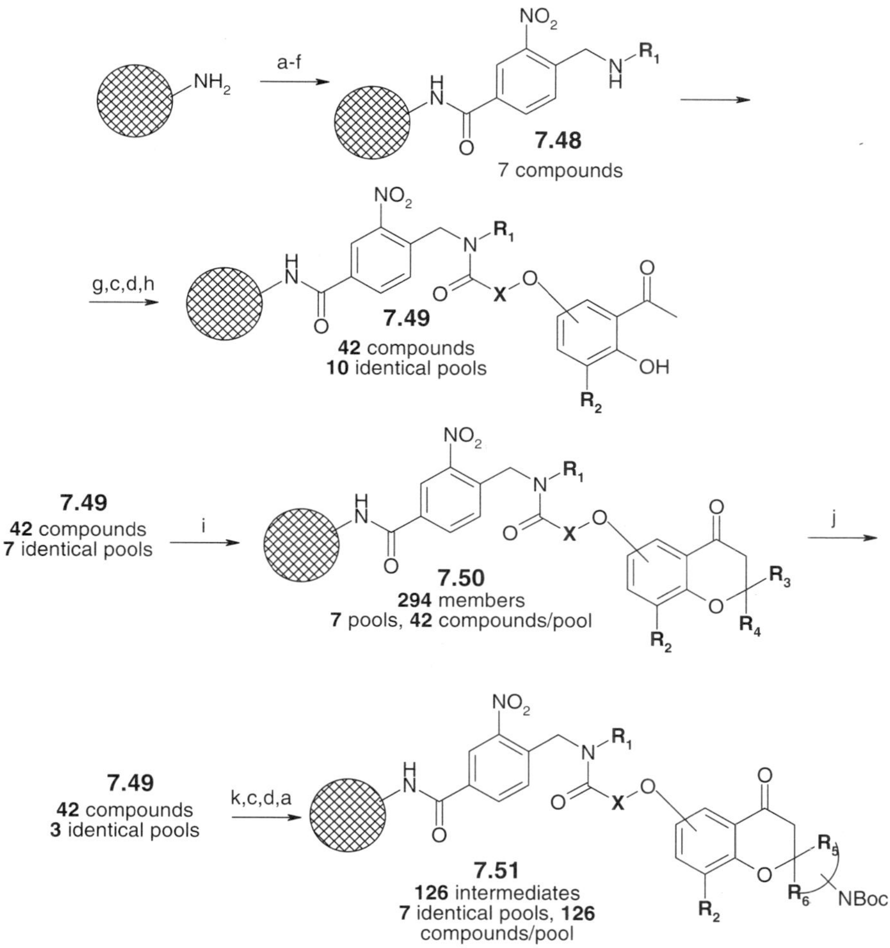

a: resin portioning (1 to 7); b: coupling with $\mathbf{M_1}$; c: encoding; d: mix in one pool; e: TFA;
f: resin portioning (1 to 6); g: coupling with $\mathbf{M_{2a,b}}$; h: resin portioning (1 to 10);
i: coupling with $\mathbf{M_{3a}}$; j: see step d, Fig. 7.30; k: coupling with $\mathbf{M_{3b}}$.

Figure 7.29 SP synthetic strategy to the SP benzopyran encoded (electrophoric tags) pool libraries **L6–L9**: synthesis of SP intermediates **7.50** and **7.51**.

The complete list of monomers and encoding tags used is reported in Figs. 7.33 and 7.34. Monomer sets $\mathbf{M_1}$, $\mathbf{M_{3b}}$, $\mathbf{M_4}$, and $\mathbf{M_5}$ were coded by 3 tags each, while monomer sets $\mathbf{M_{2a}}$, $\mathbf{M_{2b}}$, and $\mathbf{M_{3a}}$ were coded by 2 tags to give a total of 18 tags to cover the whole library synthesis. The resin aliquots leading to libraries **L6–L8** were kept divided, so that it was not necessary to code for archiving/transformations leading to **L6**, **L7**, and **L8**. Moreover, **L9** was archived as 10 pools where the monomer set $\mathbf{M_6}$ was determined, thereby avoiding its encoding (Fig. 7.34).

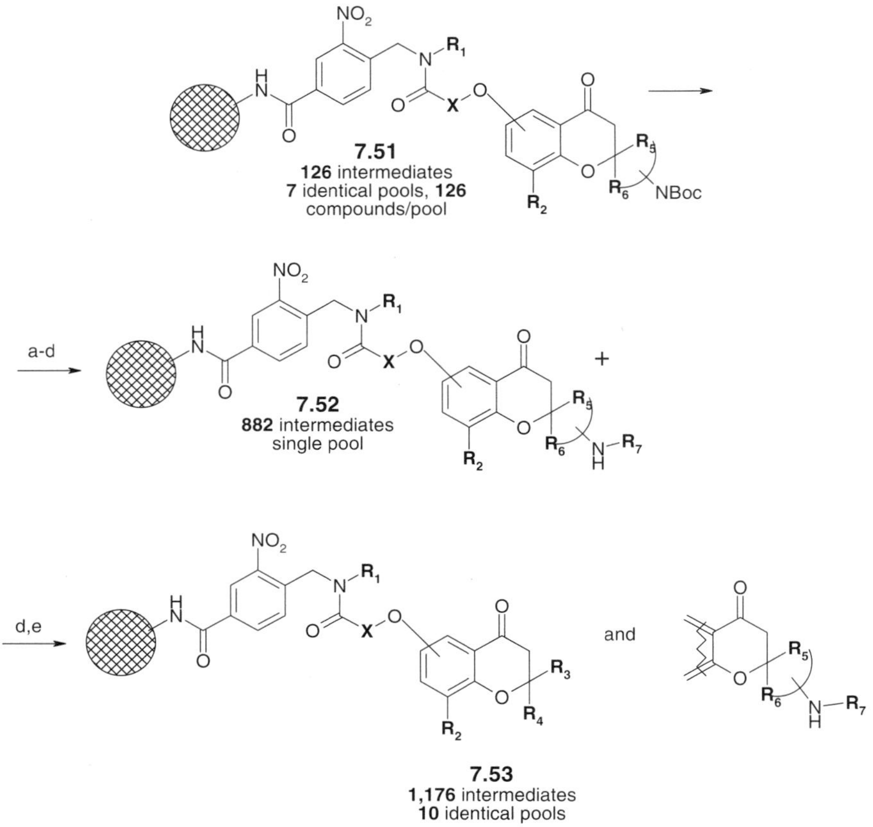

a: TFA; b: coupling with $\mathbf{M_4}$; c: encoding; d: mix in one pool; e: resin portioning (1 to 10).

Figure 7.30 SP synthetic strategy to the SP benzopyran encoded (electrophoric tags) pool libraries **L6–L9**: synthesis of SP intermediates **7.53**.

The authors did not report the biological activity of the primary libraries **L6–L9**, which were tested as a source of biologically relevant compounds. The libraries' quality was inferred from the parallel synthesis of six library individuals from **L6–L9** (**7.55–7.60**, Fig. 7.35), which were characterized by NMR, MS, HPLC, and gravimetric yield after photorelease. The encoding/decoding procedure was defined in the paper as being "greatly facilitated by encoding with electrophoric tags" and indirectly "accelerating the early phases of drug discovery" (194).

7.4.3 Nonchemical Encoding

The use of tagging methods that do not require a covalent bond between the tag and the solid support is appealing because it reduces the chemical complexity while

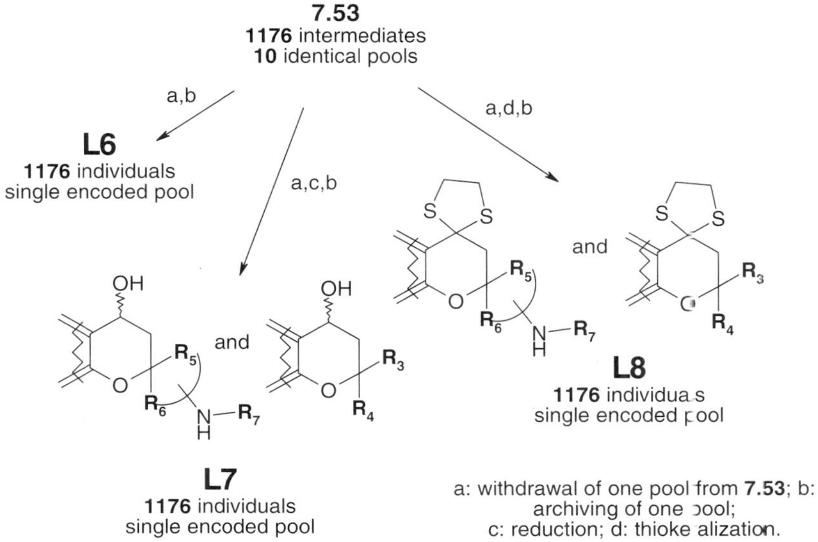

Figure 7.31 SP synthetic strategy to the SP benzopyran encoded (electrophoric tags) pool libraries **L6–L9**: synthesis of the libraries **L6–L8**.

ensuring a direct link between the code structures and each library individual. Several tagging procedures and/or moieties have been reported, including fluorophores labeling the support (218, 219), preencoded beads using intrinsically labeled solid supports (220), self-coded, mass-decipherable libraries (221), visual tagging (222), laser optical encoding (223), and non–covalently bound chemical tags (224). The most important *nonchemical encoding* technique, though, originated from two independent reports (225, 226), which presented the use of radiofrequency tags to encode chemical libraries.

Each compound is prepared in a specific reaction vessel/device where a radiofrequency tag is located, and each tag is preencoded, before the synthesis, with a readable signal that is a priori coupled with a specific final compound (*vide infra*). At any SPS step the code can be read on a coding station, and the structure of the intermediate or final compound can be determined. If a new monomer set must be added, the vessels are sorted according to the monomer representatives to be added, and the coupling takes place. Each device also contains a resin aliquot (typically a few tens of milligrams) and is permeable to the reagents in solution while preventing the beads' release. This method is particularly appealing for the organic chemist because it has the advantages of mix-and-split synthesis (high throughput, few chemical reactions) but allows the production of significant quantities of discrete compounds through so-called *directed sorting*. The principle of this technique is shown in Fig. 7.36 in a hypothetical example where 16 products are prepared using a four-step sequence and two monomer sets each composed of four representatives (**A–D** for the first set, **E–G** for the second set).

7.53
1176 intermediates
10 identical pools

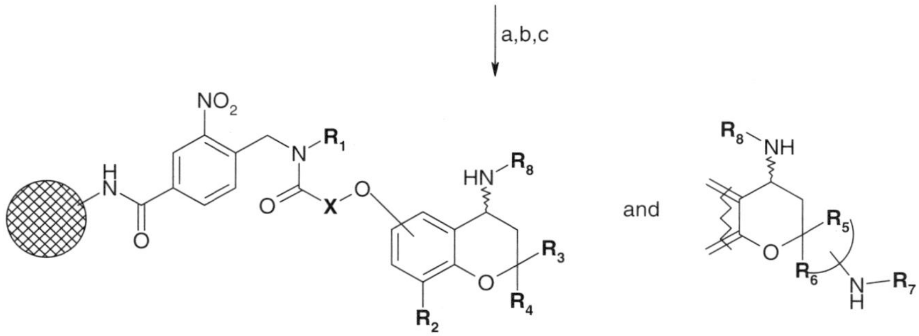

7.54
8232 intermediates
7 pools, **1,176** compounds/pool

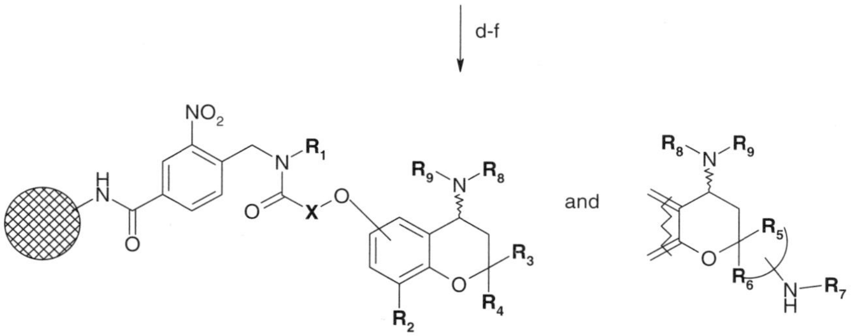

L9
82,320 individuals
10 encoded pools,
8232 compounds/pool

a: coupling with $\mathbf{M_5}$; b: reduction of imine; c: encoding; d: mix in one pool; e: resin portioning (1 to 10); f: coupling with $\mathbf{M_6}$.

Figure 7.32 SP synthetic strategy to the SP benzopyran encoded (electrophoric tags) pool libraries **L6–L9**: synthesis of the library **L9**.

The codes in each vessel are read before the first coupling, and the vessels are sorted into four flasks according to the first monomer position (the protected monomer **A** is coupled in the first flask, protected **D** in the last, step a). After the first couplings (steps b–e), the vessels are mixed and deprotected in a single reactor (steps f and g); then they are sorted according to the second monomer position (monomers **E–H**, step h)

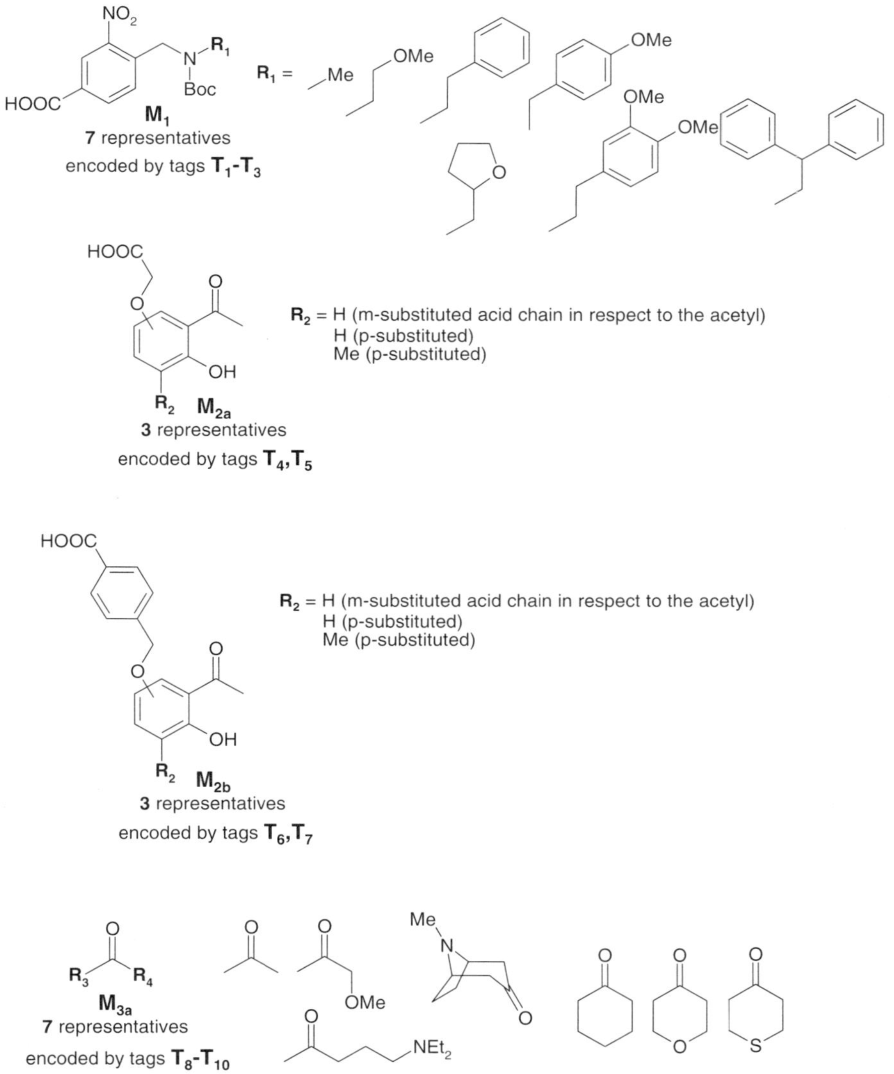

Figure 7.33 Monomer structures and tags for the SP benzopyran libraries **L6–L9**: $\mathbf{M_1}$–$\mathbf{M_{3a}}$.

and coupled with the second monomer set (steps i–l). Finally the vessels are sorted according to the individual structures and placed in a cleavage plate (step m) where pure, released library individuals are recovered after the cleavage (step n). The preparation of the same 16-member library using parallel synthesis would have required 4 (coupling with monomers **A–D**) + 4 (deprotections) + 16 (coupling with monomers **E–H**) + 16 (cleavage) = 40 reactions. Directed sorting of radiofrequency-encoded vessels required 4 (coupling with monomers **A–D**) + 1 (deprotection) + 4

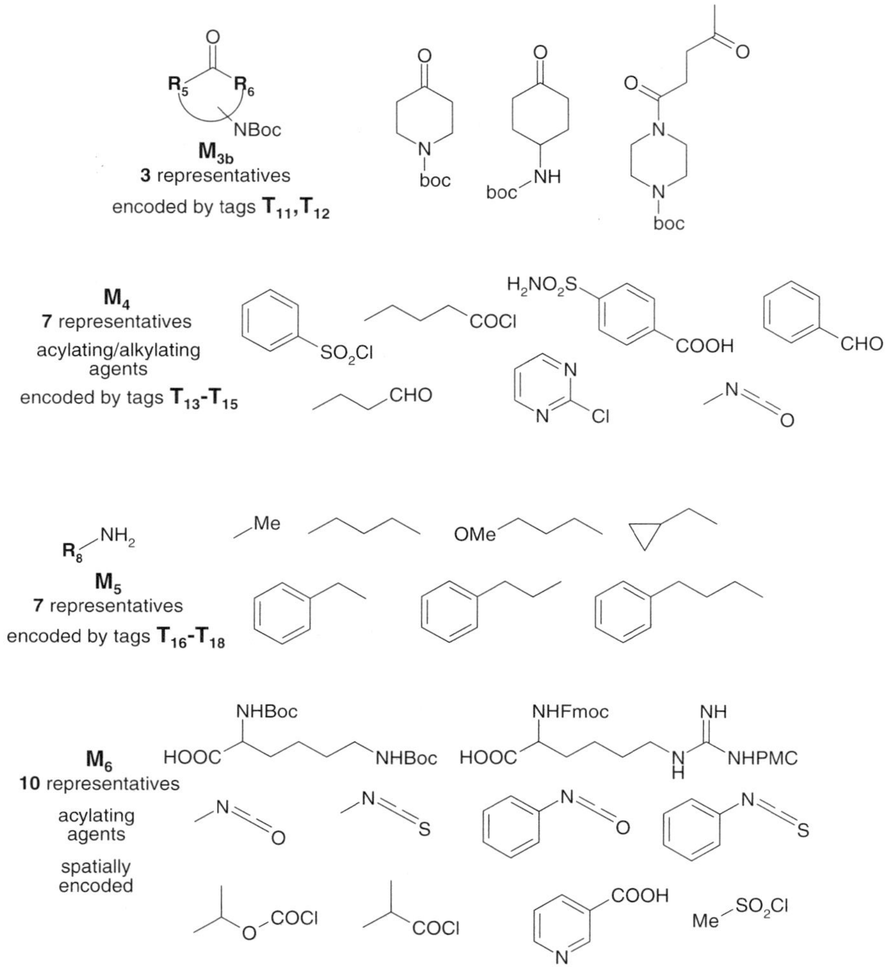

Figure 7.34 Monomer structures and tags for the SP benzopyran libraries **L6–L9**: $\mathbf{M_{3b}}$–$\mathbf{M_6}$.

(coupling with monomers **E–H**) + 16 (cleavage) = 25 reactions. This reduction in the number of reactions performed becomes more significant when the number of chemical steps or the number of monomers increase and allows an easier and faster SPS of discretes using a modified mix-and-split technique. The apparatus to perform radiofrequency-encoded SP reactions is commercially available using either resin beads in microvessels (227), grafted microtubes (228, 229) or pins as reactors (230), and the SPS of libraries with up to several thousands of individuals with good yields and purities have been reported (231–235). The interest generated by this technique has recently stimulated improvements in the handling and sorting of large amounts of microreactors (236), in the final cleavage protocols (237), and in their direct use for

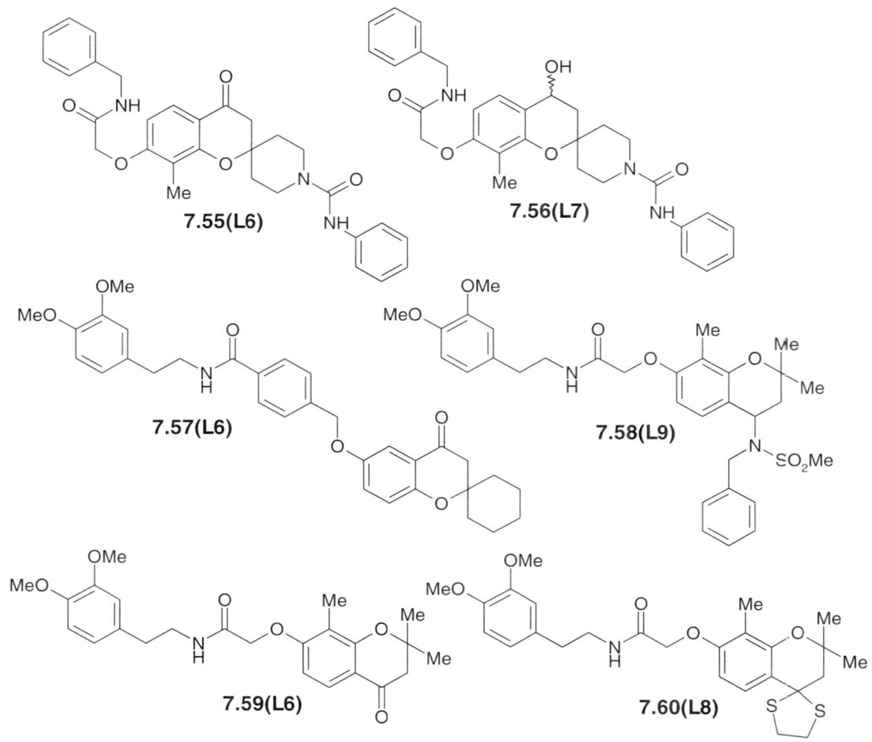

Figure 7.35 Structures of six characterized individuals **7.55–7.60** from **L6–L9**.

biological screening (238). We can easily foresee a steady and copious flow of reports of radiofrequency-encoded SP libraries in the future, especially considering the reasonable cost of the necessary equipment, which is already present in most laboratories that perform combinatorial synthesis of chemical libraries.

7.4.4 An Example: Synthesis of a Tyrphostin Radiofrequence Encoded Library

Recently Shi et al. (239) reported the synthesis of a 432-member focused library **L10** as a source of tyrosine kinase inhibitors using benzylidene malononitriles, or tyrphostins, as structural motifs to design the library. Both the general tyrphostin structure (**7.61**) and an example of an active compound on a specific tyrosine kinase (**7.62**) are reported in Fig. 7.37.

The reactors selected were 432 aminomethylated MicroTubes **7.63**, which are polystyrene-grafted polypropylene tubes with a ≈50 μ*M* loading capacity per reactor, that encapsulate a preencoded radiofrequency tag (Fig. 7.38). The tubes were reacted with the Fmoc-protected, acid-labile Knorr linker (step a), the residual amine functions

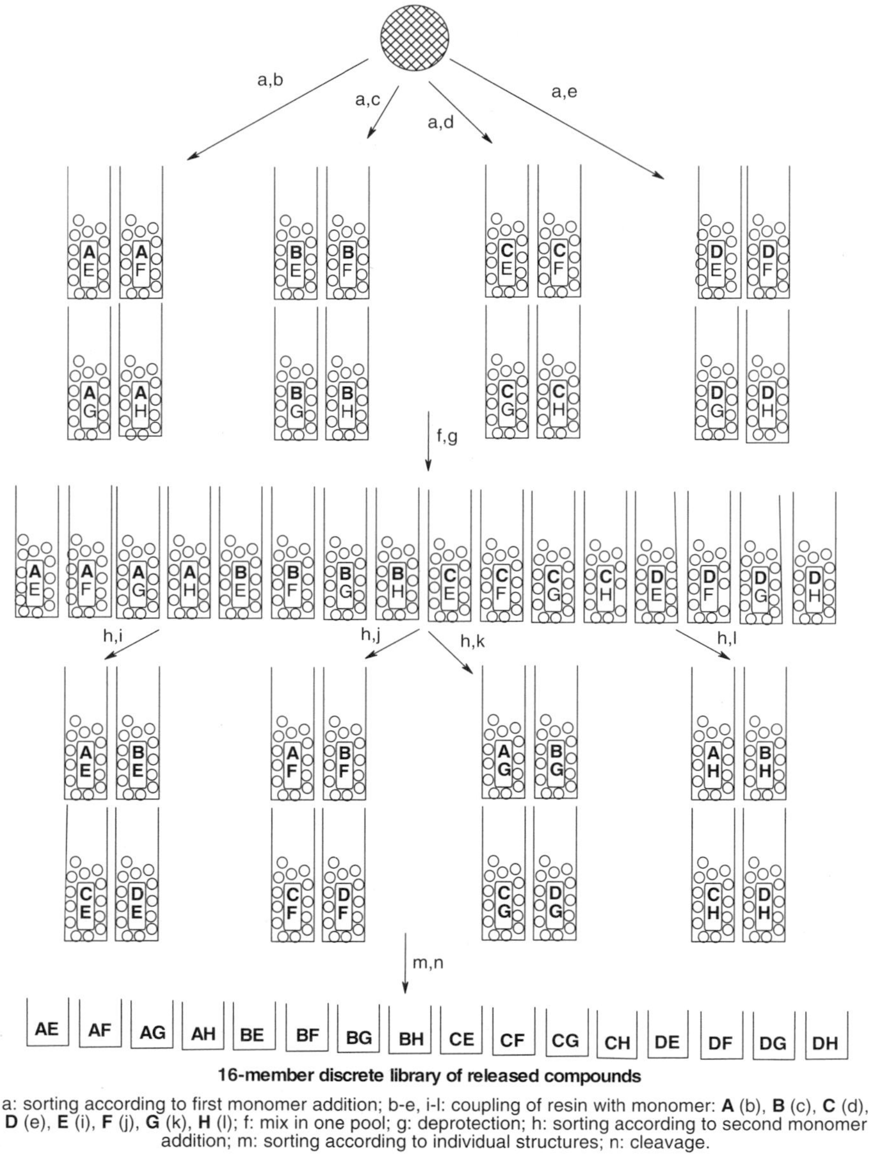

Figure 7.36 Radiofrequency encoding: directed sorting for a hypothetical 16-member discrete library.

were capped (step b), and the Fmoc group was removed (step c) to give **7.64** with a ≈50 μ*M* loading (Fmoc reading). The above reactions were performed in a single flask, while the coupling with $\mathbf{M_1}$ (18 aromatic aldehydes) required prior sorting of the preencoded tubes (step d) and their partitioning into 18 flasks, where each aldehyde

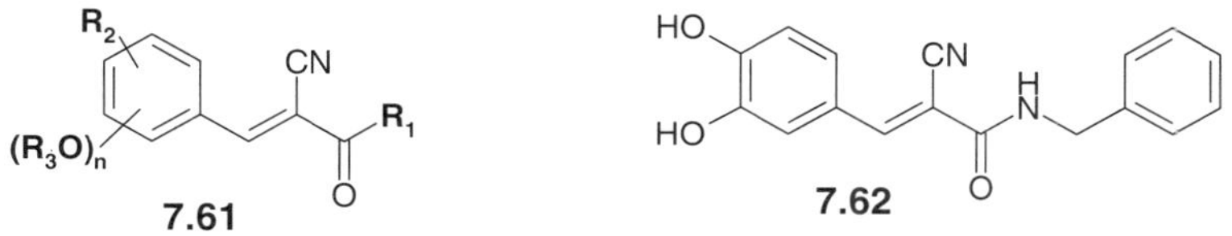

Figure 7.37 General structure of tyrphostins (**7.61**) and of an active compound on a tyrosine kinase (**7.62**).

was coupled to 24 tubes (step e), and the corresponding imines were reduced (step f) to give secondary amines **7.65**. All the reactors were then pooled (step g) and coupled with cyanoacetic acid using multiple cycles (step h) to give the amides **7.66**. The tubes were then sorted according to **M2** (eight aromatic aldehydes bearing a phenol group,

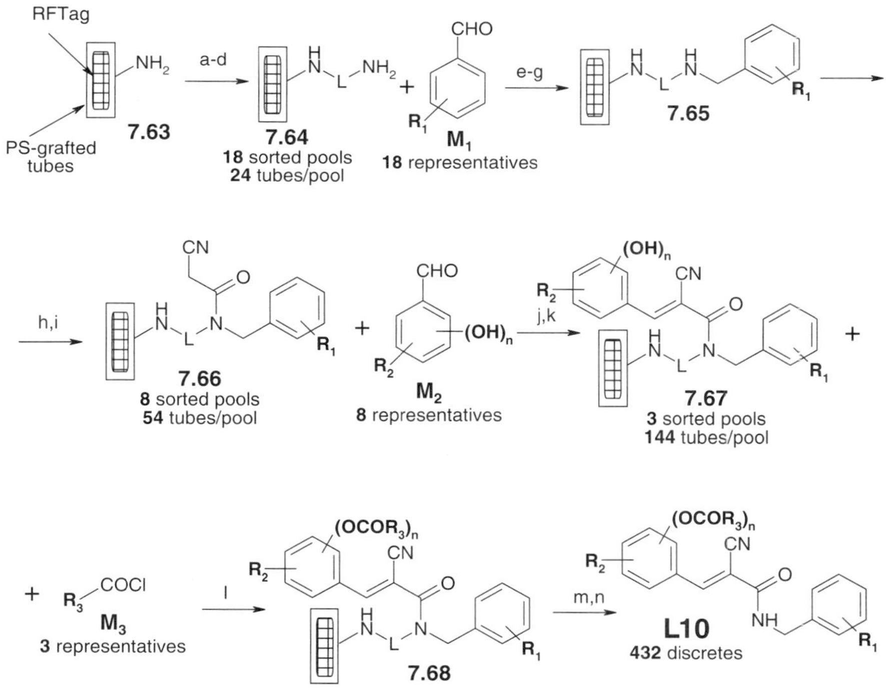

a: Knorr linker, PyBOP, DIPEA, DCM; b: Ac_2O, DIPEA; c: 20% piperidine/DMF; d: sorting according to $\mathbf{M_1}$; e: coupling with $\mathbf{M_1}$; f: $NaCNBH_3$, AcOH; g: mix in one pool; h: cyanoacetic acid, DIC, DMF, three cycles; i: sorting according to $\mathbf{M_2}$; j: coupling with $\mathbf{M_2}$; k: sorting according to $\mathbf{M_3}$; l: coupling with $\mathbf{M_3}$; m: sorting as individual compounds; n: 4% TFA/benzene.

L = Knorr linker

Figure 7.38 SP synthesis of a radiofrequency encoded, 432-member discrete tyrphostin-inspired library **L10**.

step i), and the aldol condensation was performed in eight flasks, each containing 54 tubes (step j), to give the unsaturated nitriles **7.67**. Finally, the tubes were again sorted, this time according to **M3** (two acyl chlorides and an empty position to keep the free phenol, step k), and coupled to give the final supported esters/phenols **7.68** (step l). These were sorted and identified (step m), then accordingly arranged in 432 positions of several microplates and cleaved (step n, Fig. 7.38) to give the library **L10** as 432 discretes with good yields and purities determined by TLC of all samples and by MS and NMR of 5% of library compounds selected randomly (24 samples).

The power of this encoding method to prepare discrete libraries is again summarized by the number of chemical reactions performed. If the library had been prepared by parallel synthesis, a total of 1 (step a) + 1 (step b) + 1 (step c) + 18 (step e) + 18 (step f) + 18 (step h) + 144 (step j) + 288 (step l) + 432 (step n) = 921 reactions would have been necessary. Directed sorting allowed the number to be limited to 1 (step a) + 1 (step b) + 1 (step c) + 18 (step e) + 18 (step f) + 1 (step h) + 8 (step j) + 2 (step l) + 432 (step n) = 482 reactions, while ensuring the same quality of the final discrete products.

7.5 NEW TRENDS IN SOLID-PHASE POOL LIBRARIES

7.5.1 Bead-Based Libraries: High-Throughput Synthesis

Many factors have recently contributed to a general decrease of interest for large, primary SP pool libraries. These include the increased throughput of solution and SP discrete libraries, the advent of computational tools to create virtual libraries and to rationally select smaller subsets to be prepared, and the assumed lower quality of large SP pool libraries, which originates from their more difficult analytical characterization. Despite all these factors, some recent examples of bead-based pool libraries have clearly shown how a rigorous, integrated approach may produce high-quality, very large primary or biased-targeted libraries to be tested using HTS. The amount of information generated by these libraries, together with the moderate amount of efforts necessary, should convince even the more conservative chemist of the huge potential embedded into this library format and that a rigorous chemical approach is the gateway to successful SP pool libraries.

We will present here a very recent example by Tan et al. (196, 240), who are heavily involved in the so-called *chemical genetic approach*, (see Section 9.1.4) where large numbers of compounds must be routinely screened to identify small-molecule ligands/inhibitors, which are able to influence as many gene products as possible. This group developed some miniaturized, cell-based HTS (241–244), which allow testing of bead-based libraries and detection of active, cell-permeable compounds that interact with proteins (242), or even compounds that influence posttranslational modifications of gene products (243). These miniaturized assays require large collections of meaningful small molecules to be tested, and the authors decided to use encoded (189), bead-based libraries prepared by mix-and-split methods. They introduced some limitations upfront, such as water-compatible Tentagel supports and a photolabile linker

(245) which are limiting the choice among many organic reaction conditions, and also decided to start from structures similar to known natural products and to combinatorialize them to obtain millions of small organic compounds. Even more importantly, they decided to be extremely rigorous in assessing the purity of such a large SP pool library.

The authors chose as a template to inspire their large bead-based library synthesis the compound **7.69** (Fig. 7.39) (246, 247). This polycyclic scaffold is interesting as such, representing a constrained, highly functionalized small organic molecule, but it can also be considered as a highly flexible, geometrically and stereochemically pure scaffold where to introduce chemical diversity by mild chemical transformations without the need of protection/deprotection steps. A reasonable retrosynthetic scheme was designed to provide the **7.69**-inspired SP scaffold **7.70** (Fig. 7.39). This scaffold allows a priori four primary diversifications (Fig. 7.39, full arrows): 1) nucleophilic addition to the epoxide, 2) nucleophilic addition to the lactone, 3) decoration of the **R1** substituent (either through different nitrones, or using functionalizable nitrones), and 4) reductive cleavage of the N-O isoxazoline bond. Transformations 1, 2 and 4, if successfully assessed, would open secondary diversifications (Fig. 7.39, dashed arrows): 1′) decoration of the hydroxyl function originated from epoxide opening; 2′) decoration of the hydroxyl function originated from lactone opening; 3′) decoration

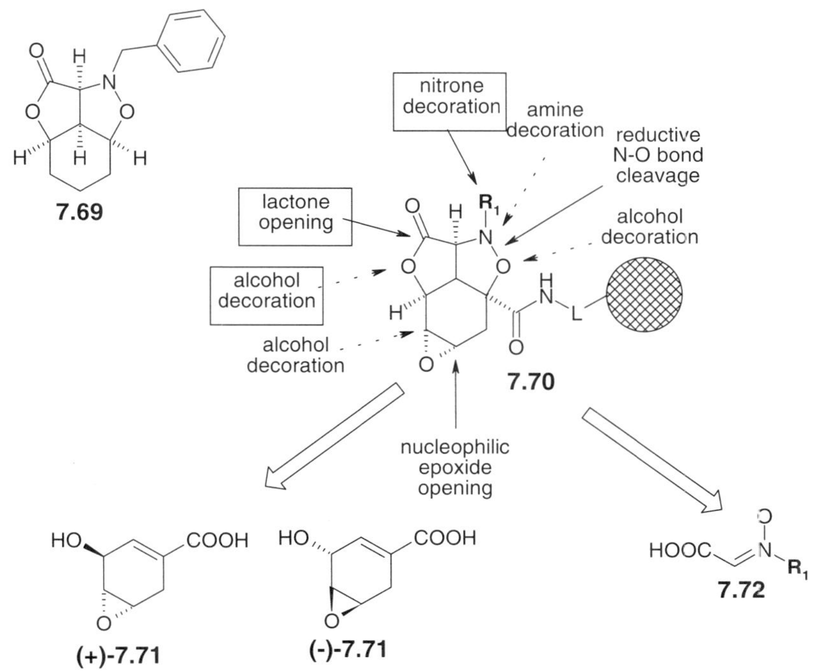

Figure 7.39 Bead-based libraries: structures of a natural product-biased scaffold **7.69** and of its adapted SP version **7.70** for the synthesis of a large bead-based library.

of the hydroxyl function originated from N-O cleavage; 3″) decoration of the amine function originated from N-O cleavage.

Having planned to synthesize a large library and to use a large number of monomers, the authors were particularly strict in the SP chemical assessment for each of the above transformations. While all of them produced the expected compounds, even after an extensive optimization of the reaction conditions several of them either resulted limited to only a few substrates (not enough flexible) or did not provide the reaction products with enough purity (cut-off >90% after photocleavage of the linker). The reader is invited to read carefully the ref. 240, which is an ideal example of rigorousness in assessing combinatorial-friendly SP transformations, to access all the details of this feasibility study. The result of this assessment focused the authors' efforts on three high quality transformations (boxed, Fig. 7.39): lactone opening with amines, esterification of the resulting OH and decoration of $\mathbf{R_1}$ via the use of bis-functionalized benzylnitrones.

The two epoxide enantiomers (+) and (−) **7.71** were selected as chemical starting points for library generation: Their synthesis from shikimic acid with reasonable overall yields was already known (248, 249), and they had a carboxylic acid handle for SPS. The other key intermediates were the three benzylnitrone carboxylic acids **7.72a–c**, which were prepared from the corresponding benzyl alcohols (250, 251) (Fig. 7.40). The two epoxycyclohexenols were supported onto a PEG-based resin, loaded with a photolabile linker, to give resin-bound **7.73a,b** (from now on only one enantiomer will be shown in the figures, but the synthetic pathway was continued with both

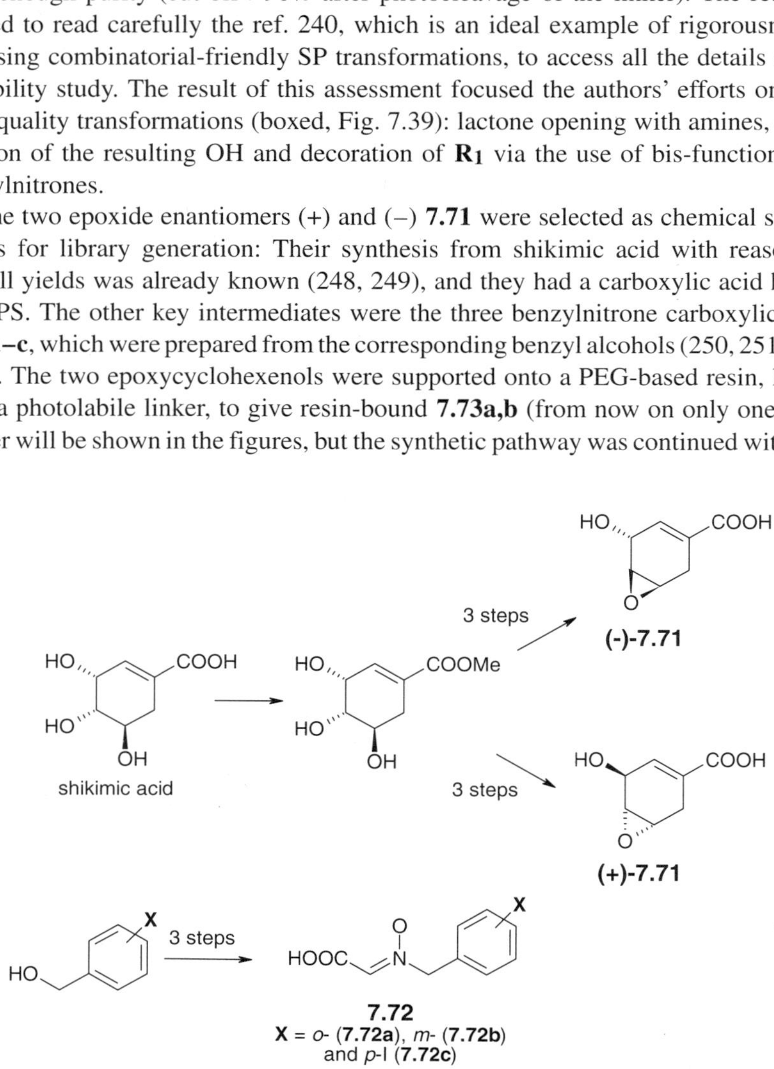

Figure 7.40 Solution synthesis of key synthons **(+)-7.71**, **(−)-7.71** and **7.72a–c**.

enantiomers of **7.71**). These two compounds were condensed with nitrones **7.72a–c**, producing six resin-bound tetracyclic benzyl iodides **7.74** (Fig. 7.41).

The authors selected a multistep synthetic pathway in accordance with the findings from chemistry assessment employing (a) an *alkyne coupling* with the iodoaromatic function, (b) the *opening of the lactone* ring with an amine, and (c) the *esterification* of the generated hydroxyl with carboxylic acids. They started from the six resin-bound compounds **7.74a–c** and **7.75a–c** (containing an ε-aminocaproic acid spacer, Fig. 7.42) and prepared six final, released compounds **7.82a–c** and **7.83a–c** using different monomers, cleaving and isolating a portion of each intermediate **7.76–7.79** in solution (Fig. 7.42). A complete off-bead analytical characterization of intermediates and final compounds (NMR, MS, TLC, HPLC) confirmed an 80–90% purity of photocleaved compounds, with good yields from the six-step SP sequence [from the loading of (+)-**7.71** onto the two resins] for all the compounds.

These encouraging results employing mild chemical conditions prompted the authors to perform a *monomer rehearsal*: The scheme of this crucial operation is reported in Fig. 7.43. Scaffold **7.74a** was selected and reacted in 50 different vials with 50 alkynes according to the Sonogashira–Castro alkyne coupling; 23 monomers produced, after cleavage, the expected coupling product with >90% yields and purity (HPLC, HPLC-MS, sometimes TLC and FAB-MS) and were selected, together with seven less optimal monomers (>70% yields, high purity), to give a first monomer set of 30 rehearsed alkynes. A large amount of scaffold **7.76a** was then prepared by coupling *t*-butylacetylene with **7.74a**, and this intermediate was reacted with 87 primary amines to give 54 high-performing monomers and 8 less optimal ones, for a

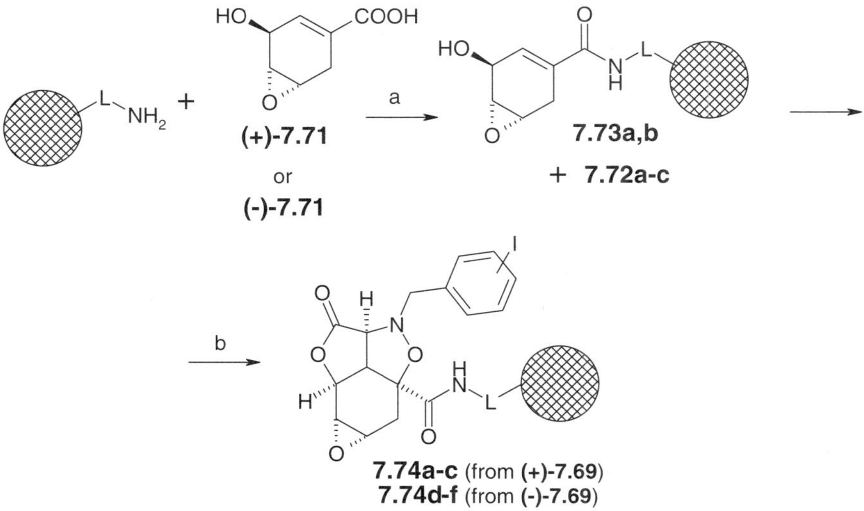

a: PyBOP, DIPEA, NMP, rt; b: HATU, DIPEA, DMAP, DCM, rt.

Figure 7.41 SP synthesis of key intermediates **7.74a–f**.

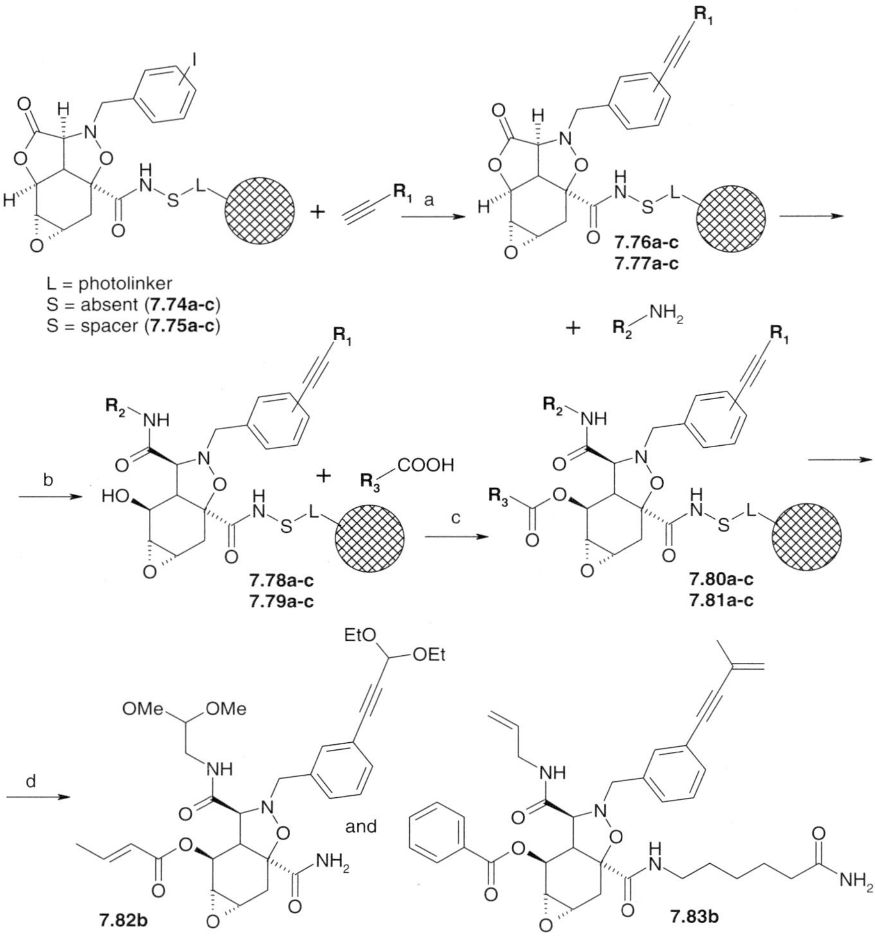

7.81a $\mathbf{R_1}$ = t-Bu (o-substituted); $\mathbf{R_2}$ = *p*-MeOBenzyl; $\mathbf{R_3}$ = i-Pr
7.81c $\mathbf{R_1}$ = n-Pr (p-substituted); $\mathbf{R_2}$ = *p*-MeOPhenethyl; $\mathbf{R_3}$ = Et
7.83a $\mathbf{R_1}$ = benzyl (o-substituted); $\mathbf{R_2}$ = cyclobutyl; $\mathbf{R_3}$ = *p*-MeOBenzyl
7.83c $\mathbf{R_1}$ = *p*-ClPhenyl (p-substituted); $\mathbf{R_2}$ = *o*-MeOBenzyl; $\mathbf{R_3}$ = i-Pr

a: CuI, $(PPh_3)_2PdCl_2$, DIPEA, DMF, rt, 15-45'; b: 2-HOPyridine, THF, rt, 12-16 hrs;
c: DIC, DIPEA, DMAP, DCM, rt, 12-16 hrs; d: cleavage.

Figure 7.42 SP chemistry assessment for the large, natural products–biased library **L12**: compounds **7.82a–c** and **7.83a–c**.

second monomer set of 62 rehearsed primary amines. A similar procedure was repeated with resin-bound **7.78a**, obtained by reaction of **7.76a** with *p*-methoxybenzyl amine, and 98 carboxylic acids were tested to eventually produce a third rehearsed monomer set made of 62 carboxylic acids (44 high performing, 18 less than optimal, Fig. 7.43).

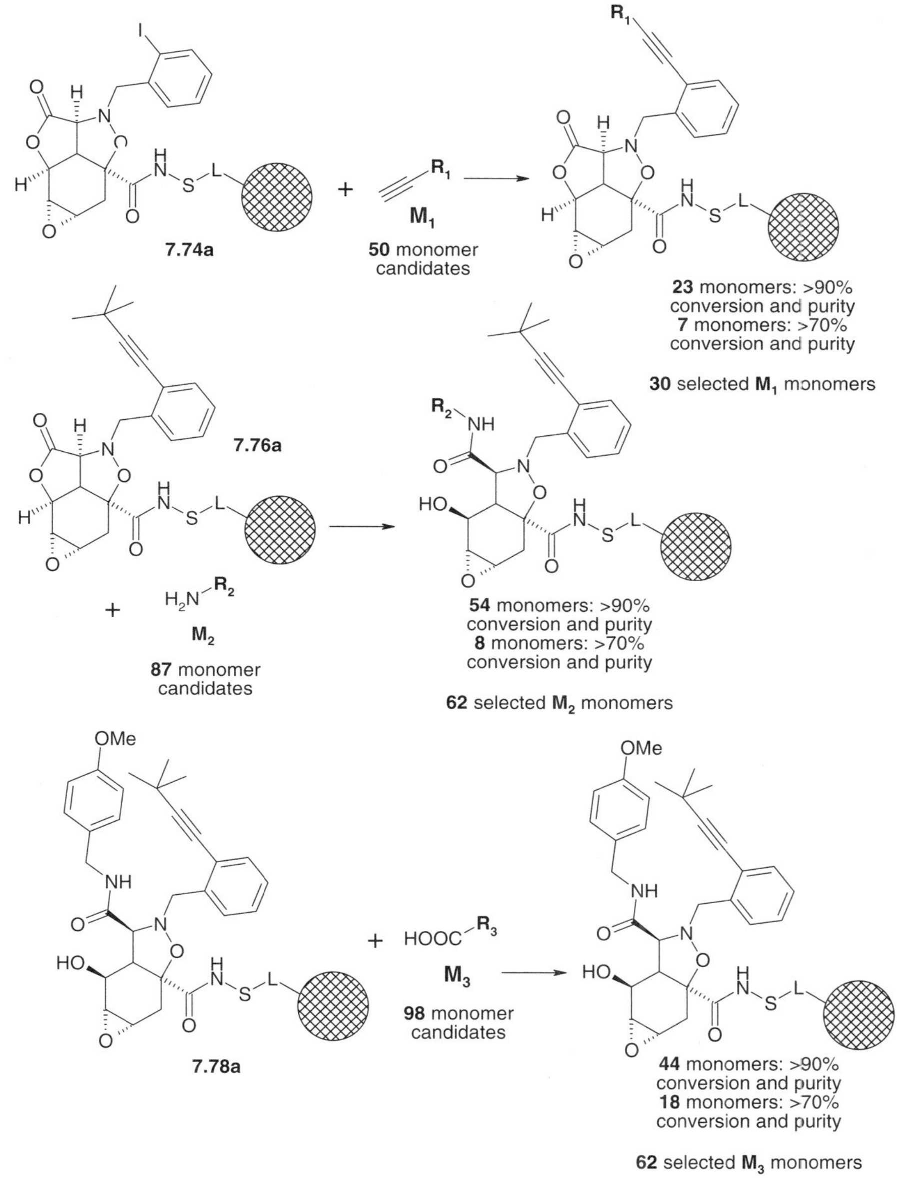

Figure 7.43 SP monomer rehearsal for the large, natural products–biased library **L12**.

Monomer rehearsal was followed by a *model library synthesis* aimed at confirming the feasibility of mix-and-split synthesis for the selected SP library synthetic scheme and to check for unwanted interactions of rehearsed building blocks at different reaction sites than the ones expected. A series of eight monomer representatives from each monomer set was selected, including the so-called skip codon (no reaction), and a total of eight pools made of 64 compounds were produced (**L11**, Fig. 7.44). The monomers sets **M1–M3** (Fig. 7.45) were selected so that each model library individual prepared had a different MW, to give 456 different masses to be detected; each pool contained the same eight lactones derived from the amine skip codon, with no opening of the lactone, and subsequently no reaction with the carboxylic acids. The eight pools were submitted to HPLC/MS analysis and all the expected MWs were spotted. From them, 88% of peaks (400 MWs) were detected at a significant intensity, and a general good quality was found except for $\mathbf{M_{2,1}}$ (skip codon) and $\mathbf{M_{2,6}}$ at the amine position. This led to the elimination of $\mathbf{M_{2,6}}$, which could have generated problems, and to the exclusion of unreacted butyrolactones (skip codon) from the last coupling step.

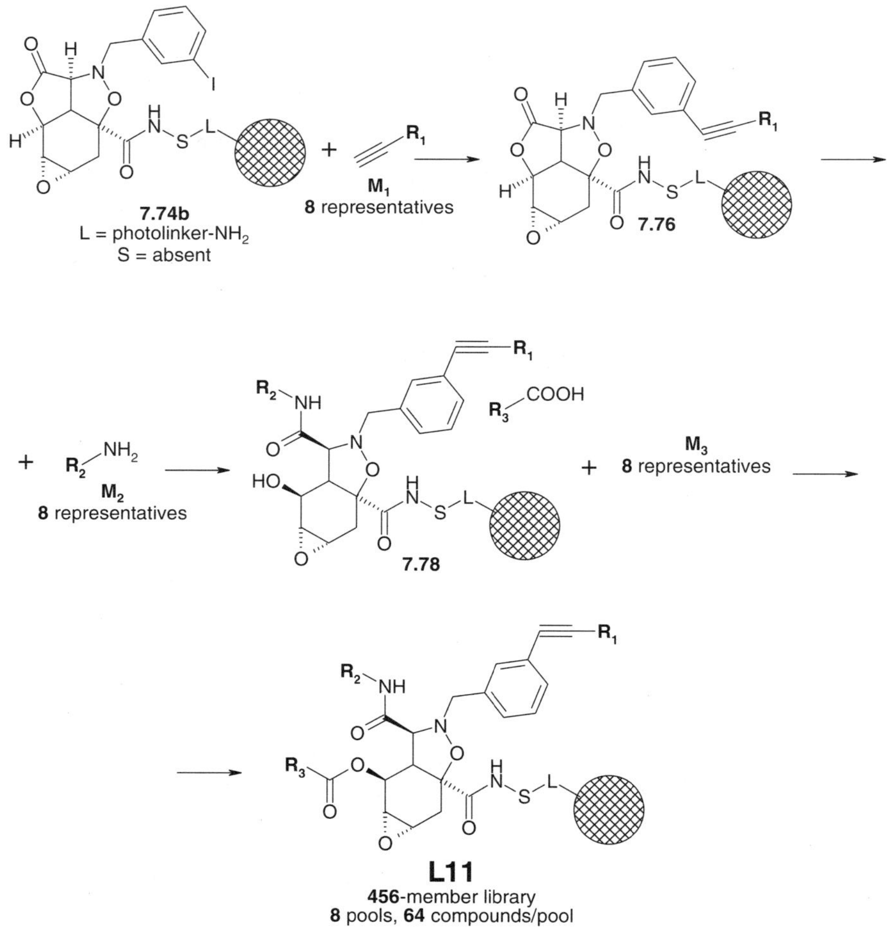

Figure 7.44 SP synthesis of the natural products-biased library model library **L11**.

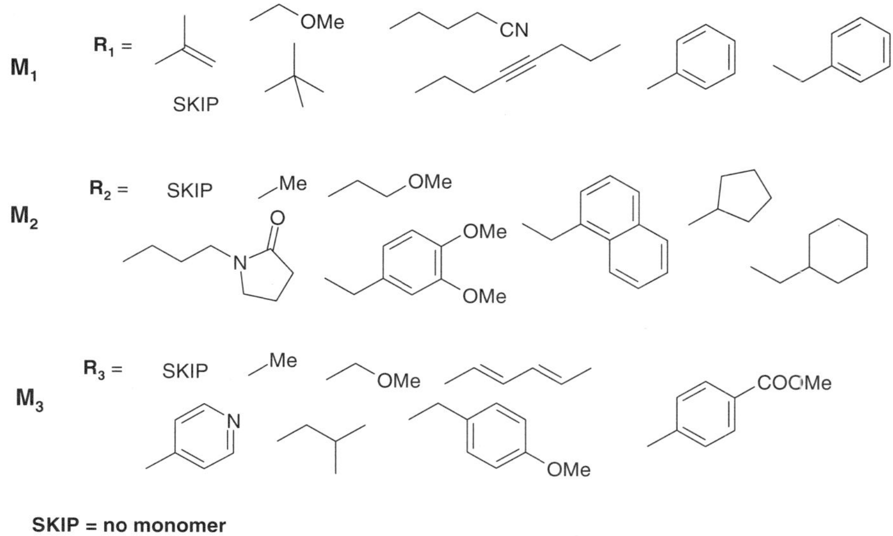

Figure 7.45 Monomer sets used for the synthesis of the natural products-biased library **L11**.

Finally, the large encoded library **L12** was prepared using three spacers on the resin (Gly, aminocaproic acid and a skip spacer), (+) and (–) **7.71** and the three nitrones **7.72a–c**, to give 18 iodobenzyl tetracyclic scaffolds [**7.74a–f**, **7.75a–f**, **7.84a–f** from (+) **7.71** (**a–c**) and from (**–**) **7.71** (**d–f**)] (Fig. 7.46). These scaffolds, using mix-and-split synthesis, were coupled to the previously rehearsed monomers: 31 alkynes (30 + the skip codon, 558 compounds, **7.76**, **7.77**, **7.85**), 63 primary amines (62 + skip, 34,596 compounds, **7.78**, **7.79**, **7.86**, + 558 from the previous skip lactone), and 63 carboxylic acids (62 + skip, 2,179,548 compounds, **L12**, Fig. 7.46). The binary encoding technique (189) required two tags for the spacers, for the epoxycyclohexenol isomers, and for the nitrones; five tags for the alkyne couplings; and six tags each for the amine and the carboxylic acid couplings; the reported encoding procedure (189) was significantly optimized to afford cleaner and more reliable decoding results. A total of 23 tags, inserted after each monomer coupling onto SP, was sufficient to encode the $3 \times 2 \times 3 \times 31 = 558$ butyrolactones and the $3 \times 2 \times 3 \times 31 \times 62 \times 63 = 2{,}179{,}548$ final library individuals, producing an ≈2,180,000-member library. Details of the analytical characterization of single beads were not given, but the decoding of single beads from each library pool produced satisfactory results (*vide infra*).

The extreme flexibility of these scaffolds would definitely allow the preparation of other large, primary libraries using chemistry-friendly synthetic schemes. The epoxide could be touched, as could the N–O function, which could be reductively cleaved and eventually used to obtain two new handles on constrained, stereo-determined novel scaffolds. The potential of such an integrated approach, where all the steps toward a large bead-based library are carefully assessed, is clearly enormous, especially if any library prepared would contain embedded biological information (starting from natu-

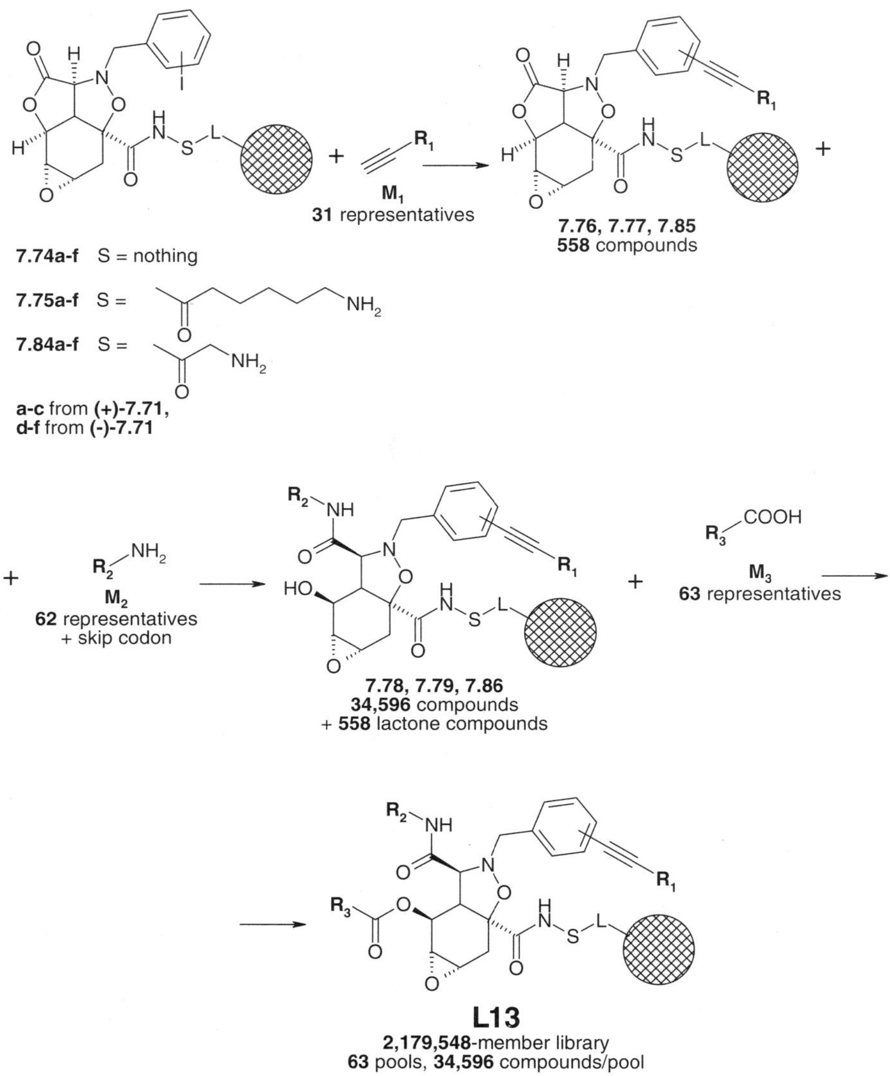

Figure 7.46 SP synthesis of the large, natural products–biased library **L12**.

ral and/or active scaffolds) as in the example (240). Possibly the library would be prepared in a >50 library equivalents-quantity to be tested on more than one specific assay. The efforts required to prepare a >2,000,000 library should be considered acceptable especially considering the developed miniaturized HTS formats (see next section); such chemistry, which should also be feasible using alternative chemical starting points and/or reaction schemes, should allow the generation of invaluable proprietary chemical diversity.

7.5.2 Bead-Based Libraries: Miniaturized High Throughput Screening

The library described in the previous section was tested on a miniaturized nanodroplet assay to spot ligand-protein interactions (242), and the authors decided to use 2.5 grams of photolinker-resin construct (around 2.7 million beads/gram, around 6,600,000 beads) to ensure the successful preparation of three library equivalents, corresponding to a >95% probability of representation for each library individual. The compounds were partitioned into polydimethylsiloxane plates (6500 assays in a 10 cm-plastic dish) by using a wetting- (pipetting a beads-cells-agar suspension into the wells) dewetting technique (pipetting off the excess liquid and leaving in the well uniform nanodroplets of 50–150 nL due to surface tension). Using a 10 mg suspension of beads in 0.006% aqueous agar resulted in a significant amount of wells (>60%) containing one to three beads in a nanodroplet, thus allowing the testing, and then the decoding, of such beads in an automated manner. As soon as the nanodroplets containing the beads, the cells and the media were deposited, the photolinker was UV-cleaved at 365 nM, releasing the library individuals from the beads. Active beads showed an effect on cells and were finally decoded (189). After the whole screening, a family of compounds from the library (**7.87**, Fig. 7.47), was identified as active for the specific biological target.

MacBeath et al. recently reported a miniaturized screening format, inspired by the DNA microarray technique and called Small Molecule Printing (SMP), to maximize the synthetic efforts to produce a bead-based large library (244). In a validation experiment, SP beads were delivered to polypropylene plates (one per well, using bead pickers) and cleaved in a small volume of solvent to afford a concentrated solution (high μ*M*). The authors used a high precision robotic instrument (252) to deliver 1 nL-aliquots from each well to the surface of many chemically derivatized microscope slides. The grafted functionality on the slides reacted with the library individuals, immobilizing each of them on its surface at a very high density (>1000 spots per cm^2); each slide carried thousands of compounds, representing a subset of the bead-based library, and could be tested with an on-slide screen format using soluble targets (a

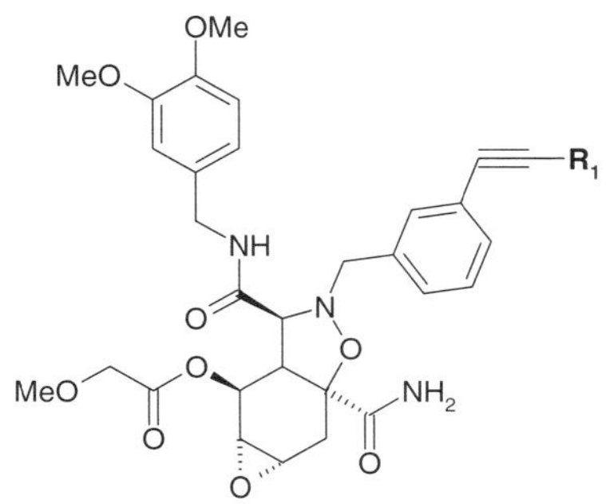

7.87

Figure 7.47 Structure of a family of protein ligands (**7.87**) from the miniaturized screening of the large, natural products-biased library **L12**.

fluorescent-tagged isolated receptor in the example). A throughput of 150 printed slides per print run of the robot could easily be obtained; up to 3,000 slide spots (thus up to 3,000 screens) could derive from the releasate of a single 425 μm diameter PS bead. The slides were then processed through a slide scanner to detect the positives (a UV scanner in ref. 244). The constraints imposed by hooking appropriate chemical groups of the library individuals on the slides (see Section 7.2.3) are more than compensated by the extremely high throughput of the miniaturized assay which requires extremely low quantities of chemicals and biological reagents often expensive or difficult to obtain. HTS campaigns for all the orphan receptors and for many recombinant proteins deriving from the expression of genetic libraries may assess their relevance via the discovery of chemicals interacting with them (high throughput chemical genetics, see also Section 9.1.4); this may become at least in part possible by using such miniaturized approaches, and may unravel novel relevant mechanisms to cure various important diseases.

Other interesting reports related to the synthesis and screening of large bead-based SP pool libraries can be found in the literature (204, 253, 254), reinforcing the concepts that were presented for the specific example reported here. In a nutshell, the saga of large, SP pool libraries of small organic molecules should gain new strength in the future, rather than being completely overpowered by other library formats and synthesis techniques.

REFERENCES

1. Furka, A., Sebestyen, F., Asgedom, M. and Dibo, G., *Int. J. Pept. Protein Res.* **37**, 487–493 (1991).
2. Lam, K. S., Salmon, S. E., Hersh, E. M., Hruby, V. J., Kazmierski, W. M. and Knapp, R. J., *Nature* **354**, 82–84 (1991).
3. Houghten, R., Pinilla, C., Blondelle, S. E., Appel, J. R., Dooley, C. T. and Cuervo, J. H., *Nature* **354**, 84–86 (1991).
4. Baldwin, J. J., Burbaum, J. J., Henderson, I. and Ohlmeyer, M. H. J., *J. Am. Chem. Soc.* **117**, 5588–5589 (1995).
5. Dankwardt, S. M., Newman, S. R. and Krstenansky, J. L., *Tetrahedron Lett.* **36**, 4923–4926 (1995).
6. Burbaum, J. J., Ohlmeyer, M. H. J., Reader, J. C., Henderson, I., Dillard, L. W., Li, G., Randle, T. L., Sigal, N. H., Chelsky, D. and Baldwin, J. J., *Proc. Natl. Acad. Sci. USA* **92**, 6027–6031 (1995).
7. Storer, R., *Drug Discovery Today* **1**, 248–254 (1996).
8. Gravert, D. J. and Janda, K. D., *Curr. Opin. Chem. Biol.* **1**, 107–113 (1997).
9. Bailey, N., Cooper, A. W. J., Deal, M. J., Dean, A. W., Gore, A. L., Hawes, M. C., Judd, D. B., Merritt, A. T., Storer, R., Travers, S. and Watson, S. P., *Chimia* **51**, 832–837 (1997).
10. Carell, T., Wintner, E. A. and Rebek, J. Jr., *Angew. Chem. Int. Ed. Engl.* **33**, 2061–2064 (1994).
11. An, H., Haly, B. D. and Cook, P. D., *J. Med. Chem.* **41**, 706–716 (1998).

12. Boger, D. L., Jiang, W. and Goldberg, J., *J. Org. Chem.* **64**, 7094–7100 (1999).
13. Lam, K. S., Lebl, M. and Krchnak, V., *Chem. Rev.* **97**, 411–448 (1997).
14. Hardin, J. H. and Smietana, F. R., *Mol. Diversity* **1**, 270–274 (1995).
15. DeWitt, S. H., *Ann. Rep. Combi. Chem. Mol. Div.* **1**, 69–77 (1997).
16. Sugarman, J. H., Rava, R. P., Kedar, H., Dower, W. J., Barrett, R. W., Gallop, M. A. and Needels, M. C., *WO Patent 9512608 A1*, May 5th, 1995.
17. Zuckermann, R. N., Kerr, J. M., Siani, M. A. and Banville, S. C., *Int. J. Pept. Protein Res.* **40**, 497–506 (1992).
18. Lashkari, D. A., Hunicke-Smith, S. P., Norgren, R. M., Davis, R. W. and Brennan, T., *Proc. Natl. Acad. Sci. USA* **92**, 7912–7915 (1995).
19. Boutin, J. A. and Fauchere, J.-L., *Am. Biotechnol. Lab.* **14**, 36–40 (1996).
20. Stankova, M. and Lebl, M., *Mol. Diversity* **2**, 75–80 (1996).
21. Needels, M. and Sugarman, J., in *Combinatorial Chemistry and Molecular Diversity in Drug Discovery*, E. M. Gordon and J. Kerwin (Eds.). Wiley-Liss, New York, 1998, pp. 339–348.
22. Stanchfield, J., Wright, D., Hsu, S., Lamsa, M. and Robbins, A., *BioTechniques* **20**, 292–296 (1996).
23. Fitch, W. L., *Ann. Rep. Combi. Chem. Mol. Div.* **1**, 59–68 (1997).
24. Fitch, W. L., Look, G. C. and Detre, G., in *Combinatorial Chemistry and Molecular Diversity in Drug Discovery*, E. M. Gordon and J. F. Kerwin, Jr. (Eds.). John Wiley and Sons, New York, 1998, pp. 349–368.
25. Sepetov, N. and Issakova, O., in *Combinatorial Chemistry and Combinatorial Technologies: Methods and Applications*, S. Miertus and G. Fassina (Eds.). Marcel Dekker, New York, 1999, pp. 169–203.
26. Fathi, R., Patel, R. and Cook, A. F., *Mol. Diversity* **2**, 125–134 (1997).
27. Whitehouse, D. L., Nelson, K. H., Jr., Savinov, S. N., Lowe, R. S. and Austin, D. J., *Bioorg. Med. Chem.* **6**, 1273–1282 (1998).
28. Richter, H. and Jung, G., *Mol. Diversity* **3**, 191–194 (1998).
29. Kurth, M. J., Randall, L. A. A. and Takenouchi, K., *J. Org. Chem.* **61**, 8755–8761 (1996).
30. Phillips, G. B. and Wei, G. P., *Tetrahedron Lett.* **37**, 4887–4890 (1996).
31. Haap, W. J., Kaiser, D., Walk, T. B. and Jung, G., *Tetrahedron* **54**, 3705–3724 (1998).
32. Amparo Lago, M., Nguyen, T. T. and Bhatnagar, P., *Tetrahedron Lett.* **39**, 3885–3888 (1998).
33. Zhu, T. and Boons, G.-J., *Angew. Chem. Int. Ed. Engl.* **37**, 1898–1900 (1998).
34. Chen, C., Ahlberg Randall, L. A., Miller, R. B., Jones, A. D. and Kurth, M. J., *Tetrahedron* **53**, 6595–6609 (1997).
35. Whitehouse, D. L., Nelson, K. H., Jr., Savinov, S. N. and Austin, D. J., *Tetrahedron Lett.* **38**, 7139–7142 (1997).
36. Dunayevskiy, Y., Vouros, P., Carell, T., Wintner, E. A. and Rebek, J. Jr., *Anal. Chem.* **67**, 2906–2915 (1995).
37. Metzger, J. W., Wiesmuller, K.-H., Kienle, S., Brunjes, J. and Jung, G., in *Peptide and Non-Peptide Libraries: A Handbook for the Search of Lead Structures*, G. Jung (Ed.). VCH, Weinheim, Germany, 1997, pp. 247–285.

38. Heerding, D. A., Takata, D. T., Kwon, C., Huffman, W. F. and Samanen, J., *Tetrahedron Lett.* **39**, 6815–6818 (1998).
39. Thompson, L. A. and Ellman, J. A., *Chem. Rev.* **96**, 555–600 (1996).
40. Balkenhohl, F., von dem Bussche-Huennefeld, C., Lansky, A. and Zechel, C., *Angew. Chem. Int. Ed. Engl.* **35**, 2288–2337 (1996).
41. Nefzi, A. and Houghten, R. A., *Chem. Rev.* **97**, 449–472 (1997).
42. Houghten, R. A., Pinilla, C., Appel, J. R., Blondelle, S. E., Dooley, C. T., Eichler, J., Nefzi, A. and Ostresh, J. M., *J. Med. Chem.* **42**, 3743–3778 (1999).
43. Pei, Y., Houghten, R. A. and Kiely, J. S., *Tetrahedron Lett.* **38**, 3349–3352 (1997).
44. Krchnak, V. and Weichsel, A. S., *Tetrahedron Lett.* **38**, 7299–7302 (1997).
45. Cao, J., Cuny, G. D. and Hauske, J. R., *Mol. Diversity* **3**, 173–179 (1998).
46. Nefzi, A., *Bioorg. Med. Chem. Lett.* **8**, 2273–2278 (1998).
47. Neustadt, B. R., Smith, E. M., Lindo, N., Nechuta, T., Bronnenkant, A., Wu, A., Armstrong, L. and Kumar, C., *Bioorg. Med. Chem. Lett.* **8**, 2395–2398 (1998).
48. Gennari, C., Longari, C., Ressel, S., Salom, B., Piarulli, U., Ceccarelli, S. and Mielgo, A., *Eur. J. Org. Chem.,* 2437–2449 (1998).
49. Kelly, M. A., Liang, H., Sytwu, I.-I., Vlattas, I., Lyons, N. L., Bowen, B. R. and Wennogle, L. P., *Biochemistry* **35**, 11747–11755 (1996).
50. Blom, K. F., Larsen, B. S. and McEwen, C. N., *J. Comb. Chem.* **1**, 82–90 (1999).
51. Davis, R. G., Anderegg, R. J. and Blanchard, S. G., *Tetrahedron* **55**, 11653–11667 (1999).
52. Dunayevskiy, Y. M., Lai, J.-J., Quinn, C., Talley, F. and Vouros, P., *Rapid Commun. Mass Spectrom.* **11**, 1178–1184 (1997).
53. Chu, Y.-H., Kirby, D. P. and Karger, B. L., *J. Am. Chem. Soc.* **117**, 5419–5420 (1995).
54. Chu, Y.-H., Dunayevskiy, Y. M., Kirby, D. P., Vouros, P. and Karger, B. L., *J. Am. Chem. Soc.* **118**, 7827–7835 (1996).
55. Cheng, C. C. and Chu, Y.-H., *Am. Lab.* **30**, 79–81 (1998).
56. Dunayevskiy, Y. M., Lyubarskaya, Y. V., Chu, Y.-H., Vouros, P. and Karger, B. L., *J. Med. Chem.* **41**, 1201–1204 (1998).
57. Chu, Y.-H. and Cheng, C. C., *Cell. Mol. Life Sci.* **54**, 663–683 (1998).
58. Sun, S., Headrick, J., Staller, T. and Sepaniak, M., *J. Microcolumn Separations* **10**, 653–660 (1998).
59. Lyubarskaya, Y. V., Carr, S. A., Dunnington, D., Prichett, W. P., Fisher, S. M., Appelbaum, E. R., Jones, C. S. and Karger, B. L., *Anal. Chem.* **70**, 4761–4770 (1998).
60. Wieboldt, R., Zweigenbaum, J. and Henion, J., *Anal. Chem.* **69**, 1683–1691 (1997).
61. Nedved, M. L., Habibi-Goudarzi, S., Ganem, B. and Henion, J. D., *Anal. Chem.* **68**, 4228–4236 (1996).
62. Wu, J., Takayama, S., Wong, C.-H. and Siuzdak, G., *Chem. Biol.* **4**, 653–657 (1997).
63. Berlin, K., Jain, R. K., Tetzlaff, C., Steinbeck, C. and Richert, C., *Chem. Biol.* **4**, 63–77 (1997).
64. Hsieh, F., Keshishian, H. and Muir, C., *J. Biomol. Screening* **3**, 189–198 (1998).
65. Siegel, M. M., Tabei, K., Bebernitz, G. A. and Baum, E. Z., *J. Mass Spectrom.* **33**, 264–273 (1998).
66. Holtzapple, C. K. and Stanker, L. H., *Anal. Chem.* **70**, 4817–4821 (1998).

67. van Breemen, R. B., Huang, C.-R., Nikolic, D., Woodbury, C. P., Zhao, Y.-Z. and Venton, D. L., *Anal. Chem.* **69**, 2159–2164 (1997).

68. Zhao, Y.-Z., van Breemen, R. B., Nikolic, D., Huang, C.-R., Woodbury, C. P., Schilling, A. and Venton, D. L., *J. Med. Chem.* **40**, 4006–4012 (1997).

69. Nikolic, D. and van Breemen, R. B., *Combi. Chem. High. Throughput Screen.* **1**, 47–55 (1998).

70. Woodbury, C. P. and Venton, D. L., *Am. Lab.* **30**, 16–19 (1998).

71. Bruce, J. E., Anderson, G. A., Chen, R., Cheng, X., Gale, D. C., Hofstadler, S. A., Schwartz, B. L. and Smith, R. D., *Rapid Commun. Mass Spectrom.* **9**, 644–650 (1995).

72. Cheng, X., Chen, R., Bruce, J. E., Schwartz, B. L., Anderson, G. A., Hofstadler, S. A., Gale, D. C., Smith, R. D., Gao, J. and Sigal, G. B., *J. Am. Chem. Soc.* **117**, 8859–8860 (1995).

73. Gao, J., Cheng, X., Chen, R., Sigal, G. B., Bruce, J. E., Schwartz, B. L., Hofstadler, S. A., Anderson, G. A., Smith, R. D. and Whitesides, G M., *J. Med. Chem.* **39**, 1949–1955 (1996).

74. Hofstadler, S. A., Sannes-Lowery, K. A., Crooke, S. T., Ecker, D. J., Sasmor, H., Manalili, S. and Griffey, R. H., *Anal. Chem.* **71**, 3436–3440 (1999).

75. Shuker, S. B., Hajduk, P. J., Meadows, R. P. and Fesik, S. W., *Science* **274**, 1531–1534 (1996).

76. Hajduk, P. J., Meadows, R. P. and Fesik, S. W., *Science* **278**, 498–499 (1997).

77. Marek, D., poster presented at *Thirty ninth Experimental Nuclear Magnetic Resonance Conference*, *P227*, Asilomar, CA (1998).

78. Hajduk, P. J., Dinges, J., Miknis, G. F., Merlock, M., Middleton, T., Kempf, D. J., Egan, D. A., Walter, K. A., Robins, T. S., Shuker, S. B., Holzman, T. F. and Fesik, S. W., *J. Med. Chem.* **40**, 3144–3150 (1999).

79. Hajduk, P. J., Sheppard, G., Nettesheim, D. G., Olejniczak, E. T., Shuker, S. B., Meadows, R. P., Steinman, D. H., Carrera, G. M., Marcotte, P. A., Severin, J., Walter, K., Smith, H., Gubbins, E., Simmer, R., Holzman, T. F., Morgan, D. W., Davidsen, S. K. and Fesik, S. W., *J. Am. Chem. Soc.* **119**, 5818–5827 (1997).

80. Hajduk, P. J., Gerfin, T., Boehlen, J.-M., Haeberli, M., Marek, D. and Fesik, S. W., *J. Med. Chem.* **42**, 2315–2317 (1999).

81. Hajduk, P. J., Dinges, J., Schkeryantz, J. M., Janowick, D., Kaminski, M., Tufano, M., Augeri, D. J., Petros, A., Nienaber, V., Zhong, P., Hammond, R., Coen, M., Beutel, B., Katz, L. and Fesik, S. W., *J. Med Chem.* **42**, 3852–3859 (1999).

82. Lin, M., Shapiro, M. J. and Wareing, J. R., *J. Am. Chem. Soc.* **119**, 5249–5250 (1997).

83. Lin, M., Shapiro, M. J. and Wareing, J. R., *J. Org. Chem.* **62**, 8930–8931 (1997).

84. Shapiro, M. J., Chin, J., Chen, A., Wareing, J. R., Tang, Q., Tommasi, R. A., Marepalli, H. R., *Tetrahedron Lett.* **40**, 6141–6143 (1999).

85. Bleicher, K., Lin, M., Shapiro, M. J. and Wareing, J. R., *J. Org. Chem.* **63**, 8486–8490 (1998).

86. Anderson, R. C., Lin, M., and Shapiro, M. J., *J. Comb. Chem.* **1**, 69–72 (1999).

87. Hajduk, P. J., Olejniczak, E. T. and Fesik, S. W., *J. Am. Chem. Soc.* **119**, 12257–12261 (1997).

88. Meyer, B., Weimar, T. and Peters, T., *Eur. J. Biochem.* **246**, 705–709 (1997).

89. Henrichsen, D., Ernst, B., Magnani, J. L., Wang, W.-T., Meyer, B. and Peters, T., *Angew. Chem. Int. Ed.* **38**, 98–102 (1999).

90. Fejzo, J., Lepre, C. A., Peng, J. W., Bemis, G. W., Ajay, Murcko, M. A. and Moore, J. M., *Chem. Biol.* **6**, 755–769 (1999).

91. Comprehensive Medicinal Chemistry (CDC) database, version CDC3D98.1, MDL Information Systems, San Leandro, CA, US.

92. Bemis, G. W. and Murcko, M. A., *J. Med. Chem.* **39**, 2887–2893 (1996).

93. Available Chemical Directory (ADC) database, version ACD98.2, MDL Information Systems, San Leandro, CA, US.

94. Chen, A. and Shapiro, M. J., *J. Am. Chem. Soc.* **120**, 10258–10259 (1998).

95. Mayer, M. and Meyer, B., *Angew. Chem. Int. Ed.* **38**, 1784–1788 (1999).

96. Klein, J., Meinecke, R., Mayer, M. and Meyer, B., *J. Am. Chem. Soc.* **121**, 5336–5337 (1999).

97. Youngquist, R. S., Fuentes, G. R., Miller, C. M., Ridder, G. M., Lacey, M. P. and Keough, T., *Adv. Mass Spectrometry* **14**, 423–448 (1998).

98. Sussmuth, R. D. and Jung, G., *J. Chromatogr., B: Biomed. Sci. Applic.* **725**, 49–65 (1999).

99. Keifer, P. A., *Curr. Opin. Biotechnol.* **10**, 34–41 (1999).

100. Moore, J. M., *Curr. Opin. Biotechnol.* **10**, 54–58 (1999).

101. Gounarides, J. S., Chen, A. and Shapiro, M. J., *J. Chromatogr., B: Biomed. Sci. Applic.* **725**, 79–90 (1999).

102. Chen, A. and Shapiro, M. J., *Anal. Chem.* **71**, 669A–675A (1999).

103. Woodbury, C. P., Jr. and Venton, D. L., *J. Chromatogr., B: Biomed. Sci. Applic.* **725**, 113–137 (1999).

104. Eliseev, A. V., *Curr. Opin. Drug Discovery Dev.* **1**, 106–115 (1998).

105. Lam, K. S. and Lebl, M., *Methods* **6**, 372–380 (1994).

106. Lam, K. S., Wade, S., Abdul-Latif, F. and Lebl, M., *J. Immunol. Methods* **180**, 219–223 (1995).

107. Lam, K. S. and Lebl, M., *Peptide and Non-Peptide Libraries: A Handbook for the Search of Lead Structures,* G. Jung (Ed.). VCH, Weinheim, Germany, 1996, pp. 173–201.

108. Lam, K. S., Lake, D., Salmon, S. E., Smith, J., Chen, M.-L., Wade, S., Abdul-Latif, F., Knapp, R. J., Leblova, Z. et al., *Methods* **9**, 482–493 (1996).

109. Lam, K. S., *Methods Mol. Biol.* **87**, 7–12 (1998).

110. Lebl, M., Krchnak, V., Sepetov, N. F., Seligmann, B., Strop, P., Felder, S. and Lam, K. S., *Biopolymers* **37**, 177–198 (1995).

111. Ohlmeyer, M. H. J., Swanson, R. N., Dillard, L., Reader, J. C., Asouline, G., Kobayashi, R., Wigler, M. and Still, W. Clark., *Proc. Natl. Acad. Sci. USA* **90**, 10922–10926 (1993).

112. Boyce, R., Li, G., Nestler, H. P., Suenaga, T. and Still, W. Clark., *J. Am. Chem. Soc.* **116**, 7955–7956 (1994).

113. Nestler, H. P., *Mol. Diversity* **2**, 35–40 (1996).

114. Yu, Z. and Chu, Y.-H., *Bioorg. Med. Chem. Lett.* **7**, 95–98 (1997).

115. Hwang, S., Tamilarasu, N., Ryan, K., Huq, I., Richter, S., Still, W. Clark and Rana, T. M., *Proc. Natl. Acad. Sci. USA* **96**, 12997–13002 (1999).

116. Needels, M. C., Jones, D. G., Tate, E. H., Heinkel, G. L., Kochersperger, L. M., Dower, W. J., Barrett, R. W. and Gallop, M. A., *Proc. Natl. Acad. Sci. USA* **90**, 10700–10704 (1993).

117. St. Hilaire, P. M., Willert, M., Juliano, M. A., Juliano, L. and Meldal, M., *J. Comb. Chem.* **1**, 509–523 (1999).

118. Kassarjian, A., Schellenberger, V. and Turck, C. W., *Pept. Res.* **6**, 129–133 (1993).

119. Nestler, H. P., Wennemers, H., Sherlock, R. and Dong, D. L. Y., *Bioorg. Med. Chem. Lett.* **6**, 1327–1330 (1996).

120. Lam, K. S. and Wu, J., *Methods* **6**, 401–403 (1994).

121. Meldal, M., Svendsen, L, Breddam, K. and Auzanneau, F.-I., *Proc. Natl. Acad. Sci. USA* **91**, 3314–3318 (1994).

122. Meldal, M. and Svendsen, I., *J. Chem. Soc., Perkin Trans. 1*, 1591–1596 (1995).

123. Lou, Q., Leftwich, M. E. and Lam, K. S., *Bioorg. Med. Chem.* **4**, 677–682 (1996).

124. Salmon, S. E., Lam, K. S., Lebl, M., Kandola, A., Khattri, P. S., Wade, S., Patek, M., Kocis, P., Krchnak, V. et al., *Proc. Natl. Acad. Sci. USA* **90**, 11708–11712 (1993).

125. Berkessel, A. and Herault, D. A., *Angew. Chem. Int. Ed.* **38**, 102–105 (1999).

126. Cho, C. Y., Liu, C. W., Wemmer, D. E. and Schultz, P. G., *Bioorg. Med. Chem.* **7**, 1171–1179 (1999).

127. St. Hilaire, P. M., Lowary, T. L., Meldal, M. and Bock, K., *J. Am. Chem. Soc.* **120**, 13312–13320 (1998).

128. Madden, D., Krchnak, V. and Lebl, M., *Persp. Drug Disc. Design* **2**, 269–285 (1995).

129. Vagner, J., Barany, G., Lam, K. S., Krchnak, V., Sepetov, N. F., Ostrem, J. A., Strop, P. and Lebl, M., *Proc. Natl. Acad. Sci. USA* **93**, 8194–8199 (1996).

130. Leon, S., Quarrell, R. and Lowe, G., *Bioorg. Med. Chem. Lett.* **8**, 2997–3002 (1998).

131. Quarrell, R., Claridge, T. D. W., Weaver, G. W. and Lowe, G., *Mol. Diversity* **1**, 223–232 (1996).

132. Marsh, I. R., Smith, H. K., Leblanc, C. and Bradley, M., *Mol. Diversity* **2**, 165–170 (1997).

133. Smith, H. K. and Bradley, M., *J. Comb. Chem.* **1**, 326–332 (1999).

134. Kapoor, T. M., Andreotti, A. H. and Schreiber, S. L., *J. Am. Chem. Soc.* **120**, 23–29 (1998).

135. Morken, J. P., Kapoor, T. M., Feng, S., Shirai, F. and Schreiber, S. L., *J. Am. Chem. Soc.* **120**, 30–36 (1998).

136. Rickles, R. J., Botfield, M. C., Zhou, X.-M., Henry, P. A., Brugge, J. S. and Zoller, M. J., *Proc. Natl. Acad. Sci. USA* **92**, 10909–10913 (1995).

137. Combs, A. P., Kapoor, T. M., Feng, S., Chen, J. K., Daude-Snow, L. F. and Schreiber, S. L., *J. Am. Chem. Soc.* **118**, 287–288 (1996).

138. Erb, E., Janda, K. D. and Brenner, S., *Proc. Natl. Acad. Sci. USA* **91**, 11422–11426 (1994).

139. Sutherlin, D. P. and Armstrong, R. W., *J. Org. Chem.* **62**, 5267–5283 (1997).

140. Gordeev, M. F., Patel, D. V. and Gordon, E. M., *J. Org. Chem.* **61**, 924–928 (1996).

141. Gordeev, M. F., Patel, D. V., England, B. P., Jonnalagadda, S., Combs, J. D. and Gordon, E. M., *Bioorg. Med. Chem.* **6**, 883–889 (1998).

142. Bellemann, P., Schade, A. and Towart, R., *Proc. Natl. Acad. Sci. USA* **80**, 2356–2360 (1983).

143. Baldwin, J. J. and Dolle, R. E., *Annu. Rep. Combi. Chem. Mol. Div.* **1**, 287–297 (1997).

144. Schriemer, D. C. and Hindsgaul, O., *Combi. Chem. High Throughput Screen.* **1**, 155–170 (1998).

145. Seneci, P., in *Combinatorial Chemistry and Combinatorial Technologies: Methods and Applications*, S. Miertus and G. Fassina (Eds.). Marvel Dekker, New York, 1999, 91–125.

146. Rohrer, S. P., Birzin, E. T., Mosley, R. T., Berk, S. C., Hutchins, S. M., Shen, D.-M., Xiong, Y., Hayes, E. C., Parmar, R. M. and Foor, F., *Science* **282**, 737–740 (1998).

147. Rohrer, S. P. and Berk, S. C., *Curr. Opin. Drug Discovery Dev.* **2**, 293–303 (1999).

148. Berk, S. C., Rohrer, S. P., Degrado, S. J., Birzin, E. T., Mosley, R. T., Hutchins, S. M., Pasternak, A., Schaeffer, J. M., Underwood, D. J. and Chapman, K. T., *J. Comb. Chem.* **1**, 388–396 (1999).

149. Heizmann, G., Hildebrand, P., Tanner, H., Ketterer, S., Pansky, A., Froidevaux, S., Beglinger, C. and Eberle, A. N., *J. Recept. Signal Transduction Res.* **19**, 449–466 (1999).

150. Lohse, A., Jensen, K. B. and Bols, M., *Tetrahedron Lett.* **40**, 3033–3036 (1999).

151. Lohse, A., Jensen, K. B., Lundgren, K. and Bols, M., *Bioorg. Med. Chem.* **7**, 1965–1971 (1999).

152. Roussel, P., Bradley, M., Kane, P., Bailey, C., Arnold, R. and Cross, A., *Tetrahedron* **55**, 6219–6230 (1999).

153. Szardenings, A. K., Antonenko, V., Campbell, D. A., DeFrancisco, N., Ida, S., Shi, L., Sharkov, N., Tien, D., Wang, Y. and Navre, M., *J. Med. Chem.* **42**, 1348–1357 (1999).

154. Ng, S., Goodson, B., Ehrhardt, A., Moos, W. H., Siam, M. and Winter, J., *Bioorg. Med. Chem.* **7**, 1781–1785 (1999).

155. Martin, E. J., Blaney, J. M., Siani, M. A., Spellmeyer, D. C., Wong, A. K. and Moos, W. H., *J. Med Chem.* **38**, 1431–1436 (1995).

156. Abato, P., Conroy, J. L. and Seto, C. T., *J. Med. Chem.* **42**, 4001–4009 (1999).

157. Pinilla, C., Appel, J. R., Blanc, P. and Houghten, R. A., *BioTechniques* **13**, 901–905 (1992).

158. Pinilla, C., Buencamino, J., Appel, J. R., Hopp, T. P. and Houghten, R. A., *Mol. Diversity* **1**, 21–28 (1995).

159. Dooley, C. T., Ny, P., Bidlack, J. M. and Houghten, R. A., *J. Biol. Chem.* **273**, 18848–18856 (1998).

160. Pinilla, C., Appel, J. R., Campbell, G. D., Buencamino, J., Benkirane, N., Muller, S. and Greenspan, N. S., *J. Mol. Biol.* **283**, 1013–1025 (1998).

161. Apletalina, E., Appel, J., Lamango, N. S., Houghten, R. A. and Lindberg, I., *J. Biol. Chem.* **273**, 26589–26595 (1998).

162. Pirrung, M. C. and Chen, J., *J. Am. Chem. Soc.* **117**, 1240–1245 (1995).

163. Pirrung, M. C., Chau, J. H. L. and Chen, J., *Chem. Biol.* **2**, 621–626 (1995).

164. Smith, P. W., Lai, J. Y. Q., Whittington, A. R., Cox, B., Houston, J. G., Stylli, C. H., Banks, M. N. and Tiller, P. R., *Bioorg. Med. Chem. Lett.* **4**, 2821–2824 (1994).

165. Appel, J. R., Johnson, J., Narayanan, V. L. and Houghten, R. A., *Mol. Diversity* **4**, 91–102 (1999).

166. Blondelle, S. E., Crooks, E., Ostresh, J. M. and Houghten, R. A., *Antimicrob. Ag. Chemother.* **43**, 106–114 (1999).

167. Berk, S. C. and Chapman, K. T., *Bioorg. Med. Chem. Lett.* **7**, 837–842 (1997).

168. Deprez, B., Williard, X., Bourel, L., Coste, H., Hyafil, F. and Tartar, A., *J. Am. Chem. Soc.* **117**, 5405–5406 (1995).

169. Carell, T., Wintner, E. A, Sutherland, A. J, Rebek, J. Jr, Dunayevskiy, Y. M and Vouros, P., *Chem. Biol.* **2**, 171–183 (1995).

170. Campian, E., Peterson, M. L., Saneii, H. H. and Furka, A., *Bioorg. Med. Chem. Lett.* **8**, 2357–2362 (1998).

171. Blake, J. and Litzi-Davis, L., *Bioconjug. Chem.* **3**, 510–513 (1992).

172. Boger, D. L., Chai, W. and Jing, Q., *J. Am. Chem. Soc.* **120**, 7220–7225 (1998).

173. Freier, S. M., Konings, D. A. M., Wyatt, J. R. and Ecker, D. J., *Bioorg. Med Chem.* **4**, 717–725 (1996).

174. Stankova, M., Strop, P., Chen, C. and Lebl, M., in *Molecular Diversity and Combinatorial Chemistry*, Chaiken, I. M. and Janda, K. D. (Eds.). ACS, Washington, DC, 1996, pp. 137–141.

175. Konings, D. A. M., Wyatt, J. R., Ecker, D. J. and Freier, S. M., *J. Med. Chem.* **39**, 2710–2719 (1996).

176. Konings, D. A. M., Wyatt, J. R., Ecker, D. J. and Freier, S. M., *J. Med. Chem.* **40**, 4386–4395 (1997).

177. Brenner, S. and Lerner, R. A., *Proc. Natl. Acad. Sci. USA* **89**, 5381–5383 (1992).

178. Kerr, J. M., Banville, S. C. and Zuckermann, R. N., *J. Am. Chem. Soc.* **115**, 2529–2531 (1993).

179. Nielsen, J. and Janda, K. D., *Methods: Meth. Enzymol.* **6**, 361–371 (1994).

180. Nielsen, J., Brenner, S. and Janda, K. D., *J. Am. Chem. Soc.* **115**, 9812–9813 (1993).

181. Needels, M. C., Jones, D. G., Tate, E. H., Heinkel, G. L., Kochersperger, L. M., Dower, W. J., Barrett, R. W. and Gallop, M. A., *Proc. Natl. Acad. Sci. USA* **90**, 10700–10704 (1993).

182. Jones, D. G., *Polym. Prepr.* **35**, 981–982 (1994).

183. Fenniri, H., Janda, K. D. and Lerner, R. A., *Proc. Natl. Acad. Sci. USA* **92**, 2278–2282 (1995).

184. Fenniri, H., *CHEMTECH* **26**, 15–25 (1996).

185. Nikolaiev, V., Stierandova, A., Krchnak, V., Seligmann, B., Lam, K. S., Salmon, S. E. and Lebl, M., *Pept. Res.* **6**, 161–170 (1993).

186. Hornik, V. and Hadas, E., *React. Polym.* **22**, 213–220 (1994).

187. Felder, E. R., Heizmann, G., Matthews, I. T., Rink, H. and Spieser, E., *Mol. Diversity* **1**, 109–112 (1996).

188. Dawson, P. E., Fitzgerald, M. C., Muir, T. W. and Kent, S. B. H., *J. Am. Chem. Soc.* **119**, 7917–7927 (1997).

189. Nestler, H. P., Bartlett, P. A. and Still, W. Clark, *J. Org. Chem.* **59**, 4723–4724 (1994).

190. Burger, M. T. and Clark Still, W., *J. Org. Chem.* **60**, 7382–7383 (1995).

191. Cheng, Y., Suenaga, T. and Still, W. Clark, *J. Am. Chem. Soc.* **118**, 1813–1814 (1996).

192. Clark Still, W., *Acc. Chem. Res.* **29**, 155–163 (1996).

193. Liang, R., Yan, L, Loebach, J., Ge, M., Uozumi, Y., Sekanina, K., Horan, N., Gildersleeve, J., Thompson, C. et al., *Science* **274**, 1520–1522 (1996).

194. Baldwin, J. J., *Mol. Diversity* **2**, 81–88 (1996).

195. Dilanni Carroll, C., Patel, H., Johnson, T. O., Guo, T., Orlowski, M., He, Z.-M., Cavallaro, C. L., Guo, J., Oksman, A., Gluzman, I. Y., Connelly, J., Chelsky, D., Goldberg, D. E. and Dolle, R. E., *Bioorg. Med. Chem. Lett.* **8**, 2315–2320 (1998).

196. Tan, D. S., Foley, M. A., Shair, M. D. and Schreiber, S. L., *J. Am. Chem. Soc.* **120**, 8565–8566 (1998).

197. Ni, Z.-J., Maclean, D., Holmes, C. P., Murphy, M. M., Ruhland, B., Jacobs, J. W., Gordon, E. M. and Gallop, M. A., *J. Med. Chem.* **39**, 1601–1608 (1996).

198. Fitch, W. L., Baer, T. A., Chen, W., Holden, F., Holmes, C. P., Maclean, D., Shah, N., Sullivan, E., Tang, M., Waybourn, P., Fischer, S. M., Miller, C. A., Snyder, L. R., *J. Comb. Chem.* **1**, 188–194 (1999).

199. Lane, S. J. and Pipe, A., *Rapid Commun. Mass Spectrom.* **12**, 667–674 (1998).

200. Lane, S. J. and Pipe, A., *Rapid Commun. Mass Spectrom.* **13**, 798–814 (1999).

201. Atuegbu, A., Maclean, D., Nguyen, C., Gordon, E. M. and Jacobs, J. W., *Bioorg. Med. Chem.* **4**, 1097–1106 (1996).

202. Ni, Z.-J., Maclean, D., Holmes, C. P. and Gallop, M. A., *Meth. Enzymol.* **267**, 261–272 (1996).

203. MacLean, D., Schullek, J. R., Murphy, M. M., Ni, Z.-J., Gordon, E. M. and Gallop, M. A., *Proc. Natl. Acad. Sci. USA* **94**, 2805–2810 (1997).

204. Silen, J. L., Lu, A. T., Solas, D. W., Gore, M. A., Maclean, D., Shah, N. H., Coffin, J. M., Bhinderwala, N. S., Wang, Y., Tsutsui, K. T., Look, G. C., Campbell, D. A., Hale, R. L., Navre, M. and DeLuca-Flaherty, C. R., *Antimicrob. Ag. Chemother.* **42**, 1447–1453 (1998).

205. Garigipati, R. S. and Adams, J. L., *WO Patent 9630337 A1*, 961003, 32 pp.

206. Geysen, H. M., Wagner, C. D., Bodnar, W. M., Markworth, C. J., Parke, G. J., Schoenen, F. J., Wagner, D. S. and Kinder, D. S., *Chem. Biol.* **3**, 679–688 (1996).

207. Wagner, D. S., Markworth, C. J., Wagner, C. D., Schoenen, F. J., Rewerts, C. E., Kay, B. K. and Geysen, H. M., *Combi. Chem. High Throughput Screen.* **1**, 143–153 (1998).

208. Rahman, S. S., Busby, D. J. and Lee, D. C., *J. Org. Chem.* **63**, 6196–6199 (1998).

209. Hochlowski, J. E., Whittern, D. N. and Sowin, T. J., *J. Comb. Chem.* **1**, 291–293 (1999).

210. Scott, R. H., Barnes, C., Gerhard, U. and Balasubramanian, S., *Chem. Commun.*, 1331–1332 (1999).

211. Guo, J., Wu, J., Siuzdak, G. and Gary, F. M. G., *Angew. Chem. Int. Ed.* **38**, 1755–1758 (1999).

212. Czarnik, A. W., *Curr. Opin. Chem. Biol.* **1**, 60–66 (1997).

213. Jacobs, J. W. and Ni, Z.-J., in *Combinatorial Chemistry and Molecular Diversity in Drug Discovery,* E. M. Gordon and J. Kerwin (Eds.). Wiley-Liss, New York, 1998, pp. 271–290.

214. Seneci, P., in *Combinatorial Chemistry and Combinatorial Technologies: Methods and Applications*, S. Miertus and G. Fassina (Eds.). Marcel Dekker, New York, 1999, pp. 127–167.

215. Terrett, N., *Combinatorial Chemistry.* Oxford University Press, Oxford, UK, 1999, pp. 77–94.

216. Baldwin, J. J. and Dolle, R., in *A Practical Guide to Combinatorial Chemistry*, A. W. Czarnik and S. H. DeWitt (Eds.). ACS, Washington DC, 1997, pp. 153–174.

217. Barnes, C,. Scott, R. H. and Balasubramanian, S., *Recent Res. Dev. Org. Chem.* **2**, 367–379 (1998).

218. Scott, R. H. and Balasubramanian, S., *Bioorg. Med. Chem.* **7**, 1567–1572 (1997).

219. Egner, B. J., Rana, S., Smith, H., Bouloc, N., Frey, J. G., Brocklesby, W. S. and Bradley, M., *Chem. Commun.* 735–736 (1997)

220. Main, B. G. and Shute, R. E., *WO Patent 9703931 A1*, June 2nd, 1997.

221. Hughes, I., *J. Med. Chem.* **41**, 3804–3811 (1998).

222. Guiles, J. W., Lanter, C. L. and Rivero, R. A., *Angew. Chem. Int. Ed.* **37**, 926–932 (1998).

223. Xiao, X., Zhao, C., Potash, H. and Nova, M. P., *Angew. Chem. Int. Ed. Engl.* **36**, 780–782 (1997).

224. Gercken, B., Felder, E., Rink, H. and Vetter, D., *Spec. Publ - R. Soc. Chem* **241**, 222–234 (1999).

225. Nicolaou, K. C., Xiao, X.-Y., Parandoosh, Z., Senyei, A. and Nova, M. P., *Angew. Chem. Int. Ed. Engl.* **34**, 2289–2291 (1995).

226. Moran, E. J., Sarshar, S., Cargill, J. F., Shahbaz, M. M., Lio, A., Mjalli, A. M. M. and Armstrong, R. W., *J. Am. Chem. Soc.* **117**, 10787–10788 (1995).

227. Mendonca, A. J., *Am. Laboratory* **30**, 46–47 (1998).

228. Li, R., Xiao, X.-Y. and Czarnik, A. W., *Tetrahedron Lett.* **39**, 8581–8584 (1998).

229. Zhao, C., Shi, S., Mir, D., Hurst, D., Li, R., Xiao, X.-Y., Lillig, J. and Czarnik, A. W., *J. Comb. Chem.* **1**, 91–95 (1999).

230. TranSort and TranStems, Chiron Technologies Pty Ltd.
http://www.chirontechnologies.com.au/product/frame2.htm

231. Xiao, X.-Y., Parandoosh, Z. and Nova, M. P., *J. Org. Chem.* **62**, 6029–6033 (1997).

232. Nicolaou, K. C., Vourloumis, D., Li, T., Pastor, J., Winssinger, N., He, Y., Ninkovic, S., Sarabia, F., Vallberg, H., Roschangar, F., King, N. P., Finlay, M. R. V., Giannakakou, P., Verdier-Pinard, P. and Hamel, E., *Angew. Chem. Int. Ed. Engl.* **36**, 2097–2103 (1997).

233. Zhu, Z. and McKittrick, B., *Tetrahedron Lett.* **39**, 7479–7482 (1998).

234. Brown, A. R., *J. Comb. Chem.* **1**, 283–285 (1999).

235. Sofia, M. J., Allanson, N., Hatzenbuhler, N. T., Jain, R., Kakarla, R., Kogan, N., Liang, R., Liu, D., Silva, D. J., Wang, H., Gange, D., Anderson, J., Chen, A., Chi, F., Dulina, R., Huang, B., Kamau, M., Wang, C., Baizman, E., Branstrom, A., Bristol, N., Goldman, R., Han, K., Longley, C., Midha, S. and Axelrod, H. R., *J. Med. Chem.* **42**, 3193–3198 (1999).

236. Andres, C. J., Swann, R. T., Severino, J., Grant-Young, K., Edinger, K., Mongillo, J. and Deshpande, M. S., *Biotechnol. Bioeng.* **61**, 93–94 (1998).

237. Andres, C. J., Swann, R. T., Grant-Young, K., D'Andrea, S. V. and Deshpande, M. S., *Comb. Chem. High Throughput Screening* **2**, 29–32 (1999).

238. Parandoosh, Z., Knowles, S. K., Xiao, X-Y., Zhao, C., David, G. S. and Nova, M. P., *Combi. Chem. High Throughput Screen.* **1**, 135–142 (1998).

239. Shi, S., Xiao, X.-Y. and Czarnik, A. W., *Biotechnol. Bioeng.* **61**, 6–12 (1998).

240. Tan, D. S., Foley, M. A., Stockwell, B. R., Shair, M. D. and Schreiber, S. L., *J. Am. Chem. Soc.* **121**, 9073–9087 (1999).

241. Borchardt, A., Liberles, S. D., Biggar, S. R., Crabtree, G. R. and Schreiber, S. L., *Chem. Biol.* **4**, 961–968 (1997).

242. You, A. J., Jackman, R. J., Whitesides, G. M. and Schreiber, S. L., *Chem. Biol.* **4**, 969–975 (1997).

243. Stockwell, B. R., Haggarty, S. J. and Schreiber, S. L., *Chem. Biol.* **6**, 71–83 (1999).

244. MacBeath, G., Koehler, A. N. and Schreiber, S. L., *J. Am. Chem. Soc.* **121**, 7967–7968 (1999).

245. Brown, B. B., Wagner, D. S. and Geysen, H. M., *Mol. Diversity* **1**, 4–12 (1995).

246. Tamura, O., Okabe, T., Yamaguchi, T., Gotanda, K., Noe, K. and Sakamoto, M., *Tetrahedron* **51**, 107–118 (1995).

247. Tamura, O., Okabe, T., Yamaguchi, T., Kotani, J., Gotanda, K., and Sakamoto, M., *Tetrahedron* **51**, 119–128 (1995).

248. McGowan, D. A. and Berchtold, G. A., *J. Org. Chem.* **46**, 2381–2383 (1981).

249. Wood, H. B. and Ganem, B., *J. Am. Chem. Soc.* **112**, 8907–8909 (1990).

250. Keirs, D. and Overton, K., *Heterocycles* **28**, 841–848 (1989).

251. Acheson, R. M. and Lee, G. C. M., *J. Chem. Soc., Perkin Trans. 1*, 2321–2332 (1987).

252. Schena, M., Shalon, D., Davis, R. W. and Brown, P. O., *Science* **270**, 467–470 (1995).

253. Oldenburg, K. R., Vo, K. T., Ruhland, B., Schatz, P. J. and Yuan, Z., *J. Biomol. Screening* **1**, 123–130 (1996).

254. Schullek, J. R., Butler, J. H., Ni, Z.-J., Chen, D. and Yuan, Z., *Anal. Biochem.* **246**, 20–29 (1997).

8 Synthetic Organic Libraries: Solution-Phase Libraries

The previous chapters have shown how combinatorial technologies have always been associated with chemistry in the SP and how this chemistry is flexible enough to provide libraries in many formats. More recently, there has been growing interest in libraries of small organic molecules prepared under homogeneous reaction conditions. These libraries are usually called *solution-phase libraries* and will be discussed in this chapter to demonstrate how they present a useful alternative to SP libraries and how they can provide the chemist with more options with regards to the choice of library format.

The first section will compare the advantages and drawbacks of SP solution-phase libraries, generically defining their applicability. This comparison will be expanded to cover the differences between libraries of *pools* and *discretes* prepared in the heterogeneous and homogeneous phases. The synthesis of solution-phase discrete libraries is covered in the second section, in which the design, assessment, monomer rehearsal, reaction monitoring, analytical characterization, automation, and synthesis of these libraries will be illustrated through several examples. The purification of intermediates and the final library components is the leitmotiv of the following three sections: The various approaches to increasing the *purity* of the final products and to decreasing the time necessary for the *purification* of the library are presented in the third section; the use of *supported reagents* in solution-phase combinatorial chemistry are reviewed in the fourth section; and *soluble supports* in library synthesis are the subject of the fifth section. Finally, *new trends* in the field of solution-phase libraries are discussed through the presentation of some more recent and perhaps controversial papers that have appeared in the literature.

8.1 SOLUTION- VERSUS SOLID-PHASE SYNTHETIC LIBRARIES: WHICH ONES TO USE?

8.1.1 Differences, Advantages, and Possible Issues

Synthetic strategies based on solution-phase (Fig. 8.1, left) or SP reactions (Fig. 8.1, right) can be compared by considering the synthesis of single-target compounds. To all intents and purposes, there is no apparent similarity between the two techniques:

- Homogeneous reactions are performed under the conditions of *classical organic chemistry* using combinations of reagents, solvents, temperature, concentration,

SOLUTION-PHASE	SOLID-PHASE
- experimental flexibility	- purification protocols
- available expertise	- excess of reagents in solution
- assessed chemical routes	- automation-friendly
- reaction monitoring	- mix and split possible
- no additional steps	- easier miniaturization

Figure 8.1 Solution phase versus solid phase: relative strengths.

and catalyst that are suited to the selected synthetic route. The use of a solid support limits the choice of reagents (interactions with the support itself must be avoided), solvents (a good swelling is needed), temperature (extremely low and high temperatures must be avoided), concentrations (diluted reagent solutions will not yield complete coupling with the support), and catalysts (heterogeneous catalysts cannot be used).

- A *knowledge* of the theoretical and experimental notions of solution-phase chemistry is possessed by any good chemist; however, a deep understanding of the implications of carrying out a reaction on a support, such as the choice of reaction vessels or SP synthesizer and of equipment needed for the handling and/or purification of samples, is acquired mostly by direct experience and is not yet widely spread among synthetic chemists in general.
- When the target molecule is similar to known structures, or when a sound synthetic scheme is drawn, the *assessment* of the best experimental conditions in solution is straightforward, while the transfer of a reaction sequence onto SP requires more theoretical (selection of the best support and linker) and experimental work to find appropriate reaction conditions.
- Homogeneous reactions are *monitored* by simple and effective methods, such as TLC, and *characterized* using a wide variety of analytical techniques, whereas the synthesis of supported intermediates and/or final products can only be monitored by sophisticated on-bead methods or after cleavage from the support.
- The use of a support and a linker requires at least *two additional chemical steps*, the anchoring to the support at the beginning of the synthesis and the final cleavage of the target compound.

The only advantages of SPS techniques for the preparation of single compounds are the *purification/work-up* procedures, which usually involve filtering off the solvents and washing the resin beads with fresh solvents and the use of a *large excess of reagents* in solution to drive the reactions to completion. Reactions in solution usually require work-up procedures including extractions, concentrations, drying, and often chromatographic purification. Nevertheless, the advantages of the synthesis of single compounds in solution override these inconveniences and generally render this the preferred format for single-compound synthesis. Solid-phase chemistry has some niche applications, for example when the support is used as a protecting group in a synthetic scheme and the loading/cleavage steps are equivalent to a protection/deprotection protocol.

The comparison is different when dealing with combinatorial libraries:

- *Simultaneous handling* of multiple reactions make simple automated purification protocols extremely important when large libraries are being considered.
- Resin beads, which can be considered as individual microreactors, allow the use of *mix-and-split* techniques, which are not applicable when using homogeneous reaction conditions.
- *Miniaturization* of reaction scale-down to the level of a single bead in SP reduces the amount of precious reagents and solvents used.
- Solid-phase is more suited to partially or totally *automated* synthetic procedures, freeing the chemist from having to perform the more repetitive operations.

Having said this, the synthesis of solution-phase libraries is possible, and indeed is more appropriate, than the corresponding SP chemistry under certain circumstances. The value of such libraries is discussed in the following sections, but it is appropriate to stress the concept of *complementarity*, rather than mutual exclusivity of the solution and SP library formats. The goal for a combinatorial chemist is always to select the best library format according to the needs of the project without exclusion of individual options due to personal preference.

8.1.2 Pool Libraries: Solution Versus Solid Phase

SP pool libraries benefit greatly from the mix-and-split technique, which can produce high-quality libraries when a fully validated and rehearsed synthetic scheme is used. The loading of a single compound on each bead, the roughly equimolar quantity of each library component in a pool, and the ease of handling and purification of resin aliquots are not matched by any library synthesis method in solution. Only the postsynthesis mix of solutions would provide comparable pools, but such a method is only useful for the pooling of large, preexisting compound collections for screening. This is a serious limitation for solution-phase pool libraries; nevertheless, several interesting approaches to the synthesis of pools in solution have been reported and will be discussed here.

Most of the early work was restricted to a few high-yielding chemical transformations typically leading to amide libraries and employing monomer sets as equimolar mixtures of all their representatives. For example, Smith et al. (1) reported a 1600-member library of amides as two 40-pool sublibraries suitable for structure determination by positional scanning (see Section 7.3.3) and Storer (2) and Bailey et al. (3) described the preparation of a 160,000-member library of diamides made by capping an acid function with a mixture of amines and obtaining 4000 pools that were subsequently deconvoluted using iterative deconvolution. The synthesis and deconvolution by positional scanning of five small pool libraries was reported recently; namely, carbamates were obtained from the reaction of alcohols with isocyanates (54 members divided into two sublibraries) (4); tetrahydroacridines from the reaction of cyclohexanones with aromatic aminonitriles (72 members, two sublibraries) (5); diamide analogues of a human neurokinin 3 receptor-antagonist from the reaction of highly functionalized amines with acylating agents (49 members, two sub-libraries) (6); polyenes from the Sonogashira coupling reaction of vinyl iodides with alkynes (42 members, two sublibraries) (7); chalcones from aldolic condensation of naphthyl ketones with benzaldehydes (20 members, two sublibraries) (8). A 700-member pool library of functionalized thiazoles prepared from a-haloketones, aromatic nitroaldehydes, and cyclic anhydrides aimed at finding leukotriene D_4 antagonists by iterative deconvolution has been described (9). Vendeville et al. (10) reported the synthesis of a 2560-member tetrahydroisoquinoline carboxylic acid (Tic)-based library as 32 pools of 80 compounds that was iteratively deconvoluted to give a low-nanomolar positive against a parasitic prolyl endopeptidase target. Decoration of orthogonally protected piperazine with aryl nitrofluorides and acyl chlorides to give a 48-member arylpiperazine pool library was described by Neuville and Zhu (11). Epoxide opening with amines to give a library of over 6000 β-aminoalcohols made as four-member pools was described by Chng and Ganesan (12). Epoxide opening was also used by van Niel et al. (13) to prepare a large pool library of 3-aryloxy-2-propanolamines which was iteratively deconvoluated to give a dual affinity ligand for 5-HT_{1A} and serotonin re-uptake receptors. Yu et al. (14) prepared 13 large β-cyclodextrin libraries (>28,000 individuals) using a common scaffold and decorating it with mixtures of amines; the libraries were tested for their phosphatase-like hydrolytic activity and iterative deconvolution was planned for the active pools.

A few other reports merit more attention. The decoration of a number of symmetrical carboxylic acid scaffolds (**8.1**–**8.3**, Fig. 8.2) with amines to give large amide or urea libraries that were characterized and deconvoluted using subtractive deconvolution (see section 7.3.3) has been reported by Rebek and co-workers (15–22). Solution-phase pool libraries of polyazapyridinocyclophane scaffolds containing macrocycles of various sizes (**8.4**–**8.6**) and the synthesis of linear pyridinopolyamine, oxyamine, and piperazine scaffolds (**8.7**–**8.9**) made by the simultaneous addition of functionalities via selective deprotection/coupling protocols followed by iterative deconvolution have also been described (23–29); the same group has also carefully monitored the synthesis of some pool libraries via capillary electrophoresis (30), and has deconvoluted some other libraries via HPLC fractionation (31). Large solution-phase pool libraries have been prepared from iminodiacetic acid–based (**8.10**, **8.11**) and tricyclic scaffolds (**8.12**, Fig. 8.2) by acylation, olefin metathesis, and reduction, with deconvolution either by positional scanning or by deletion synthesis (32–34).

The above reports have produced valuable, large solution-phase libraries taking advantage of specific structural features, such as the symmetry of common scaffolds, the wide knowledge accumulated on the reactions used, or the extreme similarity of reactivities of the monomers. Thus, these approaches cannot be generally applied to a wide diversity of scaffolds, monomers with significantly different reactivities, or difficult reaction steps that produce incomplete reactions or side-products. Consequently, the general synthesis of solution-phase pool libraries is not possible, and for

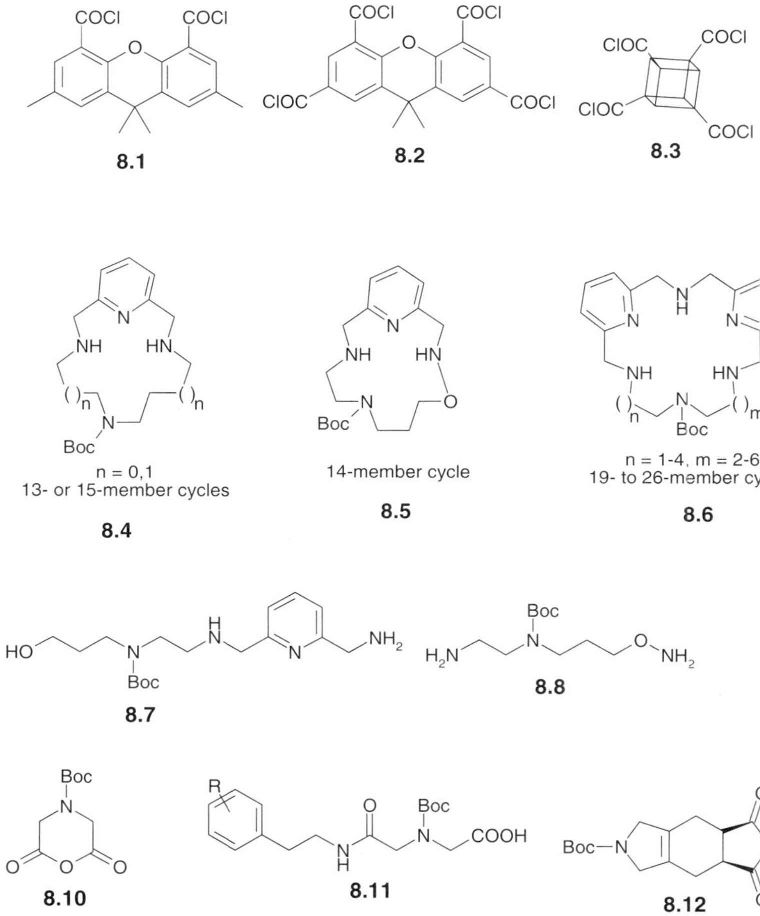

Figure 8.2 Scaffolds **8.1–8.12** for solution-phase pool library synthesis

this reason more attention has been payed to the preparation of SP pool libraries to create valuable primary libraries.

Two intriguing, recent reports took advantage of different reaction pathways from mixtures of reagents to generate some predictable pool diversity in solution. Boger et al. (35) validated the synthesis of solution pool libraries of biaryls, obtained through the coupling of equimolar quantities of trisubstituted aryl iodides **8.13a–e** to give 2- to 15-member mixtures of biaryls **8.14** in predictable relative abundancies; the HPLC traces of the reaction products confirmed the presence of all the expected compounds in roughly the predicted amounts (Fig. 8.3). Boger et al. recently exploited further this chemistry and prepared a 64,980-member solution-phase pool library containing both

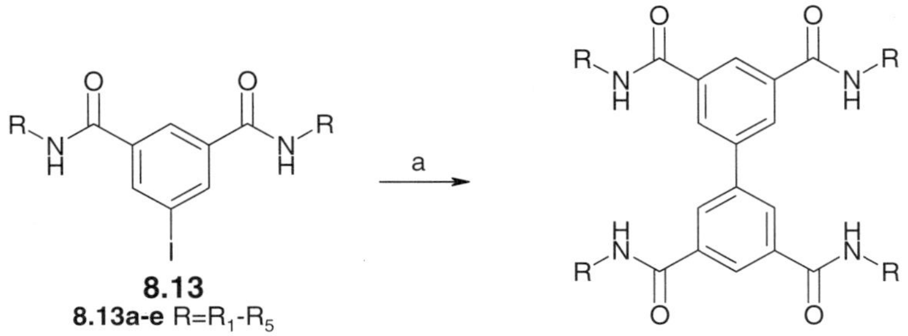

8.13
8.13a-e $R=R_1$-R_5

8.14
R=identical (R_1R_1 to R_5R_5) 1eq.
R=different (R_1R_2 to R_4R_5) 2 eqs..
15 individuals

a: equimolar mixture of **8.13a-e**, 10% Pd/C, TEA, 100°C, 16 hrs.

Figure 8.3 Chemistry assessment for the synthesis of a solution pool library of biaryls: synthesis of **8.14**.

biaryls from Pd-catalyzed iodoarene couplings and acetylene-connected biaryls from Stille couplings (36). Klumpp et al. (37) validated the synthesis of pool libraries of 3,3-diaryloxindoles in solution via the condensation of isatins **8.15** with substituted benzenes **8.16** (Fig. 8.4); as an example, the condensation between one equivalent of the isatins **8.15a,b** and two equivalents of the aromatics **8.16a–d** produced all the 20 expected isomers **8.17** (not including enantiomers) as confirmed by GC-MS. For both approaches additional building blocks would allow the synthesis of larger libraries, and the introduction of modifiable substituents on the aromatic rings would allow further diversification of the libraries. Nevertheless, substantial work has to be done

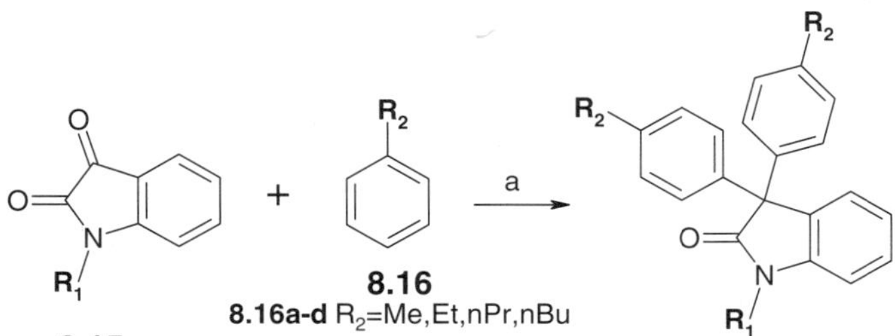

8.15
8.15a,b R_1=Me,Ph

8.16
8.16a-d R_2=Me,Et,nPr,nBu

8.17
20 isomers
(enantiomers not included)

a: mixture of **8.15** (1 eq.), mixture of **8.16** (2 eqs.), triflic acid, rt, 12 hrs.

Figure 8.4 Chemistry assessment for the synthesis of a solution pool library of 3,3-diaryloxindoles: synthesis of **8.17**.

before making such methodologies general and reliable enough to prepare large, high-quality primary solution-phase libraries.

8.1.3 Discrete Libraries: Solution Phase Versus Solid Phase

The demand for libraries varying in size from a few tens of compounds for extremely focused arrays to tens of thousands of individuals for biased-targeted or primary libraries of discretes is constantly growing in many areas of research. Given the advantages of ease of handling and purification in SP chemistry, it could be said that the more individuals making up the planned library, the more the SP format is appealing, while solution-phase chemistry is best suited to small arrays. However, even this generalization is an oversimplification.

Small focused libraries of up to 100 compounds are commonly used in the late phases of a project when previous research has already identified one or more molecules with the desired activity. The main purpose of such libraries is the final optimization of these advanced compounds, and thus individuals in these arrays must be prepared in high purity in order to maximize the quality of information obtained from testing them. If the structure of the parent compound comes from rational design, perhaps through the use of a model to simulate the target–ligand interaction, and has been prepared through classical organic chemistry in solution, the preparation of a small focused library in SP will not pay off the effort expended in transferring the synthetic scheme onto SP. Solution-phase techniques generally allow a much faster synthesis of the array and the small number of library components in this part of the project makes the necessary purification protocols to obtain high-quality compounds quite acceptable. It often happens, though, that the model structures for a small focused library derive from a primary, or biased-targeted, SP library. In this case the synthetic scheme in the SP has already been optimized, and the derived focused libraries can be more readily prepared on SP.

When medium–large libraries varying from several hundreds to thousands of members are involved, the purification steps and the ease of handling aliquots of solutions become the crucial factors driving the chemists to privilege SP chemistry for the preparation of such libraries. However, many exceptions are known and the solution phase may be better suited to specific purposes. The library synthesis may contain steps that are difficult to transfer onto SP or may require solvents in which common resins swell poorly or reaction conditions that are incompatible with the solid support. The appropriate synthetic or analytical equipment and the expertise to design and validate an SPS scheme through rehearsal of the chosen monomer sets to performing the library synthesis may not be available, and therefore many justifications for preparing medium–large discrete libraries in the solution phase can be found. Moreover, recent efforts to set up simple, automated purification procedures for large numbers of individuals in solution are currently producing promising results.

We will now focus on the preparation of discrete libraries in the solution phase using examples reported in the literature in which the usefulness of this technique is apparent.

8.2 SOLUTION-PHASE DISCRETE LIBRARIES

8.2.1 Design and Synthesis of Solution-Phase Discrete Libraries

Synthesis of solution-phase libraries is similar to the synthesis of single compounds in classical organic chemistry protocols, although the overall throughput is increased by simultaneous handling of multiple solutions that are treated with reagents, worked up, and purified according to a common reaction scheme. The majority of the steps that are relevant to prepare a high-quality SP library (see Chapter 3) are the same as for the preparation of a high-quality solution-phase library and are summarized in Fig. 8.5.

The *design* of a library that is project dependent and identical for both solution and solid phases, involves the identification of a suitable structural motif together with its randomization points, which are available for generation of diversity (Fig. 8.6, top) and the overall definition of library size. A *retrosynthetic study* follows in which a reasonable synthetic route in solution is determined, while work on SP requires the choice of a suitable support and of a linker as well (Fig. 8.6, bottom).

The next phase for SP is the *validation* of the selected route in solution. It is useful to check the outcome of the reaction on several compounds to uncover potential weaknesses. This step is obviously redundant in the solution phase, because it corresponds to the following *chemical assessment* step, even though the synthesis of a few more fully characterized discretes may be desirable (Fig. 8.7, top). Typically, the assessment of the chemistry in solution takes several weeks to check a few representative monomers with different reactivities in order to discard unsuitable monomer

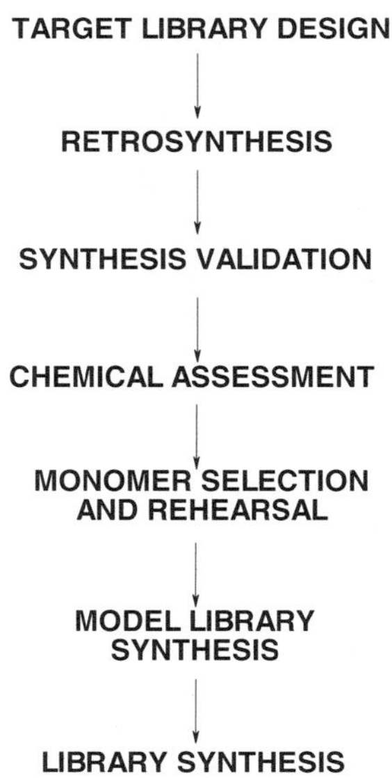

Figure 8.5 Logical steps to a successful solution-phase library synthesis.

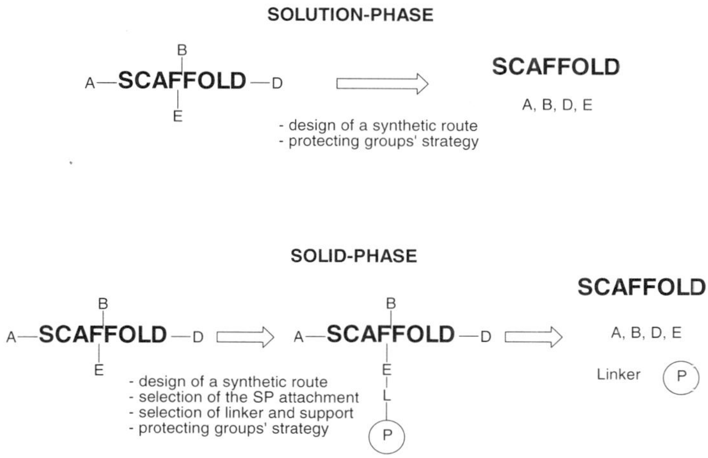

Figure 8.6 Design of a synthetic organic library and retrosynthetic studies: solution- versus solid-phase formats.

subclasses, refine the chemical route and experimental conditions leading to high-quality final compounds, and select simple handling and purification procedures for intermediates (if and when necessary) and final compounds. The same step on SP (Fig. 8.7, bottom) that has been shown in previous chapters is the most time-consuming and effort-requiring phase. Typically, the transfer and optimization of reaction conditions from solution phase to SP to give a reliable synthetic scheme requires several months to be successfully completed.

The optimized reaction conditions to make the same library in solution or on SP differ significantly in most examples because of the influence exerted by the heterogeneous support (see Chapter 1), which may lead to different purities and yields of the final compounds. When the chemistry assessment, either in solution or on solid phase, does not give satisfactory results after having exploited all the reasonable experimental conditions, the other phase should at least be considered before abandoning the library synthesis.

Potential representative monomers are then *selected* from commercially available compounds or from internal collections, possibly through the application of computational methods to select only the most significant examples, and the chemistry is *rehearsed* using these. The process is identical both in SP and in solution, but the support makes monitoring of the reactions, the precise determination of yields and purities of the reaction products, and the structure and the quantity of impurities more difficult and time consuming in the SP (Fig. 8.8). If the proper analytical equipment and expertise for working in SP are not available, the selection or rejection of a monomer candidate may be more difficult, less accurate, or even wrong.

Monomer rehearsal is strongly influenced by the selection of the phase. The higher flexibility in terms of conditions in solution may allow the successful rehearsal of more monomers, whereas the solid support or the linker may interfere with reactive mono-

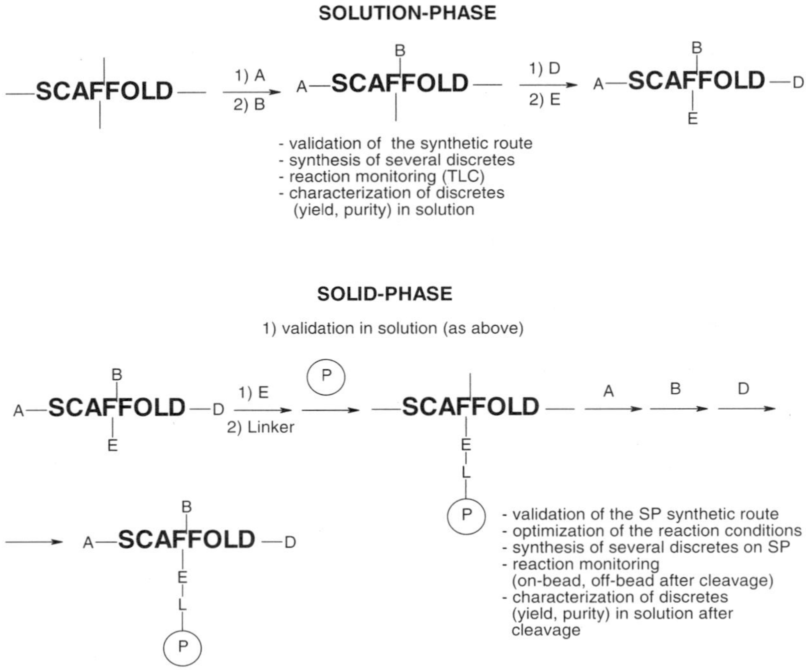

Figure 8.7 Chemical assessment: solution- versus solid-phase formats.

mers and premature cleavage from the resin may sometimes interfere. Reactions with poorly reactive monomers are driven to completion on SP by addition of a large excess of the monomer, and SP site isolation (see Section 1.1.5) may prevent double reactions of bifunctional monomers. The planning of a library synthesis should take into account these issues related to the choice of monomers and should accordingly privilege one of the phases for the synthesis of the library. A large monomer class may be split into subclasses, and two libraries in the two different phases may be designed. An example of this approach could be a subclass of stable, unreactive monomers that must be used in large excess, in which case SP would be preferred, and a subclass of reactive monomers that can be used stoichiometrically, in which case the solution is preferred.

The next step is the synthesis of a *model library* and the goal of both solid and solution phases is to use the same protocols for the library synthesis as were determined during rehearsal and to check if problems related to parallelization of the synthesis arise. This information is acquired much more easily in solution, and the overall process of solution-phase library synthesis is accelerated. However, a major additional step for the optimization of the synthesis is the testing of the handling and purification procedures in parallel, which must eventually yield a set of high-quality final discretes. This study identifies the essential purification steps, taking into account the planned size of the final library. In general, accurate methods with low throughput are not

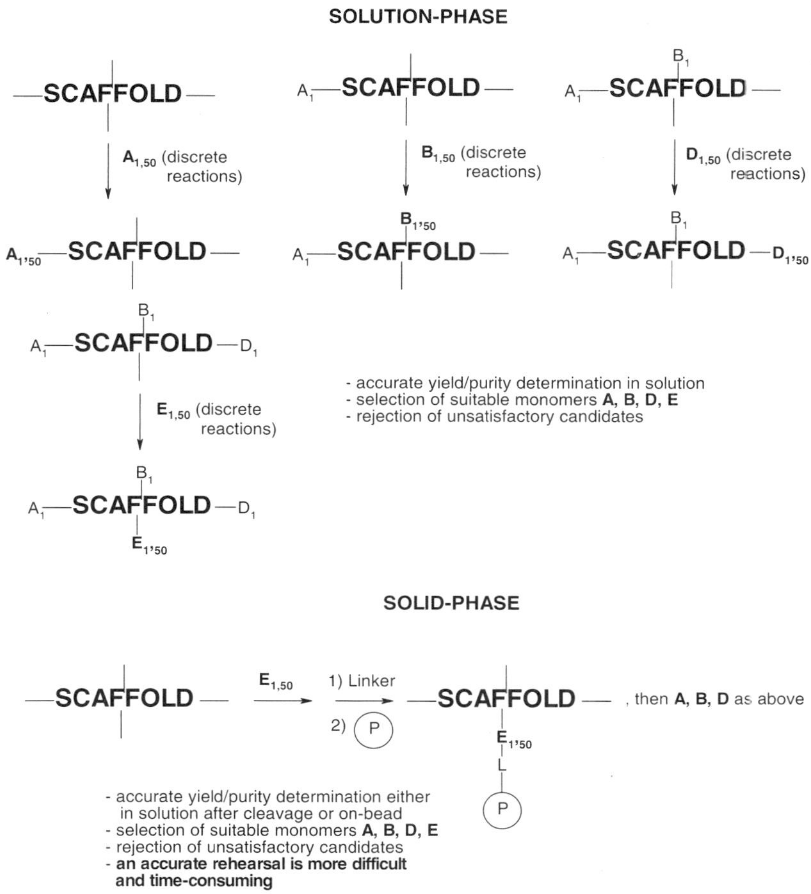

Figure 8.8 Monomer rehearsal: soluition- versus solid-phase fonnats.

suitable for medium to large solution-phase discrete libraries and must be replaced by high-throughput, less accurate but procedurally more simple protocols.

Finally, the *library synthesis* is performed and both solution-phase and SP synthesis lead to high-quality discrete libraries using the appropriate reaction conditions, monomer sets, and analytical and synthetic instrumentation. It is clear, though, that small- to medium-size libraries in solution require less effort and time for their preparation (providing that the library is not a focused expansion of a primary SP library) and the faster acquisition of information derived from screening them for activity is highly desirable.

Some recent reviews that thoroughly cover solution-phase combinatorial libraries are provided in references 2 and 38–45. Two examples will be described in details in the following sections and only a few recent reports will be cited here. Stewart et al. (46) have reported the preparation of a focused, 48-member phenothiazine amide library as a source of new cyclooxygenase-2 (COX-2) inhibitors. Other small-scale solution-phase libraries that have been reported recently include a 20-member androstane model library obtained from the decoration of dihydrotestosterone, by Maltais and Poirier (47); a 24-member library of perhydrooxazinones obtained via aza-Diels–Alder reaction, by Panunzio et al. (48); a library of dehydroamino acids made through the Passerini multicomponent condensation, by Kim et al. (49); a 288-member library of decorated piperidines with opioid antagonist activity, by Thomas et al. (50); an 192-member library of benzodiazepinediones from Ugi condensation employing a convertible isocyanide, by Hulme et al. (51); a library of himbacine-derived muscarinic antagonists obtained via Sonogashira coupling and intramolecular Diels–Alder reaction, by Doller et al. (52); and a library of pyrimidin-4-yl substituted α-amino acids obtained through the cyclocondensation between alkynyl ketones and benzamidines, by Adlington et al. (53).

8.2.2 An Example: Synthesis of an Aminoglucopyranoside Library

Recently, Wong et al. (54) have reported the synthesis of a small, 24-member focused library of aminoglucopyranosides (**L1**, Fig. 8.9) inspired by the structure of paromomycin, a known aminoglycoside antibiotic (**8.18**, Fig. 8.7). A retrosynthetic study identified the azido sugar **8.19** as the key intermediate for the library synthesis; its preparation on a multigram scale is also given in Fig. 8.9.

The synthetic scheme used to prepare the library is shown in Fig. 8.10. The reaction steps, amide coupling, ozonization, reductive amination, and catalytic reduction, are trivial for carbohydrate substrates, and the authors decided that assessment of the chemistry for the library synthesis would not have been necessary. The availability of **8.19** in multigram quantities reduced the significance of potentially low-yielding steps. The rehearsal of the monomers was also avoided because of the small size of the two monomer sets **M$_1$** (four Fmoc-protected a-amino acids, Fig. 8.10) and **M$_2$** (six amines, Fig. 8.10), which were inspired by a model for the interaction between paromomycin and RNA (55). Finally, such a small array could be considered as a model library for a much larger solution-phase library of potential RNA binding molecules.

The synthesis of the library proceeded smoothly as planned, and only two purifications were necessary. The four intermediates **8.20** were chromatographed and the final, basic library individuals were purified by ion-exchange chromatography, both steps being amenable to automation for synthesis of a larger library. The library **L1** validated the chemical route and confirmed the structural hypothesis of 1,3-hydroxyamine-containing carbohydrate scaffolds as RNA binders. The compounds were tested and showed RNA-binding activity, even if the desired sequence specificity was not observed (54).

8.2.3 An Example: Synthesis of a Thiohydantoin Library

Sim and Ganesan (56) have recently reported the partial synthesis of a 3078-member thiohydantoin library in solution (**L2**, Fig. 8.11). The initial studies were aimed at the preparation of a library of thiohydantoins in SP based on the adaptation of a previously

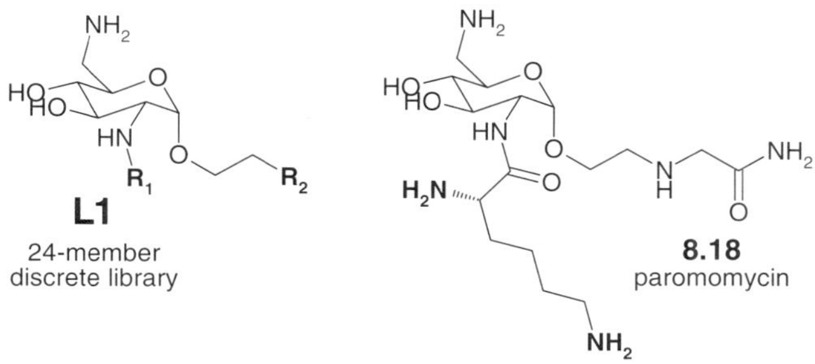

L1
24-member discrete library

8.18
paromomycin

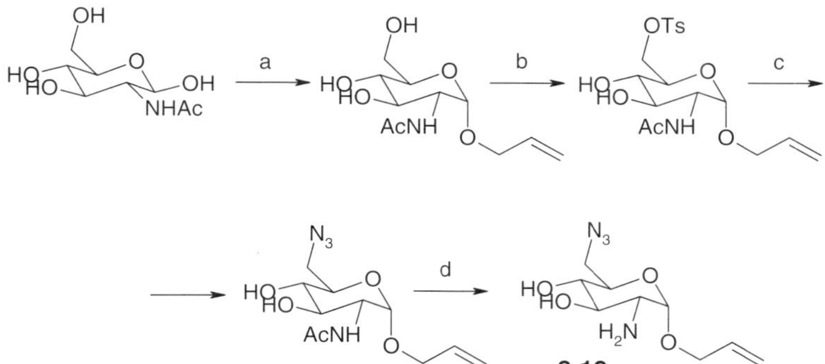

8.19

a: CH_2=CH-CH_2OH, $BF_3.Et_2O$, reflux, 2 hrs; b: TsCl, Py, 0°C, 12 hrs; c: NaN_3, DMF, 80°C, 3 hrs; d: aq. $Ba(OH)_2$, reflux, 3 hrs.

Figure 8.9 Synthesis of a key intermediate **8.19** to the solution-phase discrete aminoglucopyranoside library **L1**.

reported route (57) in which it had been observed that *N*-alkylated α-amino acid esters (obtained from α-amino acid esters as **M1** and aldehydes **M2**, Fig. 8.11) could be cyclized directly to give thiohydantoins (**L2**) when treated with isothiocyanates (**M3**) under mild basic conditions rather than forming an intermediate thiourea **8.23** (Fig. 8.11) (40). A straightforward synthetic scheme for the preparation of a solution-phase library was thus designed to produce the desired heterocycles. The authors did not consider an interesting SP alternative in which the starting α-amino ester could be bound to the support via the ester function and the thiohydantoin could be released via cyclative cleavage, sequestering the excess isothiocyanate and the base by solid-supported scavengers (see Section 8.4), and preparing even large SP pool libraries in high yields and purities.

The assessment of the chemistry and the rehearsal of the monomer set were run in parallel, with some important findings. Many natural α-amino acid esters (**M1**) gave good results, as did the aromatic aldehydes (**M2**) in general, while aliphatic and deactivated aromatic aldehydes were less satisfactory and thus rejected. Commercially

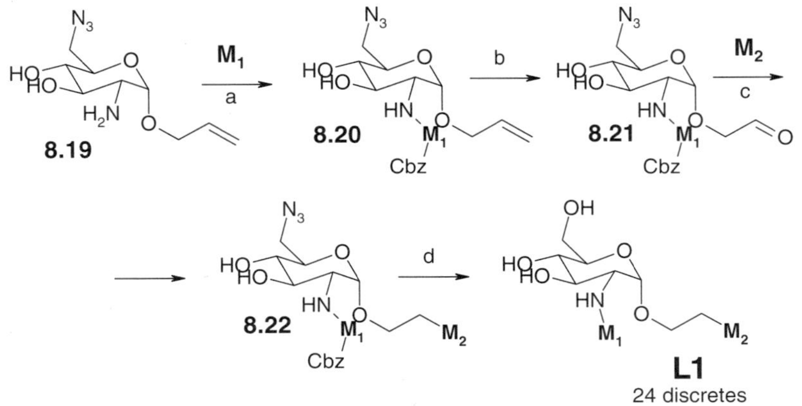

a: DMF, TEA; b: i, O_3, MeOH, -78°C, ii, DMS; c: $NaCNBH_3$, AcOH; d: H_2, $Pd(OH)_2$, AcOH, H_2O.

M_1 = 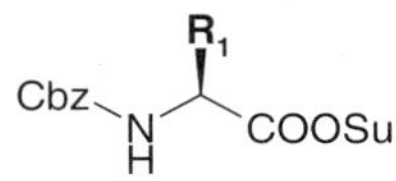

Cbz-Gly, Cbz-L-Ala, Cbz-L-Lys, Cbz-L-Arg.

4 monomers

M_2 = 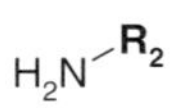

H-Gly-NH_2, H-L-Ala-NH_2, H-L-Leu-NH_2, H-L-Phe-NH_2, H_2N-CH_2-Ph, H_2N-CH_2-CH_2-NH-Cbz

6 monomers

Figure 8.10 Synthesis of the solution-phase, discrete aminoglucopyranoside library **L1**.

available isothiocyanates (**M3**) were successfully rehearsed with the exception of sterically hindered examples, which were rejected. The final selection of the three monomer sets (Fig. 8.12) produced a total of $9 \times 18 \times 19 = 3078$ library individuals sequentially prepared as small arrays of compounds.

The synthetic protocols included four extractions and an anhydrification with $MgSO_4$ followed by filtration. This nonautomated procedure for the preparation of compounds on a 0.1-mmol scale (typically tens of milligrams for each individual) as a medium-size solution library required significant effort during purification and work-up; the authors reported that only around 600 library members were prepared when the article was submitted for publication.

This example shows how the preparation of medium–large solution libraries requires automated or semiautomated synthetic and purification protocols for the rapid production of the designed library. Automation of discrete library synthesis in solution

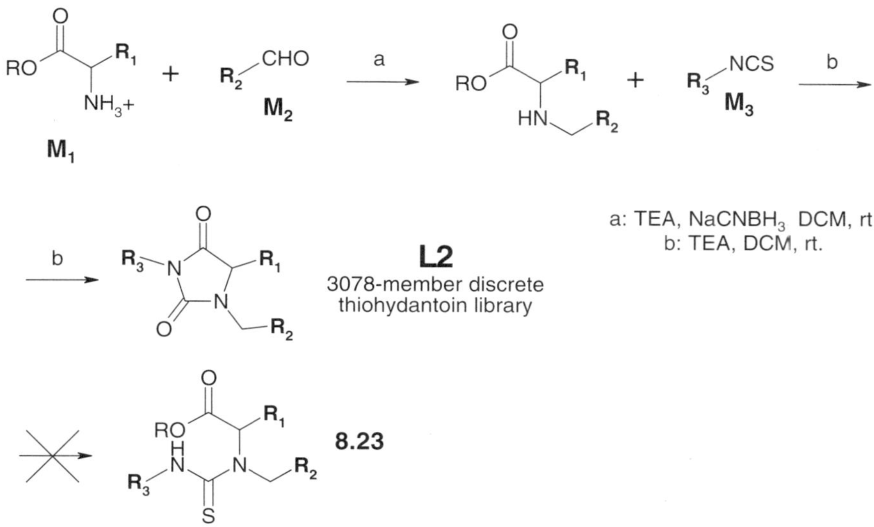

Figure 8.11 Synthetic scheme to the solution-phase, discrete thiohydantoin library **L2**.

is covered in the following sections, while automated work-up and purification procedures are covered in Sections 8.3–8.5.

8.2.4 Automation and Solution-Phase Discrete Libraries

The synthesis of small solution-phase discrete libraries does not require automation, and a set of multichannel pipettes is enough to deliver solvents and reagents into the reaction vessels. The reaction vessels are arranged in racks that spatially isolate each vessel and allow the transfer of the whole array onto an orbital stirring unit, onto a heating/cooling instrument, or into a closed system under an inert atmosphere. Even several hundred discrete reactions can be performed and processed rapidly by manual techniques. Some commercially available instruments are able to handle in parallel a few tens of compounds (stirring, heating, cooling) (58, 59), while others also automate the work-up/purification procedures allowing the manual delivery of reagents even under an inert atmosphere (60, 61).

Robotic dispensing units based on pipetting arms can be used to partially automate library synthesis in solution, which included purification steps consisting of liquid–liquid or solid–liquid extraction followed by withdrawal of the desired phase or by elution of the product from the SP (62, 63; see Section 8.3). The dispensing units of automated SP synthesizers can be used for this purpose, and sometimes manufacturers claim that it is possible to run solution-phase library synthesis on these automated instruments, even though such claims must be validated for many operations. Dispensing units are the core of fully automated modular systems specifically designed for solution-phase synthesis (64, 65). They contain stirring/vortexing and heating/cooling

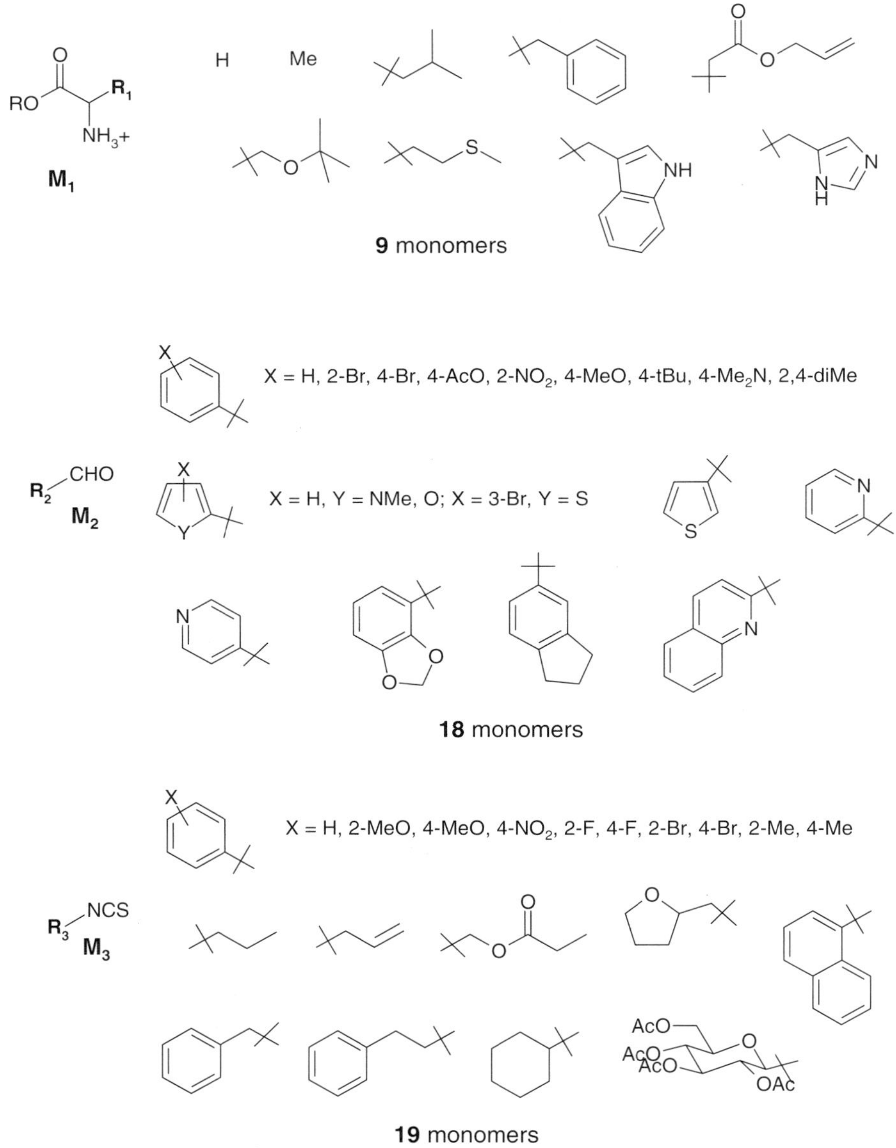

Figure 8.12 Synthesis of the solution-phase, discrete thiohydantoin library **L2**: structure of the monomer sets **M1**–**M3**.

modules and allow the fully automated, parallel handling of several arrays processed through different modules at a given time. Many of the most commonly encountered organic reaction conditions, including working under an inert atmosphere (via rubber septa that are pierced by robotically controlled needles), are compatible with these instruments. The example described in the next section is based on the use of a fully automated synthesizer to generate a small solution library. A recent report (66) described the use of several synthesizers for different operations in the same automated solution library synthesis, optimizing the performance of each step and reducing the lag times.

Work up/purification procedures have also largely benefited from the commercial availability of semiautomated or automated instrumentation. The panel of tools spans from small, 96-well based devices to increase the throughput of manual operations (67, 68) to semiautomated systems able to purify in parallel combinatorial samples (69) or to concentrate in parallel large, discrete solution libraries (70, 71). Several systems developed in-house have also been recently reported (72, 73), and a recent review covered the most recent trends in automated high-throughput purification methods for solution discrete libraries (74).

Some companies (75, 76) have developed their own automated instruments for combinatorial library synthesis in solution to produce large, purified arrays of discretes (up to several tens of thousands of individuals); the available information is obviously scarce, but an example of such a proprietary integrated synthesizer will be presented in section 8.2.6.

8.2.5 An Example: Synthesis of a Triazine Library

Whitten et al. (77) reported the synthesis of a focused triazine library **L3** (Fig. 8.13) of >350 members based on the known corticotrophin releasing factor 1 (CRF-1) antagonist, **8.24** (Fig. 8.13), to discover more potent triazine analogues. The two-step synthetic scheme leading to **L3** from dichlorotriazines **8.25a,b** using two amine monomer sets $\mathbf{M_1}$ and $\mathbf{M_2}$, is reported in Fig. 8.13.

The first chlorine displacement was carried out at rt, while the second required heating. Both reactions are well known in the literature and chemical assessment was not considered necessary. A modular automated system (64) allowed the simultaneous processing of several arrays of glass vials (1.8 mL each, typically 5–25 vials per array, each stoppered with a rubber septum) through the stirring unit, the heating unit, and the extraction/purification modules. The results of the first arrays were used as both rehearsal of the monomers and a model library synthesis. Most of the amine monomers were commercially available, while the primary amines **8.27** were also prepared on the automated instrument (Fig. 8.14). The library synthesis was quite successful, with around 70% library individuals prepared in milligram amounts and successfully QCed by HPLC/MS (purity >70%). The confirmed compounds were tested as CRF-1 antagonists, while the other samples were discarded.

The only manual operations required for the synthesis of this library were the preparation of stock solutions of **8.25a,b** (0.5 *M* THF) and the monomer representatives $\mathbf{M_1}$ (1.5 *M* DIPEA and 3 *M* THF) and $\mathbf{M_2}$ (1 *M* THF). The other operations

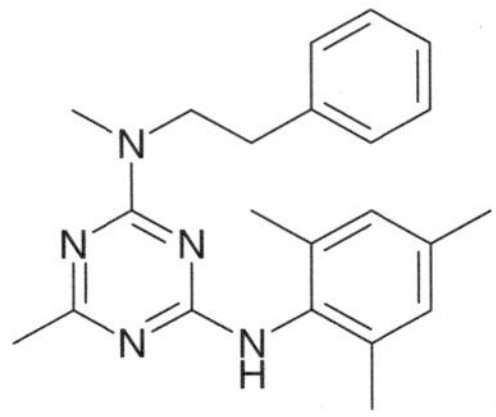

8.24

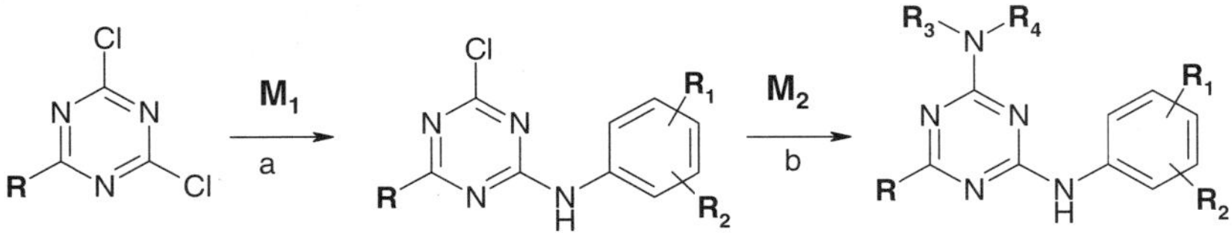

8.25a R=Me
8.25b R=Et

8.26

L3
>350-member
discrete library

a: THF, rt, 10'; b: THF, 65°C, 1 hr.

M_1 : substituted anilines

M_2 : primary and secondary alkyl amines

Figure 8.13 Synthesis of the solution-phase, discrete triazine library **L3** inspired by the CRF-1 antagonist **8.24**.

were controlled and automated by the instrument software. A brief description of the main software commands is reported in Fig. 8.15 (Method 1):

- *2STEP1/WAIT2* implied the addition of **8.25a** or **8.25b** (0.04 mL, stock solution) to septum-sealed 1.8-mL vials followed by the monomer set **M1** (0.026 mL, stock solution). The reactions were left standing for 10 min at room temperature (*WAIT2*).

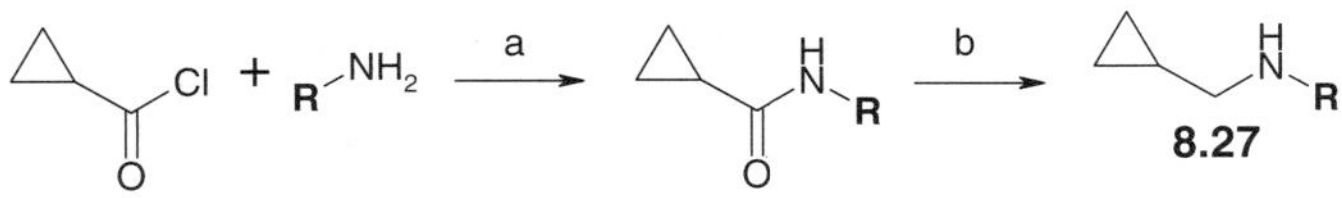

8.27

a: dry Et_2O, rt; b: BH_3, THF, rt.

Figure 8.14 Synthesis of a cyclopropane-based monomer subset for the solution-phase, discrete triazine library **L3**.

METHOD 1	METHOD 2
- 2STEP1 - WAIT2 - 2STEP2 - WAIT2 - 2STEP3 - ROBOTFAS	- CYCLOPTR - ROBOTFAS

Figure 8.15 Software automated protocols for the synthesis of the solution-phase, discrete triazine library **L3**.

- *2STEP2/WAIT2* treated the resulting solutions of **8.21** with 10 to 20 equivalents of $\mathbf{M_2}$ (stock solutions) and heated them at 65 °C for 1 h. The reactions were then cooled to rt (*WAIT2*).
- *2STEP3* dispensed EtOAc (0.5 mL) and 1 *N* HCl (0.5 mL) to each vial and transferred the organic layers to new vials after partition.
- *ROBOTFAS* analyzed all the vials by GC-MS, then evaporated their contents to dryness with nitrogen, producing crude **L3** as discretes with good average yields and purities.

The synthesis of **8.27** (Method 2) consisted of the following:

- CYCLOPTR, where the acyl chloride and the amines in THF were stirred for 1 h before partitioning the crude between ether and 1 *N* HCl. The organic phases were transferred to new vials and washed with aqueous sodium hydrogen carbonate and then to a second vial for drying with sodium sulfate. The ethereal solutions were transferred to vials containing borane and stirred for 6 h, after which time the reactions were quenched with aqueous sodium hydrogen carbonate. The organic layers were transferred to new vials and heated at 125 °C for 9 h.
- *ROBOTFAS* was performed as in the above sequence.

Fully automated instruments with low–medium throughput, typically a few hundred compounds per week for one- to three-step synthetic schemes in solutions, are suitable for small, focused libraries where more challenging reaction conditions such as heating, using reactive intermediates, or when inert atmospheres are needed. These instruments are also suitable for intermediate steps in the construction of larger libraries. For example, monomer rehearsal can be performed by reacting a common intermediate with various monomer candidates, and a small model library may be prepared. Even chemistry assessment may be tackled with these instruments.

Among some recent papers, the same researchers (78) reported the synthesis of two libraries of piperazines and piperidines containing 1086 and 835 discretes, respectively, via one-step alkylation or acylation of several scaffolds. Bhat et al. (79) used the same instrument to prepare a 26-member library of paclitaxel C7 esters via a three-step procedure. An array of twenty 1,2,4-oxadiazoles (80) was prepared through a three-step procedure from carboxylic acids and benzamidoximes using a semiautomated device (60). Frank et al. (81) reported a 107-member quinolone library prepared through a five-step protocol with a semiautomated device developed in-house.

8.2.6 An Example: Synthesis of a Spiro[Pyrrolidine-2,3′-Oxindole] Library

Powers et al. (82) reported the synthesis of a 1280-member library of discrete chalcones in solution (**L4**, Fig. 8.16) with an extremely high (96% average) compound purity. This library was used as such or through subsets to generate several primary libraries as sources of novel biologically active compounds. A 25,600-member library

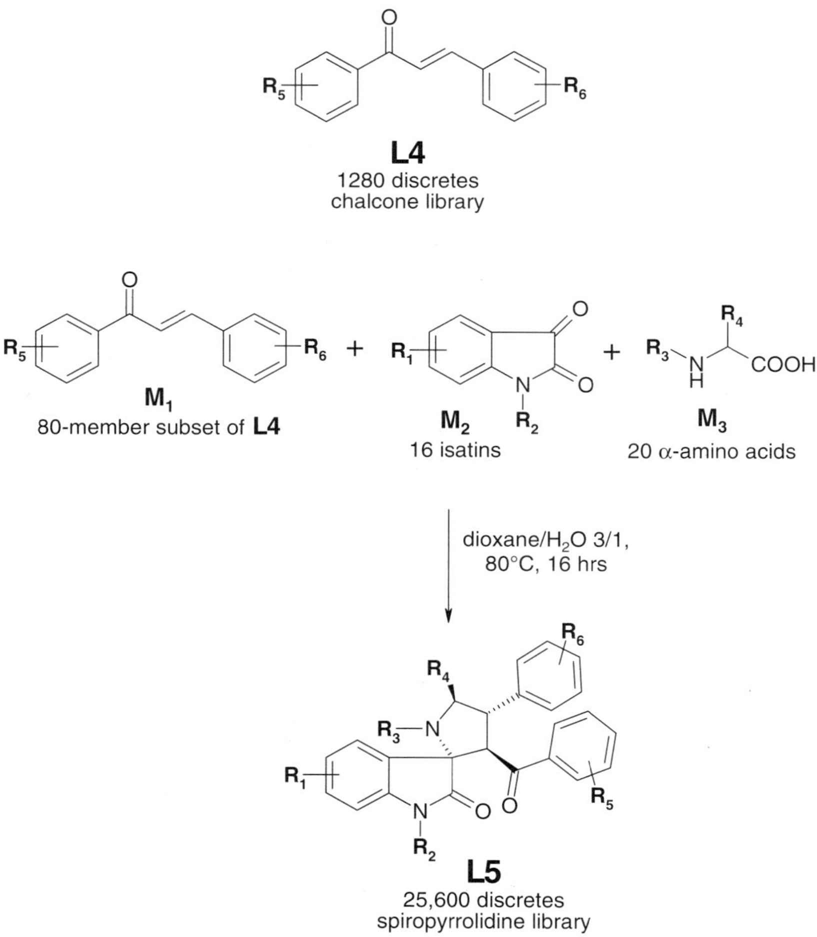

Figure 8.16 Synthesis of the solution-phase, discrete spiropyrrolidine library **L5** from a subset of the solution-phase chalcone library **L5**.

of spirotricyclic compounds (**L5**, Fig. 8.16) that was derived from the chalcone monomer subset **M1** (80 representatives) and from full monomer sets **M2** and **M3** (16 isatins and 20 α-amino acids, respectively) following the synthetic scheme reported in Fig. 8.16 was described by Fokas et al. (83).

The feasibility of the proposed chemical route was assessed, and a brief monomer rehearsal was carried out by reacting the unsubstituted chalcone **8.28** with several isatins (**8.29a–d**, Fig. 8.17) and α-amino acids (**8.30a–g**, Fig. 8.17). Most of the single reactions produced the desired spiro compounds in good yields and purities and also with excellent regio- and stereocontrol. The adverse steric interaction between the chalcone carbonyl and the **R4** substituent gave control over the regiochemistry and only one diastereomer (as a pair of enantiomers) of **8.31a–j** was generally obtained (Fig. 8.17). Two monomer candidates **M3**, **8.30f,g**, performed poorly and were discarded. The yields and purities obtained during this phase of study are reported in Table 8.1.

The large majority of the selected monomer representatives were commercially available and a few were prepared in a single reaction. The library synthesis was performed on a 50 μ*M* scale (around 20 mg/compound) using a 96-well-based, proprietary automated and integrated instrumentation called the Automated Molecular Assembly Plant (84), which is suitable for the production of medium–large libraries

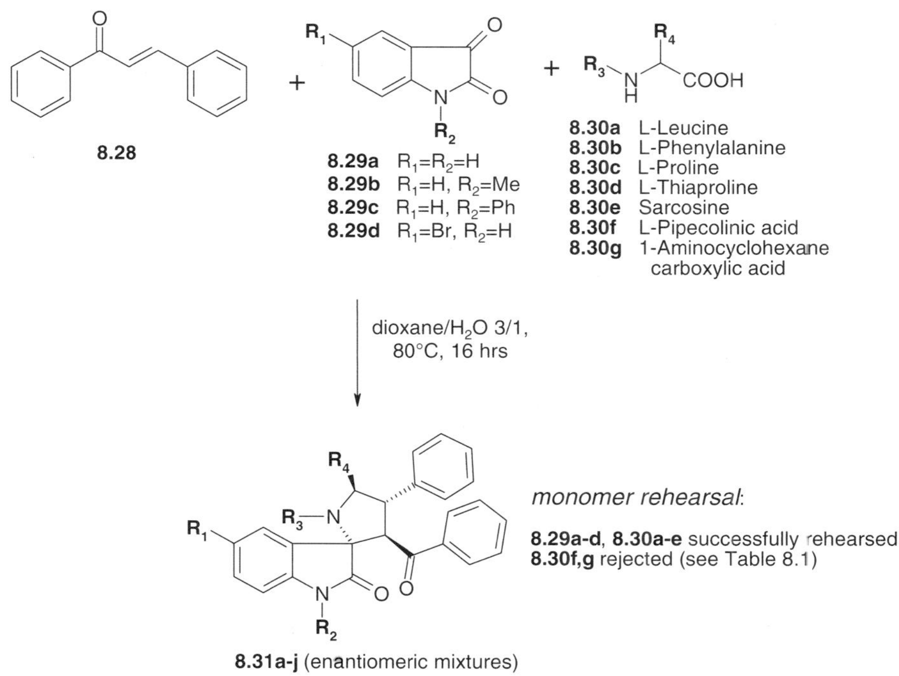

Figure 8.17 Synthesis of the solution-phase, discrete spiropyrrolidine library **L5**: monomer rehearsal.

of discretes in solution. Twenty-five percent of the resulting racemic mixtures of compounds were characterized by HPLC/MS, and generally, good purity was observed, with the exception of three amino acids and a single chalcone that gave unsatisfactory results; the corresponding products were discarded. A more extensive rehearsal of the monomers could have prevented this undesired outcome, which led to the loss of $(3 \times 1280) + (1 \times 272) = 4112$ (16% of the total library population) expected individuals from the library.

Several chalcone-based libraries (see Fig. 4.11, section 4.2.2) were combined to give a total of over 74,000 discretes (82) and, while no mention of time and resources required was made in the paper, the significant effort required to develop the automated equipment for the synthesis of medium–large libraries in solution *in-house* was surely paid back by the potential to continuously produce primary lead-seeking libraries in solution in large quantities. Only a few specialized companies may invest so heavily in combinatorial automation, but any interested party can access these companies to contract a library synthesis starting from a proprietary chemistry of interest.

The same group (85) has also reported the synthesis of a > 40,000-member triazine library from multiple decoration of cyanuric chloride. Another group that developed its own automated instrumentation reported the synthesis of a 1920-member tetrahydroquinoline library using a three-component cycloaddition between anilines, aldehydes, and alkenes (86).

TABLE 8.1 Spiro [pyrrolidine-2,3-oxindole] Discrete Library L5: Chemistry Assessment and Monomer Rehearsal

Product	R_1	R_2	R_3	R_4	R_5	Yield (%)[a]	Purity (%)[b]
8.31a	H	H	Me	H	H	65	89
8.31b	H	H	H	CH_2Ph	H	87	87
8.31c	H	H	H	$CH_2CH(Me)_2$	H	75	91
8.31b	H	H	[c]	[c]	H	83	98
8.31e	5-Br	H	[c]	[c]	H	85	90
8.31f	H	Ph	[c]	[c]	H	88	96
8.31g	H	H	[d]	[d]	H	79	81
8.31h	H	Me	[c]	[c]	H	73	81
8.31i	H	H	[e]	[e]	H	26	93
8.31j	H	H	H	[f]	[f]	ND	ND

[a]Isolated compounds.
[b]By HPLC.
[c]$R_3R_4 = (CH_2)_3$, proline.
[d]$R_3R_4 = CH_2SCH_2$, thiaproline.
[e]$R_3R_4 = (CH_2)_4$, pipecolinic acid.
[f]R_4R_5 = cyclohexyl.

8.3 PURIFICATION OF SOLUTION-PHASE LIBRARY INTERMEDIATES AND FINAL COMPOUNDS: LIQUID–LIQUID AND SOLID-PHASE EXTRACTION SYSTEMS

8.3.1 General Considerations

Organic chemistry in solution has always relied on the production of compounds in high yields, and much effort has been spent in creating, refining and applying a wide set of chemical transformations to a variety of synthetic problems. The available reactions are largely robust enough to successfully plan and execute a given chemical synthesis. Once the desired product is obtained, though, it must be purified from the reaction by-products, and typically this is done *via* a *work-up phase* where common separation techniques (extraction, evaporation, precipitation, filtration) produce a crude that is submitted to *purification* that is usually based on a chromatographic separation.

The switch from single compounds to a combinatorial library in solution increases the *complexity* of the potential issues to be addressed, as we have seen in the previous chapters. The same is true for the purification of these libraries, which cannot rely on simple filtration and washing of the resin beads, as in the SP chemistries. The separation techniques of classical chemistry are used. However, general, automated methods applicable to all members of a library have to be found. Usually, these methods are also applicable to the final purification of cleaved SP libraries, either as discretes or as pools.

While evaporation is used for the concentration and removal of solvents, usually the reaction by-products are not volatile. Similarly, filtration of precipitated or crystallized solids is not likely to be applicable to all the members of a library, and furthermore the automation of these processes is not straightforward; an interesting example of general precipitation of library members from an organic medium due to the presence of a basic ionizable group has been recently reported by Perrier and Labelle (87). *Extraction procedures* possess the desired separation properties and have been used for the purification of several solution-phase libraries; we will cover this subject in more depth in this section. An excellent review (88) has recently been published in which the interested reader will find a description of available strategies for separation and purification of single compounds and arrays.

8.3.2 Liquid–Liquid Extractions: Two-Phase Systems

The partition of a reaction between an organic and an aqueous phase is an excellent method to isolate either an organic product from water-soluble impurities or a hydrophilic product from organic impurities. In a reaction the use of water-soluble reagents, catalysts, and coupling agents that are sequestered from the reaction products by simple washing with water is often preferred. An example is 1-(3-dimethylamino-propyl)-3-ethylcarbodiimide hydrochloride (EDC), a water-soluble carbodiimide typically used for amide couplings. More exotic two-phase extraction systems using two immiscible organic solvents, while possible, have not received much attention due to their limited applicability to the purification of organic compounds, an exception being the so-called fluorous biphase systems (89), whose principle is covered in detail in Section 8.3.4.

Acidic, neutral, or basic aqueous solutions make it possible to change the phase into which a compound partitions. For example, a basic reaction product is extracted from a water-immiscible organic solvent using an acidic aqueous solution, thus removing organic impurities (step a, Fig. 8.18). Basification of the aqueous phase and extraction with an organic solvent (step b, Fig. 8.18) allows the removal of water-soluble impurities and recovery of the basic organic compound in a reasonably pure state. Phase switching can be obtained by tagging neutral organic compounds (step e, Fig.

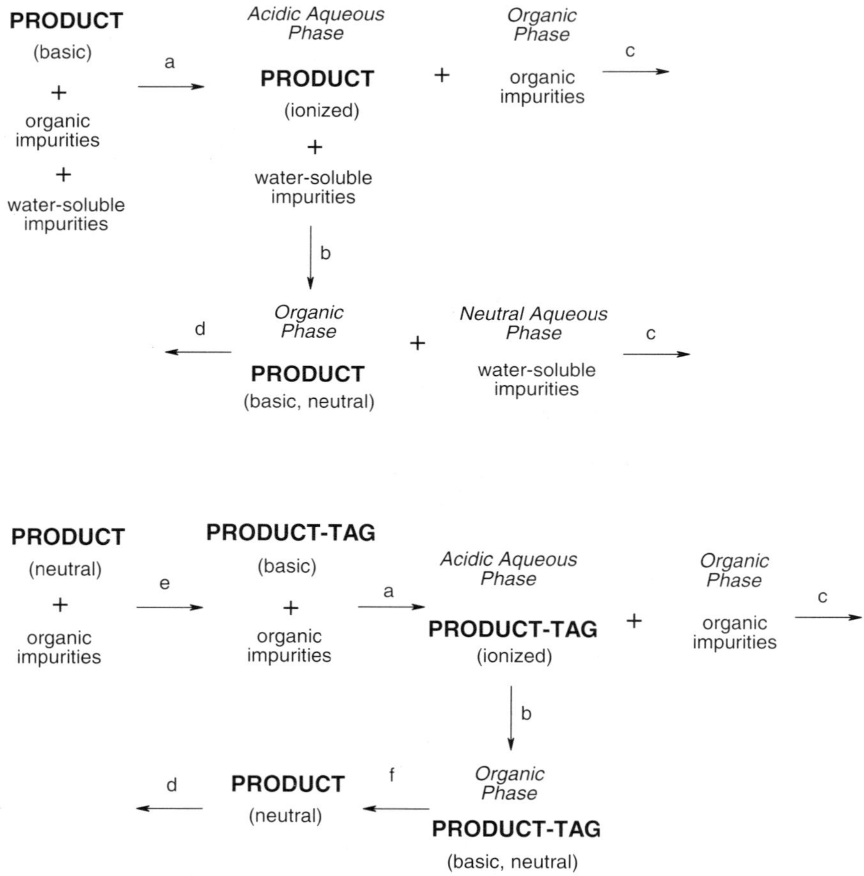

a: acidic aqueous/organic extraction; b: neutralization, then aqueous/organic extraction; c: discarded; d: product recovery; e: chemical tagging; f: de-tagging.

Figure 8.18 Liquid–liquid biphase extraction: basic principles applied to basic (top) and neutral products (bottom).

8.18); in this case, the basic tag allows the extraction of the product into aqueous acid and the same double-extraction procedure for the above example (steps a and b, Fig. 8.18) can be applied. Finally, the neutral product is detagged (step f, Fig. 8.18) and obtained as a pure sample, providing that detagging is performed without the generation of impurities. An example of a basic phase tag for liquid–liquid extraction has been recently reported (90).

Liquid–liquid two-phase extractions are sometimes suitable for parallel synthesis. Manual multichannel pipettes can handle a few tens of solutions and recover the desired liquid phase after the extraction. If numbers of individuals increase, partial or total automation is necessary and mono- or multichannel robotic pipettes deliver solutions and withdraw either the top or the bottom phase of a liquid–liquid system with extreme accuracy, transferring it to other reaction vessels. They allow the sequential preparation of medium–large libraries but are very time consuming when large libraries of > 1000 members are involved.

8.3.3 An Example: Synthesis and Purification of an Iminodiacetic-Based Triamide Library

Cheng et al. (91) have reported the manual synthesis of a 960-member, flexible triamide library **L6** made by decoration of an iminodiacetic acid–based scaffold. The structure of **L6** and the synthetic scheme leading to the library are reported in Fig. 8.19. Sequential amide coupling using well-known peptide coupling procedures involving primary amines **M1** (six representatives) and **M2** (eight representatives) and carboxylic acids **M3** (20 representatives) was considered trivial enough to avoid monomer rehearsal.

Treatment of the protected iminodiacetic acid **8.32** with EDC in DMF produced the anhydride **8.10**, which was treated in situ with **M1** to give intermediates **8.33** (steps a and b, Fig. 8.19). These were purified by washing with acid to remove the unreacted amines and any aqueous by-products (step c). Coupling with **M2** using phosphorus (1-hydroxy-1H-benzotriazolo)tri-1-pyrrolidinylhexafluoro phosphate (PyBOP) and DIPEA in DMF produced 48 intermediates **8.34** (step d) that were purified by sequential acid and basic washings to remove unreacted amines (step c) and unreacted **8.33** (step e), respectively. Deprotection of the Boc carbamate (step f) and coupling with **M3** produced 960 triamides (**L6**, step g) that were purified in the same manner as **8.34** (steps c and e, Fig. 8.19). The final compounds were obtained with overall yields varying from 10 to 71% in quantities of 30–150 mg with purity generally >95%.

Other methods are more suitable for the purification of medium–large solution-phase libraries (*vide infra*), but if small arrays are prepared and partition between aqueous and organic phases is applicable, this method rapidly produces good-quality library compounds, either alone or in conjunction with other separation techniques.

8.3.4 Liquid–Liquid Extractions: Multiphase Systems

Recently, Horvath, and Rabai (89) have reported the use of fluorous biphasic systems as an alternative to aqueous biphasic systems for phase transfer catalytic processes in

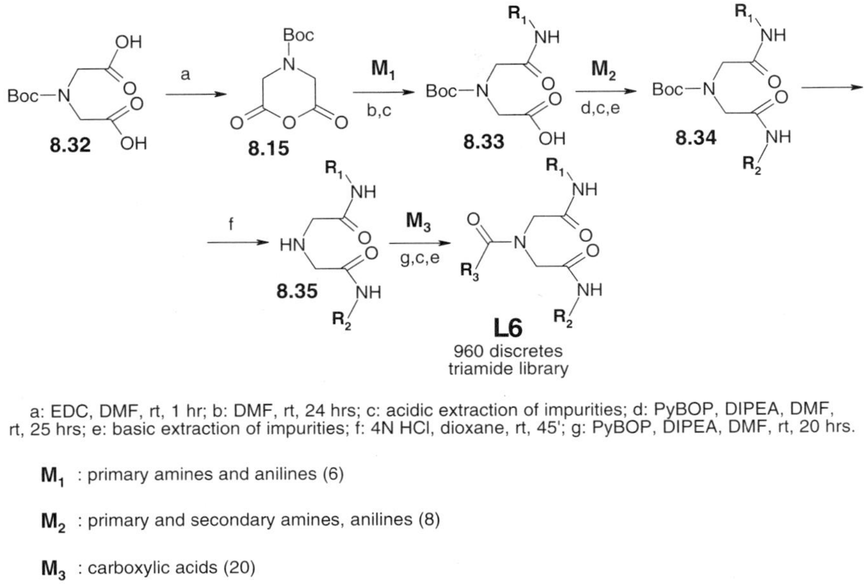

Figure 8.19 Synthesis and purification of the solution-phase, discrete triamide library **L6** using liquid–liquid extraction.

which the aqueous phase interferes with water-sensitive reactions; Barthel-Rosa and Gladysz (92) have subsequently reviewd the use of both fluorous catalysts and reagents in fluorous media. Fluorous-tagged reagents/catalysts were prepared and optimized to become miscible with the fluorous phase and immiscible with the organic phase. Ogawa and Curran (93) extended the application of this method through the use of benzotrifluoride (BTF) as a mixed solvent that dissolves organic and fluorous reagents giving a homogeneous reaction mixture, followed by a two-phase, organic–fluorous liquid extraction to isolate the desired products. The same group (97–103) and others (94–96, 106–109) have also reported the synthesis of tagged catalysts (94–97), Sn-based (98–101), Si-based (102–103), and P-based (104–107) reagents and/or protecting groups for a variety of chemical transformations and of tagged ligands for organic reactions under fluorous experimental conditions (108, 109).

The organic–fluorous system is orthogonal to the aqueous and organic phases, so that a triphase liquid–liquid–liquid system is realizable (Fig. 8.20) in which the partition between three different layers can give very pure compounds if the affinity of reagents, catalysts, and products for each phase is clearly determined. Curran (110) was the first to foresee combinatorial applications for this extraction method, which relies on fluorous tags to label reagents, monomers, and/or reaction products and the partitioning of intermediates and final discretes into a specific layer; he has also summarized the current possibilities of fluorous reagents and techniques for solution-phase parallel synthesis in a recent review (111).

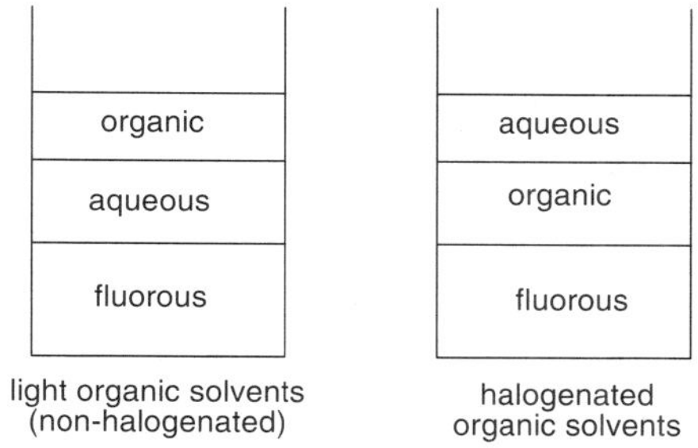

Figure 8.20 Liquid–liquid triphase extractions: aqueous/organic/fluorous mixtures.

Bi- or triphasic fluorous extraction protocols have been applied to the synthesis and purification of small arrays of isoxazolines and isoxazoles from tagged allyl alcohols and nitrile oxides (102), biaryls, and alkenylaryls from the Stille coupling of tagged tin reagents with organic halides or triflates (112, 113), using also microwave irradiation (114) and homoallylic alcohols from tagged stannanes and aldehydes (115). Recently a "fluorous switch" amine has been reacted with excess of isocyanates in an organic medium during the synthesis of an array of ureas (116); the fluorous by-product partitioned into the fluorous phase in a liquid–liquid biphasic extraction and was thus removed from the crude library discretes.

Unfortunately, this triphasic system is limited by the technical feasibility of handling single layers from a triphasic system, which is more complex than for the corresponding biphasic system, and the application of triphase extraction protocols to medium–large libraries is not practical. Curran and Luo have recently reported the so-called Light Fluorous Synthesis (LFS) (117): the lower fluorine content of fluorous tags used in LFS imparts better physico-chemical properties to the tagged compounds and makes more flexible both the reaction conditions and the purification-work up procedures. Alternative multiphase systems are also under investigation (118) and are likely to produce novel extraction phase systems. It can be expected that technological improvements in liquid handling/dispensing/separation protocols in the future will lead to wider use of aqueous/organic/fluorous extractions as a clean and effective purification method for combinatorial libraries in solution.

8.3.5 An Example: Synthesis and Purification of Amides and Dihydropyrimidines from Multicomponent Reactions

The synthesis of small arrays of organic compounds derived from multicomponent condensations was recently reported by Studer et al. (119). A 10-member amino acid amide library **L7** (Fig. 8.21) was prepared using the fluorous Ugi ("Flugi") condensation, and another 10-member dihydropyrimidine library **L8** (Fig. 8.21) was prepared using the Biginelli ("Fluginelli") condensation adapted to the fluorous phase. The key intermediates for library preparation were the silyl bromide **8.36**, prepared from a fluorous iodide (120), and the acyl bromide **8.37** and the acid **8.38**, prepared from an orthothiobenzoate (121), as shown in Fig 8.21. The structure of the fluorous tag was

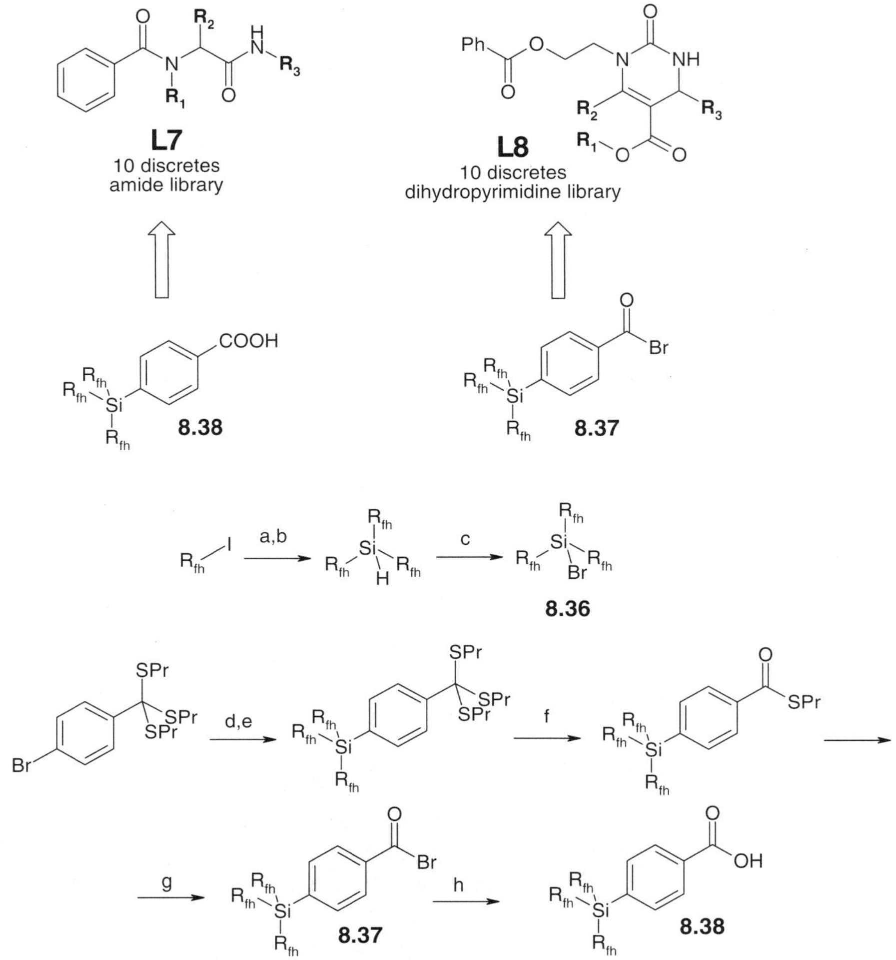

$R_{fh} = C_{10}F_{21}CH_2CH_2$-

a: Mg, Et_2O, reflux, 2 hrs; b: $HSiCl_3$, reflux, 16 hrs; c: Br_2, FC-72, rt, 12 hrs; d: t-BuLi, Et_2O, -78°C, 45'; e: **8.31**, BTF/FC-72, rt, 30'; f: $AgNO_3$, BTF/Me_2CO/THF/H_2O, rt, 12 hrs; g: Br_2, FC-72, rt, 3 hrs; h: THF/BTF/H_2O, rt, 12 hrs.

Figure 8.21 Fluorous-tagged Flugi (**L7**) and Fluginelli (**L8**) libraries : synthesis of tagged intermediates **8.36–8.38**.

optimized to have a high content of fluorine atoms, which caused the complete partition of **8.37** and **8.38** into fluorous solvents but also an ethylene spacer between the organic moiety and the fluorous tag in order to avoid changes in substrate reactivity due to the electronic effect of the fluorocarbon chain.

The synthesis of array **L7** is reported in Fig. 8.22. Compound **8.38** was reacted simultaneously with amines (**M1**, two representatives), aldehydes (**M2**, five representatives), and isonitriles (**M3**, two representatives) to give 10 compounds (not all the combinations were reacted). The reaction was performed in trifluoroethanol (TFE), another hybrid fluorous–organic solvent (step a, Fig. 8.22), and after evaporation of the TFE, the crude product **8.39** was purified by two-phase extraction between fluorous solvents and benzene (step b). After evaporation of the solvent, the fluorous tag was cleaved with TBAF (step c) and a triphasic extraction (step d, Fig. 8.22) was performed to remove the fluorosilane tag and acid **8.38**–related impurities extracted into the fluorous layer. Excess TBAF and TBAF-related impurities partitioned into the acidic aqueous layer. Yields and purities of the synthetic protocol are reported together with the structures of the library members **L7a–j** in Table 8.2.

Synthesis of the array **L8** is reported in Fig. 8.23. Bromide **8.37** was esterified with a urea alcohol (step a, Fig. 8.23) and purified by triphasic extraction (step b), giving pure **8.40** in the fluorous phase. Compound **8.40** was reacted with **M1** (β-keto esters, four examples) and with **M2** (aldehydes, three examples), as shown in step c, and then intermediates **8.41** were purified by two-phase fluorous/organic extraction (step d).

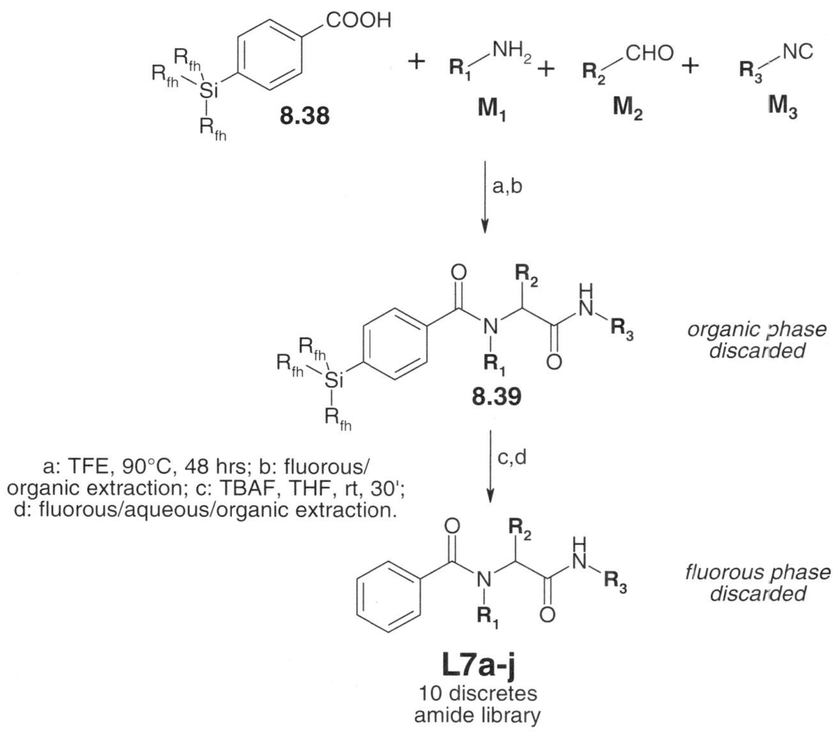

Figure 8.22 Synthesis and purification of the solution-phase, discrete amide library **L7** using fluorous aqueous/organic extraction protocols.

TABLE 8.2 Yields and Purity of the Amide Discrete Library L7

Product	R_1	R_2	R_3	Yield (%)	Purity (%)[a]
L7a	Benzyl	Phenyl	*t*-Butyl	83	85
L7b	Benzyl	4-MeOPh	*t*-Butyl	81	87
L7c	Benzyl	*c*-Hexyl	*t*-Butyl	32	89
L7d	*n*-Propyl	*c*-Hexyl	*c*-Hexyl	99	>95
L7e	Benzyl	Phenyl	*c*-Hexyl	92	80
L7f	*n*-Propyl	*i*-Propyl	*c*-Hexyl	71	>95
L7g	Benzyl	*i*-Propyl	*c*-Hexyl	61	>95
L7h	Benzyl	*c*-Hexyl	*c*-Hexyl	84	>95
L7i	Benzyl	Ethyl	*t*-Butyl	75	93
L7j	*n*-Propyl	*i*-Propyl	*t*-Butyl	78	81

[a]Determined by GC.

After removal of the solvent, the silyl tag was cleaved (step e) and the final triphasic extraction (step b, Fig. 8.23) gave the array **L8** in the organic phase, removing TBAF-related impurities and the fluorous tag by extraction into the aqueous and fluorous phases, respectively. Yields and purities are reported together with the structures of the library individuals **L8a–j** in Table 8.3. A few compounds were obtained as mixtures, probably due to lower reactivity of corresponding monomers and occurrence of side reactions.

8.3.6 Solid-Phase Extractions

The partition of compounds between two phases can also happen when one of the phases is solid, that is, using solid-phase extraction (SPE). The rationale of this method is highlighted in Fig. 8.24. Assuming a hypothetical reaction crude where our product **P** occurs together with an impurity **I**, the mixture is dissolved in a solvent (phase 1, Fig. 8.24) and eluted through a column filled with a stationary phase (phase 2) that has a high affinity for **P** and no affinity for **I**. After elution and repeated washings with fresh phase 1, **I** is eluted while **P** remains adsorbed onto phase 2 (Fig. 8.24). The final step is the elution of the column with another liquid phase (phase 3), which disrupts the noncovalent interaction between **P** and phase 2 and has a strong affinity itself for **P**, obtaining the elution of **P** into phase 3 and its quantitative recovery (Fig. 8.24).

Ion-exchange chromatography is based on this principle and has been widely used for analytical purposes, while its application in organic synthesis has been limited to scattered examples, despite its simplicity and applicability to many separation problems. When the SP is an ion-exchange resin and the product has basic or acid functions, SPE should be considered the separation technique of choice. A number of examples from the recent literature serve to illustrate the applicability of the technique to parallel purification of a number of compounds.

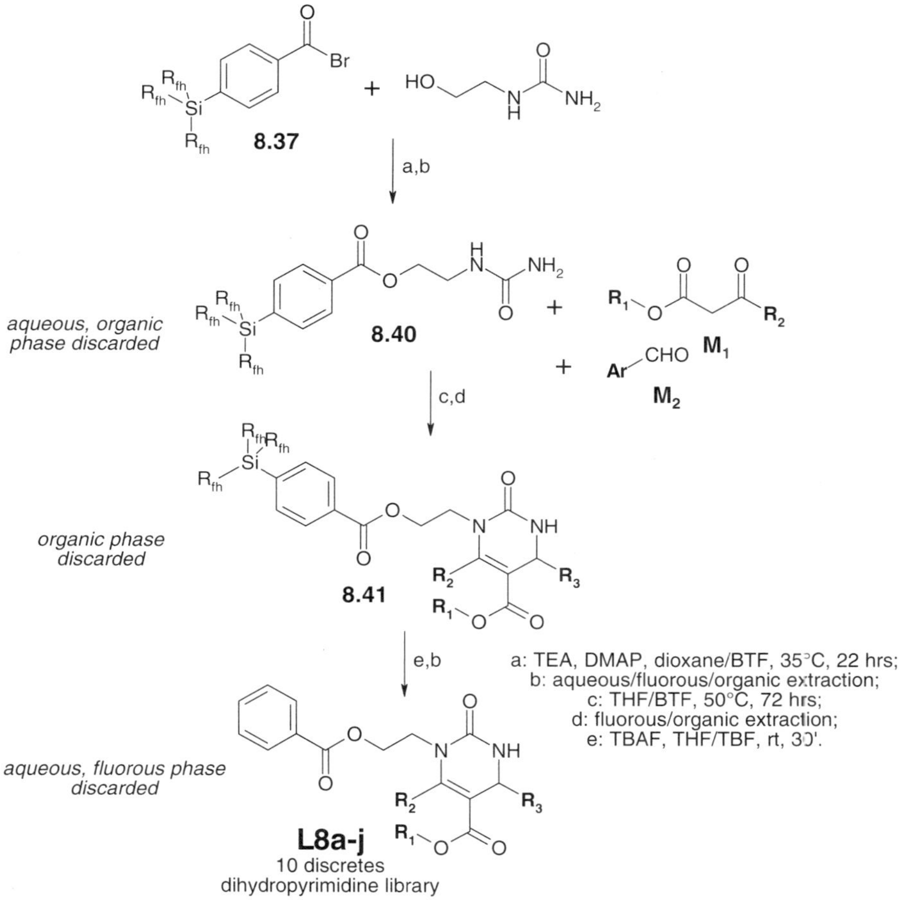

Figure 8.23 Synthesis and purification of the solution-phase, discrete dihydropyridine library **L8** using fluorous aqueous/organic extraction protocols.

Both Siegel et al. (122) and Lawrence et al. (123) have described automated systems for the purification of small arrays of amines and amides. A 48-member array of β-amino alcohols prepared from epoxides and amines was purified using SPE by Shuker et al. (124). Blackburn et al. (125) have described the purification of a 60-member 3-aminoimidazo[1,2-*a*]pyridine array obtained from a multiple-component condensation, and Bussolari et al. (126) purified a small array of phenylpropyl amines obtained from dihydrocoumarins and amines. A few applications where ion-exchange resins have been substituted with other solid phases have also recently appeared. For example, the purification of several carbohydrate arrays tagged as hydrophobic O-laurates using C_{18} silica producing up to 10–30 mg of >90% pure individuals was described by Nilsson et al. (127), and Curran et al. purified fluorous-

TABLE 8.3 Yields and Purity of the Dihydropyridine Discrete Library L8

Product	R_1	R_2	R_3	Yield (%)[a]
L8a	Ethyl	Methyl	Phenyl	71
L8b	Ethyl	Methyl	2-Naphthyl	55
L8c	Ethyl	Methyl	4-MeOPh	69
L8d	Methyl	Methyl	2-Naphthyl	70
L8e	Ethyl	Ethyl	2-Naphthyl	60
L8f	Ethyl	Ethyl	Phenyl	47
L8g	Ethyl	Ethyl	4-MeOPH	[b]
L8h	Methyl	Phenethyl	Phenyl	[b]
L8i	Methyl	Phenethyl	2-Naphthyl	[b]
L8j	Methyl	Phenethyl	4-MeOPh	[b]

[a]Determined by GC.
[b]Mixture of products.

tagged allyl alcohols (99) and aryl bromides (128) using readily available fluorous reverse-phase silica gel (129). These preliminary findings led to an expansion of SPE techniques to many more purification problems encountered in the preparation of libraries in solution.

SPE is becoming more popular for parallel synthesis/combinatorial chemistry in solution because it is more automation friendly than liquid–liquid extraction. Commercially available 96-well devices can be equipped with prepacked SPE cartridges, and automated elution and washing allow fast processing and reliable recovery of the purified compounds without the problems associated with liquid interfaces between phases found in liquid–liquid extractions. The use of SPE for the purification of libraries in solution, or for purification of cleaved libraries from a synthesis in SP, will undoubtedly take advantage from future technological advances that will increase the throughput and provide additional prepacked SP formats.

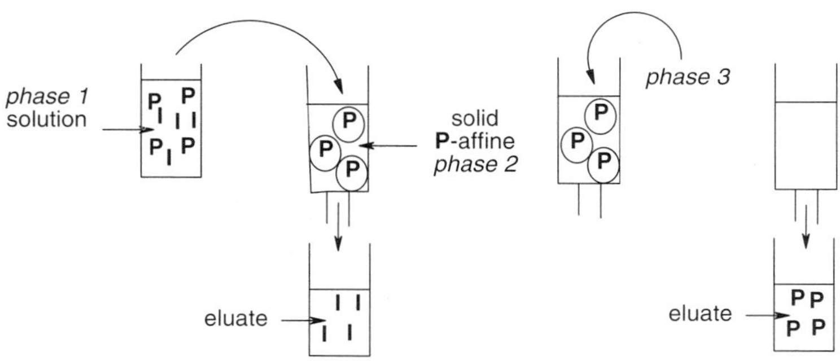

Figure 8.24 Solid-phase extractions: basic principles.

8.3.7 An Example: Synthesis and Purification of a β-Amino Alcohol Library

Siegel et al. (130) reported the synthesis and purification by parallel SPE of a focused 96-member β-amino alcohol library **L9** of potential adrenergic agonists or antagonists. The structures of known modulators (**8.42**, a β_1- and β_2-antagonist, and **8.43**, a β_2-agonist) are shown together with the synthetic scheme leading to **L9** in Fig. 8.25.

The reductive amination of monomer set $\mathbf{M_1}$ (12 N-unsubstituted β-amino alcohols) with $\mathbf{M_2}$ (8 ketones) was driven to completion by a large excess of ketone, producing a complex mixture of library individuals, excess ketones, and alcohol by-products (step a, Fig. 8.25). A strong cation exchange resin was used to separate the basic **L9** from the neutral impurities and the 96 methanolic solutions were eluted in parallel onto a block of 96 SPE cartridges. The block was then washed with fresh methanol, discarding the neutral impurities (step b). Final elution with methanolic ammonia (step c, Fig. 8.25) detached the desired products, which were characterized by HPLC, MS, and NMR, giving yields in the range of 21–99% and purities of 75–100%.

8.3.8 Chromatographic Methods

The physicochemical properties of the components of a mixture allow its separation using SPE by selective extraction of the target compound on SP, while the other

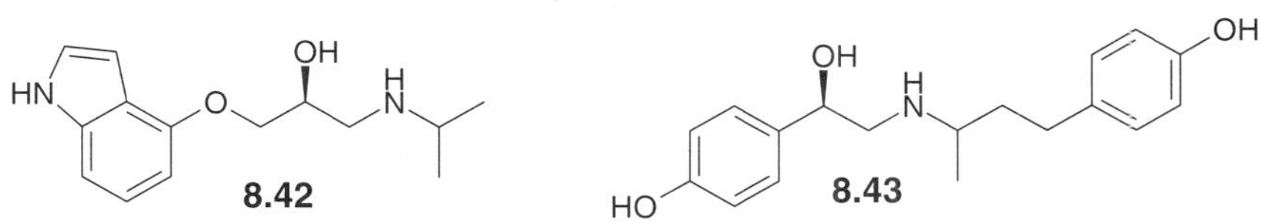

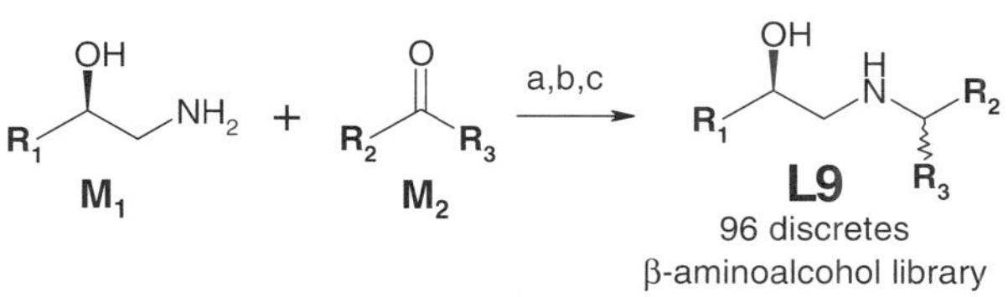

a: $NaCNBH_3$, 10% AcOH/MeOH, rt, 24 hrs; b: SPE; c: elution with NH_4OH.

$\mathbf{M_1}$: aryl-substituted β-aminoalcohols (12)

$\mathbf{M_2}$: acyclic and cyclic ketones (8)

Figure 8.25 Synthesis and purification of the solution-phase, discreteβ-amino alcohol library **L9** using SPE.

components of the mixture are eluted in the liquid phase. If all the mixture components have partial affinities for the SP and are retained with different strengths, the extraction process shifts toward more common chromatographic separations (88). This purification technique is by far the most popular in organic chemistry and direct or reversed-phase column, analytical or preparative HPLC and GC are routinely used in most chemistry laboratories.

The well-assessed reputation of chromatographic separations has stimulated their use for combinatorial purposes, and HPLC coupled with MS is the method of choice for automated, high-throughput confirmation and quality control of library individuals. Its use for automated purification of solution-phase library discretes, or cleaved SP library samples, has also been widely investigated. Preparative HPLC can easily provide significant quantities of pure compounds in conjunction with various detection techniques (diode array detection (DAD), MS, ELS, CLND). Typically, this type of system can purify and characterize from 100 to 400 compounds per day, allowing the isolation of up to 75–100 mg of each crude sample (131–134) using multiple automated instruments with high flow rates (up to 70 mL/min) and short gradients (5–10 min; see also Section 6.2).

Despite some obvious limitations as a combinatorial separation method (large volumes of solvents required, cost of multiple automated instruments), the confidence of each synthetic and analytical chemist in LC methods and the extremely high technological level of available instrumentation will keep chromatographic methods among the most popular purification techniques for small–medium solution-phase libraries, especially when large amounts of each crude material are prepared.

8.4 SOLID-PHASE ASSISTED SOLUTION-PHASE LIBRARY SYNTHESIS AND PURIFICATION

8.4.1 General Considerations

The use of supported *reagents*, *catalysts*, or *ligands* for heterogeneous reactions in organic synthesis is becoming increasingly popular and has been reviewed recently (135, 136). The switch from single to multiple reactions for the preparation of combinatorial libraries in the solution phase with the assistance of supported materials has also been studied. Insoluble supported chemicals have a major advantage that mirrors the advantage of SP synthesis; that is, the excess of a supported reagent or of a side product derived from the supported reagent is separated from the product in solution with the same automated filtering/washing procedures described for SPS. This simple concept has driven combinatorial chemists to expand the use of *solid-supported reagents* for solution-phase library synthesis and purification. Further, some of these reagents may be used to sequester excess soluble reagents or impurities from the reaction mixture (*supported scavengers*) and others can capture, via covalent bond formation, intermediates or final library compounds prepared in solution (*resin capture* reagents). The use of these supported compounds in solution-phase combinatorial chemistry is described in what follows.

8.4.2 Supported Reagents

The use of solid-supported materials in organic chemistry became significant with the pioneering work of Fetizon and Golfier in 1968 (137), and this field has been extensively reviewed (135, 138–140). The advent of combinatorial technologies and the renewed interest in solution-phase chemistry for library synthesis have also recently produced a renaissance of interest in supported reagents for application in a wider variety of modern synthetic reactions; potential advantages of supported reagents in terms of reduced pollution and cleaner, more efficient reaction routes have also been recently reviewed (141). A PS-supported distannane as a reagent for radical cyclizations has been developed by Junggebauer and Neumann (142). The well-known perruthenate oxidant for the conversion of alcohols to carbonyl compounds has been transferred to a polymer support by Hinzen and Ley (143). PS-supported HOBt has been used for ester formation and for the protection of amines as carbamates (144). Trifluoromethylaryl ketones supported on Tentagel have been used as in situ sources of polymer-bound dioxiranes for oxidations (145). The applications of PS-bound iodoso diacetate as a reusable iodination and oxidation agent (146) have been expanded. Nicolaou et al. (147) have demonstrated the use of PS–selenium resins able to perform many of the Se-promoted organic transformations common in synthesis. Sylvain et al. (148) have reported the preparation of PS-supported nitroacetate as an intermediate for novel chemical reactions. A PS-supported silyl triflate was reported by Hu and Porco to be capable of supporting enolizable allyl esters as silyl ketene acetals (149). A PS-supported cyclopentadienyl (Cp) phosphazine was prepared by Minutolo and Katzenellenbogen (150) and used as a stable, safe reagent to prepare Cp–tricarbonyl rhenium complexes. PS-supported chiral lithium amides were shown by Majewski et al. (151) to be as efficient as their soluble counterparts in the enantioselective deprotonation of ketones followed by aldol additions. PS-bound electrophilic di(acyloxy)halogenates and diazidohalogenates were prepared by Kirsching et al. and used to 1,2-functionalize glycals (152). Supported Horner-Emmons reagents on a high loading polymer resulting from ring-opening methatesis polymerization (ROMPGEL) were shown to be extremely effective by Barrett et al. (153).

These and many other reported supported reagents are applicable to combinatorial synthetic protocols in solution, and some examples have recently appeared in the literature. For example, Amberlite ion-exchange resin has been used for the construction of an array of aryl/heteroaryl ethers from phenols and alkyl bromides in solution (154) to cyclize and to purify solution arrays of hydroxy quinolinones (155) and to cyclize arrays of oxazoles from aromatic aldehydes and an isocyanide (156). PS-supported HOBt has been employed to to build amide arrays (157, 158). Another array of aryl ethers was prepared from a number of phenols and halides using a PS-supported bicyclic guanidine as a strong base (159). Poly(4-vinylpyridine) was used by Chen (160) to promote the coupling of acyl and sulfonyl chlorides with N-nucleophiles in the synthesis of amide arrays of cathepsin D inhibitors. Polystyrene-bound carbodiimide EDC has been used for the preparation of benzoxazines from anthranilates and isocyanates (161), and the same supported reagent was also employed in the synthesis of an array of acylsulfonamides from benzoic acid and sulfonamides (162). Ley's polymer-supported perruthenate together with a supported Mukayama reagent (163) was successfully employed in the synthesis of a number of pyrazoles from benzyl alcohols and enol ethers; together with a polymer-bound Wittig reagent (164) in the synthesis of an array of β-aminoalcohols from benzyl alcohols and amines; and together with supported borohydride in the synthesis of several alkaloids (165, 166). Ley reported also the use of polymer-supported (diacetoxyiodo) benzene (PSDIB) as a reusable oxidation reagent (167) for solution-phase library synthesis, and the use of supported pyridinium bromide perbromide (PS-PBP) together with supported acids and bases to provide arrys of substituted benzofurans (168), of benzodioxanes and thiazoles (169). An array of aryl ethers was made using PS-supported triphenyl phosphine in a Mitsunobu reaction (170). Arrays of cyanohydrins and highly functionalized amines from PS-supported trimethyl silyl cyanide (171) and PS-supported benzotriazoles (172), respectively, have also been described. Habermann et al. (173) have used supported DMAP, cyanoborohydride, pyridinium bromide perbromide, and various supported scavengers (see Section 8.4.6) in a complex, six-step synthesis to prepare a piperidine–thiomorpholine library without any purification step.

A somewhat neglected property of supported reagents is their mutual *isolation* in heterogeneous reactions, which allows multistep one-pot reaction schemes to be performed on a substrate in solution by *simultaneously* adding several supported reagents, which in turn react sequentially with only one of the intermediates. Rebek (174) was the first to validate this hypothesis, and subsequently Parlow (175) reported the three-step synthesis of the pyrazole array depicted in Fig. 8.26. In this example, benzylic alcohol **8.44** in solution was treated with three supported reagents. In the first step (a_1), it was oxidized to ketone **8.45** by poly(4-vinylpyridinium dichromate), which was then α-brominated to **8.46** with Amberlyst A-26-supported perbromide (step a_2). Finally, the bromine of **8.46** was displaced by 4-chloro-1-methyl-5-trifluoromethyl Amberlite IRA-900 (step a_3, Fig. 8.26) to give pyrazole **8.47** in a satisfactory 48% overall yield and good purity after filtration of the resins and evaporation of the solvent. The concept of multiple-support isolation could be exploited in combinatorial chemistry because of its attractive features, and efforts to find suitable multistep, diverse transformations coupled with several supported reagents will surely appear in the literature.

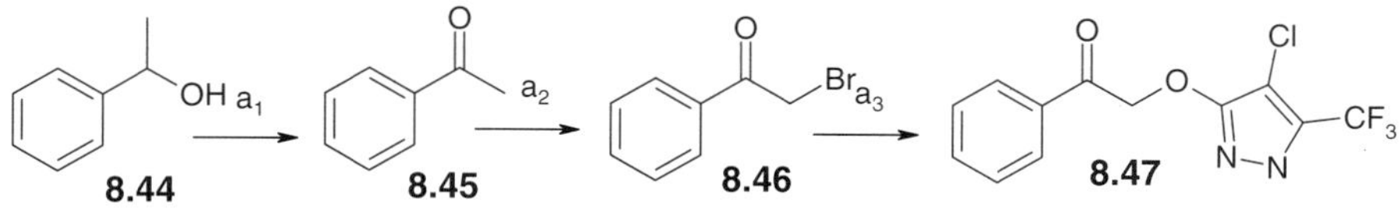

8.44 8.45 8.46 8.47

a: cyclohexane, 65°C, supported poly(4-vinylpyridinium dichromate) (oxidation a_1); supported perbromide on Amberlyst A-26 (bromination a_2); supported pyrazole on Amberlite IRA-900 (alkylation a_3).

Figure 8.26 Solid-phase site isolation: one-pot, multiple transformation of the benzylic alcohol **8.44** to the pyrazole **8.47**.

8.4.3 An Example: Synthesis and Purification of an Amine Library

Ley et al. (176) have recently reported the synthesis of a 96-member library **L10** of secondary amines (Fig. 8.27) and the assessment for its further N-functionalization via the seven-member sulfonamide array **L11** (Fig. 8.28). The synthetic scheme used employed the sets of benzylic alcohols (**M1**, 12 representatives, Fig. 8.27), amines (**M2**, 8 representatives, Fig. 8.27), and arylsulfonyl chlorides (**M3**, 7 representatives, Fig. 8.26) as monomers and is reported in Figs. 8.27 and 8.28. The authors aimed to fully automate this chemistry using an SP synthesizer and to avoid any purification step apart from filtering off the supported reagents and washing. The commercially available supported reagent **8.50** and the easily prepared reagents **8.48** and **8.49** were selected (Figs. 8.27 and 8.28).

Stock solutions of 12 monomers **M1** in dry toluene (0.052 *M*) were prepared, and 5 mL (0.26 mmol) of each were distributed by the robotic dispenser into 12 reactors. Supported **8.48** (200 mg, 0.26 mmol) was added, the reaction was stirred at 80 °C for 2 h, and after a typical SP work-up/purification protocol, each of the 12 collected organic phases containing the aldehydes was divided into eight portions. Each portion (theoretically 32 μmol) was dispensed by the robotic arm into a 96-well reaction block containing the supported reagent **8.49** (35 mg, 35 μmol). Stock solutions of eight monomers **M2** in methanol were prepared and aliquoted into the reaction block (32 μmol into each well) to obtain all of the possible combinations of final products, and the block was agitated at rt for 72 h. After the work-up/purification protocol, the organic solutions were recovered in a 96-well plate and evaporated in a centrifuge to afford the array **L10**. The analytical characterization by HPLC-MS confirmed 9 compounds with <30% purity, 28 compounds with 30–60% purity, and 59 compounds with >60% purity. Only two electron-rich monomers, $\mathbf{M_{1,8}}$ and $\mathbf{M_{2,6}}$ (Fig. 8.29), gave moderate yields of final amines, while the average quality of the library made by the unoptimized synthetic protocol was good. The two-step, polymer-assisted library

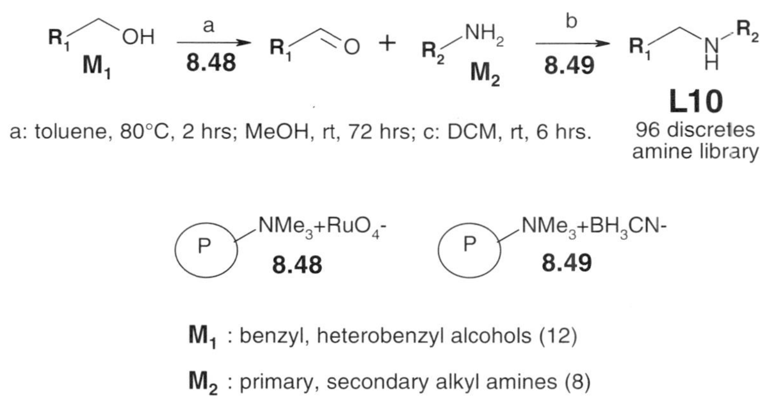

Figure 8.27 Synthesis and purification of the solution-phase, discrete amine library **L10** using polymer-supported reagents **8.48** and **8.49**.

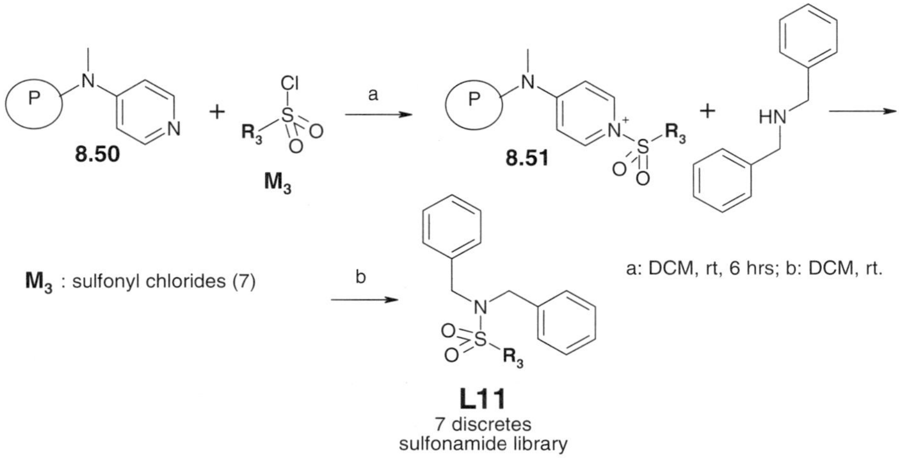

Figure 8.28 Synthesis and purification of the solution-phase, discrete sulfonamide library **L11** using the polymer-supported reagent **8.50**.

synthesis in solution was easily automated, and an expansion of the library size was proposed by the authors.

The same chemical route was further exploited by dispensing monomer set **M3** (0.4 mmol) into seven reactors and treating it with PS-supported **8.50** (0.4 mmol) to give in situ polymer-bound sulfonylpyridinium chlorides **8.51**. After stirring at rt for 30 min, stock solutions of dibenzylamine (theoretically 0.25 mmol), prepared as in **L10** from benzyl alcohol and benzylamine with >90% yield and purity, were added and stirred at rt (TLC monitoring) for 30–60 min (Fig. 8.28). After the usual work-up and purification, reasonably pure sulfonamides **L11** (>90%, HPLC-MS and NMR) were recovered.

8.4.4 Supported Catalysts and Ligands

The use of solid-supported catalysts or ligands in organic synthesis is also widespread on both a laboratory and an industrial scale, and these species have been extensively reviewed (135, 177, 178). The preparation of cheap, reusable solid-supported catalysts with high turnover numbers and similar catalytic properties to their homogeneous counterparts has proved to be a challenging goal, because supported catalysts often exhibit less effective catalytic properties. The reduced accessibility to inner catalytical sites in a support, the vicinity of catalytic sites in high-loading supports, and the disruption of symmetry in the catalyst by introduction of a bond with the solid support have all been advocated as possible explanations of the lower reactivities and/or specificities often encountered with these species. However, the advantages of recoverable, stable, solid-phase catalysts have stimulated research despite the drawbacks mentioned above, and some significant contributions have been reported over the last

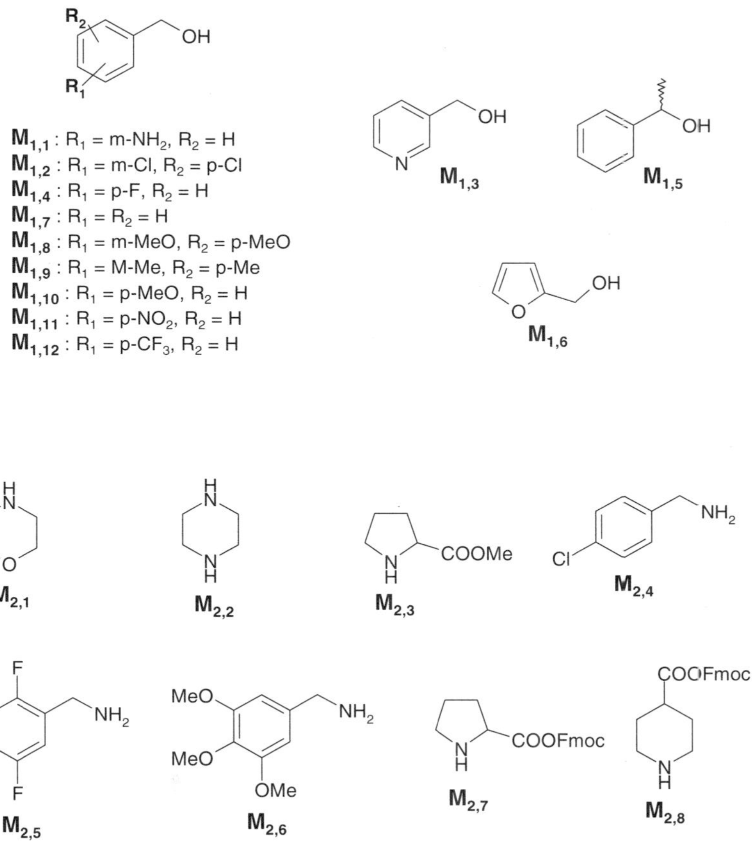

Figure 8.29 Synthesis and purification of the solution-phase, discrete amine library **L10** using polymer-supported reagents **8.48** and **8.49**: structure of the monomer sets **M_1–M_2**.

few years. Ion-exchange resin-supported lanthanides have been shown by Yu et al. (179) to be effective catalysts for a wide variety of reactions. Seebach and co-workers (180, 181) has reported a polymer-bound TADDOL (α,α,α′,α′-tetraaryl-1,3-dioxolane-4,5-dimethanol) prepared from dendritic building blocks that compared favorably in reactivity with soluble TADDOLs in the stereoselective addition of diethylzinc to benzaldehyde. A second C_2-symmetric catalyst has also been described for the same reaction (182), and Sung et al. (183) have studied the influence of various supports on the efficiency in asymmetric catalysis for supported ephedrine and camphor derivatives in the same reaction. A ROMPGEL-supported catalyst (184) for the same reaction had similar performances with its soluble counterpart in terms of efficiency and enantioselectivity. Silica-supported phenolates have been found to be efficient catalysts for Michael reactions (185). Nagayama (186) have reported a safer and cleaner PS-microencapsulated version of osmium tetraoxide. Kobayashi has reviewed the use of supported rare earth catalysis in organic synthesis (187). A copolymer between

N-phenylmaleimide and an organotin chloride was reported by Chemin et al. (188) as a suitable supported source of tin hydride catalyst with reduced tin leaking. Uozumi et al. (189), Uozumi and Watanabe (190), and Zhang and Allen (191) have supported respectively onto Tentagel, Argogel and Deloxan THP II resins several palladium catalysts that performed successfully in C–C coupling reactions; a PS-supported chiral π-allylPd catalyst for the allylation of amines with a high stability and a high turnover has also been reported recently (192). A PS-supported, recyclable ruthenium Grubbs catalyst was reported by Ahmed et al. (193) as being better than its soluble counterpart in terms of stability and lack of Ru contamination for reaction products. A PS-supported piperidine was reported as an efficient catalyst for the Knoevenagel condensation by Simpson et al. (194), with significantly reduced contamination, compatibility and side reaction problems when compared with free piperidine. Two silica-supported copper (I) complexes of imines were reported by Clark et al. (195) as smooth and effective catalysts for radical atom-transfer polymerization reactions with a variety of substrates. A comprehensive study on supported Al-based catalysts for the Diels–Alder reaction was reported by Altava et al. (196), which explored several supports and drew interesting conclusions regarding efficiency and stereoselectivity as related to the polymeric matrix. None of these examples have been applied to combinatorial library synthesis so far; two examples of successful combinatorial applications of polymer-supported catalysts are reported below.

Kobayashi and Nagayama has described a supported scandium reagent **8.52** (Fig. 8.30) as an effective catalyst for the preparation of a quinoline library **L12** in solution (Fig. 8.28, top) (197) and also for three-component reactions involving aldehydes, amines, and silyl nucleophiles (198) in the preparation of two small arrays of β-amino ketones and esters (**L13**, Fig. 8.30, bottom) or nitriles (**L14**, Fig. 8.30, bottom). The supported catalyst showed a high efficiency and turnover number and provided the products with good yields and purities.

Another recent example by Peukert and Jacobsen (199) took advantage of the first polymer supported Jacobsen's catalyst **8.53** (Fig. 8.31) comparable with the soluble catalyst in asymmetric epoxidation and its full characterization (200, 201). The supported catalyst, prepared from the activated carbonate of hydroxymethyl PS and from a soluble phenolic catalyst (201), was used to catalyze the opening of racemic alkyl epoxides (**M1**, Fig. 8.31) with substituted phenols and yielded the 50-member aryloxy alcohol library **L15** with good enantiomeric purity (average >90%, never below 80% e.e.). **8.53** was also used to produce the chiral intermediate monomer set **M3** (Fig. 8.31) which was used to make two 50-member chiral libraries **L16** (1,4-diaryloxy 2-propanols) and **L17** (3-aryloxy-2-hydroxy propanamines) with excellent enantiomeric excess following the straightforward synthetic schemes reported in Fig. 8.31.

8.4.5 Supported Scavengers

In 1996, Kaldor et al. (202) introduced a new application for functionalized resins in solution-phase library synthesis. In this method, the SP was used as a chemoselective reagent toward the excess of reagent in solution (needed to drive the reaction to completion) which was scavenged by the support and gave clean product in solution

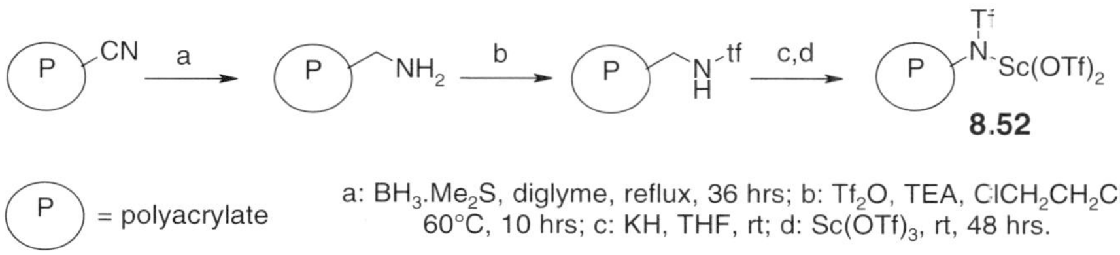

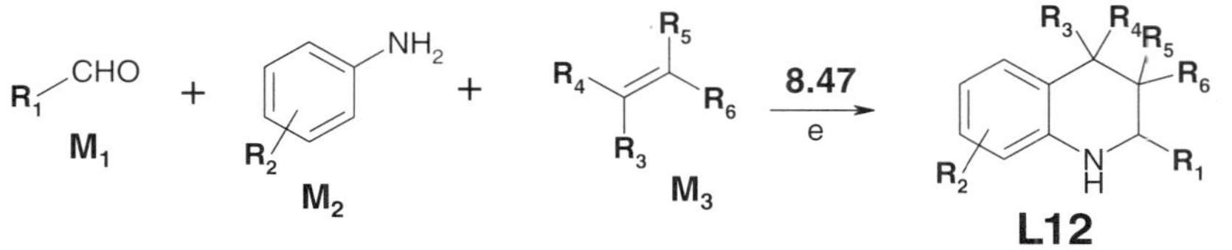

15 discretes
tetrahydroquinoline library

e: DCM/MeCN, 40°C, 15 hrs.

M_1 : aliphatic and aromatic aldehydes (8)
M_2 : anilines (3)
M_3 : alkenes, alkynes (6)

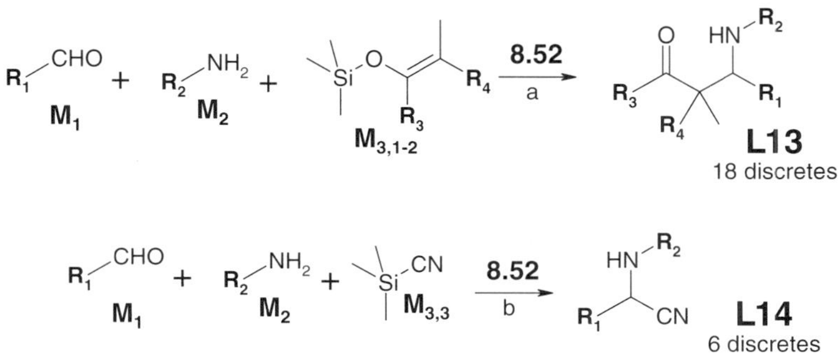

a: $MgSO_4$, DCM/MeCN, rt, 19 hrs; b: DCM/MeCN, rt, 19 hrs.

M_1 : aliphatic and aromatic aldehydes (6)
M_2 : anilines (4)
M_3 : silyl nucleophiles (3)

Figure 8.30 Synthesis and purification of the solution-phase, discrete libraries **L12–L14** using polymer-supported catalyst **8.52**.

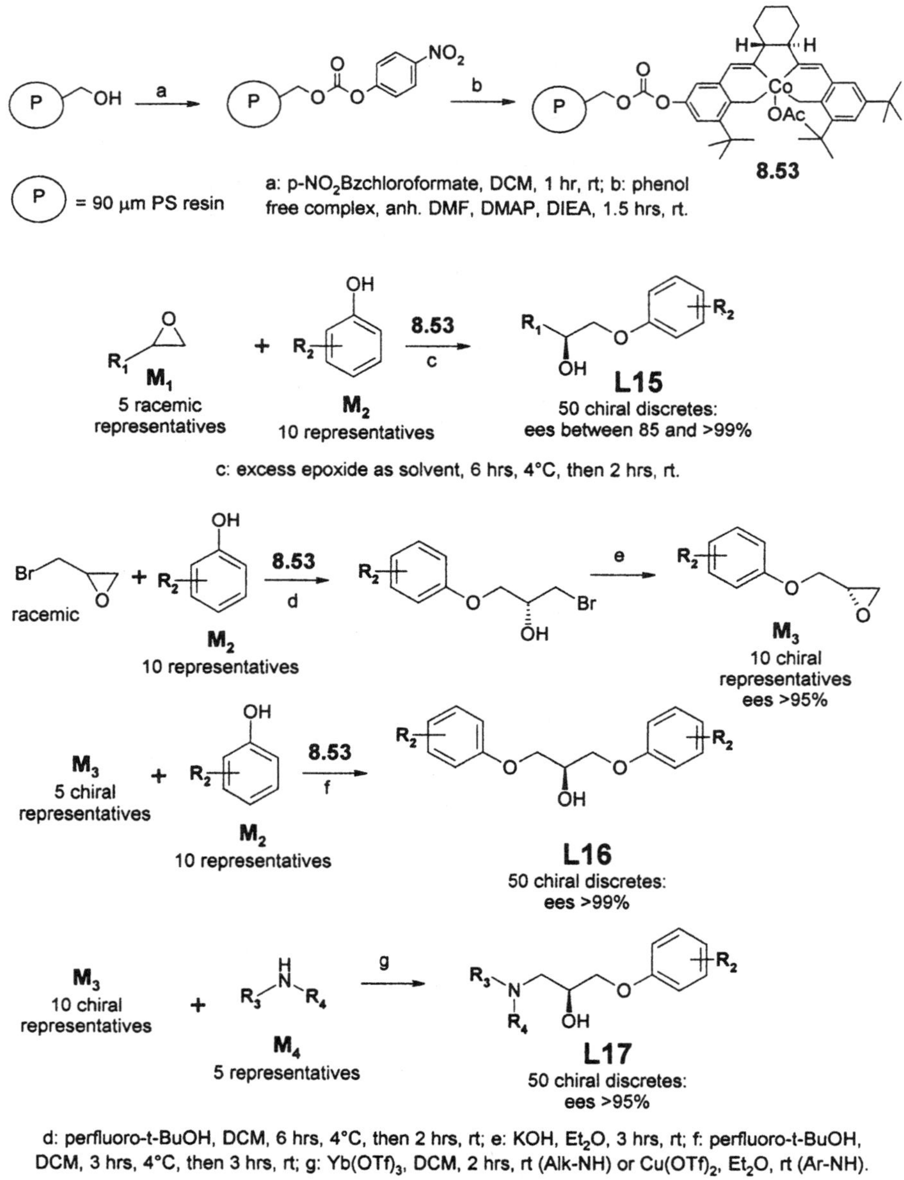

Figure 8.31 Synthesis and purification of the solution-phase discrete libraries **L15–L17** using the polymer supported catalyst **8.53**.

after filtration. Since this first report, the use of supported scavengers to sequester excess reagents, soluble impurities, or by-products through different mechanisms, described generically in Fig. 8.31, has become popular.

If the excess reagent or the reaction by-product is an acid (**AH**, Fig. 8.32, path A) or a base (**B**, Fig. 8.32, path B), addition of a complementary basic or acid support removes the soluble acid or base, respectively, leaving the product **P** in solution after filtration and washing of the support. Anionic or cationic ion-exchange resins (203–205), PS-based supports bearing basic functions (204–209), and other solid phases (210, 211) have been used to obtain good-quality intermediates, providing that these intermediates lacked a similar polar group that could bind to the supported scavenger.

Polar groups can be deliberately introduced into reagents used in combinatorial transformation to aid their eventual removal, as in path C, Fig. 8.32. In this case, the reagent **I** is tagged with a polar group **AH** and a complementary polar group on the support scavenges **I–AH** and leaves product **P** in solution. Examples of polar tagged reagents (Fig. 8.33, top) include basic carbodiimides for peptide coupling (**8.54**) (212), acidic thiols for the removal of trityls (**8.55**) (213), and protected acidic phosphines or dicarboxylates for Mitsunobu reactions (**8.56** and **8.57**) (214).

The reagent to be sequestered (**CX**, Fig. 8.32, path D) may contain a chemical functionality that is not present in the reaction product and can be reacted with a suitable reactive, supported scavenger, forming a covalent bond and again leaving pure product **P** in solution. Many supported covalent scavengers have appeared recently in the literature (202, 206, 207, 215–222), and some structures (**8.58–8.63**, Fig. 8.33,

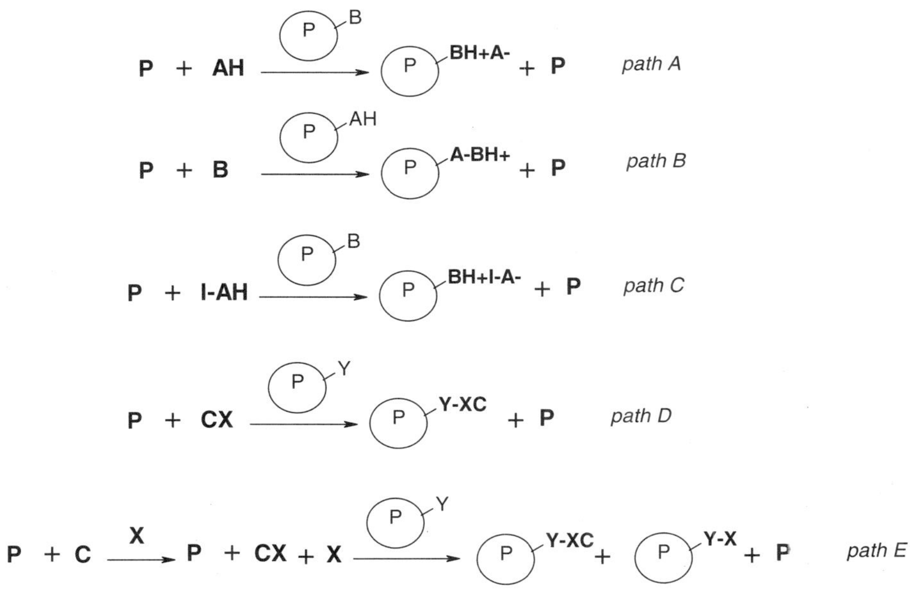

Figure 8.32 Solid-supported scavengers: basic principles.

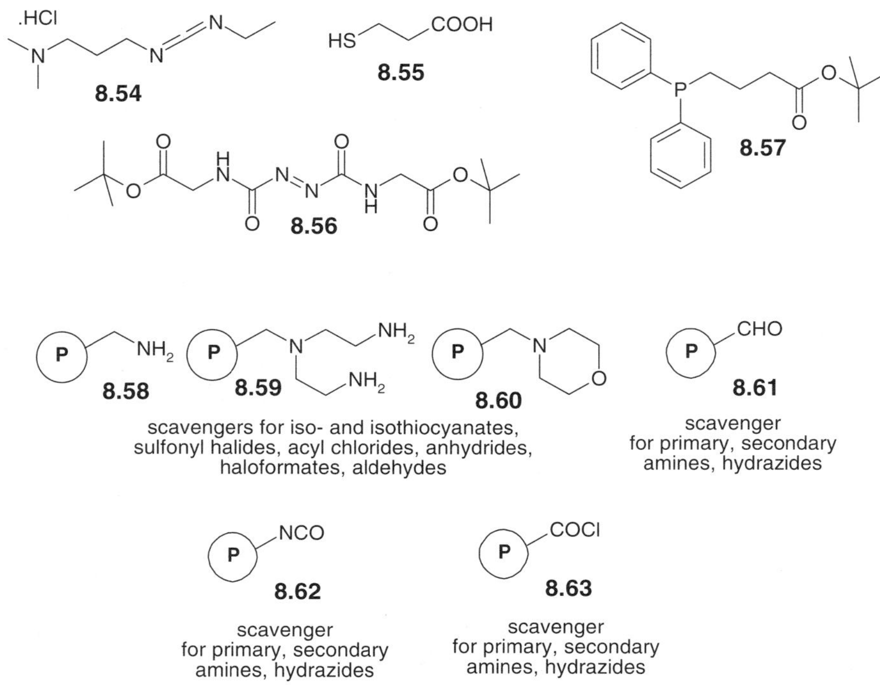

Figure 8.33 Solid-supported scavengers: tagging reagents **8.54–8.57** and supported, covalent scavengers **8.58–8.63**.

bottom) are reported along with classes of molecules that they scavenge. The limitations of the method usually depend on the absence of the scavenged chemical and similarly reactive groups in the product. However, even mixtures of compounds with similar reactivity have been successfully purified using carefully selected scavenging conditions; for example, the reaction of supported **8.62** with an excess of amine in solution did not affect the hydroxyl-containing reaction product **8.64** (Fig. 8.34) (202), which was recovered pure.

A recent approach involving in situ tagging of excess reagents or by-products that would otherwise be difficult to scavenge (Fig. 8.32, path E) has been developed. In this method the reagent **C** is tagged with a so-called sequestration-enabling reagent **X** to give **CX**, which is scavenged by a supported reagent that interacts both with **CX** and with the excess of **X** itself, leaving product **P** in solution. Examples of such reagents are provided in Fig. 8.35. The diamine **8.65** and the amino phenol **8.66** were both used to tag isocyanates, and the tagged molecules were removed with acidic and basic supports, respectively (212). The acidic thiourea **8.67** was used as a tag for α-bromoketones, which were removed by basic supports. Similarly, the sulfonyliso-cyanate **8.68** (213) and the anhydride **8.69** (223) were used to tag anilines, alcohols,

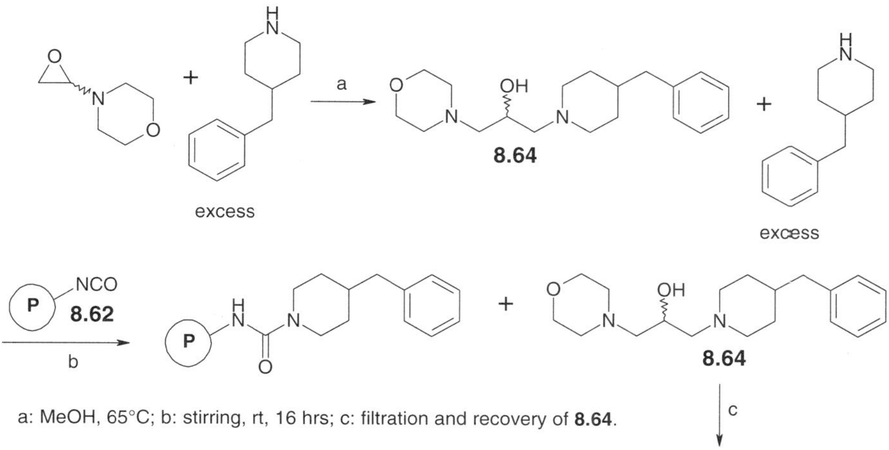

Figure 8.34 Solid-supported scavengers: purification of the β-amino alcohol **8.64** using the covalent scavenger **8.62**.

and amines, and the tagged molecules were removed from the reaction mixtures with basic supports.

The use of tetrabenzo[a,c,g,i]fluorene (Tbf) derivatives as tagging reagents, such as **8.70** (224), was reported for acids; each Tbf-tagged reagent is purified by specific adsorption on charcoal, typically stirring for 20 min in an ice bath, and subsequent washing with lipophilic solvents to recover pure intermediates, typically stirring with toluene aliquots at 40 °C until no more UV adsorbance is detected in the toluene aliquots.

These strategies provide high-quality products after simple filtration and washing, and a suitable scavenger could be applied to each combinatorial step in solution. A common feature of all the supported scavengers is their high loading (typically 1–4 mmol/g), so that their cost is significantly lower than classical PS-based SP supports with typical loadings of 0.2–0.6 mmol/g. Most of these scavengers are commercially available, and their use in combinatorial laboratories is becoming very popular. In general, they allow parallel, automated clean-up of discrete libraries and have a significantly higher throughput than a serial HPLC/MS separation. They provide chemical flexibility and typically require less assessment than a normal library of discretes prepared on SP, especially when small–medium size arrays are considered. They require only low-technology, solution-phase equipment and are becoming the preferred choice of purification method for this format of library. The area has been reviewed recently (139, 213, 225–230).

8.4.6 An Example: Synthesis and Purification of a Benzoxazinone Library

The synthesis of a 35-member, solution-phase benzoxazinone library **L18** using a five-step synthetic scheme with intermediate purifications that employed supported

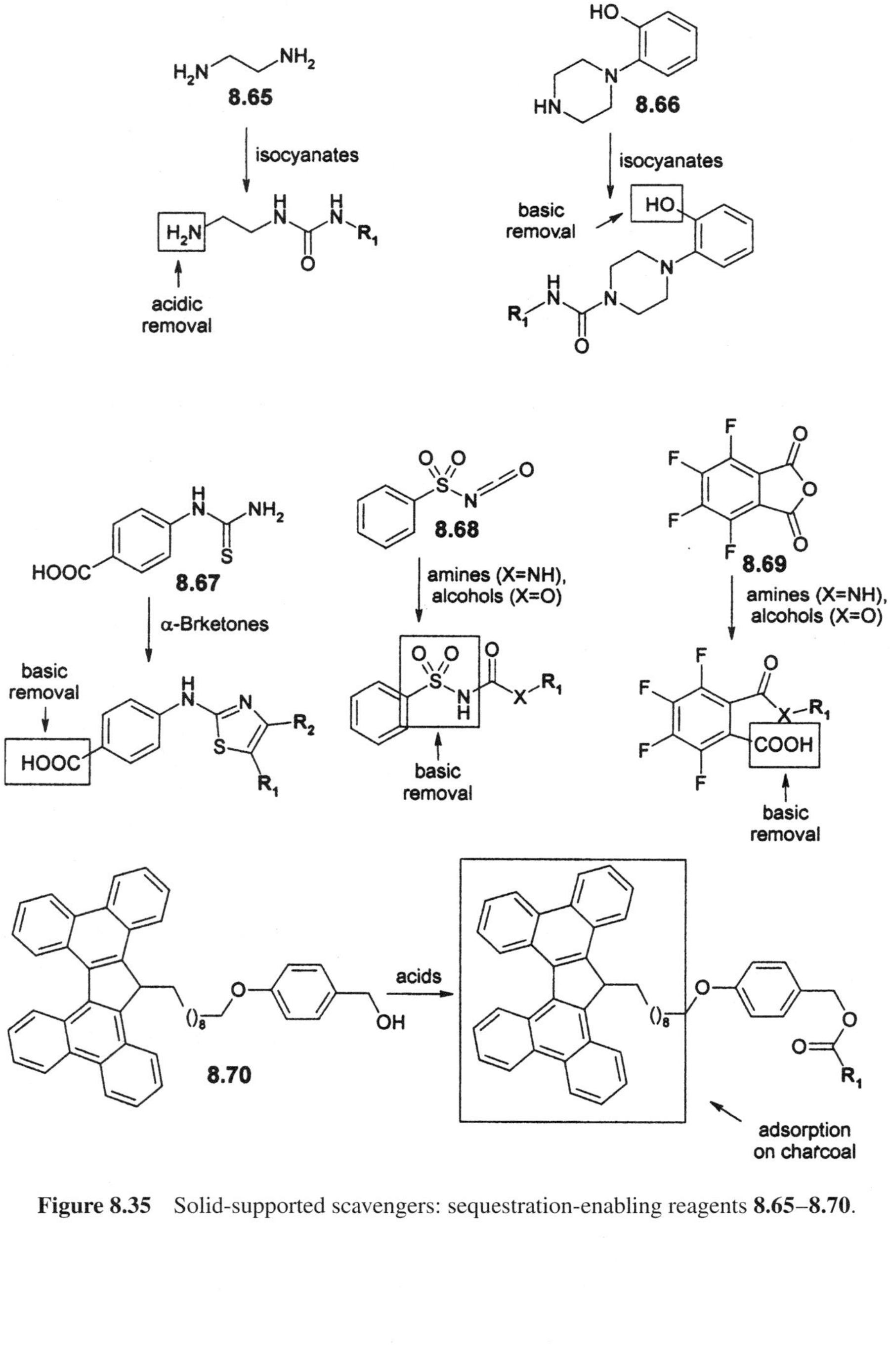

Figure 8.35 Solid-supported scavengers: sequestration-enabling reagents **8.65–8.70**.

reagents, scavengers, and sequestration-enabling reagents has been described recently (231). The synthetic scheme (Fig. 8.36, top) started from the key intermediate **8.71** (225) and used the monomer sets **M1** (five electrophiles including isocyanates, acyl chlorides, and chloroformates, Fig. 8.36) and **M2** (seven electrophiles including

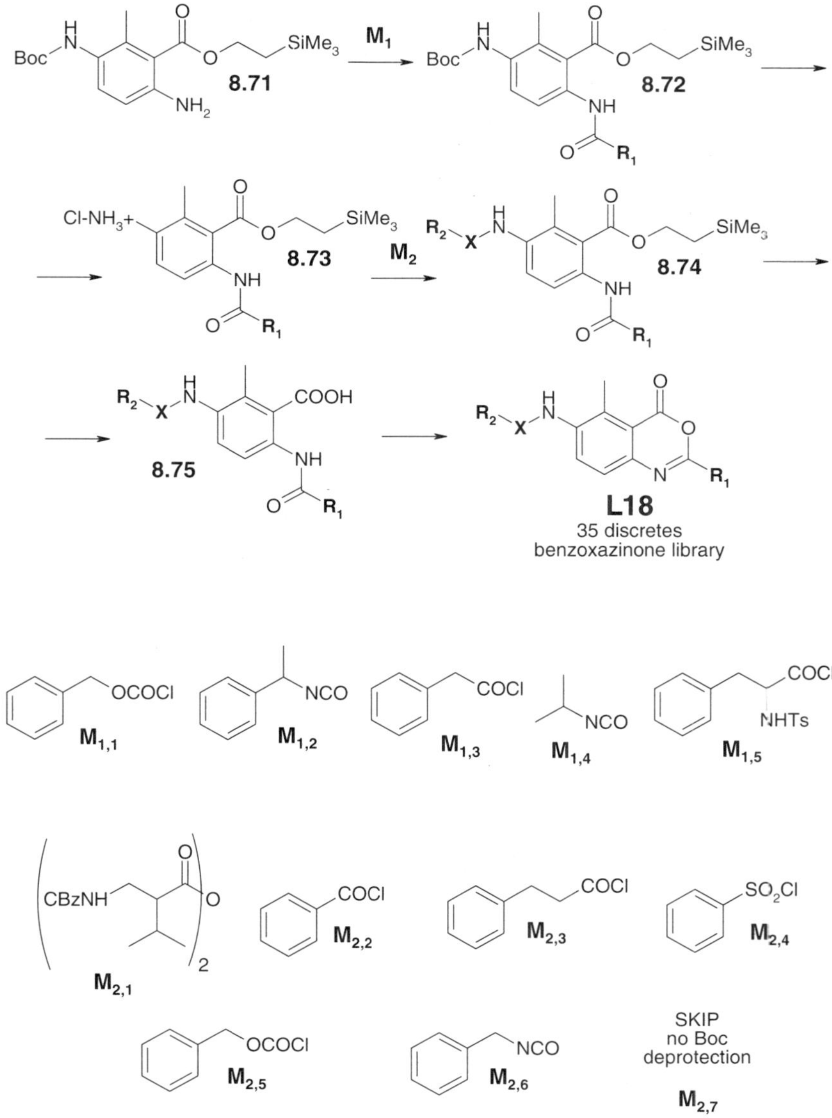

Figure 8.36 Synthesis of the solution-phase, discrete benzoxazinone library **L18** and structure of the monomer sets **M1**–**M2**.

isocyanates, acyl chlorides, sulfonyl chlorides, anhydrides, chloroformates, and the skip-Boc monomer **M2,7**, Fig. 8.36) to produce **L18**.

The first reaction (step a, Fig. 8.36) was driven to completion through the use of a 20% excess of **M1** and by pyridine as a base with acyl chlorides or chloroformates **M1,1**, **M1,3**, and **M1,5** (step a, Fig. 8.37). This protocol involved addition of the polyamine resin **8.58** to scavenge the HCl and also the excess **M1,1**, **M1,3**, and **M1,5** upon completion of the reaction, thus freeing the volatile pyridine. After filtration and washing of the resin, evaporation of the solvents gave pure **8.72**. Even two equivalents of isocyanates **M1,2** and **M1,4** and commercially available supported DMAP **8.50** in a more aggressive reaction protocol could not drive the reaction to completion (step b, Fig. 8.36). The same supported amine **8.58** was used after the second protocol, but first, an excess of the sequestration-enabling reagent **8.69** was added to transform unreacted **8.71** into an acid-tagged molecule. Supported **8.58** then scavenged the acid-tagged **8.71**, the excess **M1,2** and **M1,4**, and the excess of **8.69**, leaving products **8.72** (Fig. 8.37) after filtration and washing.

Boc deprotection (step b, Fig. 8.36) was studied, but even the optimized conditions (step a, Fig. 8.38) produced a slight amount of deprotected carboxylic acid impurity together with hydrochlorides **8.73**. Addition of **8.58** removed this side product (step b, Fig. 8.38). An aliquot of solution corresponding to monomer **M2,7** (skip Boc) was not deprotected (step c, Fig. 8.38) and was carried forward to the ester deprotection. The following coupling with an excess **M2** (step c, Fig. 8.36) was performed in pyridine (step d, Fig. 8.38), and addition of **8.58** scavenged the HCl and the excess of **M2**. The usual filtration, washing, and evaporation of solvent and pyridine gave pure products **8.74** (step e, Fig. 8.38).

The crude reaction mixture from the deprotection of the TMSE (trimethyl silyl ethanol) ester (as in step d, Fig. 8.36) contained the desired products as tetrabutylammonium salts as well as excess TBAF and the volatile trimethylsilyl fluoride and ethylene (step a, Fig. 8.39). This crude was purified by addition of a mixture of a sulfonic acid ion-exchange resin **8.76** and the same resin in its calcium sulfonate form **8.77** (206), which freed the products **8.75** from their ammonium salt and precipitated the excess fluoride as insoluble calcium fluoride (step b, Fig. 8.39). Filtration, washing, and evaporation gave pure acids **8.75**, which were finally cyclized (as in step e, Fig. 8.36) using supported EDC **8.78**, functioning here as both a supported reagent and a scavenger of unreacted **8.75** through formation of the corresponding adducts (step c, Fig. 8.39). Filtration and washing produced the array **L18**, whose components were characterized by HPLC. Table 8.4 (see Fig. 8.35 for monomer structures) indicates the overall good quality of the discrete library, considering the five-step reaction scheme.

8.4.7 An Example: Synthesis and Purification of a Library of Trisubstituted Amines

The SPS of a small array of trisubstituted amines **L19** has been recently reported (232) building on work described in two previous papers (233, 234). These reported Michael addition onto an acrylate resin **8.79**, reductive amination of amine **8.80**, quaternization

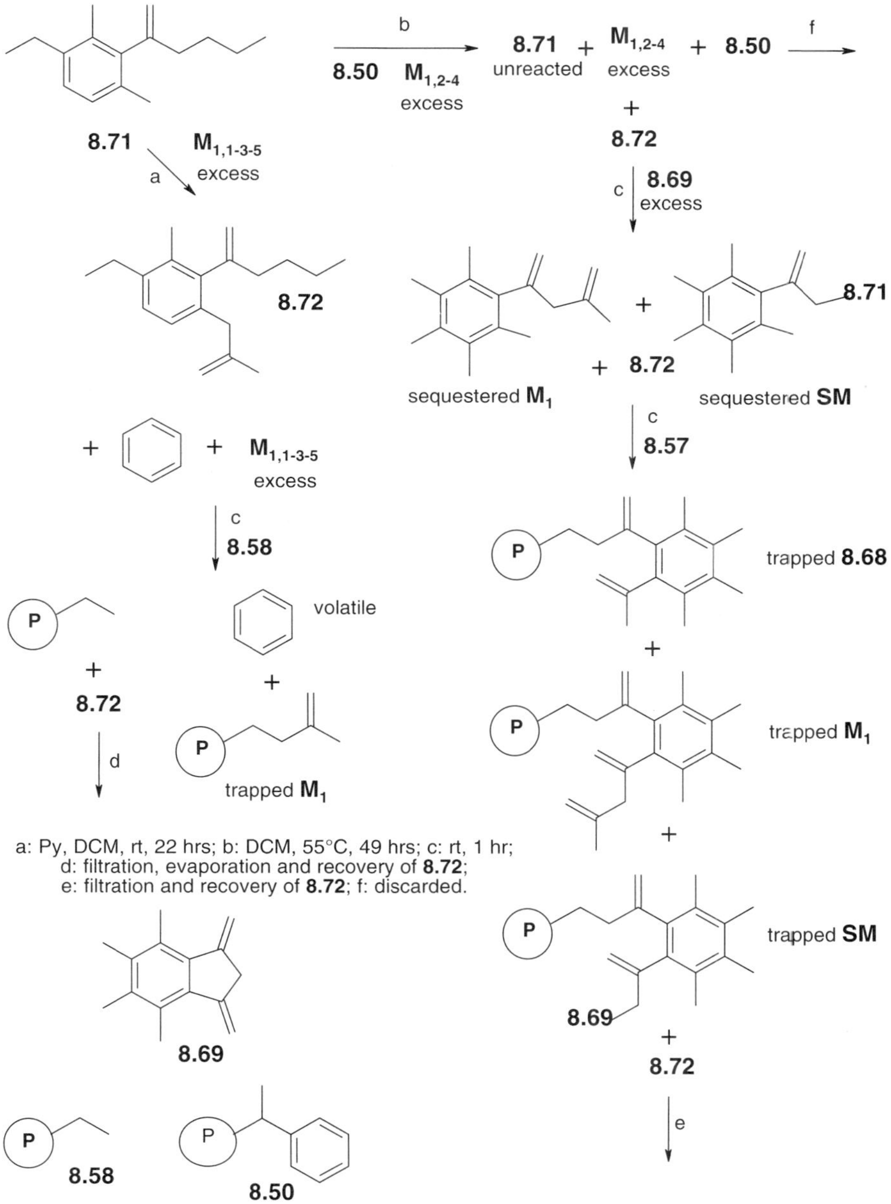

Figure 8.37 Synthesis and purification of the solution-phase discrete benzoxazinone library **L16**: amine acylation and synthesis of **8.72**.

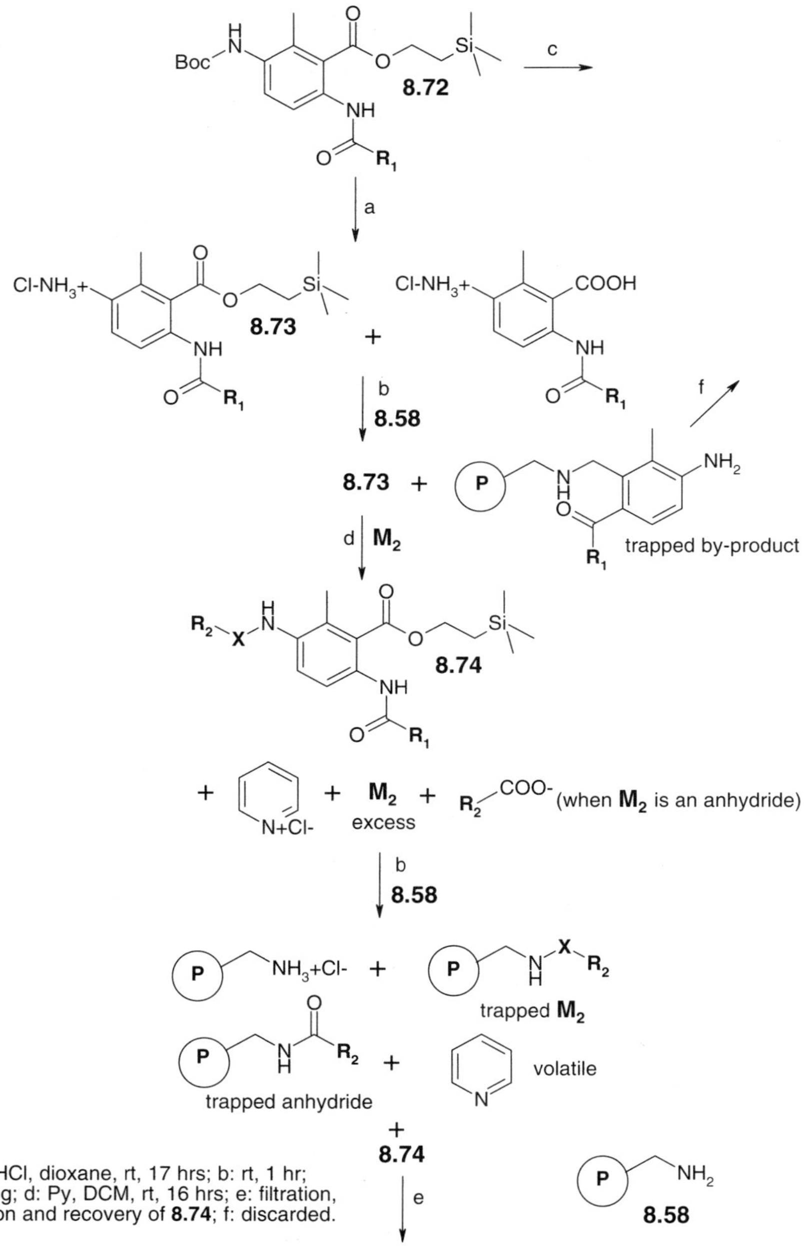

Figure 8.38 Synthesis and purification of the solution-phase discrete benzoxazinone library **L18**: Boc deprotection and amine functionalization, synthesis of **8.74**.

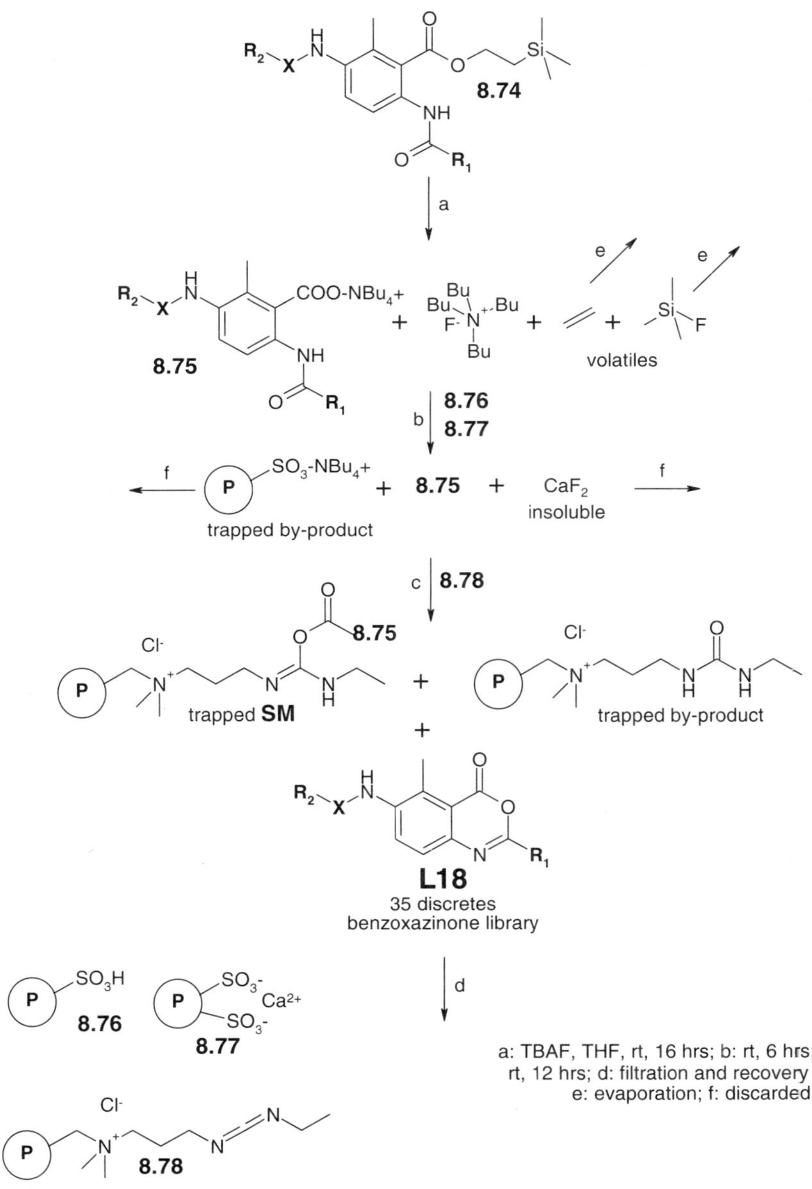

Figure 8.39 Synthesis and purification of the solution-phase discrete benzoxazinone library **L18**: silyl ester deprotection and cylcative cleavage to **L18**.

TABLE 8.4 Yields and purity of the benzoxazinone discrete library L18

	$M_{2,1}$	$M_{2,2}$	$M_{2,3}$	$M_{2,4}$	$M_{2,5}$	$M_{2,6}$	$M_{2,7}$
$M_{1,1}$	70[a]/92[b]	48/93	60/95	71/89	47/92	41/89	82/91
$M_{1,2}$	36/48	23/69	32/74	27/75	31/73	15/69	62/76
$M_{1,3}$	47/93	72/95	60/98	66/97	75/99	44/82	89/99
$M_{1,4}$	20/31	23/75	16/73	13/78	22/68	16/30	29/87
$M_{1,5}$	65/88	76/93	55/91	42/83	85/84	49/86	94/90

[a]Yields based on mass recovery.
[b]Purities, based on HPLC analysis.

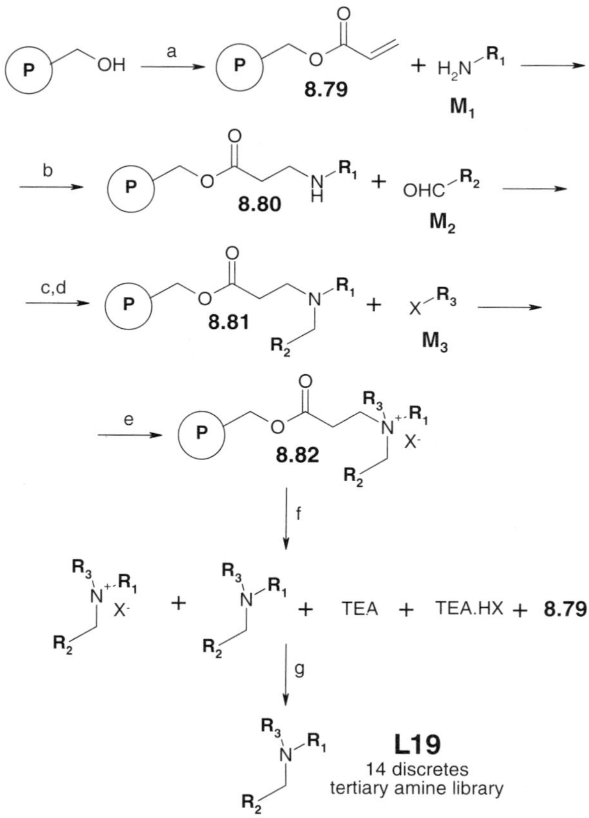

a: CH_2=CHCOCl, DIPEA, DCM, rt, 4 hrs; b: DMF, rt, 18 hrs; c: AcOH, DMF, rt, 30'; d: $NaBH(OAc)_3$, rt, 18 hrs; e: DMF, rt, 18 hrs; f: TEA, DMF, rt, 18 hrs; g: filtration, liquid-liquid extractions and SPE.

Figure 8.40 Synthesis of the SP, discrete tertiary amine library **L19**.

of the tertiary amine **8.81**, then cleavage of the ammonium salt **8.82** via Hoffman elimination, producing the library **L19** (Fig. 8.40).

An excess of TEA in the elimination step (step f, Fig. 8.40) produced significant amounts of salts as contaminants, requiring either aqueous extractions (233) or SP biphasic columns (234) for its complete sequestration from the reaction mixture. Stoichiometric amounts of TEA resulted in lower quantities of salts, but the same work-up and purification protocols were nevertheless required. The authors then used catalytic amounts of TEA in the presence of an ion-exchange resin to regenerate the soluble base, scavenging the acid moiety and allowing the total cleavage of **8.82**. This reduced the purification procedures to simple filtration, washing of the support, and evaporation of the solvent.

An intriguing finding came from the same reaction protocol with the ion-exchange resin **8.83** but in the absence of TEA. In this case, pure **L19** was isolated, albeit with longer reaction times (Fig. 8.41, top). The mechanism reported in Fig. 8.41 was postulated. Thus, if the resin-bound ammonium salt **8.82** underwent a thermal elimination (step a), even in extremely small amounts, the ammonium salt released would be converted to the amine by the ion-exchange resin **8.83** (*initiation*, Fig. 8.41). The amine would then replace TEA as a soluble base, as depicted in the *propagation* steps (Fig. 8.41). To compare the efficiency of different eliminations, three cleavage protocols were tried in parallel on **8.82**: stoichiometric TEA (**A**), an ion-exchange resin (Amberlite IRA-95, **B**), and a deprotected Rink amide resin (**C**). The yields are summarized in Table 8.5 (NMR purities were always 95% or better) and show that **B** was the best protocol, while PS–Rink resin was less satisfactory than either **A** or **B**. The structures of the library individuals **L19a–n** are reported in Fig. 8.42.

Even with SP cleavage reactions, supported reagents may be useful to directly perform or to promote clean cleavage processes for the production of high-quality libraries. Such an intriguing finding should be exploited further by other groups.

8.4.8 Resin Capture

A concept that has been repeatedly expressed is the complementarity, rather than the mutual exclusivity, of the various library formats; this also includes solution- and solid-phase libraries that are more or less suited to each planned synthetic scheme. It may well be, though, that the synthetic strategy contains steps that are more suited to homogeneous reactions and others that may benefit from the advantages of heterogeneous reactions. A hybrid approach in which the first steps are performed on SP and postcleavage combinatorial modifications in solution are used to produce the final library could be envisaged, but to date, this possibility remains largely unexploited. The opposite strategy, that is, the synthesis of advanced library intermediates in solution, their attachment onto SP, further SP transformations, and final cleavage to give a library, has been first reported by Armstrong and co-workers (235–237). This approach was named *resin capture* to stress the key event during the hybrid library synthesis. Another group has recently reported similar capture on the SP of advanced library intermediates, trapping bicyclic anhydrides derived from ring opening cross-metathesis with amine supports (238).

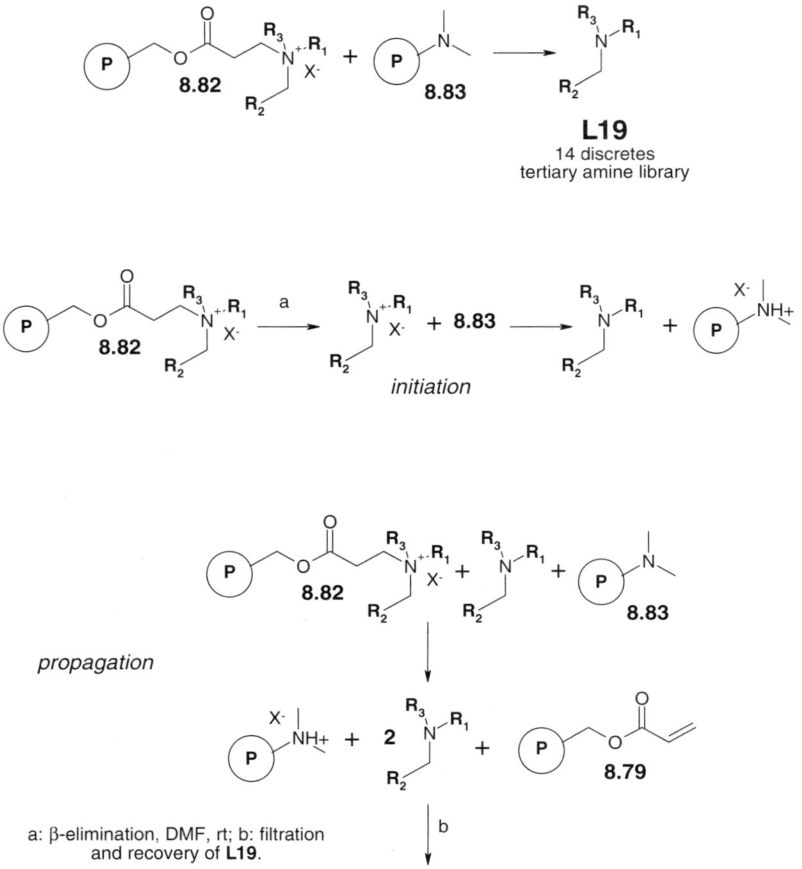

Figure 8.41 Synthesis of the SP, discrete tertiary amine library **L19**: proposed cleavage mechanism.

TABLE 8.5 Hoffman β-Elimination on 8.82: Experimental Protocols

	a[a]	b	c	d	e	f	g	h	i	j	k	l	m	n
A[b]	68	65	59	31	58	10	32	25	51	58	43	50	51	41
B[c]	70	69	61	45	65	12	37	26	40	63	55	52	58	51
C[d]	35	38	33	35	26	5	16	23	14	56	24	41	37	57

[a]**L19a–n**.

[b]Stoichiometric TEA.

[c]Ion-exchange resin.

[d]Deprotected Rink amide resin.

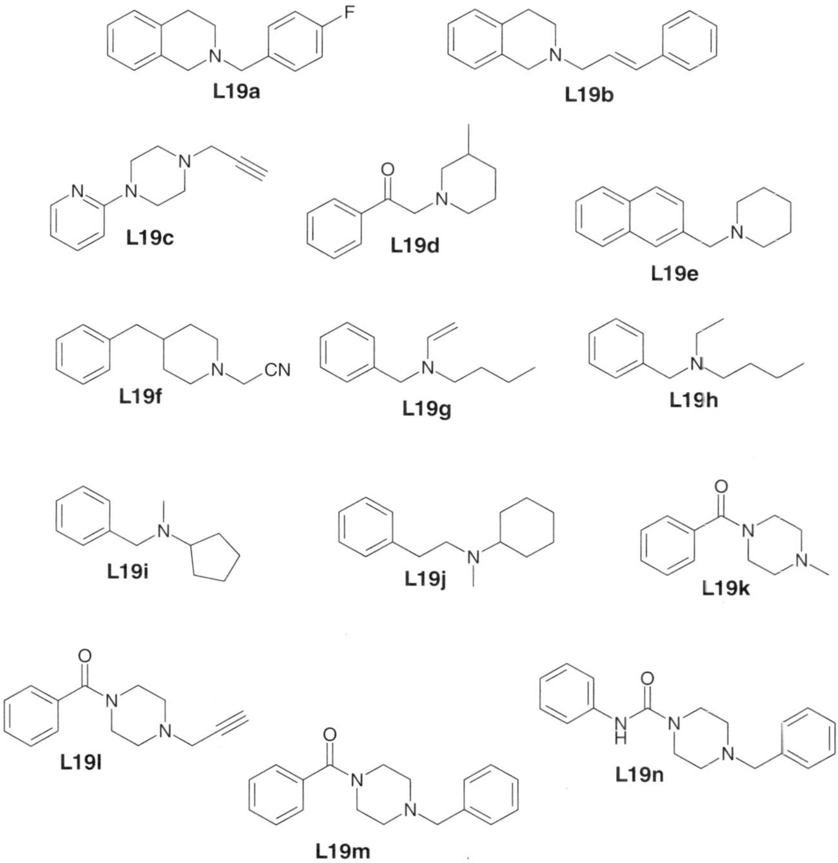

Figure 8.42 Synthesis of the SP, discrete tertiary amine library **L19**: structure of the library individuals **L19a–n**.

The synthesis of hybrid solution/solid-phase small arrays of 1,2,3-thiadiazoles, **L20** and **L21**, has also been described (239). An SP strategy was first set up (Fig. 8.43) based on commercially available PS–sulfonyl hydrazide resin that was reacted with a small set of commercially available ketones (step a) to form intermediate resin-bound sulfonyl hydrazones **8.84**, then submitted to cyclative cleavage to form the thiadiazoles **L22** (step b) with good yields (84–99%) and GC purities (94–99%).

A more diverse set of noncommercially available ketones was prepared in solution (Fig. 8.44) by the reaction of a Weinreb methoxymethyl amide (the 4-bromophenyl derivative **8.85** in the example) with a Grignard reagent (six representatives), purifying the crude mixtures using supported tosic acid **8.86** (step a). The solutions containing the ketones were transferred to other vessels containing PS–sulfonyl hydrazide resin, and the same reaction protocol described in Fig. 8.43 produced the final thiadiazoles

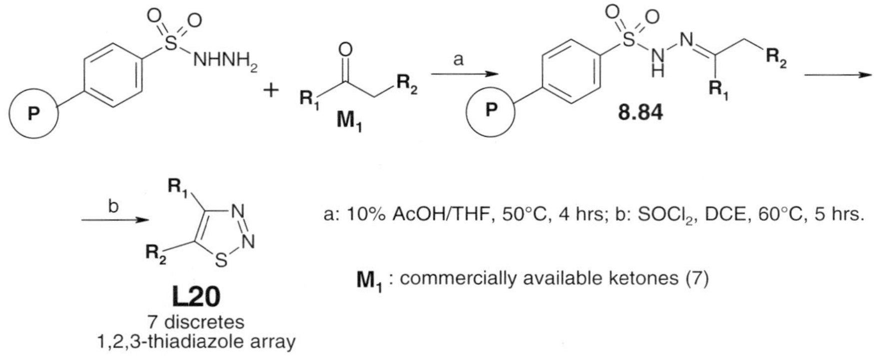

Figure 8.43 Synthesis of the SP, discrete thiadiazole library **L20**.

L21 with high yields (59–98%) and purities as measured by GC (94–100%). A hybrid approach would have benefited from further SP transformations before cleavage, and the authors proved the concept (Fig. 8.45) starting from a noncommercial ketone that was processed as in Fig. 8.44 to give a resin-bound sulfonyl hydrazone **8.88**. This compound was submitted to a Stille coupling with an aryl stannane to give **8.89**, which was submitted to cyclative cleavage to give **8.90** with 71% yield and 85% purity measured by GC (Fig. 8.45).

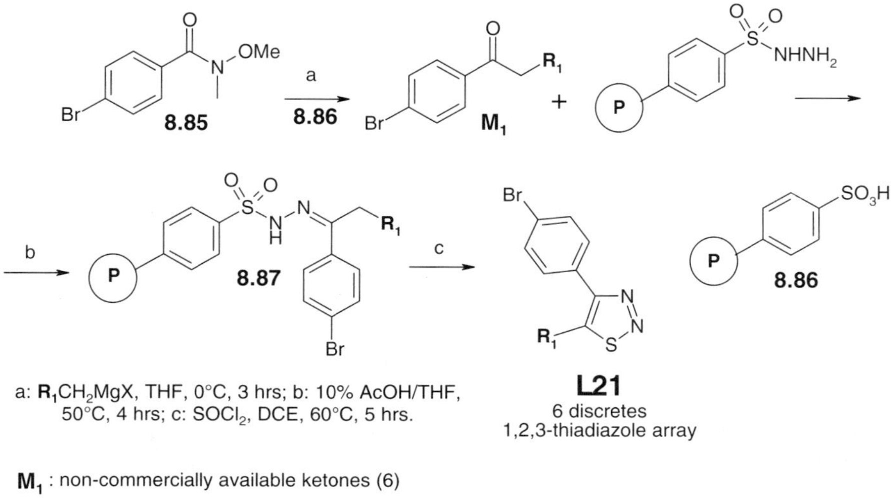

Figure 8.44 Resin capture: synthesis and purification of the discrete thiadiazole library **L21** using the supported reagent **8.86**.

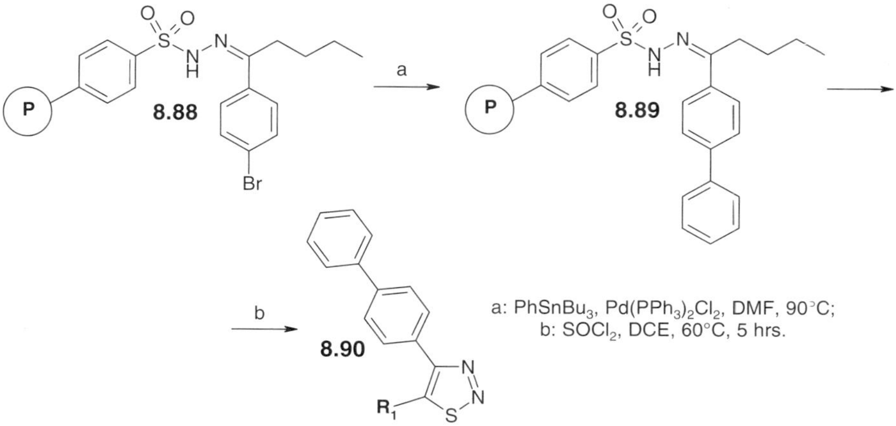

Figure 8.45 Synthesis and purification of the hybrid, discrete thiadiazole, library **L21** using resin capture: SP modification of the captured intermediate **8.88** to give the biaryl **8.90**.

8.4.9 A Comprehensive Example: Synthesis and Purification of Libraries of Tri- and Tetrasubstituted Pyrroles

Caldarelli et al. (240) have recently reported a five-step synthesis of substituted pyrrole libraries **L22** and **L23** using solid-supported reagents and scavengers. The synthesis involved oxidation of benzyl alcohols $\mathbf{M_1}$ to aldehydes (step a, Fig. 8.46). Henry reaction of aldehydes **8.91** with nitroalkanes $\mathbf{M_2}$ (step b), and acylation and elimination of nitroalcohols **8.93** (steps c and d) to give the nitrostyrenes **8.94**, which were subjected to 1,3-dipolar cycloaddition with an isocyanoacetate (step e) to give the pyrroles **8.95.** N-alkylation of these pyrroles with alkyl halides (step f) and final "library-from-a-library" hydrolysis/decarboxylation of **L22** gave a library of trisubstituted pyrroles **L23** (step g, Fig. 8.46).

The oxidation of the alcohol was performed with supported perruthenate (**8.48**, Fig. 8.46) to produce clean aldehydes **8.91** after filtration. The Henry reaction was performed in the presence of a commercially available, supported strong base **8.92** and an excess of volatile nitroalkenes, giving clean nitroalcohols **8.93** after filtration and evaporation. The reaction mixtures from the trifluoroacetylation/elimination steps were purified with commercially available amino PS resin **8.58** to scavenge the trifluoroacetates and with acidic ion-exchange resin **8.76** to remove the TEA-derived salts. Again, the nitrostyrenes **8.96** were obtained cleanly after filtration and evaporation. Cycloaddition with isocyanoacetate was promoted by the commercially available, supported guanidine base **8.95**, while the subsequent N-alkylation of the pyrroles **8.96** was performed with an excess of halide in the presence of the commercially available, supported phosphazene **8.97**. In this case, the excess halide was removed by treatment with supported **8.58**, and filtra-

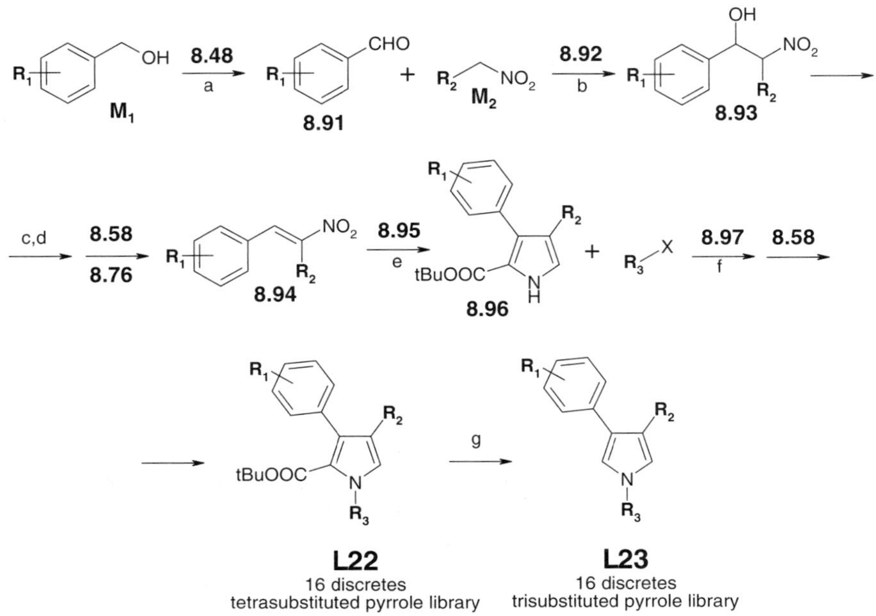

a: DCM, rt; b: neat, rt; c: TFAA, DCM, rt; d: TEA, DCM, rt; e: tBuOcoCH$_2$NC, THF/iPrOH, rt; f: DCM, rt; g: TFA, DCM, rt.

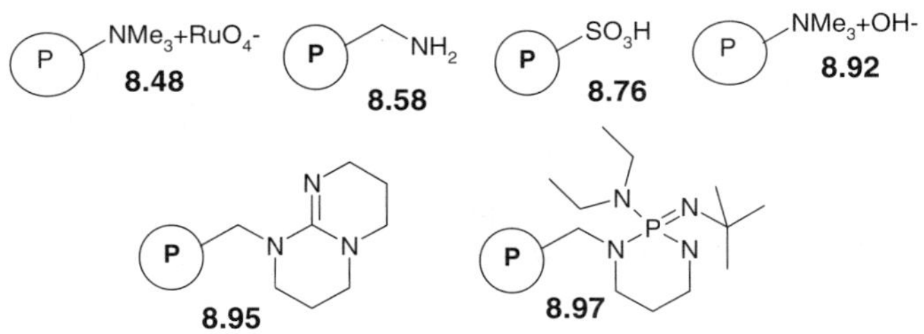

Figure 8.46 Synthesis and purification of the solution-phase, discrete pyrrole libraries **L22** and **L23** using the supported reagents **8.48, 8.58, 8.76, 8.92, 8.95, 8.97**.

tion/evaporation provided the library **L22** (Fig. 8.46). Further elaboration of **L22** to **L23** did not require any purification to remove volatile TFA and *t*-butyl alcohol. The yields of all the steps were almost quantitative (with the exception of the acetylation/elimination reactions steps c and d; Fig. 8.46) producing overall yields ranging from 30 to 80%. The purities of intermediates were checked after each step by LC-MS and were always >90%, more often being >95%.

8.5 SOLUBLE SUPPORTS IN SOLUTION-PHASE COMBINATORIAL SYNTHESIS

8.5.1 General Considerations

Solid supports for SP synthesis are an important option for the synthesis of libraries of both discretes and pools. We have extensively reviewed their properties, but we have also highlighted some critical issues such as the different reaction kinetics in heterogeneous reactions due to accessibility of inner reaction sites, the difficulty of monitoring reactions on SP, and the need for suitable expertise and instrumentation.

A few years ago, Han et al. (241) introduced the so-called *liquid-phase combinatorial chemistry* by applying the concept of liquid-phase synthesis (242) to combinatorial synthesis through the use of a soluble, linear homopolymer as a support for library synthesis. A support of this type must be soluble in most organic solvents and possibly should impart solubility onto the molecules attached to it to allow reactions to be performed and monitored in the homogeneous phase. The support must also be prone to crystallization, or at least precipitation, under well-defined experimental conditions to allow the handling of a solid material during the work-up and purification stages. Such a support would have the advantages of both solution- and solid-phase combinatorial synthesis in the different phases of a synthesis. The original work (241) used PEG-derived polymers that possessed the properties mentioned above and reported the synthesis of both a pool library of peptides and a small array of arylsulfonamides.

Since this first report, many other contributions have appeared, some of which have focused on the synthesis of libraries, whereas others describe the characterization of new polymers. The main classes of soluble supports will now be examined through examples, and their properties and use for specific applications will be reviewed.

8.5.2 PEG-Based Polymers

The use of polyethylene glycol (PEG) supports in natural oligomer synthesis has been known for some time (242–244). Polyethylene glycol polymers with average MW in the range 2000–20,000 daltons are identified as PEGs, either as bifunctional polymers with two free hydroxyls (PEG) or as monoethers capped with a methoxy function at one end (MeO-PEG). These polymers, the general structure of which is shown in Fig. 8.47, fulfill most of the criteria for an ideal soluble support:

- They are commercially available at reasonable cost in a wide range of average MWs and loadings.
- They have good solubility in most organic solvents and in water (see Fig. 8.47), but they are insoluble in *n*-hexane, diethyl ether, and cold ethanol, which induce their precipitation/crystallization, a property that can be used to advantage in purification protocols.
- They have excellent solubilizing properties that allow the dissolution of loaded insoluble molecules, providing that the loading is not high enough to overpower the influence of the PEG.

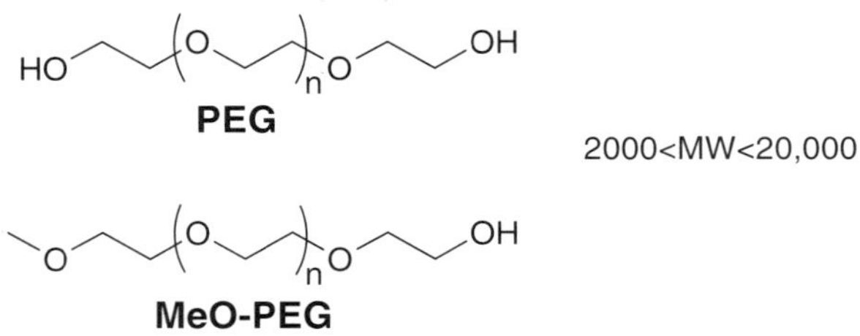

Solubility data of MeO-PEG 5000:

- Water, DCM, $CHCl_3$, DMF, Pyridine **>40%**

- MeOH, EtOH (34°C), Benzene **10%-20%**

- *Cold EtOH* ***0.1%***

- *Et_2O* ***0.01%***

Figure 8.47 Polyethylene glycol as a soluble support: structure and properties.

- Their structure does not interfere with analytical methods used to monitor the reactions (TLC of PEG-derivatized compounds can be carried out) and to characterize the reaction products (NMR signals of PEG are concentrated around 3.64 ppm, and MeO-PEG provides an internal standard at 3.38 ppm).

Soluble PEG supports have been used for different applications such as the synthesis of peptides (241, 245) and peptidomimetics (246); soluble-supported catalysts (247–249), reagents (250–253), scavengers (254), and traceless linkers (255–257); to improve purification protocols (258, 259); and high-loading PEG-derived soluble supports (260), particularly in the synthesis of arrays of small organic molecules (261–275). Several recent reviews (276–281) illustrate their usefulness but also show how additional efforts could make PEG liquid-phase combinatorial synthesis more reliable.

8.5.3 An Example: Synthesis of an Aminoimidazolinedione Array

Joon et al. (282) recently reported the synthesis of a 10-member array of 3-aminoimidazoline-2,4-diones **L24** (Fig. 8.48) using MeO-PEG-amine soluble support, the aromatic linker **8.98**, and the monomer sets $\mathbf{M_1}$ (six α-amino acid isocyanates) (283) and $\mathbf{M_2}$ (four aza α-amino acids) (247). The linker was coupled to the support via an amide bond (step a, Fig. 8.48), and deprotection of the phenolic hydroxyl (step b) gave supported **8.99**. Coupling with monomer set $\mathbf{M_1}$ (step c) and removal of the butyl ester (step b) produced **8.100**, which was coupled to monomer set $\mathbf{M_2}$ (step d) and deprotected (step b) to give **8.101**. Each deprotection step was followed by precipita-

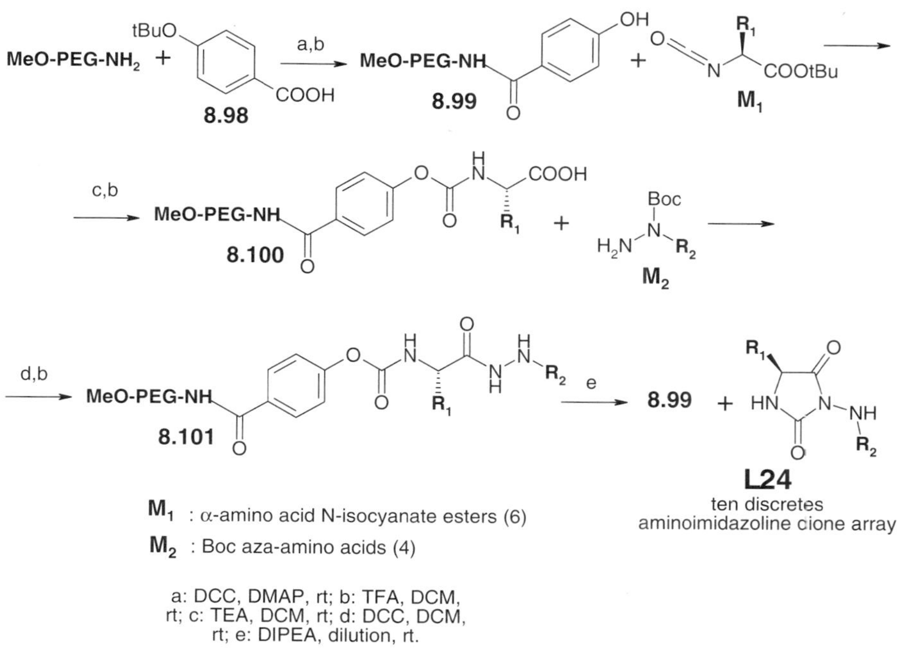

Figure 8.48 Synthesis and purification of the solution-phase discrete aminoimidazolinedione array **L24** using PEG soluble supports.

tion of the PEG-supported intermediates with diethyl ether to quickly and efficiently purify them. Monitoring of the reactions was performed by TLC and NMR spectroscopy of aliquots of the mixtures. Compounds **8.101** were easily cyclized/cleaved under weakly basic conditions (step e, Fig. 8.48) to give **L24** with yields varying from 62 to 80% and >90% purities (NMR, MS, IR). The soluble supported construct **8.99** could be recovered with good yields after the cleavage.

8.5.4 Other Soluble Polymeric Supports

While PEG-based supports are widely used for liquid-phase combinatorial chemistry, other non-PEG-based soluble polymers have also been reported for combinatorial applications. A recent review (276) contains an exhaustive list of homo- and copolymeric soluble supports used in peptide, oligonucleotide, and oligosaccharide synthesis, including combinatorial chemistry. Two of these supports have also been used for small organic molecule synthesis. Homopolymeric polyvinyl alcohol was used in conjunction with PEG for a protection/derivatization strategy in solution (284), and the copolymer between isopropylacrylamide and acrylic acid was used in the catalytic hydrogenation of a Cbz group (285).

A recent paper (286) reported the use of a commercially available soluble hybrid support prepared by copolymerization of styrene and allylic alcohol that exhibits a lipophilicity intermediate between that of PS and PEG chains. This support was grafted with a number of functional groups that were reacted with various monomer sets to produce pyrazolo[3,4-*b*]pyridines and coumarins with high yields and purities after precipitation by addition of ethanol or water. Most of the soluble support was recovered after the cleavage and retained the same efficiency in further reaction cycles. Another example is thoroughly described in the next subsection.

Further development of reliable, high-loading soluble supports with complementary solubilizing properties and with high stability to organic reaction conditions can be expected. An intriguing true combinatorial approach to the optimization of the physico-chemical properties of soluble supports has recently been reported by Gravert et al. (287) and is described in Section 11.3.2.

8.5.5 An Example: Synthesis of a Prostaglandin on a Soluble Support

Chen and Janda (288, 289) recently reported the use of non-cross-linked polystyrene (NCPS) (290) as a soluble support with complementary solubility properties with respect to PEG resins. The synthesis of a complex prostaglandin structure **8.106** (291) using the chloromethyl NCPS support **8.102**, the acid-labile THP linker, and the three synthons **8.103** (291), **8.104** (292), and **8.105** (293) is shown in Fig. 8.48.

During the complex synthetic procedure, it was necessary to use a support with a high solubility in THF even at –78 °C, in conditions in which PEG is poorly soluble (steps c, d, and f, Fig. 8.49). The extreme solubility of PEG in water would also not allow the complete removal of large quantities of salts during the aqueous removal of organometallic/inorganic salts required in steps c and f. The lipophilic, water-insoluble NCPS resin **8.102** was compatible with these requirements. Moreover, reaction conditions included solvents such as cyclohexane and benzene at temperature below 0 °C, in which NCPS is fully soluble and PEG is not and the purification protocols involved precipitation with methanol, in which NCPS is completely insoluble, in contrast to PEG.

The two single compounds prepared can be seen as a successful validation of the use of NCPS supports, which are complementary to PEG-based soluble supports. This and other similarly lipophilic supports will undoubtedly become more popular for the preparation of combinatorial libraries of small organic molecules.

8.5.6 Dendrimers as Soluble Supports

The advent of dendrimers in organic chemistry has led to many applications of these branching oligomers generated from a central core unit. Kim et al. (294) first reported their application as soluble supports in the field of combinatorial technologies, naming the approach dendrimer-supported combinatorial chemistry (DCC), then renamed by the same group combinatorial synthesis on multivalent oligomeric supports (COSMOS) (295). The elaboration of a commercially available dendrimer based on a polyamidoamine structure produced a high-loading support useful for combinatorial chemistry. For example, 7 mg of the dendrimer was equivalent in terms of loading to 100 mg of

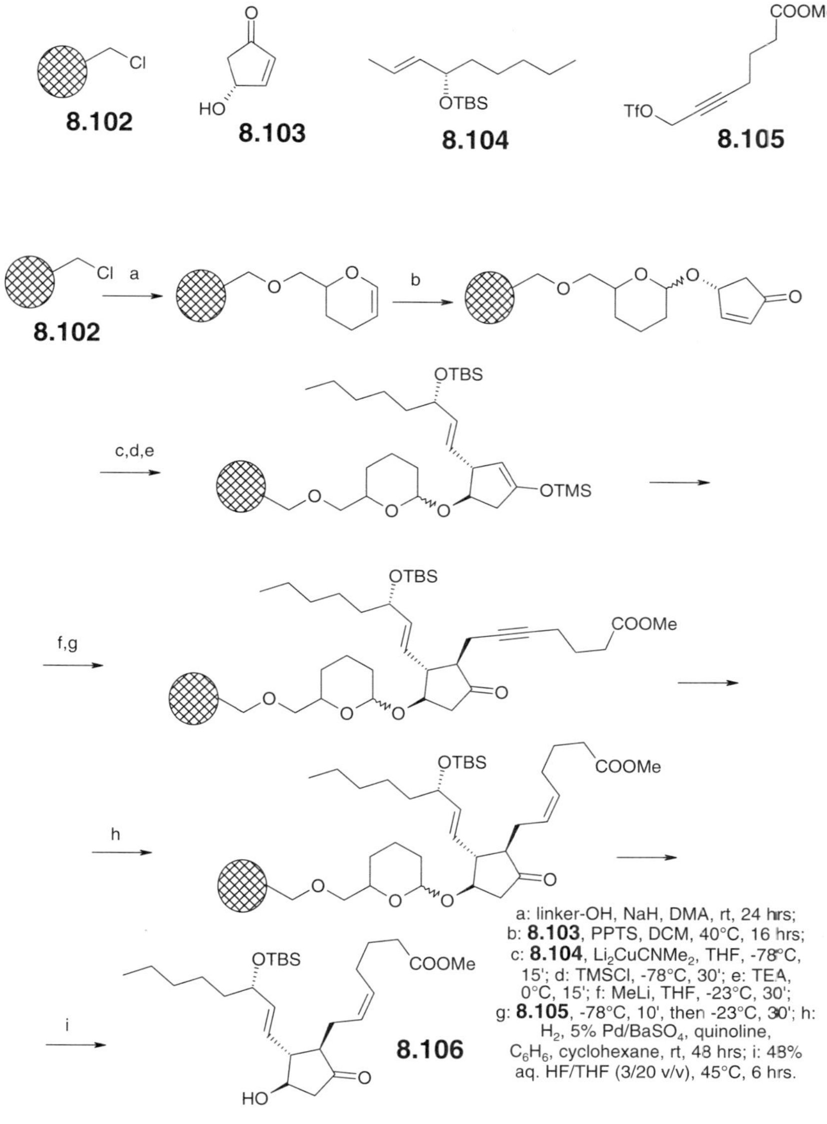

Figure 8.49 Synthesis of the prostaglandin **8.106** using a non-cross linked polystyrene (NCPS) soluble support.

a 0.23-mmol/g resin (typical capacity for a Tentagel PS). The supported intermediates were purified from reagents, salts, and by-products by size exclusion chromatography (SEC) or ultrafiltration, taking advantage of the increased MW of dendrimers supporting multiple copies of the library individuals. The compounds released after cleavage were purified by SEC, recovering the eluate solution of the small-MW library individuals and possibly recycling the high-MW dendrimers after elution from the SEC column. SEC is applicable to automated purification of small libraries, while its use for larger arrays could become somewhat rate limiting, and technological improvements would be necessary to increase its throughput.

Other reports have followed, including a paper from the same group describing the synthesis of a 48-member guanidine library (295). Gitsov et al. (296) has described the synthesis and characterization of high-loading PEG-derived dendrimers. Langer et al. (297) have reported the assembly of dendritic glycoclusters using one-pot reaction schemes starting from aliphatic polyols as core structures. Worl and Koster (298, 299) has reported the use of dendritic structures for large-scale synthesis of oligonucleotides. A hybrid approach where dendrimers are anchored onto PS resins to increase exponentially their loading has been recently presented by Mahajan et al. (300).

Cheaper, high loading hyperbranched polymers were introduced by Kantchev and Parquette and used to prepare oligosaccharides (301); their potential as solid supports is considerable. Newkome et al. have reported the combinatorial synthesis of dendritic materials with different properties (302, see also Section 11.3.2).

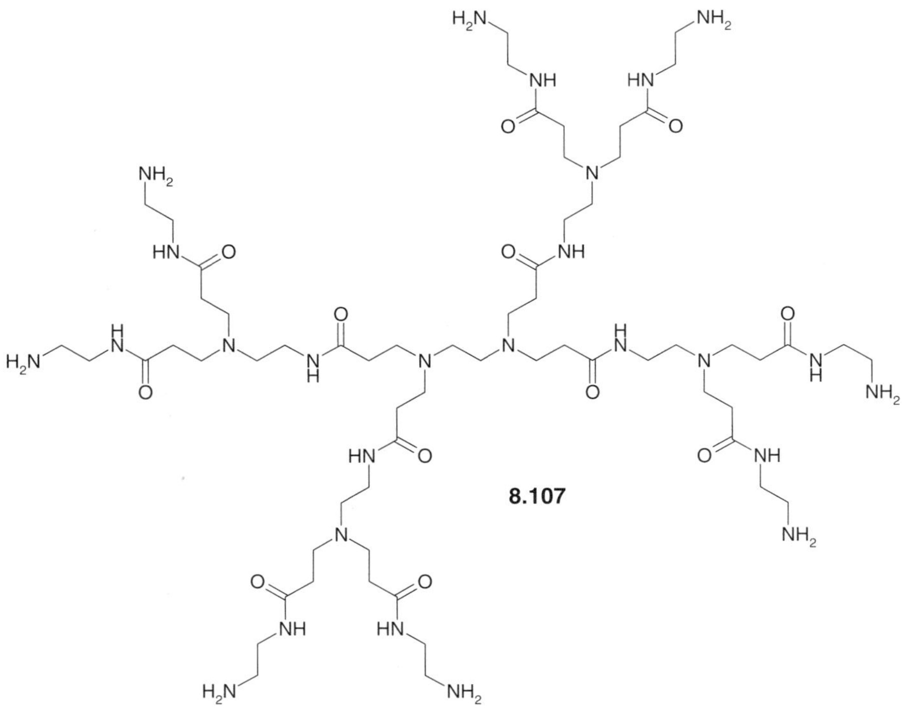

8.107

Figure 8.50 Structure of the polyamidoamine dendrimer **8.107**.

8.5.7 An Example: Synthesis of Indole Libraries on Dendrimers

Kim et al. (294) reported the synthesis of a small, 6-member array of discretes **L25** and of a small 27-member pool indole library **L26** using a commercially available, polyamide-based dendrimer support (**8.107**, Fig. 8.50) in which eight amino reaction sites per dendritic unit are present. The synthetic route for the preparation of the indole nuclei is reported in Fig. 8.51. Dendrimer **8.107** was first coupled to the base-labile

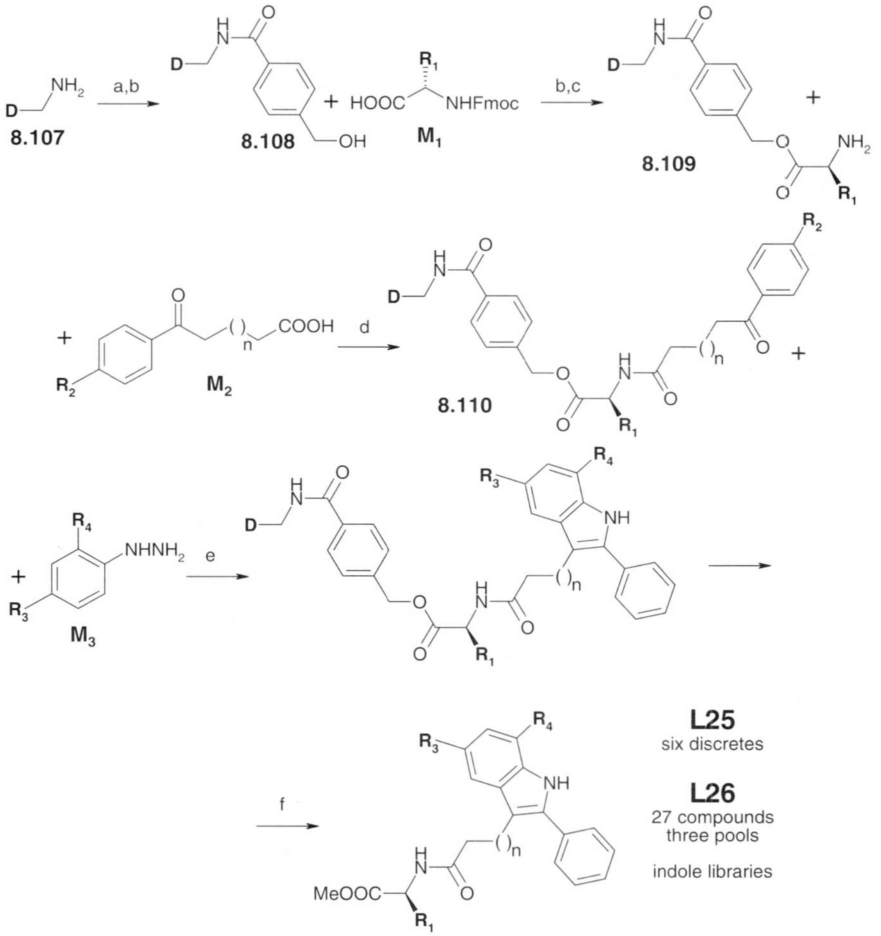

M_1 : Fmoc α-amino acids (2-**L25**, 3-**L26**)

M_2 : Subst. phenyl-γ-keto acids (1-**L25**, 3-**L26**)

M_3 : Subst. phenyl hydrazines (4-**L25**, 3-**L26**)

a: 4-HMB, EDC, DMA, rt, 18 hrs; b: EDC, DMAP, DMA, rt, 4 hrs; c: 25% piperidine, DMF, 30′; d: PyBOP, HOBt, DIPEA, DMF, rt, 3 hrs; e: $ZnCl_2$, AcOH, 70°C, 18 hrs; f: 10% TEA, MeOH, 50°C, 20 hrs.

Figure 8.51 Synthesis and purification of the discrete (**L25**) and the pool (**L26**) indole libraries using dendrimer soluble supports.

HMB (P-hydroxy methyl benzoic acid) linker (step a) to give **8.108**; then the monomer set **M1** (two α-amino acids for **L25**, three for **L26**, step b, Fig. 8.51) was coupled onto the linker followed by Fmoc deprotection (step c) to give **8.109**. The discretes were prepared in six vessels, while the pool synthesis used the mix-and-split methodology. The amine function of **8.109** was coupled to the monomer set **M2** (one ketoacid for **L25**, three for **L26**, step d) to give **8.110**; then cyclization with monomer set **M3** (four phenylhydrazines for **L25**, three for **L26**, step e) gave dendrimer-supported indoles **8.111**. Cleavage from the support produced **L25** (six discretes, Fig. 8.51) or **L26** (three pools of nine components, Fig. 8.51) as pure compounds or mixtures. SEC was used for the purification of intermediates after each synthetic step and the library members after their cleavage. In general, the discretes had a >95% purity (HPLC), while the HPLC traces of the three pools composing **L26** were in agreement with a roughly equimolar representation of each library individual.

8.6 NEW TRENDS IN SOLUTION-PHASE COMBINATORIAL SYNTHESIS

8.6.1 Dynamic Combinatorial Libraries

Synthetic organic libraries are made up of stable compounds prepared through irreversible reactions, and these compounds are screened against one or more targets using different methodologies, depending on the library format, and the active molecules are identified and further profiled to assess their usefulness. The whole process is irreversible, because a library that shows no affinity for a specific target cannot adapt its structural features to interact better with the target structure.

Dynamic interactions between a target and a molecule that lead to a reshaping of the molecule to better fit the target have been known for many years (303) and represent the basis of so-called *template-directed synthesis* (304, 305). The template acts on the macroscopic geometry of a reaction that could produce several products by shifting the equilibrium toward a single product but does not bind covalently to either the reagents or to the reaction product. Templates may act *kinetically*, operating on irreversible reactions and accelerating the formation of a product via the stabilization of its transition state. In a hypothetical example (Fig. 8.52), the reaction between **A** and **B** produces **F**, **G**, and **H** in different amounts through transition states **C**, **D**, and **E** (path A), while the template **X** binds noncovalently to the transition state **E** and leads only to the formation of **H** (path B). Other templates act *thermodynamically* when the reaction is reversible, and the noncovalent binding of the template to a specific product shifts the equilibrium toward a single product. In path C, a reversible reaction between **A** and **B** produces an equilibrium mixture of monomer **AB**, dimer **ABAB**, and cyclic dimer **cABAB**. When the template X is used, its affinity for the cyclic dimer **cABAB** shifts the equilibrium toward this compound, which is the sole reaction product (Fig. 8.52, path D). If **cABAB** is submitted to the above reversible reaction conditions without the presence of **X**, it reequilibrates, giving the same product mixture as in path C (Fig. 8.52, path E).

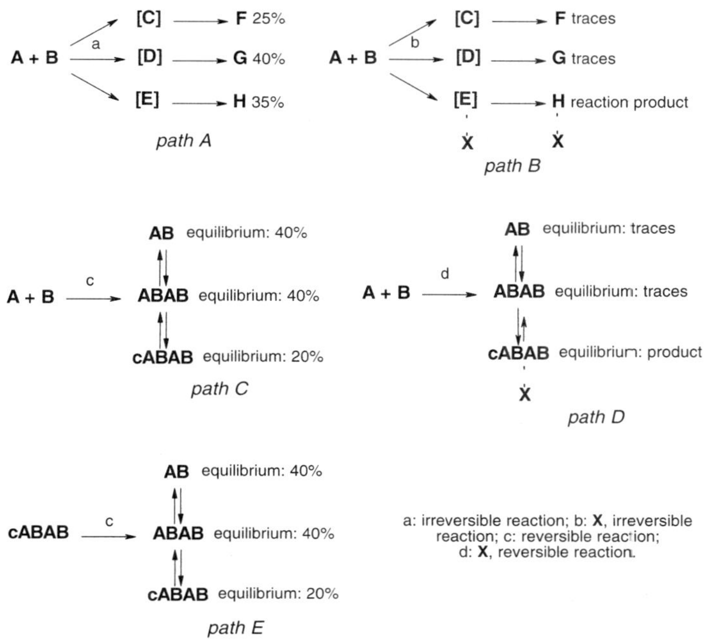

Figure 8.52 Template-assisted chemical synthesis: basic principles.

Template-directed synthesis has also been exploited for combinatorial purposes in which a reversible reaction and the use of thermodynamic templates have been used. Two different processes have been envisaged and validated, both of which consider the dynamic optimization of a receptor–ligand interaction where one of the partners is the template that drives the self-assembly of a reversible library of other partners from which the best binder for the template is selected (Fig. 8.53). If the receptor is a template, a library made using a reversible reaction is incubated with the receptor and

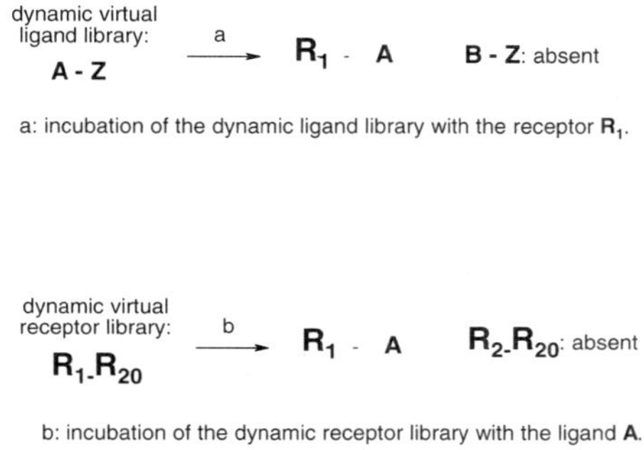

Figure 8.53 Dynamic combinatorial libraries: receptor-driven ligand library selection (top) and ligand-driven receptor library selection (bottom).

the library component(s) giving the best fit with the template is selected and produced (**A**, Fig. 8.53, top). Thus the dynamic ligand virtual library consists of a mixture of all the potential reversible combinations of the library components (**A–Z**), but only the positives (**A** in the example) are isolated and their structure is determined. If the ligand is the template, a dynamic receptor library $\mathbf{R_1}$–$\mathbf{R_{20}}$ is screened and the most active synthetic receptor ($\mathbf{R_1}$ in the example) is identified using the same approach (Fig. 8.53, bottom).

The pioneering work in this area was reported by Lehn and co-workers (306–309), both referring to ligand (imines) and to receptor libraries (barbiturate receptors and bipyridine-metal complexes-selected receptors); the principle has been further applied by other workers, for example by Eliseev and Nelen (310) to select among isomeric forms of unsaturated dicarboxylates for their affinity toward arginine receptors and to prepare a dynamic library of aryl and alkyl oximes (311); by Sakai et al. (312) to discriminate among isomeric carbohydrates as lectin binders; by Klekota and co-workers (313, 314) to select among bis(salicylaldiminato) zinc complexes for their DNA-binding efficiency; by Giger et al. (315) to prepare dynamic libraries of alkenes and by Hiraoka and Fujita (316) to select the best binders from a thermodynamic receptor library of Pd(II)-linked cages driven by the presence of 1,3,5-benzenetricarboxylic acid as a ligand; by Sanders and co-workers to make quinine macrocyclic libraries (317), to prepare reversible [2]-catenane libraries (318, 319), to produce reversible diversity by oligomerization of cinchona-based and xanthene-based building blocks (320), and to synthesize several cyclic peptidomimetic dynamic libraries (321). Six reviews (322–327) have recently covered this subject.

While the applications of dynamic libraries have so far been limited to test cases, their potential to determine and influence the best molecular arrangement of ligands by using the relevant receptor/molecule as a template is significant, providing that reliable reversible chemical reactions are developed in the future using a wide range of chemical diversity to generate large dynamic libraries.

8.6.2 An Example: Synthesis and Screening of a Reversible Imine Library

Hasenkopf et al. (306) reported the synthesis of a dynamic 12-member, template-directed imine library **L27** obtained from the reversible condensation of three aldehydes (monomer set $\mathbf{M_1}$, Fig. 8.54) with four primary amines (monomer set $\mathbf{M_2}$, Fig. 8.54) in buffered aqueous conditions followed by irreversible reduction to amines **L28** with sodium cyanoborohydride. The library was prepared in the presence of a large excess of $\mathbf{M_2}$, to prevent further condensation of an aldehyde onto the secondary amine product. A template-driven imine library **L27** was prepared in the presence of the metalloenzyme carbonic anhydrase II (CAII). After the template-assisted, reversible dynamic reaction was complete, the reducing agent was added and the amine library **L28** was produced (Fig. 8.54). Without any template the unbiased, equilibrated imine mixture **L29** was then reduced to the mixture of amines **L30** (Fig. 8.54). The different abundance of library components **L28** and **L30**, reflecting the affinity of library components for CAII, was determined by comparing the HPLC traces of the stable amine mixtures.

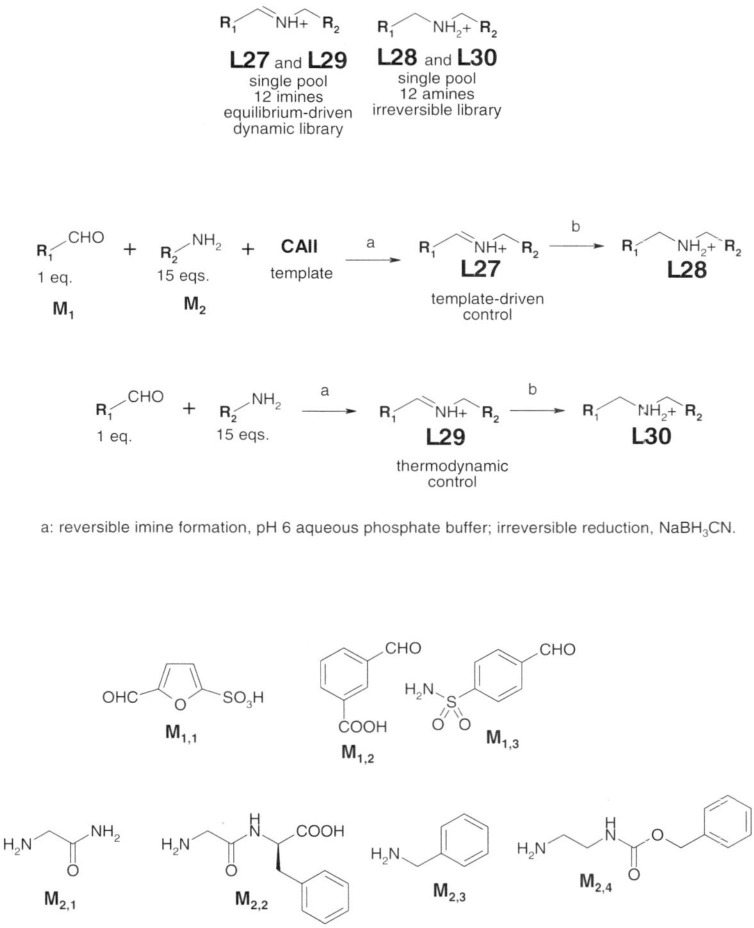

Figure 8.54 Synthesis of two template-assisted, dynamic pool libraries of Schiff bases **L27** and **L29** and structure of the monomer sets $\mathbf{M_1}$–$\mathbf{M_2}$.

The two aldehydes $\mathbf{M_{1,1}}$ and $\mathbf{M_{1,2}}$ produced the same relative amount of imines in the presence or absence of CAII, which implied that no interaction between these dynamic library members and the enzyme was observed. The relative abundance of imines from $\mathbf{M_{1,3}}$ varied in the two libraries and the two amines **8.112** and **8.113** almost disappeared from **L29** when compared to **L30**, while the formation of **8.114** and especially **8.115** was favored by the template (Fig. 8.55). These results were confirmed in four validation experiments in which the two amines were reacted with $\mathbf{M_{1,3}}$ in the presence or in the absence of the enzyme. The results reported in Table 8.6 show how **8.115** was the favored library component in the template-assisted synthesis of the mixture. Further confirmation of the specificity of **8.115** toward the template was provided by adding the known CAII inhibitor **8.116** (Fig. 8.55) to a binary mixture

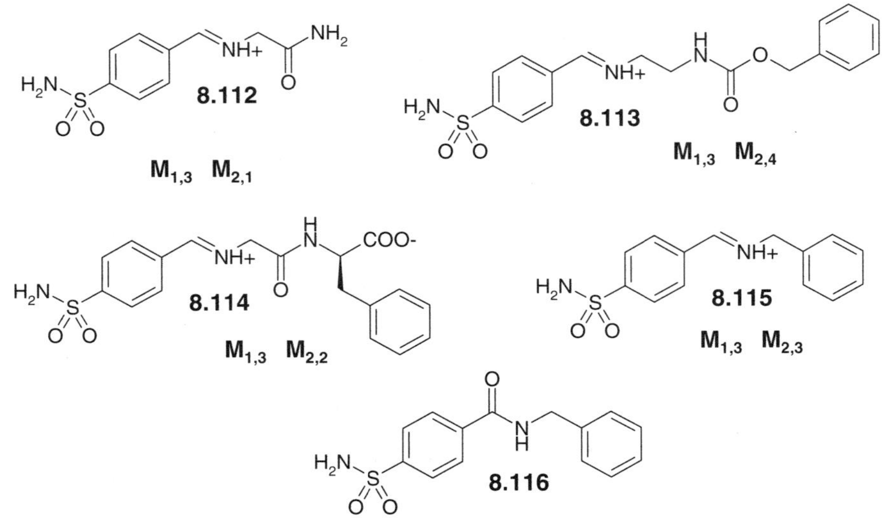

Figure 8.55 Template-favored members **8.112–8.115** from the dynamic pool library of Schiff bases **L24**.

(Table 8.6, entry d). The effect of CAII on the equilibrium of the dynamic mixture was significantly reduced in the presence of **8.116**, which bound the enzyme and reduced its template effect.

8.6.3 An Example: Synthesis of a Macrolactonic Oligocholate Dynamic Library

Brady and Sanders (328) reported the synthesis of dynamic macrolactone libraries based on the thermodynamic transesterification/cyclization of easily accessible cholate monomers **8.117–8.119** (Fig. 8.56). The transesterification conditions were first studied with **8.117**, and the mixture of potassium methoxide and a crown ether in toluene was found to be the best thermodynamic protocol. At 5 m*M* concentration of

TABLE 8.6 Template-Assisted Selection of Schiff Bases from Binary Mixtures Derived from Dynamic Library L27

Entry	Amine components	Normalized $\mathbf{M_{2,3}/M_{2,x}}$[a]
a	$\mathbf{M_{2,3}/M_{2,1}}$	15
b	$\mathbf{M_{2,3}/M_{2,2}}$	4.5
c	$\mathbf{M_{2,3}/M_{2,4}}$	21
d[b]	$\mathbf{M_{2,3}/M_{2,4}}$	2

[a]Ratio between $\mathbf{M_{2,3}}$ and the other amine component in the mixture, normalized according to relative UV responses.

[b]In presence of equimolar **8.116**.

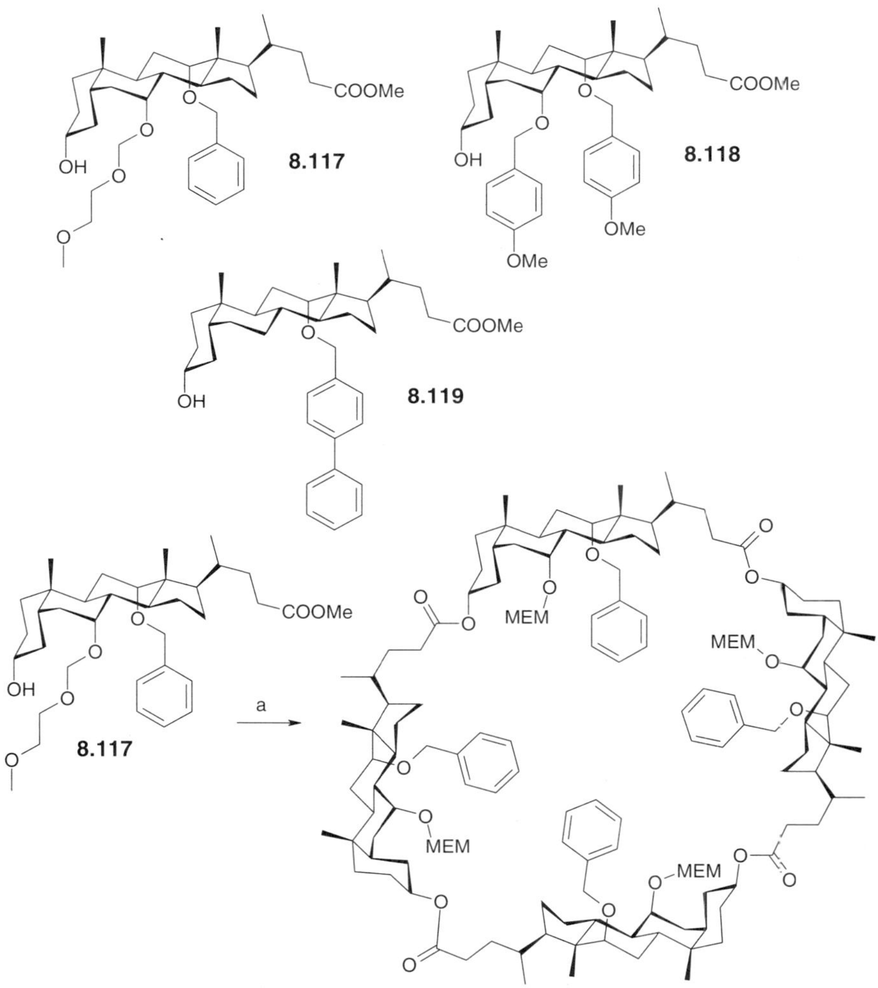

cyclic tetramer (12%) ***8.121***

+ cyclic trimer (83%) ***8.120***
+ cyclic pentamer (3%) ***8.121***

a: reversible transesterification - cyclization, 5 mM **8.117**, 0.06M MeOK/dicyclohexyl-18-crown-6, toluene, rt.

Figure 8.56 Synthesis of template-assisted, dynamic combinatorial cyclic libraries of oligocholates.

TABLE 8.7 Relative Abundance of Macrocycles from Thermodynamic Transesterification/Macrocyclization of Monomers 8.117–8.119

	8.117			**8.118**			**8.119**		
Template	Tri-	Tetra-	Penta-	Tri-	Tetra-	Penta-	Di-	Tri-	Tetra
None	83[a]	12	5	65	24	10	25	53	22
LiI	76	16	7	60	25	14	13	59	25
NaI	61	24	14	47	28	25	25	54	21
KI	75	15	8	65	23	12	32	46	17
CsI	81	12	5	65	24	10	37	48	14

[a]Expressed as % of the recoverd mass.

monomer and with 5% catalyst, the reaction produced an equilibrium mixture in which the cyclic trimer **8.120** was strongly favored versus the cyclic tetra- (**8.121**) and pentamer **8.122** (Fig. 8.56). The thermodynamic nature of the cyclization was confirmed by submitting a pure sample of tetramer **8.121** to the transesterification protocol and isolating the same mixture of products obtained from monomer **8.117**.

The cholates **8.117–8.119** were designed for the preparation of dynamic libraries with different binding affinities for alkali metal ions. The presence of a polyether chain in position 7 of **8.117** provided a recognition element for metal binding that was absent from the disubstituted *p*-methoxybenzyl substitution pattern of **8.118**, while the 7-deoxy derivative **8.119** was even less prone to metal coordination. The three monomers were submitted to transesterification/cyclization protocols, either without metal templates or using different alkali metal salts as templates. The relative abundances of cyclic dimers, trimers, tetramers, and pentamers for each experiment are reported in Table 8.7.

Monomer **8.117** produced a mixture of cyclic tri-, tetra-, and pentamers that was strongly biased toward the trimer under most conditions. Only sodium shifted the equilibrium toward the larger cycles and doubled the abundances of both the tetramer and pentamer, reflecting the increased capacity of the larger macrolactones to bind sodium. Monomer **8.118** behaved similarly to **8.117**, with a slightly lower preference for the trimer and with the same behavior in the presence of sodium salts. The theoretical recognition site provided by the 7-substituent of **8.117**, absent in **8.118**, was ruled out by these results. The deoxy monomer **8.119** provided smaller rings, including the previously unobserved dimer, but also showed a dynamic, template-assisted shift toward larger sites when Li^+ was used and toward smaller sites when larger ions (K^+ and Cs^+) were used. This unexpected finding proved the size-specific interaction of **8.119**-based macrocycles with alkali metals.

REFERENCES

1. Smith, P. W. Lai, J. Y. Q., Whittington, A. R., Cox, B., Houston, J. G., Stylli, C. H., Banks, M. N. and Tiller, P. R. *Bioorg. Med. Chem. Lett.* **4**, 2821–2824 (1994).

2. Storer, R., *Drug Disc. Today* **1**, 248–254 (1996).
3. Bailey, N., Cooper, A. W. J., Deal, M. J., Dean, A. W., Gore, A. L., Hawes, M. C., Judd, D. B., Merritt, A. T., Storer, R., Travers, S. and Watson, S. P., *Chimia* **51**, 832–837 (1997).
4. Pirrung, M. C. and Chen, J., *J. Am. Chem. Soc.* **117**, 1240–1245 (1995).
5. Pirrung, M. C., Chau, J. H. L. and Chen, J., *Chem. Biol.* **2**, 621–626 (1995).
6. Raveglia, L. F., Vitali, M., Artico, M., Graziani, D., Hay, D. W. P., Luttmann, M. A., Mena, R., Pifferi, G. and Giardina, G. A. M., *Eur. J. Med. Chem.* **34**, 825–835 (1999).
7. Andrus, M. B., Turner, T. M., Asgari, D. and Li, W., *J. Org. Chem.* **64**, 2978–2979 (1999).
8. Desphande, A. M., Argade, N. P., Natu, A. A. and Eckman, J., *Bioorg. Med Chem.* **7**, 1237–1240 (1999).
9. Maehr, H. and Yang, R., *Bioorg. Med. Chem.* **5**, 493–496 (1997).
10. Vendeville, S., Bourel, L., Davioud-Charvet, E., Grellier, P., Deprez, B. and Sergheraert, C., *Bioorg. Med. Chem. Lett.* **9**, 437–442 (1999).
11. Neuville, L. and Zhu, J., *Tetrahedron Lett.* **38**, 4091–4094 (1999).
12. Chng, B. L. and Ganesan, A., *Bioorg. Med. Chem. Lett.* **7**, 511–514 (1998).
13. van Niel, M., Beer, M. S., Castro, J. L., Cheng, S. K. F., Evans, D. C., Heald, A., Hitzel, L., Hunt, P., Mortishire-Smith, R., O'Connor, D., Watt, A. P. and MacLeod, A. M., *Bioorg. Med Chem. Lett.* **9**, 3243–3248 (1999).
14. Yu, J., Zhao, Y., Holterman, M. J. and Venton, D. L., *Bioorg. Med. Chem. Lett.* **9**, 2705–2710 (1999).
15. Carell, T. Wintner, E. A., Bashir-Hashemi, A. and Rebek, J. Jr., *Angew. Chem. Int. Ed. Engl.* **33**, 2059–2061 (1994).
16. Carell, T., Wintner, E. A. and Rebek, J. Jr., *Angew. Chem. Int. Ed. Engl.* **33**, 2061–2064 (1994).
17. Carell, T., Wintner, E. A, Sutherland, A. J, Rebek, J. Jr., Dunayevskiy, Y. M and Vouros, P., *Chem. Biol.* **2**, 171–183 (1995).
18. Shipps, G. W., Spitz, U. P. and Rebek, J. Jr., *Bioorg. Med. Chem.* **4**, 655–657 (1996).
19. Shipps, G. W. Jr., Pryor, K. E., Xian, J., Skyler, D. A., Davidson, E. H. and Rebek, J. Jr., *Proc. Natl. Acad. Sci. USA* **94**, 11833–11838 (1997).
20. Pryor, K. E., Shipps, G. W., Skyler, D. A. and Rebek, J. Jr., *Tetrahedron* **54**, 4107–4124 (1998).
21. Pryor, K. E. and Rebek, J., Jr., *Org. Lett.* **1**, 39–42 (1999).
22. Pryor, K. E. and La Clair, J. J., *Bioorg. Med. Chem. Lett.* **9**, 2291–2296 (1999).
23. An, H., Cummins, L. L., Griffey, R. H., Bharadwaj, R., Haly, B. D., Fraser, A. S., Wilson-Lingardo, L., Risen, L. M., Wyatt, J. R. and Cook, P. D., *J. Am. Chem. Soc.* **119**, 3696–3708 (1997).
24. Wang, T., An, H., Vickers, T. A., Bharadwaj, R. and Cook, P. D., *Tetrahedron* **54**, 7955–7976 (1998).
25. An, H., Haly, B. D. and Cook, P. D., *Bioorg. Med. Chem. Lett.* **8**, 2345–2350 (1998).
26. An, H., Haly, B. D., Fraser, A. S., Guinosso, C. J. and Cook, P. D., *J. Org. Chem.* **62**, 5156–5164 (1997).
27. Kung, P.-P., Bharadwaj, R., Fraser, A. S., Cook, D. R., Kawasaki, A. M. and Cook, P. D., *J. Org. Chem.* **63**, 1846–1852 (1998).

28. Kung, P.-P. and Cook, P. D., *Biotechnol. Bioeng.* **61**, 119–125 (1998).
29. Kawasaki, A. M., Casper, M. D., Gaus, H. J., Herrmann, R., Griffey, R. H. and Cook, P. D., *Nucleosides Nucleotides* **18**, 659–660 (1999).
30. Gaus, H. J., Kung, P.-P., Brooks, D., Cook, P. D. and Cummins, L. L., *Biotechnol. Bioeng.* **61**, 169–177 (1999).
31. Griffey, R. H., An, H., Cummins, L. L., Gaus, H. J., Haly, B., Herrmann, R. and Cook, P. D., *Tetrahedron* **54**, 4067–4076 (1998).
32. Boger, D. L. and Chai, W., *Tetrahedron* **54**, 3955–3970 (1998).
33. Boger, D. L., Goldberg, J., Jiang, W., Chai, W., Ducray, P., Lee, J. K., Ozer, R. S., Andersson, C.-M., *Bioorg. Med. Chem.* **6**, 1347–1378 (1998).
34. Boger, D. L., Chai, W. and Jin, Q., *J. Am. Chem. Soc.* **120**, 7220–7225 (1998).
35. Boger, D. L., Goldberg, J. and Andersson, C.-M., *J. Org. Chem.* **64**, 2422–2429 (1999).
36. Boger, D. L., Jiang, W and Goldberg, J., *J. Org. Chem.* **64**, 7094–7100 (1999).
37. Klumpp, D. A., Yeung, K. Y., Prakash, G. K. S. and Olah, G. A., *J. Org. Chem.* **63**, 4481–4484 (1998).
38. Ferritto, R., De Magistris, E.; Missio, A., Paio, A. and Seneci, P., in *Combinatorial Chemistry and Combinatorial Technologies: Methods and Applications*, S. Miertus and G. Fassina (Eds.). Marcel Dekker, Inc., New York, 1999, pp. 53–90.
39. Bunin, B., *The Combinatorial Index*. Academic Press, San Diego, CA, 1998, pp. 237–280.
40. Gayo, L. M., *Biotechnol. Bioeng.* **61**, 95–106 (1998).
41. Merritt, A. T., *Combi. Chem. High Through. Screening* **1**, 57–72 (1998).
42. Terrett, N., *Combinatorial Chemistry*. Oxford University Press, Oxford, UK, 1998, pp. 55–76.
43. Underiner, T. L. and Peterson, J. R., in *A Practical Guide to Combinatorial Chemistry*, A. W. Czamik and S. H. DeWitt (Eds.). ACS, Washington, DC, 1997, pp. 177–198.
44. Coe, D. M. and Storer, R., *Mol. Diversity* **4**, 31–38 (1999).
45. Suto, M. J., *Curr. Opin. Drug Discovery Dev.* **2**, 377–384 (1999).
46. Stewart, K. D., Loren, S., Frey, L., Otis, E., Klinghofer, V. and Hulkower, K. I., *Bioorg. Med. Chem. Lett.* **8**, 529–534 (1998).
47. Maltais, R. and Poirier, D., *Tetrahedron Lett.* **39**, 4151–4154 (1998).
48. Panunzio, M., Villa, M., Missio, A., Rossi, T. and Seneci, P., *Tetrahedron Lett.* **39**, 6585–6588 (1998).
49. Kim, S. W., Bauer, S. M. and Armstrong, R. W., *Tetrahedron Lett.* **39**, 7031–7034 (1998).
50. Thomas, J. B., Fall, M. J., Cooper, J. B., Rothman, R. B., Mascarella, S. W., Xu, H., Partilla, J. S., Dersch, C. M., McCullough, K. B., Cantrell, B. E., Zimmerman, D. M. and Carroll, F. I., *J. Med. Chem.* **41**, 5188–5197 (1998).
51. Hulme, C., Peng, J., Tang, S.-Y., Bums, C. J., Morize, I. and Labaudiniere, R., *J. Org. Chem.* **63**, 8021–8023 (1998).
52. Doller, D., Chackalamannil, S., Czamiecki, M., McQuade, R. and Ruperto, *Bioorg. Med. Chem. Lett.* **9**, 901–906 (1999).
53. Adlington, R. M., Baldwin, J. E., Catterick, D. and Pritchard, G. J., *J. Chem. Soc., Perkin Trans. I* 855–866 (1999).

54. Wong, C.-H., Hendrix, M, Manning, D. D., Rosenbohm, C. and Greenberg, W. A., *J. Am. Chem. Soc.* **120**, 8319–8327 (1998).

55. Fourmy, D., Recht, M. I., Blanchard, S. C. and Puglisi, J. D., *Science* **274**, 1367–1371 (1996).

56. Sim, M. M. and Ganesan, A., *J. Org. Chem.* **62**, 3230–3235 (1997).

57. DeWitt, S. H., Kiely, J. S., Stankovic, C. J., Schroeder, M. C., Cody, D. M. R., Pavia, M. R., *Proc. Natl. Acad. Sci. USA* **90**, 6909–6913 (1993).

58. MultiReactor, ComGenex Ltd., Hungary. *www.datanet.hu/comgenex*.

59. Reacto-Stations, Stem Corporation Ltd., US. *www.stemcorp.com*.

60. Quest 210, Argonaut Technologies, US. *www.argotech.com*.

61. Carousel Reaction Stations, Radleys, UK. *www.radleys.com*.

62. 215 Liquid Handler, Gilson Inc., US. *www.gilson.com*.

63. Iliad PS2, Charybdis Technologies, Inc., US. *www.charybtech.com*.

64. HP7686 Solution Phase Automated Synthesizer, Hewlett-Packard Inc., US. *www.hp.com*.

65. Automated Synthesis Workstation, Chemspeed Ltd., CH. *www.chemspeed.com*.

66. Brooking, P., Doran, A., Grimsey, P., Hird, N. W., MacLachlan, W. S. and Vimal, M., *Tetrahedron Lett.* **40**, 1405–1408 (1999).

67. ISOLUTE Array plates, International Sorbent Technology Ltd., UK. *www.ist-spe.com*.

68. Lollipop Phase Separator, Radleys, UK. *www.radleys.co.uk*.

69. CombiFlash, Isco, Inc., US. *www.isco.com*.

70. Atlas Evaporator, Genevac Ltd., UK. *www.genevac.co.uk*.

71. SpeedVac Systems, Savant Inc., US. *www.savec.com*.

72. Coffey, P., Ramieri, J., Garr, C. and Schultz, L., *Am. Lab.* **31**, 57–71 (1999).

73. Lebl, M., *Bioorg. Med. Chem. Lett.* **9**, 1305–1310 (1999).

74. Weller, H. N., *Mol. Diversity* **4**, 47–52 (1999).

75. Automated Molecular Assembly Plant (AMAP), ArQule Inc., US. *www.argule.com*.

76. Cerep, France. *www.cerep.fr*.

77. Whitten, J. P., Xie, Y. F., Erickson, P. E., Webb, T. R., Souza, E. B. D., Grigorladis, D. E. and McCarthy, J. R., *J. Med. Chem.* **39**, 4354–4357 (1996).

78. Xie, Y. F., Whitten, J. P., Chen, T. Y., Liu, Z. and Mccarthy, J. R., *Tetrahedron* **54**, 4077–4084 (1998).

79. Bhat, L., Liu, Y., Victory, S. F., Himes, R. H. and Georg, G. I., *Bioorg. Med. Chem. Lett.* **8**, 3181–3186 (1998).

80. Deegan, T. L., Nitz, T. J., Cebzanov, D., Pufko, D. E. and Porco, J. A. Jr., *Bioorg. Med. Chem. Lett.* **9**, 209–212 (1999).

81. Frank, K. E., Jung, M. and Mitschler, L. A., *Combi. Chem. High Through. Screen.* **1**, 73–87 (1998).

82. Powers, D. G., Casebier, D. S., Fokas, D., Ryan, W. J., Troth, J. R. and Coffen, D. L., *Tetrahedron* **54**, 4085–4096 (1998).

83. Fokas, D., Ryan, W. J., Casebier, D. S. and Coffen, D. L., *Tetrahedron Lett.* **39**, 2235–2238 (1998).

84. Rivero, R. A., in *A Practical Guide to Combinatorial Chemistry*, A. W. Czarnik and S. H. DeWitt (Eds.). ACS, Washington, DC, 1997, pp. 281–307.
85. Gustafson, G. R., Baldino, C. M., O'Donnell, M.-M. E., Sheldon, A., Tarsa, R. J., Vemi, C. J. and Coffen, D. L., *Tetrahedron* **54**, 4051–4065 (1998).
86. Baudelle, R., Melnyk, P., Deprez, B. and Tartar, A., *Tetrahedron* **54**, 4125–4140 (1998).
87. Perrier, H. and Labelle, M., *J. Org. Chem.* **64**, 2110–2113 (1999).
88. Curran, D. P., *Angew. Chem. Int. Ed.* **37**, 1174–1196 (1998).
89. Horvath, I.T. and Rabai, J., *Science* **266**, 72–75 (1994).
90. Yoshida, J., Itami, K., Mitsudo, K. and Suga, S., *Tetrahedron Lett.* **40**, 3403–3406 (1999).
91. Cheng, S., Tarby, C. M., Comer, D. D., Williams, J. P., Caporale, L. H., Myers, P. L. and Boger, D. L., *Bioorg. Med. Chem.* **4**, 727–737 (1996).
92. Barthel-Rosa, L. P. and Gladysz, J. A., *Coord. Chem. Rev.* **190–192**, 587–605 (1999).
93. Ogawa, A. and Curran, D. P., *J. Org. Chem.* **62**, 450–451 (1997).
94. Juliette, J. J. J., Rutherford, D., Horvath, I. T. and Gladysz, J. A., *J. Am. Chem. Soc.* **121**, 2696–2704 (1999).
95. Pozzi, G., Cavazzini, M., Cinato, F., Montanari, F. and Quici, S., *Eur. J. Org. Chem.* 1947–1955 (1999).
96. Kleijn, H., Rijnberg, E., Jastrzebski, J. T. B. H. and van Koten, G., *Org. Lett.* **1**, 853–855 (1999).
97. Nakamura, Y., Takeuchi, S., Ohgo, Y. and Curran, D. P., *Tetrahedron Lett.* **41**, 57–60 (2000).
98. Curran, D. P. and Hadida, S., *J. Am. Chem. Soc.* **118**, 2531–2532 (1996).
99. Curran, D. P., Luo, Z. and Degenkolb, P., *Bioorg. Med. Chem. Lett.* **8**, 2403–2408 (1998).
100. Ryu, I., Niguma, T., Minakata, S., Komatsu, M., Luo, Z. and Curran, D. P., *Tetrahedron Lett.* **40**, 2367–2370 (1999).
101. Curran, D. P., Hadida, S., Kim, S.-Y. and Luo, Z., *J. Am. Chem. Soc.* **121**, 6607–6615 (1999).
102. Studer, A. and Curran, D. P., *Tetrahedron* **53**, 6681–6696 (1997).
103. Curran, D. P., Ferritto, R. and Hua, Y., *Tetrahedron Lett.* **39**, 4937–4940 (1998).
104. Battacharyya, P., *J. Chem. Soc. Perkin Trans I*, 3609–3612 (1997).
105. Horvath, I. T., Kiss, G., Cook, R. A., Bond, J. E., Stevens, P. A., Rabai, J. and Mozeleski, E. J., *J. Am. Chem. Soc.* **120**, 3133–3143 (1998).
106. Sinou, D., Pozzi, G., Hope, E. G. and Stuart, A. M., *Tetrahedron Lett.* **40**, 849–852 (1999).
107. Moineau, J., Pozzi, G., Quici, S. and Sinou, D., *Tetrahedron Lett.* **40**, 7683–7686 (1999),
108. Quici, S., Cavazzini, M., Ceragioli, S., Montanari, F. and Pozzi, G., *Tetrahedron Lett.* **40**, 3647–3650 (1999).
109. Mathivet, T., Monflier, E., Castanet, Y., Mortreux, A. and Couturier, J.-L., *Tetrahedron Lett.* **40**, 3885–3888 (1999).
110. Curran, D. P., *Chemtracts: Org. Chem.* **9**, 75–87 (1996).
111. Curran, D. P., *Med. Res. Rev.* **19**, 432–438 (1999).
112. da Silva, L. H. M. and Loh, W., *Chem. Commun.* 787–788 (1998).
113. Larhed, M., Hoshino, M., Hadida, S., Curran, D. P. and Hallberg, A., *J. Org. Chem.* **62**, 5583–5587 (1997).

114. Hoshino, M., Degenkolb, P. and Curran, D. P., *J. Org. Chem.* **62**, 8341–8349 (1997).

115. Olofsson, K., Kim, S.-Y., Larhed, M., Curran, D. P. and Hallberg, A., *J. Org. Chem.* **64**, 4539–4541 (1999).

116. Curran, D. P. and Luo, Z., *Med. Chem. Res.* **8**, 261–265 (1998).

117. Curran, D. P. and Luo, Z., *J. Am. Chem. Soc.* **121**, 9069–9072 (1999).

118. Linclau, B., Sing, A. K. and Curran, D. P., *J. Org. Chem.* **64**, 2835–2842 (1999).

119. Studer, A., Jeger, P., Wipf, P. and Curran, D. P., *J. Org. Chem.* **62**, 2917–2924 (1997).

120. Boutevin, B., Guida-Pietrasanta, F., Ratsimihety, A., Caporiccio, G. and Gornowicz, G., *J. Fluorine Chem.* **60**, 211–223 (1993).

121. Breslow, R. and Pandey, P. S., *J. Org. Chem.* **45**, 740–741 (1980).

122. Siegel, M. G., Hahn, P. J., Dressman, B. A., Fritz, J. E., Grunwell, J. R. and Kaldor, S. W., *Tetrahedron Lett.* **38**, 3357–3360 (1997).

123. Lawrence, R. M., Biller, S. A., Fryszman, O. M. and Poss, M. A., *Synthesis* 553–558 (1997).

124. Shuker, A. J., Siegel, M. G., Matthews, D. P. and Weigel, L. O., *Tetrahedron Lett.* **38**, 6149–6152 (1997).

125. Blackburn, C., Guan, B., Fleming, P., Shiosaki, K., Tsai, S., *Tetrahedron Lett.* **39**, 3635–3638 (1998).

126. Bussolari, J. C., Rehborn, D. C. and Combs, D. W., *Tetrahedron Lett.* **40**, 1241–1244 (1999).

127. Nilsson, U. J., Fournier, E. J. L. and Hindsgaul, O., *Bioorg. Med. Chem.* **6**, 1563–1575 (1998).

128. Kainz, S., Luo, Z., Curran, D. P. and Leitner, W., *Synthesis* 1425–1427 (1998).

129. Curran, D. P., Hadida, S. and He, M., *J. Org. Chem.* **62**, 6714–6715 (1997).

130. Siegel, M. G., Shuker, A. J., Droste, C. A., Hahn, P. J., Jesudason, C. D., Mcdonald, J. H., III, Matthews, D. P., Rito, C. J. and Thorpe, A. J., *Mol. Diversity* **3**, 113–116 (1998).

131. Weller, H. N., Young, M. G., Michalczyk, S. J., Reitnauer, G. H., Cooley, R. S., Rahn, P. C., Loyd, D. J., Fiore, D. and Fischman, S. J., *Mol. Diversity* **3**, 61–70 (1997).

132. Zeng, L., Wang, X., Wang, T. and Kassel, D. B., *Comb. Chem. High Throughput Scr.* **1**, 101–111 (1998).

133. Zeng, L. and Kassel, D. B., *Anal. Chem.* **70**, 4380–4388 (1998).

134. Schultz, L., Garr, C. D., Cameron, L. M. and Bukowski, J., *Bioorg. Med. Chem. Lett.* **8**, 2409–2414 (1998).

135. *Preparative Chemistry Using Supported Reagents*, P. Laszlo (Ed.). Academic Press, London, UK, 1987.

136. *Supported Reagents and Catalysts in Chemistry*, B. K. Hodnett, A. P. Kybett. J. H. Clark and K. Smith (Eds.). Royal Chemical Society, Cambridge, UK, 1998.

137. Fetizon, M. and Golfier, M., *Comp. Rend. Acad. Sci. Ser.* **C267**, 900–903 (1968).

138. Shuttleworth, S. J., Allin, S. M. and Sharma, P. K., *Synthesis* 1217–1239 (1997).

139. Kaldor, S. W. and Siegel, M. G., *Curr. Opin. Chem. Biol.* **1**, 101–106 (1997).

140. Drewry, D. H., Coe, D. M. and Poon, S., *Med. Res. Rev.* **19**, 97–148 (1999).

141. Chesney, A., *Green Chem.* **1**, 209–219 (1999),

142. Junggebauer, J. and Neumann, W. P., *Tetrahedron* **53**, 1301–1310 (1997).

143. Hinzen, B. and Ley, S. V., *J. Chem. Soc., Perkin Trans I* 1–2 (1998).
144. Dendrinos, K. G. and Kalivretenos, A. G., *J. Chem. Soc., Perkin Trans. I* 1463–1464 (1998).
145. Boehlow, T. R., Buxton, P. C., Grocock, E. L., Marples, B. A. and Waddington, V. L., *Tetrahedron Lett.* **39**, 1839–1842 (1998).
146. Togo, H., Nogami, G. and Yokoyama, M., *Synlett* 534–536 (1998).
147. Nicolaou, K. C., Pastor, J., Barluenga, S. and Winssinger, N., *Chem. Commun.* 1947–1948 (1998).
148. Sylvain, C., Wagner, A. and Mioskowski, C., *Tetrahedron Lett.* **40**, 875–878 (1999).
149. Hu, Y. and Porco, J. A., Jr., *Tetrahedron Lett.* **40**, 3289–3292 (1999).
150. Minutolo, F. and Katzenellenbogen, J. A., *Angew. Chem. Int. Ed.* **38**, 1617–1620 (1999).
151. Majewski, M., Ulaczyk, A. and Wang, F., *Tetrahedron Lett.* **40**, 8755–8758 (1999).
152. Kirschning, A., Jesberger, M. and Monenschein, H., *Tetrahedron Lett.* **40**, 8999–9002 (1999).
153. Barrett, A. G. M., Cramp, S. M., Roberts, R. S. and Zecri, F. J., *Org. Lett.* **1**, 579–582 (1999).
154. Parlow, J. J., *Tetrahedron Lett.* **37**, 5257–5260 (1996).
155. Kulkami, B. A. and Ganesan, A., *Chem. Commun.* 785–786 (1998).
156. Kulkami, B. A. and Ganesan, A., *Tetrahedron Lett.* **40**, 5637–5638 (1999).
157. Pop, I. E., Deprez, B. P. and Tartar, A. L., *J. Org. Chem.* **62**, 2594–2603 (1997).
158. Dendrinos, K., Jeong, J., Huang, W. and Kalivretenos, A. G., *Chem. Commun.* 499–500 (1998).
159. Xu, W., Mohan, R. and Morrissey, M. M., *Tetrahedron Lett.* **38**, 7337–7340 (1997).
160. Chen, J., Dixon, B. R., Dumas, J. and Brittelli, D., *Tetrahedron Lett.* **40**, 9195–9199 (1999).
161. Buckman, B. O., Morrissey, M. M. and Mohan, R., *Tetrahedron Lett.* **39**, 1487–1490 (1998).
162. Sturino, C. F. and Labelle, M., *Tetrahedron Lett.* **39**, 5891–5894 (1998).
163. Haunert, F., Bolli, M. H., Hinzen, B. and Ley, S. V., *J. Chem. Soc., Perkin Trans. I* 2235–2237 (1998).
164. Bolli, M. H. and Ley, S. V., *J. Chem. Soc., Perkin Trans. I* 2243–2246 (1998).
165. Ley, S. V., Schucht, 0., Thomas, A. W. and Muffay, P. J., *J. Chem. Soc., Perkin Trans. I* 1251–1252 (1999).
166. Habermann, J., Ley, S. V. and Scott, J. S., *J. Chem. Soc., Perkin Trans. I* 1253–1256 (1999).
167. Ley, S. V., Thomas, A. W. and Finch, H., *J. Chem. Soc., Perkin Trans. I* 669–672 (1999).
168. Habermann, J., Ley, S. V. and Smits, R., *J. Chem. Soc., Perkin Trans. I* 2421–2423 (1999).
169. Habermann, J., Ley, S. V., Scicinski, J. J., Scott, J. S., Smits, R. and Thomas, A. W., *J. Chem. Soc., Perkin Trans. I* 2425–2427 (1999).
170. Tunoori, A. R., Dutta, D. and Georg, G. I., *Tetrahedron Lett.* **39**, 8751–8754 (1998).
171. Missio, A., Selva, S., Panunzio, M. and Seneci, P., *Biotech. Bioeng.* **71**, 38–43 (2000).
172. Paio, A., Zaramella, A., Ferritto, R., Conti, N., Marchloro, C. and Seneci, P., *J. Comb. Chem.* **1**, 317–325 (1999).

173. Habermann, J., Ley, S. V. and Scott, J. S., *J. Chem. Soc., Perkin Trans. I* 3127–3130 (1998).

174. Rebek Jr., J., *Tetrahedron* **35**, 723–731 (1979).

175. Parlow, J. J., *Tetrahedron Lett.* **36**, 1395–1396 (1995).

176. Ley, S. V., Bolli, M. H., Hinzen, B., Gervols, A.-G. and Hall, B. J., *J. Chem. Soc., Perkin Trans. I* 2239–2241 (1998).

177. Frechet, J. M. J., *Tetrahedron* **37**, 663–683 (1981).

178. Akelah, A. and Sherrington, D. C., *Chem. Rev.* **81**, 557–558 (1981).

179. Yu, L., Chen, D., Li, J. and Wang, P. G., *J. Org. Chem.* **62**, 3575–3581 (1997).

180. Rheiner, P. B., Sellner, H. and Seebach, D., *Helv. Chim. Acta* **80**, 2027–2032 (1997).

181. Sellner, H. and Seebach, D., *Angew. Chem. Int. Ed.* **38**, 1918–1920 (1999).

182. Halm, C. and Kurth, M. J., *Angew. Chem. Int. Ed. Engl.* **37**, 510–512 (1998).

183. Sung, D. W. L., Hodge, P. and Stratford, P. W., *J. Chem. Soc., Perkin Trans. I* 1463–1472 (1999).

184. Bolm, C., Dinter, C. L., Seger, A., Hocker, H. and Brozio, J., *J. Org. Chem.* **64**, 5730–5731 (1999).

185. Macquarrie, D. J., *Tetrahedron Lett.* **39**, 4125–4128 (1998).

186. Nagayama, S., Endo, M. and Kobayashi, S., *J. Org. Chem.* **63**, 6094–6095 (1998).

187. Kobayashi, S., *Top. Organomet. Chem.* **2**, 285–305 (1999).

188. Chemin, A., Deleuze, H. and Maillard, B., *J. Chem. Soc., Perkin Trans. I* 137–142 (1999).

189. Uozumi, Y., Danjo, H. and Hayashi, T., *J. Org. Chem.* **64**, 3384–3388 (1999).

190. Uozumi, Y. and Watanabe, T., *J. Org. Chem.* **64**, 6921–6922 (1999).

191. Zhang, T. Y. and Allen, M. J., *Tetrahedron Lett.* **40**, 5813–5816 (1999).

192. Bao, M., Nakamura, H. and Yamamoto, Y., *Tetrahedron Lett.* **41**, 131–134 (2000).

193. Ahmed, M., Barrett, A. G. M., Braddock, D. C., Cramp, S. M. and Procopiou, P. A., *Tetrahedron Lett.* **40**, 8657–8662 (1999).

194. Simpson, J., Rathbome, D. L. and Billington, D. C., *Tetrahedron Lett.* **40**, 7031–7033 (1999).

195. Clark, A. J., Filik, R. P., Haddleton, D. M., Radigue, A., Sanders, C. J., Thomas, G. H. and Smith, M. E., *J. Org. Chem.* **64**, 8954–8957 (1999).

196. Altava, B., Burguete, M. I., Garcia-Verdugo, E., Luis, S. V., Salvador, R. V. and Vicent, M. J., *Tetrahedron* **55**, 12897–12906 (1999).

197. Kobayashi, S. and Nagayama, S., *J. Am. Chem. Soc.* **118**, 8977–8978 (1996).

198. Kobayashi, S., Nagayama, S. and Busujima, T., *Tetrahedron Lett.* **37**, 9221–9224 (1996).

199. Peukert, S. and Jacobsen, E. N., *Org. Lett.* **1**, 1245–1248 (1999).

200. Canali, L., Cowan, E., Gibson, C. L., Sherrington, D. C. and Deleuze, H., *Chem. Commun.* 2561–2562 (1998).

201. Annis, D. A. and Jacobsen, E. N., *J. Am. Chem. Soc.* **121**, 4147–4154 (1999).

202. Kaldor, S. W., Siegel, M. G., Fritz, J. E., Dressman, B. A. and Hahn, P. J., *Tetrahedron Lett.* **37**, 7193–7196 (1996).

203. Gayo, L. M. and Suto, M. J., *Tetrahedron Lett.* **38**, 513–516 (1997).

204. Suto, M. J., Gayo-Fung, L. M., Palanki, M. S. S. and Sullivan, R., *Tetrahedron* **54**, 4141–4150 (1998).

205. Parlow, J. J., Vazquez, M. L. and Flynn, D. L., *Bioorg. Med. Chem. Lett.* **8**, 2391–2394 (1998).

206. Flynn, D. L., Crich, J. Z., Devraj, R. V., Hockerman, S. L., Parlow, J. J., South, M. S. and Woodard, S., *J. Am. Chem. Soc.* **119**, 4874–4881 (1997).

207. Booth, R. J. and Hodges, J. C., *J. Am. Chem. Soc.* **119**, 4882–4886 (1997).

208. Chen, C. and Munoz, B., *Tetrahedron Lett.* **39**, 6781–6784 (1998).

209. Yamamoto, Y., Tanabe, K. and Okonogi, T., *Chem. Lett.* 103–104 (1999).

210. Breitenbucher, J. G., Johnson, C. R., Haight, M. and Phelan, J. C., *Tetrahedron Lett.* **39**, 1295–1298 (1998).

211. Johnson, C. R., Zhang, B., Fantauzzi, P., Hocker, M. and Yager, K. M., *Tetrahedron* **54**, 4097–4106 (1998).

212. Chucholowski, A., Masquelin, T., Obrecht, D., Stadlwieser, J., and Villalgordo, J. M., *Chimia* **50**, 525–530 (1996).

213. Flynn, D. L., Devraj, R. V., Naing, W., Parlow, J. J., Weidner, J. J. and Yang, S., *Med. Chem. Res.* **8**, 219–243 (1998).

214. Starkey, G. W., Parlow, J. J. and Flynn, D. L., *Bioorg. Med. Chem. Lett.* **8**, 2385–2390 (1998).

215. Kaldor, S. W., Fritz, J. E., Tang, J. and McKinney, E. R., *Bioorg. Med. Chem. Lett.* **6**, 3041–3044 (1996).

216. Xu, W., Mohan, R. and Morrissey, M. M., *Bioorg. Med. Chem. Lett.* **8**, 1089–1092 (1998).

217. Coppola, G. M., *Tetrahedron Lett.* **39**, 8233–8236 (1998).

218. Weidner, J. J., Parlow, J. J. and Flynn, D. L., *Tetrahedron Lett.* **40**, 239–242 (1999).

219. Creswell, M. W., Bolton, G. L., Hodges, J. C. and Meppen, M., *Tetrahedron* **54**, 3983–3998 (1998).

220. Ault-Justus, S. E., Hodges, J. C. and Wilson, M. W., *Biotechnol. Bioeng.* **61**, 1722 (1998).

221. Ryder, T. R., Hu, L.-Y., Rafferty, M. F., Millerman, E., Szoke, B. G. and Tarczy-Homoch, K. *Bioorg. Med. Chem. Lett.* **9**, 1813–1818 (1999).

222. Siegel, M. G., Chaney, M. 0., Bruns, R. F., Clay, M. P., Schober, D. A., Van Abbema, A. M., Johnson, D. W., Cantrell, B. E., Hahn, P. J., Hunden, D. C., Gehlert, D. R., Zarrimnayeh, H., Omstein, P. L., Zimmerman, D. M. and Koppel, G. A., *Tetrahedron* **55**, 11619–11639 (1999).

223. Parlow, J. J., Naing, W., South, M. S. and Flynn, D. L., *Tetrahedron Lett.* **38**, 7959–7962 (1997).

224. Hay, A. M., DeWitt, S. H., MacDonald, A. A. and Ramage, R., *Synthesis* 1979–1985 (1999).

225. Parlow, J. J., Mischke, D. A. and Woodard, S. S., *J. Org. Chem.* **62**, 5908–5919 (1997).

226. Ferritto, R. and Seneci, P., *Drugs Future* **23**, 643–654 (1998).

227. Flynn, D. L., Devrai, R. V. and Parlow, J. J., *Curr. Opin. Drug Discov. Dev.* **1**, 41–50 (1998).

228. Booth, R. J. and Hodges, J. C., *Acc. Chem. Res.* **32**, 18–26 (1999).

229. Parlow, J. J., Devrei, R. V. and South, M. S., *Curr. Opin. Chem. Biol.* **3**, 320–336 (1999).

230. Flynn, D. L., *Med. Res. Rev.* **19**, 408–431 (1999).

231. Parlow, J. J. and Flynn, D. L., *Tetrahedron* **54**, 4013–4031 (1998).

232. Ouyang, X., Armstrong, R. W. and Murphy, M. M., *J. Org. Chem.* **63**, 1027–1032 (1998).

233. Morphy, J. R., Rankovic, Z. and Rees, D. C., *Tetrahedron Lett.* **37**, 3209–3212 (1996).

234. Brown, A. R., Rees, D. C., Rankovic, Z. and Morphy, J. R., *J. Am Chem. Soc.* **119**, 3288–3295 (1997).

235. Keating, T. A. and Armstrong, R. W., *J. Am. Chem. Soc.* **118**, 2574–2583 (1996).

236. Brown, S. D. and Armstrong, R. W., *J. Am. Chem. Soc.* **118**, 6331–6332 (1996).

237. Brown, S. D. and Armstrong, R. W., *J. Org. Chem.* **62**, 7076–7077 (1997).

238. Cuny, G. D., Cao, J., Sidhu, A. and Hauske, J. R., *Tetrahedron* **55**, 8169–8178 (1999).

239. Hu, Y., Baudart, S. and Porco, J. A. Jr., *J. Org. Chem.* **64**, 1049–1051 (1999).

240. Caldarelli, M., Habermann, J. and Ley, S. V., *J. Chem. Soc., Perkin Trans. I* 107–110 (1999).

241. Han, H., Wolfe, M. M., Brenner, S. and Janda, K. D., *Proc. Natl. Acad. Sci. USA* **92**, 6419–6423 (1995).

242. Bayer, E. and Mutter, M., *Nature* **237**, 512–513 (1972).

243. Bonora, G. M., Scremin, C. L., Colonna, F. P. and Garbesi, A., *Nucl. Acid Res.* **18**, 3155–3159 (1990).

244. Douglas, S. P., Whitfield, D. M. and Krepinsky, J. J., *J. Am. Chem. Soc.* **113**, 5095–5097 (1991).

245. Vandersteen, A. M., Han, H. and Janda, K. D., *Mol. Diversity* **2**, 89–96 (1996).

246. Han, H. and Janda, K. D., *J. Am. Chem. Soc.* **118**, 2539–2544 (1996).

247. Han, H. and Janda, K. D., *J. Am. Chem. Soc.* **118**, 7632–7633 (1996).

248. Han, H. and Janda, K. D., *Tetrahedron Lett.* **38**, 1527–1530 (1997).

249. Han, H. and Janda, K. D., *Angew. Chem. Int. Ed. Engl.* **36**, 1731–1733 (1997).

250. Wentworth Jr., P., Vandersteen, A. M. and Janda, K. D., *Chem. Commun.*, 759–760 (1997).

251. Brain, C. T., Paul, J. M., Loong, Y. and Oakley, P. J., *Tetrahedron Lett.* **40**, 3275–3278 (1999).

252. Sieber, F., Wentworth, P., Jr., Toker, J. D., Wentworth, A. D., Metz, W. A., Reed, N. N. and Janda, K. D., *J. Org. Chem.* **64**, 5188–5192 (1999).

253. Rafai Far, A. and Tidwell, T. T., *J. Comb. Chem.* **1**, 458–460 (1999).

254. Hori, M. and Janda, K. D., *J. Org. Chem.* **63**, 889–894 (1998).

255. Zhao, X.-y., Jung, K. W. and Janda, K. D., *Tetrahedron Lett.* **38**, 977–980 (1997).

256. Zhao, X.-y. and Janda, K. D., *Tetrahedron Lett.* **38**, 5437–5440 (1997).

257. Zhao, X. -y. and Janda, K. D., *Bioorg. Med. Chem. Lett.* **8**, 2439–2442 (1998).

258. Zhao, X.-y., Metz, W. A., Sieber, F. and Janda, K. D., *Tetrahedron Lett.* **39**. 8433–8436 (1998).

259. Benaglia, M., Cinquini, M. and Cozzi, F., *Tetrahedron Lett.* **40**, 2019–2020 (1999).

260. Benaglia, M., Annunziata, R., Cinquini, M., Cozzi, F. and Ressel, S., *J. Org. Chem.* **63**, 8628–8629 (1998).

261. Molteni, V., Annunziata, R., Cinquini, M., Cozzi, F. and Benaglia, M., *Tetrahedron Lett.* **39**, 1257–1260 (1998).

262. Blettner, C. G., Koenig, W. A., Stenzel, W. and Schotten, T., *Synlett* 295–297 (1998).

263. Pan, P.-C. and Sun, C.-M., *Tetrahedron Lett.* **39**, 9505–9508 (1998).

264. Shey, J.-Y. and Sun, C.-M., *Synlett* 1423–1425 (1998).

265. Shey, J.-Y. and Sun, C.-M., *Bioorg. Med. Chem. Lett.* **9**, 519–522 (1999).

266. Blettner, C. G., Koenig, W. A., Stenzel, W. and Schotten, T., *Tetrahedron Lett.* **40**, 2101–2102 (1999).

267. Blettner, C. G., Koenig, W. A., Ruehter, G., Stenzel, W. and Schotten, T., *Synlett* 307–310 (1999).

268. Ho, K.-C. and Sun, C.-M., *Bioorg. Med. Chem. Lett.* **9**, 1517–1520 (1999).

269. Pan, P.-C. and Sun, C.-M., *Bioorg. Med. Chem. Lett.* **9**, 1537–1540 (1999).

270. Pan, P.-C. and Sun, C.-M., *Tetrahedron Lett.* **40**, 6443–6446 (1999).

271. Yeh, C.-M. and Sun, C.-M., *Synlett* 810–812 (1999).

272. Sun, C.-M. and Shey, J.-Y., *J. Comb. Chem.* **1**, 361–363 (1999).

273. Cheng, S., Comer, D. D., Myers, P. L. and Saunders, J., *Tetrahedron Lett.* **40**, 8975–8978 (1999).

274. Yeh, C.-M. and Sun, C.-M., *Tetrahedron Lett.* **40**, 7247–7250 (1999).

275. Sieber, F., Wentworth, P. Jr. and Janda, K., *J. Comb. Chem.* **1**, 540–546 (1999).

276. Gravert, D. J. and Janda, K. D., *Chem. Rev.* **97**, 489–509 (1997).

277. Gravert, D. J. and Janda, K. D., *Curr. Opin. Chem. Biol.* **1**, 107–113 (1997).

278. Gravert, D. J. and Janda, K. D., *Drugs Future* **22**, 1147–1150 (1997).

279. Harwig, C. W., Gravert, D. J. and Janda, K. D., *Chemtracts* **12**, 1–26 (1999).

280. Wentworth, P. Jr. and Janda, K. D., *Chem. Commun.* 1917–1924 (1999).

281. Wentworth, P. Jr., *Trends Biotechnol.* **17**, 448–452 (1999).

282. Yoon, J., Cho, C.-W., Han, H. and Janda, K. D., *Chem. Commun.* 2703–2704 (1998).

283. Goldschmidt, S. and Wick, M., *Liebigs Ann. Chem.* **575**, 217–223 (1952).

284. Huang, J. and Hu, Y., *J. Appl. Polym. Sci.* **47**, 1503–1511 (1993).

285. Bergbreiter, D. E. and Caraway, J. W., *J. Am. Chem. Soc.* **118**, 6092–6093 (1996).

286. Vanden Eynde, J. J. and Rutot, D., *Tetrahedron* **55**, 2687–2694 (1999).

287. Gravert, D. J., Datta, A., Wentworth, P., Jr. and Janda, K. D., *J. Am. Chem. Soc.* **120**, 9481–9495 (1998).

288. Chen, S. and Janda, K. D., *J. Am. Chem. Soc.* **119**, 8724–8725 (1997).

289. Chen, S. and Janda, K. D., *Tetrahedron Lett.* **39**, 3943–3946 (1998).

290. Narita, M., *Bull. Soc. Chem. Jpn.* **51**, 1477–1480 (1978).

291. Paquette, L. A., *Org. Synth.* **73**, 44–49 (1996).

292. Corey, E. J., Niimura, K., Konishi, Y., Hashimoto, S. and Hamada, Y., *Tetrahedron Lett.* **27**, 2199–2202 (1986).

293. Casy, G., Patterson, J. W. and Taylor, R. J. K., *Org. Synth.* **67**, 193–201 (1989).

294. Kim, R. M., Manna, M., Hutchins, S. M., Griffin, P. R., Yates, N. A., Bernick, A. M. and Champman, K. T., *Proc. Natl. Acad. Sci. USA* **93**, 10012–10017 (1996).

295. Chang, J., Oyelaran, O., Esser, C. K., Kath, G. S., King, G. W., Uhrig, B. G., Konteatis, Z., Kim, R. M. and Chapman, K. T., *Tetrahedron Lett.* **40**, 4477–4480 (1999).

296. Gitsov, I., Wu, S. and Ivanova, P. T., *Polym. Mater. Sci. Eng.*, **77**, 214–215 (1997).

297. Langer, P., Ince, S. J. and Ley, S. V., *J. Chem. Soc., Perkin Trans. I* 3913–3916 (1998).

298. Worl, R. and Koster, H., *Tetrahedron* **55**, 2941–2956 (1999).

299. Worl, R. and Koster, H., *Tetrahedron* **55**, 2957–2972 (1999).

300. Mahajan, A., Chhabra, S. R. and Chan, W. C., *Tetrahedron Lett.* **40**, 4909–4912 (1999).

301. Kantchev, A. B. and Parquette, J. R., *Tetrahedron Lett.* **40**, 8049–8053 (1999).

302. Newkome, G. R., Weis, C. D., Moorefield, C. N., Baker, G. R., Childs, B. J. and Epperson, J., *Angew. Chem. Int. Ed.* **37**, 307–310 (1998).

303. Thompson, M. C. and Busch, D. H., *J. Am. Chem. Soc.* **86**, 213–217 (1964).

304. Anderson, S., Anderson, H. L. and Sanders, J. K. M., *Acc. Chem. Res.* **26**, 469–475 (1993).

305. Hoss, R. and Vogtle, F., *Angew. Chem. Int. Ed. Engl.* **33**, 375–384 (1994).

306. Hasenknopf, B., Lehn, J.-M., Kneisel, B. O., Baum, G. and Fenske, D., *Angew. Chem. Int. Ed. Engl.* **35**, 1838–1840 (1996).

307. Huc, I and Lehn, J.-M., *Proc. Natl. Acad. Sci. USA* **94**, 2106–2110 (1997).

308. Huc, I., Krische, M. J., Funeriu, D. P. and Lehn, J.-M., *Eur. J. Inorg. Chem.* 1415–1420 (1999).

309. Berl, V., Huc, I., Lehn, J.-M., DeCian, A. and Fischer, J., *Eur. J. Org. Chem.* 3089–3094 (1999).

310. Eliseev, A. V. and Nelen, M. I., *J. Am. Chem. Soc.* **119**, 1147–1148 (1997).

311. Polyakov, V. A., Nelen, M. I., Nazarpack-Kandlousy, N., Ryabov, A. D. and Eliseev, A. V., *J. Phys. Org. Chem.* **12**, 357–363 (1999).

312. Sakai, S., Shigemasa, Y. and Sasaki, T., *Tetrahedron Lett.* **38**, 8145–8148 (1997).

313. Klekota, B., Hammond, M. H. and Miller, B. L., *Tetrahedron Lett.* **38**, 8639–8642 (1997).

314. Klekota, B. and Miller, B. L., *Tetrahedron* **55**, 11687–11697 (1999).

315. Giger, T., Wigger, M., Audetat, S. and Benner, S. A., *Synlett* 688–691 (1998).

316. Hiraoka, S. and Fujita, M., *J. Am. Chem. Soc.* **121**, 10239–10240 (1999).

317. Rowan, S. J. and Sanders, J. K. M., *Chem. Commun.* 1407–1408 (1997).

318. Try, A. C., Harding, M. M., Hamilton, D. G. and Sanders, J. K. M., *Chem. Commun.* 723–724 (1998).

319. Hamilton, D. G., Feeder, N., Teat, S. J. and Sanders, J. K. M., *New J. Chem.* **22**, 1019–1021 (1998).

320. Rowan, S. J., Lukeman, P. S., Reynolds, D. J. and Sanders, J. K. M., *New J. Chem.* **22**, 1015–1018 (1998).

321. Cousins, G. R. L., Poulsen, S.-A. and Sanders, J. K. M., *Chem. Commun.* 1575–1576 (1999).

322. Brady, P. A. and Sanders, J. K. M., *Chem. Soc. Rev.*, 327–336 (1997).

323. Ganesan, A., *Angew. Chem. Int. Ed.* **37**, 2828–2831 (1998).

324. Klekota, B. and Miller, B. L., *Trends Biotechnol.* **17**, 205–209 (1999).

325. Eliseev, A. V. and Lehn, J.-M., *Curr. Top. MicrobioL Immunol.* **243**, 159–172 (1999).

326. Reinhoudt, D. N., Timmerman, P., Cardullo, F. and Crego-Calama, M., *NATO ASI Ser., Ser. C* **527**, 181–195 (1999).

327. Lehn, J.-M., *Chem. Eur. J.* **5**, 2455–2463 (1999).

328. Brady, P. A. and Sanders, J. K. M., *J. Chem. Soc., Perkin Trans. I* 3237–3253 (1997).

9 Applications of Synthetic Libraries

This chapter will try to justify the existence of combinatorial technologies and in particular of small organic molecule libraries, explaining the major impact they have had on the scientific community, especially in the last decade. Both the applications of synthetic libraries and the research fields that have benefited from combinatorial technologies are cited and discussed.

Our analysis naturally starts from *pharmaceutical applications*, which have been the historical driver of combinatorial technologies and still represent the most relevant field of application for any type of combinatorial library. A brief, but complete, description of the drug discovery process will help in highlighting the areas where combinatorial technologies have had a significant impact, and a few examples will be given to support the utility of synthetic organic libraries for pharmaceutical applications. Related issues that arise during the drug discovery process, such as *data management* of the chemical and biological information generated and *patenting* of synthetic libraries and *physicochemical or metabolic screens* to profile drug candidates from combinatorial libraries, which are often relevant for the whole field of combinatorial technologies, are also covered in this section. The application of combinatorial technologies to similar fields, such as agricultural and veterinary research, will then be briefly mentioned, highlighting the almost complete absence of reports in these fields up to now.

Catalysis is another popular field of application for combinatorial technologies. Discrete libraries of catalysts and ligands, or libraries of combinations of catalysts and ligands, will be described and exemplified. More intriguing encoded pool libraries of catalysts and ligands will be mentioned. Combinatorial optimization both of catalytic protocols and reaction conditions will also be covered. Finally, *molecular recognition* will be presented. Among others, libraries of synthetic receptors and libraries of coordination complexes are described, together with the adopted screening procedures to determine their activity.

9.1 PHARMACEUTICAL APPLICATIONS

9.1.1 General Considerations

Pharmaceutical research has constantly evolved during the years, and this evolution in turn has brought significant ameliorations to the patient population in curing various diseases. Evolution of the pharmaceutical industry, though, has also led to detrimental consequences for the companies active in this field. The most significant effects are increased competition, changes in the patients' needs in terms of acceptable drug profiles (including costs), and an increase in regulatory requirements to obtain new drug approvals. The macroscopic effects of these and other factors are summarized in two major phenomena. First, the cost to launch a new drug on the market has increased eight times since 1976, going from around \$50 million to an impressive average of \$400 million, which is sustainable only by major pharmaceutical companies. Second, development time has almost doubled since 1960, reaching the current average figures of 14–15 years and reducing the time window to capitalize on sales before patent expiration of a successful drug and insurgence of generic competition.

The need to counteract this tendency has led to the search for faster and/or less expensive processes for the discovery and development of drug candidates. Biological and chemical techniques to synthesize and screen early or late drug candidates in a high-throughput fashion (HTS, combinatorial technologies) are impacting positively on these factors, as already mentioned (see Section 4.1.2). In the following sections we cover the so-called *drug discovery process*, which is schematized in Fig 9.1, describing and discussing the impact of combinatorial technologies on each of the steps involved. A recent review has covered exhaustively the economic impact of combinatorial technologies in drug discovery (1).

9.1.2 From Gene to Function

Disease selection is the first step of drug discovery. The choice of one or more disease areas in which to research for new targets is based on the so-called opportunity-mapping process, which ranks the disease areas according to the unmet medical need and the risk involved. Ideally efforts are concentrated in areas of high unmet medical need, and the disease portfolio of a pharmaceutical company will include some risky areas, where little, or nothing, is known, as well as more assessed diseases for which significant knowledge is available. Combinatorial technologies are not involved at this level of the drug discovery process.

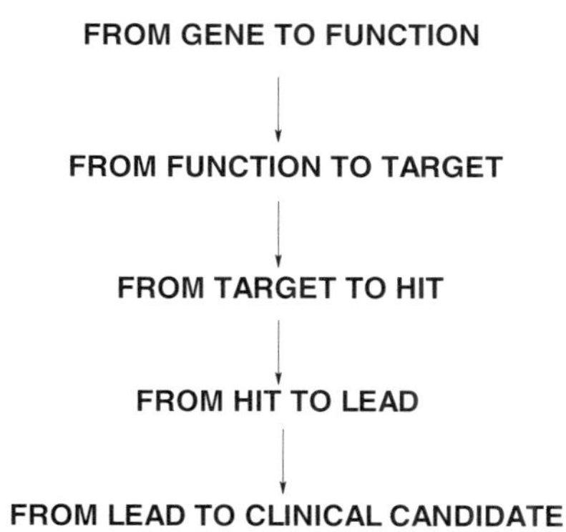

Figure 9.1 Key steps of the drug discovery process.

Target identification is the next, crucial phase. Currently marketed drugs interact with about 400 known targets (genes or gene products), which are mostly receptors, enzymes, or ion channels. The completion of the Human Genome Project, which is expected in the next five years, will unravel all of the approximately 100,000 human genes and will provide their detailed genetic and physical maps. Of these, the estimated number of important genes to cure diseases ranges between 3000 and 10,000. All these potential targets will be known to everybody, and the competitive advantage will be to identify, faster than the competition, the pharmaceutically relevant targets and to associate the significant ones to a given disease. The conventional sources for target identification (literature, in-house research, competitor information) are already and will be more heavily overwhelmed in the future by susceptibility genetics (2, 3), which establishes an association between a gene and a disease by studying a patient population. The link may be determined by differential gene expression (DGE) analysis (4, 5), which analyzes the alterations of gene transcription comparing healthy versus diseased tissues, by proteomics (6, 7), which profile the proteins from a cellular or tissue source, again comparing normal versus diseased samples, or by bioinformatic data mining (8, 9), which browses the genetic databases and identifies potential new targets by comparison of their sequence with known genes of interest. The identification of these targets may take advantage, at some level, of biological libraries (see Chapter 10) as tools, but synthetic libraries are not significantly involved in this phase.

9.1.3 From Function to Target

The identified target is first *validated*, discriminating between relevant and nonrelevant targets for a given disease. Even if a target is related to a disease, it may not be essential, and to interfere with it may not lead to novel, useful drugs. Functional genomics, or functional gene analysis (FGA) (10, 11), is fundamental in linking genomics research with the discovery of disease-relevant targets, addressing the question of their function or dysfunction in disease states. Current technologies in FGA rely on gene under- (12) or overexpression (13), on knock-out and interaction studies in cell cultures or in transgenic animals (14, 15), and on tissue distribution of new genes or on in vivo pharmacological studies in animals. These technologies use a wide variety of techniques, including combinatorial technology–related tools such as ribozymes, antisense oligonucleotides, aptamers, and antibodies. Synthetic libraries, though, are not involved in this phase of the drug discovery process; the only exception is a novel approach, termed chemical genetics and described in the next subsection (see also Sections 7.5.1 and 7.5.2), which will provide significant assistance to elucidate biological pathways relevant to diseases and to validate novel targets via their interaction with small molecules.

The consequence of validation is the *selection* of a target that is promising; that is, affecting it will likely result in curing the disease of interest. An important criterion is target tractability, related to the ease of screening chemical diversity for activity on the target. In fact, there is no therapeutic use for a target where one or preferably more screening assays cannot be designed and realized (see also next section).

9.1.4 Chemical Genetics: From Chemical Entities to Valuable Targets

The use of chemical tools to elucidate complex biological pathways is not a recent discovery; many past examples determined the identity and the function of a target through a chemical entity interacting with the macromolecule. This pharmacological approach has been largely obscured by the advent of genetics and genomics (see Sections 9.1.2 and 9.1.3), but has still the power to unravel, or to provide information about complex pathways for which genetic manipulation protocols are not yet assessed.

A higher throughput version of this approach takes advantage of combinatorial technologies and has been recently developed and exploited with the name of *chemical genetics* (16–18). More specifically, two complementary approaches have been defined and will be exemplified:

- *forward chemical genetics* (FCG), where a large collection of drug-like compounds is tested for its modulation of a complex biological pathway. If one, or more, modulators are found, their molecular target (either novel, or not known to interact with the pathway) is identified and its role in the pathway is elucidated. This provides both useful chemical entities to be progressed and unvaluable proprietary information regarding novel targets for relevant diseases;
- *reverse chemical genetics* (RCG), where a small molecule which targets a specific pathway or which causes a specific phenomenon is incubated with a large genetic collection, to spot if the expression of any of the gene products is affected. Knowing that the whole human genome will be soon available for such experiments in microarrays, and that most of the sequenced genes will not be immediately linked to a function, this approach will be important for the ambitious goal "to identify a small molecule partner for every gene product" (16).

Mayer et al. (19) reported the use of FCG to identify compounds that affect the cellular mitotic process through a screening cascade reported in Fig. 9.2. A commercially available, diverse collection of 16,320 compounds was screened on a whole cell primary cytoblot assay (20) measuring the phosphorylation level of a nucleolar protein called nucleolin (step a). This protein is phosphorylated when cells enter mitosis, and inhibitors of the mitotic process are expected to increase the level of phosphonucleolin. 139 positive compounds were able both to penetrate the cell and to increase the level of phosphorylated nucleolin.

As tubulin is a major target for antimitotics, the positives were screened for their *in vitro* effects on tubulin polymerization (step b). 52 compounds inhibited tubulin polymerization, and one stimulated it; these structures (see a few representatives, **9.1–9.5**, in Fig. 9.3), which could represent a starting point for an optimization program, were archived. The remaining 86 positives not interacting with tubulin were progressed and tested in mammalian epithelial kidney cells (BS-C-1) stained with various fluorescent reagents to visualize microtubules, actin and chromatin, i.e. the essential structural/mechanochemical mitiotic spindle (step c). 27 compounds, although increasing the number of normal mitotic cells and thus confirming the result of primary screening, did not have any immediate effect and were deemed to act on

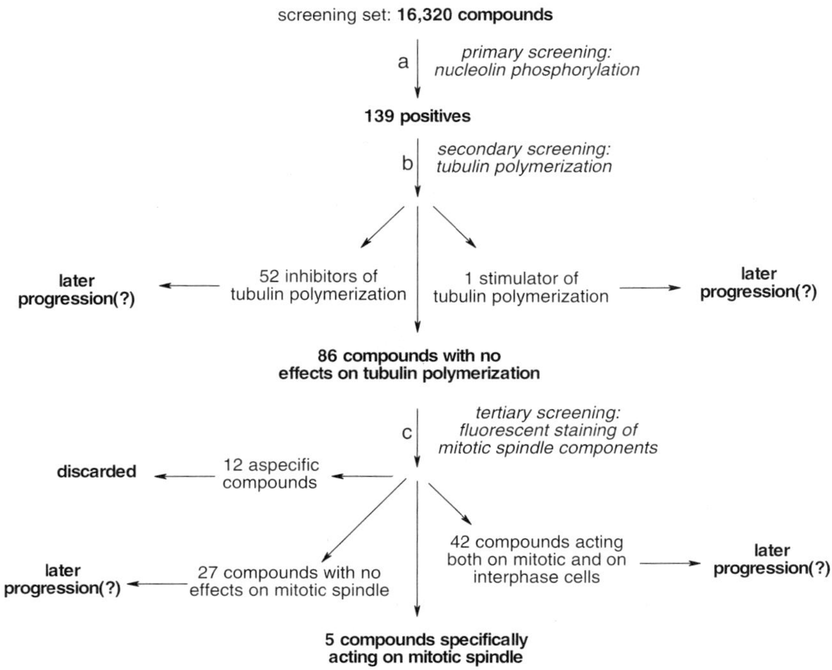

Figure 9.2 Forward Chemical Genetics (FCG): screening aimed to compounds interacting with the mitotic spindle components during mitosis.

other components of the mitotic process. These compounds were archived. 42 compounds affected the mitotic spindle (chromatin and microtubules, no effect on actin cytoskeleton) but affected microtubules also in interphase cells not yet entered into the mitotic phase. These compounds were also archived. 12 compounds showed multiple, aspecific effects and were thus discarded (Fig. 9.2).

The five remaining compounds were specifically acting on the mitotic spindle, and did not alter its components in interphase cells. One of them in particular prevented the formation of the spindle in most mitotic cells, replacing it with a monoastral microtubule formation surrounded by chromosomes; the compound (**9.6**, Fig. 9.3) was thus named monastrol. The authors compared monastrol effects with several published effects (21–23) related to inhibition of Eg5, a member of the BimC kinesin family, and showed monastrol to be the α selective Eg5 inhibitor (Eg5-driven microtubule motility inhibition = 14 μM). This compound is both the first permeable and selective inhibitor of a specific kinesin, and may have many possible applications as a tool or as the starting point for a chemical optimization program.

The same compound collection was screened by Stockwell et al. (24) on a primary FCG assay directed towards activators of a regulatory reporter gene (p3TPLux)

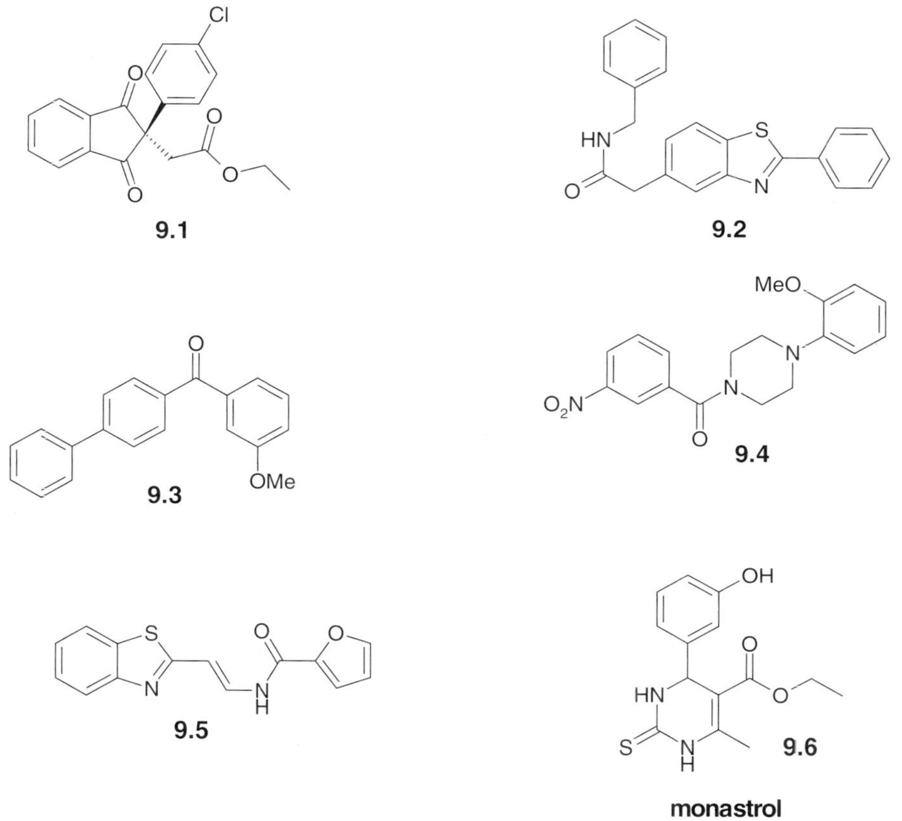

Figure 9.3 Structures of inhibitors of tubulin polymerization (**9.1–9.5**) and of monastrol **9.6**, an inhibitor of mitotic spindle assembly, found through FCG.

sensitive to TGF-β (25) (step a, Fig. 9.4). The determination of the mechanism of action for the activators of p3TPLux should have provided insights into the TGF-β signalling pathway. Four positive compounds (**9.7–9.10**, Fig. 9.4) were identified and submitted to a secondary cytoblot screen for 5-bromodeoxyuridine incorporation (20) to determine their inhibition of DNA synthesis (step b, Fig. 9.4). All of them inhibited DNA synthesis, as happens for TGF-β signalling.

Compounds **9.8** and **9.10** were submitted to an RCG screen using the vast majority of *Saccaromyces cerevisiae* genome (around 5800 genes) on a microarray format (step a, Fig. 9.5). The high similarity between the human and fungal genome was expected to ensure useful indications for the molecular targets of the two molecules. The gene expression of the whole microarray, checked with the use of fluorescent dyes, did not change in the presence of **9.10**. The expression of five genes (Fig. 9.5) was significantly increased by **9.8**. Two out of the five genes are involved in metal transport, respectively

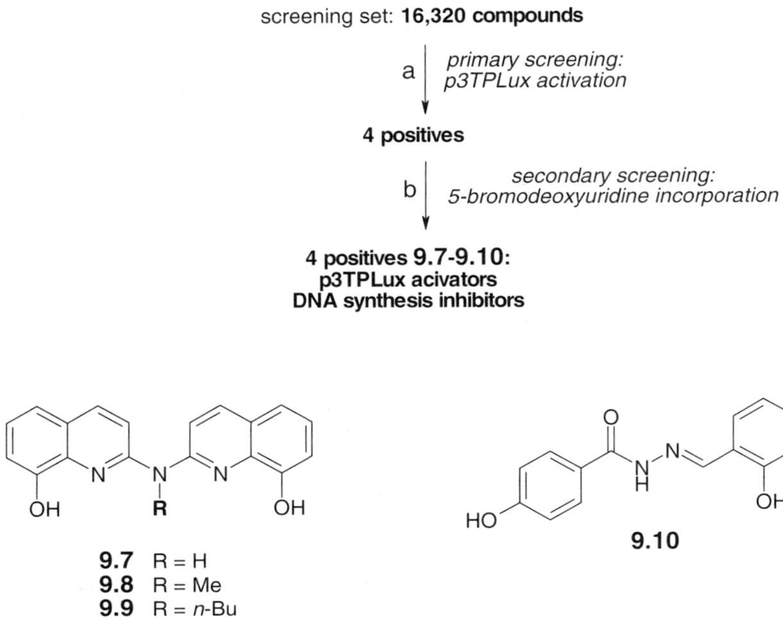

Figure 9.4 Forward Chemical Genetics (FCG): screening aimed to compounds activating the reporter gene p3TPLux and interfering with TGF-β signalling, and structures of activators **9.7–9.10**.

for Zn (*ZRT1*) (26) and Cu-depended Fe transport (*FET3*) (27); *HSP26* is a heat shock protein activated by osmotic stress (28), and the two uncharacterized genes are likely to be involved in similar, metal-related processes.

Compounds **9.8** and **9.10** were further characterized for their metal binding capacity; **9.8** confirmed a high and quite selective affinity for Zn^{2+}, Fe^{3+} and Cu^{2+}, and its activation of TGF-β signaling was reversed (Fe^{3+}) or largely reduced (Zn^{2+}, Cu^{2+}) by addition of metal chloride salts. RCG thus provided insight on TGF-β signalling (several potential genes involved) and on the relevance of metal ions for this process.

9.1.5 From Target to Hit

This is the drug discovery phase where the accumulated biological knowledge is converted into relevant, *novel chemical entities* that will be progressed along the line to hopefully become new drugs on the market. The interdependent progression of implied biological and chemical activities is thoroughly described below, with the assistance of several examples.

This phase requires the initial development of an assay able to measure significant biological interactions of a molecule with the target. This is the foundation of pharmaceutical research, and traditionally there was no real need for close interaction

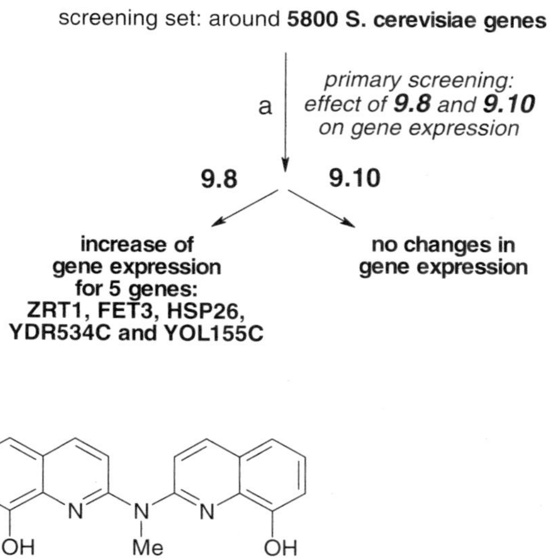

9.8
metal binder
with high affinity
for Zn^{2+}, Fe^{3+} and Cu^{2+}

Figure 9.5 Reverse Chemical Genetics (RCG): screening aimed to elucidate the mode of action and the main features of compounds **9.8** and **9.10**.

at the assay development stage between chemists and biologists. Medicinal chemists, in fact, prepared large amounts of molecules, which were then tested by the biologists on a variety of assays and were fully characterized. Eventually, the acquired structure–activity information was used to orient further chemical efforts. While this classical approach was accurate and successful on many occasions, the low throughput of both synthesis and screening was not appropriate for a quick and economical drug discovery process. This is the engine that gave rise first to high-throughput screening (HTS) and then to its chemical counterpart, high-throughput chemical synthesis or, as it is better known, combinatorial chemistry (see Section 4.1.2 for the historical background).

The significant change is represented by a quick and inexpensive biological filter to rapidly discard nonactive molecules and to focus further efforts on a few confirmed and attractive positives. The assay, thus, is automated to test even hundreds of thousands of compounds in a relatively short time, and this introduces some limitations. The following are among the most relevant:

- Only *simple assays can be automated*, while complex protocols will cause problems and will not produce reliable results.
- Automation must not affect the *robustness* of the assay, because automated protocols cannot afford last-minute adjustments or modifications.

- Automated assays must be *miniaturized*, partly because they will typically receive small quantities of compound from collections or chemical libraries, but also to reduce the costs of biological reagents.
- Automated assays must be developed using *several formats*, on an assay-specific base, and it is essential to possess the necessary expertise to select and automate the best choice for a given target.
- *Automated instrumentation* and *data management software* are necessary to analyze and archive the large number of data produced by a screening campaign.

In principle, any type of biological assay can be automated, and several excellent reviews have covered this topic (29–38). Assay development and automation are not covered further here, but their influence on chemistry at the early stages of drug discovery will be examined.

At the very start of the so-called *hit identification phase*, the project is driven by the available information. If the biological target is already well-characterized, either a pharmacophoric/structural model or even structures of known ligands/inhibitors are available. This structural information will determine the class(es) of compounds to be tested and the size of the screening set. This knowledge-based, or focused, approach calls for similarity with a model/structure, rather than for diversity, and the screening sets are typically smaller (see Sections 5.4.2 and 5.4.3). More focused approaches are dealt with in Section 9.1.7. A more frequent situation, due to the input of novel genomic-derived targets, deals with unknown or, at best, poorly characterized targets, which are less tractable but also highly rewarding if novel, active structures are identified. The goal here is to test significant chemical diversity on a single, automated primary assay to fish out positives. These are then profiled through a screening cascade of secondary assays, and one or more structures are eventually selected as starting points for an exploratory chemical project. Ideally, large sets (hundreds of thousands) of compounds from collections and from synthetic libraries are screened, providing that a reliable and robust primary screening assay has been set up. Pooling strategies to reduce the number of wells per screening can also be adopted (39).

The screening of a large, diversity-based set of compounds/libraries implies the *selection of candidates*. Selected libraries should be built around a relevant (possibly proprietary) common core scaffold, maximize their chemical diversity, and provide novel, druglike individuals according to well-known computational filters (see Section 5.3). Each pharmaceutically biased-targeted primary library is tested on many biological targets; thus a large number of library equivalents are prepared and stored to be used when needed. High-quality SP pool libraries are usually the preferred format, but also medium–large discrete libraries are becoming very popular as hit-seeking libraries. Each screening set contains library individuals or compounds from collections in very small amounts to fully exploit the synthetic efforts and maximize the compound availability for multiple screening.

The positive compounds from the primary screening are further profiled to check their usefulness. This characterization includes preliminary physicochemical property determination, toxicity data, and specificity/selectivity data when possible. The secondary screening cascade will restrict the screening outcome to a small number of confirmed hits, to be used as starting points for chemical modification projects (see Section 9.1.8).

9.1.6 Management of Combinatorial/Screening Data and Samples

Combinatorial technologies and HTS in drug discovery generate a huge amount of biological and chemical information. This section follows the flow of generated data from the design and synthesis of a library to its analytical characterization and its screening, ending with the extraction of activity data, the assessment of SAR, and the storage of the information.

The design of a primary or biased-targeted, hit-seeking library is aided by computational tools in selecting the most diverse set of library individuals according to the desired library size (see Section 5.4). The designed library is then prepared, and the individual structures are registered. The structures of hundreds of thousands of compounds, though, would require enormous efforts for their input into a database and a huge computing space for them to be stored and searched. The structures are thus stored via the reaction scheme and the building blocks (BBs) used to prepare the library. The example shown in Fig. 9.6 (40) shows how a 64,000,000-member library of hexapeptides, where 20 α-amino acids are randomized in each position, can be

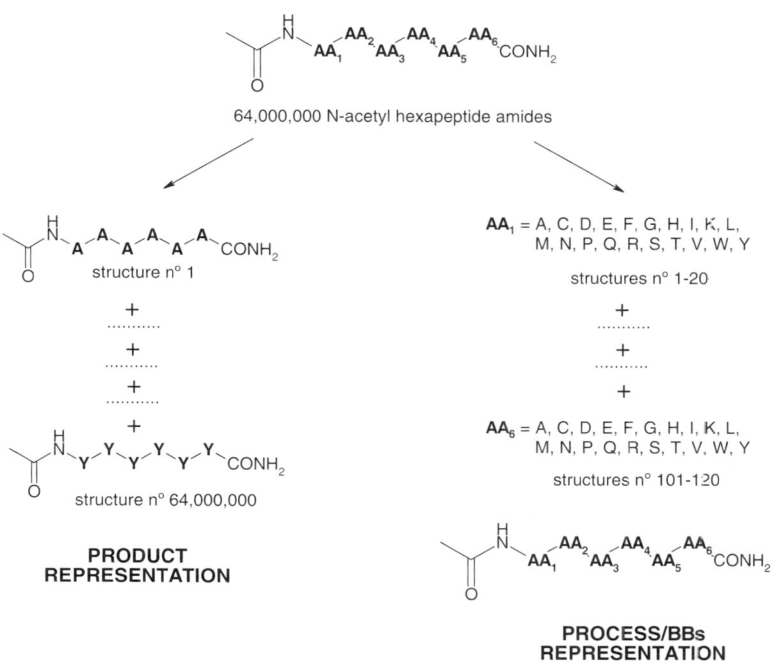

Figure 9.6 Product representation versus process/building block representation of a hypothetical 64,000,000-member peptide library.

condensed from 64,000,000 represented structures (*product representation*) to 120 monomers and a reaction scheme (*process/BB representation*). The combination of building blocks can be used at any time to extract and visualize each library component, providing that an unequivocal match can be made from the pathway and the building block structures. The example in Fig. 9.7 (41) shows how several alternatives in terms of regio- and stereochemistry may prevent the assignment of an unequivocal structure from a process/BB representation. The structures of amino acids (**M_1**), aldehydes (**M_2**), olefins (**M_3**), and acyl chlorides (**M_4**) do not allow the generation of the correct reaction products related to the olefin cycloaddition step (regiochemistry and stereochemical preference; see Fig. 9.7) via software programs.

The structural database must be linked to a *relational database* (42), which contains all the generated information for a discrete or a pool library sample. The relational database includes structural information (MW, molecular formula), physicochemical properties (log P, solubility, p*K*a), analytical results (peak values or spectra IDs from which the spectra can be found), and activity data from primary and secondary

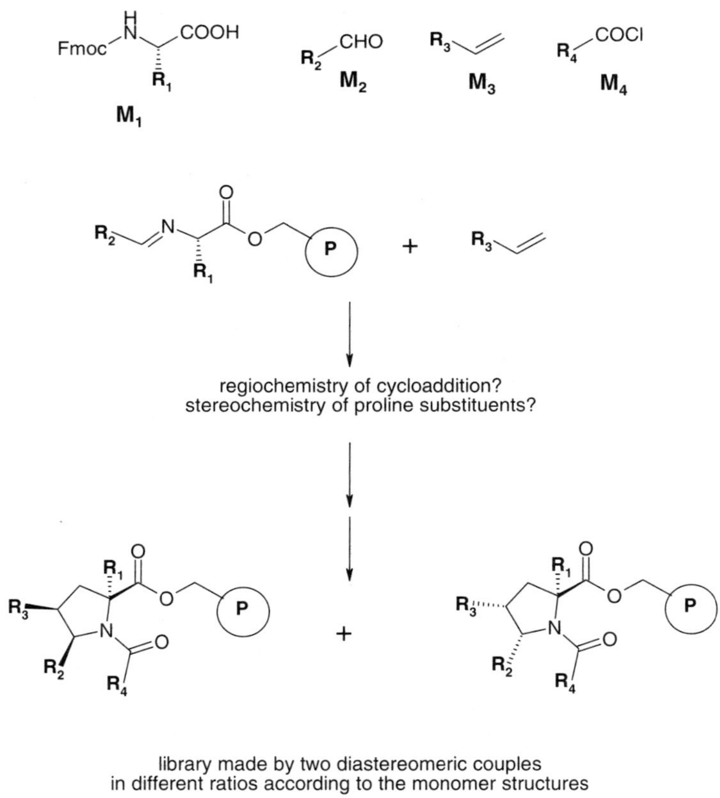

Figure 9.7 Process/BB representation of a hypothetical pyrrolidine library: regio- and stereochemical issues.

screenings plus any other available data. These data come from the library quality control, performed after the synthesis, from the HTS of a high-quality pharmaceutical-focused library, and from further profiling of some of its members. When the library is tested on many biological targets, the stored data become invaluable in judging the library usefulness and maybe also in performing related combinatorial efforts.

A *search* on the integrated structural/relational database system downloads each stored piece of information. Browsing the library, extracting SAR data related to structural subclasses, and pulling out library subsets to be tested in future screenings are the main goals of such a search. Proper data management of combinatorial and screening results is, and will become ever more, crucial to maximize the return of synthetic and biological efforts toward the discovery of novel drugs. A few excellent reviews have covered the issues related to data management of combinatorial libraries and HTS (40–46), and the reader should consult them for a more detailed picture of this relevant topic.

9.1.7 An Example: Synthesis of Chalcone-Based, Druglike Libraries

Powers (47) reported the synthesis in solution of a set of modular, discrete libraries containing >74,000 members based on the combinatorialization of a common intermediate scaffold (see also Sections 4.2.2 and 8.2.6). A core chalcone library **L1** was modified (libraries from libraries) (48), producing libraries **L2–L10** and expanding both the number and diversity of the compounds due to the smaller sets of different scaffold-derived libraries (*vide infra*).

Aldol condensation of acetophenones (**M$_1$**, 32 representatives) with aldehydes (**M$_2$**, 40 representatives) produced a high-quality core chalcone array **L1** (Fig. 9.8), with an excellent 96% average yield. The monomer structures are reported in Fig. 9.9. Combination of aromatic or heteroaromatic mono-, di-, or trisubstituted methyl ketones **M$_1$** with aromatic or heteroaromatic mono- or disubstituted aldehydes **M$_2$** produced 1280 chalcones with suitable, druglike MWs (452 > MWs > 186) and lipophilicity.

Condensation of **L1** with various reagents to produce disubstituted five-member heterocycles was envisaged (Fig. 9.10). Reaction with hydroxylamine **9.11** provided

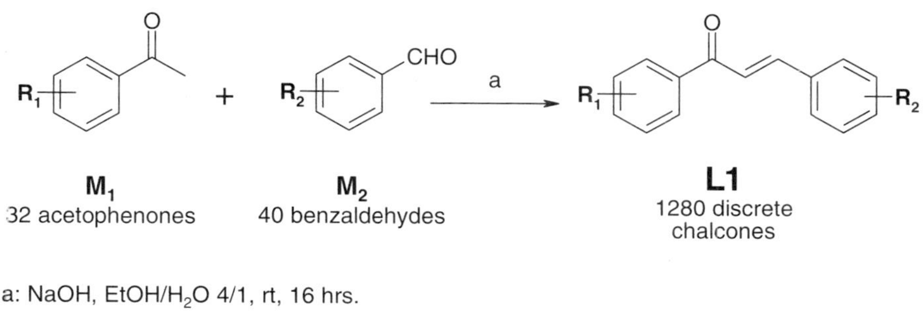

Figure 9.8 Synthesis of the solution-phase discrete chalcone library **L1**.

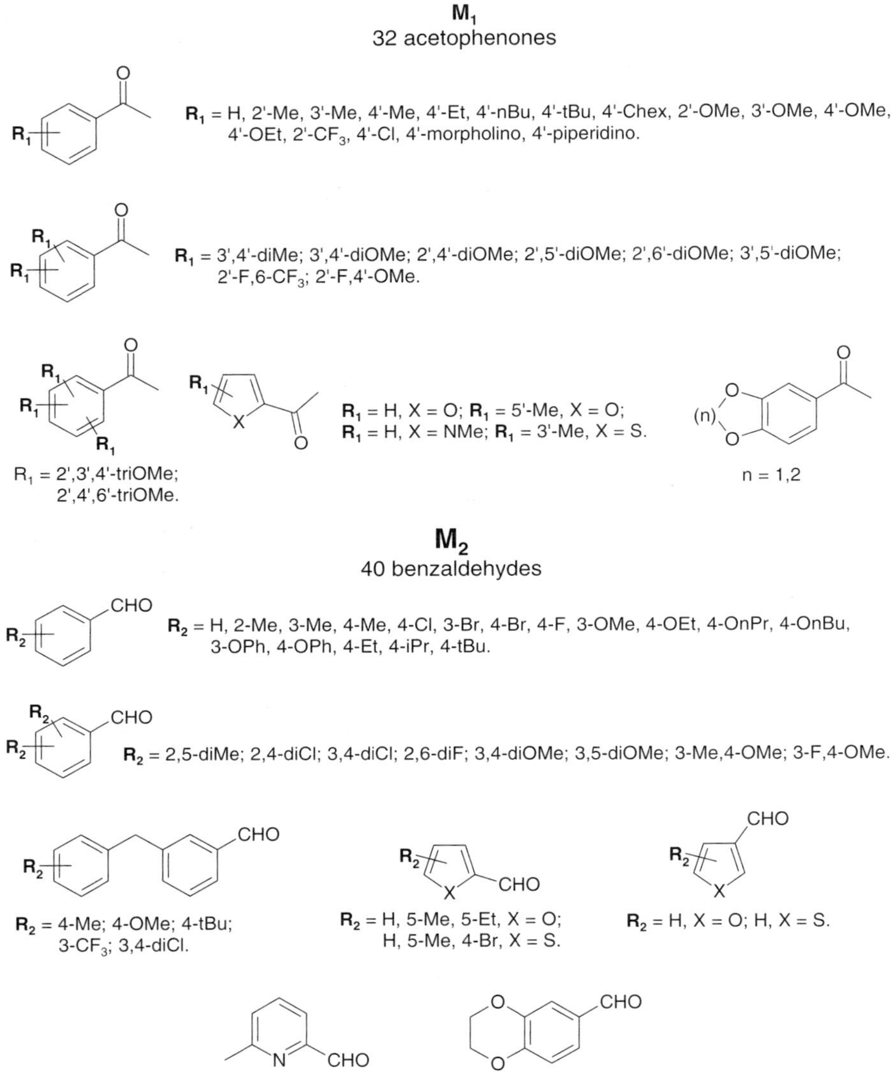

Figure 9.9 Monomer sets **M1–M2** used for the synthesis of the solution-phase discrete chalcone library **L1**.

the daughter isoxazoline library **L2** (1280 members), while condensation with hydrazine hydrate **9.12** was abandoned as it did not produce a clean set of compounds during the chemistry assessment. Phenylhydrazines **9.13a–f** performed better, and the trisubstituted pyrazoline library **L3** (7680 members) was obtained.

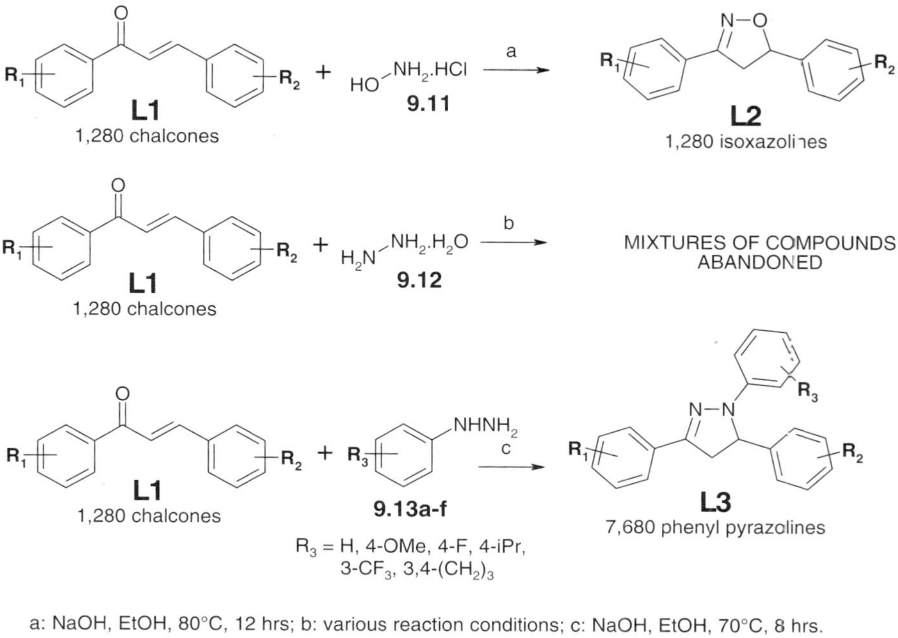

Figure 9.10 Solution-phase discrete five-membered heterocyclic libraries **L2–L3** obtained from the solution-phase discrete chalcone library **L1**.

Trisubstituted six-member carbocycles were also prepared. Condensation of **L1** with commercial acetoacetanilides **9.14a–f** produced, by one-pot Michael addition and Robinson annulation, the cyclohexenone library **L4** (7680 members, Fig. 9.11). Noncommercial acetoacetamides **M1**, prepared by condensation of diketene **9.15** with 40 primary and secondary amines (structures not shown in the paper), were used as monomers and condensed with a 320-member **L1** subset to give the expanded cyclohexenone library **L5** (12,800 members). Difficulties in preparing extremely large discrete libraries, even using fully automated robotic workstations, and the need to diversify the screening set obliged the authors to limit the number of monomers and chalcones for any specific library/scaffold (see also **L10** below).

Polysubstituted dihydropyridines were also considered as targets. Condensation of **L1** with enamino ester **9.16** and enamino nitrile **9.17** under standard Hantsch conditions only proved successful for the latter, furnishing the pyridine library **L6** (1280 members). Oxidation of the initially formed dihydropyridines is presumably favored due to the highly conjugated system formed (Fig. 9.12). Hantsch condensation with cyclic enamino esters **9.18** and **9.19a–f** was also successful, providing respectively the bicyclic libraries **L7** (1280 members) and **L8** (7680 members, Fig. 9.12).

Finally, two polycyclic systems were built. First, the cyclic guanidine embedded into the 2-aminobenzimidazole nucleus **9.20a–f** was used to build the tricyclic library **L9** (7680 members, Fig. 9.13); then a complex spiro-polyheterocycle library **L10**

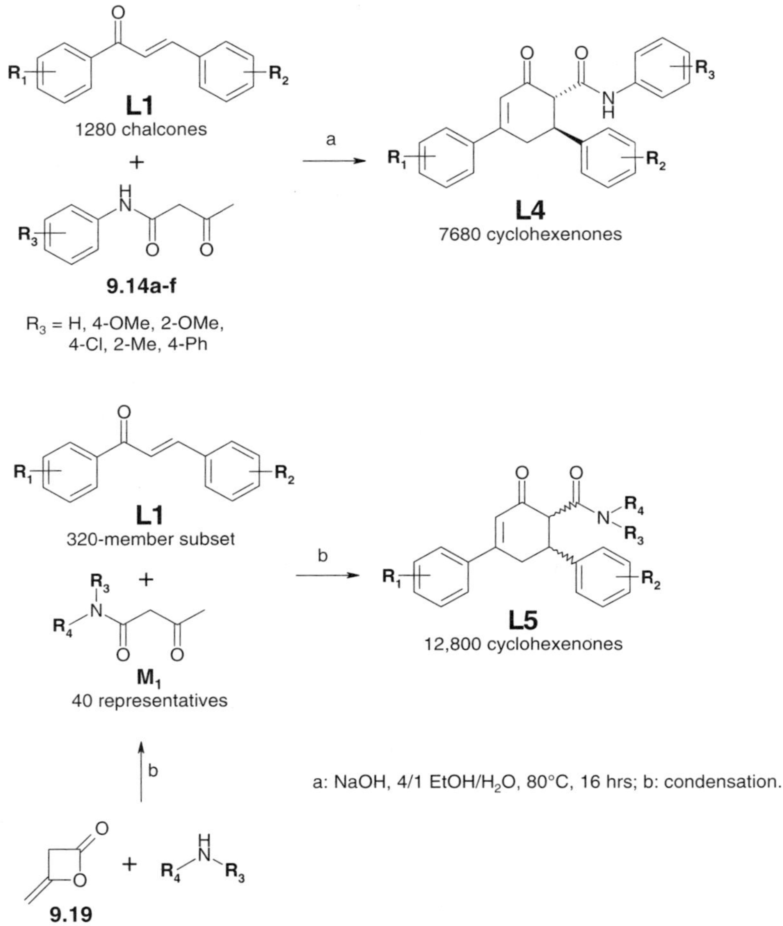

Figure 9.11 Solution-phase discrete six-membered heterocyclic libraries **L4–L5** obtained from the solution-phase discrete chalcone library **L1**.

(25,600 members) was prepared by one-pot condensation of isatins (**M1**, 16 representatives), α-amino acids (**M2**, 20 representatives), and a chalcone subset (**M3**, 80 representatives from **L1**, Fig. 9.13). The monomers for the synthesis of **L10** were again selected to maximize the library diversity and to produce a druglike profile for the library individuals to be considered as drug candidates.

The screening set of roughly 75,000 compounds was characterized analytically. Twenty-five percent of the compounds were submitted to HPLC/MS using ELSD, producing average purities of >85% for each sample. The generic libraries **L1–L10** and their assessed synthetic schemes could easily be expanded using larger sets of diversifying monomers in a more focused attempt, if one or more hits were found from a specific library, without significant efforts.

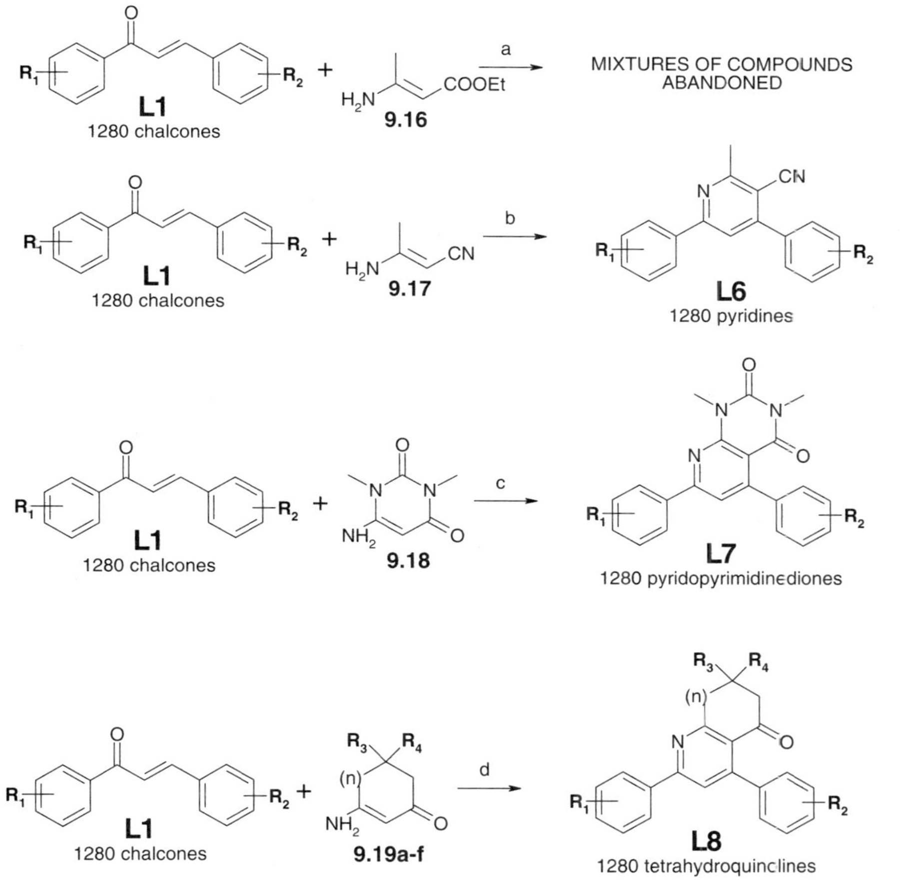

R_3 = H, Me, iPr, Ph, R_4 = H, n = 1;
R_3,R_4 = Me, n = 1; R_3,R_4 = H, n = 2

a: various reaction conditions; b: NaOH, EtOH, 70°C, 6 hrs; c: NaOH, EtOH, 80°C, 16 hrs;
d: NaOH, EtOH, 80°C, 12 hrs.

Figure 9.12 Solution-phase discrete pyridine-based libraries **L6–L8** obtained from the solution-phase discrete chalcone library **L1**.

9.1.8 From Hit to Lead

The selected hit must be *exploited rapidly* and thoroughly in this drug discovery phase. It must be chemically tractable to allow its selective derivatization/modification and the fast preparation of diverse analogues. These analogues are prepared in larger amounts (typically a few milligrams per compound) as discretes. Their thorough characterization on several assays (*vide infra*) establishes a reliable SAR for the modification of the hit nucleus and selects the most promising class of derivatives

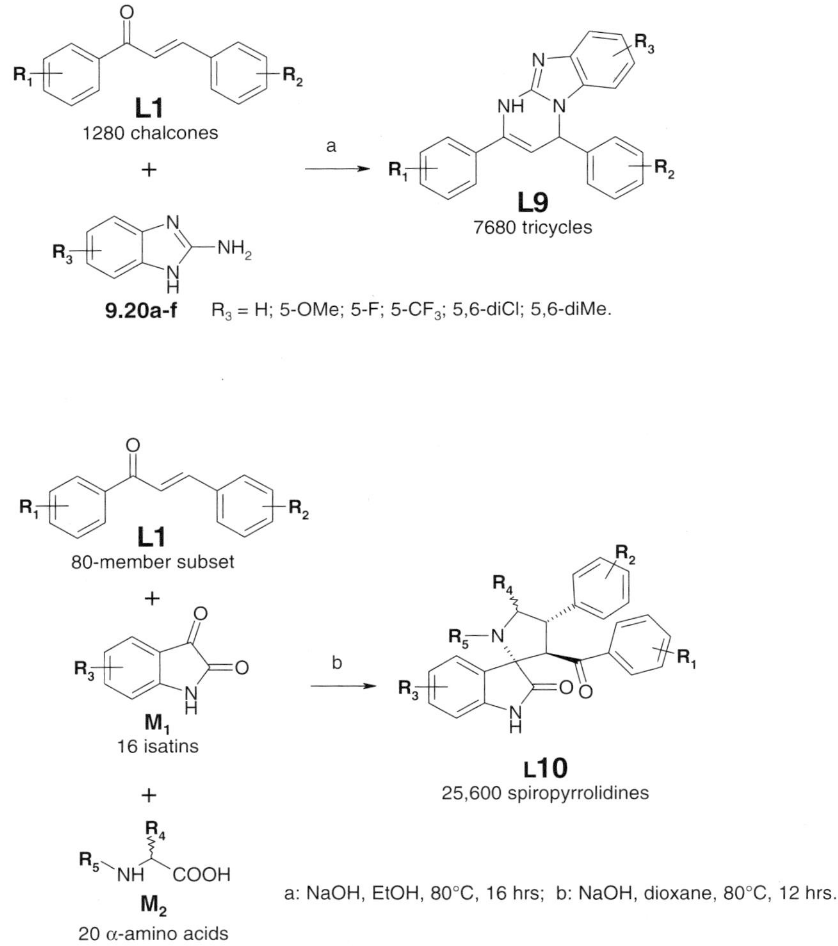

Figure 9.13 Solution-phase discrete polycyclic libraries **L9–L10** obtained from the solution-phase discrete chalcone library **L1**.

(*leads*), which will be further optimized with even more focused efforts. The same process is also applied when the hit comes from more conventional sources, such as literature searching or structural information generated in-house (e.g., X-ray structure of the target active site).

The synthetic routes available to prepare diverse analogues include the classical synthesis of single target molecules to check the feasibility of some synthetic schemes and to explore noncombinatorializable routes, as well as the parallel synthesis of several small arrays of compounds, expanding the diversity around a chemical modification. The typical size of a focused screening set designed to select a lead decreases to several hundreds to a few thousands of derivatives. These libraries are submitted to medium–high throughput biological assays measuring their potency on the target along with selectivity, toxicity, stability, and physicochemical and pharmacokinetic properties. The increased, and more stringent, set of requirements to progress a hit to the next drug discovery phase causes a significant drop in the number of potential lead candidates along the process. The parallel progression of multiple hits coming from the same primary screening campaign is, when possible, desirable.

9.1.9 Patenting Issues

Patenting a class of chemical entities is, here as in many other fields, the key that eventually leads to the return of investments and prevents the insurgence of competition. The change in the pharmaceutical market, though, has also had a strong impact on patenting policies; the increased time taken for a drug to reach the market has reduced the profitability time window for companies. In fact, given that patent protection expires after 20 years, if it is filed very early and 15 years are required to reach the market, only five years of sales without generic competition are granted. Thus, filing a patent in the late phases of drug discovery becomes appealing as only the more assessed and promising drug candidates are patented, reducing the substantial patent costs, and a larger profitability window is available.

High-throughput chemistry and biology have introduced an additional variable to the patent protection equation in drug discovery. Chemical libraries (i.e., large collections of chemical compounds) can be patented and may either represent prior art to hinder competitive research on the same structural class or even a way to claim large collections/libraries of compounds for specific applications. A conservative patenting approach can be severely damaged if competitors are actively exploiting the same biological target using high-throughput chemical and biological strategies and patent their results early. A careful evaluation of the risks versus benefits of waiting to file patents during the drug discovery process should be made, and competitor activity should be monitored regularly to facilitate implementation of the best patenting strategy for a specific project.

Combinatorial technologies–related patents have appeared since the early 1990s, and their number is growing steadily. They can be divided mostly into *structure-based* patents and *technology-based patents*. The first are broad patents claiming chemical classes of compounds and/or their screening on large families of targets (enzymes, receptors, whole cells, etc.). Some claimed generic or specific structures are reported in Fig. 9.14 together with the patent number, the claimed biological activities, and the existence of prior art as determined by the International Search Report, which could endanger some of the patent claims or even the whole patent. Technology-based patents span a wide range of applications, including methods for library synthesis, tagging methods, and synthetic and analytical combinatorial instrumentation. A sampling of these patents is reported in Fig. 9.15 together with their main claims and contents and with relevant findings provided by the International Search Report.

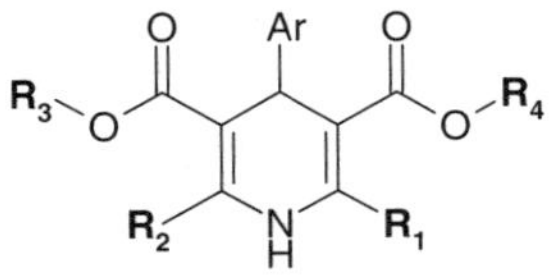

WO9633972 (1996)
Affymax/Glaxo

Generic dihydropyridine libraries for lead generation

Broad pharmaceutical activity claimed

X = NH_2, NCO, CHO, COR

EP-816309-A1 (1997)
EP-816310-A1 (1997)
EP-818431-A1 (1997)
Eli Lilly

Polymer-supported scavengers for solution-phase library synthesis

Prior Art documents found by the International Search Report

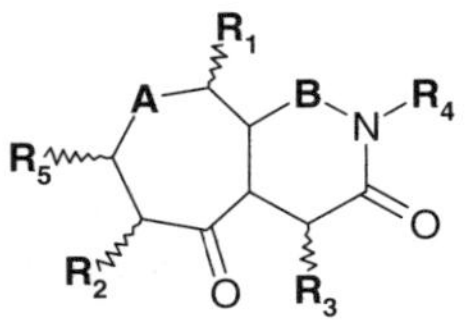

WO9715577 (1997)
Molecumetics Ltd.

Generic scaffold for peptide reverse turn mimics claimed

Libraries based on claimed scaffolds were claimed

Biological activity of claimed libraries were claimed

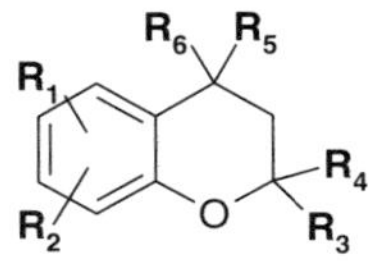

WO9530642 (1995)
Pharmacopeia

Generic dihydrobenzopyran libraries for lead generation

Activity on carbonic anhydrase

Treatment of glaucomas

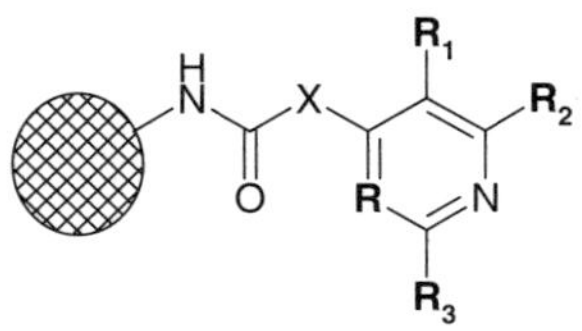

R = C-R_4, N

and other 5-, 6- and 7-member cycles

WO9815532 (1998)
Novartis

Generic solid-phase synthesis of heterocycles was claimed

A common precursor to all libraries was described and claimed

Several relevant documents found by the International Search Report

Figure 9.14 Patents regarding scaffold-based combinatorial libraries: main features and claims.

A few additional comments can be made. Many combinatorial patents have been filed by small biotechnology companies to let the scientific world appraise their libraries (or technologies), rather than to really protect them. To date, major pharmaceutical companies have not patented to a large extent in the field. The more conservative approach of patenting well-characterized, more downstream compounds is still prevalent. The time lag between first filing and patent publication may disprove such precepts, though, in the near future. Most accessible patent applications have prior art,

WO9309668 (1993)
Affymax

Light-directed masking/unmasking strategies for biooligomer library synthesis

Automated instrumentation for reagent delivery

WO9320242 (1993)
The Scripps Institute

Oligonucleotide encoded SP pool libraries

PCA deconvolution

WO9408051 (1994)
University of Columbia

Chemically encoded, bead-based SP pool libraries

Several chemical tagging methods exemplified

Several tagged SP libraries exemplified

WO9512608 (1995)
Affymax

Automated instrumentation (vessels, manifolds, flow-lines, agitation, etc.) and software to perform and control mix and split encoded SP library synthesis

Methods to transfer resin slurries using the above mentioned instrumentation

US5463564 (1995)
3-D Pharmaceutical, Inc.

Computer-based processes to define chemical libraries with selected properties

Pharmaceutical exploitation of the rational design

WO9624061 (1996)
Ontogen Corp.

Radio frequency encoded microchips

Application to SP combinatorial synthesis

WO9740383 (1997)
Glaxo Group Ltd.

A robotic instrument for the withdrawal and transfer of one, or sets of single bead from one array of reaction vessels to another

WO9811036 (1998)
Abbott Lab

A chemical encoding method based on nitrile- or acetylene-containing tagging molecules identified by IR or Raman spectroscopy

WO9851393 (1998)
Glaxo Group Ltd.

A manual, 96-well based instrument for the parallel removal of the aqueous phase from a two-phase extraction medium was claimed through freezing and removal of the ice phase

Prior art was identified by the International Search Report

WO9841534 (1998)
Biosepra, Inc. and Sepracor, Inc.

A novel family of stable and chemically inert porous ceramic supports were claimed for solid-phase combinatorial and parallel synthesis

Figure 9.15 Patents regarding technology-based combinatorial libraries: main features and claims.

which was discovered during the International Examiners' Search Report, and their relevance should be strongly affected. No major litigations concerning claimed libraries or technologies have appeared yet, but such events are likely to appear in the near future (see, e.g., the structures and claims of Figs. 9.14 and 9.15).

Some recent articles (49–55) have reviewed the field of combinatorial technologies patenting. Any combinatorial scientist must know that current and future intellectual property is going to be influenced by technological breakthroughs and by the increase in popularity of combinatorial technologies in many other fields of application.

9.1.10 An Example: Exploration of Structure–Activity Relationship of a Peptidomimetic Hit via Combinatorial/Medicinal Chemistry

Bhandari et al. (56) modified a hit structure derived from the primary screening of various libraries (>300,000 compounds) on the zinc metalloenzyme phosphomannose isomerase from the yeast *Candida albicans* (CaPMI) to find enzyme inhibitors as potential antifungal agents. During primary screening only a 1296-member SP dipeptide pool library (**L11**, Fig. 9.16) showed activity on the enzyme. Its deconvolution and analytical characterization led to the discovery of a by-product, derived from incomplete coupling, that showed activity on the enzyme. This compound (**9.21**, Fig. 9.16) showed a weak inhibitory activity on CaPMI (IC_{50} = 40 μ*M*) and was selected for further chemical profiling.

At first, substituted phenoxybutyric, phenoxyacetic, and phenylacetic acids (**M1**, 66 representatives, Fig. 9.17) were used to cap the aminoindane carboxamide **9.22** in solution to expand the capping residue diversity, producing the discrete array **L12**. Only three chlorosubstituted phenoxybutyric acids produced active compounds **9.23**–**9.25**, with the 3,4-diclorophenoxy substitution pattern as in **9.25** being found to be optimal (Fig. 9.17). This initial exploration was enhanced by using the SP synthetic route reported in Fig. 9.18. Rink-supported aminoindane carboxamide **9.26** was coupled with bromobutyric acid; then the resulting resin-bound bromide **9.27** was

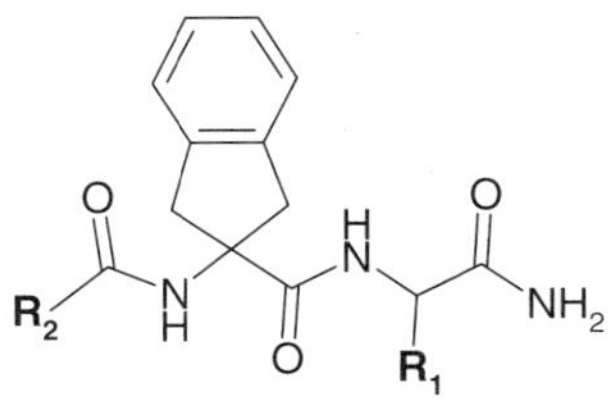

L11
1296 capped dipeptides
SP pool library
36 pools containing 36 compounds

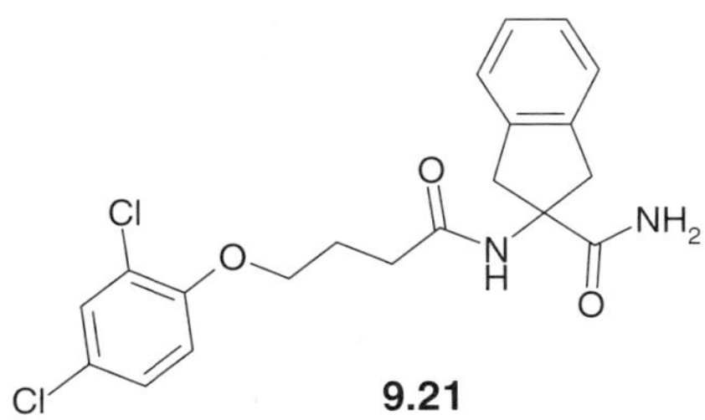

9.21
CaPMI IC_{50} = 40 μM

Figure 9.16 Structure of the primary SP peptidomimetic library **L11** and of the deconvoluted hit **9.21** active on *C. albicans* phosphomannose isomerase (CaPMI).

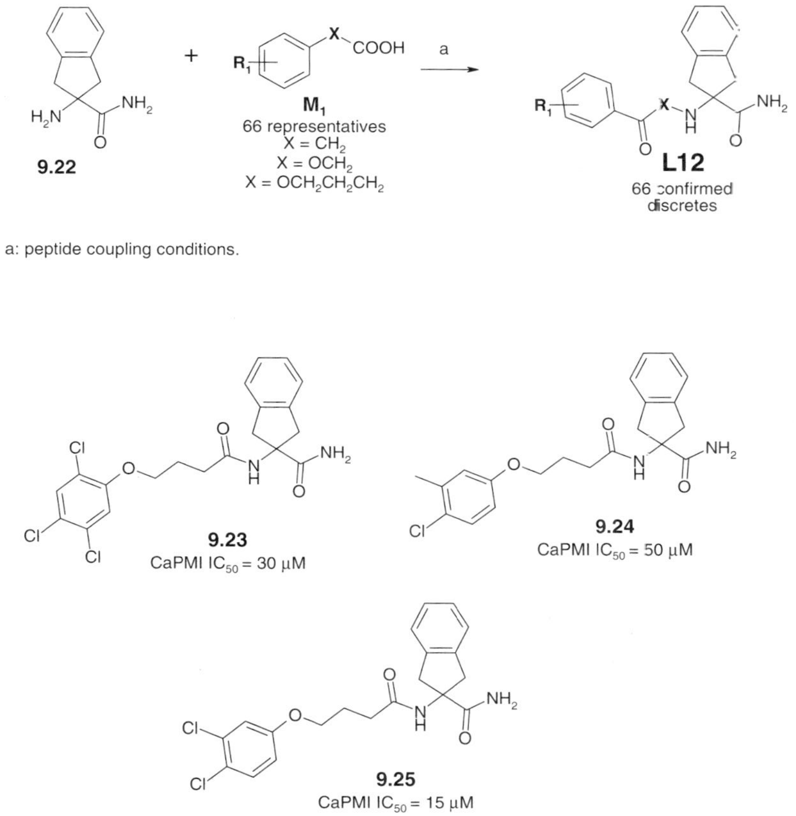

Figure 9.17 SAR from N-capping modifications of **9.21**: structures of the solution-phase peptidomimetic discrete library **L12** and of hits **9.23–9.25** obtained from its screening.

displaced with **M1** (phenols, thiophenols, anilines, and aminopyridines). While both thiophenols and anilines provided the desired compounds during the assessment, the reactivity of phenols and aminopyridines was not satisfactory. A 51-member discrete array **L13** was prepared and confirmed after quality control by HPLC. The compounds were screened, confirming the importance of the dichloro-substituted phenyl ring. A few reprepared discretes showed the general potency order thiophenol > phenol > aniline (compounds **9.28–9.33**, Fig. 9.18).

Modification of the butyl linker moiety was then studied on SP, treating **9.26** with **M1** (symmetrical anhydrides or diacids, 10 representatives, Fig. 9.19) to give the resin-bound acids **9.34**, which were pooled, reduced to alcohols **9.35**, and brominated to give the alkyl bromides **9.36**. The resin was split into 60 portions and treated with **M2** (the same 60 thiophenol and aniline nucleophiles as for **L13**) to give the 600-member pool library **L14** made of 60 pools of 10 individuals (Fig. 9.19). The library quality

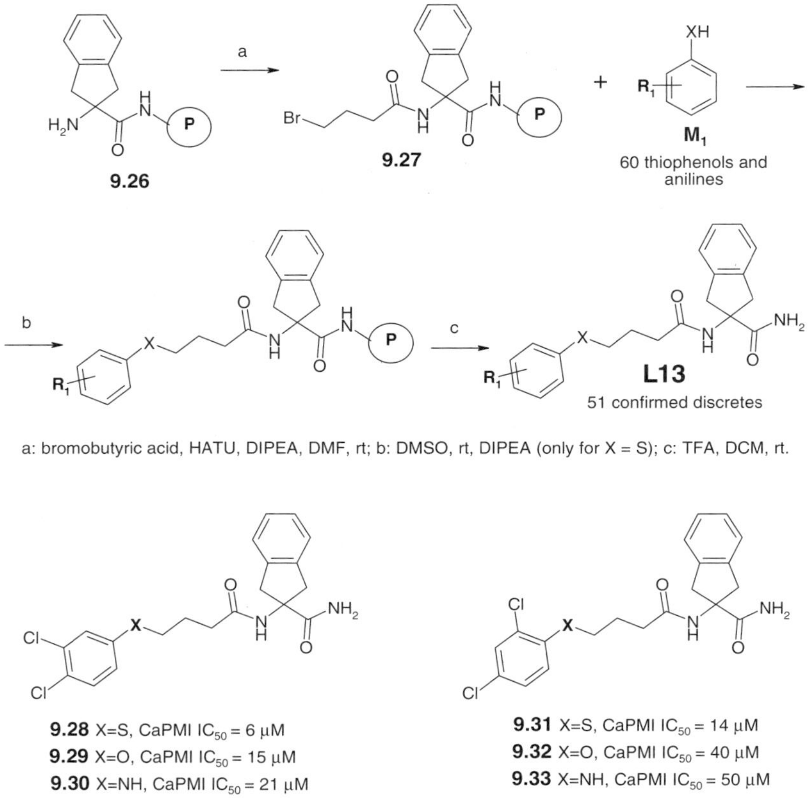

Figure 9.18 SAR from N-capping modifications of **9.21**: structures of the solution-phase peptidomimetic discrete library **L13** and of hits **9.28–9.33** obtained from its screening.

was good (HPLC/MS), but none of its components showed significant activity on CaPMI.

The carboxamide moiety was then examined, preparing several 2,4-dichlorophenoxy compounds in solution (**9.37–9.43**, Fig. 9.20). Replacement of the primary amide with small N-nucleophile-derived groups (**9.41–9.43**) maintained activity, as did the methyl ester–substituted **9.39** while the free acid **9.38**, the deletion compound **9.37**, and more complex secondary amide analogues lost inhibitory activity. The hydroxamate function significantly increased the solubility profile of **9.43**; thus it was considered relevant for the optimization of the chemical series (Fig. 9.20).

A small discrete library **L15** explored the replacement of the amino indane scaffold with aromatic, mono- and dialkylated linear or cyclic amino acids. Even small modifications were found to destroy the activity (**9.44–9.46**, Fig. 9.21). Only the amino tetralone replacement (**9.47**) afforded a weakly active compound (Fig. 9.21). Finally, a three-member discrete set of substituted amino indane–based compounds **9.52a–c**

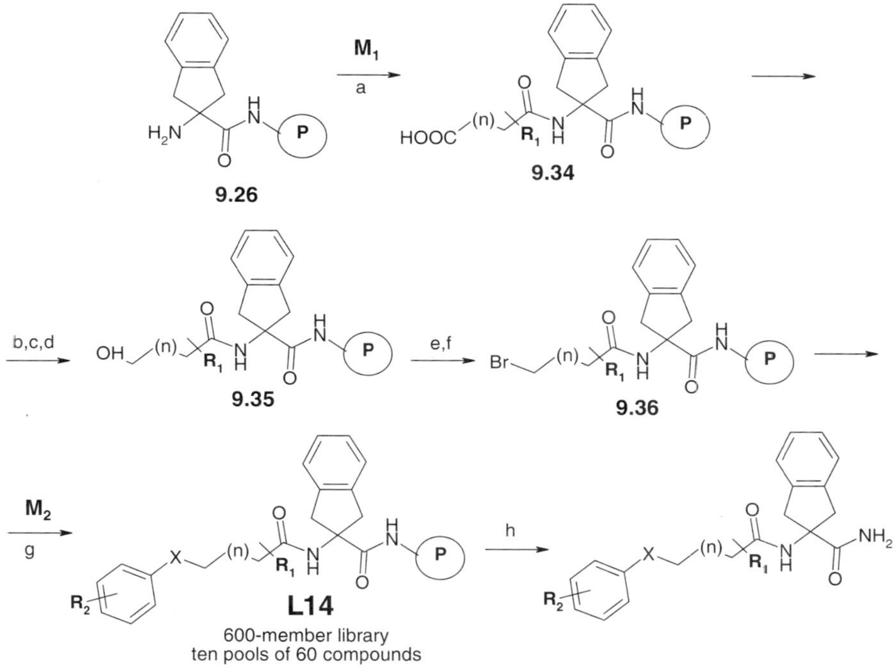

a: acylation; b: pooling of resin aliquots; c: iBuOCOCl, TEA, THF, rt; d: $NaBH_4$, H_2O; e: Ph_3PBr_2, DCM, rt; f: resin portioning (1 to 60); g: DMSO, rt, DIPEA (for X=S); h: TFA, DCM, rt

M_1: 10 representatives

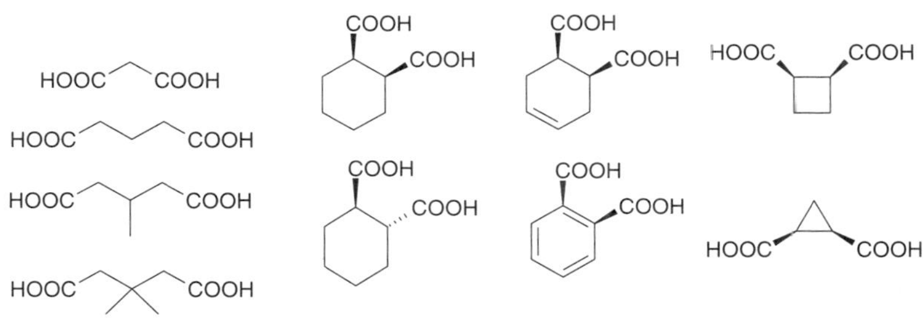

M_2: 60 representatives (30 thiophenols, 25 anilines, 5 aminopyridines)

Figure 9.19 SAR from N-capping modifications of **9.21**: structures of the solution-phase peptidomimetic pool library **L14** and of the monomer sets **M1–M2**.

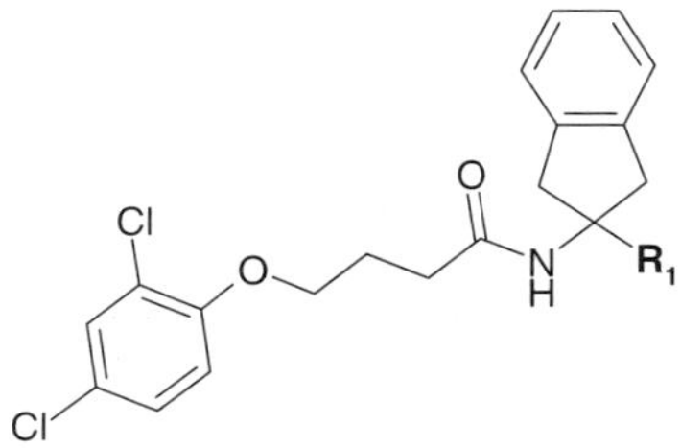

9.37 R_1=H, CaPMI IC_{50} = >200 μM
9.38 R_1=COOH, CaPMI IC_{50} = >500 μM
9.39 R_1=COOMe, CaPMI IC_{50} = 34 μM
9.40 R_1=CH_2OH, CaPMI IC_{50} = 140 μM
9.41 R_1=CONHMe, CaPMI IC_{50} = 34 μM
9.42 R_1=$CONHNH_2$, CaPMI IC_{50} = 40 μM
9.43 R_1=CONHOH, CaPMI IC_{50} = 39 μM

Figure 9.20 SAR from amide replacement in **9.21**: structures of compounds **9.37–9.43**.

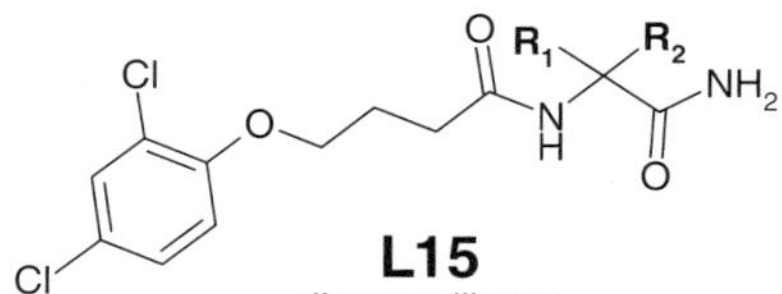

L15
discrete library

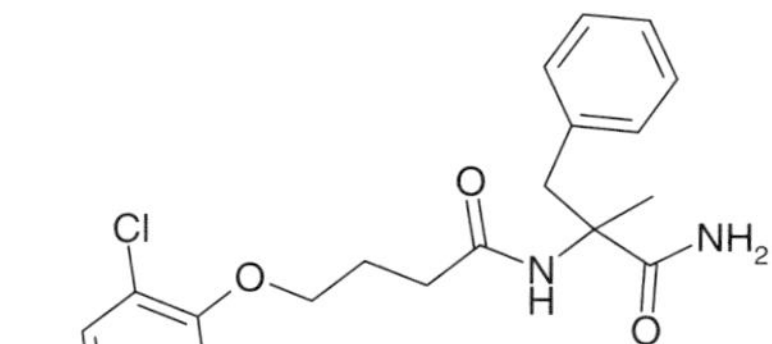

9.44
CaPMI IC_{50} > 200 μM

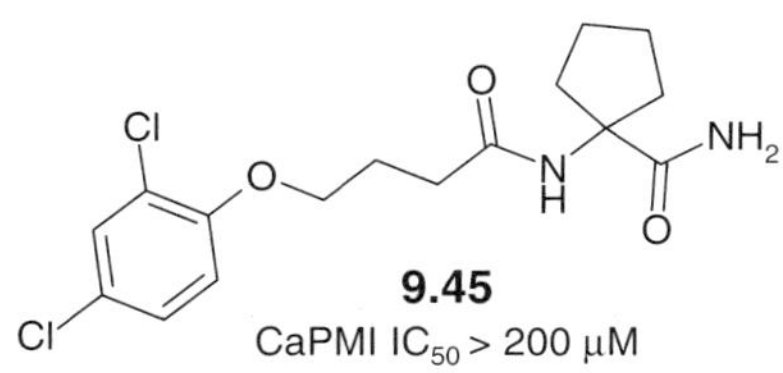

9.45
CaPMI IC_{50} > 200 μM

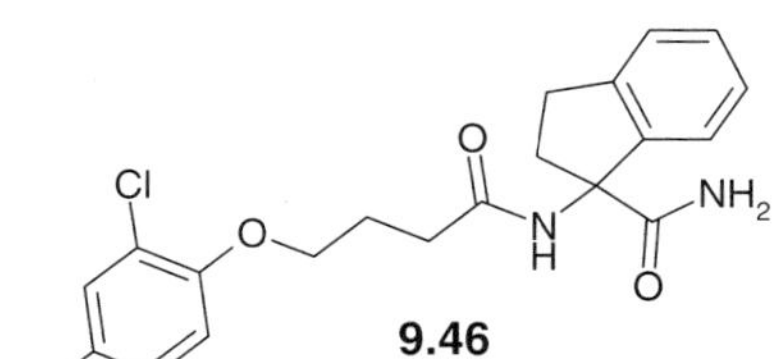

9.46
CaPMI IC_{50} > 200 μM

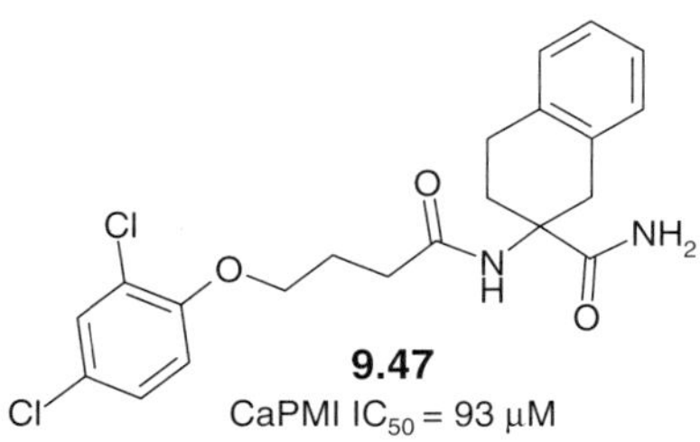

9.47
CaPMI IC_{50} = 93 μM

Figure 9.21 SAR from amino indane replacement in **9.21**: structures of the solution-phase peptidomimetic discrete library **L15** and of selected library individuals **9.44–9.47**.

was prepared following the SP strategy depicted in Fig. 9.22. The biological activity of these esters was completely lost, confirming the extremely strict structural requirements of the scaffold portion of these CaPMI inhibitors (compare with **9.39**, Fig. 9.20).

The combined information acquired by the above-mentioned efforts produced the novel compound **9.53** (Fig. 9.23) as a moderately potent inhibitor of CaPMI with good physicochemical properties, modest selectivity versus the human enzyme, and some in vivo activity against several fungal strains.

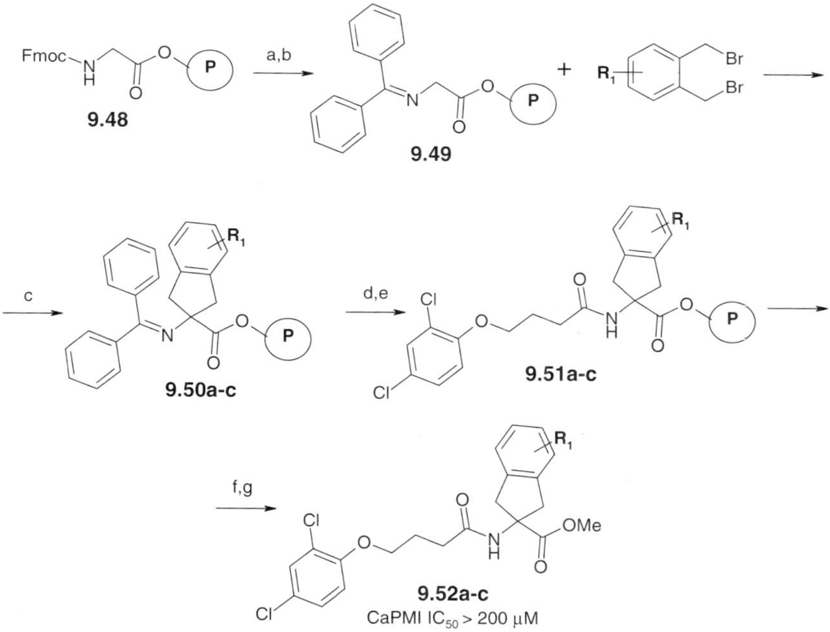

a: piperidine, DMF, rt; b: benzophenone imine, AcOH, NMP, rt; c: NaHDMS, THF, -78°C to rt; d: $NH_2OH.HCl$, THF, rt; e: acid, HATU, DIPEA, DMF, rt; f: TFA, DCM, rt; g: Me_3SiCHN_2, THF, rt.

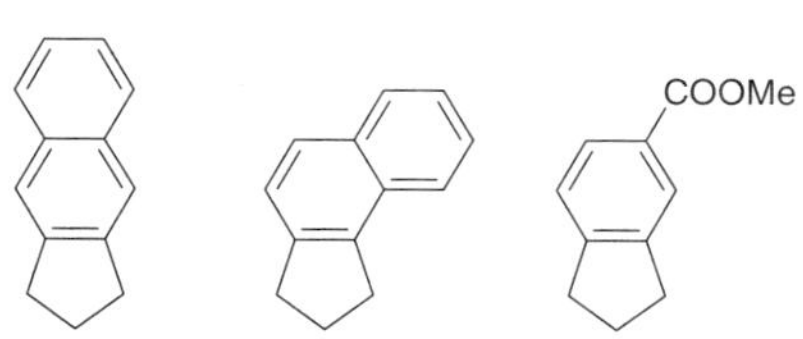

9.52a **9.52b** **9.52c**

Figure 9.22 SAR from amino indane replacement in **9.21**: structures of compounds **9.52a–c**.

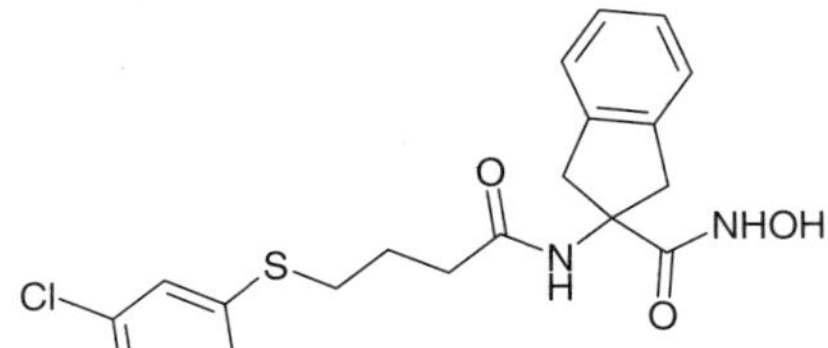

9.53
CaPMI IC_{50} = 4 μM
HumanPMI IC_{50} = 26μM
S. cerevisiae MIC = 80μM

Figure 9.23 Structure and properties of the optimized lead compound **9.53**.

9.1.11 An Example: Synthesis of a κ Opioid Receptor–Focused Piperidine Library

Thomas et al. (57) recently reported the synthesis of a 288-member discrete solution library **L16** of piperidines (Fig. 9.24), inspired by the structures of known piperidine-

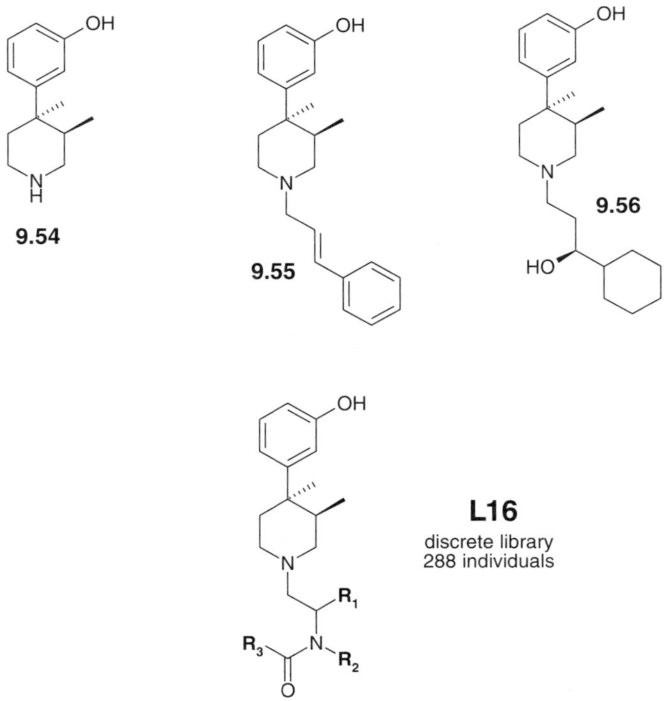

Figure 9.24 Structures of the known opioid antagonists **9.54–9.56** and of the solution-phase discrete focused library **L16** inspired by their structures.

based opioid antagonists **9.54–9.56** (Fig. 9.24) (58, 59). These structures showed a subtype selectivity toward the μ opioid receptor. The authors intended to pursue the identification of selective opioid antagonists, possibly active on the κ receptor subtype, as novel and more effective agents in drug abuse therapy. Knowing that N-substitution did not alter the antagonistic nature of the piperidine-based derivatives, a synthetic scheme based on decoration of the N-unsubstituted scaffold **9.54** was designed (Fig. 9.25).

The scaffold was first coupled with monomer set **M1** (11 N-protected α-amino acids, from which seven validated monomers were used for the library synthesis, Fig. 9.26) to give **9.57**; then the N-protecting group was removed (steps a and b, Fig. 9.25). The amide bond was reduced (step c), and the intermediates **9.58** were purified by chromatography or by crystallization. Finally, monomer set **M2** (171 carboxylic acids, from which 116 validated monomers were used for the library synthesis, Fig. 9.26) was coupled to **9.58** to give **L16** (step d, Fig. 9.25). The monomers **M1** were chosen to avoid μ-orienting cyclic lipophilic substituents at a distance of three carbon atoms from the piperidine nitrogen atom, while various aryl-, alkyl-, or alkenylaryl carboxylic acids were used as the **M2** set to explore the substitution pattern allowed at the amidic nitrogen atom. The 288 confirmed library individuals contained 7 different **M1** monomers (from 1 to 116 recurrences of the same monomer) and 116 different **M2** monomers (from 1 to 7 recurrences of the same monomer). The unexplained imbalance in **L16** composition between monomer representatives might be due to various factors, such as the successful characterization of final compounds (purity cutoff) or the decision to report only selected structures and biological data.

The 288 library individuals were tested as κ-opioid binders in a radioligand binding assay. The percent inhibition of the most relevant compounds at a concentration of 100 n*M* is listed in Table 9.1, and their structures are reported in Fig. 9.27. Among them, compound **9.56** displayed an extremely interesting in vitro κ-opioid binding activity

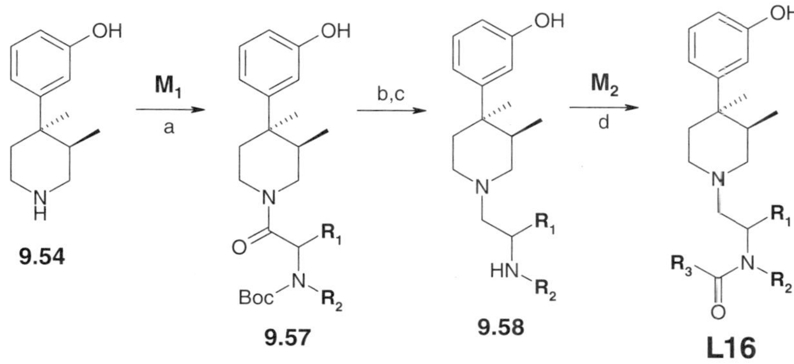

a: BOP, TEA, THF, rt; b: TFA, DCM, rt; c: $BH_3.SMe_2$, rt; d: BOP, TEA, THF, rt.

Figure 9.25 Synthetic scheme to the solution-phase discrete focused piperidine library **L16**.

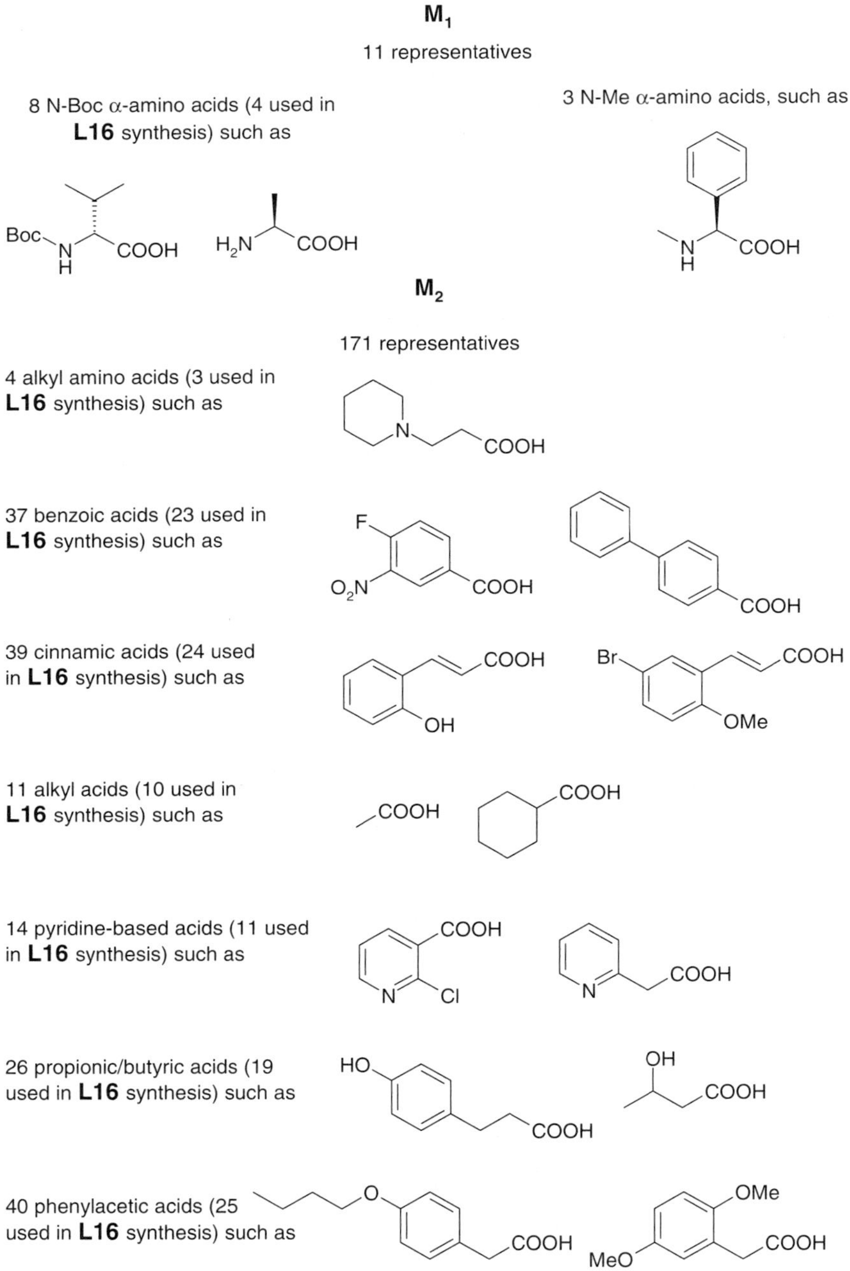

Figure 9.26 Monomer sets **M1–M2** used in the synthesis of the solution-phase discrete focused piperidine library **L16**.

TABLE 9.1 κ-Opioid Inhibition of Active Individuals from Screening of the Solution-Phase Discrete Focused Piperidine Library L16

Compound	**9.59**	**9.60**	**9.61**	**9.62**	**9.63**	**9.64**	**9.65**	**9.66**
Inhibition[a]	42	43	43	50	50	52	61	70

[a]Percent inhibition of 100 n*M*.

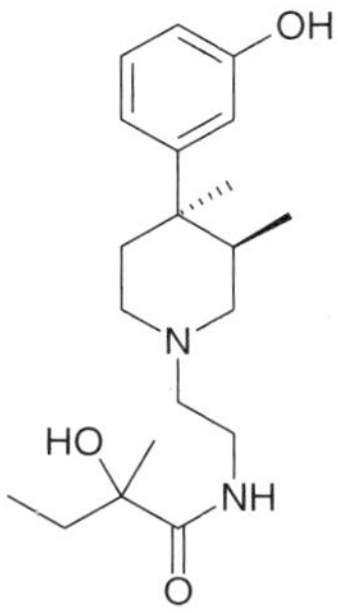

9.59

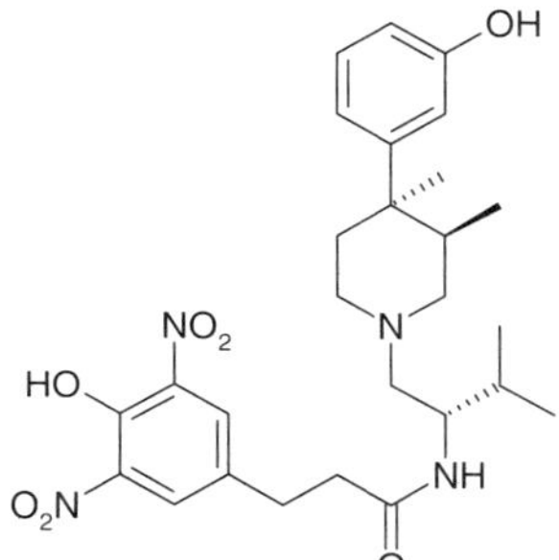

9.60

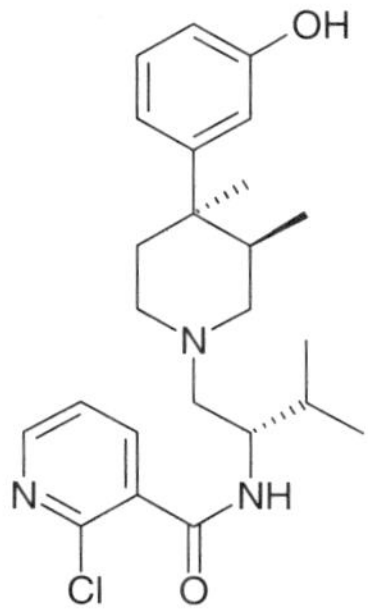

9.61

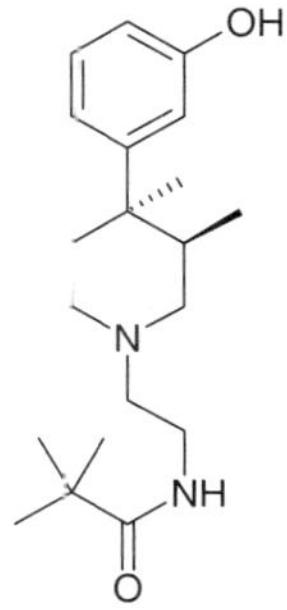

9.62

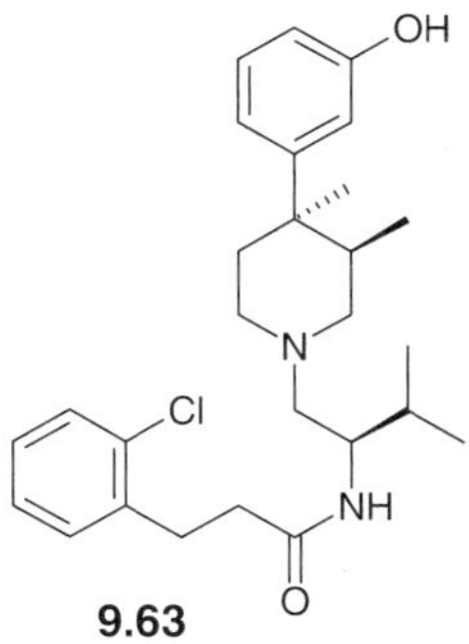

9.63

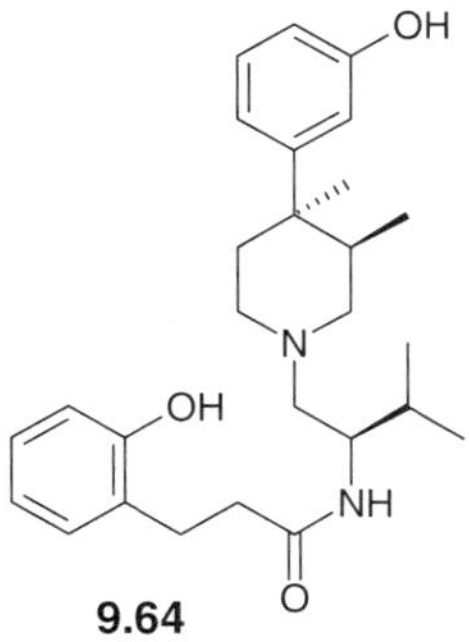

9.64

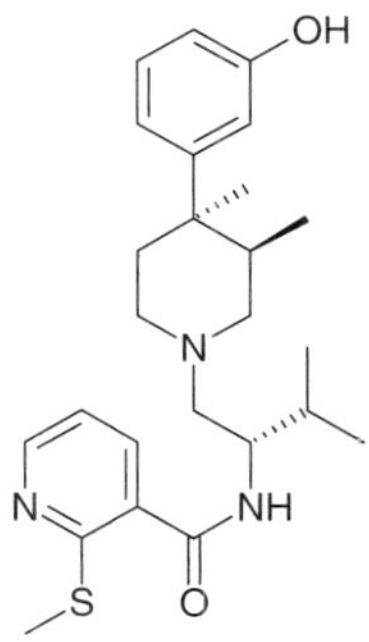

9.65

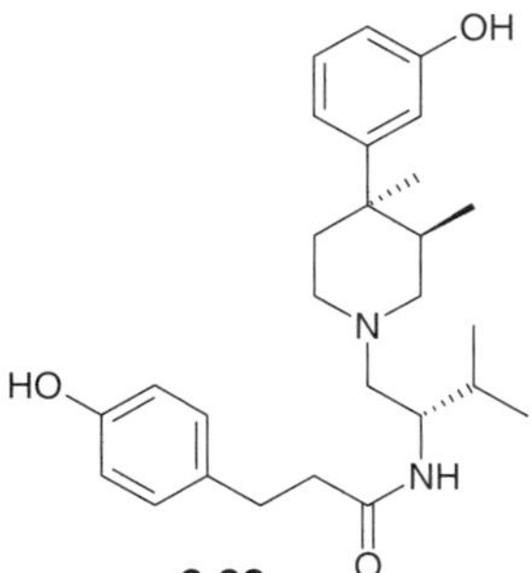

9.66

Figure 9.27 Active library individuals **9.59–9.66** from screening of **L16**.

that, unfortunately, was not confirmed in another in vitro assay (59). This result suggested a lower antagonist potency than predicted from the radiolabeled binding assay. The compound, though, remained a useful tool to further investigate the role of κ-receptor subtypes in drug abuse, and the set of acquired data represented a valuable SAR, which was useful for the authors to refine their knowledge of this structural class of opioid antagonists and to orient their future efforts in this area.

9.1.12 From Lead to Clinical Candidate

A lead structure needs refinement to be moved to the status of development candidate, which is then progressed further beyond the research phase of drug discovery. Many noncombinatorial parameters are studied in this phase, but still the synthesis of extremely focused arrays of discretes takes place. The prepared arrays undergo a thorough developability profiling, being characterized in terms of in vitro and in vivo potency, selectivity versus other, similar targets, safety issues related to toxicity and mutagenicity, physicochemical parameters including solubility and lipophilicity, and pharmacokinetic profiles, including ADME properties (see next section). The compounds are prepared in large amounts (hundred milligrams to grams), and issues related to cost of goods and chemical process routes are important; the characterization of compounds is rigorous, but nevertheless an increased chemical and biological throughput is highly beneficial. Purity is also an essential requisite, as most of the above-mentioned assays require highly pure samples to be reliable.

The drug candidates obtained are processed using classical development techniques and disciplines such as chemical and pharmaceutical development and are eventually profiled in clinical studies. Combinatorial technologies do not play a highly recognized role in these late phases, even if areas such as chemical route identification and process development optimization may largely benefit from combinatorial approaches and from SP chemistry (60–63). The overall impact of combinatorial technologies, though, significantly reduces the time required to progress molecules through the drug discovery process along with the cost of the process by focusing, as early as possible, on the most likely candidates while dropping the less promising ones.

9.1.13 HTS ADME and Physicochemical Assays

The development process to take a candidate drug to the market is historically highly influenced by the so-called ADME (absorption, distribution, metabolism, and excretion) properties of a molecule. Many candidates that exhibit otherwise good activity profiles are dropped due to unavoidable ADME failures. ADME screens have always been considered time-consuming and labor-intensive, low-throughput processes that were carried out during the late phases of the drug discovery process. More in general, an early, albeit approximate, evaluation of the physicochemical, pharmacokinetic, and toxicity properties of compounds/combinatorial arrays would be extremely useful to focus chemical efforts on druglike molecules. Such a process would eventually build large, systematic, and coherent databases to help the rationalization of ADME-influencing principles and the prediction of ADME and physicochemical properties.

In vivo screening methods to determine a pharmacokinetic profile are more difficult to automate but are more significant in that their result is the synthesis of ADME properties and can safely be used to judge the compounds' likeliness to become an ADME-friendly drug. Recent reports (64–69) have introduced cassette dosing methods to simultaneously administer mixtures of compounds. To date, a maximum of 10 components have been administered together, and their p*K* profiles were determined by analyzing the MS spectra of biological samples; a recent paper (70) reported a higher, 64-component-based cassette dosing study. Even though the screening throughput is increased, significant technological improvements are needed to obtain true HTS in vivo ADME methods; a review has recently summarized the state of the art of cassette dosing (71) and should be consulted by the interested reader. Currently, their use in lead optimization is possible, while their use in lead identification would require a much larger throughput.

In vitro studies have mostly been concerned with absorption and metabolism. The use of CACO2 monolayer cells to mimic intestinal absorption of drugs is known and validated (72), and methods to automate the assay have been reported using either CACO2 cells (73, 74) or immobilized artificial membranes (75). Papers dealing with small mixtures of peptoids (76) and with large mixtures of peptides (77) have been presented using LC (76) or MS (77) techniques to detect the permeable compounds. Improvements are to be expected in the selection of new, more user-friendly cell lines, such as MDCK (madin darby canine kidney), in the identification of reliable and sensitive detection methods and in the validation of artificial, stable lipid membrane models to allow an easier automation of the screening. Metabolism-based HTS could be realized using either fresh tissue preparations from human or animal liver or purified enzymes responsible for metabolic effects. With drug metabolism being a multivariate effect, the former format would be more useful but more difficult to set up considering the instability of tissue preparations and their fast and irreversible loss of enzymatic activity (64). Nevertheless, medium-throughput screenings using an LC-MS separation–identification technique for metabolites capable of processing 96 samples in 13 h have been reported (78) together with other relevant examples (79, 80). The use of stable, isolated cytochrome P450 xenobiotic-metabolizing enzymes is much more automation friendly, and the screening results can be used to safely rationalize and predict metabolic effects. Some examples and validation studies have been reported recently (81–83); an intriguing report based on ultrafiltration mass spectrometry claimed a throughput of 20–60 compounds per hour using rat liver microsomes (84).

ADME HTS screens have also been covered in several recent reviews (72, 85–90) that extensively comment on future trends and possibilities to improve the impact and the performance of such tools. The use of such screening as a primary filter, rather than a late optimizing tool, can be envisaged for most of the in vitro or in vivo screening formats. This will contribute to focus significant efforts in drug discovery on valuable candidates, with higher chances to be successfully developed as novel drugs, at an ever earlier stage.

High-throughput toxicological assays to determine the issues of molecules still in a relatively early phase of the drug discovery process are extremely important, and several academic and private groups are active in this field to develop predictive models, new high throughput tests and to rationalize the available information in a toxicity database. Several in vitro (91–93) and in vivo (94) toxicological HTSs have been recently reported, and even an in silico approach where the predicted toxicity of druglike substances was used together with HTS has appeared (95, 96). This field was exhaustively covered by several reviews (97–100).

The determination of the physicochemical properties of a molecule, or of an array of molecules, is also extremely important in drug discovery; in fact, the so-called developability of a drug cannot take place if the compound has problems related to its solubility or to its lipophilicity at physiological pH. High-throughput screening assays able to provide an early estimation of these properties for discrete, focused libraries or collections of compounds were reported using several assay formats and detection techniques. Turbidimetry was used to determine solubility (101), while flow injection–capillary electrophoresis (FI-CE) was used to determine the dissolution process of compounds with a throughput of 60 compounds per hour (102); the chromatographic hydrophobicity index of compounds was determined and validated as a higher throughput good alternative (50–100 compounds per day) to log *P*/log *D* (103). Penetration through biological barriers, and especially through the blood-brain barrier, has been the subject of both theoretical (104) and experimental research (69). A recent review (105) summarizes most of the current trends in physicochemical HTS and illustrates the use of computational tools to predict and to rationalize the data from HTS.

9.2 AGROCHEMICAL AND FOOD-RELATED APPLICATIONS

9.2.1 General Considerations

The areas where combinatorial technologies impact strongly in the drug discovery process (i.e., the phase of lead identification and optimization) are essentially the same for agricultural and food research. Opportune, relevant targets must be isolated and the necessary adjustments, in terms of desired properties of the active compounds (e.g., physicochemical properties and toxicity), must be made on the basis of the selected agrochemicals or food-related application. It is thus surprising that very few reports in these areas have appeared in the literature. Apart from the example reported in the next section about agrochemicals, Wong et al. (106) isolated peptide ligands for barley α-amylase using phage display libraries; peptide libraries were used to identify anti-phytopathogenic fungi (107) and were designed to identify novel pesticides (108); researchers from FMC Corporation (109) reported the synthesis and the screening of a discrete library of pyrazolopyridines as insecticidals; researchers from Zeneca (110, 111) reviewed the future trends in fungicide/agricultural research, mentioning the synthesis and screening of primary, synthetic organic libraries without divulging any structural information; Jansson et al. (112) reported an evaluation of available HTS assays to screen for novel insecticides; Petsko et al. (113) reviewed the theoretical opportunities provided by combinatorial technologies in the search of novel pesticides. Only two recent papers (114, 115) introduced the use of combinatorial technologies to prepare a flavor library and to characterize its components in terms of their odor characteristics and a review has highlighted the food-related areas of major interest for combinatorial applications (116).

Due to the substantial research efforts in these fields, the appearance of significant undertakings in the near future is to be expected for any application where a target and an activity assay can be identified. A library to be screened on an HTS assay based on an agrochemical/food-related target will be as beneficial as it has proven to be for pharmaceutical research. We can safely assume that the increased awareness of scientists for combinatorial technologies and the growing number of successful examples in drug discovery applications will eventually lead to the extension of these approaches to disciplines that have not yet been significantly touched (117).

9.2.2 An Example: Synthesis of a Primary Amide/Ester Library as a Source of Herbicidal Compounds

Parlow and Normansell (118) reported the synthesis of a pool library **L17** of amides and esters in solution, made of around 8000 members, and its screening as a source of

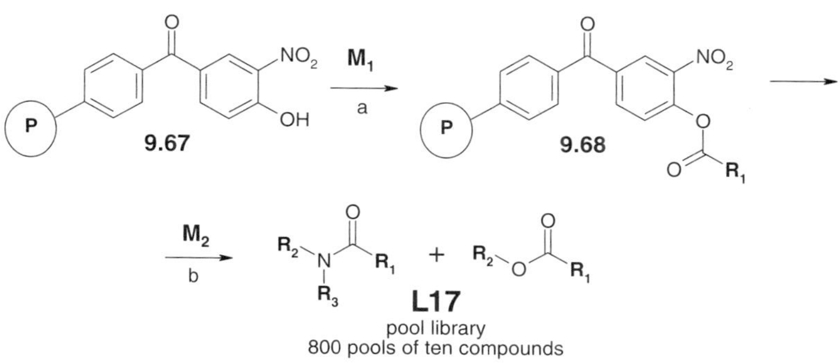

a: pyridine, DCM, rt; b: TEA, MeCN, 70°C.

$\mathbf{M_1}$: 6 sets of 10 carboxylic acids, 60 representatives

$\mathbf{M_2}$: primary and secondary alkyl amines, anilines, primary alcohols

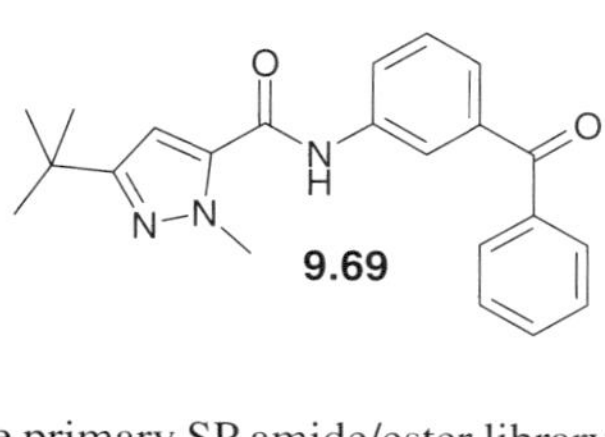

Figure 9.28 Synthesis of the primary SP amide/ester library **L17** tested for herbicidal activity and structure of the hit **9.69** derived from its screening.

herbicidal leads on a whole-plant assay. The synthesis of **L17** and its structure are reported in Fig. 9.28.

A solid-supported reagent, **9.67**, was used to activate the monomer set **M1** (60 carboxylic acids, step a, Fig. 9.28) added to the solid support as 6 mixtures of 10 representatives. The activated ester pools **9.68** were reacted with the monomer set **M2** (amines and primary alcohols, step b, Fig. 9.28) and the library **L17** was recovered, after filtration, as a set of 10-member pools with yields varying from 60 to 90% (mass recovery). Screening and deconvolution produced **9.69**, an active compound of interest (Fig. 9.28, bottom). Its structure made it suitable for further optimization via focused library efforts, while its similarity to known bleachers also pointed toward a specific mechanism of action (118). No further efforts were reported by the authors.

9.3 APPLICATIONS TO COMBINATORIAL REACTION OPTIMIZATION

9.3.1 General Considerations

The paradigm of many compounds (a library) prepared simultaneously (combinatorially) and tested rapidly (HTS format) to find a specific activity is the core of combinatorial technologies. This same paradigm can also be applied when a single compound is prepared but *a set of reaction variables* (a library) *are simultaneously* (combinatorially) *modified* and the reaction outcome is rapidly measured (HTS format) to find the optimal reaction conditions. This multivariate, simultaneous optimization of reaction conditions can be used to accelerate the SP chemistry assessment aimed toward the synthesis of an SP library as recently proved using a limited example (118) (see also Sections 6.1.2 and 6.3.5). A typical example of a biased/focused approach involves tens/hundreds of reaction conditions. Theoretically, though, the exploration of a larger set of new, unprecedented, diverse reaction conditions on a common set of reagents to promote novel reactions could represent the primary/diversity-based approach aimed at more fundamental chemistry research. A primary approach has not been exploited as of today, but it could be instrumental in creating a sort of chemical reaction database allowing the chemist to be more predictive in selecting experimental conditions for a given transformation. Complex encoding strategies could even be used to optimize reaction protocols, as foreseen previously (120).

9.3.2 An Example: Rapid Optimization of the Synthesis of an ICE Inhibitor

Warmus et al. (121) recently reported the multivariate optimization of reaction conditions for the synthesis of a known class of ICE (interleukin-1β converting enzyme) inhibitors, represented by compound **9.70**. The structure of the compound and the synthetic route, starting from the key bromide intermediate **9.71**, are reported in Fig. 9.29 (the acetylvaline moiety of **9.70** was replaced during the optimization with an Fmoc carbamate).

Coupling of **9.71** with carboxylic acid **9.72** (step a, Fig. 9.29) was performed in a set of 82 reaction conditions (parallel reaction library **L18**, Fig. 9.30). The base and

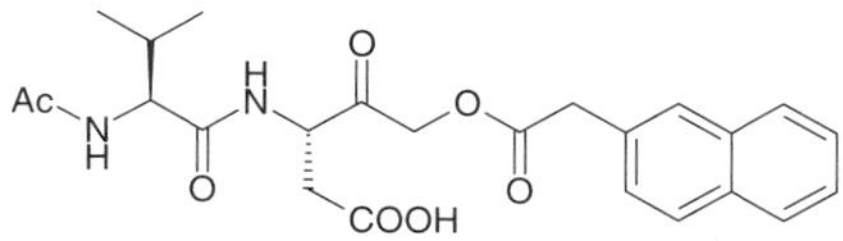

9.70

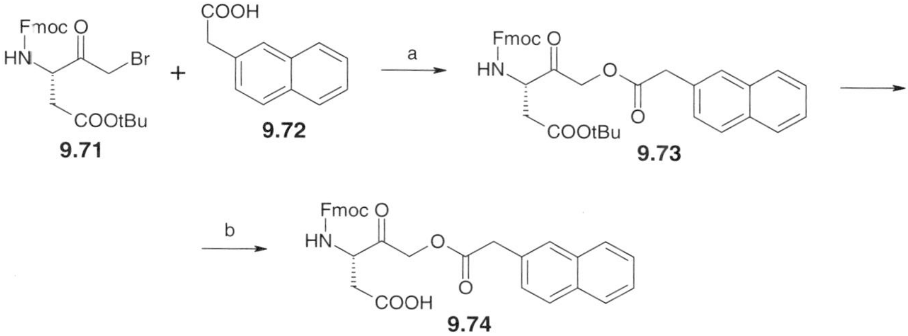

a: nucleophilic displacement; b: t-butyl ester hydrolysis.

Figure 9.29 Structure of the known ICE inhibitor **9.70** and synthesis of its simplified analogue **9.74**.

L18

82-member reaction library
optimization of nucleophilic displacement (step a, Fig. 9.25)
used *basic reagents*: KF, KF on alumina, KF on celite,
F on Amberlyst A-26, K_2CO_3, BaF, $AgCO_3$,
DIPEA, TEA, DBU, pyridine, N-Me morpholine,

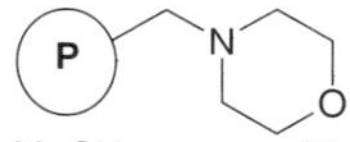

used *solvents*: DMF, MeCN, acetone, THF, H_2O/DCM, dioxane, $CHCl_3$
used *adjuvants*: NaI, Et_4NBr, $CaCO_3$

L19

39-member reaction library
optimization of t-butyl ester hydrolysis (step b, Fig. 9.25)
used *acid reagents*: TFA, HCl, TsOH, CSA,
IRP-64 resin, IRC-58 resin, DOWEX 50 resin,
Amberlite IR 120 resin, Nafion resin
used *solvents*: DCM, EtOAc, toluene, dioxane, THF, acetone
used *adjuvants*: PhSMe, Et_3SiH, m-cresol, H_2O

Figure 9.30 Parallel reaction libraries for the optimization of the nucleophilic displacement (**L18** and of the *t*-butyl ester hydrolysis (**L19**).

the solvent were changed, and some adjuvants were added while the other reaction parameters were kept constant. The reactions were run in glass vials, and the library was screened for the disappearance of **9.71** (reaction yield) and for its purity (HPLC traces). Several entries from **L18** are reported in Table 9.2. A somewhat refined ROR (reagent–outcome relationship) could be obtained, pointing toward DIPEA, TEA, and

TABLE 9.2 Yields and Purity of 9.73 from Parallel Reaction Library L18: Selected Entries

Entry	Base	Solvent	Adjuvant	Reaction Outcome[a]
1	KF	DMF	[b]	95/60
4	KF	Acetone	NaI (stoichiometric)	97/20
5	KF/alumina	DMF	[b]	66/37
8	KF/alumina	Acetone	NaI (stoichiometric)	100/27
12	KF/Celite	Acetone	NaI (stoichiometric)	85/38
13	F/Amberlyst A-26	DMF	[b]	38/26
16	K_2CO_3	DMF	[b]	98/71
22	K_2CO_3	MeCN	[b]	95/90
24	K_2CO_3	MeCN	Et_4NBr	68/42
26	BaF	DMF	$CaCO_3$	39/22
32	$AgCO_3$	MeCN	[b]	90/84
33	DIPEA	DMF	[b]	95/95
34	DIPEA	DMF	NaI (catalysis)	98/90
41	DIPEA	Dioxane	[b]	97/35
42	DIPEA	$CHCl_3$	[b]	97/67
43	**TEA**	**DMF**	[b]	**99/88**
44	TEA	DMF	NaI (catalysis)	99/46
45	TEA	THF	[b]	97/46
46	TEA	THF	NaI (catalysis)	98/13
47	MeCN	DMF	[b]	87/13
48	MeCN	DMF	NaI (catalysis)	99/46
49	Acetone	THF	[b]	99/59
50	Acetone	THF	NaI (catalysis)	99/59
51	TEA	dioxane	[b]	98/40
52	TEA	$CHCl_3$	[b]	97/22
53	DBU	DMF	[b]	Decomposition
61	DBU	Acetone	NaI (catalysis)	62/8
63	PS-morpholine	DMF	NaI (catalysis)	98/36
69	PS-morpholine	Acetone	NaI (catalysis)	86/45
76	*N*-Me morpholine	MeCN	[b]	97/49
78	Pyridine	DMF	[b]	Decomposition

[a]Reaction completion (disappearance of **9.71**)/purity.

[b]Absent.

KF as the best basic reagents and towards DMF as the best solvent. An accurate determination of yields and purity of **9.73** in all reaction vessels selected entry 43 (TEA–DMF) as the best compromise for step a, Fig. 9.29. A similar optimization was performed for the *t*-butyl ester hydrolysis (step b, Fig. 9.29), creating the 39-member reaction library **L19** by permutations of acidic reagents and solvents, and addition of adjuvants (Fig. 9.30). The screening outcome (Table 9.3) highlighted the poor performances of ion-exchange resins (no reaction) and TFA (unclean product) to prepare **9.74**, and selected entry 7 (HCl/EtOAc) as the best reaction conditions to obtain clean **9.74** (CSA (camphor sulfonic acid) actually performed slightly better, but the reaction work-up was less automation friendly). The whole manual optimization process required three to four days, and the best reaction conditions were used directly to produce a 590-member discrete library, which met the >75% purity cutoff (78).

Among other similar reports, Bray et al. (122) optimized the SP reductive amination of a model ketone with three primary amines using a 56-member reaction library with simultaneous variation of amines, reducing agent concentrations, pH, and solvents. Gayo and Suto (123) optimized the coupling of an acyl chloride with benzylamine in solution using a 30-member reaction library with simultaneous variation of the

TABLE 9.3 Yields and Purity of 9.74 from Parallel Reaction Library L19: Selected Entries

Entry	Acid	Solvent	Adjuvant	Concentration	Reaction Outcome[a]
1	TFA	DCM	[b]	20%	100/88
4	TFA	DCM	PhSMe	50%	80/65
5	TFA	DCM	Et_3SiH	50%	100/65
6	TFA	DCM	*m*-Cresol	50%	100/49
7	**HCl**	**AcOEt**	[b]	**1*M***	**48/98**
8	HCl	Toluene	[b]	1 *M*	37/97
9	HCl	Dioxane	[b]	1 *M*	29/97
10	HCl	THF	[b]	1 *M*	24/98
11	HCl	Dioxane	10% H_2O	1 *M*	NR
12	TsOH	Toluene	[b]	Solid	70/63
14	TsOH	DCM	PhSMe	Solid	65/88
16	CSA	THF	[b]	Solid	74/98
18	CSA	THF	H_2O	Solid	32/99
23	IRP-64 resin	AcOEt	[b]	Solid	NR
27	IRC-58 resin	AcOEt	[b]	Solid	NR
31	DOWEX 50 Resin	AcOEt	[b]	Solid	NR
35	Amberlite IR 120 resin	AcOE	[b]	Solid	NR
39	Nafion resin	AcOEt	[b]	Solid	NR

[a]Reaction completion (disappearance of **9.73**)/purity; NR = no reaction.
[b]Absent.

basic ion-exchange resin and solvent. Kuo et al. (124) and Hsu et al. (125) have reported the use of two automated workstations to perform a combinatorial reaction optimization. Lindsey (126) has reviewed the use of automation for combinatorial reaction optimization using also statistical tools to design the automated experiments.

9.4 APPLICATIONS TO CATALYSIS

9.4.1 General Considerations

Catalysis in organic chemistry has always occupied a preminent position in allowing access to otherwise inaccessible compounds and to unfavored chemical transformations. The quest for new catalysts or ligands, their optimization in terms of catalytic efficiency, and their application to novel chemistries has been, is, and will remain a major topic for synthetic organic chemists. Having already covered the use of supported catalysts or ligands in solution-phase library synthesis (see Section 8.4.4), here we will deal with the high-throughput synthesis and evaluation of solid- and solution-phase libraries of catalysts or ligands and chiral auxiliaries for specific chemical reactions. As in the previous section, rather than compound libraries, we will present either catalyst libraries or ligand libraries or catalytic system libraries to catalyze the chemical transformation of a single substrate. For these catalytic applications, screening is represented by measurement of the reaction outcome on a single substrate in terms of yields (or conversion), purity, and, when stereocenters are involved, the enantiomeric excess obtained.

The use of combinatorial technologies in this field was introduced by two papers. Liu and Ellman (127), in 1995, reported the SP synthesis, cleavage, and screening of 2-pyrrolidinemethanol derivatives **9.75** and **9.76** as stereoselective ligands for dialkylzinc addition to aldehydes (Fig. 9.31), proving the usefulness of combinatorial technologies in assisting the discovery of novel ligands for asymmetric catalysis.

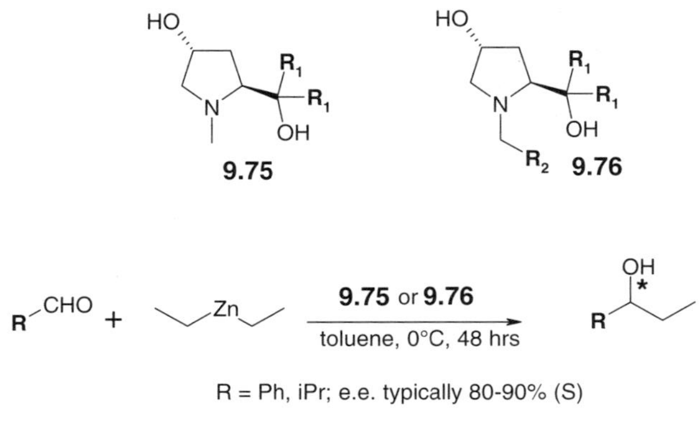

Figure 9.31 Structures of pyrrolidine-based ligands **9.75** and **9.76** for the catalyzed addition of zinc to aldehydes.

Subsequently, in 1996, Burgess et al. (128) reported a high-throughput catalyst screening for a C–H insertion reaction (see next section), widening the application of combinatorial technologies to the fast optimization of catalytic conditions for a given reaction. Since then, many papers and reports have appeared regarding synthetic organic libraries aimed toward the discovery of new catalyst/ligand systems. These are covered in this section through several examples and are thoroughly referenced. Other approaches regarding inorganic libraries of catalysts, catalytic antibodies, or molecularly imprinted polymers to catalyze organic reactions are among the main topics of the next two chapters.

The most obvious approach utilizes *parallel synthesis* to prepare arrays containing different catalytic systems, testing them on a single chemical transformation and measuring their efficiency. This is done varying the catalyst, or the ligand, or both of them simultaneously. Three examples of this approach are reported in Sections 9.4.2, 9.4.3 (catalytic system libraries), and 9.4.4 (ligand library). A specular approach employs an array of chemical substrates to determine the specificity and the general efficiency of a given catalytic system and is exemplified in Section 9.4.5. Finally, two intriguing examples of SP pool libraries, screened via either deconvolutive methods (Section 9.4.6) or chemical encoding (Section 9.4.7), determine the activity of a catalytic system using high-throughput mix and split synthetic methodologies. These examples are followed by an exhaustive list of references, which are also summarized in several recent reviews (129–137).

9.4.2 An Example: Screening of a Library of Catalytic Systems for a C–H Insertion Reaction

Burgess et al. (128) reported the catalyst screening of a 96-member array of catalytic systems **L20** on a C–H insertion reaction of substrate **9.77** (Fig. 9.32), a transformation usually catalyzed by rhodium (138) or copper salts (139) in the presence of chiral ligands (140). The stereochemical outcome was measured on the diastereomeric couple **9.79–9.80**, obtained following uncatalyzed oxidation of **9.78** (Fig. 9.32), to simplify the determination of the chiral products while evaluating the stereoselectivity of the tricycle formation. The stereoselectivity of the C–H insertion was not significantly influenced by the presence of the (L)-methyl ester (128).

The composition of **L20** is reported in Fig. 9.33. Five chiral ligands were used, including three chiral bis(oxazolidines) **9.81** (141), **9.82** (142), and **9.83** (143); sparteine **9.84** and a salen-type compound **9.85** (144). Seven metal ions were used with the different ligands, silver, and scandium as well as the assessed copper and rhodium salts were trialed with each ligand, while gold, ytterbium, and lanthanium salts were employed more conservatively with only one or two ligands each. The 24 combinations of catalyst and ligand were tested in four solvents. Each catalytic system was tested on a 10-mg sample of **9.77** using small quantities of ligands and catalysts (10% in moles, 0.6–1.5 mg) in a 200-μL solution stirred at rt or at 10 °C. The reaction outcome was measured by HPLC, the diastereomeric couple **9.79–9.80** being sufficiently distinguishable. The whole process of catalytic reactions, work-up, and purification of samples and analytical determination of reaction products was completed in a few days for the 96-member array **L20**.

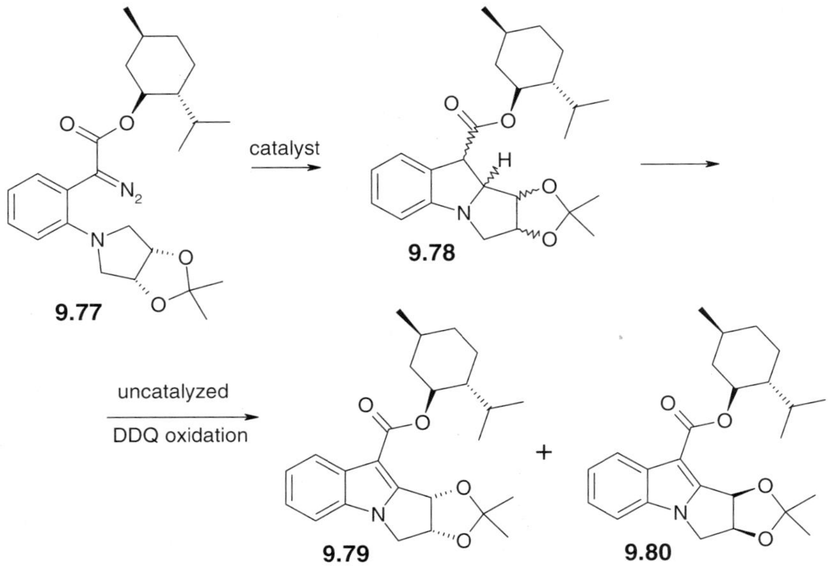

Figure 9.32 Catalyzed C–H insertion on azo compound **9.77** and uncatalyzed oxidation of **9.78** to tricyclic structures **9.79** and **9.80**.

The observed stereochemical outcome of the C–H insertion varied from 1 : 2.4 (**9.80** favored) to 5.9 : 1 (**9.79** favored), with many examples of scarce asymmetric induction. The use of THF generally led to higher stereoselection than the other solvents. The most significant results of the screening experiments are reported in Table 9.4. Seventeen catalytic combinations gave encouraging results and were repeated on a larger scale to reconfirm the observed activities. Several data were not reconfirmed, probably because of experimental errors and heterogeneities in the small-scale primary screening. The already known catalytic system (entry 18) was surpassed in terms of diastereomeric ratio (entry 19) and of both diastereomeric ratio and reaction yield (entry 20). The use of silver ion as C–H insertion catalyst was novel and highlighted the potential of combinatorial catalyst screening in affording relevant, unpredictable new catalytic systems.

An increased reliability of the results generated in the primary screening, an increased throughput, and the use of larger sets and libraries of catalytic systems were the main issues emerging from this pivotal contribution. The following examples will include more recent reports that partially address these concerns.

9.4.3 An Example: Screening of a Library of Catalytic Systems for a Hydrosilylation Reaction

Cooper (145) recently reported the use of dye-containing substrates as screening tools for the rapid evaluation of different catalytic systems. Hydrosilylation was chosen as

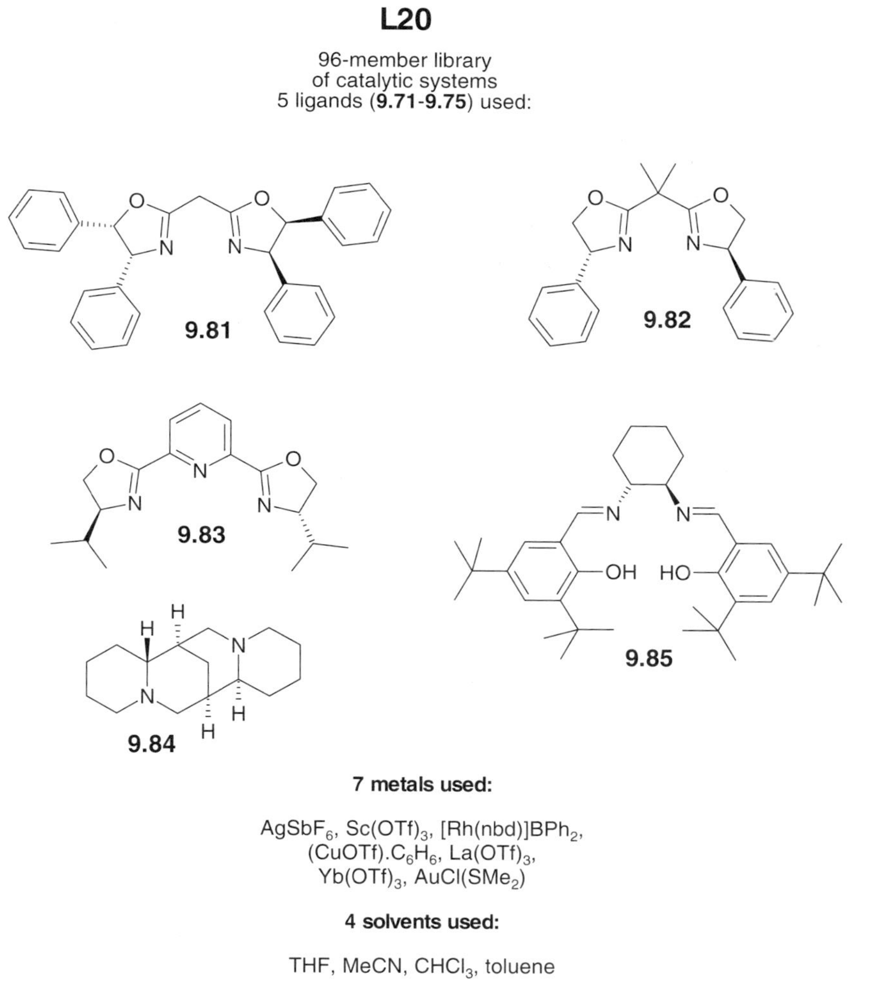

Figure 9.33 Catalyzed hydrosilylation: structure of the solution-phase catalytic system discrete library **L20**.

a reaction, and two compounds, **9.88** and **9.89**, containing an electron-donor ferrocenyl group and an electron-acceptor pyridinium group linked respectively by a C=C (**9.88**) or C=N (**9.89**) bond were selected as substrates. Their synthesis from intermediates **9.86** and **9.87** is reported in Fig. 9.34. These compounds had strong absorption maxima in the UV–visible area and were respectively deep purple and dark blue, while their hydrosilylation produced compounds **9.90** and **9.91**, respectively, which lost the original color and became light yellow (Fig. 9.35).

A 12-member catalytic system library **L21** (Fig. 9.35) was tested using the two substrates, and the color change in each reaction vessel containing a different catalytic

TABLE 9.4 Diastereomeric Ratios and Yields of 9.79 and 9.80 Using the Solution-Phase Catalytic System Discrete Library L20: Selected Entries

Entry	Ligand	Metal Source	Solvent	Reaction Outcome[a]
1	**9.81**	$AgSbF_6$	$CHCl_3$	2.3/1.0 (6.7)
2	**9.81**	$(CuOTf)_2{\cdot}C_6H_6$	THF	3.4/1.0 (44)
3	**9.82**	$\mathbf{AgSbF_6}$	**THF**	**2.7/1.0 (44)**
4	**9.82**	$[Rh(nbd)]BPh_4$	Toluene	1.0/2.1 (36)
5	**9.82**	$\mathbf{(CuOTf)_2{\cdot}C_6H_6}$	**THF**	**4.4/1.0 (47)**
6	**9.82**	$(CuOTf)_2{\cdot}C_6H_6$	$CHCl_3$	4.0/1.0 (13)
7	**9.82**	$(CuOTf)_2{\cdot}C_6H_6$	Toluene	2.9/1.0 (44)
8	**9.83**	$AgSbF_6$	THF	2.0/1.0 (10)
9	**9.83**	$[Rh(nbd)]BPh_4$	Toluene	1.0/2.3 (17)
10	**9.83**	$(CuOTf)_2{\cdot}C_6H_6$	THF	2.0/1.0 (7.0)
11	**9.83**	$(CuOTf)_2C_6H_6$	MeCN	2.0/1.0 (5.0)
12	**9.84**	$Sc(OTf)_3$	THF	3.4/1.0 (0.91)
13	**9.84**	$[Rh(nbd)]BPh_4$	Toluene	1.0/2.4 (29)
14	**9.84**	$(CuOTf)_2{\cdot}C_6H_6$	THF	5.9/1.0 (23)
15	**9.84**	$(CuOTf)_2{\cdot}C_6H_6$	MeCN	2.1/1.0 (5.1)
16	**9.85**	$[Rh(nbd)]BPh_4$	Toluene	1.0/2.3 (15)
17	**9.85**	$(CuOTf)_2{\cdot}C_6H_6$	THF	3.2/1.0 (4.1)
18[b]	**9.82**	$(CuOTf)_2{\cdot}C_6H_6$	DCM	2.3/1.0 (18)
19[b]	**9.82**	$(CuOTf)_2{\cdot}C_6H_6$	THF	3.9/1.0 (61)
20[b]	**9.82**	$AgSbF_6$	THF	3.5/1.0 (75)

[a]Diastereomeric ratio **9.79/9.80** (yield after DDQ oxidation).
[b]Repeated as discretes on a larger scale.

system–substrate mixture was monitored. Care was taken to minimize any interference due to catalysts or solvents in the color change, and blank control vessels were added. Images were recorded with a digital camera on the occasion of two distinct events: when the first sign of bleaching appeared (t_1), corresponding to around 40% of hydrosilylation, and when the original color disappeared completely (t_2), corresponding roughly to 95% of hydrosilylation. The latter measurement was more susceptible to color interferences, but both measurements were reliable enough (see the original paper for more details).

The screening results are reported in Table 9.5. The performance of the catalytic systems with the two substrates was similar, a significant change in the reaction times being observed in only one case (entry 2). Many results, such as the good performance of Wilkinson's catalyst (entry 4), were expected, but a palladium-based catalyst (entry 10) with no previous history as a hydrosilylation catalyst provided the best results and validated the usefulness of dye-labeled substrates for the rapid screening of catalytic systems. Correlation of bleaching with hydrosilylation was confirmed for a specific case (entry 4), isolating the expected hydrosilylation product and characterizing it by

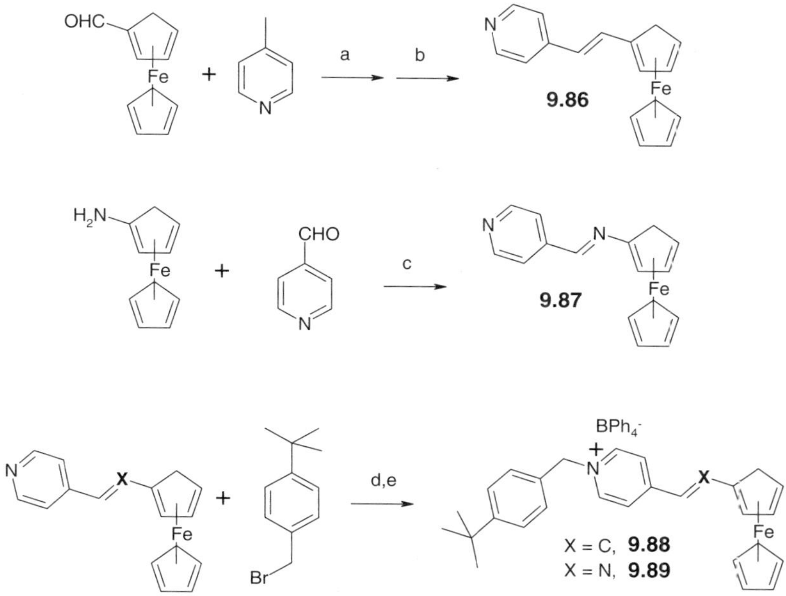

a: LDA, THF, rt, 16 hrs; b: $POCl_3$, pyridine, rt, 3 hrs;
c: $MgSO_4$, C_6H_6, reflux, 2 hrs; d: DMF, 80°C, 15'; e: $NaBPh_4$, acetone, rt, 12 hrs.

Figure 9.34 Catalyzed hydrosilylation: solution synthesis of the dye-labeled substrates **9.88** and **9.89**.

NMR. The significance of the acquired data was checked by running several reactions using cyclooctene as a substrate and isolating the reaction products. Entries 4 and 10 proved the best, in accordance with the results obtained through rapid screening.

Several parameters required further optimization. The color of the dyes was sometimes masked by colored catalysts, especially with the imine **9.89**, which has a lower ε value, large substrate–catalyst ratios were necessary, and solubility problems were encountered (see entry 12, Table 9.5). The use of dye-labeled substrates, though, proved to be both much faster than classical LC/GC detection methods for unlabeled substrates and reliable enough to screen catalytic systems. Its usefulness to assay polymer-bound substrates, or even to screen SP libraries using soluble dye-labeled reagents, was claimed by the authors.

Several other reports of discrete libraries of catalytic systems have appeared recently. Among them, Reetz et al. (146) validated the use of time-resolved IR-thermographic screening to find enantioselective, transition metal–based catalytic systems for epoxide hydrolysis in homogeneous conditions; Sigman and Jacobsen (147) reported the selection of enantioselective metal complexes as catalysts for addition of hydrogen cyanide to imines in homogeneous conditions; Berg et al. reported the synthesis and screening of hydrolytic metal complexes based on hydroxy

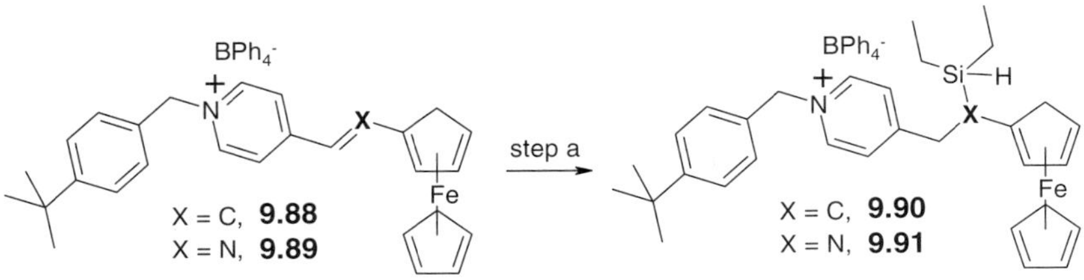

step a: single catalyst or **L21**, Ph_2SiH_2, dry THF, rt.

L21

12-member library of catalytic systems

12 catalysts used:

$[Ir(cod)(PPh_3)_2]BF_4$, $[Rh(cod)(PPh_3)_2]PF_6$, $[Rh(nbd)(PPh_3)_2]PF_6$, $RhCl(PPh_3)_3$, $[Rh(octanoate)_2]_2$, $RuCl_2(PPh_3)_3$, $NiCl_2(PPh_3)_2$, $[Ni(tss)]_2$, Cp_2ZrClH, $[Pd(Ar_2PC_6H_4CH_2)OAc]_2$, $[(nbd)Rh(triphos)]SbF_6$, $PtCl_2(NH_3)_2$

Figure 9.35 Hydrosilylation of **9.88** and **9.89** with the solution-phase-system discrete library **L21**.

aza crown ligands (148), using similar structures also for the cleavage of phospho di- and triesters and even DNA plasmids (149); Moye-Sherman et al. (150) selected optimal enantioselective metal complexes for the cyclopropanation of dehydrophenyl alanine derivatives; Bromidge (151) reported the identification of metal complexes that catalyzed an asymmetric aza Diels–Alder reaction producing a substituted pyridone; Huffman and Reider (152) optimized the stereoselectivity of the diastereoselective reductive amination leading to the angiotensin converting enzyme (ACE) inhibitor enalapril using high-throughput parallel screening of catalytic systems; Reetz et al. (153) reported a novel HTS method for enantioselective catalytic system libraries applied to several lipase-catalyzed reactions; Lavastre and Morken (154) presented a high throughput, visual assay to rapidly screen catalytic systems for allylic alkylation using various metal and ligands.

9.4.4 An Example: High-Throughput Screening of a Ligand Library for Heck C–C Coupling

Shaughnessy et al. (155) reported the rapid, sequential screening of three phosphine ligand libraries **L22–L24** in a Heck coupling reaction involving a fluorescent substrate **9.92** and the supported aryl bromide **9.93**, whose structure and synthesis are reported in Fig. 9.36.

TABLE 9.5 Hydrosilylation Kinetics for the Individuals from the Solution-Phase Discrete Catalytic System Library L21[a]

Catalyst		**9.88/t_1**	9.88/t_2	9.88/$t_2 - t_1$	9.89/t_1	9.89/t_2	9.89/$t_2 - t_1$
1	$[Ir(cod(PPh_3)_2]BF_4$	5 s	>45 min[b]	>45 min	5 s	>45 min	>45 min
2	$[Rh(cod)(PPh_3)_2]PF_6$	3 s	35 s	32 s	2 s	9 min	8.58 min
3	$[Rh(nbd)(PPh_3)_2]PF_6$	4 s	1 min	56 s	15 s	1.45 min	1.30 min
4	$RhCl(PPh_3)_3$	3 s	2 min	1.57 min	3 s	3 min	2.57 min
5	$[Rh(octanoate)_2]_2$	NR	NR	NR	NR	NR	NR
6	$RuCl_2(PPh_3)_3$	10 s	>45 min	>45 min	10 s	>45 min	>45 min
7	$NiCl_2(PPh_3)_2$	NR	NR	NR	NR	NR	NR
8	$[Ni(tss)]_2$	NR	NR	NR	NR	NR	NR
9	Cp_2ZrClH	NR	NR	NR	NR	NR	NR
10	$[Pd(Ar_2PC_6H_4CH_2)OAc]_2$	1 s	1.15 min	1.14 min	1 s	1.15 min	1.14 min
11	$[(nbd)Rh(triphos)]SbF_6$	NR	NR	NR	NR	NR	NR
12	$PtCl_2(NH_3)_2$	sol	sol	sol	sol	sol	sol

[a] t_1 = initial bleaching time, t_2 = final bleaching time; NR means no reaction in 45 min; sol means insoluble catalyst in the assay conditions.

[b] Color interference preventing the observation of final bleaching time.

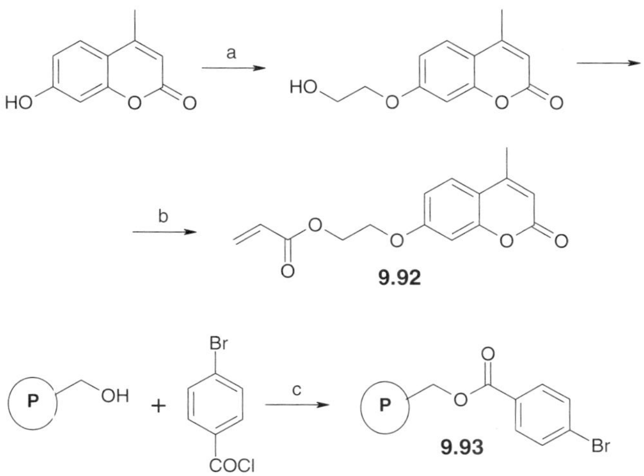

a: bromoethanol, K_2CO_3, DMF, 100°C, 15 hrs; b: acryloyl chloride, TEA, DCM, rt, 2 hrs; c: pyridine, DCM.

Figure 9.36 Ligand libraries for the catalyzed Heck reaction: synthesis of reagents **9.92** in solution and **9.93** on SP.

To begin with, the reactivity of **9.92** was compared to a typical substrate, butyl acrylate, in solution. Reaction with a common aryl halide produced similar reaction yields of **9.94** and **9.95**, respectively (Fig. 9.37, top). The reactivity of **9.92** and butyl acrylate were then compared on solid phase, using **9.93**, standard coupling conditions, and various reaction times (Fig. 9.37, bottom). The coupling to give **9.96** was complete after 4 h, with the fluorescence of the beads treated with **9.92** being either weak (55% conversion by fluorescence measurement, 2 h) or strong (quantitative conversion, 4 h).

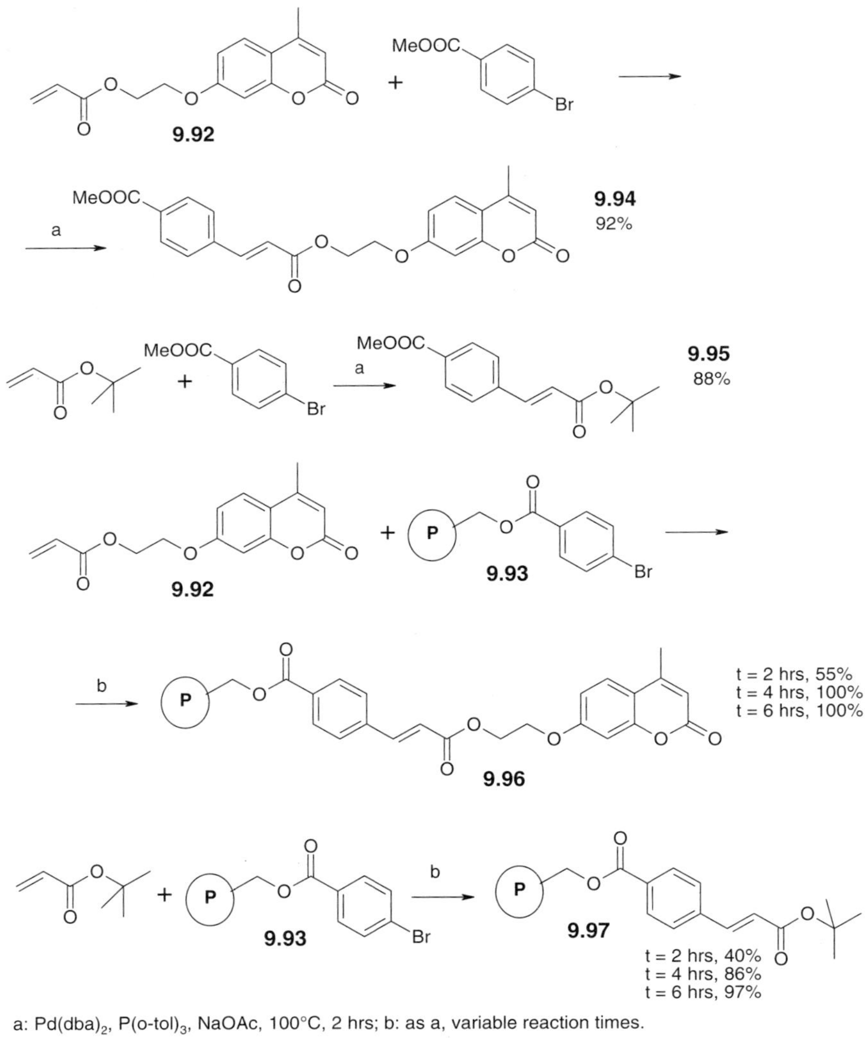

Figure 9.37 Fluorescent determination of catalytic activity in the Heck coupling of the labeled substrate **9.92** with supported **9.93**: validation studies versus Heck couplings in solution.

These results were slightly better than those obtained employing *t*-butyl acrylate to produce **9.97** (GC determination after cleavage). When **9.93** was treated with fluorescent substrate **9.92** without catalyst, the washed resin beads showed no fluorescence, proving the overall reliability of the fluorescent screening and the absence of interferences.

Two 20-member libraries of mono- and diphosphine ligands (**L22** and **L23**, Fig. 9.38) were then screened in two parallel arrays, containing **9.92**, **9.93**, $Pd(dba)_2$ and NaOAc, heated at 100°C for 4 h. The ligand structures of **L22** included hindered *o*-substituted arylphosphines, arylphosphines with different electronic properties or with weak coordination properties, and alkylphosphines. **L23** included diphosphines with various steric and electronic properties as well as with different backbones (Fig. 9.38). The level of bead fluorescence was measured roughly as being absent (**F_1**), moderate (**F_2**), or strong (**F_3**), and the data for the best 12 ligands are reported in Table 9.6 (all the reactions were repeated twice, giving the same results). The same 40 reactions were also performed in solution, replacing **9.93** with a soluble aryl bromide, and their outcome was determined by GC. The results reported in Table 9.6 highlighted a general accordance between the homogeneous and heterogeneous reaction systems, with strongly fluorescent beads corresponding to >80% conversion in solution and nonfluorescent beads corresponding generally to poor ligands in solution (only one example produced >60% conversion in solution and did not produce any SP fluorescence; see Table 9.6). Most of the 11 ligands, 8 from **L22** and 3 from **L23**, that showed moderate or strong fluorescence were sterically hindered structures (Table 9.6 and Fig. 9.39).

The best performers from **L22** and **L23** were tested using three more demanding reaction protocols, either decreasing the reaction temperature to 75°C, or to 50°C, or using a less reactive supported aryl chloride (replacing Br with Cl in **9.93**). An additional 5 ligands (Fig. 9.39) were added to the 11 library-generated active ligands, producing the 16-member ligand library **L24**, whose screening results are reported in Table 9.7. Only 2 ligands produced strongly fluorescent beads in all the reaction conditions. Confirmation of these results in solution was successful, in that a best performer was selected (**$L24_{12}$**). This ferrocene ligand (Fig. 9.39) was as effective as a known, optimized dimeric palladacycle ligand (156) when tested on activated substrates but was significantly better on unactivated substrates (155).

Fluorescent detection was successfully validated in this specific example. Its higher throughput, when compared to more classical LC/GC detection methods (2–3 h versus 16 h), makes it appealing to screen larger ligand libraries using fluorescent substrates. The known phenomenon of fluorescence quenching on resin beads (157) was not relevant in this example, but a rigorous assessment should always be made, as it was here, to determine the reliability of the fluorescent screen to predict real catalytic efficiency for any given reaction.

Many other reports of ligand libraries for specific catalytic applications have been reported. Among them, Gilbertson and co-workers reported a chiral phosphine library, tested in the rhodium-catalyzed asymmetric hydrogenation of an enamide (158, 159), and a similar library for the palladium-catalyzed allylation of malonates (160, 161); Hoveyda and co-workers (162, 163) reported a chiral Schiff base library, screened in the titanium-catalyzed opening of epoxides with (TMSCN) (trimethyl silyl cyanide);

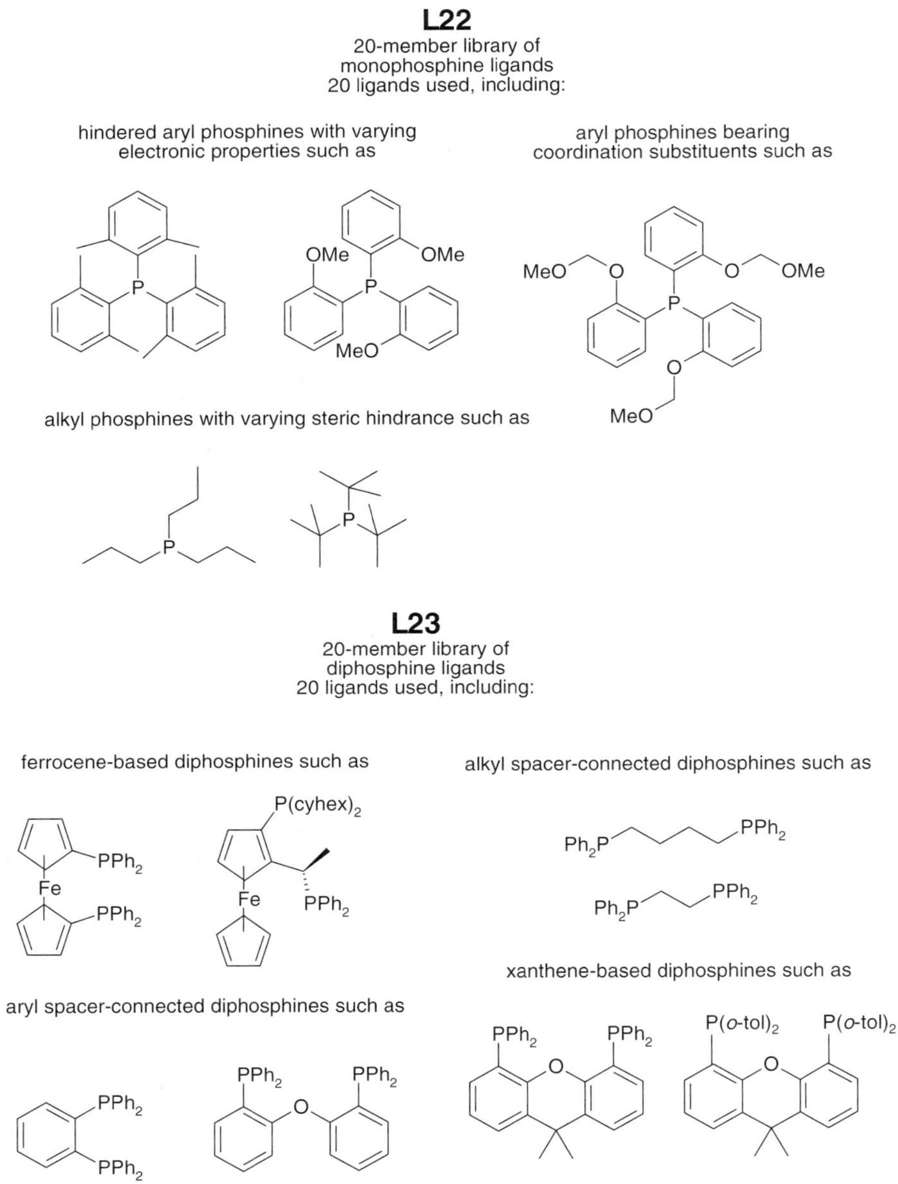

Figure 9.38 Mono- (**L22**) and diphosphine (**L23**) solution-phase discrete ligand libraries for the catalyzed Heck reaction of the labeled substrate **9.92** with supported **9.93**.

TABLE 9.6 Ligands from Solution-Phase Discrete Mono- and Diphosphine Libraries L22 and L23 for the Catalyzed Heck Reaction of Labeled Substrate 9.92 with Supported 9.93: Screening Results

Entry	Ligand	Fluorescent Yield[a]	GC Yield[b]
1	Di(2,4-xylyl)PPh	F_2	82
2	P(*o*-tolyl)$_3$	F_3	99
3	P(2,4-xylyl)$_3$	F_3	97
4	P(α-naphthyl)$_3$	F_3	96
5	P(*o*-anisyl)$_3$	F_2	20
6	(2-CF$_3$Ph)P(*o*-anisyl)$_2$	F_1	64
7	(DPPh)EtOMe	F_3	66
8	(2-MOMPh)$_3$P	F_2	85
9	P(tBu)$_3$	F_3	95
10	DPPDPE	F_2	42
11	DTPDPE	F_3	65
12	DTPX	F_3	100

[a]Reaction on solid phase (**9.93**); F_1 = no fluorescence, F_2 = weak fluorescence, F_3 = strong fluorescence.
[b]Reaction in solution (soluble aryl bromide).

Sigman and Jacobsen (164) reported a Schiff base library, screened for the asymmetric Strecker reaction; Porte et al. (165) presented the screening of a chiral phosphine oxazoline library in the palladium-catalyzed allylation of malonates; Gennari et al. (166) reported a chiral disulfonamide library, screened in the titanium-catalyzed addition of Et_2Zn to aldehydes; Buck et al. (167) reported a library of novel carboxylate ligands, screened for the rhodium-catalyzed asymmetric carbenoid Si–H insertion; Ding et al. (168) reported a library of diol ligands–chiral nitrogen activators for the zinc-catalyzed addition of Et_2Zn to aldehydes; Hinderling and Chen (169) reported the MS screening of a ligand diimine library made for the Pd^{II}-catalyzed olefin polymerization; Havranek et al. (170) reported the TLC/GC screening of 1,2-phenylene diamine amide peptidomimetics for the Mn-catalyzed oxidation of alkanes to carbonyl compounds and alkenes to carbonyl compounds and epoxides; Altava et al. (171) presented a library of supported, chiral β-aminoalcohols as chiral auxiliaries reacted with $LiAlH_4$ to give reducing agents for the enantioselective reduction of acetophenone.

9.4.5 An Example: Screening of a Substrate Library of Ketones for Their Asymmetric Reduction

Gao and Kagan (172) reported the screening of several mixtures of ketones **L25–L29**, composed of compounds **9.98–9.111** (Fig. 9.40), as substrates for asymmetric catalytic reduction using Corey's oxazaborolidine **9.112** (173), establishing its general applicability to the generic reaction shown in Fig. 9.40, top.

The substrates were discernible using chiral HPLC as a detection method for measuring the presence of residual ketones and formed alcohol stereoisomers. In several

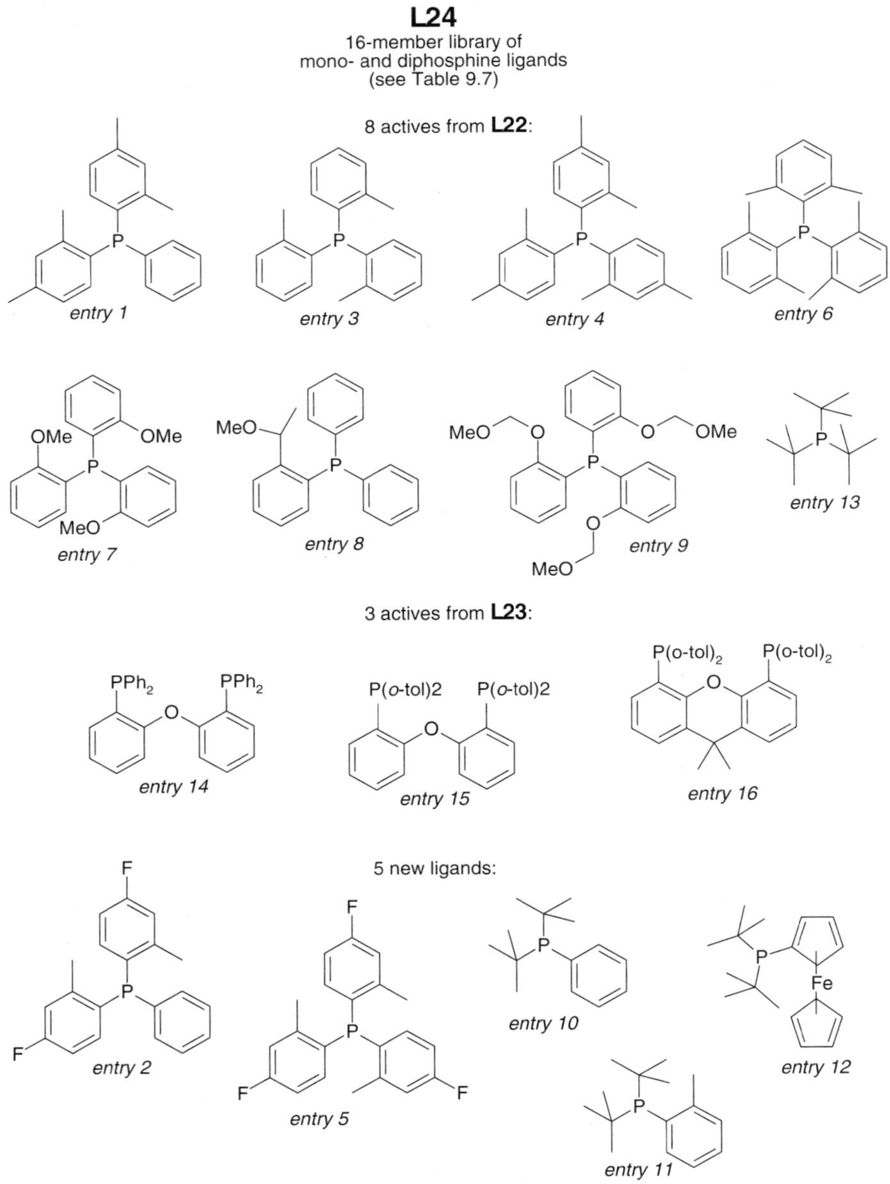

Figure 9.39 Mono- and diphosphine focused solution-phase discrete ligand library **L24** for the catalyzed Heck reaction of the labeled substrate **9.92** with supported **9.93**.

TABLE 9.7 Ligands from Solution-Phase Discrete Mono- and Diphosphine Library L24[a] for the Catalyzed Heck Reaction of Labeled Substrate 9.92 with Supported 9.93: Screening Results

Entry	Ligand	**9.93**/75°C	**9.93**/50°C	SP-ArCl 100°C
1	Di(2,4-xylyl)PPh	F_1	F_1	F_1
2	Di(2-Me,4-FPh)PPh	F_3	F_3	F_1
3	P(*o*-tolyl)$_3$	F_3	F_1	F_1
4	P(2,4-xylyl)$_3$	F_3	F_3	F_1
5	P(2-Me,4-FPh)$_3$	F_1	F_1	F_1
6	P(2,6-xylyl)$_3$	F_1	F_1	F_1
7	P(*o*-anisyl)$_3$	F_1	F_1	F_1
8	(DPPh)EtOMe	F_1	F_1	F_1
9	(2-MOMPh)$_3$P	F_1	F_1	F_1
10	(tBu)$_2$PPh	F_3	F_1	F_3
11	(tBu)$_2$P(*o*-tolyl)	F_1	F_1	F_1
12	(tBu)$_2$P-ferrocenyl	F_3	F_3	F_3
13	P(tBu)$_3$	F_3	F_3	F_3
14	DPPDPE	F_1	F_1	F_1
15	DTPDPE	F_1	F_1	F_1
16	DTPX	F_3	F_1	F_1

[a] F_1 = no fluorescence, F_3 = strong fluorescence.

cases (libraries **L26–L28**) a preliminary flash chromatography was necessary to separate the library members. The reduction of the mixtures was compared with the reduction of isolated ketones, and an almost complete reproducibility of the results obtained in the two protocols was observed (Table 9.8). Thus, the reduction of ketone mixtures allowed determination of the chiral preference (*S* or *R*) and the enantiomeric excess produced by the catalyst in the presence of substrates with different electronic or steric properties.

The rapid assessment of the catalytic profiles of several catalytic systems, tested on substrate libraries, should become a useful tool to speed the selection of optimally catalyzed reaction conditions, providing that faster detection methods affording reliable results in shorter times will be available. Eventually, the parallel use of substrate and catalytic system libraries should become routine to thoroughly assess the catalytic properties and the specificity of a chemical transformation.

9.4.6 An Example: Synthesis and Screening of a Catalytic System Library as Alkene Epoxidation Catalysts

Francis and Jacobsen (174) reported a 5760-member pool library of ligand–metal complexes **L30**, which is shown in Fig. 9.41 together with its synthetic scheme. First, the monomer set **M1** (five α-amino acids, Fig. 9.41) was coupled onto five portions of aminomethyl PS resin using typical peptide coupling conditions to give, after Fmoc

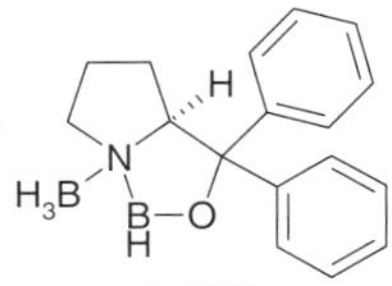

9.102

L25
4-member substrate library

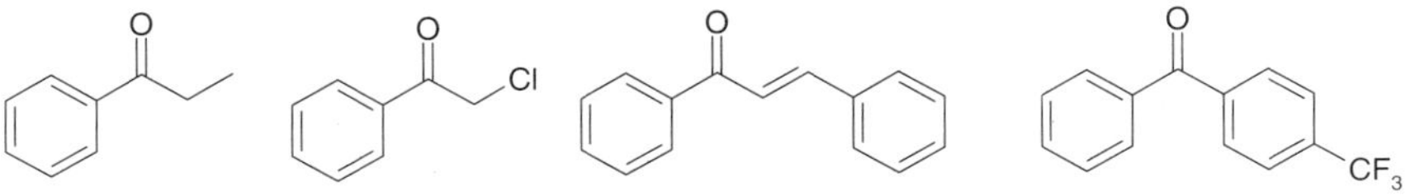

9.98 9.99 9.100 9.101

L26
7-member substrate library

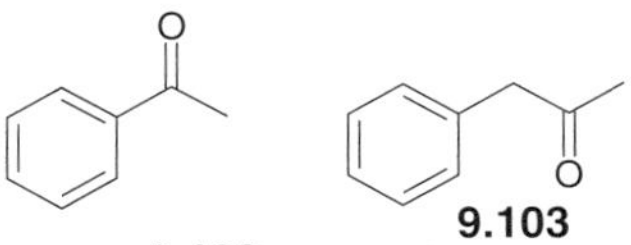

9.102 9.103

9.98 9.100
9.99 9.101

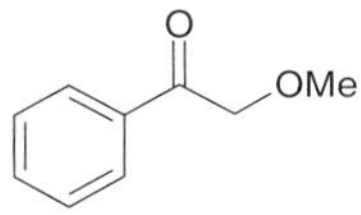

9.104

L27
5-member substrate library

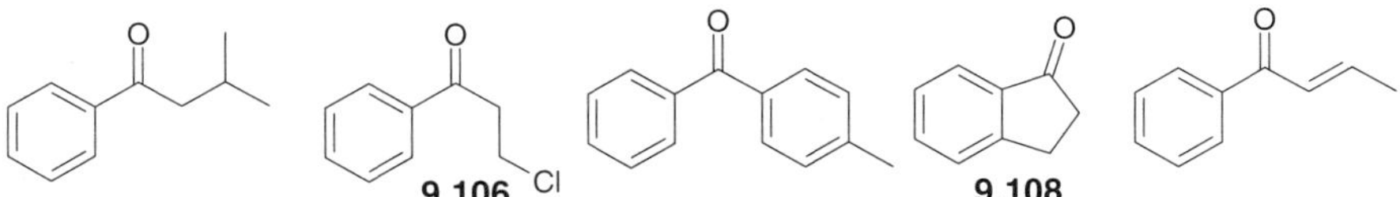

9.105 9.106 9.107 9.108 9.109

L28
6-member substrate library

9.99 9.102 9.104
9.101 9.103 9.105

L29
6-member substrate library

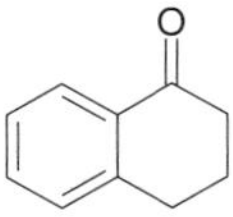

9.110

9.98 9.106
9.100 9.107

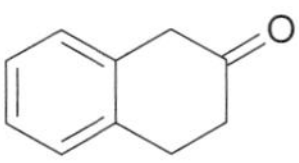

9.111

Figure 9.40 Solution-phase pool substrate ketone libraries **L25–L29**.

TABLE 9.8 Enantiomeric Excess of Alcohols 9.98–9.111 Obtained from Reductions of Libraries L25–L29

Entry	Ketone	Library	Mixture e.e.[a]	Isolated e.e.[b]	Configuration
1	**9.98**	**L25,26,29**	90	92	*R*
2	**9.99**	**L25,26,28**	96	95	*S*
3	**9.100**	**L25,26,29**	29	35	*S*
4	**9.101**	**L25,26,28**	35	39	*S*
5	**9.102**	**L26,28**	97	97	*R*
6	**9.103**	**L26,28**	32	31	*R*
7	**9.104**	**L26,28**	94	95	*S*
8	**9.105**	**L27,28**	46	24	*S*
9	**9.106**	**L27,29**	87	90	*R*
10	**9.107**	**L27,29**	19	18	*S*
11	**9.108**	**L27**	92	93	*R*
12	**9.109**	**L27**	76	71	*R*
13	**9.110**	**L29**	96	96	*R*
14	**9.111**	**L29**	5.6	4.8	*S*

[a]Average enantiomeric excess from different libraries (entries 1–10) or from a single library (entries 11–14).
[b]Enantiomeric excess from discrete ketone reduction.

deprotection, compounds **9.113** (steps a and b). After pooling of the resin and splitting into three aliquots (step c), the resin was coupled with the monomer set $\mathbf{M_2}$ (three compounds, including a β-aminoalcohol $\mathbf{M_{2,1}}$, an α,β-diamine $\mathbf{M_{2,2}}$, and an α-amino acid $\mathbf{M_{2,3}}$, steps d_1–d_3, Fig. 9.41) and deprotected (steps e_1–e_3) to give resin-bound amines **9.114a–o**. A structure similar to known epoxidation catalysts was also added, starting from Wang PS resin and coupling it sequentially with an aldehyde to give **9.115** and with building block $\mathbf{M_{2,2}}$ to give **9.114p** (Fig. 9.41) as a positive control for screening. The 16 resin aliquots were pooled, split into 12 portions (step f), and coupled with monomer set $\mathbf{M_3}$ (12 capping agents, including aldehydes, acids, and a skip codon, steps g_1–g_2, Fig. 9.41) to give resin-bound imines **9.116** and amides **9.117**. The resin aliquots were then pooled (step h) and exposed to monomer set $\mathbf{M_4}$ (30 metal ion sources where the metal, the oxidation state, or the counterion was changed) in solution for 1 h (step i) to produce **L30** as 30 pools of 192 compounds after filtration and rinsing of the beads (Fig. 9.41). The quality of the library was checked using both visual detection of the expected colors of the ligand–metal complexes and qualitative detection of the complexes using inorganic staining reagents. Approximately 80% of the expected library individuals were detected, and the library was progressed to screening.

The selected reaction was the epoxidation of *trans*-β-methylstyrene (**9.118**, Fig. 9.42). At first, various potential reaction conditions were tested for the production of epoxide **9.119** as an enantiomeric couple (chiral GC detection). Among them, hydrogen peroxide and *tert*-butylhydroperoxide gave good results. The use of the former reagent

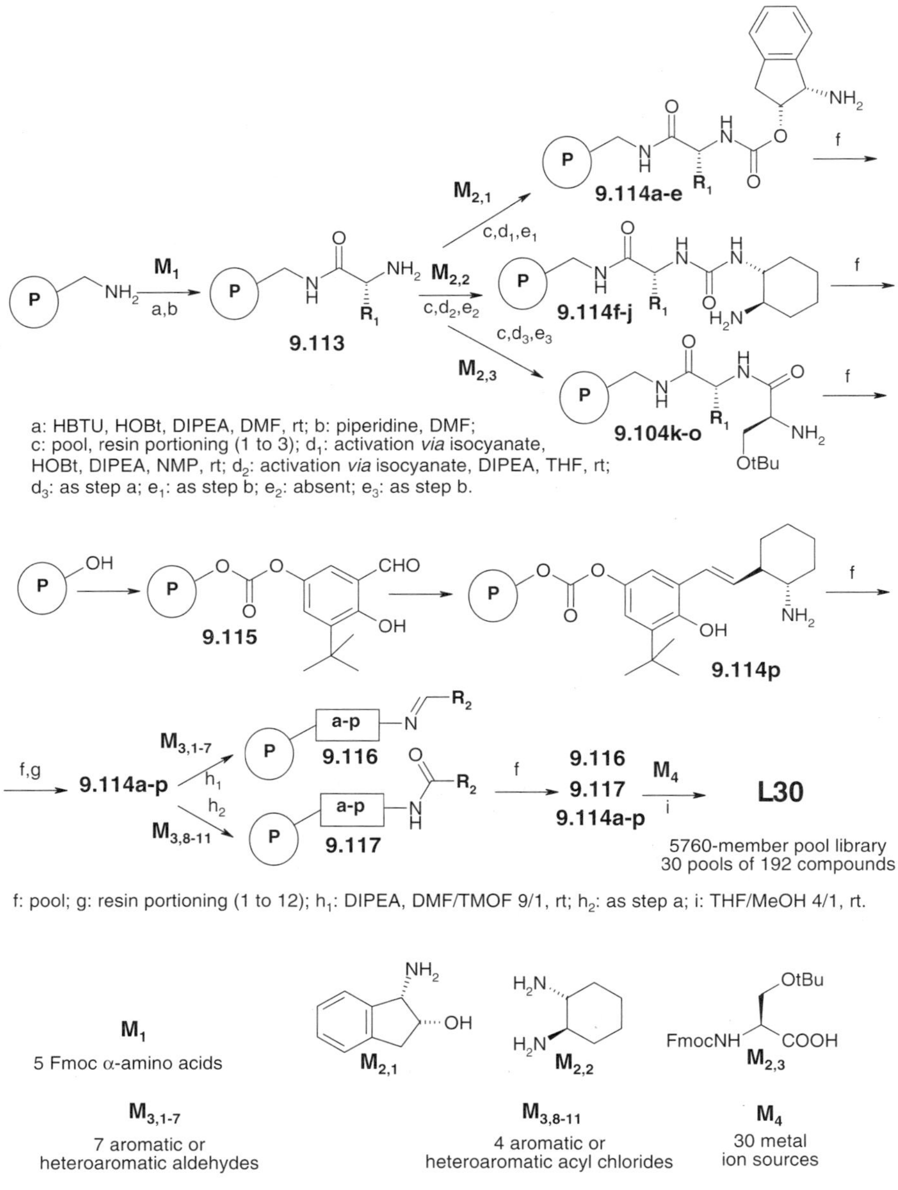

Figure 9.41 Catalyzed alkene epoxidation: synthesis and structures of the SP pool peptidomimetic catalytic system library **L30** and of the monomer sets **M1**–**M4**.

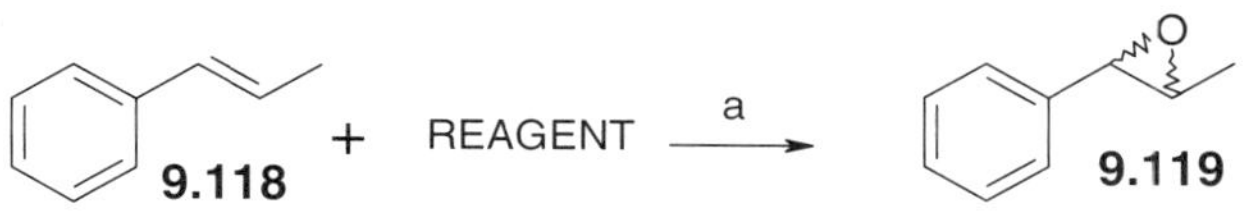

a: catalyst, DCM/tBuOH 1/1, rt.

tested reagents: O_2, $NaIO_4$, 4-PPNO, NMO, tBuOOH, H_2O_2

selected: **30% aq. H_2O_2**

Figure 9.42 Catalyzed epoxidation of **9.118**: selection of the oxidant.

was described in Ref. 174, while the other is the subject of another report. Deconvolution of the library **L30** in the hydrogen peroxide epoxidation of **9.118** was then started.

The 30 **L30** pools, varying the $\mathbf{M_4}$ constituent, were first screened in the epoxidation reaction (Fig. 9.43). As a control, 30 parallel reactions containing only the metal source were also performed to determine the metal–ligand-dependent catalysis. The results, expressed as relative yields to an internal standard (from 0 = no reaction to 3 = maximum observed), are reported in Table 9.9. Seventeen of the 30 pools showed either ligand-dependent (7) or ligand-independent (10) catalytic epoxidation activity. The most significant ligand-dependent catalytic activity was displayed by Fe ions, with Fe^{2+} ion and Cl^- counterion as the best combination (compare entries 4–6, Table 9.9).

Having determined the best $\mathbf{M_4}$ component as $\mathbf{M_{4,7}}$, library **L31** was prepared as 12 sublibraries of 16 compounds containing different capping monomers $\mathbf{M_3}$ and subsequently exposed to a solution of $FeCl_2$ (Fig. 9.43). The screening results are presented in Table 9.10. Two pyridine-based capping monomers ($\mathbf{M_{3,1}}$ and $\mathbf{M_{3,8}}$) were roughly equivalent to each other and significantly more active than the other 16-member pools in promoting the epoxidation reaction.

The synthesis of the 32 discretes composing the most active pools mentioned above should have produced the most active catalytic system. The authors, though, prepared **L32** as 192 discretes, corresponding to all the pools tested above in **L31**, using the radiofrequency encoding approach with directed sorting (see Sections 7.4.3 and 7.4.4). The identification of three active ligands **9.120–9.122** (Fig. 9.43) followed this third iterative screening. The deconvolutive approach was validated by these results, where all the best ligands contained the expected monomers $\mathbf{M_{3,1}}$ (**9.120**) and $\mathbf{M_{3,8}}$ (**9.121** and **9.122**), identified in the previous iterative round. The best monomer in position $\mathbf{M_2}$ was also clearly selected as L-serine, while position $\mathbf{M_1}$ was represented by two similar monomers, L-serine (**9.120** and **9.121**) and L-cysteine (**9.122**).

The three selected ligands were excellent in promoting the epoxidation, but the stereoselection was negligible, varying from 4 to 7%. An additional 96-member parallel library **L33** was built to improve the stereochemical outcome of the reaction, building on the SAR acquired from **L30–L32**. Its structure and the composition of the monomer sets $\mathbf{M_1}$–$\mathbf{M_3}$ are reported in Fig. 9.44. Three best ligands **9.123–9.125** (Fig. 9.44) were selected using the same screening protocol on **L33** as described previously.

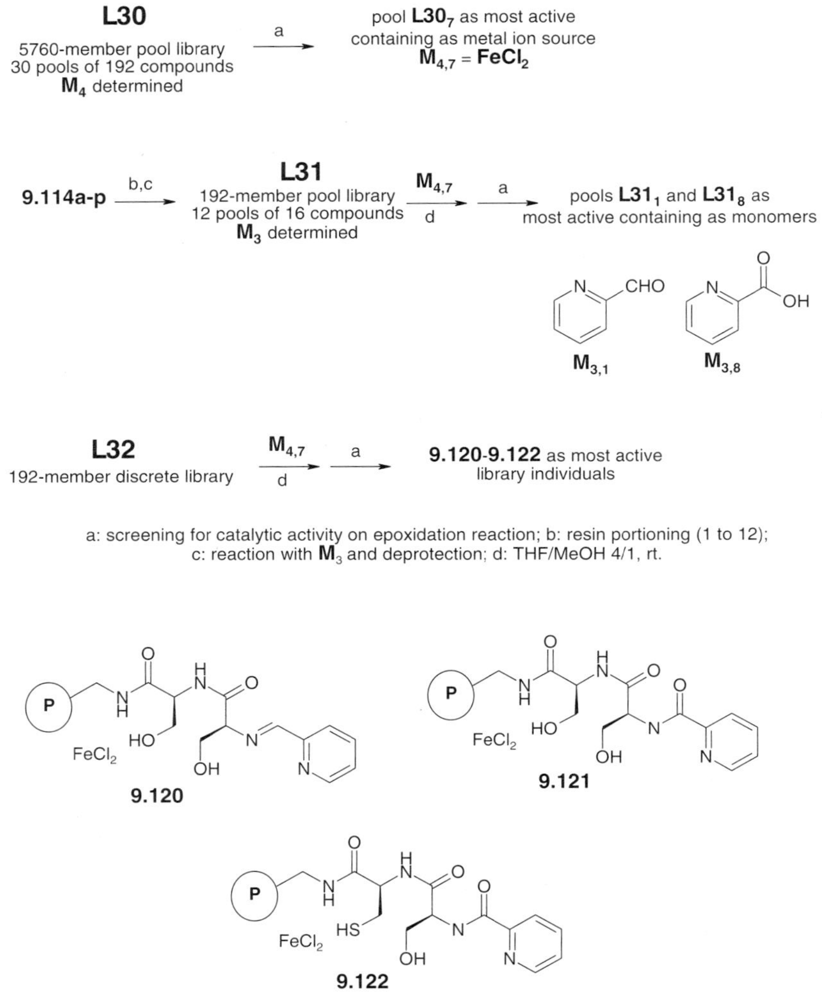

Figure 9.43 Deconvolution of the SP pool peptidomimetic library of catalytic systems **L30** through lower complexity libraries **L31** and **L32** and identification of catalytic composites **9.120–9.122** obtained from its screening.

A clean conversion of **9.118** to **9.119** was obtained, with moderate stereoselection (15–20% enantiomeric excess e.e.) for the *R,R* (**9.123**, **9.124**) or the *S,S* enantiomer (**9.125**). Resynthesis of the unbound analogues of **9.120–9.125** was planned to test their effectiveness in the same reaction and to evaluate the influence of the solid support on the reaction outcome. The deconvolution-prone, pooled format for catalyst discovery has been used by Venton (175) for the synthesis of a >28,000-member β-cyclodextrin library (13 pools) as a source of Zn-based catalytic systems with phosphatase-like activity.

TABLE 9.9 Screening of the SP Pool Peptidomimetic Encoded Catalytic Library L30 for Epoxidation of 9.118

Entry	Metal Sources	Yield with **L30**[a]	Yield with Metals Only[a]
1	$Ti(OiPr)_4$	0.5–0.75	0.25–0.5
2	$VOCl_3$	0–0.25	0
3	$VOSO_4$	1.5–1.75	2.75–3
4	**$FeCl_2$**	**2.5–2.75**	**0–0.25**
5	$FeCl_3$	1.25–1.5	0–0.25
6	$Fe(acac)_3$	0.25–0.5	0.5–0.75
7	$CoCl_2$	0–0.25	0
8	$CuCl_2$	0–0.25	0
9	$Cu(Oac)_2$	0–0.25	0
10	$[RuCl_2(cymol)]_2$	0–0.25	0
11	$RuCl_3$	0.25–0.5	0–0.25
12	$Ru(acac)_2$	0–0.25	0
13	$[Rh(OAc)_2]_2$	0	0.25–0.5
14	$MeReO_3$	0	2.75–3
15	$IrCl_4$	0	0–0.25
16	$YbCl_3$	0	0–0.25

[a]Calculation based on the amount of **9.119** compared with an internal standard.

TABLE 9.10 Activity Screening of the SP Encoded Pool Peptidomimetic Catalytic System Library L31: Deconvolution of M_3

Entry	Monomer	Catalytic Activity[a]
1	$\mathbf{M_{3,1}}$	3.25–3.50
2	$\mathbf{M_{3,2}}$	0.75–1
3	$\mathbf{M_{3,3}}$	0.75–1
4	$\mathbf{M_{3,4}}$	0.75–1
5	$\mathbf{M_{3,5}}$	1–1.25
6	$\mathbf{M_{3,6}}$	0.50–0.75
7	$\mathbf{M_{3,7}}$	0.50–0.75
8	$\mathbf{M_{3,8}}$	3.50–3.75
9	$\mathbf{M_{3,9}}$	1.75–2
10	$\mathbf{M_{3,10}}$	0.5–0.75
11	$\mathbf{M_{3,11}}$	0.25–0.5
12	$\mathbf{M_{3,12}}$	0.5–0.75

[a]Calculated as the yield of **9.119** compared to an internal standard.

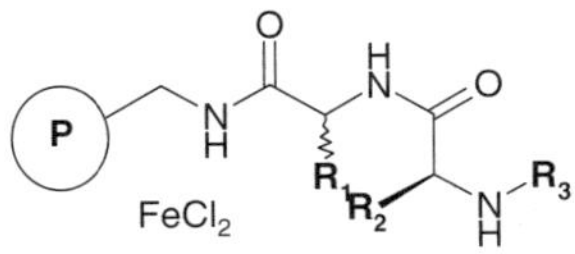

L33

96-member discrete library

$\mathbf{M_1}$: L- or D-Ser, L- or D-Thr;
$\mathbf{M_2}$: L-Ser, L-Leu, L-Phe;
$\mathbf{M_3}$: 8 pyridyl-containing aldehydes (3) or acyl chlorides (5).

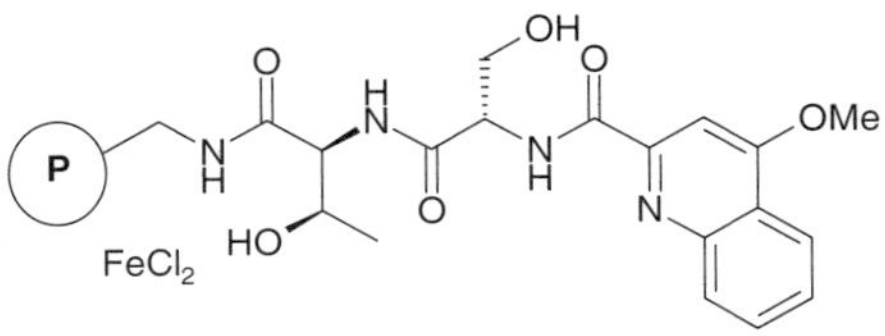

9.123

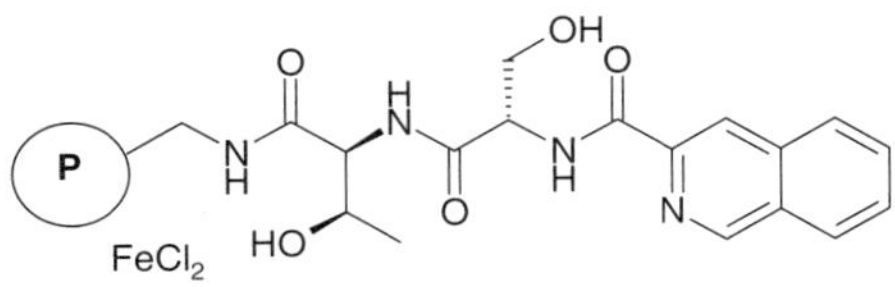

9.124

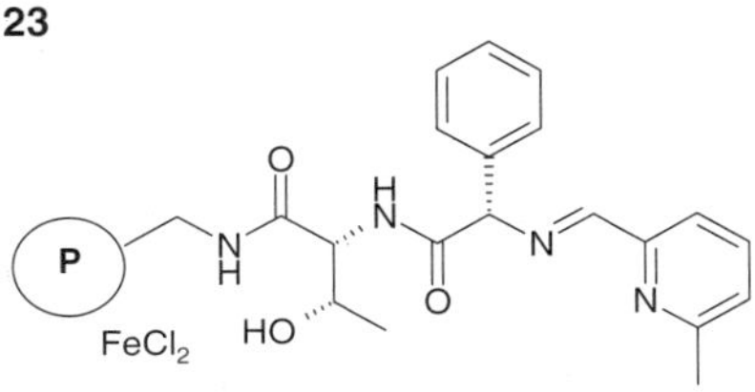

9.125

Figure 9.44 Structure of the SP peptidomimetic focused catalytic library **L33** and of the enantioselective catalytic composites **9.123–9.125** obtained from its screening.

9.4.7 An Example: Synthesis and Screening of an Encoded Acylation Catalytic Library

Taylor and Morken (176) reported the synthesis and on-bead screening of a 3150-member encoded SP library of potential acylation catalysts **L34**, whose synthesis and structure are reported in Fig. 9.45. Resin-supported bromide **9.126**, obtained by treatment of amino Tentagel resin with bromoacetic acid (step a), was reacted (step b) with the monomer set $\mathbf{M_1}$ (primary amines, 15 representatives, Fig. 9.46); then the secondary amine function of intermediates **9.127** was coupled (step c) with monomer set $\mathbf{M_2}$ (15 representatives, N-protected α-amino acids, carboxylic acids, and a skip codon, Fig. 9.46). Deprotection of the nitrogen (step d) and capping (step e) with monomer set $\mathbf{M_3}$ (carboxylic acids, 15 representatives, and a skip codon, Fig. 9.46) produced **L34** as 15 pools, each made of 210 compounds (Fig. 9.45). Each monomer representative was encoded using a popular encoding method (177). The library structure captured the concept of a basic center and a nucleophilic center mutually arranged in different orientations to detect a suitable bifunctional acylation catalyst. A known acylation catalyst (supported monomer $\mathbf{M_{2,12}}$) was added among the monomer set $\mathbf{M_2}$ to validate the applicability of the screening and its ability to detect catalysts in an encoded SP pool format.

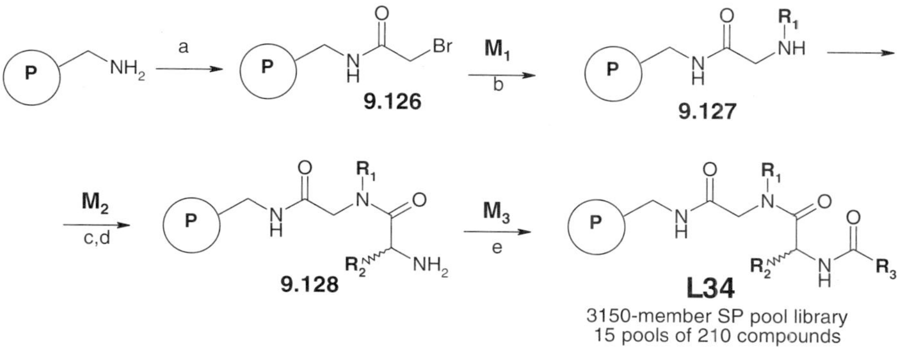

a: amide coupling; b: nucleophilic displacement; c: acylation; d: TFA, DCM, rt; e: amide coupling.

Figure 9.45 Catalyzed acylations synthesis of the SP catalytic peptidomimetic encoded library **L34**.

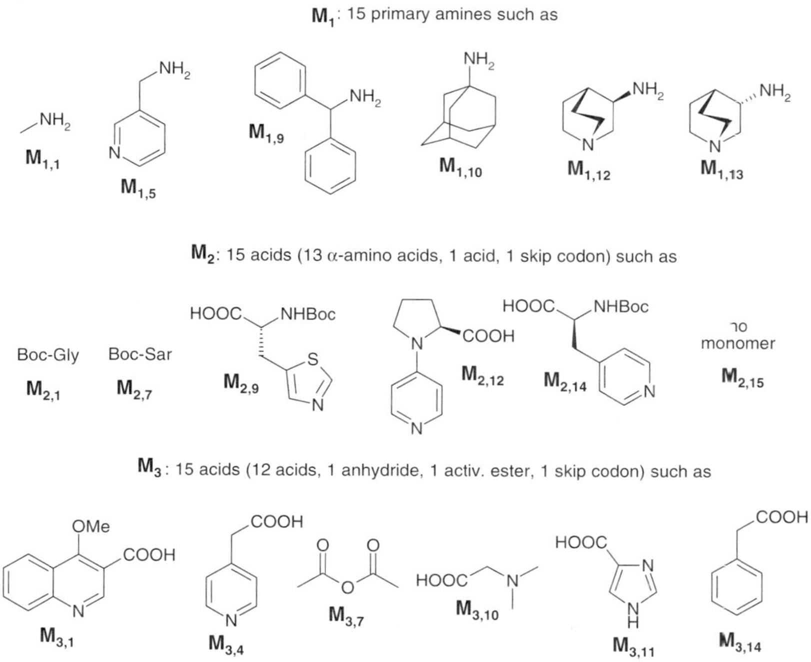

Figure 9.46 Monomer sets **M1–M3** used in the synthesis of the SP peptidomimetic encoded library of catalysts **L34**.

The catalyst library was screened on an acylation reaction between ethanol and acetic anhydride (Fig. 9.47, equation 1). A thermographic IR assay was chosen as the detection method. Previous studies (178) have shown the applicability of this technique to materials science libraries, where ignition temperatures of pellets with various compositions could be measured. With a given heat of reaction, the progress of a chemical reaction can be monitored by measuring the temperature increase of the reaction medium, and for catalyzed reactions this phenomenon is correlated with the efficiency of the catalytic system. Thus, the goal was to measure the temperature of each resin bead loaded with a potential catalyst, spotting any temperature increase, which should have been related to catalytic activity of the supported library individual in the acylation reaction. Larger 300-μm beads were chosen to maximize the temperature increase, and chloroform was used as a cosolvent so that the beads floated on top of the solution, thereby avoiding solvent interference with IR transmission. To validate this method, *N*-4-pyridylproline (monomer **M2,12**) was linked to amino Tentagel

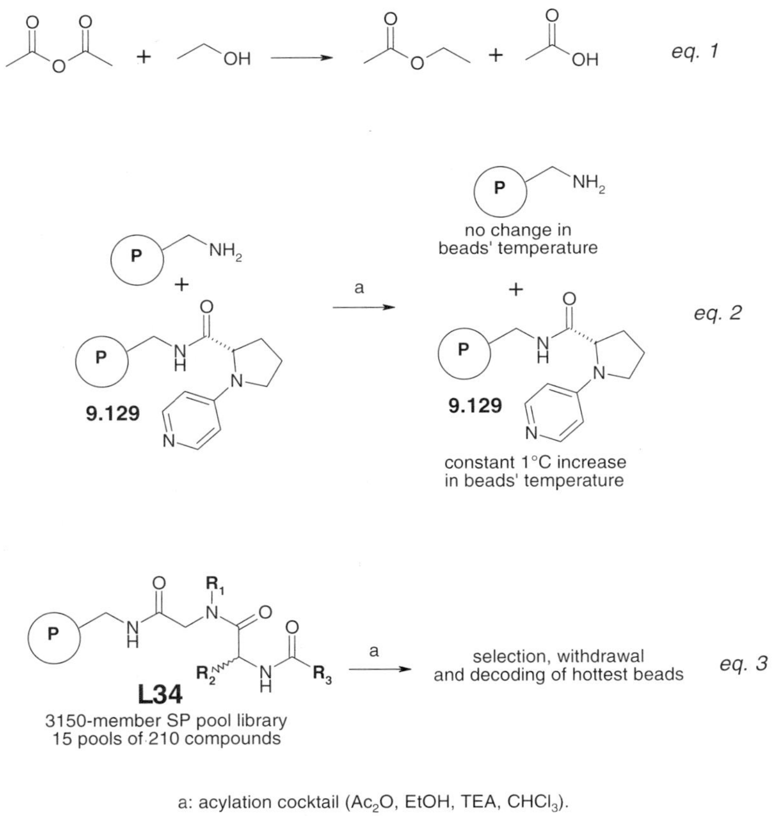

Figure 9.47 On-bead IR thermography screening of the SP peptidomimetic encoded library of catalysts **L34** tested in the acylation of ethanol with acetic anhydride.

300-μm beads to give the supported catalyst **9.129**, and several beads were added to an acylation cocktail (Fig. 9.47, equation 2) together with nonfunctionalized beads. IR thermography consistently spotted a 1 °C increase of temperature for each bead loaded with the catalyst with respect to the other beads, thus validating the detection method in this demanding screening protocol.

Library screening involved around 7000 beads (610 mg, around 2.3 library equivalents) to represent the large majority of library individuals. The beads were added to an acylation cocktail (Fig. 9.47, equation 3), and the hottest beads were spotted, maintaining a large excess of acylation reagents in the solution (see the original work for more technical details). Several beads were identified, and 23 of them were decoded (the number of hot beads that were not decoded was not reported) to give the four structures **9.130–9.133** (Fig. 9.48). Structures **9.130–9.132** contained the capping monomer **$M_{2,12}$**. This monomer prevented the coupling of **M_3** representatives, thus producing 15 copies of **$M_{2,12}$**-containing library individuals where only one copy of

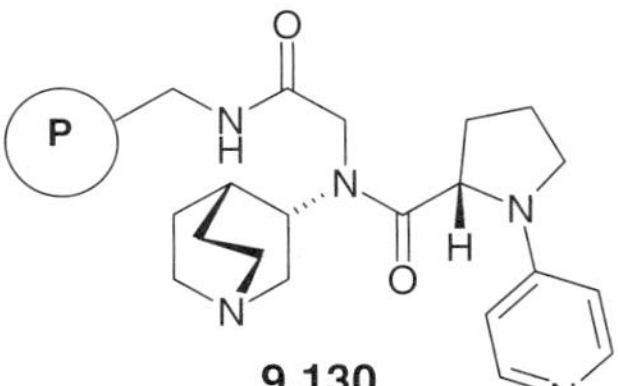

9.130
11 beads

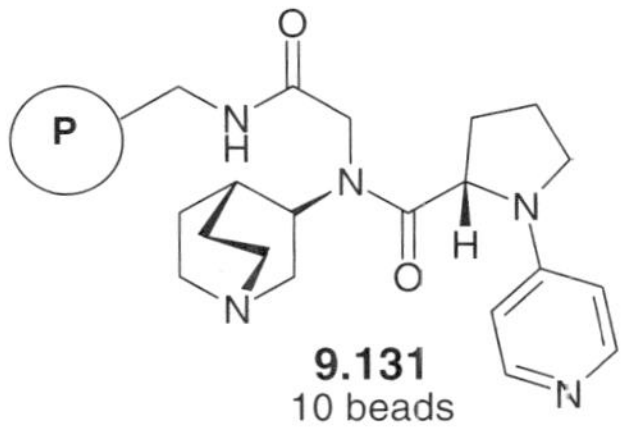

9.131
10 beads

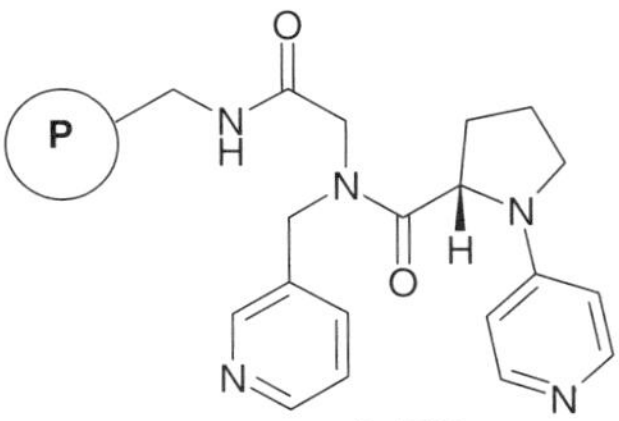

9.132
1 bead

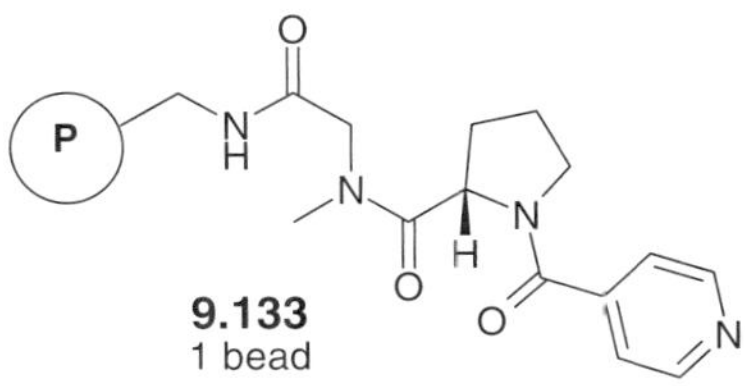

9.133
1 bead

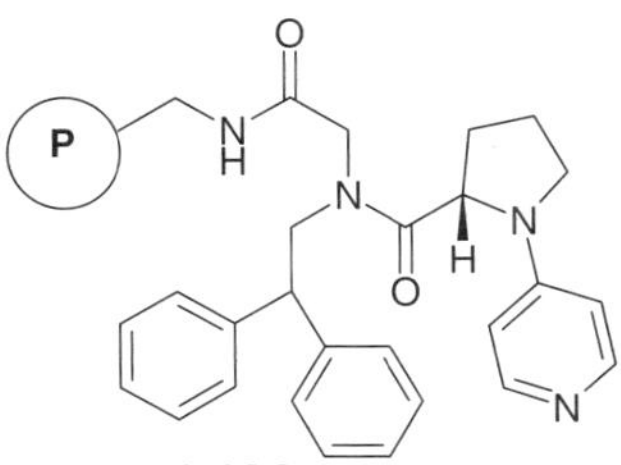

9.134

Figure 9.48 Structure of the decoded acylation catalysts **9.130–9.133** from screening of **L34** and of the less active catalyst **9.134**.

TABLE 9.11 Efficiency of Library Individuals 9.129–9.134 from Screening of the SP Pool Encoded Peptidomimetic Library of Catalysts L34 as Acylation Catalysts

Entry	Catalyst	Percent Conversion[a]
1	**9.130**	39
2	**9.131**	24
3	**9.132**	23
4	**9.133**	ND[b]
5	**9.134**	9
6	**9.129**	14

[a]Calculated after 9 min incubation with the acylation cocktail as an average of three runs.
[b]Not detected.

M3-containing library members was formed. The frequency of **9.130** and **9.131** among the hot beads roughly paralleled their frequency in the pooled library (Fig. 9.48), while **9.132** was spotted only once for unknown reasons. Compound **9.133** belonged to a different chemical class, but its catalytic activity was not reconfirmed when tested as a discrete. Compounds **9.130–9.132** all resulted more active catalysts than **9.129** (Table 9.11), with **9.130** being the most active compound. A less active catalyst from the library, **9.134** (Fig. 9.48), was also reprepared and tested as a discrete (Table 9.11). Its weaker activity was not spotted by IR thermography, confirming the ability of the detection method to select the most active catalysts during the screening.

The pooled encoded format was also used by Boussie and co-workers (179, 180) to prepare and screen Pd- and Ni-diimine ligand complexes as olefin polymerization catalytic systems. The successful application of deconvolutive or encoding methods to medium–large SP catalytic system pool libraries should become routine in the future. Careful validation of the synthetic scheme, the screening, and the detection methods will ensure the applicability of HTS of SP catalytic pool libraries to each specific transformation investigated.

9.5 APPLICATIONS TO MOLECULAR RECOGNITION

9.5.1 General Considerations

The existence of biomolecules, such as proteins, peptides, carbohydrates, and nucleotides, able to recognize a ligand with extremely high selectivity and specificity through a series of noncovalent interactions has long been a challenge for synthetic chemists. The access, via synthetic routes, to artificial receptors able to tightly bind specific substrates for various applications would produce, among others, artificial catalytic systems to catalyze known chemical transformations, synthetic target mimicks for setting up relevant pharmaceutical screening assays, and synthetic affinity receptors useful for the isolation, purification, and/or identification of active principles. All these applications would largely benefit from the higher stability and tractability of synthetic recognition systems when compared to their biological counterparts, as well as from the possibility to tailor their structure and to reduce extremely complex biomolecules to smaller and simpler, yet efficient artificial models. Another advantage would be the lower cost and ease of an assessed synthetic process to produce a synthetic supramolecular recognition element, rather than overproducing and isolating large quantities of the corresponding biomolecule. Even more importantly, a full understanding of supramolecular phenomena, and transfer of this knowledge to synthetic chemistry routes, would allow the design and synthesis of artificial elements, with no biological counterpart, to catalyze new reactions, isolate new compounds, and set up novel relevant assays.

Implications and applications of supramolecular chemistry have been reviewed recently (181–183). Synthetic receptors derived from cyclodextrins (184), calixarenes (185), fullerenes (186), crown ethers (187), and dendrimers (188, 189) have been reported. Unfortunately, the elusive balance of obtaining high affinity by means of a large number of weak, noncovalent interactions, as between biomolecules and their ligands, has proved extremely difficult to translate onto synthetic counterparts, and synthetic receptors matching the prowess of enzymes in detecting their substrates and processing them are still far from reality.

Combinatorial technologies can help the search for supramolecular entities, and several significant reviews have thoroughly illustrated the applications of libraries in molecular recognition (190–196). Synthetic peptide libraries have been used to refine the selectivity of synthetic receptors and to determine the tolerability for substitutions in a given position of di- and tripeptide structures (197–203), speeding up the characterization of these entities. Even more importantly, libraries of synthetic receptors have been prepared using rigid scaffolds decorated with amino acidic building blocks to modulate their recognition properties. Examples from Clark Still and co-workers (204, 205) and from Nestler and co-workers (206, 207) presented different peptidosteroid-based rigid libraries and more flexible pentamethylene linker libraries to bind enkephalin-related peptides. Fessmann and Kilburn (208) reported a 2197-member receptor library based on a trisubstituted pyridine scaffold tested in an on-bead fluorescent assay for their binding with a dansylated tripeptide. Goodman et al. (209) dealt with metal-template strategies to assemble synthetic receptor libraries (209), in analogy to template-assisted library synthesis, which was discussed in Section 8.6. Jung et al. (210) reported an application for cyclopeptide libraries as synthetic receptors for enantiomeric resolution of racemic amino acid mixtures. The next section introduces an example reporting both the synthesis and screening of a ligand library to determine the specificity of a synthetic receptor and then the use of this information to design a synthetic receptor library to optimize the recognition properties.

Another example described in Section 9.5.3 reports the synthesis and screening of a synthetic receptor library aimed at transition metal binding. Other similar examples have been reported recently. Burger and Clark Still (211) prepared ionophoric, cyclen-based libraries decorated by amino acid units and screened them for their ability to complex copper and cobalt ions; Malin et al. (212) identified novel hexapeptidic technetium-binding sequences from the screening of cellulose-bound libraries; and

Shibata et al. (213) prepared and screened an SP oligopeptide library for its ability to complex cobalt ions, isolating several high-binding sequences.

Finally, libraries aimed to chiral resolution of racemates will be covered here; in particular, the use of chiral stationary phases (CSPs) has recently been reported for the identification of materials to be used for chiral separation of racemates by HPLC. The group of Frechet reported the selection of two macroporous polymethacrylate-supported 4-aryl-1,4-dihydropyrimidines (DHPs) as CSPs for the separation of amino acid, anti-inflammatory drugs, and DHP racemates from an 140-member discrete DHP library (214, 215) as well as a deconvolutive approach for the identification of the best selector phase from a 36-member pool library of macroporous polymethacrylate-grafted amino acid anilides (216, 217). Welch and co-workers (218, 219) reported the selection of the best CSP for the separation of a racemic amino acid amide from a 50-member discrete dipeptide *N*-3,5-dinitrobenzoyl amide library and the follow-up, focused 71-member library (220). Wang and Li (221) reported the synthesis and the Circular Dichroism- (CD) based screening of a 16-member library of CSPs for the HPLC resolution of a leucine ester. Welch et al. recently reviewed the field of combinatorial libraries for the discovery of novel CSPs (222). Dyer et al. (223) reported an automated synthetic and screening procedure based on Differential Scanning Calorimetry (DSC) for the selection of chiral diastereomeric salts to resolve racemic mixtures by crystallization. Clark Still reported another example which is discussed in detail in Section 9.5.4.

9.5.2 An Example: Ligand and Receptor Libraries Based on Guanidinium Tweezer Receptors

Davies et al. (224) reported the synthesis of a 1000-member SP autoencoded, cyclic peptide library **L35** (Fig. 9.49) using a recently reported synthetic strategy and standard mix-and-split peptide conditions. Each monomer set **M_1–M_3** was composed of ten α-amino acids (Fig. 9.49). Such a library could expose the C-terminus inverted tripeptide library (**L35c**, 80% of sites, screening portion) by mild acidic cleavage (step a, Fig. 9.49), allowing an on-bead screening protocol (*vide infra*) and selection of positives (step b). A subsequent strong acidic cleavage (step c, Fig. 9.49) released the tripeptide library **L35r** in solution while exposing the N-terminus sequenceable portion of the same resin-bound library (**L35s**, 20% of sites, on-bead Edman sequencing of positive beads, step d).

The C-exposed library **L35c** was screened for its binding affinity for the guanidinium-based, fluorescent tweezer receptor **9.135** (225), whose structure is reported in Fig. 9.50. The dansylated chains should have assured the fluorescent detection of beads containing ligand tripeptidic sequences, and a series of control experiments ruled out the possibility of significant, nonspecific interaction of the beads with the receptor or with the biological assay constituents. The assay conditions were carefully optimized, and around 7000 beads (seven library equivalents, complete representation of the library individuals) were incubated with **9.135**.

Microscope observation of the beads after incubation spotted the brilliantly fluorescent ones from the background, containing both nonfluorescent and slightly fluo-

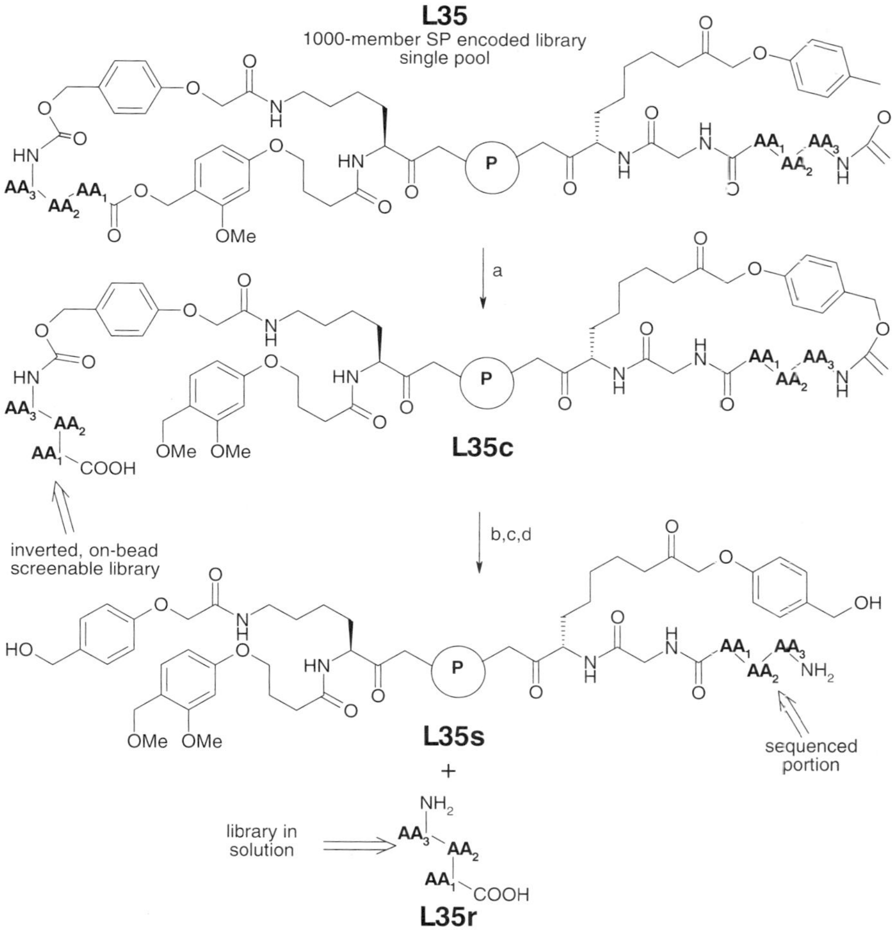

a: 1% TFA in DCM, rt; b: on-bead screening, positive beads selection and withdrawn
c: 100% TFA, rt; d: Edman sequencing.

M_1-M_3 = AA_1-AA_3 = L-Ala, Gly, L-Glu(OtBu), L-Leu, L-Met, L-Phe, D-Phgly, L-Pro, L-Ser(OtBu), L-Val.

Figure 9.49 On-bead screening and decoding protocol for the inverted SP peptide library of artificial receptors **L35**.

rescent beads. Around 3% of beads were positive, corresponding to around 200 of the 7000 beads tested, or to around 30 active library individuals (if all active beads were reliably spotted by the screening). Twenty beads were manually removed, cleaved with strong acid conditions, and sequenced. Fifteen different sequences were determined, showing a strong conservation in the C-terminus position (two residues, Val strongly preferred with 95% recurrence), some preferences in the N-terminus position (five

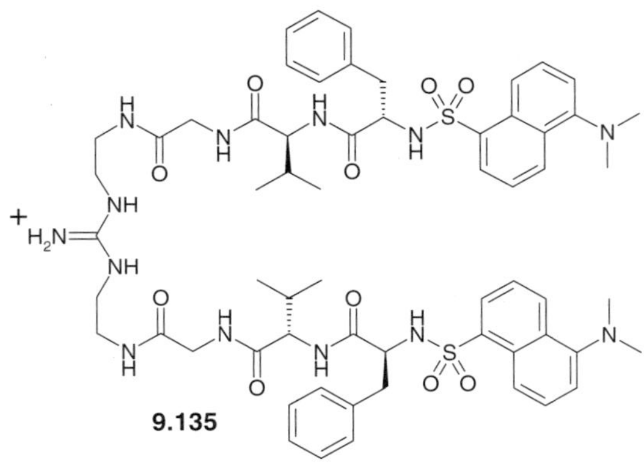

9.135

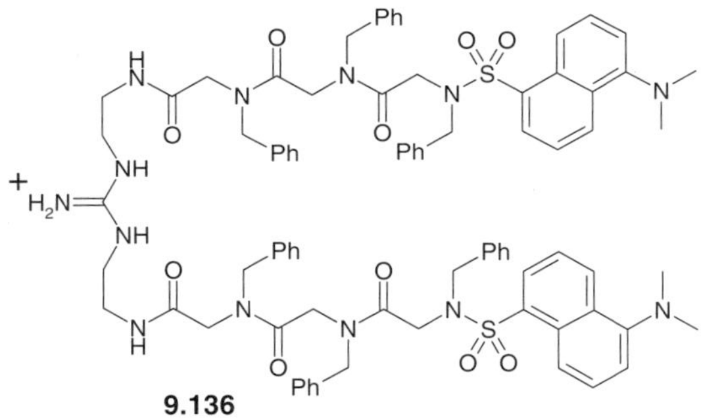

9.136

Figure 9.50 Structure of the guanidinium-based tweezer receptors **9.135** and **9.136**.

residues, Glu, Pro, and Phe strongly preferred with 40, 25, and 20% recurrence, respectively), and less specificity in the middle (seven residues, Glu and Ser preferred with 25 and 20% recurrence, respectively). The binding affinities of active library components were not determined in solution, nor was the library screened in solution to validate the results obtained from on-bead screening.

A similar guanidinium-based, peptoid tweezer (**9.136**, Fig. 9.50) was prepared with a somewhat similar synthesis (226). The library **L35c** was incubated with this synthetic receptor, but no brilliantly fluorescent beads were observed. The guanidinium arm alone was apparently not enough to elicit a binding, which probably also required the presence of hydrogen bonding from peptidic NH groups, as for the peptide tweezer **9.135**.

The feasibility of on-bead screening of an SP tweezer library with soluble tripeptides was then evaluated. An 125-member SP pool library **L36** was prepared (Fig. 9.51) from the key resin-bound intermediate **9.137** using mix-and-split protocols, three monomer sets **M1**–**M3**, each containing five α-amino acid representatives, and the

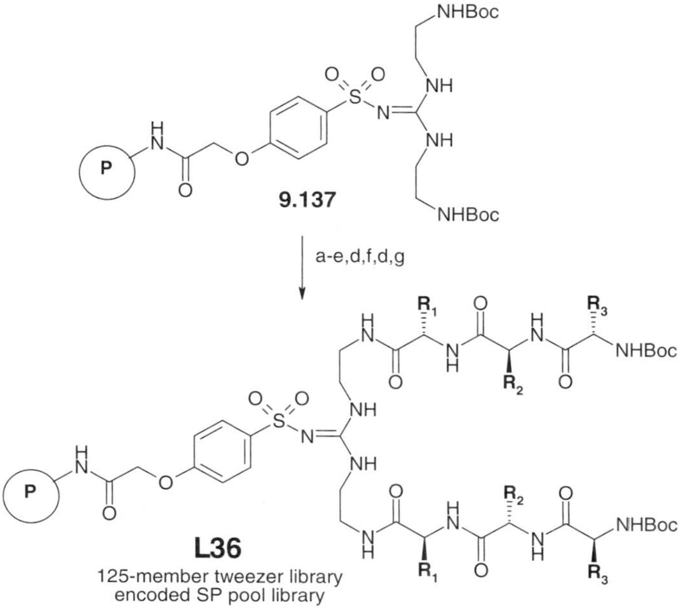

a: TFA/DCM 1/4, rt; b: DIPEA; c: coupling with $\mathbf{M_1}$; d: piperidine, DMF, rt; e: coupling with $\mathbf{M_2}$; f: coupling with $\mathbf{M_3}$; g: $(Boc)_2O$, DIPEA, rt.

$\mathbf{M_1}$-$\mathbf{M_3}$ = L-Ala, Gly, L-Phe, L-Ser(OtBu), L-Val

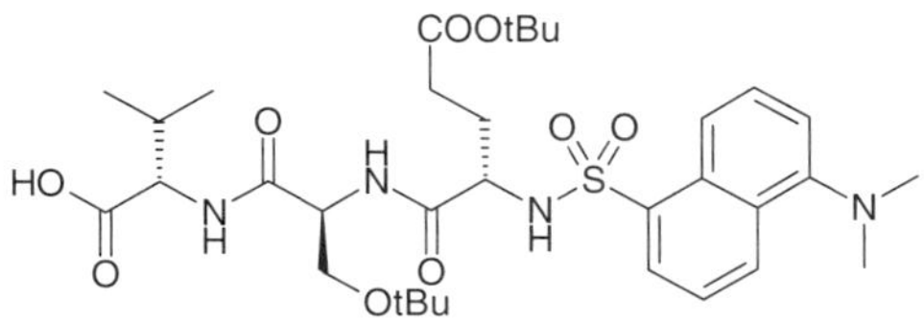

9.138

Figure 9.51 Synthesis of the SP-encoded pool, guanidinium-based tweezer receptor library **L36** and structure of the selection substrate **9.138**.

same synthetic strategy used for the synthesis of **9.135**. Compound **9.135** was included in the library structure as a positive control. The library was incubated with the tripeptide **9.138** (Fig. 9.51), selected among the active sequences obtained from the previous screening, but no brilliantly fluorescent beads were obtained even after prolonged incubation times. Probably the transition from a strongly basic, unprotected guanidinium group (as in **9.135**) to a weakly basic, supported sulfonylated guanidinium group (as for **L36** individuals) affected the binding affinity of the synthetic tweezers. The same library should be made with a different linker, preserving the guanidinium basicity. This could determine the importance of steric hindrance (represented by the resin bead) around the core guanidinium, with eventual testing of the library after release into solution to cross-check the on-bead screening results.

9.5.3 An Example: Synthesis and Screening of a Capped Peptidomimetic Library for Metal Binding Activity

Francis et al. (227) reported the synthesis of a >10,000-member encoded SP pool library **L37** characterized by a turn element (monomers **M2**, prepared using simple synthetic routes) surrounded by two α-amino acids (monomers **M1** and **M3**) and capped on the N-terminus (monomers **M4**). The structure of the library and the monomer sets are reported in Fig. 9.52. Mix-and-split protocols and a popular encoding method (177) were used to prepare and encode the library.

Library **L37** was tested on-bead for its coordination of Ni^{2+} and Fe^{3+} ions, and suitable experimental incubation protocols were set up for both metal ion solutions (Figs. 9.53 and 9.54, respectively). As the preliminary screenings at high metal concentration detected a large number of stained beads, two library equivalents (around 24,000 beads, representing >90% of the library population) were screened in the presence of decreasing metal ion concentration until only a small number of stained beads remained. With a 2.5×10^{-4} *M* Ni^{2+} concentration, only six beads were positive, and their isolation and decoding produced compounds **9.139–9.142** (Fig. 9.53). Protected histidine as **M1** and **M3** and only two **M2** and **M4** monomers were observed, indicating marked structural preferences for Ni^{2+} coordination. Solution synthesis and screening of **9.139** and **9.140** confirmed their coordination with Ni^{2+}. With a 5×10^{-6} *M* Fe^{3+} concentration 64 beads were spotted, isolated, and decoded. All of them had a fixed monomer in positions **M3** and **M4** (see generic structure **9.143**, Fig. 9.54), half of them had L-methionine as preferred **M1** monomer, but no preference was detectable for the turn element. Distinct preferences for Ni^{2+} and Fe^{3+} coordination were observed, and simultaneous on-bead library screening with solutions of Ni^{2+} and Fe^{3+} confirmed this tendency. Other metal ions (Cu^{2+}, Pt^{4+}, Sn^{4+}, Pd^{2+}) were also used as screening probes, producing results that, albeit less rationalizable, again highlighted the different preferences of various metal salts toward **L37** individuals.

9.5.4 An Example: Selection of an Enantioselective Resolving Resin Using On-Bead Screening

Weingarten et al. (228) reported the synthesis of an encoded SP 60-member pool library **L38** that was screened to find the best chiral selector for the resolution of racemate mixtures of dye-containing amino acid Pfp esters **9.144a** (L-enantiomer, blue dye) and **9.144b** (D-enantiomer, red dye, Fig. 9.55). The structures of the library **L38** and of the three monomer sets **M1** (15 representatives, *N*-Fmoc α-amino acids), **M2**

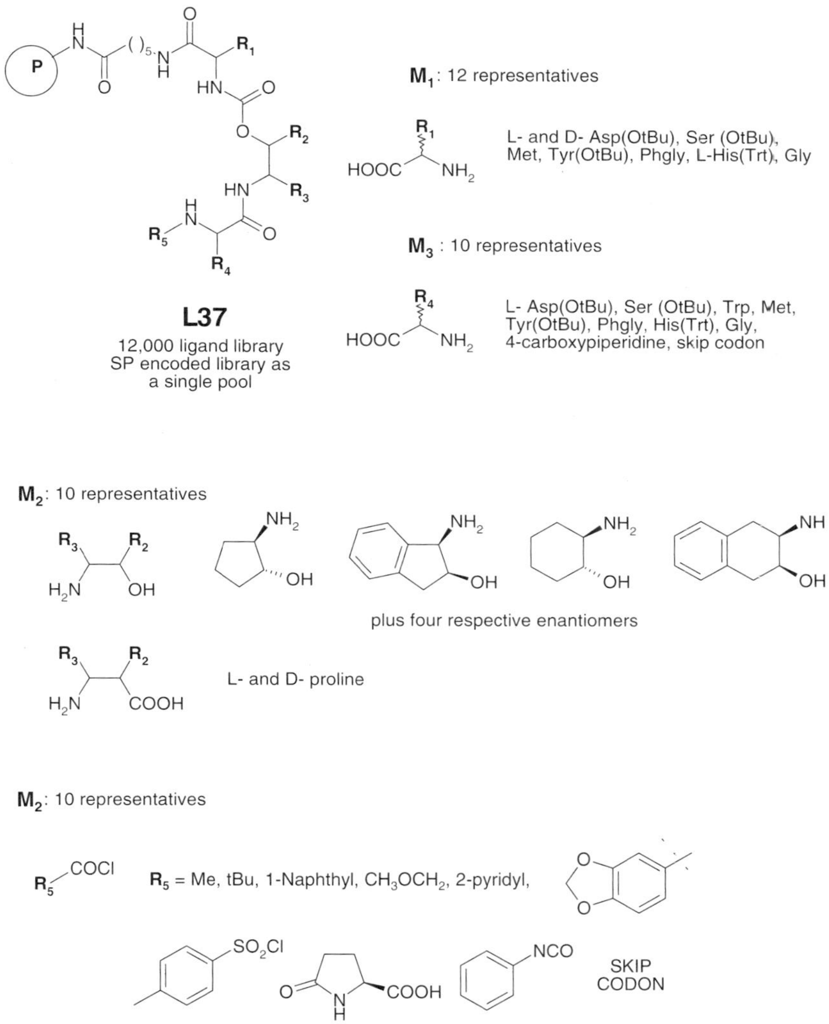

Figure 9.52 Structure of the SP-encoded primary peptidomimetic library of metal complexing agents **L37** and of the monomer sets **M1**–**M4** used for its synthesis.

(two representatives, pyrrolidine diamine linkers), and **M3** (two representatives, cyclic polyamides) are also shown in Fig. 9.55.

While the monomer set **M1** was made up of commercially available materials, the components of the **M2** and **M3** sets were prepared following the reported synthetic schemes (Fig. 9.56) starting from commercial (**M3**) or easily obtainable precursors (**9.145**, **M2**) (229). The four monomers **M2,1–2** and **M3,1–2** were then assembled and

L37 24,000 beads, two library equivalents —a,b→ 6 positive beads for Ni^{++} complexation; after decoding, 4 structures **9.139-9.142**

a: 2.5×10^{-4} $Ni(OAc)_2$, MeOH, rt; b: dimethylglioxime, MeOH, rt.

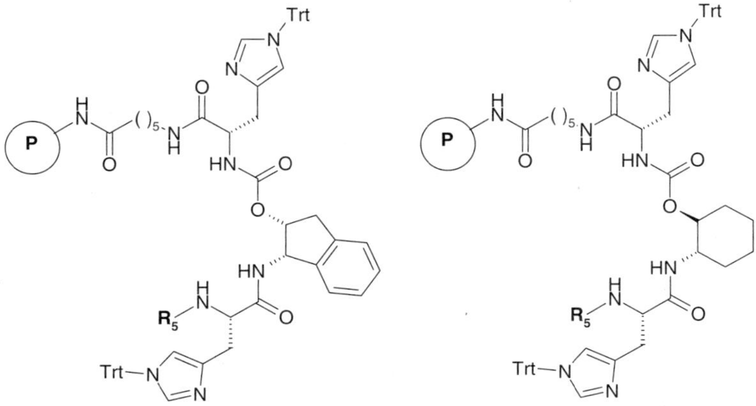

R_5 = Me 9.139
R_5 = 1-Naphthyl 9.140

R_5 = Me 9.141
R_5 = 1-Naphthyl 9.142

Figure 9.53 Screening of the SP encoded primary peptidomimetic library of metal complexing agents **L37** for Ni^{2+} complexation and structures of the hits **9.139–9.142** obtained from its screening.

L37 24,000 beads, two library equivalents —a,b→ 64 positive beads for Fe^{+++} complexation; after decoding, $\mathbf{M_3}$ and $\mathbf{M_4}$ determined (see **9.143**)

a: 5×10^{-6} $FeCl_3$, AcOH, NaOAc, MeOH, rt; b: KSCN, AcOH, MeOH, rt.

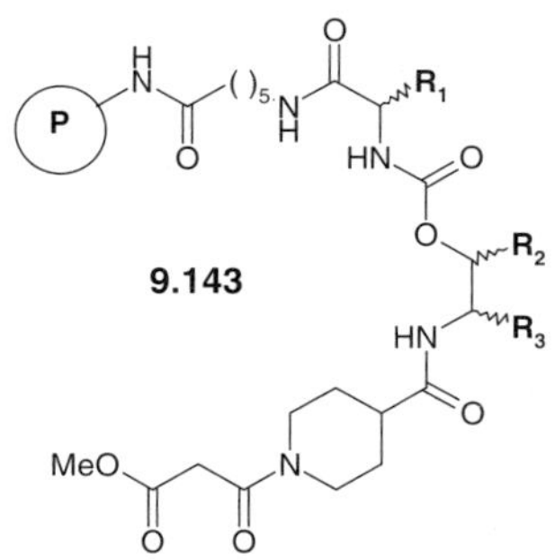

9.143

Figure 9.54 Screening of the SP encoded primary peptidomimetic library of metal complexing agents **L37** for Fe^{3+} complexation and structures of the hit **9.143** obtained from its screening.

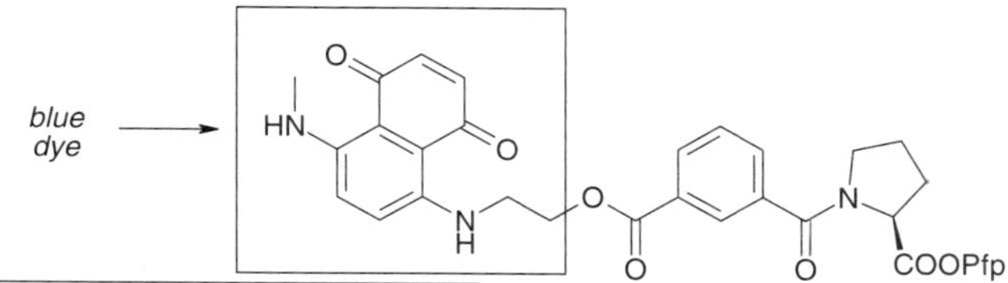

9.144a

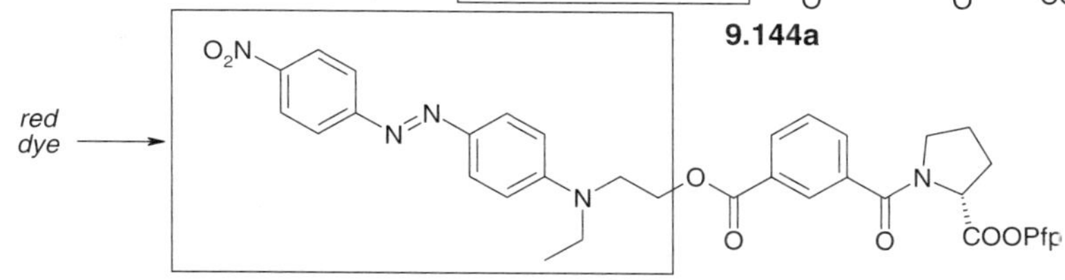

9.144b

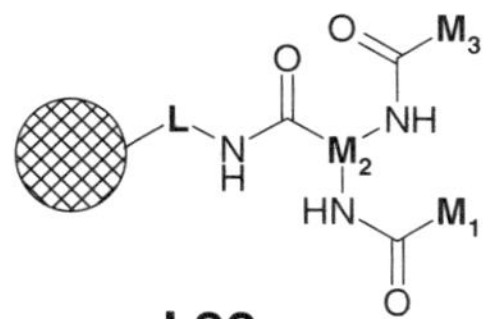

L38
60-member
encoded pool library

M_1 Gly, L- and D- Val, L- and D-Pro, L- and D-Phe, L- and D-Asn(Trt), L- and D-His (Trt), L- and D-Asp(OtBu), L- and D-Ser(OtBu); all N-Fmoc protected.

M_2

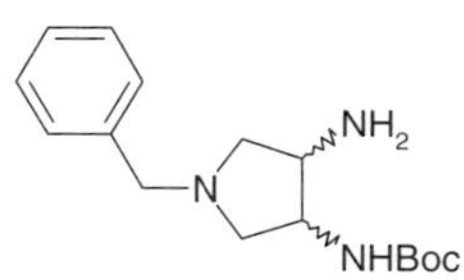

$M_{2,1}$: RR enantiomer
$M_{2,2}$: SS enantiomer

M_3

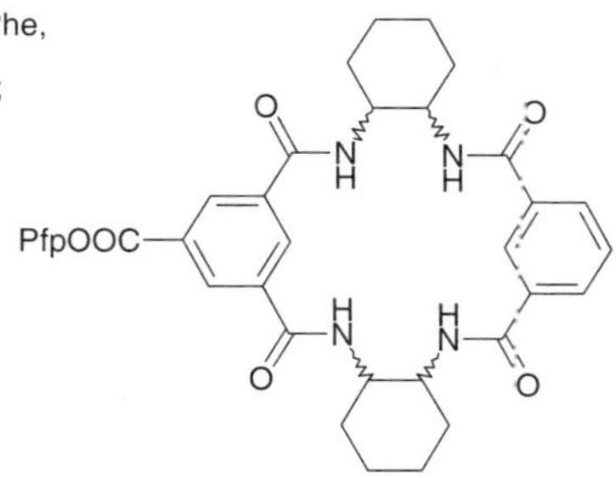

$M_{3,1}$: RRRR enantiomer
$M_{3,2}$: SSSS enantiomer

Figure 9.55 Structure of the SP peptidomimetic encoded pool chiral selector library **L38** and of the monomer sets **M_1–M_3** used for its synthesis.

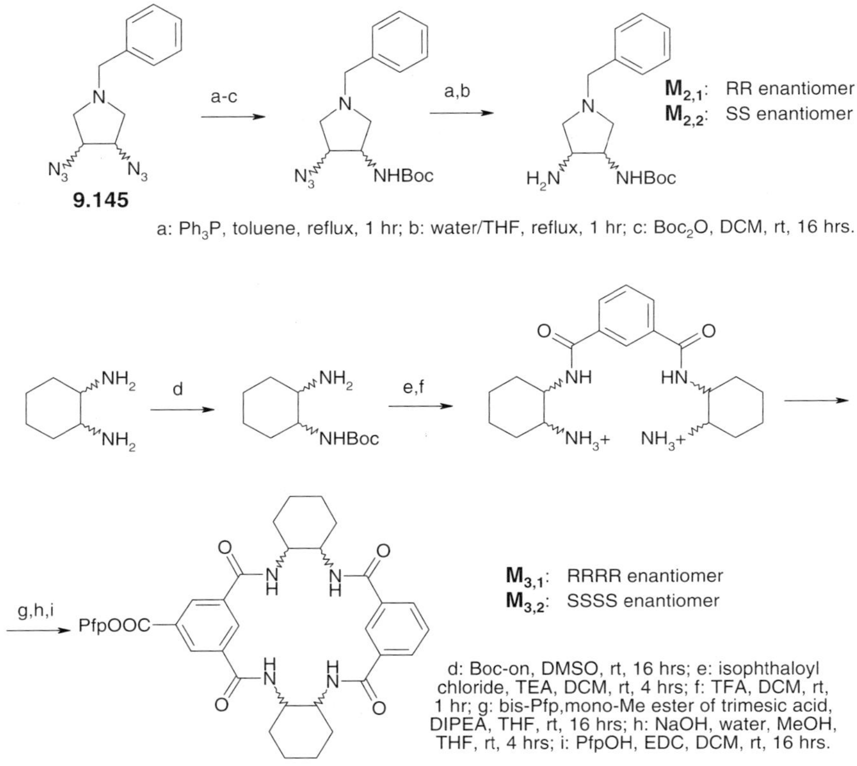

Figure 9.56 Synthesis of the monomer sets **M2–M3** used for the synthesis of the SP peptidomimetic encoded pool chiral selector library **L38**.

elaborated to give four Boc-protected isomeric constructs **9.146a–d** (steps a–c, Fig. 9.57). The library **L38** was then prepared as shown in Fig. 9.58. Aminomethyl PS resin was acylated with an Fmoc-protected amino acid linker (step a), and after Fmoc deprotection (step b) the resin was divided into four aliquots and treated with four tags, according to a popular encoding method (step c) (177). The portions of encoded resin were then acylated with **9.146a–d** to give the four encoded discretes **9.147a–d** (step d), and, after pooling, the resin was deprotected, split into 15 aliquots, and tagged (steps e and f). Finally, the encoded intermediates were acylated with **M1**, pooled, and deprotected (steps g and b, Fig. 9.58) to give the encoded library **L38**.

The library was screened simply by incubating an excess of **9.144a,b** (equimolar amounts of the two colored enantiomers) with a library aliquot in chloroform for 4 h. The library was then washed and dried and finally observed with a microscope to determine the color of each bead. The most selective beads showed either blue

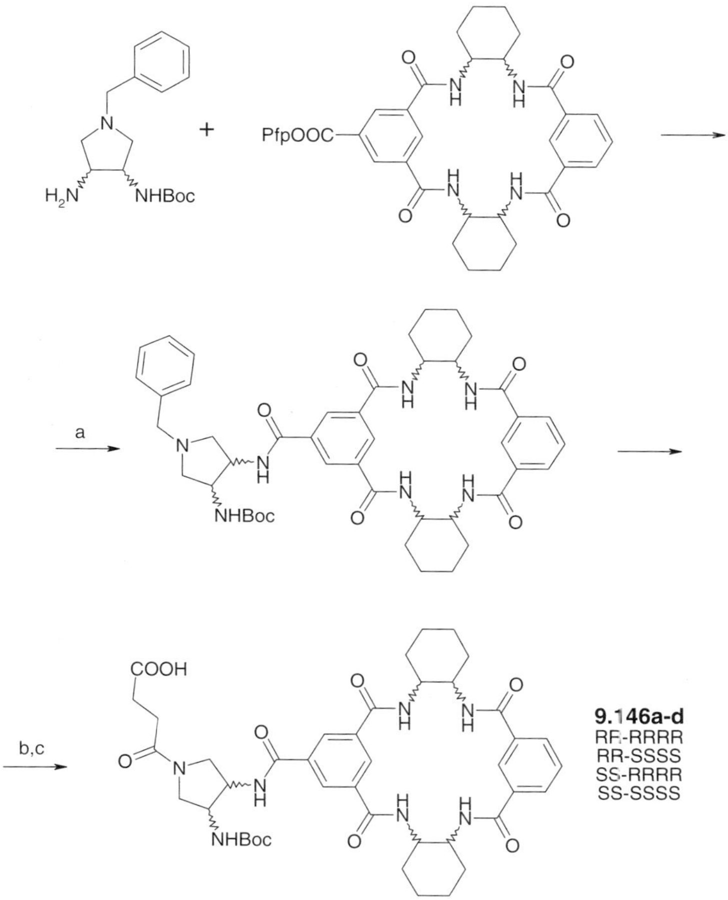

a: TEA, THF, rt, 4 hrs; b: H_2, Pearlman catalyst, MeOH, HCl, rt, 26 hrs; c: succinic anhydride, TEA, DMAP, DCM, rt, 16 hrs.

Figure 9.57 Synthesis of the SP peptidomimetic encoded pool chiral selector library **L38**: preparation of the key intermediates **9.146a–d** in solution.

(L-selective) or red color (D-selective), while beads with low selectivity assumed a brown coloration. Among the selected beads, two enantiomeric structures **9.148a,b** (Fig. 9.59) showed a significant enantioselectivity that was confirmed upon resynthesis. Their enantioselectivity was also confirmed with the separation of *N*-acyl proline racemates lacking the dye moieties, thus highlighting the possible usefulness of **9.148a,b** as chiral selectors.

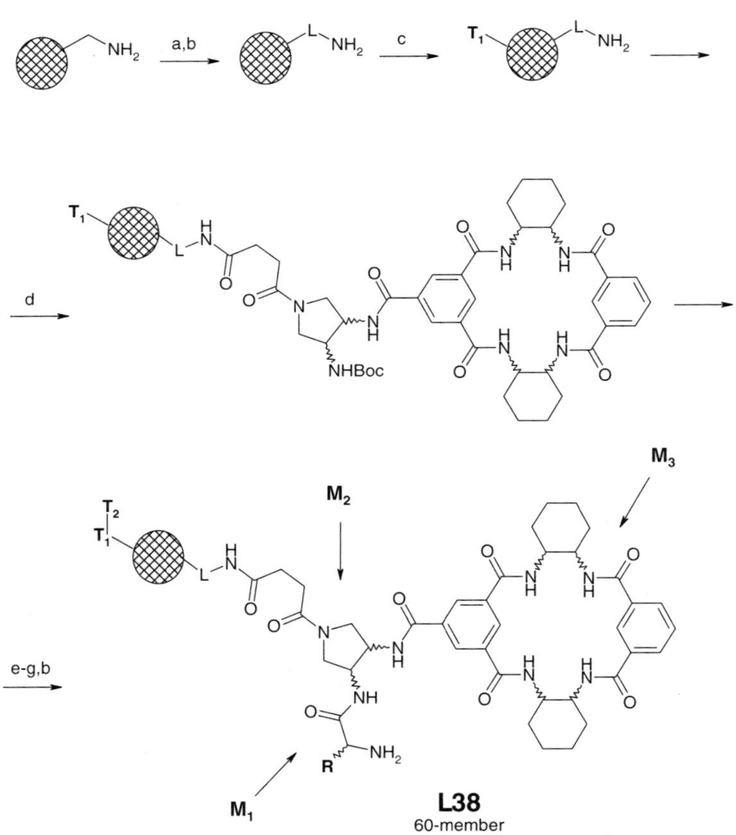

L38
60-member
encoded pool library

a: N-Fmoc caproic acid, HOBt, DIC, DCM, 2 hrs; b: piperidine, DMF, rt, 30'; c: resin portioning (1 to 4), tagging with $\mathbf{T_1}$; d: **9.146a-d**, HOBt, EDC, DCM, DMF, rt, 16 hrs; e: pooling, then TFA, DCM, rt, 5'; f: resin portioning (1 to 15), tagging with $\mathbf{T_2}$; g: $\mathbf{M_1}$, DIC, DCM, DMF, rt, 2hrs, repeated twice.

Figure 9.58 SP Elaboration of **9.146a–d** into the SP peptidomimetic encoded pool chiral selector library **L38**.

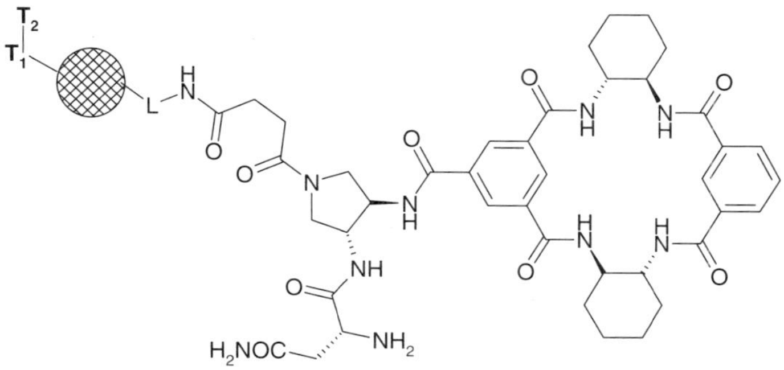

9.148a
D-Asn-RR-RRRR
81% selection for
9.144a

9.148a
other enantiomer
L-Asn-SS-SSSS
73% selection for
9.144b

Figure 9.59 Chiral selectors individuated from screening of the SP peptidomimetic encoded pool chiral selector library **L38**: compounds **9.148a** and **9.148b**.

REFERENCES

1. Persidis, A., in *Combinatorial Chemistry and Combinatorial Technologies: Methods and Applications*, S. Miertus and G. Fassina (Eds.). Marcel Dekker, Inc., New York, 1999, pp. 405–422.
2. Tucker, G. T., *J. Pharm. Pharmacol.* **46**, 417–424 (1994).
3. Heward, J. and Gough, S. C. L., *Clin. Sci.* **93**, 479–491 (1997).
4. Pirtle, R. M. and Pirtle, I. L., in *Transfer RNA in Protein Synthesis*, D. L. Hatfield, B. J. Lee and R. M. Pirtle (Eds.). CRC, Boca Raton, FL, 1992, pp. 141–155.
5. Liang, P., Wang, F., Zhu, W., O'Connell, R. P. and Averboukh, L., *Methods Mol. Med.* **13**, 3–13 (1998).
6. Anderson, N. L. and Anderson, N. G., *Electrophoresis* **19**, 1853–1861 (1998).
7. Stubberfield, C. R. and Page, M. J., *Expert Opin. Invest. Drugs* **8**, 65–70 (1999).
8. Andrade, M. A. and Sander, C., *Curr. Opin. Biotechnol.* **8**, 675–683 (1997).
9. Smith, T. F., *Trends Genet.* **14**, 291–293 (1998).
10. Koonin, E. V., Tatusov, R. L. and Galperin, M. Y., *Curr. Opin. Struct. Biol.* **8**, 355–363 (1998).
11. Schena, M., Heller, R. A., Theriault, T. P., Konrad, K., Lachenmeier, E. and Davis, R. W., *Trends Biotechnol.* **16**, 301–306 (1998).
12. Thakker, R. V., *Princ. Bone Biol.*, 837–851 (1996).

13. Guerre-Millo, M., *Biomed. Pharmacother.* **51**, 318–323 (1997).
14. Van Cott, K. E. and Velander, W. H., *Expert Opin. Invest. Drugs* **7**, 1683–1690 (1998).
15. Heddle, J. A., *Environ. Mol. Mutagen.* **32**, 110–114 (1998).
16. Schreiber, S. L., *Bioorg. Med. Chem.* **6**, 1127–1152 (1998).
17. Crews, C. M. and Splittgerber, U., *Trends Biol. Sci.* **24**, 317–320 (1999).
18. http://www-schreiber.chem.harvard.edu/
19. Mayer, T. U., Kapoor, T. M., Haggarty, S. J., King, R. W., Schreiber, S. L. and Mitchison, T. J., *Science* **286**, 971–974 (1999).
20. Stockwell, B. R., Haggarty, S. J. and Schreiber, S. L., *Chem. Biol.* **6**, 71–83 (1999).
21. Hagan, I. and Yanagida, M., *Nature* **356**, 74–76 (1992).
22. Blangy, A., Lane, H. A., d'Herin, P., Harper, M., Kress, M. and Nigg, E. A., *Cell* **83**, 1159–1169 (1995).
23. Sawin, K. E., *Nature* **359**, 540–542 (1992).
24. Stockwell, B. R., Hardwick, J. S., Tong, J. K. and Schreiber, S. L., *J. Am. Chem. Soc.* **121**, 10662–10663 (1999).
25. Carcamo, J., Zentella, A. and Massague, J., *Mol. Cell. Biol.* **15**, 1573–1581 (1995).
26. Zhao, H. and Eide, D., *Proc. Natl. Acad. Sci. USA* **93**, 2454–2458 (1996).
27. Askwith, C., Eide, D., Van Ho, A., Bernard, P. S., Li, L., Davis-Kaplan, S., Sipe, D. M. and Kaplan, J., *Cell* **76**, 403–410 (1994).
28. Varela, J. C., van Beekvelt, C., Planta, R. J. and Mager, W. H., *Mol. Microbiol.* **6**, 2183–2190 (1992).
29. Mattheakis, L. C., Olivan, S. E., Dias, J. M. and Horthrop, J. P., *Chem. Biol.* **6**, 835–844 (1999).
30. Sarwar Nasir, M. and Jolley, M. E., *Comb. Chem. High Throughput Screening* **2**, 177–190 (1999).
31. Zuck, P., Lao, Z., Skwish, S., Glickman, J. F., Yang, K., Burbaum, J. and Inglese, J., *Proc. Natl. Acad. Sci. USA* **96**, 11122–11127 (1999).
32. Beutel, B. A., in *Combinatorial Chemistry and Molecular Diversity in Drug Discovery* E. M. Gordon and J. Kerwin (Eds.). Wiley-Liss, New York, 1998, pp. 421–432.
33. Hill, D. C., *Curr. Opin. Drug Discovery Dev.* **1**, 92–97 (1998).
34. Fernandes, P. B., *Curr. Opin. Chem. Biol.* **2**, 597–603 (1998).
35. Kenny, B. A., Bushfield, M., Parry-Smith, D. J., Fogarty, S. and Treherne, J. M., *Prog. Drug Res.* **51**, 245–269 (1998).
36. Venton, D. L. and Woodbury, C. P., *Chemometr. Intell. Lab. Systems* **48**, 131–150 (1999).
37. Fox, S., Farr-Jones, S. and Yund, M. A., *J. Biomol. Screen.* **4**, 183–186 (1999).
38. Carrano, L. and Donadio, S., in *Combinatorial Chemistry and Combinatorial Technologies: Methods and Applications*, S. Miertus and G. Fassina (Eds.). Marcel Dekker, Inc., New York, 1999, pp. 233–250.
39. Hann, M., Hudson, B., Lewell, X., Lifely, R., Miller, L. and Ramsden, N., *J. Chem. Inf. Comput. Sci.* **39**, 897–902 (1999).
40. Leland, B. A., *J. Chem. Inf. Comput. Sci.* **37**, 62–70 (1997).
41. Muskal, S. M., in *A Practical Guide to Combinatorial Chemistry,* A. W. Czarnik and S. H. Dewitt (Eds.). ACS, Washington, DC, 1997, pp. 357–397.

42. Bauer, B. E., in *High Throughput Screening*, J. P. Devlin (Ed.). Marcel Dekker, Inc., New York, 1997, pp. 551–581.

43. Weininger, D., in *Combinatorial Chemistry and Molecular Diversity in Drug Discovery*, E. M. Gordon and J. Kerwin (Eds.). Wiley-Liss, New York, 1998, pp. 483–498.

44. Heuer, K. I., *Med. Res. Rev.* **19**, 209–221 (1999).

45. Lee, M. S., Nakanishi, H. and Kahn, M., *Curr. Opin. Drug Discovery Dev.* **2**, 332–341 (1999).

46. Kuzniar, E., *Comb. Chem. High Throughput Screening* **2**, 237–245 (1999).

47. Powers, D. G., Casebier, D. S., Fokas, D., Ryan, W. J., Troth, J. R. and Coffen, D. L., *Tetrahedron* **54**, 4085–4096 (1998).

48. Ostresh, J. M., Husar, G. M., Blondelle, S., Dorner, B., Weber, P. A. and Houghten, R. A., *Proc. Natl. Acad. Sci. USA* **91**, 11138–11142 (1994).

49. Newton, C. G., *Exp. Opin. Ther. Patents* **6**, 827–835 (1996).

50. Bozicevic, K., *Annu. Rep. Combi. Chem. Mol. Diversity* **1**, 298–313 (1997).

51. Newton, C. G., *Exp. Opin. Ther. Patents* **7**, 1183–1194 (1997).

52. Green, G. D., *Mol. Diversity* **3**, 247–252 (1998).

53. Jakobsen, P. H., Kurtzhals, P. and Poulsen, F., *Exp. Opin. Ther. Patents* **8**, 1157–1165 (1998).

54. Mulligan, L. and Steele, P., *Exp. Opin. Ther. Patents* **9**, 507–514 (1999).

55. Harper, M. F. and Newton, C. G., *Exp. Opin. Ther. Patents* **9**, 583–624 (1999).

56. Bhandari, A., Jones, D. G., Schullek, J. R., Vo, K., Schunk, C. A., Tamanaha, L. L., Chen, D., Yuan, Z., Needels, M. C. and Gallop, M. A., *Bioorg. Med. Chem. Lett.* **8**, 2303–2308 (1998).

57. Thomas, J. B., Fall, M. J., Cooper, J. B., Rothman, R. B., Mascarella, S. W., Xu, H., Partilla, J. S., Dersch, C. M., McCullough, K. B., Cantrell, B. E., Zimmerman, D. M. and Carroll, F. I., *J. Med. Chem.* **41**, 5188–5197 (1998).

58. Thomas, J. B., Mascarella, S. W., Rothman, R. B., Partilla, J. S., Xu, H., McCullough, K. B., Dersch, C. M., Cantrell, B. E., Zimmerman, D. M. and Carroll, F. I., *J. Med. Chem.* **41**, 1980–1990 (1998).

59. Mitch, C. H., Leander, J. D., Mendelsohn, L. G., Shaw, W. N., Wong, D. T., Cantrell, B. E., Johnson, B. G., Reel, J. K., Snoddy, J. D. et al., *J. Med. Chem.* **36**, 2842–2850 (1993).

60. Harre, M., Nickisch, K. and Tilstam, U., *React. Funct. Polym.* **41**, 111–114 (1999).

61. Raillard, S. P., Ji, G., Mann, A. D. and Baer, T. A., *Org. Proc. Res. Dev.* **3**, 177–183 (1999).

62. Edwards, P. J., Gardner, M., Klute, W., Smith, G. F. and Terrett, N. K., *Curr. Opin. Drug Discovery Dev.* **2**, 321–331 (1999).

63. Harre, M., Tilstam, U. and Weinmann, H., *Org. Proc. Res. Dev.* **3**, 304–318 (1999).

64. Berman, J., Halm, K., Adkison, K. and Shaffer, J., *J. Med. Chem.* **40**, 827–829 (1997).

65. Olah, T. V., Mcloughlin, D. A. and Gilbert, J. D., *Rapid Commun. Mass Spectrom.* **11**, 17–23 (1997).

66. Kuo, B.-S., Van Noord, T., Feng, M. R. and Wright, D. S., *J. Pharm. Biomed. Anal.* **16**, 837–846 (1998).

67. Frick, L. W., Higton, D. M., Wring, S. A. and Wells-Knecht, K. J., *Med. Chem. Res.* **8**, 472–477 (1998).

68. Corcoran, O., Nicholson, J. K. and Wilson, I. D., *Anal. Commun.* **36**, 259–261 (1999).
69. Atherton, J. P., Van Noord, T. J. and Kuo, B.-S., *J. Pharm. Biomed. Anal.* **20**, 39–47 (1999).
70. Beaudry, F., Le Blanc, J. C. Y., Coutu, M. and Brown, N. K., *Rapid Commun. Mass. Spectrom.* **12**, 1216–1222 (1998).
71. Bayliss, M. K. and Frick, L. W., *Curr. Opin. Drug Discovery Dev.* **2**, 20–25 (1999).
72. Tarbit, M. H. and Berman, J., *Curr. Opin. Chem. Biol.* **2**, 411–416 (1998).
73. Mandagere, A. K., Correll, M. A., Mooney, J. P., Poole, J. C., Hwang, K.-K. and Cheng, L. K., *Pharmaceut. Res.* **13**, S237 (1996).
74. Rubas, W., Cromwell, M. E. M., Mrsny, R. J., Ingle, G. and Elias, K. A., *Pharmaceut. Res.* **13**, 23–26 (1996).
75. Kansy, M., Senner, F. and Gubernator, K., *J. Med. Chem.* **41**, 1007–1010 (1998).
76. Taylor, E. W., Gibbons, J. A. and Braeckman, R. A., *Pharmaceut. Res.* **14**, 572–577 (1997).
77. Stevenson, C. L., Augustijns, P. F. and Hendren, R. W., *Int. J. Pharmaceut.* **177**, 103–115 (1999).
78. Lim, H. K., Stellingweif, S., Sisenwine, S. and Chan, K. W., *J. Chromatog. A* **831**, 227–241 (1999).
79. Li, A. P., Gorycki, P. D., Hengstler, J. G., Kedderis, G. L., Koebe, H. G., Rahmani, R., de Sousas, G., Silva, J. M. and Skett, P., *Chem.-Biol. Interact.* **121**, 17–35 (1999).
80. Caldwell, G. W., Masucci, J. A. and Chacon, E., *Comb. Chem. High Throughput Screening* **2**, 39–51 (1999).
81. Crespi, C. L., Miller, V. P. and Penman, B. W., *Anal. Biochem.* **248**, 188–190 (1997).
82. Crespi, C. L., Miller, V. P. and Penman, B. W., *Med. Chem. Res.* **8,** 457–471 (1998).
83. Onderwater, R. C. A., Venhorst, J., Commandeur, J. N. M. and Vermeulen, N. P. E., *Chem. Res. Toxicol.* **12**, 555–559 (1999).
84. Van Breemen, R. B., Nikolic, B. and Bolton, J. L., *Drug Metab. Dispos.* **26**, 85–90 (1998).
85. Gibbons, J. A., Taylor, E. W. and Braeckman, R. A., in *Combinatorial Chemistry and Molecular Diversity in Drug Discovery*, E. M. Gordon and J. Kerwin (Eds.). Wiley-Liss, New York, 1998, pp. 453–474.
86. Rodrigues, A. D., *Pharmaceut. Res.* **14**, 1504–1510 (1997).
87. Rodrigues, A. D., *Med. Chem. Res.* **8**, 422–433 (1998).
88. Smith, D. A. and Van der Waterbeemd, H., *Curr. Opin. Chem. Biol.* **3**, 373–378 (1999).
89. Darvas, F. and Dorman, G., *Chim. Oggi* **17**, 10–13 (1999).
90. Watt, A. P., Morrison, D. and Evans, D. C., *Drug Disc. Today* **5**, 17–24 (2000).
91. Spraul, M., Hofmann, M., Ackermann, M., Nicholls, A. W., Damment, S., J. P., Haselden, J. N., Shockcor, J. P., Nicholson, J. K. and Lindon, J. C., *Analyst* **122**, 339–341 (1997).
92. Afshari, C. A., Nuwaysir, E. F. and Barrett, J. C., *Cancer Res.* **59**, 4759–4760 (1999).
93. Todd, M. D., Lin, X., Stankowski, L. F. Jr., Desai, M. and Wolfgang, G. H. I., *J. Biomol. Screen.* **4**, 259–268 (1999).
94. Shapiro, M. A., Sesnie, J. C., Joannides, E. T., Malta, T. E., Dever, J. A. and Heifetz, C. L., *J. Pharmacol. Toxicol. Methods* **36**, 185–187 (1996).
95. Cronin, M. T., *Pharm. Pharmacol. Commun.* **4**, 157–163 (1998).
96. Wang, J., Lai, L. and Tang, Y., *J. Chem. Inf. Comput. Sci.* **39**, 1173–1189 (1999).
97. Kramer, P. J., *J. Pharm. Pharmacol.* **50**, 395–405 (1998).

98. Gibson, G. G., *Biomed. Health Res.* **25**, 13–23 (1998).

99. Johnson, D. E., Braeckman, R. A. and Wolfgang, G. H., *Curr. Opin. Drug Discovery Dev.* **2**, 49–57 (1999).

100. Todd, M. D. and Ulrich, R. G., *Curr. Opin. Drug Discovery Dev.* **2**, 58–68 (1999).

101. Lipinski, C. A., Bissett, B. D., Campos, G. andy Feeney, P. J., *Book of Abstracts, MEDI-161.* 217th ACS, Washington, DC Anaheim, CA, March 21st–25th 1999.

102. Chen, H.-W. and Fang, Z.-L., *Anal. Chim. Acta* **376**, 209–220 (1998).

103. Valko, K., Bevan, C. and Reynolds, D., *Anal. Chem.* **69**, 2022–2029 (1997).

104. Clark, D. E., *J. Pharm. Sci.* **88**, 815–821 (1999).

105. Lipinski, C. A., Lombardo, F., Dominy, B. W. and Feeney, P. J., *Adv. Drug Delivery Rev.* **23**, 3–25 (1997).

106. Wong, D. W. S., Robertson, G. H., Tillin, S. J. and Wong, C., *J. Agric. Food Chem.* **46**, 3852–3857 (1998).

107. Reed, J. D., Edwards, D. L. and Gonzales, C. F., *Mol. Plant-Microbe Interact.* **10**, 537–549 (1997).

108. Mee, R. P., Auton, T. R., Eldridge, M. D. and Brooks, A. G., *Pestic. Sci.* **51**, 91–92 (1997).

109. Lyga, J. W., Ali, S. F., Webster, C. A., Simmons, K. A., Plummer, J. and Galietti, M. E., *Electron. Conf. Heterocycl. Chem. Proc.* (1997).

110. Ridley, S. M., Elliott, A. C., Yeung, M. and Youle, D., *Pestic. Sci.* **54**, 327–337 (1998).

111. Corran, A. J., Renwick, A. and Dunbar, S. J., *Pestic. Sci.* **54**, 338–344 (1998).

112. Jansson, R. K., Halliday, W. R. and Argentine, J. A., *J. Econ. Entomol.* **90**, 1500–1507 (1997).

113. Petsko, G. A., Ringe, D. and Hogan, J., *Spec. Publ.-R. Soc. Chem.* **233**, 66–70 (1999).

114. Khan, J. A., Gijs, L., Berger, C., Martin, N., Piraprez, G., Spinnler, H. E., Vulfson, E. N. and Collin, S., *J. Agric. Food Chem.* **47**, 3269–3273 (1999).

115. Berger, C., Martin, N., Collin, S., Gijs, L., Khan, J. A., Piraprez, G., Spinnler, H. E. and Vulfson, E. N., *J. Agric. Food Chem.* **47**, 3274–3279 (1999).

116. Wong, D. W. S. and Robertson, G. H., *Adv. Exp. Med. Biol.* **464**, 91–105 (1999).

117. Strauss, C. R., *Aust. J. Chem.* **52**, 83–96 (1999).

118. Parlow, J. J. and Normansell, J. E., *Mol. Diversity* **1**, 266–269 (1996).

119. Pilipauskas, D. R., *Med. Res. Rev.* **19**, 463–474 (1999).

120. Geysen, H. M., Wagner, C. D., Bodnar, W. M., Markworth, C. J., Parke, G. J., Schoenen, F. J., Wagner, D. S. and Kinder, D. S., *Chem. Biol.* **6**, 679–688 (1996).

121. Warmus, J. S., Ryder, T. R., Hodges, J. C., Kennedy, R. M. and Brady, K. D., *Bioorg. Med. Chem. Lett.* **8**, 2309–2314 (1998).

122. Bray, A. M., Chiefari, D. S., Valerio, R. M. and Maeji, N. J., *Tetrahedron Lett.* **36**, 5081–5084 (1995).

123. Gayo, L. M. and Suto, M. J., *Tetrahedron Lett.* **38**, 513–516 (1997).

124. Kuo, P. Y., Du, H., Corkan, A., Yang, K. and Lindsey, J. S., *Chemom. Intell. Lab. Systems* **48**, 219–234 (1999).

125. Hsu, L. C., Ping, L.-J., Webb, E. C. and Wrzosek, T. J., *Lab. Robot. Autom.* **11**, 186–192 (1999).

126. Lindsey, J. S., in *A Practical Guide to Combinatorial Chemistry*, A. W. Czarnik and S. H. DeWitt (Eds.). ACS, Washington, DC, 1997, pp. 309–326.

127. Liu, G. and Ellman, J. A., *J. Org. Chem.* **60**, 7712–7713 (1995).

128. Burgess, K., Lim, H.-J., Porte, A. M. and Sulikowski, G. A., *Angew. Chem., Int. Ed. Engl.* **35**, 220–222 (1996).

129. Borman, S., *Chem. Eng. News* **74**, 37–39 (1996).

130. Burgess, K. and Porte, A. M., *Adv. Catalyt. Proc.* **2**, 69–82 (1997).

131. Weinberg, W. H., Jandeleit, B., Self, K. and Turner, H., *Curr. Opin. Solid State Mater. Sci.* **3**, 104–110 (1998).

132. Hoveyda, A. H., *Chem. Biol.* **5**, R187–R191 (1998).

133. Jandeleit, B. and Weinberg, W. H., *Chem. Ind.*, 795–798 (1998).

134. Jandeleit, B., Turner, H. W., Uno, T., Van Beek, J. A. M. and Weinberg, W. H., *CATTECH* **2**, 101–123 (1998).

135. Bein, T., *Angew. Chem., Int. Ed. Engl.* **38**, 323–326 (1999).

136. Kuntz, K. W., Snapper, M. L. and Hoveyda, A. H., *Curr. Opin. Chem. Biol.* **3**, 313–319 (1999).

137. Newsam, J. M. and Schuth, F., *Biotech. Bioeng.* **61**, 203–216 (1999).

138. Doyle, M. P., *Chem. Rev.* **86**, 919–940 (1986).

139. Lim, H.-J. and Sulikowski, G. A., *J. Org. Chem.* **60**, 2326–2327 (1995).

140. Doyle, M. P., Dyatkin, A. B., Roos, G. H. P., Canas, F., Pierson, D. A., van Basten, A., Mueller, P. and Polleux, P., *J. Am. Chem. Soc.* **116**, 4507–4508 (1994).

141. Lowenthal, R. E., Abiko, A. and Masamune, S., *Tetrahedron Lett.* **31**, 6005–6008 (1990).

142. Fritschi, H., Leutenegger, U., Siegmann, K., Pfaltz, A., Keller, W. and Krathy, C., *Helv. Chim. Acta* **71**, 1541–1552 (1988).

143. Pfaltz, A., *Acc. Chem. Res.* **26**, 339–345 (1993).

144. Jacobsen, E. N., Zhang, W., Muci, A. R., Ecker, J. R. and Deng, L., *J. Am. Chem. Soc.* **113**, 7063–7064 (1991).

145. Cooper, A. C., *J. Am. Chem. Soc.* **120**, 9971–9972 (1998).

146. Reetz, M. T., Becker, M. H., Kuhling, K. M. and Holzwarth, A., *Angew. Chem., Int. Ed. Engl.* **37**, 2647–2650 (1998).

147. Sigman, M. S. and Jacobsen, E. N., *J. Am. Chem. Soc.* **120**, 5315–5316 (1998).

148. Berg, T., Vandersteen, A. M. and Janda, K. D., *Bioorg. Med. Chem. Lett.* **8**, 1221–1224 (1998).

149. Berg, T., Simeonov, A. and Janda, K. D., *J. Comb. Chem.* **1**, 96–100 (1999).

150. Moye-Sherman, D., Welch, M. B., Reibenspies, J. and Burgess, K., *Chem. Commun.*, 2377–2378 (1998).

151. Bromidge, S., Wilson, P. C. and Whiting, A., *Tetrahedron Lett.* **39**, 8905–8908 (1998).

152. Huffman, M. A. and Reider, P. J., *Tetrahedron Lett.* **40**, 831–834 (1999).

153. Reetz, M. T., Becker, M. H., Klein, H.-W. and Stockigt, D., *Angew. Chem., Int. Ed. Engl.* **38**, 1758–1761 (1999).

154. Lavastre, O. and Morken, J. P., *Angew. Chem., Int. Ed. Engl.* **38**, 3163–3165 (1999).

155. Shaughnessy, K. H., Kim, P. and Hartwig, J. F., *J. Am. Chem. Soc.* **121**, 2123–2132 (1999).

156. Herrmann, W. A., Brossmer, C., Reisinger, C.-P., Riermeier, T. H., Ofele, K. and Beller, M., *Chem. Eur. J.* **3**, 1357–1364 (1997).

157. Scott, R. H. and Balasubramanian, S., *Bioorg. Med. Chem. Lett.* **7**, 1567–1572 (1997).

158. Gilbertson, S. R. and Wang, X., *Tetrahedron Lett.* **36**, 6475–6478 (1996).

159. Gilbertson, S. R. and Wang, X., *Tetrahedron* **55**, 11609–11618 (1999).

160. Gilbertson, S. R. and Chang, C.-W. T., *Chem. Commun.*, 975–976 (1997).

161. Gilbertson, S. R. and Chang, C.-W. T., *J. Org. Chem.* **63**, 8424–8431 (1998).

162. Cole, B. M., Shimizu, K. D., Krueger, C. A., Harrity, J. P. A., Snapper, M. L. and Hoveyda, A. H., *Angew. Chem., Int. Ed. Engl.* **35**, 1668–1671 (1996).

163. Shimizu, K. D., Cole, B. M., Krueger, C. A., Kuntz, K. W., Snapper, M. L. and Hoveyda, A. H., *Angew. Chem., Int. Ed. Engl.* **36**, 1704–1707 (1997).

164. Sigman, M. S. and Jacobsen, E. N., *J. Am. Chem. Soc.* **120**, 4901–4902 (1998).

165. Porte, A. M., Reibenspies, J. and Burgess, K., *J. Am. Chem. Soc.* **120**, 9180–9187 (1998).

166. Gennari, C., Ceccarelli, S., Piarulli, U., Montalbetti, C. A. G. N. and Jackson, R. F. W., *J. Org. Chem.* **63**, 5312–5313 (1998).

167. Buck, R. T., Coe, D. M., Drysdale, M. J., Moody, C. J. and Pearson, N. D., *Tetrahedron Lett.* **39**, 7181–7184 (1998).

168. Ding, K., Ishii, A. and Mikami, K., *Angew. Chem., Int. Ed. Engl.* **38**, 497–501 (1999).

169. Hinderling, C. and Chen, P., *Angew. Chem., Int. Ed. Engl.* **38**, 2253–2256 (1999).

170. Havranek, M., Singh, A. and Sames, D., *J. Am. Chem. Soc.* **121**, 8965–8966 (1999).

171. Altava, B., Burguete, M. I., Garcia-Verdugo, E., Luis, S. V., Pozo, O. and Salvador, R. V., *Eur. J. Org. Chem.*, 2263–2267 (1999).

172. Gao, X. and Kagan, H. B., *Chirality* **10**, 120–124 (1998).

173. Corey, E. J., Wess, G., Xiang, Y. B. and Singh, A. K., *J. Am. Chem. Soc.* **109**, 4717–4718 (1987).

174. Francis, M. B. and Jacobsen, E. N., *Angew. Chem., Int. Ed. Engl.* **38**, 937–941 (1999).

175. Yu, J., Zhao, Y., Holterman, M. J. and Venton, D. L., *Bioorg. Med. Chem. Lett.* **9**, 2705–2710 (1999).

176. Taylor, S. J. and Morken, J. P., *Science* **280**, 267–270 (1998).

177. Ohlmeyer, M. H. J., Swanson, R. N., Dillard, L., Reader, J. C., Asouline, G., Kobayashi, R., Wigler, M. and Still, W. Clark., *Proc. Natl. Acad. Sci. USA* **90**, 10922–10926 (1993).

178. Moates, F. C., Somani, M., Annamalai, J., Richardson, J. T., Luss, D. and Willson, R. C., *Ind. Eng. Chem. Res.* **35**, 4801–4803 (1996).

179. Boussie, T. R., Coutard, C., Turner, H., Murphy, V. and Powers, T. S., *Angew. Chem., Int. Ed. Engl.* **37**, 3272–3275 (1998).

180. Boussie, T. R., Murphy, V., Hall, K. A., Coutard, C., Dales, C., Petro, M., Carlson, E., Turner, H. W. and Powers, T. S., *Tetrahedron* **55**, 11699–11710 (1999).

181. Lehn, J. M., *Supramolecular Chemistry: Concepts and Perspectives.* VCH, Weinheim, Germany, 1995.

182. Inoue, Y. and Wada, T., *Adv. Supramol. Chem.* **4**, 55–96 (1997).

183. Cantrill, S. J., Fyfe, M. C. T., Raymo, F. M. and Stoddart, J. F., *NATO Sci. Ser., Ser. C* **519**, 17–35 (1999).

184. Harada, A., *Adv. Polym. Sci.* **133**, 141–191 (1997).

185. Casnatt, A., *Gazz. Chim. Ital.* **127**, 637–649 (1997).
186. Diederich, F. and Gomez-Lopez, M., *Chimia* **52**, 551–556 (1998).
187. Fyfe, M. C. T. and Stoddart, J. F., *Adv. Supramol. Chem.* **5**, 1–53 (1999).
188. Zeng, F. and Zimmerman, S. C., *Chem. Rev.* **97**, 1681–1712 (1997).
189. Narayanan, V. V. and Newkome, G. R., *Top. Curr. Chem.* **197**, 19–77 (1998).
190. Clark Still, W., *Acc. Chem. Res.* **29**, 155–163 (1996).
191. Fenniri, H., *Curr. Med. Chem.* **3**, 343–378 (1996).
192. Gennari, C., Nestler, H. P., Piarulli, U. and Salom, B., *Liebigs Ann./Recueil*, 637–647 (1997).
193. de Miguel, Y. R. and Sanders, J. K. M., *Curr. Opin. Chem. Biol.* **2**, 417–421 (1998).
194. Timmermann, P. and Reinhoudt, D. N., *Adv. Mater.* **11**, 71–74 (1999).
195. Linton, B. and Hamilton, A. D., *Curr. Opin. Chem. Biol.* **3**, 307–312 (1999).
196. Reinhoudt, D. N., Timmerman, P., Cardullo, F. and Crego-Calama, M., *NATO ASI Ser., Ser. C* **527**, 181–195 (1999).
197. Yoon, S. S. and Clark Still, W., *Tetrahedron* **51**, 567–578 (1995).
198. Gennari, C., Nestler, H. P., Salom, B. and Still, W. C., *Angew. Chem., Int. Ed. Engl.* **34**, 1765–1768 (1995).
199. Shao, Y. F. and Clark Still, W., *J. Org. Chem.* **61**, 6086–6087 (1996).
200. Tomeiro, M. and Clark Still, W., *Tetrahedron* **53**, 8739–8750 (1997).
201. Burger, M. T. and Clark Still, W., *J. Org. Chem.* **62**, 4785–4790 (1997).
202. Lowik, D. W. P. M., Weingarten, M. D., Broekema, M., Brouwer, A. J., Still, W. Clark and Liskamp, R. M. J., *Angew. Chem., Int. Ed. Engl.* **37**, 1846–1850 (1998).
203. Iorio, E. J. and Clark Still, W., *Bioorg. Med. Chem. Lett.* **9**, 2145–2150 (1999).
204. Boyce, R., Li, G., Nestler, H. P., Suenaga, T. and Still, W. Clark., *J. Am. Chem. Soc.* **116**, 7955–7956 (1994).
205. Cheng, Y., Suenaga, T. and Still, W. Clark, *J. Am. Chem. Soc.* **118**, 1813–1814 (1996).
206. Nestler, H. P., *Mol. Diversity* **2**, 35–40 (1996).
207. Dong, D. L., Liu, R., Sherlock, R., Wigler, M. H. and Nestler, H. P., *Chem. Biol.* **6**, 133–141 (1999).
208. Fessmann, T. and Kilburn, J. D., *Angew. Chem. Int. Ed.* **38**, 1993–1996 (1999).
209. Goodman, M. S., Jubian, V., Linton, B. and Hamilton, A. D., *J. Am. Chem. Soc.* **117**, 11610–11611 (1995).
210. Jung, G., Hofstetter, H., Feiertag, S., Stoll, D., Hofstetter, O., Wiesmueller, K.-H. and Schurig, V., *Angew. Chem., Int. Ed. Engl.* **35**, 2148–2150 (1996).
211. Burger, M. T. and Clark Still, W., *J. Org. Chem.* **60**, 7382–7383 (1995).
212. Malin, R., Steinbrecher, R., Janssen, J., Semmler, W., Noll, B., Johannsen, B., Froemmel, C., Hoehne, W. and Schneider-Mergener, J., *J. Am. Chem. Soc.* **117**, 11821–11822 (1995).
213. Shibata, N., Baldwin, J. E. and Wood, M. E., *Bioorg. Med. Chem. Lett.* **7**, 413–416 (1997).
214. Lewandowski, K., Murer, P., Svec, F. and Frechet, J. M. J., *Chem. Commun.*, 2237–2238 (1998).
215. Lewandowski, K., Murer, P., Svec, F. and Frechet, J. M. J., *J. Comb. Chem.* **1**, 105–112 (1999).

216. Murer, P., Lewandowski, K., Svec, F. and Frechet, J. M. J., *Chem. Commun.*, 2559–2560 (1998).

217. Murer, P., Lewandowski, K., Svec, F. and Frechet, J. M. J., *Anal. Chem.* **71**, 1278–1284 (1999).

218. Welch, C. J., Bhat, G. and Protopopova, M. N., *Enantiomer* **3**, 463–469 (1998).

219. Welch, C. J., Protopopova, M. N. and Bhat, G., *Enantiomer* **3**, 471–476 (1998).

220. Welch, C. J., Bhat, G. and Protopopova, M. N., *J. Comb. Chem* **1**, 364–367 (1999).

221. Wang, Y. and Li, T., *Anal. Chem.* **71**, 4178–4182 (1999).

222. Welch, C. J., Protopopova, M. N. and Bhat, G., *Spec. Publ.-R. Soc. Chem.* **235**, 129–138 (1999).

223. Dyer, U. C., Henderson, D. A. and Mitchell, M. B., *Org. Proc. Res. Dev.* **3**, 161–165 (1999).

224. Davies, M., Bonnat, M., Guillier, F., Kilburn, J. D. and Bradley, M., *J. Org. Chem.* **63**, 8696–8703 (1998).

225. Davies, M. and Bradley, M., *Angew. Chem., Int. Ed. Engl.* **36**, 1097–1099 (1997).

226. Bonnat, M., Bradley, M. and Kolburn, J. D., *Tetrahedron Lett.* **37**, 5409–5412 (1996).

227. Francis, M. B., Finney, N. S. and Jacobsen, E. N., *J. Am. Chem. Soc.* **118**, 8983–8984 (1996).

228. Weingarten, M. D., Sekanina, K. and Still, W. Clark, *J. Am. Chem. Soc.* **120**, 9112–9113 (1998).

229. Reddy, D. R. and Thornton, E. R., *J. Chem. Soc., Chem. Commun.*, 172–173 (1992).

10 Biosynthetic Combinatorial Libraries

The wide range of combinatorial libraries that are produced by biological or biochemical methods will be discussed briefly in this chapter. The aim here is to simply provide a description of the main features of such libraries together with several examples highlighting selected applications.

In the first section, *biosynthetic polypeptide libraries* produced by in vitro or in vivo biological methods are covered, with particular emphasis on their versatility for applications in the fields of pharmaceutical research, molecular recognition, and catalysis. An extensive coverage of display techniques in phage clones is provided through detailed consideration of several examples. *Biological oligonucleotide libraries* and their selection and amplification to give strong ligands (aptamers) or oligonucleotidic enzymes (ribozymes) are described in the second section.

Manipulation of the biosynthetic pathways leading to natural compounds, so-called *combinatorial biosynthesis*, is presented in the third section, with particular attention paid to the opportunities arising from polyketide biosynthesis. Finally, *combinatorial biotransformation* of natural or synthetic compounds by means of isolated enzymes or whole microorganisms is presented in the fourth section.

10.1 BIOSYNTHETIC POLYPEPTIDE LIBRARIES

10.1.1 General Considerations

Biological sources possess several appealing features related to the evolutionary process as potential producers of combinatorial libraries. The two most important bio-oligomers, nucleic acids and proteins, are strictly connected because the genetic information inherent in the oligonucleotidic chain, that is, the nucleic acid sequence of the coding gene, is translated into the oligomeric polypeptide, that is, the gene product that carries out a specific function within the biological system. These oligomeric structures have evolved in every living organism through time to produce libraries of nucleic acids and proteins based on the ability of the system to mutate its components and the ability to pass the mutated, favorable structure to its progeny.

Although many factors are involved in the evolution of an organism, its life-cycle plays an important part regulating the pace of evolution and can be compared to the chemical synthesis of a library. Higher organisms evolve very slowly and can be compared to classical, information-rich low-throughput chemistry methods to prepare a few complex compounds by sequential methods whereas prokaryotic and lower eukaryotic organisms have a much faster life-cycle and can mutate and replicate at high rates under suitable conditions, which is comparable to the high-throughput combinatorial synthesis of extremely large primary pool libraries as sources of positives. These latter organisms are extremely appealing for use as *vectors* for the preparation of libraries, providing that an appropriate piece of randomized genetic information (cDNA) can be inserted into the genome for eventual expression by the biological machinery to eventually obtain the corresponding randomized gene products (polypeptide sequences) as a peptide library.

Biological methods for the generation of peptide libraries (Fig. 10.1) can be divided into two classes using either in vitro or in vivo techniques, the former being based on the in vitro translation/transcription (IVTT) machinery and the latter on the expression of peptide libraries through the introduction of suitable genetic material into either prokaryotic or eukaryotic cells. Several excellent reviews that provide an overview of these methods have been published recently (1–5). Examples of in vitro systems are ribosome display (6–9), where the production units are the ternary complexes formed between the randomized messenger RNAs (genetic information), the ribosomes (the machinery to transcribe the information and to translate it into peptides) and the polypeptide library products, and the peptide–RNA system assembled during IVTT by the action of puromycin (10). Miscellaneous systems are described in references 11 and 12.

More work has been carried out on the related in vivo systems. These techniques are known as *surface display*, in that a virus particle, or a bacterial cell, is used to express the polypeptide library components and then to export it onto the surface of the cell to be displayed and, hence, to be available for screening. By far the most used method is phage display, in which peptide libraries are displayed on the surface of a phage, amplified in iterative rounds using *Escherichia coli* as the infected host organism, and selected for their biological properties using target-based selection protocols. This technique is described in detail in the next section. Other viruses such

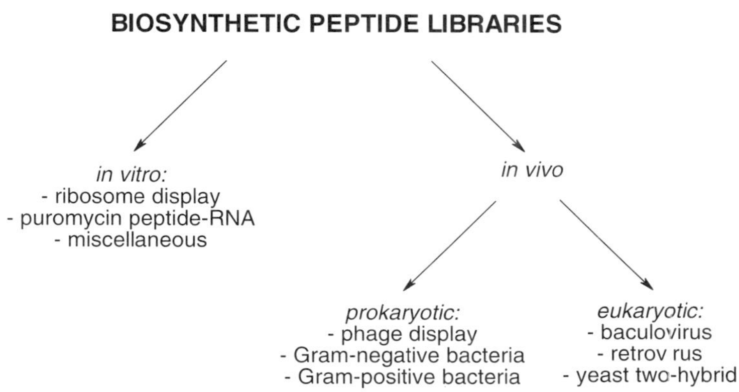

Figure 10.1 Biosynthetic peptide library sources.

as mammalian retroviruses (13), baculovirus (14), and modified adenovirus (15), which are able to infect mammalian cells and are thus amplifiable in these hosts, have also been reported.

Both gram-negative and, more recently, gram-positive bacteria (1, 16, 17) have been used to display various peptide libraries that were screened to find ligands (18–20), antibodies (21, 22), and vaccines (23). Surface display on yeasts has produced the very popular yeast–two hybrid system and some of its variants; several recent papers and reviews are referenced here (24–27). These methods have been used to prepare and select polypeptides and proteins for various applications, including the identification of binding partners in protein–protein interactions where this is the technique of choice (28–31).

10.1.2 Phage Display of Polypeptide Libraries

In 1985, Smith (32) reported the insertion of foreign DNA sequences into phage genes with the resultant peptide expressed or *displayed* on the surface of the phage capsid. A hybrid phage displaying a foreign peptide could be isolated from wild-type phages by affinity purification using a receptor with affinity for the peptide anchored to a solid phase and washing away the unbound phage capsids. The infectivity of the phage and its ability to propagate in a suitable bacterial host were maintained by insertion of the nucleotide sequence in selected gene regions such that the isolated hybrid phage was amplified in the host to produce a large population of phages displaying the same peptide. Slightly later, Parmley and Smith applied this principle to the selection and affinity purification of different gene products (33), opening the route to several phage-displayed polypeptide libraries in 1989 (34) and 1990 (35–37). Phage display has become popular in recent years with many hundreds of papers describing a number of applications that will be discussed below after a brief description of the basic principles behind the technique.

Standard recombinant DNA techniques allow the insertion of a foreign piece of DNA into a recombinant vector, which in phage display is the wild-type phage DNA. When this phage infects its standard host, the gram-negative bacterium *E. coli*, the foreign DNA insert is replicated together with the phage DNA vector. Moreover, being an expression vector, the foreign insert is converted into the corresponding polypeptide sequence through translation. A peculiarity and at the same time an advantage of phage display originates from the location of the gene insert, which is introduced into the gene sequence coding for phage coat proteins and thus is expressed with them as a hybrid protein on the surface of the phage. Careful selection of the insertion loci allows the display of the foreign polypeptide as part of the phage capsid (Fig. 10.2).

Filamentous phages are typically used for phage display because of their properties, although other bacterial phages have also been used to a lesser extent (38–40). The infection of *E. coli* starts when one of the phage coat proteins (*vide infra*) connects with the pilus of the bacterial cell (Fig. 10.3, step a). The coat proteins start to dissolve and the single-stranded phage DNA (ssDNA) penetrates the cell and enters the cytoplasm while the whole virion disappears (step b). The ssDNA is replicated by the biological machinery of the host to give a double-stranded form suitable for replication

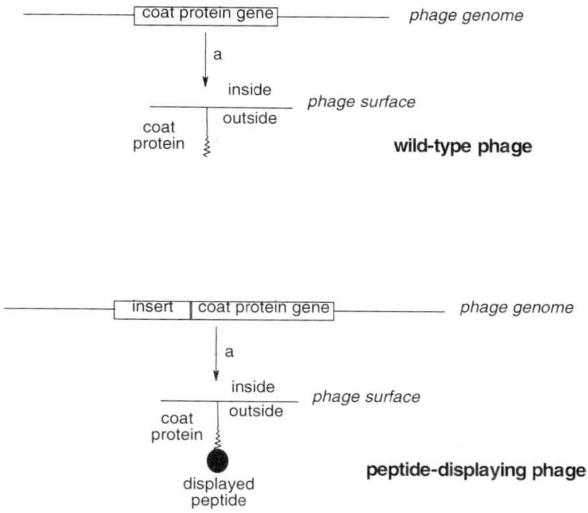

a: translation of the genomic information.

Figure 10.2 Display of a peptide sequence on the phage surface via fusion onto phage coat proteins.

(step c) that is used by the host to produce multiple copies of the foreign ssDNA phage strand and to transcribe phage genes such as the coat proteins (step d). The progeny ssDNA strands are surrounded by the coat proteins produced and are externalized from the cytoplasm, eventually emerging from the bacterial surface as whole virions (step e, Fig. 10.3). This process, which makes on average several hundred phage particles per cell at each division cycle, continues indefinitely without significantly affecting the bacterial cell life-cycle while producing extremely large populations of phages. If the foreign DNA insert is represented by a random mixture of oligonucleotides with each sequence being recombined in a phage vector, then phage infection will amplify each ssDNA and eventually produce a population of phages each displaying a single polypeptide chain. The result of this process is a true library of displayed peptides, as shown in Fig. 10.4, where n library individuals, each displayed on x copies of phage clones, are represented.

An intriguing comparison can be made between phage display libraries and synthetic SP, pool polypeptide libraries:

- A microunit bearing multiple copies of a single library individual exists for both formats (the resin bead versus the phage virion); that is, the one-bead, one-compound concept is paralleled by the one-virion, one-compound construct.
- The location of each peptide molecule is defined on the microunit (the resin loading sites versus the phage coat protein sites) as is the loading per microunit (number of sites on a bead or on the phage surface).

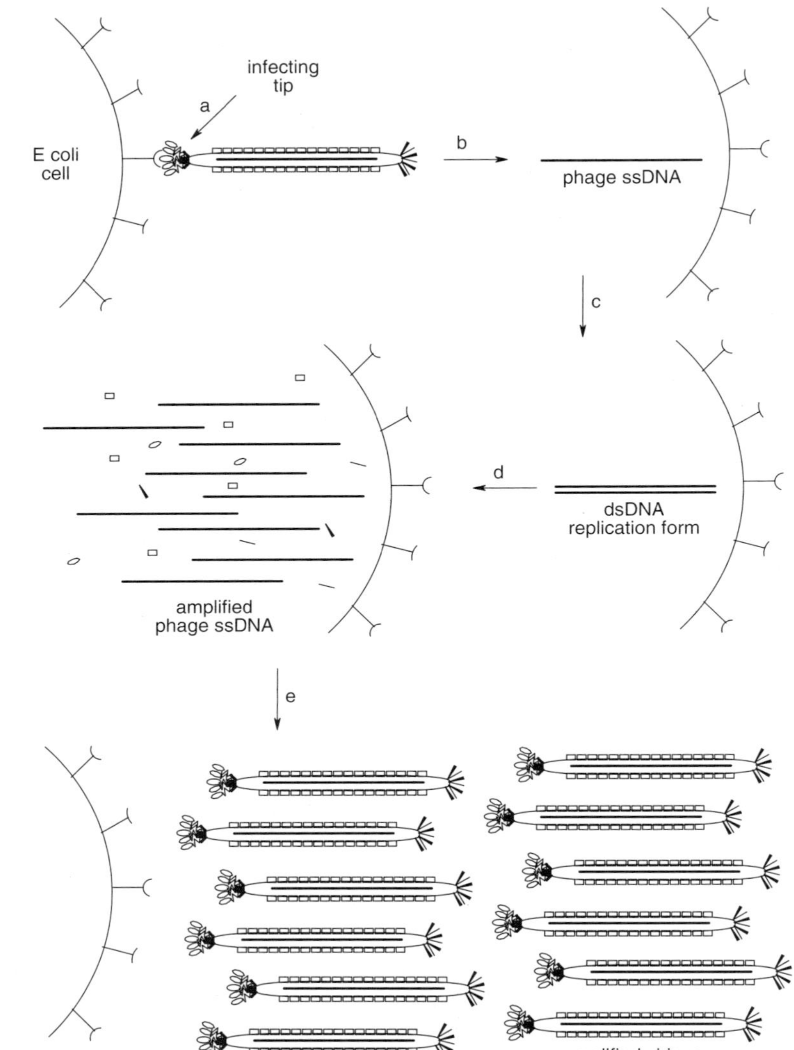

a: *E. coli* infection *via* connection of pIII coat protein and the pilus; b: internalization of ssDNA and disappearance of the virion; c: dsDNA formation; d: DNA replication/amplification; e: virion assembly and externalization of the amplified phage population.

Figure 10.3 Infection and replication of phages: the whole cycle.

- Multiple copies of each library individual are easily obtained by controlling the library production steps (mix-and-split synthesis versus DNA recombination and amplification of phage populations).
- Library individuals can be screened as microunit-bound entities (on-bead screening versus on-phage screening).

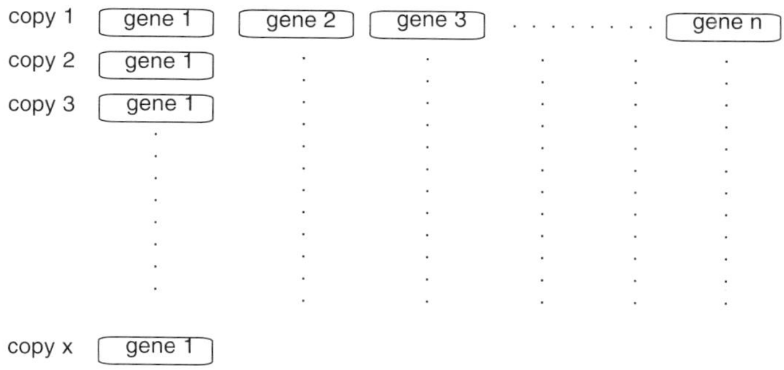

Figure 10.4 Schematic representation of a phage display library: multiple copies of many genomes.

- The structure(s) of active sequences can be extracted from the whole library (deconvolution or encoding versus target-assisted selection and isolation of displayed polypeptides).
- Second-generation focused libraries can be designed using the results from the primary screening to optimize the activity of library individuals.

Each of these features is briefly presented and discussed for phage display, starting with the morphology of the microunit and the sites of attachment of the peptide chains. The phage capsid is made of various coat proteins, two of which are important for fusion with foreign peptides. The schematic representation of a phage, highlighting the position of the coat proteins, is shown in Fig. 10.5. The vast majority of the phage coat (87% by mass) is made up of around 2700 copies of a small, 50-residue coat protein named pVIII and each copy is encoded by a single phage gene VIII. The copies are helically arranged and make up the 1-μm-long, 6-nm-wide filamentous phage

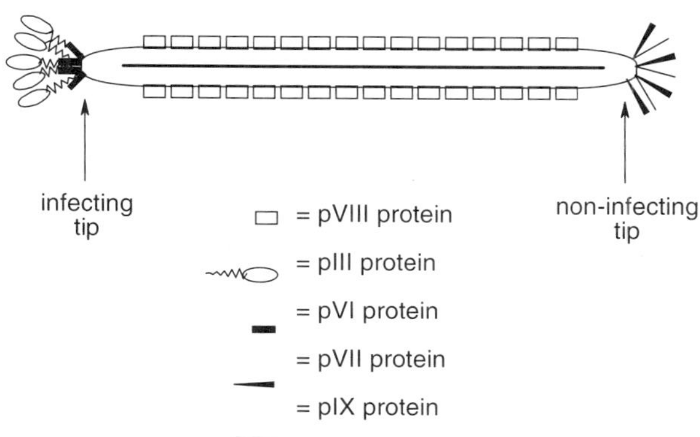

Figure 10.5 Structure of philamentous phages: phage coat proteins.

capsid. The two tips of the rod-shaped phage bear five copies of four coat proteins, two at the infecting tip (pIII and pVI, genes III and VI) and two at the other tip (pVII and pIX, genes VII and IX). The pIII proteins (200 amino acids) are responsible for the infection of bacterial hosts and, together with the abundant pVIII proteins, have been used to display polypeptides on filamentous phages. Both proteins have the N-terminus end displayed on the capsid surface, and the foreign peptide is expressed in proximity to this area. Careful selection of the insertion junctions of the foreign DNA into the genes III or VIII displays the polypeptide onto the phage surface and maintains the virion infectivity for the host and its ability to reconstruct its structure correctly while being externalized by the host. A recent report (41) validated also the use of pVII and pIX coat proteins to display, through a phagemid format, combinatorial heterodimeric arrays of antibody structures (*vide infra*).

The use of pIII and pVIII as supports for the polypeptide is exploitable in several ways. The virion genome may contain a single copy of either recombined gene III or VIII, displaying a peptide chain fused with five copies of pIII protein (type 3, Fig. 10.6) or around 2700 copies of pVIII (type 8, Fig. 10.6). It may contain two copies of the selected coat protein gene, one as a wild-type sequence and one as a recombinant gene, to produce a mosaic virion that displays 25–100 copies (type 88, Fig. 10.6) or even one single copy of a polypeptide chain (type 33, Fig. 10.6). Finally, the wild-type virion may be coupled with a special phage plasmid (phagemid) bearing a hybrid copy of gene III (type 3+3, Fig. 10.6) or gene VIII (type 8+8, Fig. 10.6). The presence of wild-type and hybrid phages and phagemids reduces the number of peptides displayed per phage population.

The importance of the loading per particle (from 2700 copies per phage for type VIII to even 1 copy per 100 phage virions with 3+3 constructs) is related to the activity of the displayed peptides. The presence of many peptide copies per phage particle increases the probability of spotting weakly active sequences through the additive effect of each peptide–receptor interaction. On the contrary, if only high-affinity binders are desired, a lower number of copies, or even monovalent phages, are desirable. The size of the foreign peptide sequence is also important. Small peptides may be displayed and accommodated even on the surface of a type 8 phage, but larger ones require less dense environments to maintain their flexibility as displayed sequences and to preserve the essential phage characteristics.

The structure of the phage library is determined by the sequence of the foreign DNA inserted into the coat protein phage genes. A single structure per virion is derived from the unique nature of the genetic information contained in each single phage recombined ssDNA. Library synthesis starts with the synthesis of the recombinant ssDNA strands bearing the foreign peptide coding sequences. Standard recombinant DNA techniques allow the careful control of the structure of the phage library as derived from the genetic information, and the reduced dimensions of these ssDNA chains make the production of 10^8–10^9-member libraries a reasonable task. Once prepared, the phage vectors are introduced as naked DNA into *E. coli* cells using electroporation (42), and replication/amplification steps immediately start following the processes depicted in Fig. 10.3. The phage population generated is freed from the *E. coli* cells, purified, and submitted to a selection process aimed at identifying binders for the

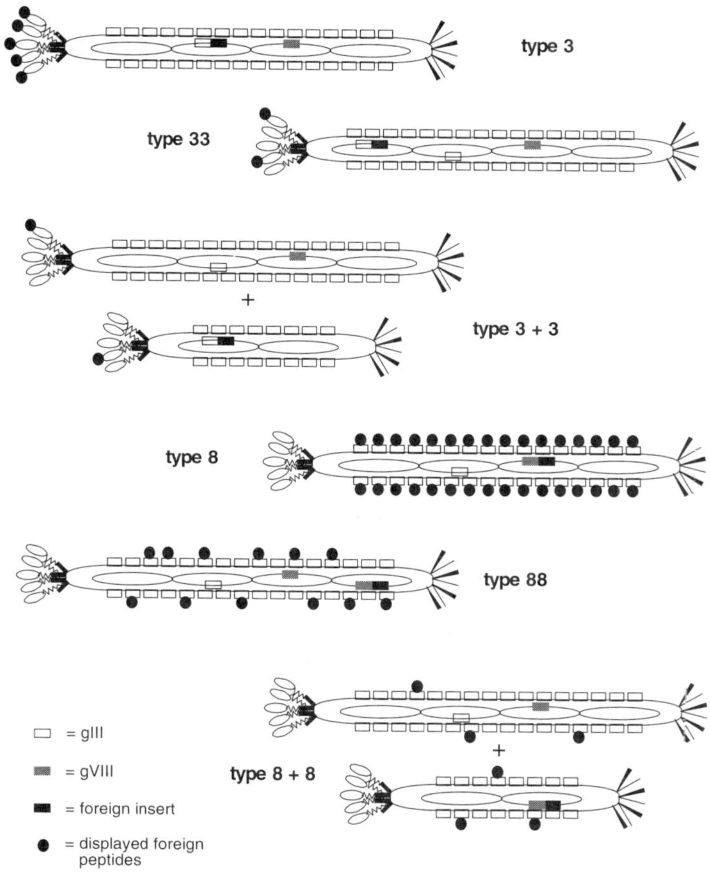

Figure 10.6 Common display formats of filamentous phage libraries.

relevant target (*vide infra*). Selected phages are isolated, and further amplification/selection cycles (typically lasting one day) are performed until the required structural information and/or potency toward the target are obtained. The structure of displayed peptides is obtained from sequencing the DNA of the selected clones.

Amplification of a phage library by the biological machinery of the host is the major advantage of phage display in that large numbers of copies of peptides are obtained starting from cheap precursors (filamentous phage, *E. coli* cells, and oligonucleotide strands), and library copies and/or selected individuals can be replicated indefinitely. Among the few *caveats* to this technique is the fact that the abundance of each library member will necessarily be unbalanced because some of the 20 amino acids are coded only by one of the 64 DNA coding triplets, while others are coded by multiple codons and thus will be more represented in the peptide chains. A large redundancy must be used to represent all the library individuals in at least one copy. Display of large peptides in a type VIII phage may stress the *E. coli* machinery, forcing it to prepare too many copies of oligonucleotide and peptide oligomers, slowing down the life-cycle of the cell and again privileging the amino acids coded by multiple triplets in the library population.

When a library is prepared and displayed, it is screened and deconvoluted to find active structures from the library population. The phage capsid is stable enough to be purified from the biological medium while retaining its viability and infectivity to start a new selection/amplification cycle. The bonds between the inserted foreign peptides and the phage coat proteins, though, cannot usually be broken to release the peptides in solution. The fusion of foreign peptides onto pIII or pVIII coat proteins allows a large degree of conformational freedom for the displayed chains, making them solutionlike for the binding to any type of receptor. The screening therefore takes place "on-phage" and allows the recovery of viable phages to start a new amplification/selection cycle. It is flexible and sensitive, can spot even weakly binding activities for the selected receptor, and allows the fast, automated separation and isolation of target-bound phages from unbound phages. All of these properties are embedded in the concept of target-assisted screening (see Sections 7.2.2–7.2.4), providing that the selected target is bound onto a solid support to facilitate the separation of active phages from inactive library individuals. Suitable solid supports for large receptors include plastic surfaces, tubes, Petri dishes, and microtiter wells to which the target is adsorbed or bound nonspecifically to ensure a good percentage of adsorbed targets with accessible binding sites for the display library. If the target is a smaller entity, specific covalent linkages or interactions with supports (resin beads) or with supported molecules (target-specific antibodies) may be used to anchor it to the support. The phage library is incubated with the supported target and the desired, noncovalent interaction with phages displaying active peptide sequences takes place (step a, Fig. 10.7). The incubation medium containing unbound phages is separated and the solid support is washed thoroughly (step b); then the absorbed ligand is eluted from the support, thus recovering the first selected phage population (step c, Fig. 10.7). Either this population can be used for further rounds of amplification and selection if the observed activity is low or negligible or it can be structurally deconvoluted by DNA sequencing to determine the active polypeptide(s). Several rounds of selection and amplification are usually necessary to select peptides with the required levels of activity.

Two strictly related criteria are important for a selection protocol and must be carefully adjusted to obtain positive results from phage display. The first is the stringency of measurement, that is, the cutoff level at which a peptide sequence is recognized as active and the experimental conditions under which the activity is measured; the other is the yield, that is, the percentage of phage particles surviving the selection process. In the first selection round for a phage library, it is reasonable to look for sequences with weak activity so that a high loading (e.g., as found in a type VIII clone) and a loose stringency may produce starting points that can be optimized in further rounds of amplification and selection. These iterative cycles will progressively raise the cutoff level to allow the eventual isolation of populations containing a large number of phages bearing several highly active peptide

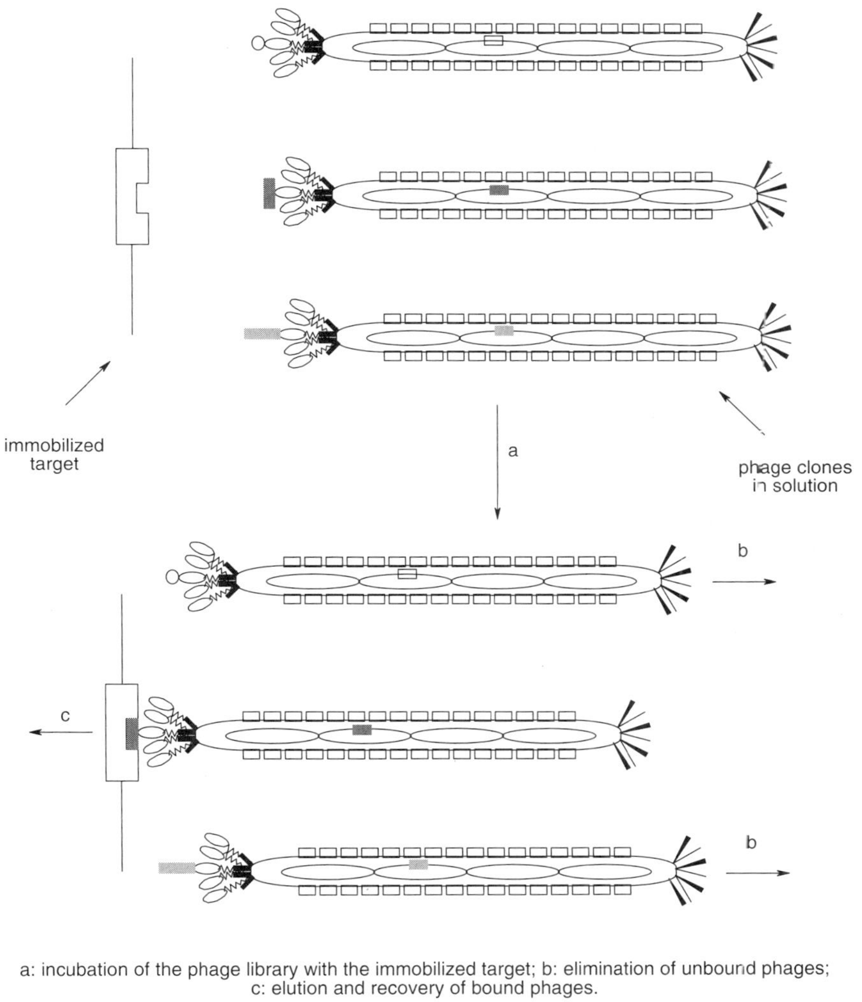

Figure 10.7 Selection of phages carrying binding sequences via target-assisted screening.

sequences. Further increase in affinity may come from using lower loading clones such as a type 33 or 3+3. Structural deconvolution by DNA sequencing may lead to the design of focused libraries where the DNA inserts are biased toward the positives obtained with the aim of optimizing the active structures from the primary phage library.

The next two sections describe several examples, providing both an overview of the main applications of phage display libraries and a brief bibliography. More detailed descriptions of the various aspects of a phage display can be found in several recent reviews (3, 4, 43–52).

10.1.3 Phage Display Libraries: Looking for Small Peptide Recognition Motifs

The synthesis of peptide libraries to identify more or less refined ligand recognition motifs for the binding to a receptor is one of the main applications for phage display libraries. A recent example by Kraft et al. (53) reported the screening of a 7-mer (**L1**) and a 12-mer (**L2**) random commercial display library for binding to two different integrins (Fig. 10.8). The libraries were incubated on receptor-coated plates for 1 h, and the bound phages were eluted at acidic pH and amplified after removal of phages in solution. After three rounds of amplification and selection, the active phages were sequenced and a series of hepta- and dodecapeptide structures were obtained (Fig. 10.9).

The known binding motif RGD was observed in many of the 12-meric bound sequences to αvβ6 integrin (51%, italic, Fig. 10.9), but a significant amount of peptides (27%, bold, Fig. 10.9) contained the unexpected $X_1X_2DLX_3X_4LX_5$ motif. The X_1–X_5 positions showed marked preferences for specific amino acids (Fig. 10.9). The heptameric library **L1** contained only 5% of the truncated DLXXL motifs, showing the importance of the full eight-residue recognition module found from **L2** in defining specificity for the αvβ6 integrin. A specific clone (**10.1**, Fig. 10.10) was used to determine the biological properties of this class of binders. This clone **10.1** inhibited the RGD-dependent binding to αvβ6 integrin in vitro and in whole cells, and the importance of the eight residues was confirmed in a deletion study (Fig. 10.10). Many reports of binding motifs isolated from phage display libraries have appeared recently in the literature, and a few are referenced here for the interested reader (54–68).

Phage libraries have also been used to study the substrate specificity of enzymes by finding an improved artificial substrate. Coombs et al. (69) reported the detailed assessment of specificity for a serine protease belonging to the α-chymotrypsin family, the prostate specific antigen (PSA). They used both substrate optimization by single-point mutations and phage display libraries. The sequence of the 14-member substrate **10.2** (70) was used to start the iterative optimization process (Fig. 10.11) in which substitution or exchange of the P1, P2, or P2′ residues increased the substrate affinity

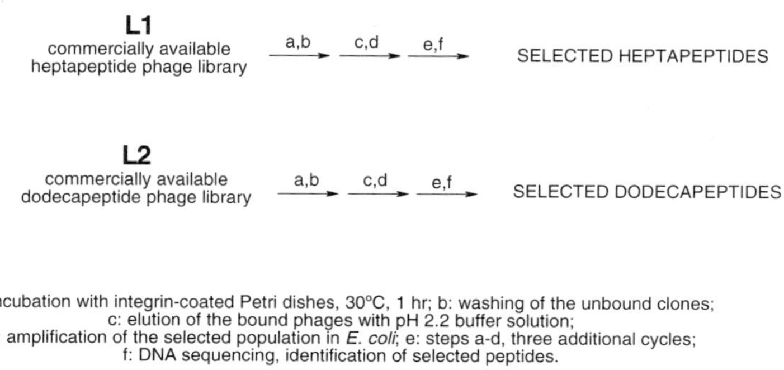

Figure 10.8 Screening of the phage libraries **L1** and **L2** for integrin binding: the selection/amplification iterative process.

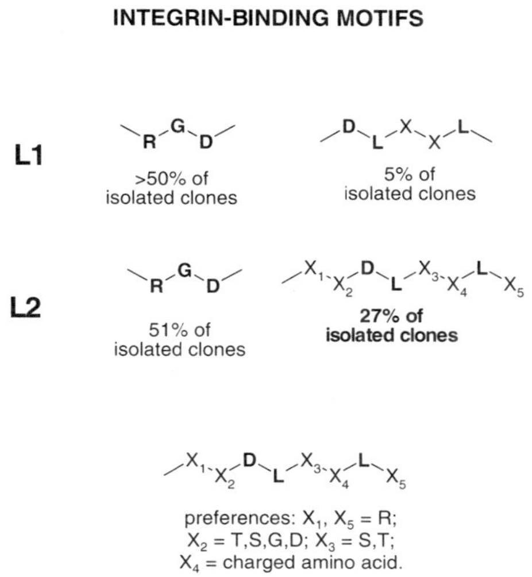

Figure 10.9 Screening of the phage libraries **L1** and **L2** for integrin binding: selected binding motifs.

and consequently the catalytic efficiency from 46 (**10.2**) to 2200 for **10.3**, which was the best substrate (Fig. 10.11).

Alternatively, a random phage library of octapeptides **L3** was built through an elaborate construct to allow the members of the library to be assayed for protease activity (Fig. 10.12). The library was supported on the pIII coat protein by inserting a flexible four-amino-acid linker sequence between the coat protein and the random library. The 8-mer library insert was followed by another six-amino-acid linker sequence and by a known epitope for a monoclonal antibody. The stability of the construct to PSA digestion was carefully checked to prevent nonspecific cleavage of the linkers or the pIII coat protein sequence. The phage library **L3** was incubated with PSA for 4 h (step a, Fig. 10.12), and the phages bearing PSA substrate sequences were processed by cleaving the epitope for 3-E7 mAb (step b). The mixture was treated with the antibody, and the antibody-bound phages were removed from the solution by absorption and discarded (step c). The soluble phages that produced PSA substrate sequences were amplified (step d) and used for further amplification/selection cycles (step e, Fig. 10.12). After five iterative cycles, 88 phage clones were isolated and sequenced and a refined binding motif for the P5–P3′ region was obtained. Several positions showed a marked preference for a single amino acid (P4, P3, P2, P1, and P2′) while other positions (P5, P1′, and P3′) privileged several amino acids (Fig. 10.13). The relative efficiency of the 88 clones as substrates was measured by kinetic experiments (step a, Fig. 10.13), and nine preferred sequences were selected to define a more precise recognition/cleavage motif. This information was conveyed into the 14-mer consensus peptide **10.4** containing the SSYYSG P4–P2′ sequence, which showed better properties than **10.3**, the result of iterative substrate optimization. Other

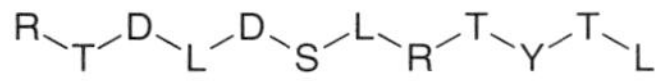

10.1

in vitro inhibition of fibronectins (IC_{50}):

$\alpha v\beta 6$: 20 nM

$\alpha v\beta 3$, $\alpha v\beta 5$, $\alpha IIb\beta 3$: >50 μM

in vivo: active in cell attachment assays

DELETION STUDY:

in vitro inhibition of immobilized $\alpha v\beta 6$

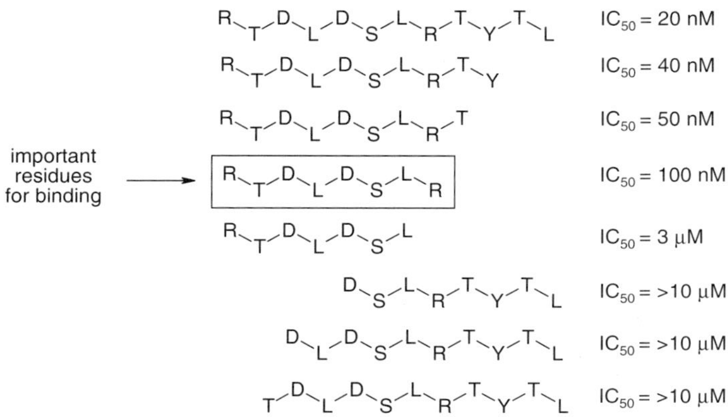

Figure 10.10 Screening of the phage libraries **L1** and **L2** for integrin binding: characterization and specificity of the most active clone **10.1**.

workers have reported the optimization of substrates for five serine proteases (71), stromelysin and matrilysin (72), and granzyme B (73).

Several examples of constrained phage libraries have been aimed at reducing the conformational freedom of potential ligands in order to increase affinity for a receptor: for example, Gee et al. (74) identified cyclic peptide ligands for a PDZ domain of syntrophins, a family of membrane proteins that mediate protein–protein interactions. Two libraries were used in this work: a 12-mer linear library **L4** and an 11-mer library **L5** made up of 10 random amino acids and a final C-terminal cysteine (Fig. 10.14). Screening was carried out by coating microtiter wells with three fusion proteins containing three PDZ domains (α_1, β_1, and β_2 syntrophins) and measuring the binding affinity of peptide sequences. Incubation of the library **L4** for 2 h did not lead to the selection of any sequence, even after three rounds of selection and amplification, whereas library **L5** gave three positives that were sequenced and found to have the structures **10.5–10.7** (Fig. 10.14). The three sequences showed a common cysteine residue in position 7, hinting at a cyclic structure caused by the disulfide bond between the two cysteines present in the peptide. This hypothesis was validated by incubation

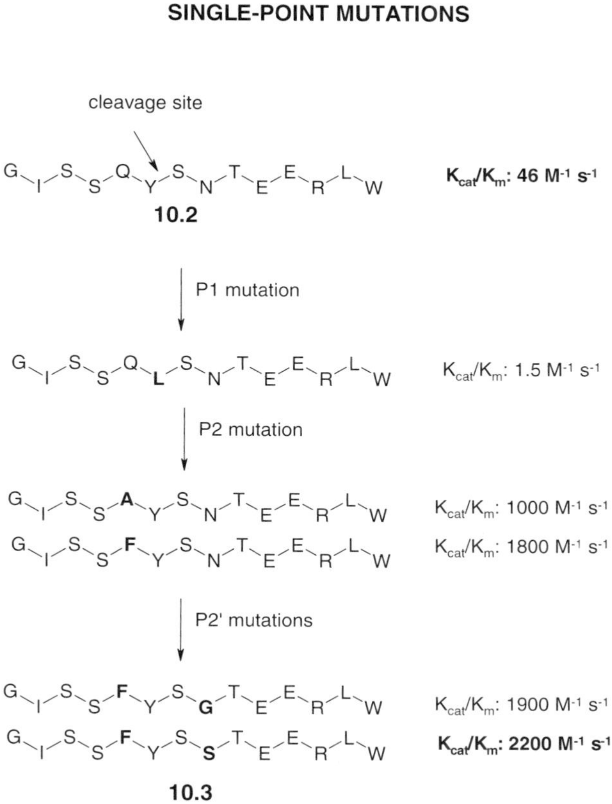

Figure 10.11 Single-point mutations to improve the specificity of the prostate specific antigen (PSA) substrate **10.2**.

of the three peptides with the receptors in the presence of the reducing agent dithiothreitol (DTT), known to inhibit disulfide bond formation in peptides. The binding affinity of compounds **10.5–10.7** was significantly decreased proportionally to the increase of DTT. Other examples of constrained, cyclic peptide libraries displayed on phage surfaces have been reported recently (75–79).

Small peptide phage libraries have been used to screen for binding to organic molecules to identify consensus motifs that are then compared to known protein sequences to eventually determine biologically relevant interactions. Rodi et al. (80) screened a commercially available 12-mer library **L6** fused to phage pIII coat protein for its binding with a biotinylated derivative of taxol supported onto streptavidin-coated petri dishes (Fig. 10.15). The isolated sequences after one, two, and three rounds of selection and amplification were compared for common structural motifs. Two 5-mer motifs were identified, and a search in protein databases found six binding candidates for taxol (Fig. 10.15). More sophisticated

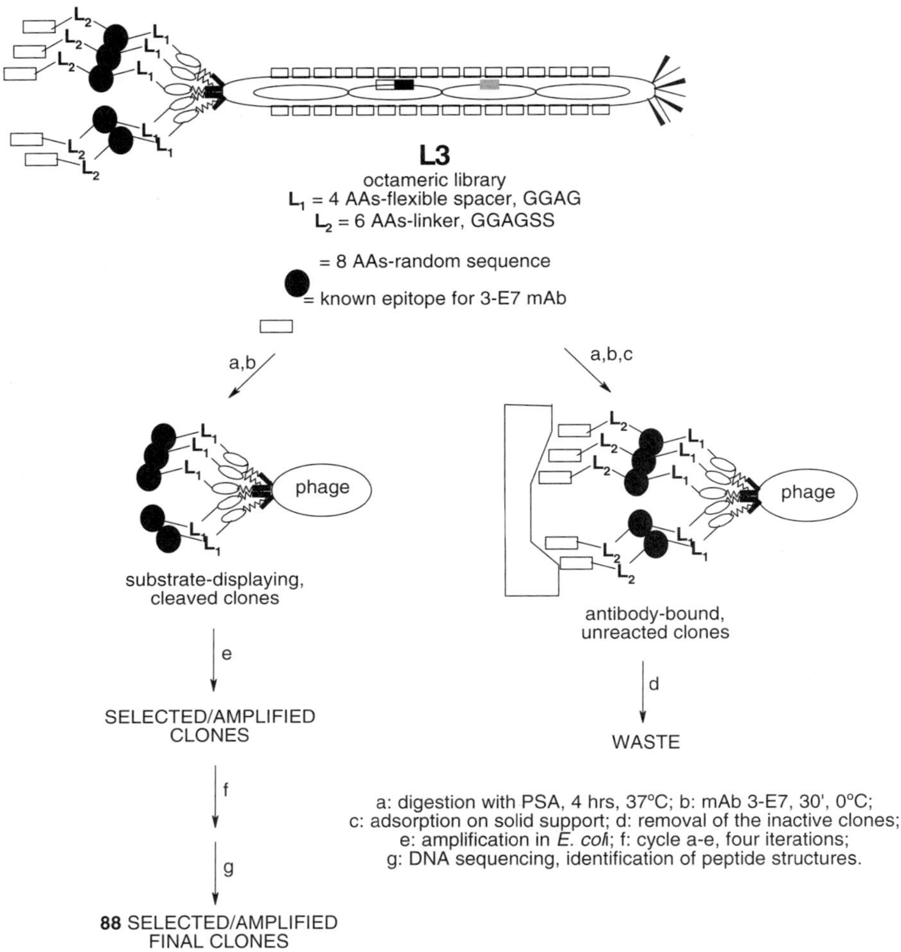

Figure 10.12 Screening of the phage library **L3** to improve the specificity of PSA substrates: the selection/amplification process.

similarity analyses of the selected peptides and of the whole sequence of the six proteins confirmed only the anti-apoptotic protein Bcl-2 as a binding partner for taxol. Further investigations elucidated the location of the peptide binding region in Bcl-2, the structural changes deriving from the binding with taxol and the nanomolar binding affinity of library individuals. The absence of homology between the selected peptides and tubulin, the primary binding target for taxol, was attributed to constraints imposed by the phage structure on the dodecapeptides, which could not assume any tubulinlike conformation for the binding.

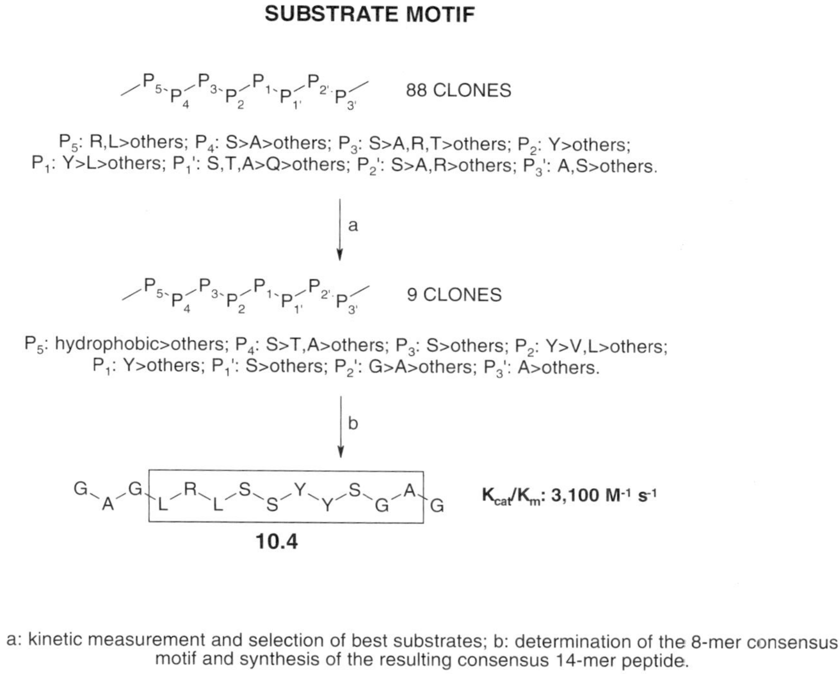

Figure 10.13 Screening of the phage library **L3** to improve the specificity of PSA substrates: characterization of selected clones and structure of the most active substrate **10.4**.

10.1.4 Phage Display Libraries: Large Peptide Sequences as Receptors, Antibodies, or Enzymes

The number of hexapeptides obtained by complete randomization of the 20 natural α-amino acids is 6.4×10^7, whereas the population of a large phage display library is made up of 10^8–10^9 clones and cannot exceed this number to allow the handling of the library during screening, selection, and amplification. For this reason the complete randomization of long peptide chains is not realizable, and alternative methods for the display and optimization of large peptides, antibodies, or protein structures on phages have been developed.

In analogy with constrained peptide libraries, several reports have described the use of small proteins, protein domains, or antibodies as scaffolds for the display of random polypeptide sequences to obtain novel binding proteins or antibodies. Koide et al. (81) used the tenth FN3 sequence, a 94-amino-acid fibronectin domain (82, 83) known to be involved in molecular recognition, as a scaffold to build a phagemid 3+3 library **L7** (Fig. 10.16) where less than a copy of modified FN3 was present on each phage capsid. The $\approx 10^8$-member library was screened using plates coated with ubiquitin, a small protein for which native FN3 does not have any affinity. The library was made by randomizing five amino acids in positions 26–30 (BC) and five amino acids in

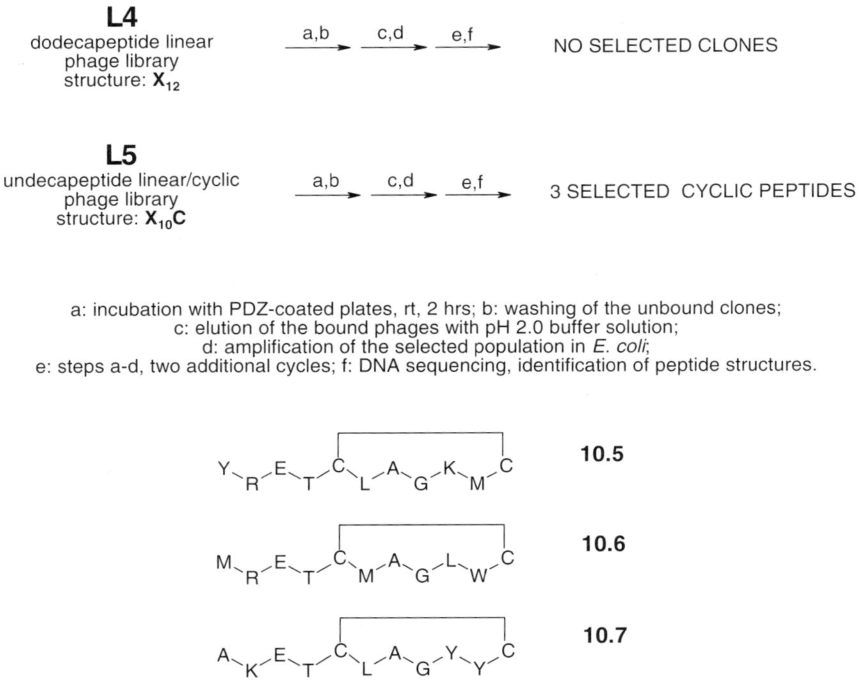

Figure 10.14 Screening of the cyclic phage libraries **L4** and **L5** for PDZ domains (PDZ) syntrophin binding: the selection/amplification process and the structures of the best binders **10.5–10.7**.

positions 77–81 and deleting the 82–84 Pro–Ala–Ser sequence of FN3 (FG, Fig. 10.16) sequences, which corresponded to two nonconserved loops of FN3 and the absence of which was not likely to reduce the protein stability. After five selection/amplification cycles 11 clones were selected (steps a–f, Fig. 10.15) among which the sequence **10.8** (Ubi4) was dominant. This novel motif (only the residue in position 30 of wild-type FN3 was conserved; Fig. 10.16) was confirmed as interacting with ubiquitin with an IC_{50} of 5 μM. An alanine scan, in which each randomized amino acid was replaced in turn with L-alanine, showed a general decrease of affinity between ubiquitin and the mutated protein domains. Other examples of scaffolded phage libraries from antibodies (84–90) or proteins (91–96) with small randomization sequences have been reported recently. This subject has also been reviewed extensively (97, 98).

Phage display libraries of more heavily randomized antibodies aimed at improving the binding affinity for an antigen using a library DNA encoding method schematized in Fig. 10.16 have been reported (99). Recombinant phage libraries of antibodies have been assembled from two distinct vector libraries, one containing a repertoire of V_H (heavy-chain gene fragments) and another containing a repertoire of V_L (light-chain

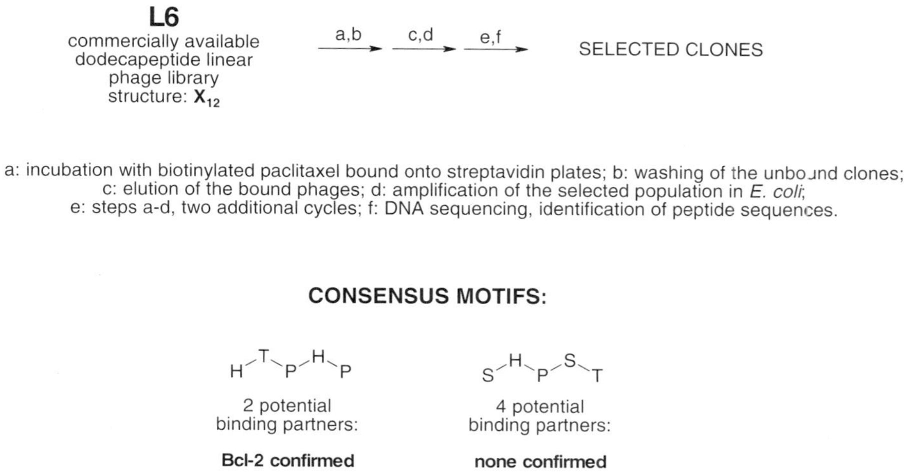

Figure 10.15 Screening of the phage library **L6** to find macromolecular binding partners for paclitaxel: identification of Bcl-2 as a binding partner.

gene fragments). Both vector libraries have been processed, then ligated to assemble a V_H–V_L genetic library, which was eventually translated and displayed on a phage surface. The careful construction of the vectors and their ligation (C_1–C_4, linker structure, Fig. 10.17) ensured that only full mAbs-displaying phage particles maintained their infectivity, thus allowing a degree of self-control through the discarding of nonfunctional products of the biological machinery. Phagemids containing foreign DNA sequences fused onto the pIII gene and expressing one copy of mAbs on each phage capsid have been commonly used as genetic vectors.

Steinberger et al. (100) assembled a light-chain gene library and ligated it onto the pIII coat gene, then processed the construct and ligated a library of V_H chains genes onto it (Fig. 10.18) to produce a focused library **L8** of $\approx 5 \times 10^7$ individual antibody-clones. The light- and heavy-chain libraries were assembled into **L8**, ensuring the presence of a cleavable site between the recombinant antibody and the phage coat protein (Fig. 10.18). The encoding nucleotide sequences were derived from and biased toward IgE (immunoglobulin E) obtained from a grass pollen allergic patient. **L8** was incubated onto coated plates with several purified recombinant grass pollen allergens for 2 h, binding phages were selected, and the amplification/selection cycle was repeated four times (steps a–e, Fig. 10.18). The binding clones were solubilized by cleavage of the junction to the pIII protein (step f) and 20 of them tested for their specificity for group 5 grass pollen allergens, and four specific sequences were identified and structurally characterized after DNA sequencing (steps g and h, Fig. 10.18). The V_H fragment for all the isolated clones was identical, showing a selection consensus during the iterative cycles, and, although the V_L fragments were extremely similar, several different patterns were observed, especially in CDR1 and CDR3

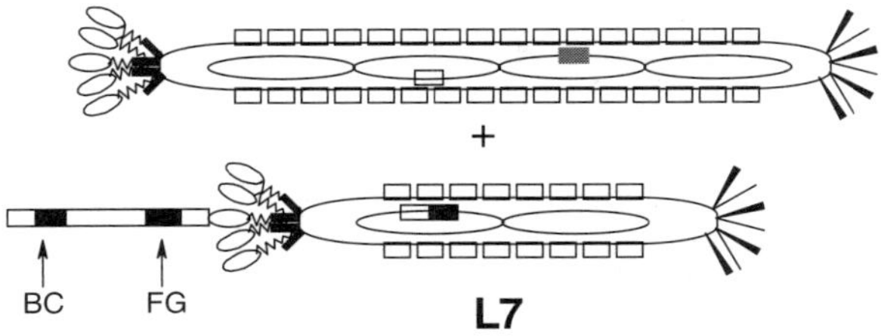

L7
FN3-based linear phage library
structure: $\mathbf{D_{25}X_5D_{46}X_5D_{10}}$
5 AAs-randomizations in loops **BC** (AAs 26-30) and **FG** (AAs 77-81)

$\xrightarrow{a,b}$ $\xrightarrow{c,d}$ $\xrightarrow{e,f}$ 11 SELECTED CLONES $\xrightarrow{g}$ **10.8**

a: incubation with ubiquitin-coated plates 2h, rt; b: washing of the unbound clones; c: elution of the bound phages with soluble ubiquitin; d: amplification of the selected population in *E. coli*; e: steps a-d, five additional cycles; f: DNA sequencing, identification of peptide structures; g: selection of the best clone.

SEQUENCE COMPARISON

	26-30	**77-81**
FN3 wild-type	A V T V R	G R G E S
10.8 (Ubi4 clone)	S R L R R	P P W R V

Figure 10.16 Screening of the constrained phage library **L7** for FN3 fibronectin binding: the selection/amplification process and the structure of the best binder **10.8**.

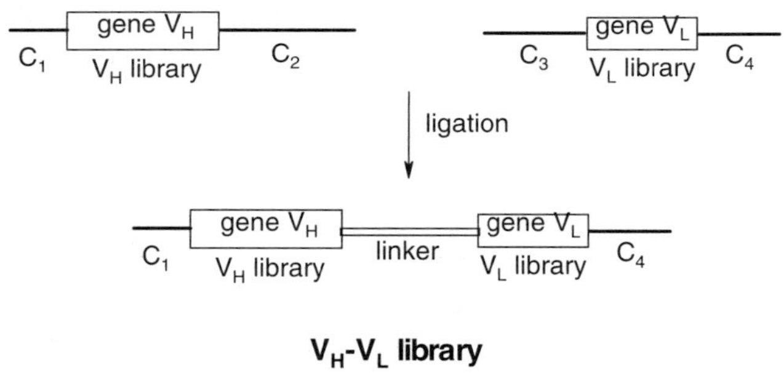

Figure 10.17 Phage display of antibodies: construction of the genetic information for the displayed V_H–V_L antibody library.

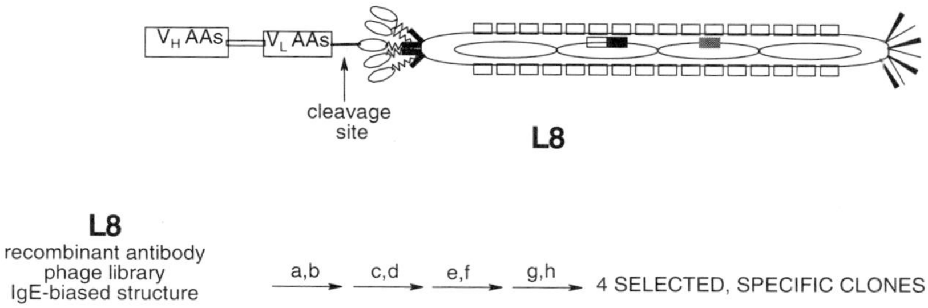

a: incubation with grass pollen allergens-coated plates, rt, 2 hrs; b: washing of the unbound clones; c: elution of the bound phages, pH 2.2; d: amplification of the selected population in *E. coli*; e: steps a-d, four additional cycles; f: cleavage of the library; g: check of mAb specificity; h: DNA sequencing, identification of peptide structures.

Figure 10.18 Screening of the IgE-biased phage library **L8** of antibodies: the selection/amplification process.

(complementarity-determining regions). The four clones were tested on a panel of clinically relevant group 5 allergens from different grass species, their group 5 specificity and their recognition of allergens from different grass species were observed, and their potential use as therapeutic tools in allergology was forecasted. Many other reports of combinatorial antibody libraries, originating either from the optimization of known antibody structures (101–110) or from random assembly of heavy- and light-chain fragments (111–120), have been reported together with several comprehensive reviews (121–125).

A specific class of antibodies selected to perform chemical transformations, the *catalytic antibodies* (126), has also been the target of phage display libraries. Fujii et al. (127) reported the affinity maturation of the esterolytic antibody Mab 17E11, generated by immunization with a transition-state analogue (TSA) **10.9** to allow the isolation of regioselective esterolytic activities at C_4 (128). Mab 17E11 regiochemically deacylated the 4-position of the sugar **10.10** to give compound **10.11** (Fig. 10.19). The same catalytic antibody showed a greatly reduced activity when the C_6-hindered analogue **10.12** (Fig. 10.19) was used as a substrate. An improvement of catalytic properties was sought via phage display of catalytic antibody libraries because a previous site-directed mutagenesis approach based on the antibody modeling in the presence of the substrate did not produce better catalytic antibodies for the hydrolysis of **10.12**. The library **L9** (10^7 clones) was generated by randomizing six amino acid positions in the heavy-chain CDR3 loop, known by molecular modeling to interact with the C_6 region of the TSA. The library was screened on plates coated with bovine serum albumin–coupled TSA **10.13** (Fig. 10.20). The selected population was iterated for four rounds; then 124 randomly picked clones were rescreened and checked for their specific binding to **10.13** (steps a–h, Fig. 10.20). The 24 resulting clones were sequenced, and six of them showed different amino acid sequences compared with Mab 17E11 (three to six-residues changed). Two of these clones were found to be better catalysts than mAb 17E11 for the regiochemical deprotection of **10.12**, the best

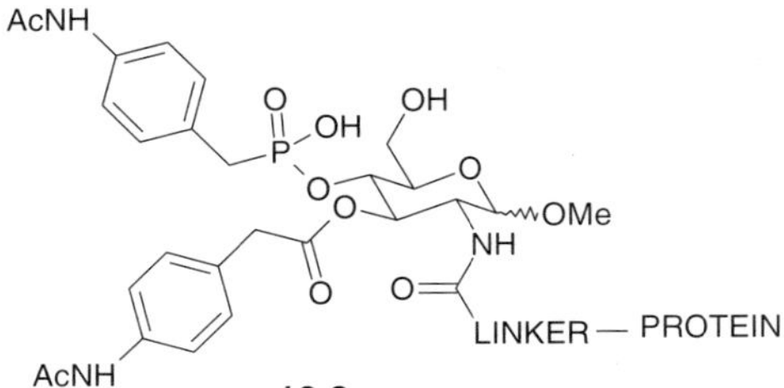

10.9

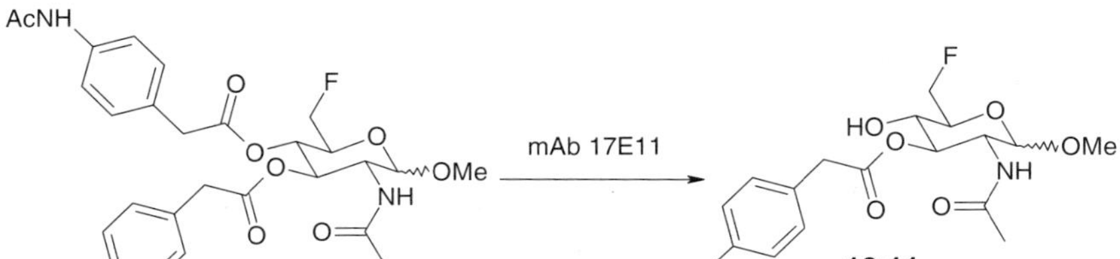

10.10 **10.11**

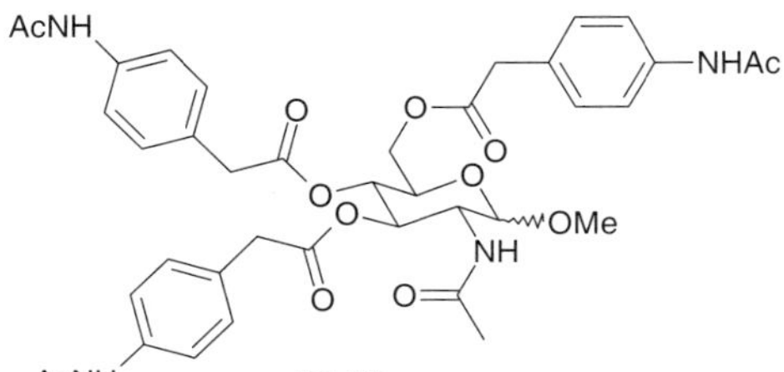

10.12
poor substrate
for mAb 17E11

Figure 10.19 Catalytic antibodies: structures of the transition state analogue/selection substrate (**10.9**), the catalyzed reaction substrate (**10.10**) and product (**10.11**), and a modified poor substrate for the esterolytic antibody Mab 17E11.

being roughly 12 times more efficient than the parent antibody. Four other reports describing libraries of catalytic antibodies have been published recently (86, 129–131), including an intriguing report on the synthesis of selection reagents for the discovery of catalytic metallo- antibodies.

Selection of a catalytic antibody via its binding to a TSA assumes that this binding will be predictive of the catalytic efficiency of the selected antibody. An intriguing application in which the selection of an antibody via a process based directly on catalytic efficiency was reported by Gao et al. (132). The acylated tripeptide amides **10.14a,b** and esters **10.15a,b** were used as substrates for hydrolysis to acids **10.16a,b** (Fig. 10.21), and a catalytic antibody was looked for through the generation a display

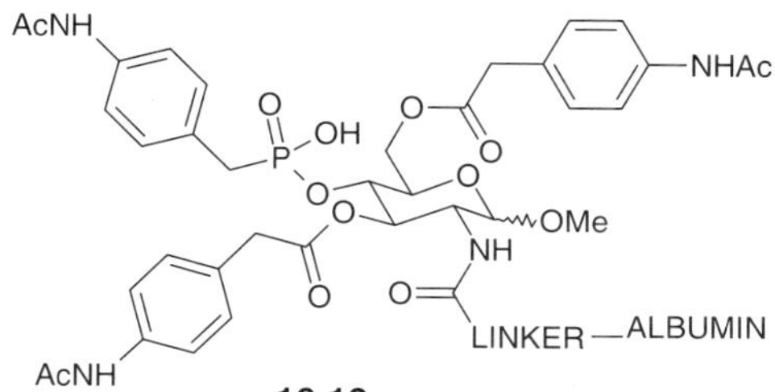

10.13

L9
recombinant antibody
V_H-V_L phage library
mAb 17E11-biased structure
6-mer random sequence
in the CDR3 domain

$\xrightarrow{a,b}$ $\xrightarrow{c,d}$ $\xrightarrow{e,f}$ $\xrightarrow{g,h}$ 6 SELECTED CLONES

a: incubation with conjugated albumin-**10.13**-coated plates, rt, 2 hrs; b: washing of the unbound clones; c: elution of the bound phages, pH 2.2; d: amplification of the selected population in *E. coli*; e: steps a-d, three additional cycles; f: affinity measurement with free **10.13;** g: DNA sequencing, identification of peptide structures; h: selection of the most active mutated clones.

Figure 10.20 Screening of the mAb 17E11–biased antibody phage library **L9** to find optimized catalytic antibodies for the transition state analogue/selection substrate **10.13**: the selection/amplification process.

library **L10** (2×10^8 members) starting with cDNA sequences from mouse spleen cells immunized with the boronic acid substrate analogue **10.17** (Fig. 10.21). Four rounds of selection/amplification produced several positive clones (steps a–f, Fig. 10.21) from which one was isolated, sequenced, and purified having full catalytic specificity for the amide **10.14a** (4×10^4 enhancement of the reaction rate), with no effect on the hydrolysis of the stereoisomer **10.14b** or the ester analogue **10.15a**. The same authors reported two other examples of joint catalysis and infectivity in the selection of catalytic antibodies by phage display (133, 134); other researchers used similar approaches where the selected phage clones and the catalytic reagents and products were connected (135, 136).

Phage libraries of proteins have also been produced by insertion of DNA sequences from various species into the phage genome with expression of the foreign proteins as single copies per phage clone fused onto the pIII coat protein. The proteins displayed on phage capsids have been used to study potentially new protein–protein interactions to clarify relevant biological processes or to identify unknown homologue proteins in different organisms. Yamabhai et al. (137) screened a phage-expressed library of frog cDNA **L11** for its binding to a biotinylated peptide **10.18** (Fig. 10.22) known to bind to Src homology 3 (SH3) domains of proteins. Identification of positive clones should have led to novel frog proteins involved in signal transduction (see the original paper for more details). This approach, named cloning of ligand targets (COLT) had been

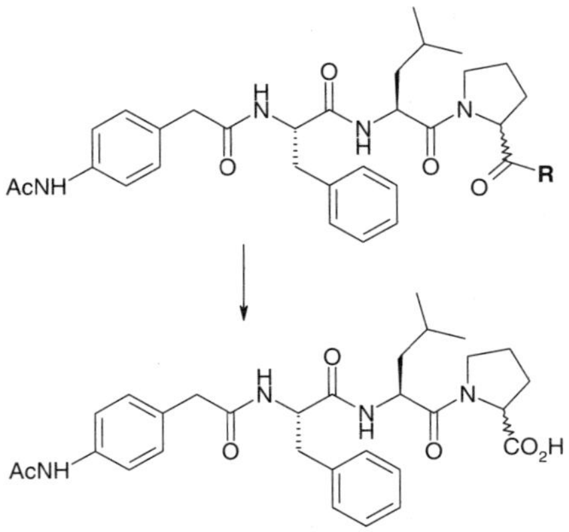

10.14a (S), **b** (R) R = NH_2

10.15a (S), **b** (R) R = OM_e

10.16a (S), **b** (R)

L10

recombinant antibody
V_H-V_L phage library
mouse spleen-derived structure

a,b → c,d → e,f → 1 SELECTED CLONE

a: incubation with conjugated albumin-**10.17**-coated plates, 37°C, 2 hrs; b: washing of the unbound clones; c: elution of the bound phages, pH 2.2; d: amplification of the selected population in *E. coli*; e: steps a-d, three additional cycles; f: DNA sequencing, identification of peptide structures.

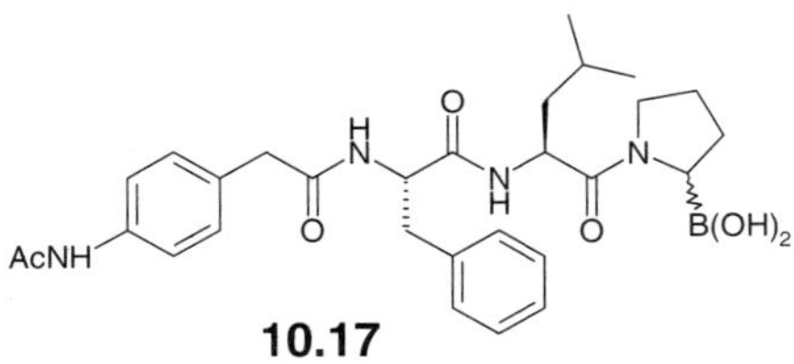

10.17

Figure 10.21 Direct selection of catalytic antibodies from phage libraries: the selection/amplification process to an improved stereoselective hydrolytic antibody from the library **L10** using the transition state analogue/selection substrate **10.17**.

previously used to identify other SH3-binding proteins (138). Screening of **L11** (steps a–d, Fig. 10.22) produced a single clone that eventually led to the sequence of a novel protein named intersectin by the authors. Structural analysis highlighted the presence of five SH3 domains as well as two Eps15 homology (EH) domains in the structure of intersectin. These latter domains are known to be involved in protein–protein interaction.

The binding specificity of these EH modules was checked with two phage libraries **L12** (linear 9-mers on pVIII coat protein) and **L13** (11-mers with a terminal cysteine on pIII coat protein) that were incubated with plates coated with constructs containing the intersectin EH domains EH_a and EH_b (Fig. 10.23). After selection and amplification of **L12** and **L13**, many positives containing the known NPF (asparagine–proline–

a: incubation with **10.18**; b: washing of the unbound library individuals; c: hybridization; d: DNA sequencing, identification of the peptide structures.

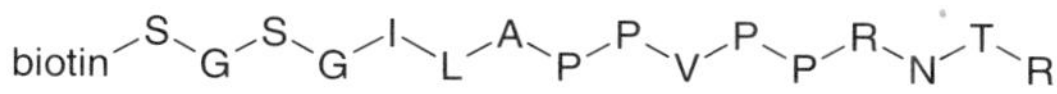

10.18

Figure 10.22 Screening of the frog cDNA phage-expressed library **L11** to find macromolecular binding partners for SH3 protein domains: the selection/amplification process using the biotinylated selection substrate **10.18**.

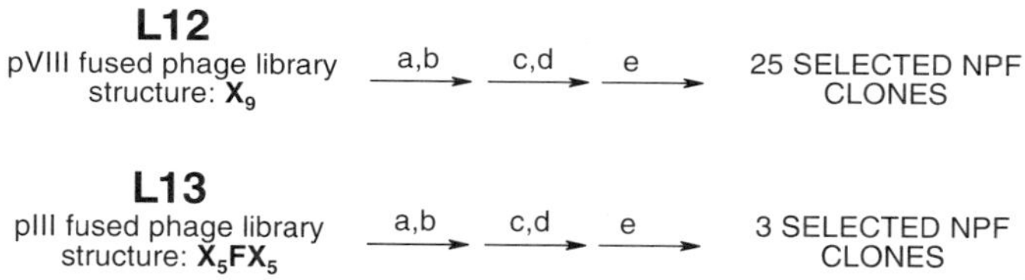

a: incubation with EH_a and EH_b intersectin domains-coated plates; b: washing of the unbound clones; c: elution of the bound phages; d: amplification of the selected population in *E. coli*; e: DNA sequencing, identification of the peptide structures.

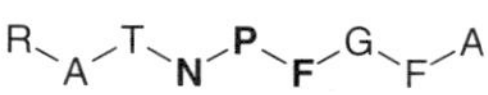

10.19

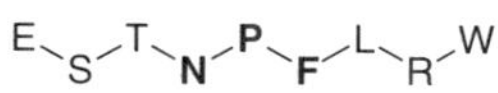

10.20

CLONES FROM L12
BINDING WITH INTERSECTIN: **NOT CONFIRMED**

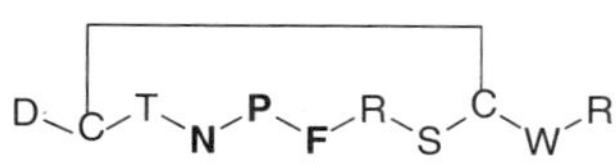

10.21

CLONE FROM L13
BINDING WITH INTERSECTIN: **CONFIRMED**

Figure 10.23 Screening of the phage display libraries **L12** and **L13** to find macromolecular binding partners for intersection: the selection/amplification process and the structure of selected binding sequences **10.19–10.21** for the EH domains of intersectin.

phenylalanine) binding motif for EH domains were obtained. Further refinements of binding specificities and affinities showed the constrained peptide **10.21** from **L13** to be a strong binder, while two linear examples from **L8** (**10.19** and **10.20**) failed to show significant binding to intersectin (Fig. 10.23). These findings confirmed both the importance of constrained sequences (e.g., **10.21** completely lost its affinity for EH upon substitution of the cysteines with serines) and the amplification of weak affinities by pVIII display (2700 copies of pVIII-displayed, weakly actives **10.19** and **10.20** versus five copies of **10.21** by pIII display). The bound intersectin EH domains were also screened, albeit unsuccessfully, using the phage-expressed frog cDNA library **L11** in the search for intersectin protein binders. However, a different cDNA library **L14** from mouse embryo produced four positive clones. These were in turn connected to three protein sequences out of which two novel proteins, named intersectin-binding proteins (Ibp), were characterized. The characterization of SH3 and EH domains of intersectin was then completed using other phage libraries. Homology searches found similar proteins from other organisms and other candidate proteins potentially able to interact with the SH3 domains of frog intersectin. Two recent reviews (139, 140) covered extensively the use and applications of phage displayed cDNA libraries.

10.2 BIOSYNTHETIC OLIGONUCLEOTIDE LIBRARIES

10.2.1 General Considerations

Biosynthetic polypeptide libraries are the translation of a genetic, oligonucleotide-based diversity information into a biologically relevant pool of potential binders, catalysts, or receptors. The previous section illustrated how the availability of complex DNA sequences and of the appropriate techniques for their synthesis and manipulation allows the biological production of diverse polypeptide libraries. However, the intrinsic properties of nucleotide sequences have themselves stimulated researchers into building a variety of oligonucleotide (ON) libraries per se for a variety of applications.

ON libraries have been employed in two main fields. First, the ability of nucleotides to bind strongly to a variety of partners either as small oligomers or as larger DNA or RNA sequences is well known. Thus, various examples of ON libraries targeted toward the discovery of DNA, or more often RNA, ligands (*aptamers*) (141, 142) for other nucleotide sequences, small molecules, drugs, or proteins are described below. Second, the discovery of naturally occurring catalytic RNA sequences (*ribozymes*) (143, 144) has led both to the optimization of natural ribozyme sequences and to the quest for novel artificial ribozymes and deoxyribozymes via combinatorial ON libraries capable of catalyzing reactions such as oligonucleotide cleavage and ligation, peptide bond formation, ester hydrolysis, and C–C bond formation.

The first part of this section is devoted to the description of the in vitro selection process, which enables the iterative selection/amplification cycles to reliably and accurately furnish ON sequences.

10.2.2 In Vitro Selection of ON Sequences

The state of the art in the SPS of oligonucleotides (see Section 2.2) has reached excellent levels of reliability and performance, which can be summarized as follows:

- Sequences of >100 oligomers can be routinely synthesized with extremely high purity and yield.
- Libraries of up to 10^{16} individuals can be made and handled using various techniques.
- Four commercially available building blocks with similar reactivity and with orthogonal protection allow the construction of any target ON sequence.
- The reduced number of building blocks allows the complete randomization of long ONs ($4^{26} \approx 10^{16}$ individuals).
- Efficient, fully automated SP ON synthesizers are commercially available.

Comparing the SP synthesis of ONs versus peptides, the latter does not exceed the complete randomization of octamers ($20^8 \approx 2.6 \times 10^{10}$ individuals) and cannot attain the same yield and purity in each coupling step, thus preventing even the synthesis of single >100-mer target sequences. Therefore, the synthesis of large, randomized peptides is better carried out using biological tools, whereas the SPS of ONs produces large, high-quality libraries that are not biased by any of the potential problems of using the biosynthetic machinery of a cell, such as viability, infectivity, and depletion of amino acid pools. Only when the randomized sequence is extremely long, as is the case for ribozymes, are the DNA strands ligated using biochemical tools or prepared by genetic mutations of preexisting sequences (*vide infra*).

The main steps involved in the synthesis, selection, and amplification of ON libraries are reported in Fig. 10.24. Automated synthesis of libraries of DNA fragments of a given length usually uses the phosphoramidite SP strategy (Section 2.2.1) with equimolar quantities of each nucleotide at each elongation step to give equally represented ON libraries (step a). The random ssDNA fragments, which usually vary from 30 to 300 nucleotides, are flanked on both sides by constant regions that allow their amplification (step c) and transcription into the corresponding RNA sequences using RNA polymerase (step d) after conversion to double-strand DNA (dsDNA, step b). The RNA sequences are submitted to a target-assisted screening, which selects the active library components through their immobilization with the support-bound target (step e) and discards the unbound library individuals by washing (step f), as already seen for biosynthetic peptide libraries. The main principles encountered in the selection of phage libraries such as stringency and yields are also important here, but the increased quality of the library due to the use of equimolar quantities of library individuals and the lack of constraints related to the anchorage onto the phage capsid make the process extremely reliable and also allow further modification of the stringency conditions by fine tuning the selection parameters in vitro. After recovery of the selected RNAs by disruption of the target–RNA complex (step g), they are transcribed into ssDNA using reverse transcriptase (step h) and then converted to dsDNA (step b). This material is then amplified to allow iterative selection cycles to

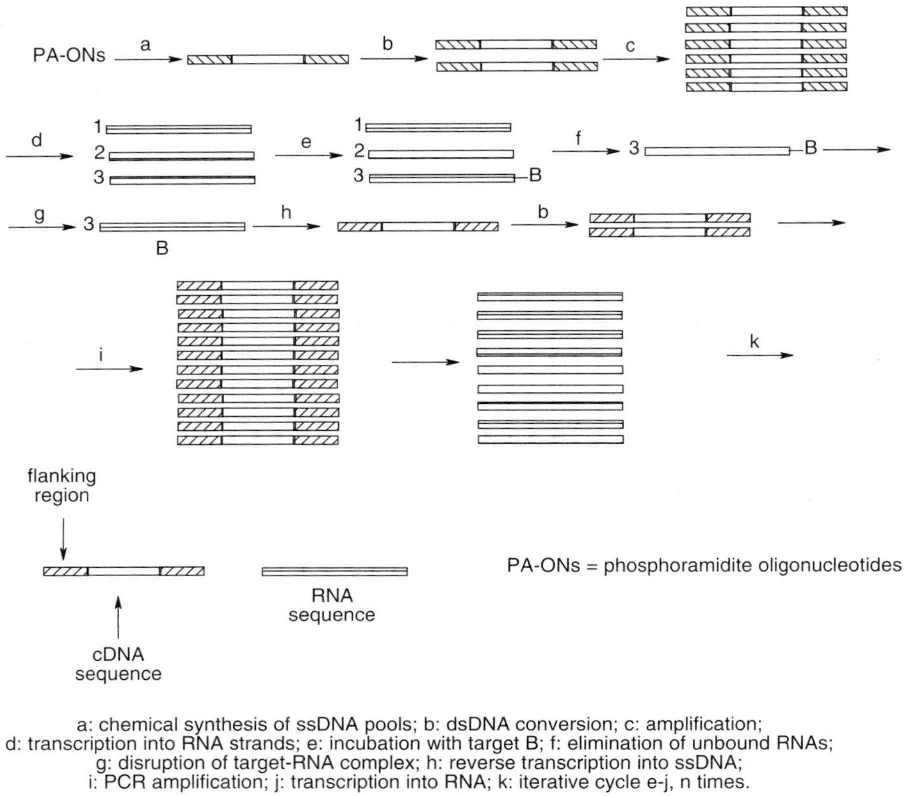

Figure 10.24 Synthesis, selection, and amplification of biosynthetic ON libraries: the whole process.

arrive at the best library individuals. At first, amplification was carried out using in vivo biological tools (145), as already described for the phage display techniques, but later more efficient in vitro systems were developed and among them the polymerase chain reaction (PCR, step i) (146) is by far the most used because of its fidelity and efficiency. The flanking regions of the DNA strand are kept constant and are used to drive the amplification of the random portion of the DNA. The amplified dsDNA sequences are then transformed into the corresponding RNA strands (step j), and a second iterative cycle is carried out (step k, Fig. 10.24). Typically, up to 15 cycles with increasing stringency are used to select up to 1000 best sequences.

The length of the random ON sequence of a library must be carefully considered in order to extract meaningful results from the screening as longer sequences may be the optimal binding partners for a target but their relative abundance in an ON library may be too low. In a hypothetical example of a 10^{12}-member library of 20-mers, only one library individual contains a given 20-mer motif ($4^{20} = 10^{14}$), but around

10,000,000 library members contain a specific 10-mer insert ($4^{10} = 10^7$, $10^{14}/10^7 = 10^7$). Even if the shorter sequence is a weaker binder than the former, the relative abundance of the shorter motif will drive the selection and amplification cycles toward suboptimal, more abundant, shorter ONs. To overcome this limitation, extremely long randomized sequences are considered for ON libraries to allow the selection of long binders (with a high price to pay in terms of synthetic complexity); usually, the cutoff level is raised during the final iterative cycles in order to isolate a few long molecules with extremely high affinity for the target and to discard the weak, shorter binders.

The in vitro selection/amplification strategy has also been applied to modified ONs, especially where the 2′-ribose position has been changed and where phosphorothioates or other phosphate replacements have been used (Fig. 10.25, top). Several structures of modified ON chains that have been synthetically produced to obtain constrained sequences or sequences with higher stability to nucleases have also been reported (Fig. 10.25, bottom). Examples of biosynthetic modified ON libraries are covered in the next section.

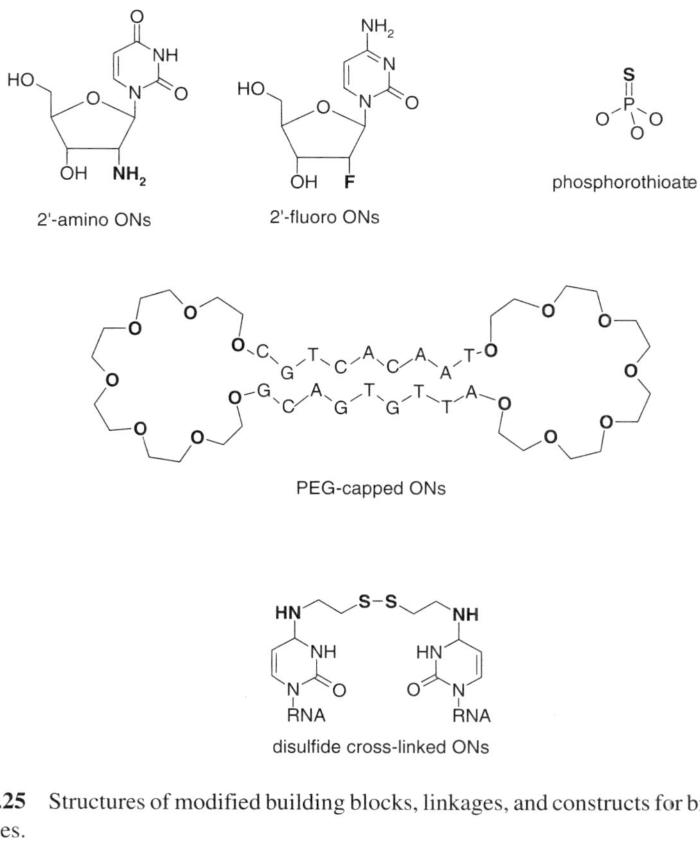

Figure 10.25 Structures of modified building blocks, linkages, and constructs for biosynthetic ON libraries.

10.2.3 ON Libraries: Binding to Small Molecules, Drugs, and Proteins

In vitro selection of ON sequences from random libraries as binding motifs for a series of molecular targets ranging from small molecules up to macromolecules is a well-studied area that has been extensively covered in several recent reviews (147–155).

Small molecules have often been the targets for aptamer selection. Haller and Sarnow (156) have reported the screening of a 10^{13}-member, ON 90-mer library **L15** that incorporated randomization of 40 inner residues to select for specific binders of 7-methyl guanosine **10.22** (Fig. 10.26) and of **10.22**-containing oligonucleotides (capped oligonucleotides). An iterative cycle of the screening strategy is also depicted in Fig. 10.26. Library **L15** was incubated with sepharose-bound m^7-GTP (7-methyl guanosine tri phosphate) (step a) for 1 h at 4°C; then the unbound library fraction was washed with fresh assay buffer (step b), and the bound library members were eluted with an excess of free m^7-GTP (step c). The selected RNAs were reverse transcribed to the corresponding cDNA, and the ODN sequences were amplified by PCR (steps d and e). Transcription using RNA polymerase (step f) produced an enriched ON library that was submitted to a second iterative cycle that was identical to the first, except for a preliminary elution of the bound library fraction with GTP to discard RNA binders acting on features of m^7-GTP different from the methylated site (step g, Fig. 10.26). The amount of bound RNA was low after the first cycle (less than 1% of the library population) but steadily increased in the following cycles. After the eighth cycle, the PCR-amplified cDNA pool was cloned and several of these clones were characterized. Among them, the aptamer **10.23** showed significant specificity for **10.22** containing

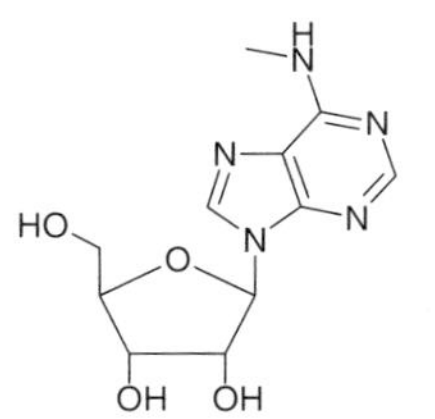

10.22

L15
40-mer randomized sequence
10^{13} members $\xrightarrow{\text{a-f}}$ SELECTED/AMPLIFIED FIRST GENERATION CLONES $\xrightarrow{\text{g}}$ FINAL CLONES $\xrightarrow{\text{h}}$

a: incubation with sepharose-supported **10.22**; b: washing of the unbound RNAs; c: elution of the bound RNAs with an excess of free **10.22**; d: reverse transcription into DNA; e: PCR amplification; f: transcription into RNA; g: a-f, seven cycles with preliminary elution with free GTP after step b; h: cloning and sequencing of selected clones.

BEST BINDER: **10.23** IC_{50} = **500 nM**

Figure 10.26 Screening of the biosynthetic ON library **L15** for aptamers binding to m^7-GTP (**10.22**): the selection/amplification process and the structure of the most active aptamer **10.23**.

nucleotides versus their desmethyl analogues (Table 10.1) and proved to be a useful tool in inhibiting capping-dependent translation in eukaryotic cells in vivo (157). Many other reports of in vitro selection of aptamers binding to small molecules have been reported, including coenzyme A (158), vitamin B_{12} (159), flavin adenine dinucleotide (FAD) (160), theophylline (161), and adenosine triphosphate (ADP) (162).

Burke et al. (163) reported two libraries of 10^{14}–10^{15}-member ON pools **L16** and **L17** containing 118-mers and 134-mers with inner randomization of 70 and 80 nucleotides, respectively. Both of these libraries were screened using agarose-bound chloramphenicol **10.25** (Cam, Fig. 10.27) to find families of aptamers able to shed light on the mode of binding of Cam (**10.24**, Fig. 10.27) to the bacterial ribosome. Library synthesis, transcription and reverse transcription from and to DNA sequences, selection, and amplification of selected sequences were performed using standard procedures (Fig. 10.26). Elution of resin-bound RNAs was performed with **10.24** with increased stringency for 12 cycles. The selected library population was low for the first seven cycles (<0.2%) but increased to 2–6% in the eighth round, eventually reaching 50% in the final cycle. Seventy-four aptamers were isolated after cloning of the individual sequences with various frequencies from both **L16** and **L17** (Fig. 10.27). It was found that, while several individuals shared common sequence elements, there were no sequences common to all of the selected aptamers. The most active aptamers showed a low micromolar dissociation constant, representing a 1000-fold increase of the original affinity of the random **L16** and **L17** sequences. The structural information acquired was then used to build a shorter 50-mer **10.26**, retaining full Cam affinity, which was used to elucidate the characteristics of the ribosome–aptamer complex

TABLE 10.1 Specificity of Aptamer 10.23 Toward m^7-GTP-Containing Nucleosides and Nucleotides

Free Competitor	Binding potency, IC_{50} (μM)
m^7-GTP, **10.22**	0.5
GTP	1200
m7GpppG	1
GpppG	>2800
m7GpppA	2.2
m7GDP	1.5
dGDP	2900
GMP	1800
Guanosine	90
ITP	4700
UTP	6500
CTP	6000
ATP	4800

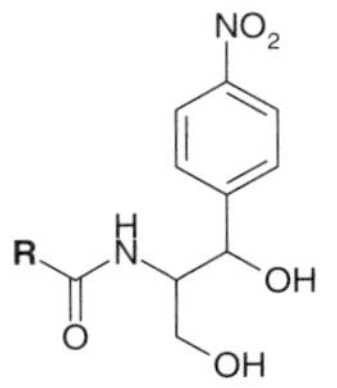

10.24 R = $CHCl_2$
10.25 R = $(CH_2)_n$-RESIN

L16
134-mers, 80-mer randomized sequence; 10^{14}-10^{15} members —a-f→ SELECTED/AMPLIFIED FIRST GENERATION CLONES —g,h→ 45 SELECTED CLONES

L17
118-mers, 70-mer randomized sequence; 10^{14}-10^{15} members —a-f→ SELECTED/AMPLIFIED FIRST GENERATION CLONES —g,h→ 29 SELECTED CLONES

SELECTED CLONES —i→ **10.26** 50-mer

a: incubation with agarose-supported **10.25**; b: washing of the unbound RNAs; c: elution of the bound RNAs with an excess of free **10.24**; d: reverse transcription into DNA; e: PCR amplification; f: transcription into RNA; g: a-f, 11 cycles; h: cloning and sequencing of selected clones; i: truncation of ON sequences.

Figure 10.27 Screening of the biosynthetic ON libraries **L16** and **L17** for aptamers binding to chloramphenicol **10.24**: the selection/amplification process using the supported selection substrate **10.25**.

(163). Other reports regarding the selection of antibiotic-binding aptamers, including aminoglycosides, have also recently appeared (164–167).

The preparation of aptamers binding to RNA, DNA, or proteins has been the subject of many reports in the literature. Lebruska and Maher (168) has described the synthesis and screening of a library of 10^{14} ON 100-mers, **L18**, in which each member was made of 60 randomized bases flanked by two constant sequences. **L18** was used to select aptamers for the proteic transcription factor NF-κB that binds to duplex DNA strands. The search for RNA strands with comparable affinity for the transcription factor was designed to produce potential tools, or even antisense therapeutics for antiviral and anticancer therapy. The library was incubated with the protein homodimer ($p50_2$, Fig. 10.28) for 2 h at 37°C (step a), then filtered through nitrocellulose membranes and washed (steps b and c) to remove the unbound members of the library (>99%). Elution with a mild denaturating buffer recovered the bound library fraction (step d), which was then reverse transcribed and amplified using standard procedures (steps e–g). The new RNA pool was submitted to several iterative cycles (step h), and the population obtained after 14 cycles was cloned to give two related sequences (step i), **10.27** (21 recurrences) and **10.28** (3 recurrences, Fig. 10.28). The two sequences displayed a low nanomolar affinity for the receptor, similar to the natural ligand. The sequences of

L18
100-mers, 60-mer randomized sequence; 10^{14} members —a-g→ SELECTED/AMPLIFIED FIRST GENERATION CLONES

—h→ FINAL CLONES —i→ **10.27-10.28** —j→ **10.29** 26-mer
10.30 31-mer

a: incubation with $p50_2$; b: filtration on nitrocellulose; c: washing of the unbound RNAs; d elution of the bound RNAs; e: reverse transcription into DNA; f: PCR amplification; g: transcription into RNA; h: a-g, 13 cycles; i: cloning and sequencing of selected clones; j: truncation of ON sequences.

Figure 10.28 Screening of the biosynthetic ON library **L18** for aptamers binding to the transcription factor NF-κB: the selection/amplification process and the structures of the most active aptamers **10.29** and **10.30**.

these two aptamers were used to identify the essential regions for binding to NF-κB, and two reduced ON sequences **10.29** (26-mer) and **10.30** (31-mer) were prepared (Fig. 10.28). An excess of **10.27** and **10.29** displaced the duplex DNA strand from NF-κB thus demonstrating the potential usefulness of these aptamers in modifying or influencing the process of transcription in a disease state. Other examples of aptamer selection for proteins (169–173), for receptors (174), or for nucleic acids (175, 176) have appeared in the literature recently.

Several reports have described the use of an in vivo selection method for the detection of biologically relevant interactions with macromolecules. In this work, the difficulties encountered in the in vivo selection/amplification cycle were compensated by the effectiveness of the selected sequences on the target in vivo. Ferber and Maher (177) screened ON libraries in *E. coli* cells and found positive effectors of amplification of an expressed plasmid. The selection of high-affinity RNA aptamers that could be used as markers or even as drugs targeted against the living cells of *Trypanosoma* protozoan parasites have been described by Homann and Goringer (178). Bruno and Kiel has reported the selection of DNA aptamers using anthrax spores (179) as biosensors in biological warfare detection. Davis et al. (180), Smith and co-workers (181, 182) have used aptamer sequences as staining agents and markers to track disease-related targets in vivo.

Therapeutic applications of selected RNA aptamers are usually prevented by the considerable drawbacks of such molecules, including their poor stability in biological media. Modified biosynthetic ON libraries, which are still accepted by the biological machinery of the cell during transcription, translation, and amplification but are considered to be more "druglike," have been the subject of several reports. For example, Bridonneau et al. (183) have reported the synthesis and selection in vitro of a 3×10^{14}, ≈100-mer modified ON library **L19** for the detection of high-affinity aptamers for human nonpancreatic secretory phospholipase A_2 (hnps-PLA_2). The chemical synthesis of **L19** employed the purine nucleotides A and G, and the 2′-NH_2 pyrimidine nucleobases **10.31** and **10.32** (Fig. 10.29), a modification that is known to

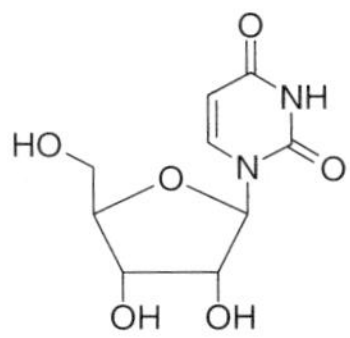

10.31

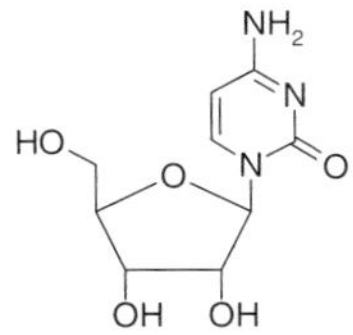

10.32

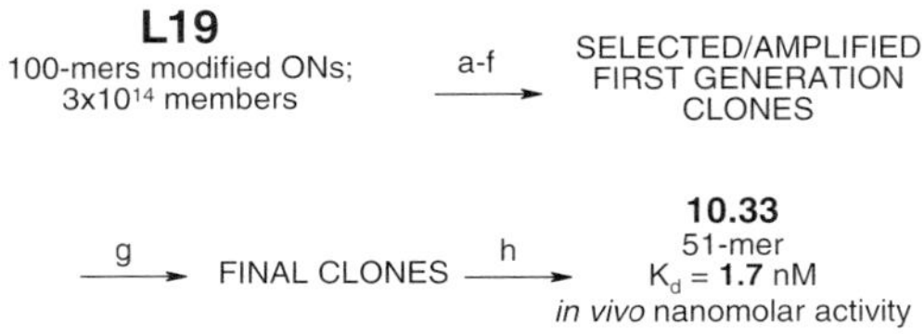

a: incubation with hnps-PLA_2 immobilized onto agarose with pAbs; b: washing of the unbound RNAs; c: elution of the bound RNAs; d: reverse transcription into DNA; e: PCR amplification; f: transcription into RNA; g: a-f, 13 cycles, cloning and sequencing of selected clones; h: truncation of ON sequence.

Figure 10.29 Screening of the biosynthetic ON library **L19** for aptamers binding to hnps-PLA_2: the selection/amplification process and the structure of the most active aptamer **10.33**.

increase the stability of the ONs (184) without affecting the iterative cycles of selection/amplification. The library was selected and amplified through 11 cycles following standard protocols (steps a–g, Fig. 10.29). The affinity binding was performed with hnps-PLA_2 polyclonal antibodies supported on agarose beads (cycles 1–5, 8, and 9, step a, Fig. 10.29) or with nitrocellulose filtering using free hnps-PLA_2 (cycles 6, 7, 10, and 11, step b, Fig. 10.29). With both procedures, incubation of the RNA pools with bound or free enzymes lasted 5 min at 37°C, and the final clones isolated and individually amplified to generate several low-affinity binders (which were discarded) together with a family of high-affinity binders that contained a number of aptamers with picomolar affinity (183). Refinement of the aptamer structure and detailed mechanistic studies resulted in a 51-mer sequence **10.33** (step h, Fig. 10.29) with specific and significant in vitro and in vivo affinity for hnps-PLA_2 and with similar potency to some recently described small organic molecule inhibitors (185). Further examples of modified amplifiable ON libraries have also been presented recently (186, 187).

An intriguing expansion of this concept is represented by the so-called selection–reflection strategy. If a chiral molecule is the binding target, its enantiomer is prepared and in vitro selection is used to select a high-affinity natural D-RNA or D-DNA aptamer (188). The mirror image of this aptamer/ligand, that is, an unnatural L-RNA, L-DNA, or a D-peptide, is then prepared that has by definition the same binding/inhibitory activity on the mirror image of the chiral selector (i.e., the target). However, the L-ON or D-peptide nature of the selected oligomer makes it resistant to degradation by nucleases or proteases. This concept has been exploited by several groups to select and characterize L-RNA aptamers (189, 190), L-DNA aptamers (191), and D-peptide ligands (188).

10.2.4 ON Libraries: the Ribozyme World

The fascinating concept of a primordial RNA world (192) in which all the functions encoded into and performed by polypeptides were originally carried out by oligonucleotides cannot be proven, but the discovery of naturally occurring RNA sequences possessing catalytic activity, the ribozymes, is another indication of the multiple functions that may be embedded into an ON sequence. The characterization and optimization of naturally occurring ribozymes as well as the creation of synthetic ribozymes and DNAzymes via in vitro selection from biosynthetic ON libraries have been the subject of many reports. Many catalytic RNA activities have been found in addition to the naturally occurring functions of RNA and DNA cleavage and splicing. This field has been widely covered in several recent reviews (149, 151, 152, 153, 193–199). Some examples will be briefly illustrated in this section with particular attention to synthetic ribozymes with specificity for organic reactions.

Joyce and co-workers (200–202) reported the evolution of a biased library **L20** inspired by the structure of the *Tetrahymena* ribozyme, which in nature cleaves or ligates specific RNA substrates (203). A library of 10^{13} 413-mers, **L20**, was prepared by random mutagenesis of the catalytic core of the natural ribozyme. The library population was obtained by introducing 5% of randomized nucleotides in each of 35 positions of the ribozyme, obtaining a starting library population bearing from one to more than seven point mutations (Table 10.2). Diversification in the following iterative cycles was obtained via the 0.1% mutation rate embedded in the PCR amplification process. The library was screened (200) in iterative cycles for its DNA-cleaving activity, which was negligible at 37°C for the natural ribozyme. The structural bias guaranteed for sequence-specific cleavage or ligation properties of selected ribozymes, that is, the lack of the 2′-OH in ODN substrates, required new substrate–ribozyme

TABLE 10.2 Composition of Library L20: Mutations and Frequencies

Mutation Number[a]	Recurrence[b]	Sequences in **L20**[c]	Copies/Sequence
0	0.1	1	9×10^9
1	0.6	420	2×10^8
2	2.1	90,000	3×10^6
3	5.0	1×10^7	50,000
4	9.0	1×10^9	900
5	12.8	1×10^{11}	15
6	15.2	7×10^{12}	0.3
7+	55.4		

[a]0 = wild-type, up to seven modifications.

[b]Percentage of population representing the mutation number.

[c]Number of sequences containing the mutation number.

interactions to compensate for the 2′-OH-based RNA interactions and possibly to switch the cleaving preference toward DNA strands.

The selection mechanism is shown schematically in Fig. 10.30. The members of the library that catalyzed the cleavage of the DNA substrate **10.34** ligated its 3′-portion onto the mutated ribozyme molecule (steps a and b). This 3′-portion was carefully selected to be recognized by a primer (step c) that initiated cDNA synthesis in the presence of reverse transcriptase leading to the amplification of the DNA-cleaving ribozymes through synthesis of dsDNA (step d) and transcription/amplification of the corresponding ribozymes (step e). The inactive library individuals were not recognized by the primer, thus preventing their amplification (step f, Fig. 10.30). Incubation of the library with **10.34** (10 μ*M*) was done at 37°C for 1 h, and nine selection/amplification cycles were performed. Selected representatives of round 9 such as **10.35** showed a 100-fold increased DNA cleavage activity, which became appreciable at 37 °C; however, their efficiency as RNA-cleaving enzymes remained largely prevalent (Fig. 10.31). In a further report Tsang and Joyce (201) applied more stringent selection parameters for 18 additional selection/amplification cycles on the final population from the first paper (round 9, **L21**, Fig. 10.32); the concentration of substrate **10.34** was reduced to 0.2 μ*M* and from round 19 the incubation period was reduced to 5 min. This stringency caused an increase in both the catalytic rate and the binding affinity for the substrate, producing selected representatives from round 27 with comparable DNA- and RNA-cleaving performances (Fig. 10.32, top). The high degree of selection for DNA cleavage (10,000-fold increase from **L20**), though, was obtained together with a slight improvement in the RNA cleavage, which led to ribozymes such as **10.36** with broader specificity including DNA strands, rather than DNA-specific ribozymes (Fig. 10.32). In a following paper Tsang and Joyce (202) conceived a modified selection procedure (Fig. 10.32, bottom) in which an RNA inhibitor was added together

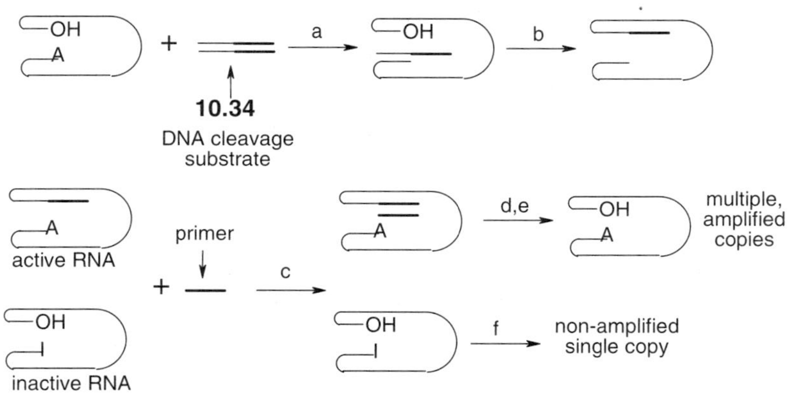

a: DNA substrate recognition; b: cleavage and ligation of the 3' portion; c: treatment with a primer for cDNA synthesis; d: DNA amplification; e: transcription/amplification into RNA; f: no reaction.

Figure 10.30 Evolution of a natural, RNA-cleaving ribozyme into a DNA-cleaving enzyme: the selection/amplification process using the selection/amplification substrate **10.34**.

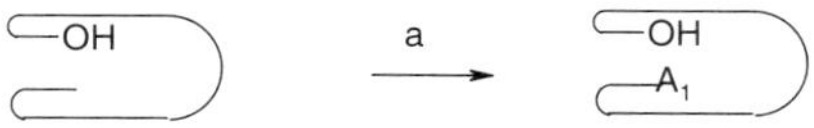

L20
413-mers
10^{13}-member ON library
randomization of 35 positions
in the catalytic domain

L21
30-fold improvement
in substrate cleavage

a: nine selection/amplification cycles as in Fig. 10.30; incubation with **10.34** (10 μM) for 1 hr at 37°C.

SELECTED INDIVIDUAL FROM L21:
10.35
K_{cat}/K_m for the cleavage of **10.34** = 3,600 M^{-1} min^{-1} at 37° with 10μM Mg^{++}
K_{cat}/K_m for the cleavage of natural RNA substrate = 10^7 M^{-1} min^{-1} at 37° with 10μM Mg^{++}

WILD-TYPE Tetrahymena RIBOZYME:
K_{cat}/K_m for the cleavage of **10.34** = 36 M^{-1} min^{-1} at 50° with 10μM Mg^{++}
K_{cat}/K_m for the cleavage of natural RNA substrate = 10^7 M^{-1} min^{-1} at 37° with 10μM Mg^{++}

RESULTS:
>100-fold increase of DNA-cleavage activity;
still $3x10^3$ higher RNA-cleavage activity.
RNA-SPECIFIC CLEAVING RIBOZYME

Figure 10.31 Evolution of a natural, RNA-cleaving ribozyme into a DNA-cleaving enzyme: screening of the biosynthetic ON ribozyme library **L20** and selection of the biosynthetic ON ribozyme library **L21**.

with the DNA substrate. In this case, selection of DNA-cleaving ribozymes was not changed (step b), but if a comparable or even higher activity on RNA was embedded into the ribozyme, it bound to the RNA inhibitor and was sequestered from the amplifiable library pool, which was then processed as described above. Thirty-six additional cycles were performed starting from the population of round 27 (**L22**, Fig. 10.33). Selected representatives from round 63 such as **10.37** showed a fivefold increase in DNA cleavage and a twofold decrease in RNA cleavage activity (Fig. 10.32). For the first time higher DNA cleavage activity was detected even though the specificity was not yet significant. Analysis of the mutation history from round 0 (**L20**) to round 63 allowed an understanding of the major interactions driving toward DNA specificity (200, 202). Other examples of focused ON libraries derived from natural sequences have been reported by Ekland and Bartel (204), Schmitt and Lehman (205) and Ordoukhanian and Joyce (206) who described the optimization of natural RNA ligases; by Tusoul et al. (207), reporting the synthesis of small nuclear RNA-inspired libraries; by Zarrinkar and Sullenger (208) and Pierce and Ruffner (209), who reported the synthesis of libraries inspired by the so-called hammerhead ribozymes; by Putlitz et al., who elaborated by combinatorial methods the structure of the so-called hairpin ribozymes (210); by Cole and Dorit (211) who evolved the M1 RNA natural ribozyme into a DNA-cleaving enzyme.

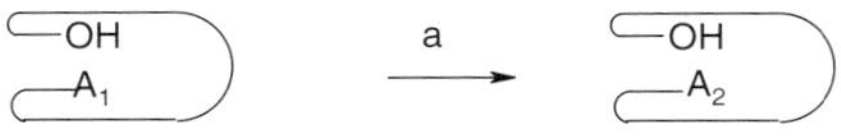

L21 413-mers **L22**

a: 18 selection/amplification cycles as in Fig. 10.30; incubation with **10.34** (0.2 μM) for 1 hr at 37°C in rounds 10-18, for 5' in rounds 19-27.

SELECTED INDIVIDUAL FROM L22:
10.36

K_{cat}/K_m for the cleavage of **10.34** = $2.9 \times 10^6\ M^{-1}\ min^{-1}$ at 37° with 10μM Mg^{++}

K_{cat}/K_m for the cleavage of natural RNA substrate = $2.1 \times 10^7\ M^{-1}\ min^{-1}$ at 37° with 10μM Mg^{++}

RESULTS:

*around 1000-fold increase of DNA-cleavage activity with respect to **10.35**;*
around 10 times higher RNA-cleavage activity.
BROAD (RNA > DNA) SPECIFICITY RIBOZYME

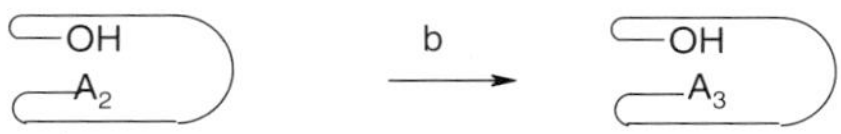

L22 413-mers **selected ribozymes**

a: 36 selection/amplification cycles as in Fig. 10.30; incubation with **10.34** (0.2 μM) for 5' at 37°C in presence of an RNA inhibitor of the natural RNA cleavage reaction.

SELECTED INDIVIDUAL:
10.37

K_{cat}/K_m for the cleavage of **10.34** = $1.5 \times 10^7\ M^{-1}\ min^{-1}$ at 37° with 10μM Mg^{++}

K_{cat}/K_m for the cleavage of natural RNA substrate = $6.9 \times 10^6\ M^{-1}\ min^{-1}$ at 37° with 10μM Mg^{++}

RESULTS:

*5-fold increase of DNA-cleavage activity with respect to **10.36**;*
2 times lower RNA-cleavage activity.
BROAD (DNA > RNA) SPECIFICITY RIBOZYME

Figure 10.32 Evolution of a natural, RNA-cleaving ribozyme into a DNA-cleaving enzyme: screening of the biosynthetic ON ribozyme libraries **L21** and **L22** and selection of the DNA-selective ribozyme **10.37**.

Ribozymes acting on nucleotide sequences have often been selected from random libraries, and many artificial sequences have been obtained. Landweber and Porovskaya (212) recently reported the selection of a family of small ribozymes from a 1.6×10^{15}-member, 132-mer library **L23** able to ligate multiple RNA substrates. In vitro selection was performed with three different RNA substrates for six cycles in the presence of Mg^{2+} (steps a–c, Fig. 10.33), and multiple substrates were used to select

L23
132-mers
1.6×10^{15}-member ON library
100 inner randomized positions

→ a → FIRST POPULATION → b → SELECTED CLONES

a: 6 selection/amplification cycles using substrate 1 (cycles 1,4), 2 (cycles 2,5), 3 (cycles 3,6) ;
b: 7 cycles with heavily mutagenic PCR conditions.

CONSENSUS SEQUENCE:

29-mer sequence (from final clones)
UGUGGU**GAAA**AAAUAGAAUCGAUCUUGUC

containing also the GAAA-based Mn^{++}-dependent
RNA-cleavage motif.

↓

10.38
small 29-mer ribozyme
dual catalytic activity
pH- and divalent ion-dependent activity switch

Figure 10.33 Selection of artificial ribozymes with dual, switchable RNA cleavage and ligation activity from the biosynthetic ON ribozyme library **L23**: the selection/amplification process and the structure of the optimized ribozyme **10.38**.

for ribozymes with broad specificity. Seven further, more focused cycles were performed using heavily mutagenic PCR amplification of selected sequences to produce a family of related catalytic RNA molecules. Comparison of their sequences highlighted a highly conserved, 29-mer catalytic core (**10.38**, Fig. 10.33) that was prepared separately and showed similar, Mg^{2+}-dependent activity to the integral sequences selected from round 63. Surprisingly, this 29-mer sequence also contained a 7-mer that was already known as a Mn^{2+}-dependent RNA cleavage motif (Fig. 10.33) previously observed in naturally occurring ribozymes and that also occurred in the longer sequences (213, 214). Further studies highlighted the dual nature of **10.38**, which, according to the pH and the presence of different divalent cations, was able to switch from ligation of RNA to self-cleavage.

The existence of a coordination site for divalent cations and the possibility for the ribozyme to adopt two different conformations, one suited to ligation, the other to self-cleavage, were possible explanations for this dual activity. The occurrence of a parallel, catalytic activity that was selective for Mn^{2+}, even though this cation was never used in the in vitro selection experiments, hinted at the flexibility of ribozymes able to adapt their properties to different conditions and that could allow the evolution of new catalytic activities. Other artificial ribozymes selected in vitro have been reported by Vaish et al. (hammerhead-like) (215), Yu et al. (hairpin-like) (216), Robertson and Ellington (allosteric ligase activated by ON effectors) (217), and

Santoro and Joyce (218). The last authors cited reported the selection of multipurpose RNA-cleaving DNAzymes with potential applications in molecular biology. The elusive RNA replicases, which should be able to replicate RNA structures including their own to sustain the RNA world hypothesis, are a common target for research; a recent review (219) summarized the efforts in this field.

A primordial RNA world would have required the efficient assembly of single nucleotides into ON chains. Sugar phosphates (220) and nucleobases (221) have been prepared under conditions that supposedly mimic those found in this primordial world, but the assembly of the two nucleotide constituents remained a challenge. Unrau and Bartel (222) studied the nucleophilic attack of activated ribose (pRpp, **10.39**) on uracil (**10.40**) to give uridine 5′-phosphate (UMP, **10.41**, Fig. 10.34), a reaction that is catalyzed by the enzyme uracylphosphoribosyl transferase (UPRT). A 1.5×10^{15}-member library of 294-mer ONs, **L24**, with 228 random positions was prepared using standard synthetic protocols and was submitted to the in vitro selection strategy depicted in Fig. 10.35. Activated ribose **10.39** was first condensed with activated adenosine **10.42** (step a); then the resulting compound **10.43** was condensed onto **L24** (step b) to give **10.44**. Compound **10.44** was incubated in the presence of 4-thiouracil **10.45** (step c, 8 m*M*) (223) for 18 h in the first six cycles, gradually decreasing the concentration of **10.45** to 40 μ*M* and the incubation time to 7.5 min for five additional rounds. The presence of a 4-thio group on **10.46** allowed the separation of the reacted RNA structures from the inactive components (step a, Fig. 10.36). The reaction of **10.46** with iodoacetyl biotin to give **10.47** (step b) allowed the separation of the reacted RNA by capture with supported streptavidin (step c, Fig. 10.36). The bound sequences were eluted and submitted to standard reverse transcription, PCR, and transcription protocols (steps d–g). The new pools of RNA were submitted to further selection cycles (step h, Fig. 10.36). Error-prone PCR (224) was used for amplification of individuals from rounds 4–6 to introduce mutations that were absent from **L24** and to include additional structural diversity. After 11 selection cycles 35 clones were randomly selected, and three families of ribozymes were identified. Representative examples of these families showed significant catalytic activity, with up to 10^7 times enhancement compared with the noncatalyzed reaction and selectivity toward **10.45**, as only uracil was slowly condensed with 3′-derivatized **10.44** while the other thionucleobases did

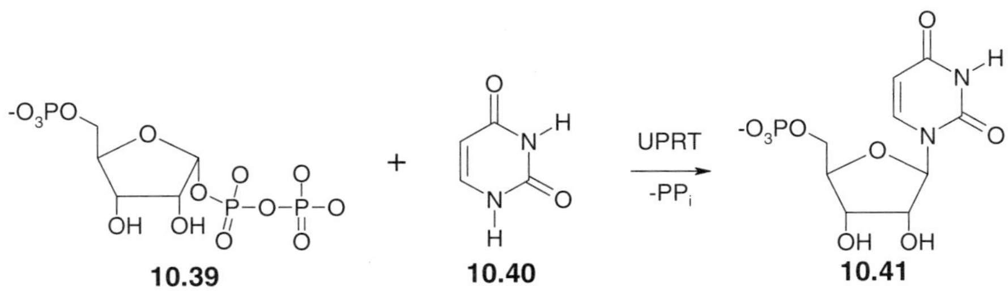

Figure 10.34 Uracylphosphoribosyltransferase (UPRT)-catalyzed synthesis of uridine-5-phosphate **10.41**.

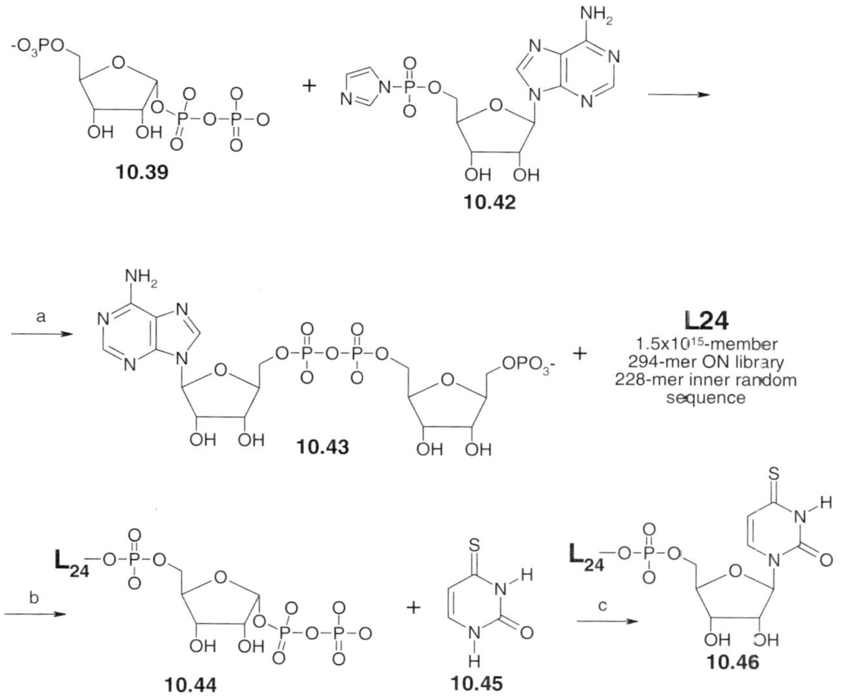

a: aq. $MgCl_2$, 2 hrs, 50°C; b: T4 RNA ligase, aqueous buffer, rt, 4 hrs; c: incubation at 23°C for 12 cycles using decreasing concentrations of **10.45** (from 8 mM to 40 μM) and incubation times (from 18 hrs to 7.5').

Figure 10.35 Selection of artificial ribozymes with nucleotide-assembling properties from the biosynthetic ON ribozyme library **L24**: the selection/amplification process to **10.45** using the slection substrate **10.45**.

not react. This work is of great significance because it assessed both the feasibility of nucleotide synthesis by RNA catalysis and the use of small molecules, such as **10.45**, as substrates for the selection of ribozymes.

Zhang and Cech (225) have reported the in vitro selection of ribozymes catalyzing peptide bond formation from a 1.3×10^{15}-member library of 222-mer modified ONs, **L25**, containing 142 randomized positions. The library was prepared from the corresponding DNA sequence, but translation was performed in the presence of guanosine-5-monophosphorothioate (GMPS, **10.48**). The final 5′-GMPS-RNA pool was reacted with a phenylalanine–linker construct (**10.49**, step a, Fig. 10.37) to form **L25** as a 5′-Phe-S-S-RNA pool. This library, bearing a free amino group at its 5′-end, was designed to mimic the aminoacyl acceptor bound to the aminoacyl (A) site of the ribosome where a growing peptide chain is built. The peptidyl donor tRNA bound to the ribosomal site P was mimicked by the simple, methionine-containing labeled

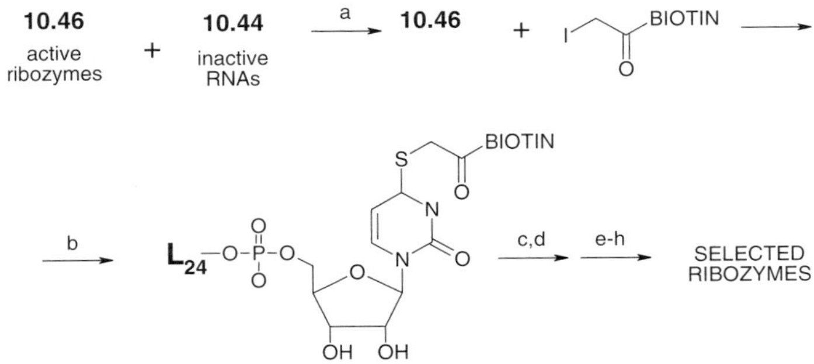

10.47

a: separation by two-dimensional TLC; b: aq. buffer, DMF, rt, 3 hrs; c: capture of **10.47** with streptavidin-coated beads; d: elution of ribozymes; e: reverse transcription to DNA; f: PCR amplification; g: transcription into RNA; h: 11 iterative cycles and final selection.

3 REPRESENTATIVE RIBOZYME FAMILIES:

up to 10^7 **enhancement** of nucleotide synthesis compared to uncatalyzed reaction; **specificity for 10.46**, as no reaction was observed with 2-thiouracil, 2,4-thiouracil, 2-thiocytosine, 2-thiopyrimidine, 2-thiopyridine or 5-carboxy-2-thiouracil.

Figure 10.36 Selection of artificial ribozymes with nucleotide-assembling properties from the biosynthetic ON ribozyme library **L24**: the selection/amplification process from **10.46** to three ribozyme families with nucleotide synthesis properties.

selection substrate **10.50** (AMP-Met-biotin, Fig. 10.37). The in vitro selection strategy adopted is shown in Fig. 10.38. Incubation of **L25** with **10.50** (8 m*M*) for 20 h at 25°C (step a) in the presence of Mg^{2+} (50 m*M*) induced the nucleophilic attack of the free NH_2 onto the ester bond of **10.50**, releasing AMP and linking the biotinyl-Met moiety to the active RNAs **10.51**. The library was eluted through a streptavidin column that complexed the biotin-containing ribozymes (step b). Treatment with dithiothreitol (DTT, step c) reduced the disulfide bonds and eluted the selected free 5′-GMPS-RNAs **10.52** that had been selected (Fig. 10.38). After standard reverse transcription and PCR amplification, the transcription of dsDNA was again performed in the presence of GMPS (step d). The RNA pool was converted to 5′-Phe-S-S-RNA as described in Fig. 10.37, and a new selection cycle was started. The first nine selection rounds used the same reaction conditions (step e), and then the concentration of **10.50** (up to 200 μ*M*) and the incubation time (up to 1 h) were reduced to increase the stringency (step f, Fig. 10.38). The RNA population selected after 19 cycles, corresponding to a >10,000-fold increase in catalytic efficiency, was processed individually, and 75 sequences were cloned. Of these, nine sequences were found to be more active than the selection pool and could be divided into two main structural families. The most active RNA sequence (**10.53**, Fig. 10.38) was fully characterized and showed a markedly enhanced catalytic activity (10^6 times the uncatalyzed peptide bond formation). This activity was lost

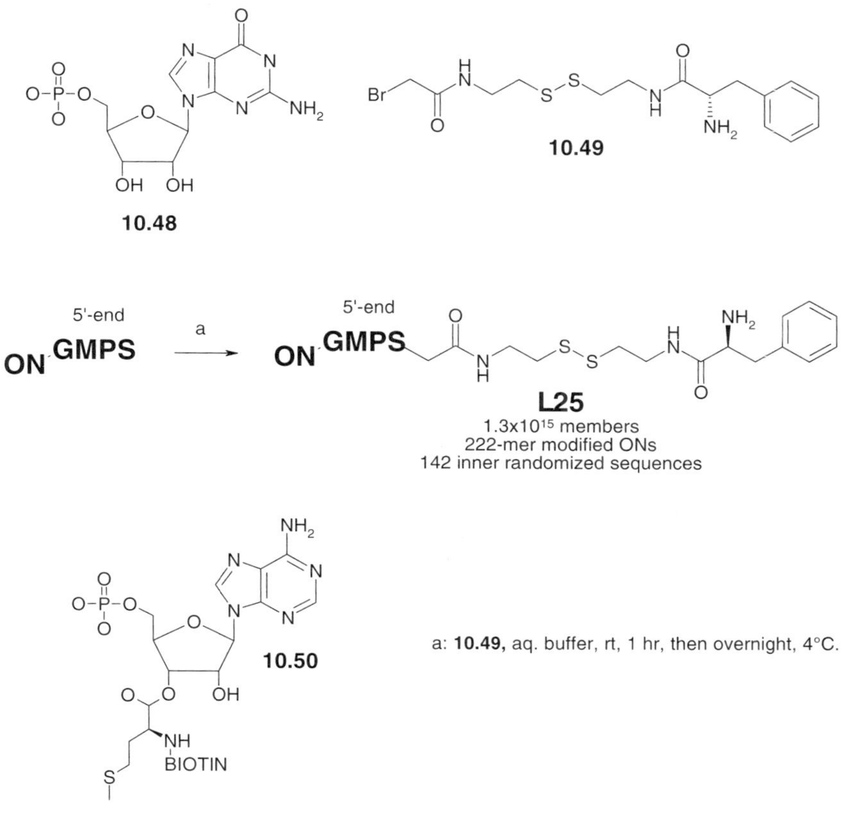

Figure 10.37 Selection of artificial ribozymes with peptide-assembling properties from the biosynthetic ON riboyzyme labeled library **L25**: structure of the key intermediates **10.49** and **10.50**.

upon deletion of both the 3′- and 5′-ends and removal of Mg^{2+} ions and the amide link between **10.50** and 5′-Phe-S-S-RNA. The activity was also diminished when the linker between Phe and GMPS was shortened. The ribozyme **10.53** also showed a broader substrate specificity and was able to use AMP-Leu-biotin and, to a lesser extent, phenylalanine and lysine-containing substrates as donors. Even though the activity of **10.53** was significantly lower than the natural peptide–peptide bond-forming machinery of the ribosome, **10.53** successfully demonstrated the feasibility of peptide bond-forming ribozymes able to act on multiple substrates. A few similar reports (226–229) have confirmed the activity of families of ribozyme in acylations and peptide bond formation.

Naturally occurring and synthetic ribozymes often require metal ion cofactors (usually Mg^{2+}) to be effective, as demonstrated in the previous examples. However, the involvement of nucleotide-like cofactors (230) and allosteric regulatory mecha-

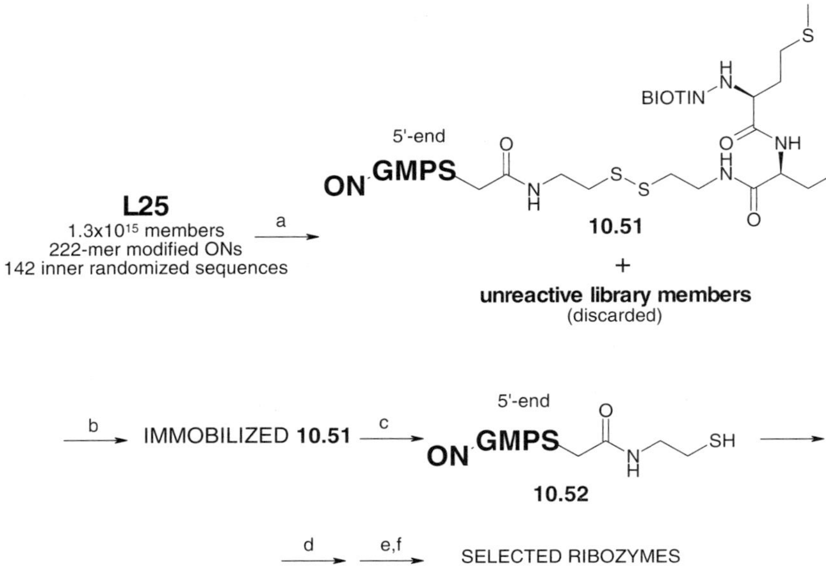

a: incubation with **10.50** (8 mM), 25°C, 20 hrs; b: absorption on a streptavidin column; c: DTT, rt; d: standard reverse transcription, PCR amplification, transcription protocols; e: eight iterative selection/amplification cycles; f: ten iterative cycles with decreasing **10.50** (up to 200 μM) and decreasing incubation time (up to 1 hr).

MOST ACTIVE RIBOZYME:
10.53
$>10^6$ increase of catalytic efficiency compared to the uncatalyzed reaction; broad substrate specificity.

Figure 10.38 Selection of artificial ribozymes with peptide-assembling properties from the biosynthetic ON ribozyme labeled library **L25**: the selection/amplification process using **10.50** as the selection substrate to the most active ribozyme **10.53**.

nisms (217, 231) has also been reported and illustrates the high degree of versatility found among the ribozymes. Roth and Breaker (232) have described the selection of L-histidine-dependent RNA cleavage DNAzymes in vitro from a 2×10^{13}-member library of >100-mer modified ODNs, **L26**, in which a biotin moiety was attached onto the 5′-position. The library contained a single oligonucleotide as the self-cleavage site and a 40-mer random sequence from which to select the active DNAzymes (Fig. 10.39). **L26** was absorbed onto a streptavidin-derivatized column matrix (step a) via its biotinylated end, so that self-cleavage of the active library individuals caused their release during incubation (*vide infra*) while the unreactive DNAs remained anchored onto the column. The column was first washed with aqueous buffer to discard non-histidine-dependent DNAzymes (step b) and was then eluted with three portions

L26
2×10^{13} members
>100-mer modified ODNs
1 ON at the cleavage site
40-mer inner randomized sequences
5'-biotinylated individuals

a,b ⟶ c-e ⟶ f,g ⟶ 6 SELECTED DNAzymes
1 STRUCTURAL FAMILY

a: adsorption onto a streptavidin column; b: elution with buffers, to waste; c: elution with L-histidine (50 mM) and EDTA, pH 7.5, 23°C, 1 hr; d: standard protocols for PCR amplification; e: reconstruction of biotinylated end; f: six repeated a-e cycles; g: 11 iterative cycles with shorter elution times (up to 15').

REPRESENTATIVE DNAzyme:
10.54
39-mer selected sequence (deletion)
significant but low His-dependent, RNA-cleavage activity

↓ h

L27
10.54-focused library
10^{13}-member ODNs
up to 7 mutated residues
in the 39-mer catalytic sequence

↓ i

10.55 and **10.56**
100-fold increased activity in respect to **10.54**
strong dependence from L-His
strong cofactor selectivity:
lack of activity for D-His, modified histidines and other AAs.

h: synthesis of **L27** (0.21 degeneracy of each position in the 39-mer sequence); i: standard selection/amplification protocols, five cycles.

Figure 10.39 Selection of artificial ribozymes with amino acidic cofactors from the biosynthetic modified on DNAzyme libraries **L26** and **L27**: the selection/amplification process to the L-his dependent ribozymes **10.55** and **10.56**.

of 50 m*M* L-histidine solutions for 1 h at pH 7.5 and 23°C (step c) to select for the L-His-dependent ribozymes. The solutions also contained the metal chelator ethylenediaminetetraacetic acid (EDTA) to prevent the selection of metal cofactor-dependent DNAzymes. Selected DNA sequences were PCR amplified and the proper biotinylated 5′-ends were reconstructed to provide a new pool of DNA for another selection cycle (steps d and e). Six cycles were performed as above (step f); then an additional four followed with shorter elution times (15–25 min) to increase stringency (step g). The DNA population after the 11th cycle was characterized, and a family of six sequences of L-histidine-dependent DNAzymes was obtained. These sequences contained 39 ONs in the randomized sequence, probably due to a single deletion event during in vitro selection. Their activity, although significant, was ~1,000-fold lower than most natural self-cleaving ribozymes. For this reason, the structure of the ribozyme **10.54** (Fig. 10.39) was used to prepare a focused library of ODNs **L27** in which all the individuals containing up to seven modifications with respect to **10.54** were represented. Five rounds of reselection with lower L-histidine concentrations (5 m*M*) produced a further 100-fold increase in catalytic efficiency (step i, Fig. 10.39). Two isolated DNAzymes, **10.55** and **10.56** (Fig. 10.39), were fully characterized and found to have a strict requirement for L-His, as similar amino acids showed little or no cofactor activity apart from L-histidine methyl ester. The selection of cofactor-dependent ribozymes could increase the versatility and possible applications for catalytic ON sequences.

Ribozyme-catalyzed reactions involving C–C bond formations have also been reported. Seelig and Jaschke (233) presented the in vitro selection of ribozyme catalysts for the Diels–Alder reaction between maleimide and anthracene, employing a 2×10^{14}-member library of 160-mer modified ONs (**L28**) with 120 randomized positions. The selection strategy used is shown in Fig. 10.40. Library **L28** was prepared from the corresponding dsDNA sequences, and transcription initiation was performed in the presence of ternary complexes between guanosine monophosphate (**10.57**), PEG (**10.58**), and anthracene (**10.59**, step a, Fig. 10.40). The library obtained contained a 5′-anthracene–PEG appendage and was incubated with biotin-modified maleimide **10.60** (25 μ*M*) as the Diels–Alder substrate in the presence of various metal cations at 25°C for 1 h (step b). The reaction caused the attachment of the biotin-bearing adduct **10.61** onto the active RNAs, and immobilization with streptavidin-functionalized agarose (step c) selected the Diels–Alder catalysts from **L28** (Fig. 10.40). Their elution using standard protocols freed the selected ribozymes (step d, Fig. 10.40). Reverse transcription, PCR amplification, and modified transcription (234) yielded another population of 5′-modified RNAs (step a, Fig. 10.41) that were submitted to a new round of selection. The first five cycles were performed as above (step b), while stringency was then increased between the 6th and 10th cycles by reducing the concentration of **10.61** to 2.5 μ*M* and the incubation time to 1 min, step d. After round 10, the selected population was characterized, and 35 different sequences were obtained, among which **10.62** (Fig. 10.41) showed an ~20,000-fold increase in efficiency with respect to the uncatalyzed Diels–Alder reaction. Further studies identified the minimal structural requirements for catalysis and allowed the preparation of several active truncated ribozymes (**10.63–10.66**, from 39- to 57-mers, Fig. 10.41) that maintained a similar level of activity as the original ribozymes. The specificity for this reaction was high, as several related Diels–Alder substrates did not react with the ribozyme **10.66**. The length of the tether between the anthracene and RNA was not important, as three PEGs with various lengths (7, 10, and 16 ethylene glycol units compared with the 13-unit construct originally employed) were inserted into **L28** and gave similar reactions with **10.61**.

The results from this and another similar study (235) could be the precursors to the identification of novel, potent ribozymes able to catalyze organic reactions and to

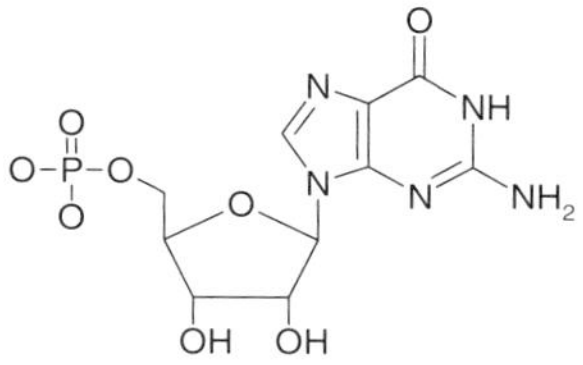

10.57

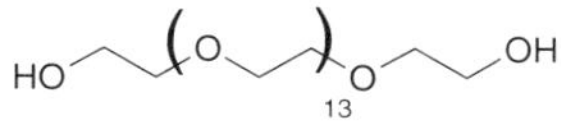

10.58

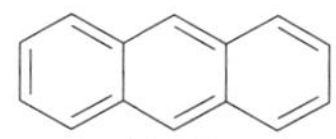

10.59

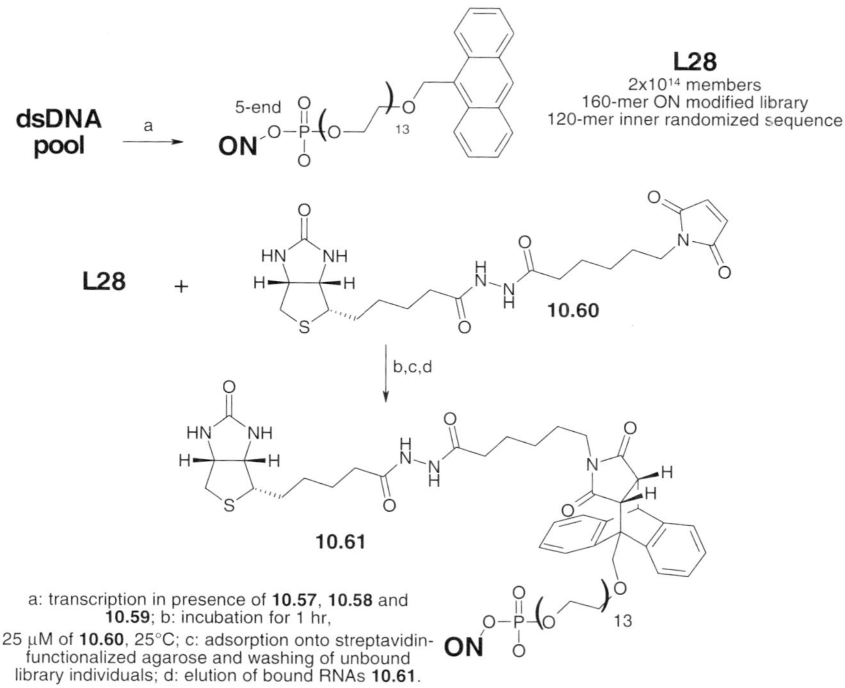

Figure 10.40 Selection of artificial ribozymes with catalytic properties for a Diels–Alder reaction from the biosynthetic ON modified ribozyme library **L28**: library synthesis and the selection process using the biotinylated selection dienophile **10.60**.

provide useful tools for biochemical and chemical applications; the same is true for a report highlighting the selection of a ribozyme with cholesterol esterase activity (236). Even if current examples suffer many drawbacks and cannot be applied for practical purposes, the concept of RNA-based enzymes has been thoroughly validated, and major improvements driven by technological advancements can be expected in this area in the near future.

An intriguing review has recently dealt with in vitro selection of nucleotide- and peptide-related bio-oligomers ranging from small to large sequences (237); the inter-

10.61
2×10^{14} members
160-mer ON modified library
120-mer inner randomized sequence

→ a,b → c → SELECTED RIBOZYMES

a: reverse transcription, PCR amplification and transcription as in standard protocols; b: four iterative cycles; c: five additional cycles with reduced concentrations of **10.60** (up to 2.5 μM) and incubation times (up to 1').

BEST SELECTED INDIVIDUAL FROM L28
10.62
around 20,000-fold increase of catalytic efficiency compared to the uncatalyzed reaction

↓ d

10.63-10.66
four truncated (39-mer to 57-mer) ribozymes
with efficiency similar to **10.62**
and high specificity for **10.61**

d: truncation of ON sequences unnecessary for catalytic activity.

Figure 10.41 Selection of artificial ribozymes with catalytic properties for a Diels–Alder reaction from the biosynthetic ON modified ribozyme library **L28**: the selection/amplification process from **10.61** to the optimized ribozymes **10.63–10.66**.

ested reader could further expand the knowledge of this area by consulting this relevant paper.

10.3 COMBINATORIAL BIOSYNTHESIS OF NATURAL PRODUCTS

10.3.1 General Considerations

Natural products (NPs) have long been a source of biologically active compounds, and their extraction and synthetic modification, especially for pharmaceutical purposes (238, 239), have been well-studied. Their structures exhibit a wide degree of diversity and levels of complexity that are seldom attained in totally synthetic structures, and this is reflected in the wide range of biological properties they display.

Recently, combinatorial technologies and libraries have somewhat replaced NPs as a source of diversity in pharmaceutical research. A comparison of combinatorial technologies with the fermentation of an NP-producing organism show that they both produce a library of compounds. While the former library is structurally determined a priori either by the selected synthetic scheme or by recombinant genetic information in the case of biosynthetic peptide and ON libraries, the latter is the result of the metabolic complexity of the producing strain and has to be deconvoluted in order to determine the exact structure of an active compound. Whereas in combinatorial libraries the members of the libraries are effectively present in roughly equimolar amounts and their complexity can be resolved with relative ease, natural extracts contain many chemical entities including macromolecules, inorganic materials, and small organic molecules in different amounts, such that an active component may well be present at a very low concentration and the work-up procedures to purify, separate, and structurally characterize any individual from the extract can be long and laborious. Finally, the large majority of extracts (>99%) showing an interesting biological activity are known and therefore unexploitable compounds, while combinatorial libraries may be designed to contain only novel, druglike individuals. Over time a number of efficient methods have been introduced to reduce the time required to purify and isolate a natural product and determine the novelty of its structure. Nevertheless, the diversity embedded in NPs is mostly applied in specific therapeutic areas or where combinatorial methods have failed to produce active compounds.

The biosynthesis of a natural product is an extremely complex event that is carried out by multienzyme systems showing a high specificity for the sequential elaboration of simple precursors into complex end products. Often these systems perform iterative processes to synthesize the final NP as an assembly of units, with each cycle being performed by a multifunctional protein containing several active sites responsible for each transformation in a given cycle (e.g., polyketides, *vide infra*). The cluster of active sites in a multifunctional protein is referred to as a module (240). Genetic manipulation of these modules has allowed the full characterization of several of these pathways in order to understand the features of the many enzymes involved and to test their structural specificity. Combinatorial alteration of naturally occurring modules, usually referred to as *combinatorial biosynthesis* (241), involves either modification of the order of individual enzymes in the module complex or deletion or duplication of an enzyme activity through deletion or addition of a component to the module. This process can be completely controlled by genetic manipulation and results in the production of modified analogues of the parent NP. Libraries of modified NPs obtained by combinatorial biosynthesis are structurally defined, in that the design of several module modifications determines the expected structure of any member of the library in any specific library well (Fig. 10.42). The final compounds are novel, because only modifications leading to unprecedented compounds are considered and can be detected, purified, and isolated from the biological mixture using well-known protocols. These modified NP libraries ensure access to biologically relevant NP diversity that is constantly increased by the elucidation and characterization of novel biosynthetic pathways. The following sections will focus on the most studied biosynthetic pathways and on some specific examples of their combinatorial modification.

10.3.2 Combinatorial Biosynthesis of Polyketides

Polyketides (PKs) are a typical example of a large and diverse class of NPs that derive from several related biosynthetic pathways. Their structures contain repeating units iteratively assembled into a range of diverse chemical structures (Fig. 10.43). PKs can be taken as an example of the application of combinatorial biosynthesis as both the

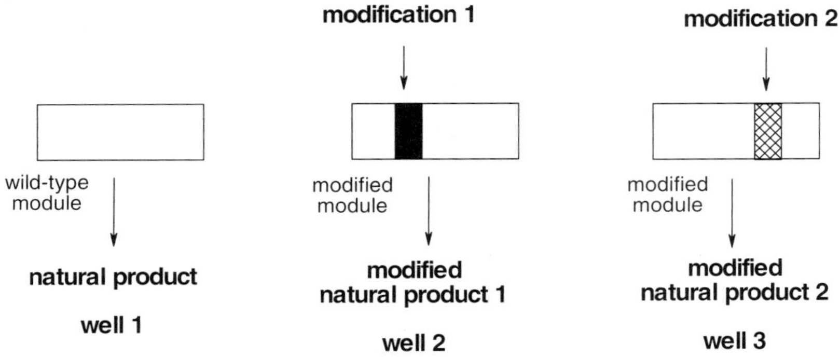

Figure 10.42 Combinatorial biosynthesis: structure of a hypothetical three-member discrete library of modified natural products.

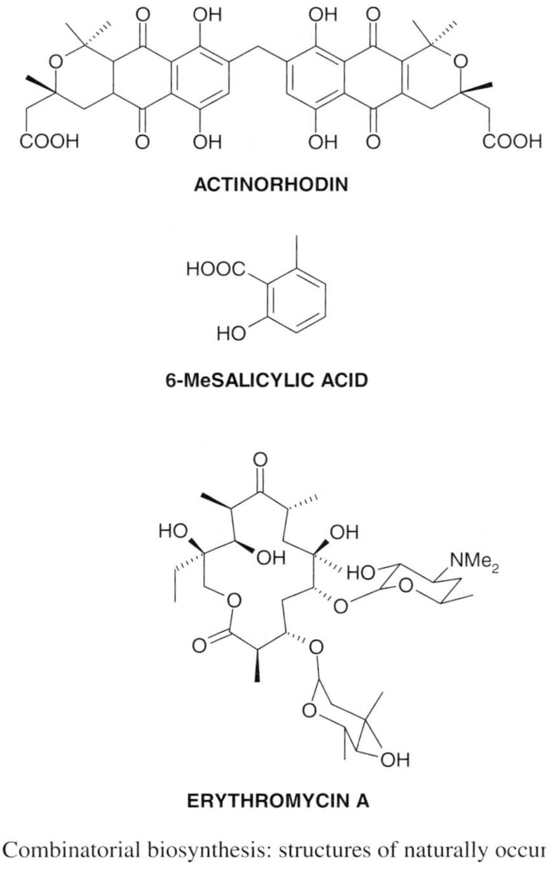

Figure 10.43 Combinatorial biosynthesis: structures of naturally occurring polyketides.

chemistry and the biology of their biosynthesis are well known (242–245) and they are suitable for extensive manipulation to produce libraries of modified PKs. Standard protocols for the genetic manipulation of PK producers and for the expression of engineered, novel PKs (246, 247) are now routinely used.

Two major classes of multifunctional polyketide synthase (PKS) enzymes can be considered. Iterative PKS enzymes are made up of a single module that iteratively processes biosynthetic units for a synthetic cycle, adding a new monomeric unit and repeating the whole cycle until the assembly of the final PK is finished. Modular PKS enzymes are made up of several modules embedded into multifunctional proteins that are specifically responsible for one of the several cycles necessary to produce the final PK. The biosynthetic intermediates remain anchored to an individual module until the cycle is successfully terminated; then they are shunted to the next module in order to continue the PK biosynthesis. Among iterative PKS enzymes, actinorhodin synthase produces actinorhodin (**10.69**, Fig. 10.44) by the iterative action of a minimal PKS enzyme composed of a ketosynthase (KS), which carries the growing PK chain and couples it with a carboxylate extender unit **10.67** loaded onto an acyl carrier protein (ACP) and transported by an acyltransferase (AT). Iteration of this cycle is sometimes punctuated by other activities such as a ketoreductase (KR), an aromatase (ARO), and a cyclase (CYC). The advanced intermediate **10.68** is converted into **10.69** by several tailoring enzymes (Fig. 10.44). The combinatorial potential of these pathways is high, including different starter and extender units, manipulation of ARO and CYC activities, the use of KR activities from several different producing organisms, and the use of tailoring enzymes from other sources. Several reviews (248–252) and papers (253–264) have extensively covered the subject including an example of iterative PKS from plants (265).

Modular PKS enzymes are responsible for the synthesis of a wide diversity of structures and seem to have more relaxed specificities in several of the enzymatic steps. Their enormous appeal for combinatorial purposes, though, derives from the presence of multiple modules that can be manipulated independently, allowing the production of rings of different sizes and with potential stereochemical variation at each PK carbon. The higher complexity of these pathways has somewhat hindered their exploitation, but recently, several have been fully characterized. Among them, by far the most studied modular multienzyme complex is 6-deoxyerythronolide B synthase (DEBS; 240, 266, 267), which produces the 14-member macrolide 6-deoxyerythronolide B (**10.70**, Fig. 10.45). DEBS contains three large subunits each of which contains two PKS enzyme modules. Each module contains the minimal PKS enzyme (*vide supra*) and either none (**M3**), one (ketoreductase KR; **M1**, **M2**, **M5**, and **M6**), or three (dehydratase DH–enoyl reductase ER–ketoreductase KR, **M4**) catalytic activities that produce a keto (**M3**), an hydroxy (**M1**, **M2**, **M5** and **M6**), or an unsubstituted methylene (**M4**) on the last monomeric unit of the growing chain (Fig. 10.45). A final thioesterase (TE) activity catalyzes lactone formation with concomitant release of **10.70** from the multienzyme complex. Introduction of TE activity after an upstream module allows various reduced-size macrolides (**10.71–10.73**, Fig. 10.45) to be obtained.

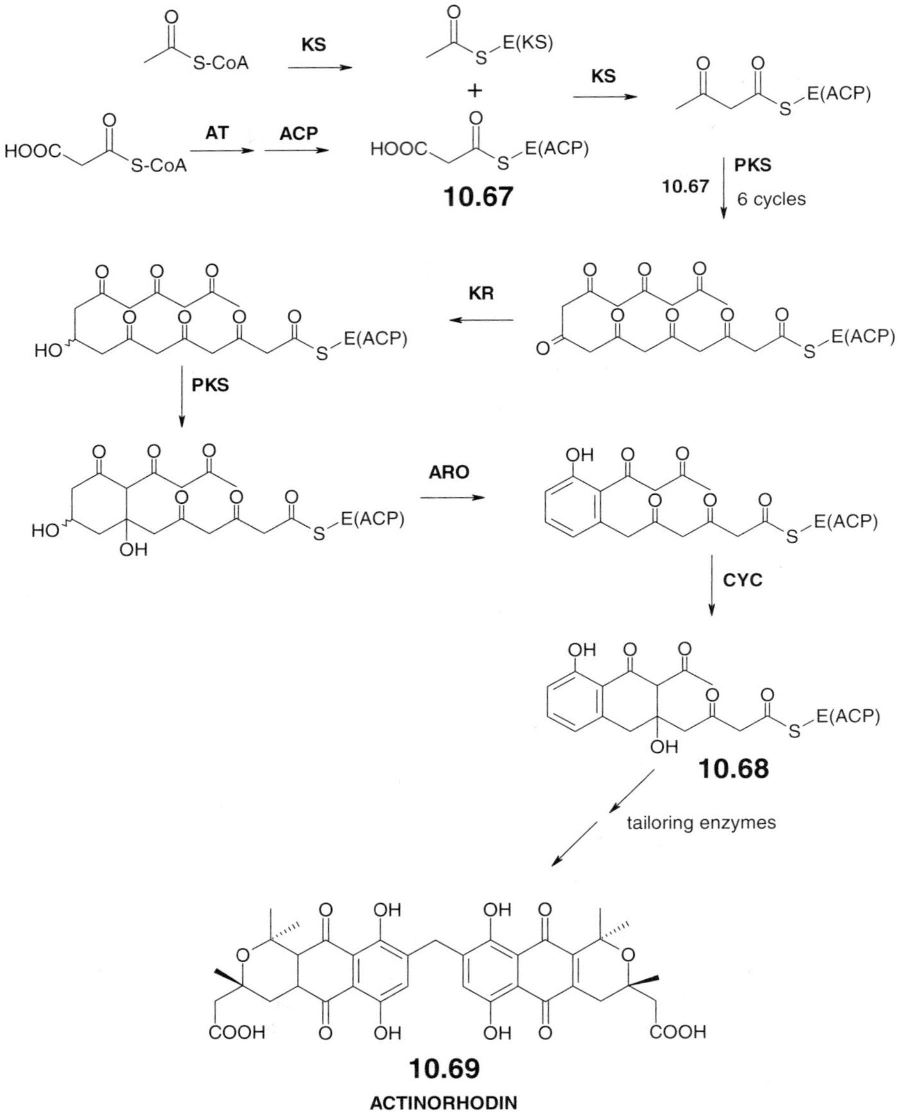

Figure 10.44 Iterative PKS enzymes: biosynthesis of actinorhodin (**10.69**).

The specificity of many components of DEBS has been thoroughly studied to allow its careful manipulation (268–272), and its potential for combinatorial biosynthesis including several combinatorial applications via rational modifications of DEBS has been thoroughly reviewed (251, 252, 273–279). Other large modular PKS enzymes have also been characterized (280–288) and will eventually be used to design and

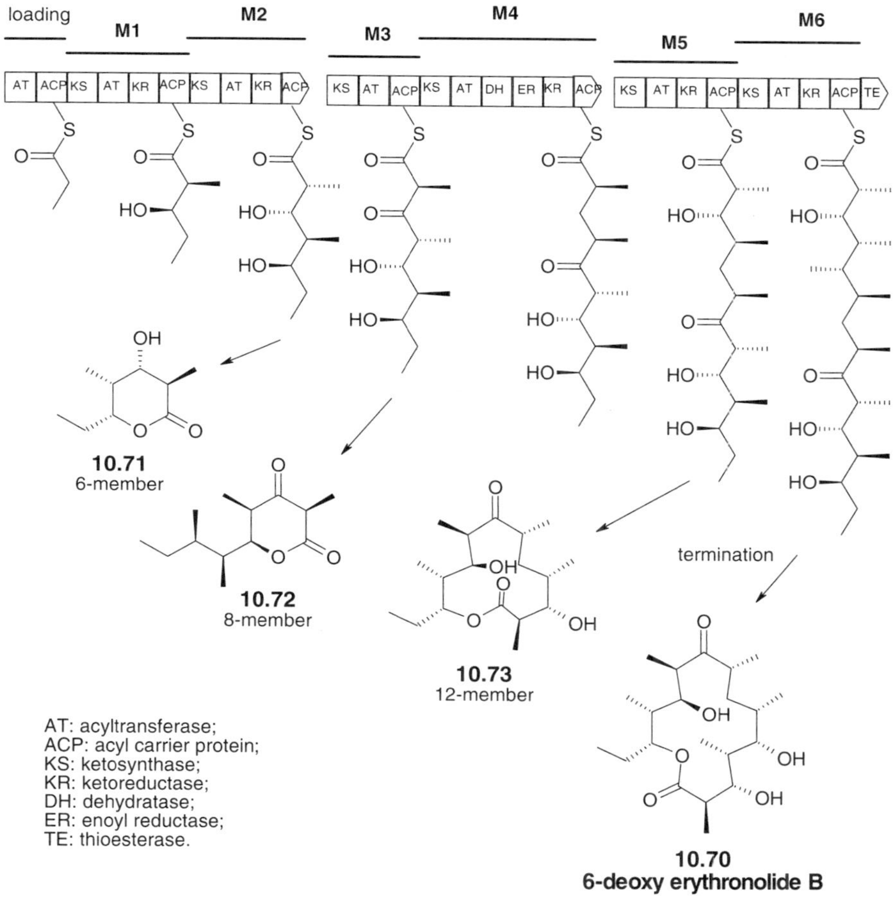

Figure 10.45 Modular PKS enzymes: biosynthesis of 6-deoxy erythronolide B (**10.70**) and modified biosynthetic products **10.71–10.73**.

execute novel combinatorial biosynthetic approaches. The same is true for modular biosynthetic pathways other than PKS enzymes, such as nonribosomal peptide synthetases (276, 289, 290), which have also been reviewed recently (241, 291, 292), and deoxy sugar biosynthetic pathways (293, 294).

Jacobsen et al. (295) reported the biosynthesis of a 16-membered macrolide from a modified DEBS multienzyme in which KS in **M1** was inactivated and the unnatural di- (**10.74**) and triketides (**10.75** and **10.76**, Fig. 10.46) were used to feed **M1** or **M2** (Fig. 10.46) (295). The products **10.70**, **10.77**, and **10.78** show how not only decreased ring sizes but also larger ones (path 3, Fig. 10.46) are available by simple genetic manipulations, and they also demonstrate the selectivity of the modules even for the chirality of a single atom (compare paths 2 and 3, Fig. 10.46). Jacobsen also reported

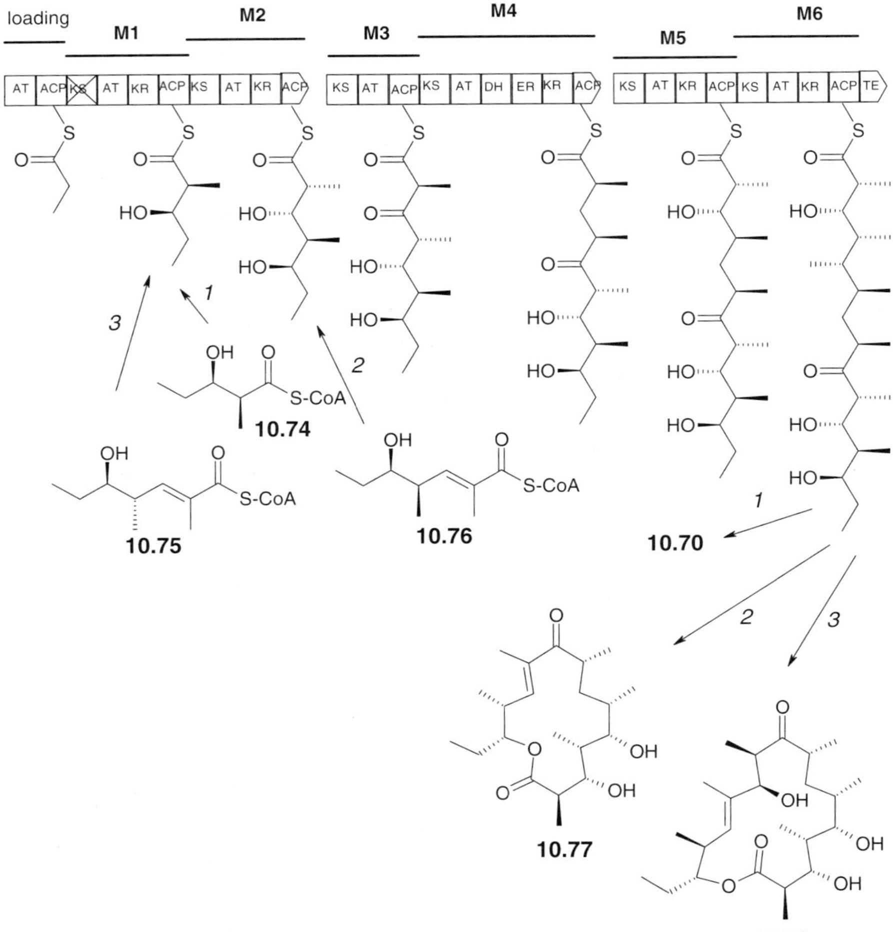

Figure 10.46 Combinatorial biosynthesis: manipulation of the ring size and the stereochemical transformations of DEBS modules to produce analogues **10.77** and **10.78**.

the first successful shuffling of entire modules, rather than single enzymatic activities, to increase the diversity of PKs obtained. The small macrolide **10.79** was produced by hybrid bimodular subunits **M1–M3** or **M1–M6**, where a polypeptide linker connected the two modules and allowed the processing of the substrates, as for the natural bimodular **M1–M2** (Fig. 10.47). More significantly, though, the replacement of **M2** in DEBS with a module from rifamycin PKS (*rap*M5) containing the same activities of **M2** gave the natural PK **10.70** with reasonable yield (Fig. 10.47) (296).

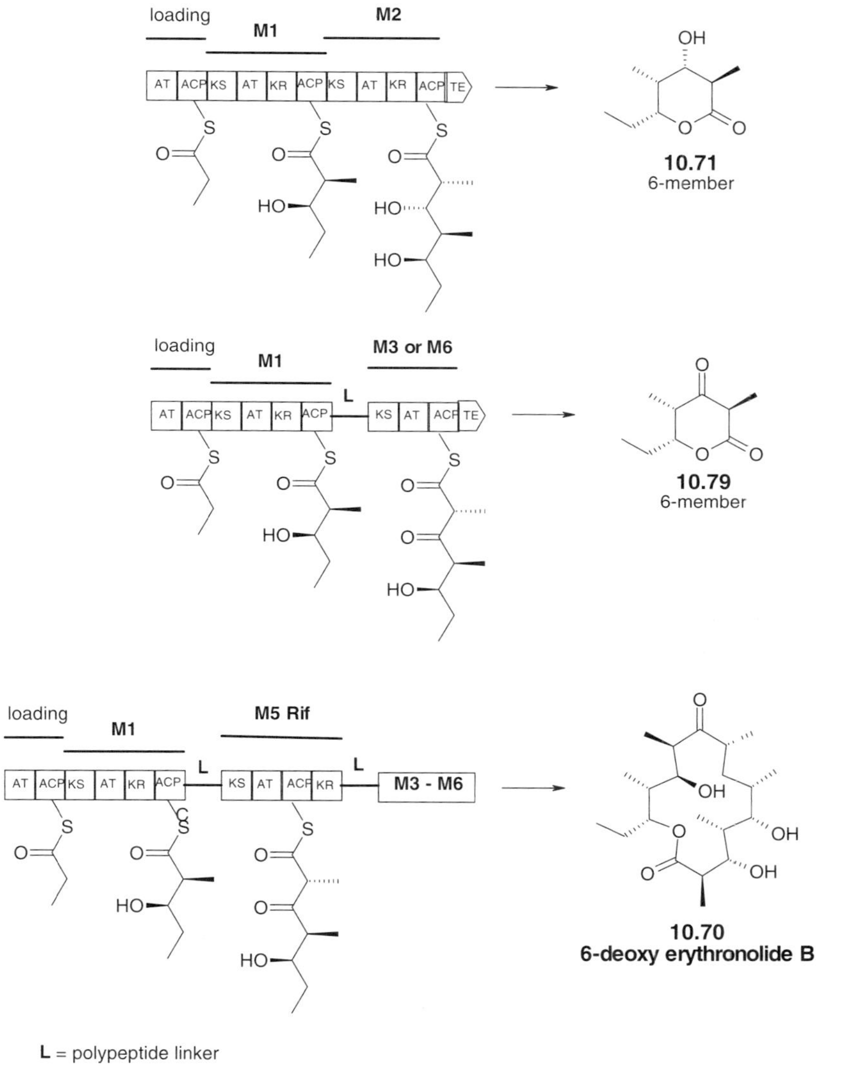

Figure 10.47 Combinatorial biosynthesis: manipulation and substitution of whole DEBS modules to produce **10.70**, **10.71** and **10.79**.

10.3.3 An Example: Synthesis of a Library of 14-Member Macrolides

McDaniel et al. (297) reported the synthesis of a >100-member library of macrolides, **L29**, obtained by manipulation of DEBS. All the modules **M1**–**M6** were altered with a single domain substitution, and some of the resulting PKs are shown in Fig. 10.48, with the modified structural fragment highlighted in bold. For example, the replacement of DEBS AT enzymatic domains with the corresponding AT of

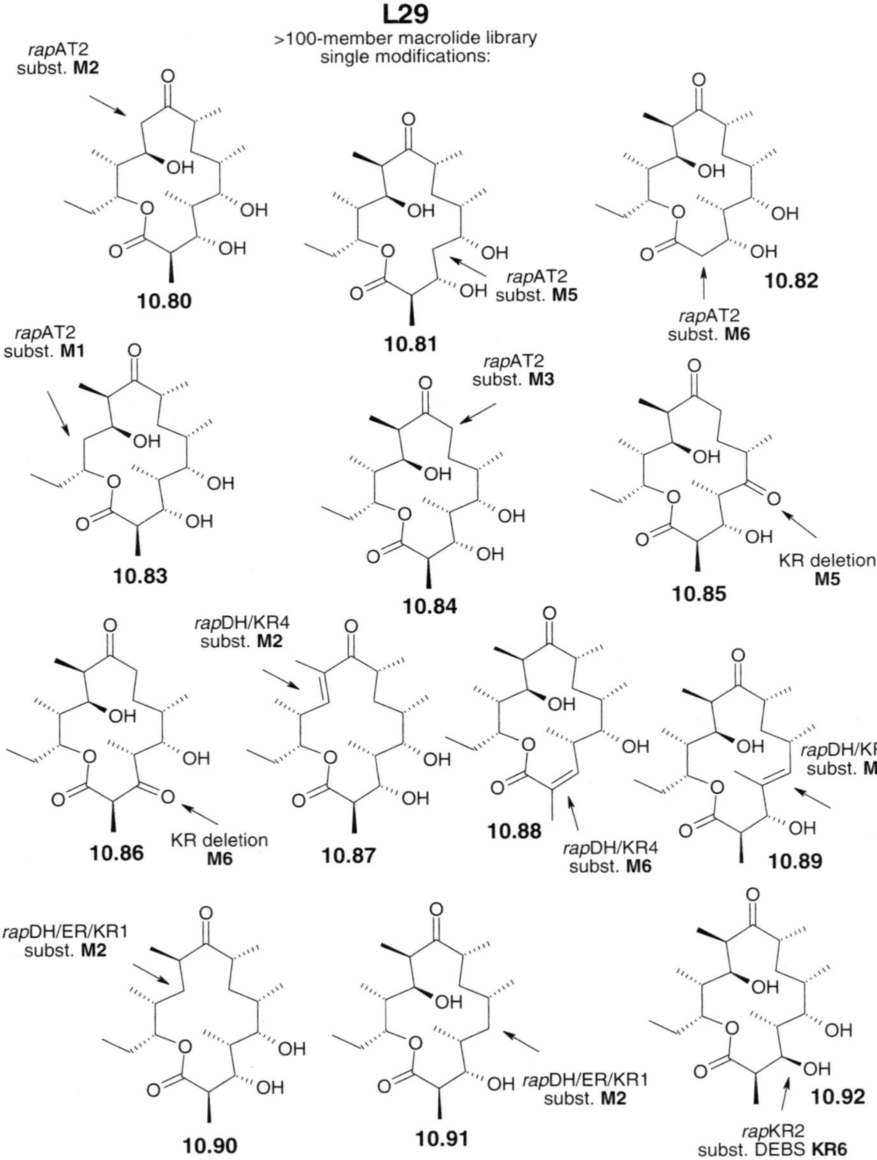

Figure 10.48 Combinatorial biosynthesis of the PK library **L29** from modified DEBS modules: single-domain substitution products.

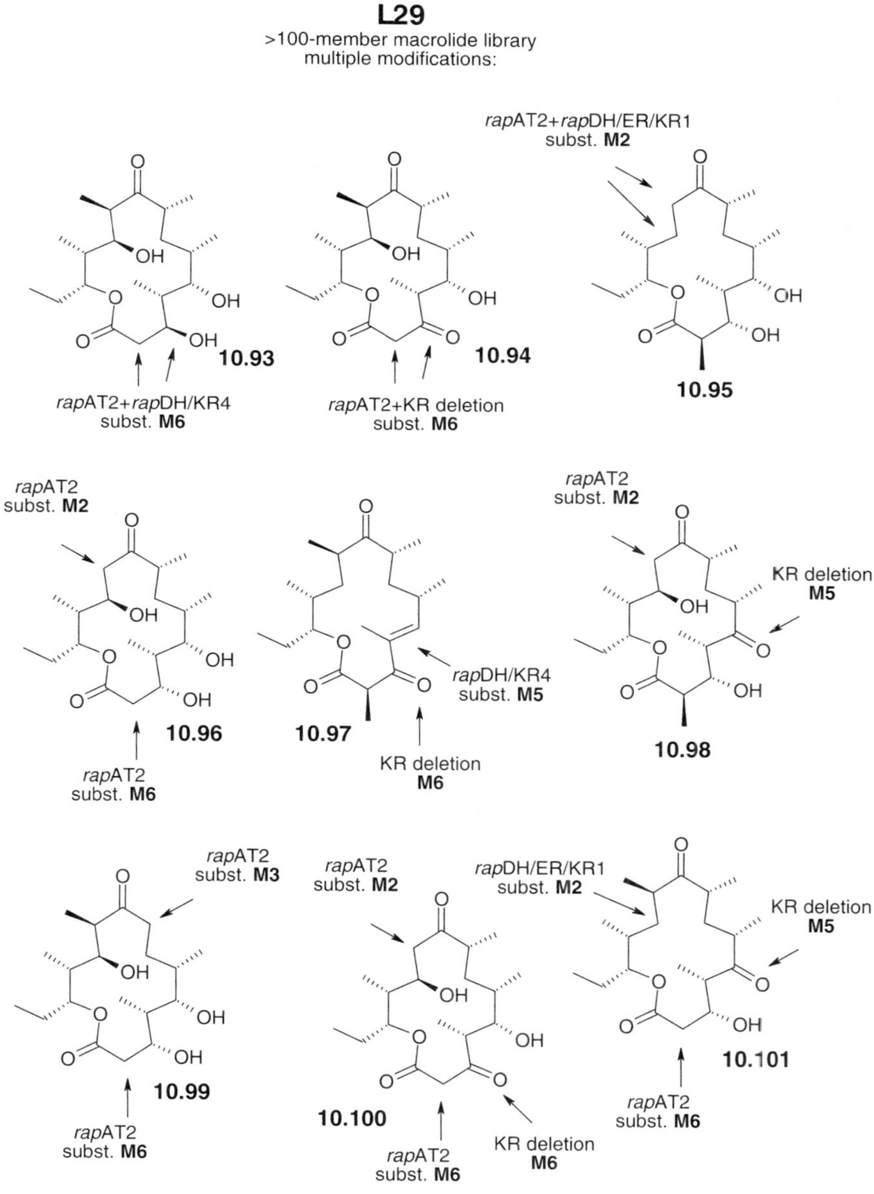

Figure 10.49 Combinatorial biosynthesis of the PK library **L29** from modified DEBS modules: double- and triple-domain substitution products.

rifampicin **M2**, *rap*AT2, gave various desmethyl erythronolide B derivatives **10.80–10.84**, whereas the replacement of their KR domains with a synthetic amino acidic linker to suppress ketone reduction gave keto erythronolides **10.85–10.86**. Anhydro erythronolides **10.87–10.89** were obtained by introducing the *rap*DH/KR4 domains of rifampicin **M4**, and inclusion of the *rap*DH/ER/KR4 **M4** domains led to deoxy erythronolides **10.90–10.91**. The epimeric erythronolide **10.92** was formed when the *rap*KR2 was incorporated. Double or triple mutations were also inserted into DEBS (Fig. 10.49), in the same module (**10.93–10.95**), in two modules (**10.96–10.99**), or even in three modules (**10.100–10.101**), to generate a total of >100 novel PKs, including several by-products obtained from each of the manipulated multienzyme proteins, to give the library **L29**.

Another report by Xue et al. (298) presented a multiple-plasmid strategy employed to increase exponentially the number of polyketides obtainable from a limited number of experiments using DEBS as a test PKS. A library of 43 fully characterized polyketides (6dEB, 11 single mutations, 26 double mutations, 5 triple mutations) was obtained.

The flexibility of each DEBS module was proven by this work, which will almost certainly be followed in the near future by further work aimed at the mutation of this and other macrolide pathways. The major obstacle encountered during this work was the marked reduction in productivity (ranging from 1 to 70% of the wild-type **10.70**), especially when several modules were modified. The use of higher yielding replacements and alterations along with careful optimization of productivity by genetic means could overcome this problem.

10.4 COMBINATORIAL BIOCATALYSIS

10.4.1 General Considerations

Combinatorial chemistry has proven its usefulness for the synthesis of chemical libraries with different degrees of complexity embedded in the scaffolds and the building blocks used. The synthesis of polyfunctionalized molecules in a combinatorial format, though, often requires the careful adjustment of experimental conditions and the selection of orthogonal protecting groups to prevent side reactions, degradation, or problems with regioselectivity. The synthesis of libraries of complex, chiral compounds has mostly been an unattainable target for combinatorial chemists.

Enzymes have often been used as reagents in organic reactions (299, 300). However, several new directions in the development of biocatalysts such as the utilization of enzymes from extremophiles (301, 302), nonaqueous enzyme technology (303, 304), and directed evolution (305, 306) now ensure the wider applicability of purified enzymes, and even whole cells, to organic biotransformations (307). Many of these enzymes are commercially available, inexpensive, and able to perform a wide range of chemical transformations, including the introduction of new functional groups on a scaffold, the modification of existing functionalities, and addition onto functional groups. Their most appealing features as reagents in combinatorial chemistry are the conversion of substrates under mild conditions, the absence of side-products, their complete regio- and stereoselectivity, and the broad substrate specificity that is common for many enzymes. Their use for combinatorial purposes, especially on complex substrates bearing functional groups susceptible to enzymatic reactions, could allow a series of modifications of the original scaffold, including sequential multiple transformations to be carried out. A general scheme for so-called *combinatorial biocatalysis* (308) is depicted in Fig. 10.50, showing the parallel with natural evolution in living cells (left) and illustrating the selection of drug properties from chemical leads (right). Combinatorial biocatalysis has been applied to both the unbiased derivatization of small organic scaffolds and to the focused decoration of complex natural products, and both approaches are described through consideration of two examples in the following sections.

10.4.2 An Example: Synthesis of a Small Molecule–Derived Biocatalysis Library

Khmelnitsky et al. (309) reported the synthesis of solution-phase libraries via combinatorial biocatalysis using a number of different substrates. One of these, bicyclo[2.2.2]oct-5-ene-2,3-*trans* dimethanol (BOD, **10.102**), is shown in Fig. 10.51 together with the biotransformation strategy that was applied. The selected reactions

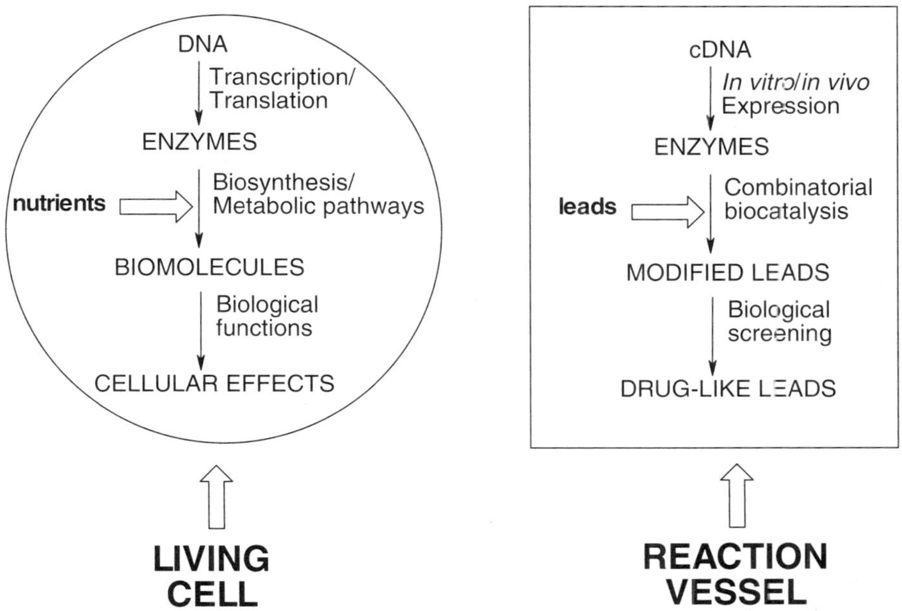

Figure 10.50 Combinatorial biocatalysis: comparison with living cell processes.

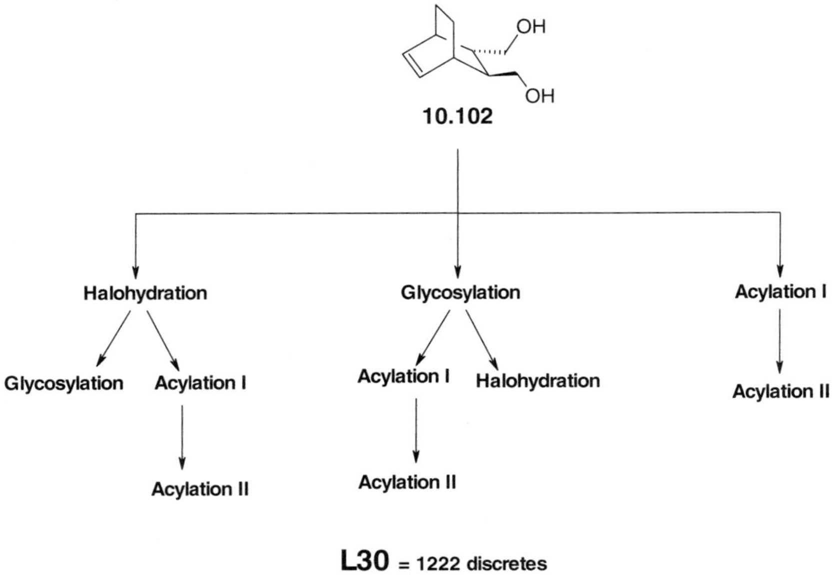

Figure 10.51 Combinatorial biocatalysis: synthesis of the BOD-focused biocatalytic discrete library **L30**.

introduced new functional groups on BOD (halohydration) or selectively functionalized one of the existing alcoholic functions (glycosylation, acylation). The reactions were carried out either in organic or in aqueous solvents, and the experimental conditions were adjusted to take account of the stability of the enzyme and the solubility of the substrate. Each biotransformation was monitored using chromatographic methods, and the final products were characterized by MS and NMR. The different sequence of enzymes used and the specificity of each of them produced a large, diverse library of BOD analogues. Some selected structures of this biotransformation library **L30** (1222 derivatives in total) are shown in Fig. 10.52 to illustrate the variety of functional groups, products, and physicochemical properties that can be obtained by such an approach. It is noteworthy that all the isolated compounds are also optically active.

Khmelnitsky also reported using adenosine and 2,3-(methylene dioxy)benzaldehyde as substrates to create libraries of 92 and 457 compounds, respectively. Taxol has been acylated to give a library of 200 compounds, among which two showed a significant improvement in water solubility compared to the parent compound.

10.4.3 An Example: Synthesis of a Bergenin-Derived Library By Combinatorial Biocatalysis

Mozhaev et al. (310) have reported the synthesis of an 167-member, focused library **L31** obtained by biocatalytic manipulation of bergenin, a polyhydroxylated flavonoid

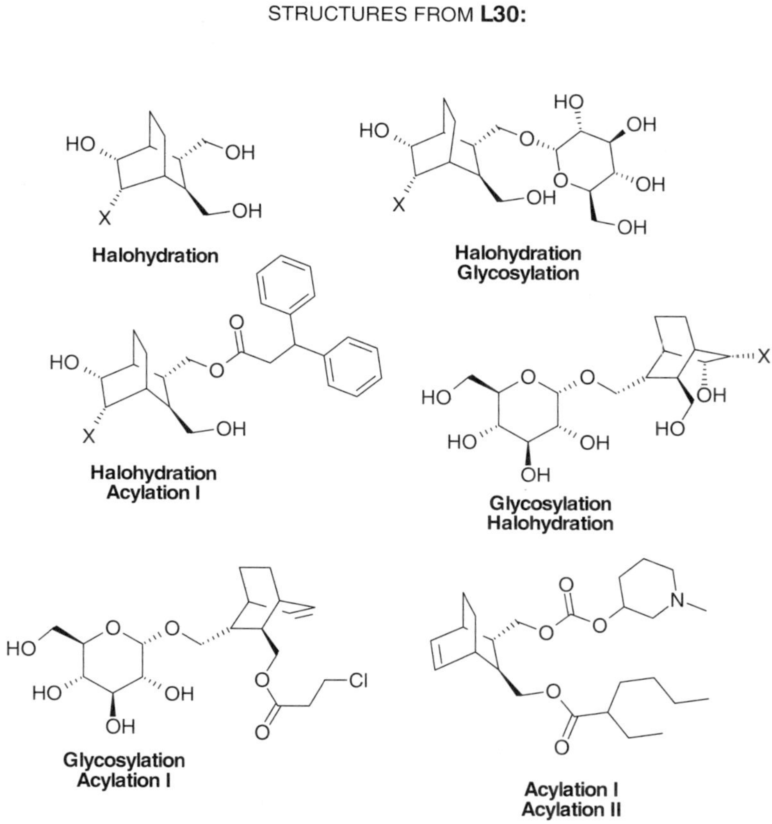

Figure 10.52 Combinatorial biocatalysis: structures of several individuals from the biocatalytic focused discrete library **L30**.

(**10.103**, Fig. 10.53). A thorough search among commercially available acylating enzymes was carried out to select those able to regioselectively acylate any of the five OH groups. However, although many 11-regioselective enzymes able to discriminate the primary OH from the others were found, only one 4,11-diacylating enzyme, subtilisin Carlsberg, was found that was able to acylate any of the other hydroxyls. The enzymes were purchased and immobilized (311) or lyophilized in the presence of KCl (312) to facilitate handling and to increase their catalytic activity. Various chemical conditions for acylation of **10.103** were also attempted, but invariably only mixtures of mono-, di-, and triacylated bergenins were obtained.

The library synthetic scheme is reported in Fig. 10.53. The synthesis was performed in 96-well plates using an automated liquid dispensing unit for sampling liquid aliquots. Bergenin was first submitted to 11-regioselective acylation with a mixture of four immobilized lipases (step a) and the acyl donor monomer set $\mathbf{M_1}$ (12 representatives, Fig. 10.54) in organic solvents, using 168 reaction wells. Purification of the

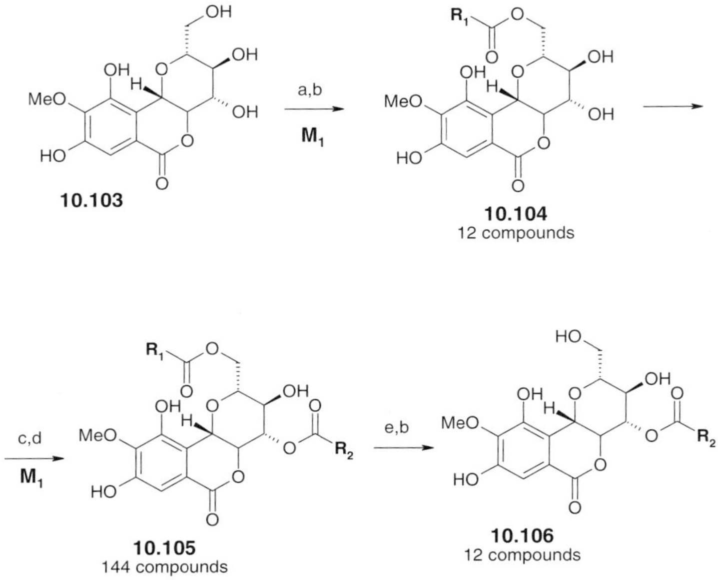

L31 = **10.104**+**10.105**+**10.106** = **167** discretes
(one non-confirmed diacylated library member)

a: lipase catalytic mixture (PS30, FAP-15, Chirazyme L-2, Chirazyme L-9), acetonitrile, 45°C;
b: separation of the immobilized enzyme mixture by filtration; c: Subtilisin Carlsberg/95% KCl, acetonitrile, toluene, DMSO, 45°C; d: separation of liophilized subtilisin by centrifugation; e: lipase mixture as in step a, acetonitrile/water 98/2, 45°C, 96 hrs.

Figure 10.53 Synthesis of the bergenin-focused discrete library **L31** using combinatorial biocatalysis.

products (step b) included removal of the immobilized enzymes by filtration of the solutions through the filter of the plates, evaporation of solvents and extraction of the excess of the acyl donor with *n*-hexane. Twelve wells containing twelve 11-acyl bergenins **10.104** were analyzed and archived, while the monoacylated residues in the other wells were treated with lyophilized subtilisin Carlsberg and the same acyl donor set $\mathbf{M_1}$ (step c) followed by centrifugation to remove the enzyme (step d). One hundred and forty-four wells containing all the possible combinations of homo- and hetero-4,11-diacylated bergenins **10.105** were characterized and archived, while the remaining 12 were treated with the same enzyme cocktail used in step a but in presence of water (step e). In this case, the 11-acyl group was regioselectively hydrolyzed and twelve 4-acylated bergenins **10.106** were obtained after the usual purification procedure (step b, Fig. 10.53).

Only one diacylated derivative out of the 168 reaction products in **L31** was not confirmed after HPLC/MS quality control, probably due to the steric hindrance of the acyl donor, whereas all the other library individuals were obtained in 60–90% yields

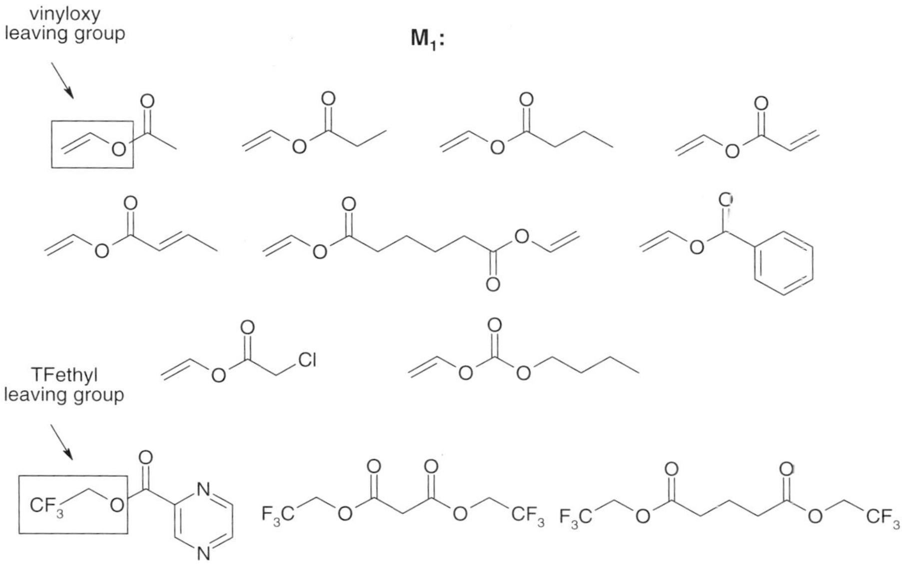

Figure 10.54 Monomers $\mathbf{M_1}$ used to prepare the bergenin-focused discrete library **L31**.

and good purities. The synthesis of **L31** by conventional organic synthesis would have required a complex multistep protection–deprotection strategy, and the final products would undoubtedly have been obtained with lower yields.

A larger 600-member bergenin-based library using acylations, oxidations, halogenations, and glycosylations and a 24-member 3,6-dihydroxytropane library have been prepared by the same group (313); a library of acylated taxol derivatives was also reported (314). A survey of theoretical applications for combinatorial biocatalysis has recently appeared (315).

REFERENCES

1. Stahl, S. and Uhlen, M., *Trends Biotechnol.* **15**, 185–192 (1997).
2. Miller, G. P., Zhong, W., Smiley, J. and Benkovic, S. J., in *Combinatorial Chemistry: Synthesis and Application*, S. R. Wilson and A. W. Czarnik (Eds.). Wiley, New York, 1997, pp. 241–259.
3. Kay, B. K., Kurakin, A. V. and Hyde-DeRuyscher, R., *Drug Discovery Today* **3**, 370–378 (1998).
4. Rodi, D. J. and Makowski, L., *Curr. Opin. Biotechnol.* **10**, 87–93 (1999).
5. Shusta, E. V., VanAntwerp, J. and Wittrup, K. D., *Curr. Opin. Biotechnol.* **10**, 117–122 (1999).

6. Mattheakis, L. C., Bhatt, R. R. and Dower, W. J., *Proc. Natl. Acad. Sci. USA* **91**, 9022–9026 (1994).
7. He, M. and Taussig, M. J., *Nucleic Acids Res.* **25**, 5132–5134 (1997).
8. Hanes, J. and Pluckthun, A., *Proc. Natl. Acad. Sci. USA* **94**, 4937–4942 (1997).
9. Makeyev, E. V., Kolb, V. A. and Spirin, A. S., *FEBS Letters* **444**, 177–180 (1999).
10. Roberts, R. W. and Szostak, J. W., *Proc. Natl. Acad. Sci. USA* **94**, 12297–12302 (1997).
11. Ohuchi, S., Nakano, H. and Yamane, T., *Nucleic Acids Res.* **26**, 4339–4346 (1998).
12. Tawfik, D. S. and Griffiths, A. D., *Nat. Biotechnol.* **16**, 652–656 (1998).
13. Buchholz, C. J., Peng, K.-W., Morling, F. J., Zhang, J., Cosset, F.-L. and Russell, S. J., *Nat. Biotechnol.* **16**, 951–954 (1998).
14. Ernst, W., Grabherr, R., Wegner, D., Borth, N., Grassauer, A. and Katinger, H., *Nucleic Acids Res.* **26**, 1718–1723 (1998).
15. Douglas, J. T., Miller, C. R., Kim, M., Dmitriev, I., Mikheeva, G., Krasnykh, V. and Curiel, D. T., *Nat. Biotechnol.* **17**, 470–475 (1999).
16. Samuelson, P., Cano, F., Robert, A. and Stahl, S., *FEMS Microbiol. Lett.* **179**, 131–139 (1999).
17. Gunneriusson, E., Samuelson, P., Ringdahl, J., Gronlund, H., Nygren, P.-A. and Stahl, S., *Appl. Environm. Microbiol.* **65**, 4134–4140 (1999).
18. Lu, Z. K., Murray, K. S., Van Cleave, V., LaVallie, E. R., Stahl, M. L. and McCoy, J. M., *Biotechnology* **13**, 366–372 (1995).
19. Wentzel, A., Christmann, A., Kratzner, R. and Kolmar, H., *J. Biol. Chem.* **274**, 21037–21043 (1999).
20. Christmann, A., Walter, K., Wentzel, A., Kratzner, R. and Kolmar, H., *Prot. Eng.* **12**, 797–806 (1999).
21. Daugherty, P. S., Olsen, M. J., Iverson, B. L. and Georgiou, G., *Protein Eng.* **11**, 825–832 (1998).
22. Dhillon, J. K., Drew, P. D. and Porter, A. J. R., *Lett. Appl. Microbiol.* **28**, 350–354 (1999).
23. Georgiou, G., Stathopoulos, C., Daugherty, P. S., Nayak, A. R., Iverson, B. L. and Curtiss, R. III, *Nat. Biotechnol.* **15**, 29–34 (1997).
24. Boder, E. T. and Wittrup, K. D., *Nat. Biotechnol.* **15**, 553–557 (1997).
25. Phizicky, E. M. and Fields, S., *Microbiol. Rev.* **59**, 94–123 (1995).
26. Brachmann, R. K. and Boeke, J. D., *Curr. Opin. Biotechnol.* **8**, 561–568 (1997).
27. Cereghino, G. P. L. and Cregg, J. M., *Curr. Opin. Biotechnol.* **10**, 422–427 (1999).
28. Johnson, N. and Varshavsky, A., *Proc. Natl. Acad. Sci. USA* **91**, 10340–10344 (1994).
29. Aronheim, A., Zandi, E., Hennemann, H., Elledge, S. J. and Karin, M., *Mol. Cell. Biol.* **17**, 3094–3102 (1997).
30. Broder, Y. C., Katz, S. and Aronheim, A., *Curr. Biol.* **8**, 1121–1128 (1998).
31. Enright, A. J., Iliopoulos, I., Kyrpides, N. C. and Ouzounis, C. A., *Nature* **402**, 86–90 (1999).
32. Smith, G. P., *Science* **228**, 1315–1317 (1985).
33. Parmley, S. F. and Smith, G. P., *Gene* **73**, 305–318 (1988).
34. Ilyichev, A. A., *Proc. Acad. Sci. USSR* **307**, 196–203 (1989).
35. Scott, J. K. and Smith, G. P., *Science* **249**, 386–390 (1990).

36. Devlin, J. J., Panganiban, L. C. and Devlin, P. E., *Science* **249**, 404–406 (1990).
37. Cwirla, S. E., Peters, E. A., Barrett, R. W. and Dower, W. J., *Proc. Natl. Acad. Sci. USA* **87**, 6378–6382 (1990).
38. Ren, Z.-J. and Black, L. W., *Gene* **215**, 439–444 (1998).
39. Rondot, S., Anthony, K. G., Diubel, S., Ida, N., Wiemann, S., Beyreuther, K., Frost, L. S., Little, M. and Breitling, F., *J. Mol. Biol.* **279**, 589–603 (1998).
40. Santini, C., Brennan, D., Mennuni, C., Hoess, R. H., Nicosia, A., Cortese, R. and Luzzago, A., *J. Mol. Biol.* **282**, 125–135 (1998).
41. Gao, C., Mao, S., Lo, C.-H. L., Wirsching, P., Lerner, R. A. and Janda, K. D., *Proc. Natl. Acad. Sci. USA* **96**, 6025–6030 (1999).
42. Dower, W. J., Miller, J. A. and Ragsdale, C. W., *Nucleic Acids Res.* **16**, 6127–6145 (1988).
43. Collins, J., *Ann. Rep. in Combi. Chem. and Mol. Diversity* **1**, 210–262 (1997).
44. Smith, G. P. and Petrenko, V. A., *Chem. Rev.* **97**, 391–410 (1997).
45. Levitan, B., *J. Mol. Biol.* **277**, 893–916 (1998).
46. Thompson, J. E. and Williams, A. J., in *Molecular Biomethods Handbook*, R. Rapley and J. M. Walker (Eds.). Humana, Totowa,. NJ, 1998, pp. 581–594.
47. Zwick, M. B., Shen, J. and Scott, J. K., *Curr. Opin. Biotechnol.* **9**, 427–436 (1998).
48. Cesareni, G., Castagnoli, L. and Cestra, G., *Combi. Chem. High Throughput Synth.* **2**, 1–17 (1999).
49. Yip, Y. L. and Ward, R. L., *Comb. Chem. High Throughput Screening* **2**, 125–138 (1999).
50. Forrer, P., Jung, S. and Pluckthun, A., *Curr. Opin. Struct. Biol.* **9**, 514–520 (1999).
51. Dani, M., in *Combinatorial Chemistry and Combinatorial Technologies: Methods and Applications*, S. Miertus and G. Fassina (Eds.). Marcel Dekker, Inc., New York, 1999, pp. 275–299.
52. Dani, M., in *Combinatorial Chemistry and Combinatorial Technologies: Methods and Applications*, S. Miertus and G. Fassina (Eds.). Marcel Dekker, Inc., New York, 1999, pp. 301–322.
53. Kraft, S., Diefenbach, B., Mehta, R., Jonczyk, A., Luckenbach, G. A. and Goodman, S. L., *J. Biol. Chem.* **274**, 1979–1985 (1999).
54. Sato, A., Ida, N., Fukuyama, M., Miwa, K., Kazami, J. and Nakamura, H., *Biochemistry* **35**, 10441–10447 (1996).
55. Szardenings, M., Tornroth, S., Mutulis, F., Muceniece, R., Keinanen, K., Kuusinen, A. and Wikberg, J. E. S., *J. Biol. Chem.* **272**, 27943–27948 (1997).
56. Renschler, M. F., Dower, W. J. and Levy, R., *Methods Mol. Biol.* **87**, 209–234 (1998).
57. Pierce, H. H., Schachat, F., Brandt, P. W., Lombardo, C. R. and Kay, B. K., *J. Biol. Chem.* **273**, 23448–23453 (1998).
58. Krook, M., Lindbladh, C., Eriksen, J. A. and Mosbach, K., *Molecular Diversity* **3**, 149–159 (1998).
59. Bartoli, F., Nuzzo, M., Urbanelli, L., Bellintani, F., Prezzi, C., Cortese, R. and Monaci, P., *Nat. Biotechnol.* **16**, 1068–1073 (1998).
60. Reineke, U., Sabat, R., Misselwitz, R., Welfle, H., Volk, H.-D. and Schneider-Mergener, J., *Nat. Biotechnol.* **17**, 271–275 (1999).
61. Chirinos-Rojas, C. L., Steward, M. W. and Partidos, C. D., *Immunology* **96**, 109–113 (1999).

62. Blank, M., Shoenfeld, Y., Cabilly, S., Heldman, Y., Fridkin, M. and Katchalski-Katzir, E., *Proc. Natl. Acad. Sci. USA* **96**, 5164–5168 (1999).
63. Cwirla, S. E., Dower, W. J. and Li, M., *Methods Mol. Biol.* **128**, 131–141 (1999).
64. Houimel, M., Mach, J.-P., Corthesy-Theulaz, I., Corthesy, B. and Fisch, I., *Eur. J. Biochem.* **262**, 774–780 (1999).
65. McConnell, S. J., Thon, V. J. and Spinella, D. G., *Comb. Chem. High Throughput Screening* **2**, 155–163 (1999).
66. Koivunen, E., Restel, B. H., Rajotte, D., Lahdenranta, J., Hagedorn, M., Arap, W. and Pasqualini, R., *Meth. Mol. Biol.* **129**, 3–17 (1999).
67. Wang, B., Yang, H., Liu, Y.-C., Jelinek, T., Zhang, L., Ruoslahti, E. and Fu, H., *Biochemistry* **38**, 12499–12504 (1999).
68. Pasqualini, R., *Quart. J. Nucl. Med.* **43**, 159–162 (1999).
69. Coombs, G. S., Bergstrom, R. C., Pellequer, J.-L., Baker, S. I., Navre, M., Smith, M. M., Tainer, J. A., Madison, E. L. and Corey, D. R., *Chem. Biol.* **5**, 475–488 (1998).
70. Lilja, H., Abrahamsson, P. A. and Lundwall, A., *J. Biol. Chem.* **264**, 1894–1900 (1989).
71. Kiczak, L., Koscielska, K., Otlewski, J., Czerwinski, M. and Dadlez, M., *Biol. Chem.* **380**, 101–105 (1999).
72. Smith, M. M., Shi, L. and Navre, M., *J. Biol. Chem.* **270**, 6440–6449 (1995).
73. Harris, J. L., Peterson, E. P., Hudig, D., Thornberry, N. A. and Craik, C. S., *J. Biol. Chem.* **273**, 27364–27373 (1998).
74. Gee, S. H.,Sekely, S.A., Lombardo, C., Kurakin, A., Froehner, S. C. and Kay, B. K., *J. Biol. Chem.* **273**, 21980–21987 (1998).
75. Pasqualini, R., Koivunen, E. and Ruoslahti, E., *J. Cell. Biol.* **130**, 1189–1196 (1995).
76. Wrighton, N. C., Farrell, F. X., Chang, R., Kashyap, A. K., Barbone, F. P., Mulcahy, L. S., Johnson, D. L., Barrett, R. W., Jolliffe, L. K. and Dower, W. J., *Science* **273**, 458–463 (1996).
77. Oligino, L., Lung, F.-D. T., Sastry, L., Bigelow, J., Cao, T., Curran, M., Burke, T. R. B. Jr.; Wang, S., Krag, D., Roller, P. P. and King, C. R., *J. Biol. Chem.* **272**, 29046–29052 (1997).
78. Pierce, H. H., Adey, N. and Kay, B. K., *Molecular Diversity* **1**, 259–265 (1996).
79. Ladner, R. C., *Quart. J. Nucl. Med.* **43**, 119–124 (1999).
80. Rodi, D. J., Janes, R. W., Sanganee, H. J., Holton, R. A., Wallace, B. A. and Makowski, L., *J. Mol. Biol.* **285**, 197–203 (1999).
81. Koide, A., Bailey, C. W., Huang, X. and Koide, S., *J. Mol. Biol.* **284**, 1141–1151 (1998).
82. Leahy, D. J., Hendrickson, W. A., Aukhil, I. and Erickson, H. P., *Science* **258**, 987–991 (1992).
83. Dickinson, C. D., Veerapandian, B., Dai, X. P., Hamlin, R. C., Nguyen, H. X., Ruoslahti, E. and Ely, K. R., *J. Mol. Biol.* **236**, 1079–1092 (1994).
84. Ward, E. S., Gussow, D., Griffiths, A. D., Jones, P. T. and Winter, G., *Nature* **341**, 544–546 (1989).
85. McCafferty, J., Griffiths, A. D., Winter, G. and Chiswell, D. J., *Nature* **348**, 552–554 (1990).
86. Barbas, C. F. III, Bain, J. D., Hoekstra, D. M., and Lerner, R. A., *Proc. Natl. Acad. Sci. USA* **89**, 4457–4461 (1992).

87. Davies, J. and Riechmann, L., *Biotechnology* **13**, 475–479 (1995).

88. Arkin, M. R. and Wells, J. A., *J. Mol. Biol.* **284**, 1083–1094 (1998).

89. Mulligan-Kehoe, M. J. and Russo, A., *J. Mol. Biol.* **289**, 41–55 (1999).

90. Lamminmaki, U., Pauperio, S., Westerlund-Karlsson, A., Karvinen, J., Virtanen, P. L., Lovgren, T. and Saviranta, P., *J. Mol. Biol.* **291**, 589–602 (1999).

91. Pessi, A., Bianchi, E., Crameri, A., Venturini, S., Tramontano, A. and Sollazzo, M., *Nature* **362**, 367–369 (1993).

92. Martin, F., *EMBO J.* **13**, 5303–5309 (1994).

93. McConnell, S. J. and Hoess, R. H., *J. Mol. Biol.* **250**, 460–470 (1995).

94. Ku, J. and Schultz, P. G., *Proc. Natl. Acad. Sci. USA* **92**, 6552–6556 (1995).

95. Nord, K., Nilsson, J., Nilsson, B., Uhlen, M. and Nygren, P.-A., *Protein Eng.* **8**, 601–608 (1995).

96. Rudgers, G. W. and Palzkill, T., *J. Biol. Chem.* **274**, 6963–6971 (1999).

97. Nygren, P.-A. and Uhlen, M., *Curr. Opin. Struct. Biol.* **7**, 463–469 (1997).

98. Hoess, R. H., in *Combinatorial Chemistry and Molecular Diversity in Drug Discovery*, E. M. Gordon and J. F. Kerwin Jr. (Eds.). Wiley, New York, 1998, pp. 389–399.

99. Huse, W. D., Sastry, L., Iverson, S. A., Kang, A. S., Alting-Mees, M., Burton, D. R., Benkovic, S. J. and Lerner, R. A., *Science* **246**, 1275–1281 (1989).

100. Steinberger, P., Kraft, D. and Valenta, R., *J. Biol. Chem.* **271**, 10967–10972 (1996).

101. Andersen, P. S., Stryhn, A., Hansen, B. E., Fugger, L., Engberg, J. and Buus, S., *Proc. Natl. Acad. Sci. USA* **93**, 1820–1824 (1996).

102. Pini, A., Viti, F., Santucci, A., Carnemolla, B., Zardi, L., Neri, P. and Neri, D., *J. Biol. Chem.* **273**, 21769–21776 (1998).

103. Osbourn, J. K., Earnshaw, J. C., Johnson, K. S., Parmentier, M., Timmermans, V. and McCafferty, J., *Nat. Biotechnol.* **16**, 778–781 (1998).

104. Rader, C., Cheresh, D. A. and Barbas, C. F. III, *Proc. Natl. Acad. Sci. USA* **95**, 8910–8915 (1998).

105. Huls, G. A., Heijnen, I. A. F. M., Cuomo, M. E., Koningsberger, J. C., Wiegman, L., Boel, E., Van Der Vuurst De Vries, A.-R., Loyson, S. A. J., Helfrich, W., Van Berge Henegouwen, G. P., Van Meijer, M., De Kruif, J. and Logtenberg, T., *Nat. Biotechnol.* **17**, 276–281 (1999).

106. de Bruin, R., Spelt, K., Mol, J., Koes, R. and Quatrocchio, F., *Nat. Biotechnol.* **17**, 397–399 (1999).

107. Beste, G., Schmidt, F. S., Stibora, T. and Skerra, A., *Proc. Natl. Acad. Sci. USA* **96**, 1898–1903 (1999).

108. Kirkham, P. M., Neri, D. and Winter, G., *J. Mol. Biol.* **285**, 909–915 (1999).

109. Tang, Y., Beuerlein, G., Pecht, G., Chilton, T., Huse, W. D. and Watkins, J. D., *J. Biol. Chem.* **274**, 27371–27378 (1999).

110. Wu, H., Nie, Y., Huse, W. D. and Watkins, J. D., *J. Mol. Biol.* **294**, 151–162 (1999).

111. Carnemolla, B., Neri, D., Castellani, P., Leprini, A., Neri, G., Pini, A., Winter, G. and Zardi, L., *Int. J. Cancer* **68**, 397–405 (1996).

112. Fan, J., Griffiths, A. D., Lockhart, A., Cross, R. A. and Amos, L. A., *J. Mol. Biol.* **259**, 325–330 (1996).

113. Stausbol-Gron, B., Wind, T., Kjaer, S., Kahns, L., Hansen, N. J. V., Kristensen, P., Clark, B. F. C., *FEBS Lett.* **391**, 71–75 (1996).

114. Vaughan, T. J., Williams, A. J., Pritchard, K., Osbourn, J. K., Pope, A. R., Earnshaw, J. C., McCafferty, J., Hodits, R. A., Wilton, J. and Johnson, K. S., *Nat. Biotechnol.* **14**, 309–314 (1996).

115. Schmaljohn, C., Custer, D., VanderZanden, L., Spik, K., Rossi, C. and Bray, M., *Virology* **258**, 189–200 (1999).

116. Reiter, Y., Schuck, P., Boyd, L. F. and Plaksin, D., *J. Mol. Biol.* **290**, 685–698 (1999).

117. Qiu, J., Kai, M., Padlan, E. A. and Marcus, D. M., *J. Neuroimmunol.* **97**, 172–181 (1999).

118. Gao, C., Bruemmer, O., Mao, S. and Janda, K. D., *J. Am. Chem. Soc.* **121**, 6517–6518 (1999).

119. Jung, S., Honegger, A. and Pluckthun, A., *J. Mol. Biol.* **294**, 163–180 (1999).

120. Harper, K., Toth, R. L., Mayo, M. A. and Torrance, L., *J. Virol. Meth.* **81**, 159–168 (1999).

121. Burton, D. R. and Barbas, C. F. III, *Adv. Immunol.* **57**, 191–280 (1994).

122. Hoogenboom, H. R., *Trends Biotechnol.* **15**, 62–70 (1997).

123. Rader, C. and Barbas, C. F. III, *Curr. Opin. Biotechnol.* **8**, 503–508 (1998).

124. Griffiths, A. D. and Duncan, A. R., *Curr. Opin. Biotechnol.* **9**, 102–108 (1998).

125. Bradbury, A., Sblattero, D., Hoogenboom, H. R. and Hufton, S., in *Combinatorial Chemistry and Combinatorial Technologies: Methods and Applications*, S. Miertus and G. Fassina (Eds.). Marcel Dekker, Inc., New York, 1999, pp. 323–364.

126. Lerner, R. A., Benkovic, S. J. and Schultz, P. G., *Science* **252**, 659–667 (1991).

127. Fujii, I., Fukuyama, S., Iwabuchi, Y. and Tanimura, R., *Nat. Biotechnol.* **16**, 463–467 (1998).

128. Iwabuchi, Y., Miyashita, H., Tanimura, R., Kinoshita, K., Kikuchi, M. and Fujii, I., *J. Am. Chem. Soc.* **116**, 771–772 (1994).

129. Baca, M., Scanlan, T. S., Stephenson, R. C. and Wells, J. A., *Proc. Natl. Acad. Sci. USA* **94**, 10063–10068 (1997).

130. Tanaka, F., Almer, H., Lerner, R. A. and Barbas, C. F. III, *Tetrahedron Lett.* **40**, 8063–8066 (1999).

131. Brummer, O., Gao, C., Mao, S., Weiner, D. P. and Janda, K. D., *Lett. Pept. Sci.* **6**, 295–302 (1999).

132. Gao, C., Lavey, B. J., Lo, C.-H. L., Datta, A., Wentworth, P. Jr. and Janda, K. D., *J. Am. Chem. Soc.* **120**, 2211–2217 (1998).

133. Janda, K. D., Lo, L.-C., Lo, C.-H. L., Sim, M.-M., Wang, R., Wong, C.-H. and Lerner, R. A., *Science* **275**, 945–948 (1997).

134. Gao, C., Lin, C.-H., Lo, C.-H. L., Mao, S., Wirsching, P., Lerner, R. A. and Janda, K. D., *Proc. Natl. Acad. Sci. USA* **94**, 11777–11782 (1997).

135. Demartis, S., Huber, A., Viti, F., Lozzi, L., Giovannoni, L., Neri, P., Winter, G. and Neri, D., *J. Mol. Biol.* **286**, 617–633 (1999).

136. Jestin, J.-L., Kristensen, P. and Winter, G., *Angew. Chem., Int. Ed. Engl.* **38**, 1124–1127 (1999).

137. Yamabhai, M., Hoffman, N. G., Hardison, N. L., McPherson, P. S., Castagnoli, L., Cesareni, G. and Kay, B. K., *J. Biol. Chem.* **273**, 31401–31407 (1998).

138. Sparks, A. B., Hoffman, N. G., McConnell, S. J., Fowlkes, D. M. and Kay, B. K., *Nat. Biotech.* **14**, 741–744 (1996).

139. Crameri, R. and Walter, G., *Comb. Chem. High Throughput Screening* **2**, 63–72 (1999).

140. Sche, P. P., McKenzie, K. M., White, J. D. and Austin, D. J., *Chem. Biol.* **6**, 707–716 (1999).

141. Ellington, A. D., *Curr. Biol.* **4**, 427–429 (1994).

142. Uphoff, K. W., Bell, S. D. and Ellington, A. D., *Curr. Opin. Struct. Biol.* **6**, 281–288 (1996).

143. Kruger, K., Grabowski, P. J., Zaug, A. J., Sands, J., Gottschling, D. E. and Cech, T. R., *Cell* **31**, 147–157 (1982).

144. Guerrier-Takada, C., Gardiner, K., Marsh, T., Pace, N. and Altman, S., *Cell* **35**, 849–857 (1983).

145. Mills, D. R., Peterson, R. G. and Spiegelman, S., *Proc. Natl. Acad. Sci. USA* **58**, 217–224 (1967).

146. Conrad, R. C., Giver, L., Tian, Y. and Ellington, A. D., *Methods Enzymol.* **267**, 336–367 (1996).

147. Osborne, S. E. and Ellington, A. D., *Chem. Rev.* **97**, 349–370 (1997).

148. Eaton, B. E., *Curr. Opin. Chem. Biol.* **1**, 10–16 (1997).

149. Mei, H.-Y. and Czarnik, A. W., in *Combinatorial Chemistry and Molecular Diversity in Drug Discovery*, E. M. Gordon and J. F. Kerwin Jr. (Eds.). Wiley, New York, 1998, pp. 165–178.

150. Bacher, J. M. and Ellington, A. D., *Drug Disc. Today* **3**, 265–273 (1998).

151. Famulok, M. and Jenne, A., *Curr. Opin. Chem. Biol.* **2**, 320–327 (1998).

152. Famulok, M., *Ernst Schering Res. Found. Workshop* **26**, 135–149 (1998).

153. Famulok, M., *Curr. Opin. Struct. Biol.* **9**, 324–329 (1999).

154. Jaschke, A., Frauendorf, C. and Hausch, F., *Synlett*, 825–833 (1999).

155. Jayasena, S. D., *Clin. Chem.* **45**, 1628–1650 (1999).

156. Haller, A. A. and Sarnow, P., *Proc. Natl. Acad. Sci. USA* **94**, 8521–8526 (1997).

157. Burke, D. H. and Hoffman, D. C., *Biochemistry* **37**, 4653–4663 (1998).

158. Lorsch, J. R. and Szostak, J. W., *Biochemistry* **33**, 973–982 (1994).

159. Burgstaller, P. and Famulok, M., *Angew. Chem., Int. Ed. Engl.* **33**, 1084–1087 (1994).

160. Zimmermann, G. R., Jenison, R. D., Wick, C. L., Simorre, J.-P. and Pardi, A., *Nat. Struct. Biol.* **4**, 644–649 (1997).

161. Sassanfar, M. and Szostak, J. W., *Nature* **364**, 550–553 (1993).

162. Ulrich, H., Ippolito, J. E., Pagan, O. R., Eterovic, V. A., Hann, R. M., Shi, H., Lis, J. T.; Eldefrawi, M. E. and Hess, G. P., *Proc. Natl. Acad. Sci. USA* **95**, 14051–14056 (1998).

163. Burke, D. H., Hoffman, D. C., Brown, A., Hansen, M., Pardi, A. and Gold, L., *Chem. Biol.* **4**, 833–843 (1997).

164. Lato, S. M., Boles, A. R. and Ellington, A. D., *Chem. Biol.* **2**, 291–303 (1995).

165. Wang, Y. and Rando, R. R., *Chem. Biol.* **2**, 281–290 (1995).

166. Famulok, M. and Huttenhofer, A., *Biochemistry* **35**, 4265–4270 (1996).

167. Hofstadler, S. A., Sannes-Lowery, K. A., Crooke, S. T., Ecker, D. J., Sasmor, H., Manalili, S. and Griffey, R. H., *Anal. Chem.* **71**, 3436–3440 (1999).

168. Lebruska, L. L. and Maher, L. J., *Biochemistry* **38**, 3168–3174 (1999).

169. O'Connell, D., Koenig, A., Jennings, S., Hicke, B., Han, H.-L., Fitzwater, T., Chang, Y.-F., Varki, N., Parma, D. and Varki, A., *Proc. Natl. Acad. Sci. USA* **93**, 5883–5887 (1996).

170. Brown, D., Brown, J., Kang, C., Gold, L. and Allen, P., *J. Biol. Chem.* **272**, 14969–14974 (1997).

171. Gal, S. W., Amontov, S., Urvil, P. T., Vishnuvardhan, D., Nishikawa, F., Kumar, P. K. R. and Nishikawa, S., *Eur. J. Biochem.* **252**, 553–562 (1998).

172. Klug, S. J., Huttenhofer, A. and Famulok, M., *RNA* **5**, 1180–1190 (1999).

173. Shi, H., Hoffman, B. E. and Lis, J. T., *Proc. Natl. Acad. Sci. USA* **96**, 10033–10038 (1999).

174. Hirao, I., Spingola, M., Peabody, D. and Ellington, A. D., *Mol. Diversity* **4,** 75–89 (1999).

175. Paillart, J. C., Marquet, R., Skripkin, E., Ehresmann, C. and Ehresmann, B., *Biochimie* **78**, 639–653 (1996).

176. Boiziau, C., Dausse, E., Yurchenko, L. and Toulme, J.-J., *J. Biol. Chem.* **274**, 12730–12737 (1999).

177. Ferber, M. J. and Maher, L. J., *J. Mol. Biol.* **279**, 565–576 (1998).

178. Homann, M. and Goringer, H. U., *Nucleic Acids Res.* **27**, 2006–2014 (1999).

179. Bruno, J. G. and Kiel, J. L., *Biosens. Bioelectron.* **14**, 457–464 (1999).

180. Davis, K. A., Lin, Y., Abrams, B. and Jayasena, S. D., *Nucleic Acids Res.* **26**, 3915–3924 (1998).

181. Smith, D., Kirschenheuter, G. P., Charlton, J., Guidot, D. M. and Repine, J. E., *Chem. Biol.* **2**, 741–750 (1995).

182. Charlton, J., Sennello, J. and Smith, D., *Chem. Biol.* **4**, 809–816 (1997).

183. Bridonneau, P., Chang, Y.-F., O'Connell, D., Gill, S. C., Snyder, D. W., Johnson, L., Goodson, T. Jr., Herron, D. K. and Parma, D. H., *J. Med. Chem.* **41**, 778–786 (1998).

184. Lee, S. and Sullenger, B. A., *Nat. Biotechnol.* **15**, 41–45 (1997).

185. Draheim, S. E., Bach, N. J., Dillard, R. D., Berry, D. R., Carlson, D. G., Chirgadze, N. Y., Clawson, D. K., Hartley, L. W., Johnson, L. M. et al., *J. Med. Chem.* **39**, 5159–5175 (1996).

186. King, D. J., Ventura, D. A., Brasier, A. R., Gorenstein, D. G., *Biochemistry* **37**, 16489–16493 (1998).

187. Jhaveri, S., Olwin, B. and Ellington, A. D., *Bioorg. Med. Chem. Lett.* **8**, 2285–2290 (1998).

188. Schumacher, T. N. M., Mayr, L. M., Minor, D. L. Jr., Milhollen, M. A., Burgess, M. W. and Kim, P. S., *Science* **271**, 1854–1857 (1996).

189. Klussmann, S., Nolte, A., Bald, R., Erdmann, V. A. and Fuerste, J. P., *Nat. Biotechnol.* **14**, 1112–1115 (1996).

190. Nolte, A., Klussmann, S., Bald, R., Erdmann, V. A. and Fuerste, J. P., *Nat. Biotechnol.* **14**, 1116–1119 (1996).

191. Williams, K. P., Liu, X.-H., Schumacher, T. N. M., Lin, H. Y., Ausiello, D. A., Kim, P. S. and Bartel, D. P., *Proc. Natl. Acad. Sci. USA* **94**, 11285–11290 (1997).

192. Joyce, G. F. and Orgel, L. E., *The RNA World*, R. F. Gesteland and J. F. Atkins (Eds.). Cold Spring Harbor Lab., Cold Spring Harbor, NY, 1993, pp. 1–25.

193. Breaker, R. R., *Chem. Rev.* **97**, 371–390 (1997).

194. Lorsch, J. R. and Szostak, J. W., *Acc. Chem. Res.* **29**, 103–110 (1996).

195. Yarus, M., *Curr. Opin. Chem. Biol.* **3**, 260–267 (1999).

196. Li, Y. and Breaker, *Curr. Opin. Struct. Biol.* **9**, 315–323 (1999).
197. Tanner, N. K., *FEMS Microb. Rev.* **23**, 257–275 (1999).
198. Carola, C. and Eckstein, F., *Curr. Opin. Chem. Biol.* **3**, 274–283 (1999).
199. Soukup, G. A. and Breaker, R. R., *Trends Biotechnol.* **17**, 469–476 (1999).
200. Beaudry, A. A. and Joyce, G. F., *Science* **257**, 635–641 (1992).
201. Tsang, J. and Joyce, G. F., *Biochemistry* **33**, 5966–5973 (1994).
202. Tsang, J. and Joyce, G. F., *J. Mol. Biol.* **262**, 31–42 (1996).
203. Zaug, A. J., Been, M. D. and Cech, T. R., *Nature* **324**, 429–433 (1986).
204. Ekland, E. H. and Bartel, D. P., *Nucleic Acids Res.* **23**, 3231–3238 (1995).
205. Schmitt, T. and Lehman, N., *Chem. Biol.* **6**, 857–869 (1999).
206. Ordoukhanian, P. and Joyce, G. F., *Chem. Biol.* **6**, 881–889 (1999).
207. Tuschl, T., Sharp, P. A. and Bartel, D. P., *EMBO J.* **17**, 2637–2650 (1998).
208. Zarrinkar, P. P. and Sullenger, B. A., *Biochemistry* **38**, 3426–3432 (1999).
209. Pierce, M. L. and Ruffner, D. E., *Nucleic Acids Res.* **26**, 5093–5101 (1998).
210. Putlitz, J. Z., Yu, Q., Burke, J. M. and Wands, J. R., *J. Virol.* **73**, 5381–5387 (1999).
211. Cole, K. B. and Dorit, R. L., *J. Mol. Biol.* **292**, 931–944 (1999).
212. Landweber, L. F. and Pokrovskaya, I. D., *Proc. Natl. Acad. Sci. USA* **96**, 173–178 (1999).
213. Kazakov, S. and Altman, S., *Proc. Natl. Acad. Sci. USA* **89**, 7939–7943 (1992).
214. Dange, V., Van Atta, R. B. and Hecht, S. M., *Science* **248**, 585–588 (1990).
215. Vaish, N. K., Heaton, P. A., Fedorova, O. and Eckstein, F., *Proc. Natl. Acad. Sci. USA* **95**, 2158–2162 (1998).
216. Yu, Q., Pecchia, D. B., Kingsley, S. L., Heckman, J. E. and Burke, J. M., *J. Biol. Chem.* **273**, 23524–23533 (1998).
217. Robertson, M. P. and Ellington, A. D., *Nat. Biotechnol.* **17**, 62–66 (1999).
218. Santoro, S. W. and Joyce, G. F., *Proc. Natl. Acad. Sci. USA* **94**, 4262–4266 (1997).
219. Bartel, D. P., *Cold Spring Harbor Monogr. Ser.* **37**, 143–162 (1999).
220. Pitsch, S., Eschenmoser, A., Gedulin, B., Hui, S. and Arrhenius, G., *Origin Life Evol. Biosphere* **25**, 297–334 (1995).
221. Robertson, M. P. and Miller, S. L., *Nature* **375**, 772–774 (1995).
222. Unrau, P. J. and Bartel, D. P., *Nature* **395**, 260–263 (1998).
223. Mizuno, Y., *Chem. Pharm. Bull.* **10**, 647–652 (1962).
224. Zakour, R. A. and Loeb, L. A., *Nature* **295**, 708–710 (1982).
225. Zhang, B. and Cech, T. R., *Nature* **390**, 96–100 (1997).
226. Illangasekare, M., Sanchez, G., Nickles, T. and Yarus, M., *Science* **267**, 643–647 (1995).
227. Lohse, P. A. and Szostak, J. W., *Nature* **381**, 442–444 (1996).
228. Wiegand, T. W., Janssen, R. C. and Eaton, B. E., *Chem. Biol.* **4**, 675–683 (1997).
229. Jenne, A. and Famulok, M., *Chem. Biol.* **5**, 23–34 (1998).
230. Breaker, R. R. and Joyce, G. F., *J. Mol. Evolut.* **40**, 551–558 (1995).
231. Tang, J. and Breaker, R. R., *Chem. Biol.* **4**, 453–459 (1997).
232. Roth, A. and Breaker, R. R., *Proc. Natl. Acad. Sci. USA* **95**, 6027–6031 (1998).

233. Seelig, B. and Jaschke, A., *Chem. Biol.* **6**, 167–176 (1999).

234. Seelig, B. and Jaschke, A., *Bioconjugate Chem.* **10**, 371–378 (1999).

235. Tarasow, T. M., Tarasow, S. L. and Eaton, B. E., *Nature* **389**, 54–57 (1997).

236. Chun, S.-M., Jeong, S., Kim, J.-M., Chong, B.-O., Park, Y.-K., Park, H. and Hu, J., *J. Am. Chem. Soc.* **121**, 10844–10845 (1999).

237. Roberts, R. W. and Ja, W. W., *Curr. Opin. Struct. Biol.* **9**, 521–529 (1999).

238. Waterman, P. G., *Adv. Drug Discovery Tech.*, 13–23 (1998).

239. Fallis, A. G., *Can. J. Chem.* **77**, 159–177 (1999).

240. Donadio, S., Staver, M. J., McAlpine, J. B., Swanson, S. J. and Katz, L., *Science* **252**, 675–679 (1991).

241. Khosla, C., in *Combinatorial Chemistry and Molecular Diversity in Drug Discovery*, E. M. Gordon and J. F. Kerwin Jr. (Eds.). Wiley, New York, 1998, pp. 401–417.

242. O'Hagan, D., *The Polyketide Metabolites.* Ellis Horwood, Chichester, UK, 1991.

243. Katz, L. and Donadio, S., *Annu. Rev. Microbiol.* **47**, 875–912 (1993).

244. O'Hagan, D., *Nat. Prod. Rep.* **12**, 1–32 (1995).

245. Hutchinson, C. R. and Fujii, I., *Annu. Rev. Microbiol.* **49**, 201–238 (1995).

246. McDaniel, R., Ebert-Khosia, S., Hopwood, D. A. and Khosla, C., *Science* **262**, 1546–1550 (1993).

247. Kao, C. M., Katz, L. and Khosla, C., *Science* **265**, 509–512 (1994).

248. Tsoi, C. J. and Khosla, C., *Chem. Biol.* **2**, 355–362 (1995).

249. Rohr, J., *Angew. Chem., Int. Ed. Engl.* **34**, 881–885 (1995).

250. Rohr, J., Mendez, C. and Salas, J. A., *Bioorg. Chem.* **27**, 41–54 (1999).

251. Hutchinson, C. R., *Proc. Natl. Acad. Sci. USA* **96**, 3336–3338 (1999).

252. Khosla, C., Gokhale, R. S., Jacobsen, J. R. and Cane, D. E., *Annu. Rev. Biochem.* **68**, 219–253 (1999).

253. McDaniel, R., Ebert-Khosia, S., Hopwood, D. A. and Khosla, C., *Nature* **375**, 549–554 (1995).

254. Yu, T.-W., Shen, Y., McDaniel, R., Floss, H. G., Khosla, C., Hopwood, D. A. and Moore, B. S., *J. Am. Chem. Soc.* **120**, 7749–7759 (1998).

255. Wohlert, S.-E., Blanco, G., Lombo, F., Fernandez, E., Brana, A. F., Reich, S., Udvarnoki, G., Mendez, C., Decker, H., Frevert, J., Salas, J. A. and Rohr, J., *J. Am. Chem. Soc.* **120**, 10596–10601 (1998).

256. Prado, L., Lombo, F., Brana, A. F., Mendez, C., Rohr, J. and Salas, J. A., *Mol. Gen. Genet.* **261**, 216–225 (1999).

257. Shen, Y., Yoon, P., Yu, T.-W., Floss, H. G., Hopwood, D. and Moore, B. S., *Proc. Natl. Acad. Sci. USA* **96**, 3622–3627 (1999).

258. Bangera, M. G. and Thomashow, L. S., *J. Bacteriol.* **181**, 3155–3163 (1999).

259. Kennedy, J., Auclair, K., Kendrew, S. G., Park, C., Vederas, J. C. and Hutchinson, C. R., *Science* **284**, 1368–1372 (1999).

260. Hendrickson, L., Davis, C. R., Roach, C., Nguyen, D. K., Aldrich, T., McAda, P. C. and Reeves, C. D., *Chem. Biol.* **6**, 429–439 (1999).

261. Zawada, R. J. X. and Khosla, C., *Chem. Biol.* **6**, 607–615 (1999).

262. Kunnari, T., Kantola, J., Ylihonko, K., Klika, K. D., Mantsala, P. and Hakala, J., *J. Chem. Soc., Perkin Trans. 2*, 1649–1652 (1999).

263. Olano, C., Lomovskaya, N., Fonstein, L., Roll, J. T. and Hutchinson, C. R., *Chem. Biol.* **6**, 845–855 (1999).

264. Petkovic, H., Thamchaipenet, A., Zhou, L.-H., Hranueli, D., Raspor, P. , Waterman, P. G. and Hunter, I. S., *J. Biol. Chem.* **274**, 32829–32834 (1999).

265. Ferrer, J.-L., Jez, J. M., Bowman, M. E., Dixon, R. A. and Noel, J. P., *Nat. Struct. Biol.* **6**, 775–784 (1999).

266. Cortes, J., Haydock, S. F., Roberts, G. A., Bevitt, D. J. and Leadlay, P. F., *Nature* **348**, 176–178 (1990).

267. Caffrey, P., Bevitt, D. J., Staunton, J. and Leadlay, P. F., *FEBS Lett.* **304**, 225–228 (1992).

268. Weissman, K. J., Bycroft, M., Cutter, A. L., Hanefeld, U., Frost, E. J., Timoney, M. C., Harris, R., Handa, S., Roddis, M., Staunton, J. and Leadlay, P. F., *Chem. Biol.* **5**, 743–754 (1998).

269. Gokhale, R. S., Hunziker, D., Cane, D. E. and Khosla, C., *Chem. Biol.* **6**, 117–125 (1999).

270. Holzbaur, I. E., Harris, R. C., Bycroft, M., Cortes, J., Bisang, C., Staunton, J., Rudd, B. A. M. and Leadlay, P. F., *Chem. Biol.* **6**, 189–195 (1999).

271. Bisang, C., Long, P. F., Cortes, J., Westcott, J., Crosby, J., Matharu, A.-L., Cox, R. J., Simpson, T. J., Staunton, J. and Leadlay, P. F., *Nature* **401**, 502–505 (1999).

272. Ranganathan, A., Timoney, M., Bycroft, M., Cortes, J., Thomas, I. P., Wilkinson, B., Kellenberger, L., Hanefeld, U., Galloway, I. S., Staunton, J. and Leadlay, P. F., *Chem. Biol.* **6**, 731–741 (1999).

273. Khosla, C. and Zawada, R. J. X., *Trends Biotechnol.* **14**, 335–341 (1996).

274. Leadlay, P. F., *Curr. Opin. Chem. Biol.* **1**, 162–168 (1997).

275. Katz, L., *Chem. Rev.* **97**, 2557–2576 (1997).

276. Khosla, C., *Chem. Rev.* **97**, 2577–2590 (1997).

277. Staunton, J., *Curr. Opin. Chem. Biol.* **2**, 339–345 (1998).

278. Cane, D. E., Walsh, C. T. and Khosla, C., *Science* **282**, 63–68 (1998).

279. Katz, L. and McDaniel, R., *Med. Res. Rev.* **19**, 543–558 (1999).

280. MacNeil, D. J., Occi, J. L., Gewain, K. M., MacNeil, T., Gibbons, P. H., Ruby, C. L. and Danis, S. J., *Gene* **115**, 119–125 (1992).

281. Swan, D. G., Rodriguez, A. M., Vilches, C., Mendez, C. and Salas, J. A., *Mol. Gen. Genet.* **242**, 358–362 (1994).

282. Aparicio, J. F. and Leadlay, P. F., *Gene* **169**, 9–16 (1996).

283. Xue, Y., Zhao, L., Liu, H.-W. and Sherman, D. H., *Proc. Natl. Acad. Sci. USA* **95**, 12111–12116 (1998).

284. Aparicio, J. F., Colina, A. J., Ceballos, E. and Martin, J. F., *J. Biol. Chem.* **274**, 10133–10139 (1999).

285. Tang, L., Fu, H., Betlach, M. C. and McDaniel, R., *Chem. Biol.* **6**, 553–558 (1999).

286. Zhao, L., Ahlert, J., Xue, Y., Thorson, J. S., Sherman, D. H. and Liu, H.-W., *J. Am. Chem. Soc.* **121**, 9881–9882 (1999).

287. Yu, T., Shen, Y., Doi-Katayama, Y., Tang, L., Park, C., Moore, B. S., Hutchinson, C. R. and Floss, H. G., *Proc. Natl. Acad. Sci. USA* **96**, 9051–9056 (1999).

288. Hu, Z., Hunziker, D., Hutchinson, C. R. and Khosla, C., *Microbiology* **145**, 2335–2341 (1999).

289. Shaw-Reid, C. A., Kelleher, N. L., Losey, H. C., Gehring, A. M., Berg, C. and Walsh, C. T., *Chem. Biol.* **6**, 385–400 (1999).

290. Stachelhaus, T., Mootz, H. D. and Marahiel, M. A., *Chem. Biol.* **9**, 493–505 (1999).

291. Hutchinson, C. R., *Curr. Opin. Microbiol.* **1**, 319–329 (1998).

292. Cane, D. E. and Walsh, C. T., *Chem. Biol.* **6**, R319–R325 (1999).

293. Trefzer, A., Bechthold, A. and Salas, J. A., *Nat. Prod. Rep.* **16**, 283–299 (1999).

294. Hallis, T. M. and Liu, H.-W., *Acc. Chem. Res.* **32**, 579–588 (1999).

295. Jacobsen, J. R., Cane, D. E. and Khosla, C., *J. Am. Chem. Soc.* **120**, 9096–9097 (1998).

296. Gokhale, R. S., Tsuji, S. Y., Cane, D. E. and Khosla, C., *Science* **284**, 482–485 (1999).

297. McDaniel, R., Thamchaipenet, A., Gustafsson, C., Fu, H., Betlach, M., Betlach, M. and Ashley, G., *Proc. Natl. Acad. Sci. USA* **96**, 1846–1851 (1999).

298. Xue, Q., Ashley, G., Hutchinson, C. R. and Santi, D. V., *Proc. Natl. Acad. Sci. USA* **96**, 11740–11745 (1999).

299. Dordick, J. S., Patil, D. R., Parida, S., Ryu, K. and Rethwisch, D. G., *Chem. Ind.* **47**, 267–292 (1992).

300. Sutherland, A. G., *Annu. Rep. Prog. Chem., Sect. B* **89**, 281–298 (1993).

301. Madigan, M. T. and Marrs, B. L., *Sci. Am.* **276**, 82–87 (1997).

302. Hough, D. W. and Danson, M. J., *Curr. Opin. Chem. Biol.* **3**, 39–46 (1999).

303. Okahata, Y. and Mori, T., *Trends Biotechnol.* **15**, 50–54 (1997).

304. Khmelnitsky, Y. L. and Rich, J. O., *Curr. Opin. Chem. Biol.* **3**, 47–53 (1999).

305. Arnold, F. H. and Volkov, A. A., *Curr. Opin. Chem. Biol.* **3**, 54–59 (1999).

306. Reetz, M. T. and Jaeger, K.-E., *Top. Curr. Chem.* **200**, 31–57 (1999).

307. Dordick, J. S., Khmelnitsky, Y. L. and Sergeeva, M. V., *Curr. Opin. Microbiol.* **1**, 311–318 (1998).

308. Altreuter, D. H. and Clark, D. S., *Curr. Opin. Biotechnol.* **10**, 130–136 (1999).

309. Khmelnitsky, Y. L., Michels, P. C., Dordick, J. S. and Clark, D. S., in *Molecular Diversity and Combinatorial Chemistry: Libraries and Drug Discovery*, I. M. Chaiken and K. D. Janda (Eds.). ACS, Washington, DC, 1996, pp. 144–157.

310. Mozhaev, V. V., Budde, C. L., Rich, J. O., Usyatinsky, A. Y., Michels, P. C., Khmelnitsky, Y. L., Clark, D. S. and Dordick, J. S., *Tetrahedron* **54**, 3971–3982 (1998).

311. Svensson, I., Wehtje, E., Adlercreutz, P. and Mattiasson, B., *Biotechnol. Bioeng.* **44**, 549–556 (1994).

312. Khmelnitski, Y. L., Welch, S. H., Clark, D. S. and Dordick, J. S., *J. Am. Chem. Soc.* **116**, 2647–2648 (1994).

313. Michels, P. C., Khmelnitsky, Y. L., Dordick, J. S. and Clark, D. S., *Trends Biotechnol.* **16**, 210–215 (1998).

314. Khmelnitsky, Y. L., Budde, C., Arnold, J. M., Usyatinsky, A., Clark, D. S. and Dordick, J. S., *J. Am. Chem. Soc.* **119**, 11554–11555 (1997).

315. Krstenansky, J. L. and Khmelnitsky, Y.L., *Bioorg. Med. Chem.* **7**, 2157–2162 (1999).

11 Materials and Polymeric Combinatorial Libraries

Last, but not least, in terms of importance are the emerging fields of materials science and polymeric combinatorial libraries, which are extensively covered in this chapter. The methodologies and the protocols that are connected with these areas (synthesis, characterization, screening) are extremely different from the corresponding steps for synthetic organic libraries. However, this book does not pretend to cover extensively areas such as the synthetic methods for obtaining composite inorganic materials or quality polymer composites. These methodologies are briefly described here, as related to several examples of library preparation, and references are provided for the interested reader to expand his or her knowledge in inorganic/polymer sciences.

A survey of *solution-* and *solid-phase synthetic methods* for producing materials science libraries introduces the subject of this chapter, briefly mentioning their main features and highlighting the usefulness of each to specific combinatorial applications. A second section is devoted to the *characterization* and *high-throughput screening* of materials science libraries, with a number of examples illustrating the throughput, reliability, and user-friendliness of each screening/detection technique. Several recent reviews (1–11) covering the whole field of materials libraries should be consulted by the interested reader to enrich the information provided in this chapter.

Polymer libraries are covered according to their numerous applications, each described through a specific example. The reported examples include libraries of copolymers as liquid/solid supports with different compositions, libraries of biodegradable materials for clinical applications, libraries of stationary phases for GC/LC separations, libraries of polymeric reagents or catalysts, libraries of artificial polymeric receptors or molecularly imprinted polymers, and libraries of polymeric biosensors. The opportunities that could arise in the near future from novel applications of polymer libraries are also briefly discussed.

11.1 SYNTHESIS OF MATERIALS SCIENCE LIBRARIES

11.1.1 General Considerations

Inorganic SP chemistry is an important branch of chemistry that has been the subject of extensive research and has also been extensively reviewed (12–15). The development of new materials with specific properties is a major endeavor for chemical research. New catalysts, or superconductors, or photoluminescent materials, or ferroelectrics/dielectrics, or liquid crystals, or polymeric materials are among the most relevant targets. The possibilities offered by the combination of stable elements are almost infinite, in that tens of elements can be combined in almost any relative proportion. Each resulting material is by definition different and is characterized by properties that can be exploited and optimized for a specific application.

Unfortunately, the ability to rationally predict the properties of a specific material composition is currently poor, and thus the rational design of enhanced materials is, at best, extremely difficult and long. Moreover, the synthesis and the characterization of such materials using classical inorganic SP chemistry methods is complex and time consuming. Full exploitation of material composites thus requires significantly higher throughput synthetic and screening methods, both to prepare large numbers of materials and to construct better predictive models to help drive the rational selection of elements and relative abundances in a composite. In this context combinatorial technologies appear ideally suited to boost the discovery of new materials via their synthesis, characterization, and screening.

Organic synthesis of a target molecule requires the design of a synthetic route, the selection of suitable, commercially available precursors, and the optimization of reaction conditions. It requires reaction monitoring and product characterization with various analytical techniques as well as work-up procedures to purify and isolate the target. The large amount of available knowledge, in terms of organic reaction mechanisms and the reactivity and stability of organic molecules, allows chemists to plan and carry out the above-mentioned steps, often even optimizing reported protocols according to target-specific needs. The transfer of classical protocols to solution- or solid-phase combinatorial protocols is also becoming an assessed field, as reported in previous chapters.

Inorganic solid-state chemistry is much simpler, in that only a few general synthetic methods exist to prepare a material of virtually any composition. The main issue is the preparation of a homogeneous material where all the components have completely diffused in the mixture to obtain the desired composition. Commonly encountered diffusion barriers could produce nonhomogeneous mixtures with varying compositions and thus prevent the synthesis of the desired material. Classical solid-state synthetic methods are based on intimate mixing and heating of finely powdered inorganic solids to create homogeneous new composites. They are hampered by the macroscopic size of the particles and often do not provide high-quality materials. A more promising technique is based on the sequential deposition of thin films of each component of the desired composite (16–18). The reduced thickness of the resulting film, typically in the range of several atomic layers, allows the total diffusion of each component in the film with no resistance. As of today, all the reported efforts in combinatorial materials science involving solid reagents have used thin-film deposition techniques, which are described in more detail in the next section. Liquid-phase techniques have also been used with success for combinatorial applications in a few reports and are thus also reported.

11.1.2 Synthesis of Materials Libraries by Thin-Film Deposition

Thin-film deposition of materials has been known for many years. It was originally described (19) as the first combinatorial method to produce mixtures of three components with varying relative percentages by simultaneous deposition of three films from the three corners of a triangle (components A, B, and C, Fig. 11.1). Subsequent improvements of the technique currently allow the deposition of several components by using sophisticated devices such as electronic guns and emission jets and the sequential rather than simultaneous application of the film layers. The use of an indefinite number of components is thus possible, providing that the thickness of each sequential deposition (usually from tens to hundreds of angstroms) is controlled to prevent the nucleation and crystallization of intermediate composites at the interfaces between precursor layers. Generally, amorphous materials are obtained from the deposition. Their crystallization is eventually promoted using high-temperature standard solid-state protocols.

In order to obtain combinatorial materials science libraries, sequential deposition must be coupled with a method to diversify the composition of small areas of the deposition surface. A moving-mask system, originally designed to obtain compositional gradients (20) and then used later for the synthesis of organic libraries (21), has been successfully and repeatedly used for this purpose. The first reported materials library **L1** (22) used this technique employing eight binary masks **M0–M7**, as shown in Fig. 11.2.

Four metal oxides/carbonates (Bi, Sr, Ca, and Cu) were used to prepare an 128-member magnetoresistant library **L1**. Each library component was assembled on a 1-mm-wide, 2-mm-long site inserted in a crystal substrate (Fig. 11.2). The sequential thin-film deposition was arranged according to the following scheme:

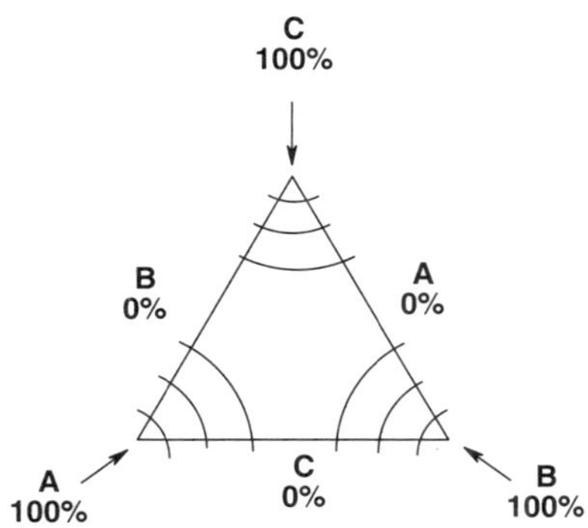

Figure 11.1 Simultaneous deposition of three materials as thin films with varying compositions on a triangular substrate.

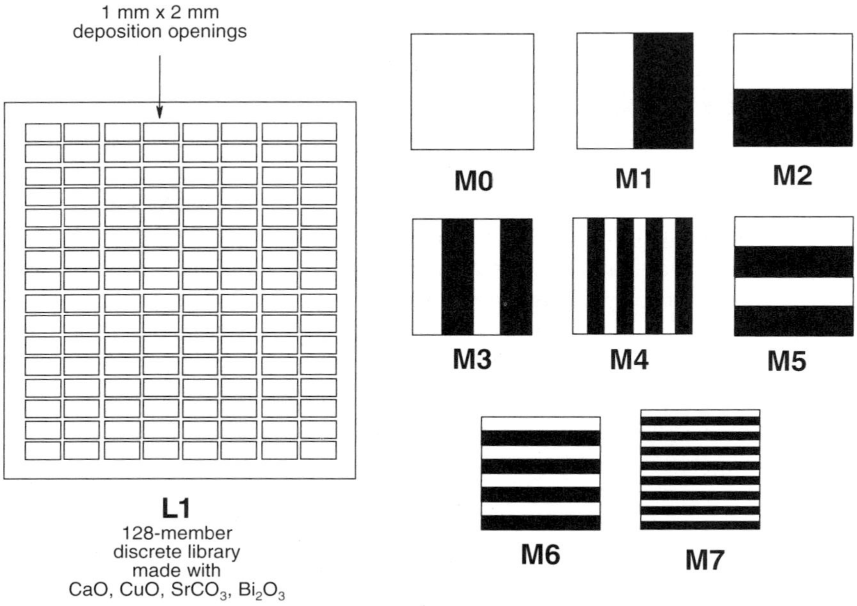

Figure 11.2 Structure of the magnetoresistant materials discrete library **L1** and of the moving masks **M1**–**M7** used for its synthesis.

1 Bi_2O_3, 300 Å, **M0**;
2. Bi_2O_3, 300 Å, **M1**;
3. CuO, 150 Å, **M0**;
4. CuO, 300 Å, **M2**;
5. CuO, 150 Å, **M3**;
6. $SrCO_3$, 300 Å, **M0**;
7. $SrCO_3$, 300 Å, **M5**;
8. CaO, 300 Å, **M6**;
9. CuO, 300 Å, **M4**; and
10. CaO, 300 Å, **M7**.

The resulting library individuals showed various magnetoresistance profiles and also provided a crude SAR related to magnetoresistance for these metal oxide mixtures (22). The use of deposition/masking solid-state techniques was subsequently reported using, among others, specific thin-film deposition protocols such as radiofrequency sputtering (23–25), physical vapor deposition (26–28), electrochemical deposition (29–31), electron beam evaporation (32, 33), and pulsed laser ablation (34–36). The library size was significantly increased (up to 25,000-member libraries; see the next section) by increasing the moving-masks complexity and by adopting deposition protocols with different properties (37). The exponential increase of publications related to combinatorial materials science libraries generated by thin-film deposition ensures a steady technological improvement that will soon allow the preparation of high-quality, extremely complex discrete libraries.

11.1.3 An Example: Synthesis of Primary and Focused Libraries of Luminescent Materials

Danielson et al. (38) reported the synthesis of an ≈25,000-member materials library **L2**, which was used as a source of luminescent red phosphor materials. The composition of **L2** and its synthesis are reported in Fig. 11.3.

Several metals, metal oxides, and metal carbonates were selected to include cations from groups IIA and IIIA (La_2O_3, Y_2O_3, MgO, $SrCO_3$, SnO_2, V, and Al_2O_3) and activator ions (Eu_2O_3, Tb_4O_7, Tm_2O_3, and CeO_2). They were deposited on an ≈ 8-cm-diameter silicon wafer that was predivided into 230-μm^2 elements, each hosting

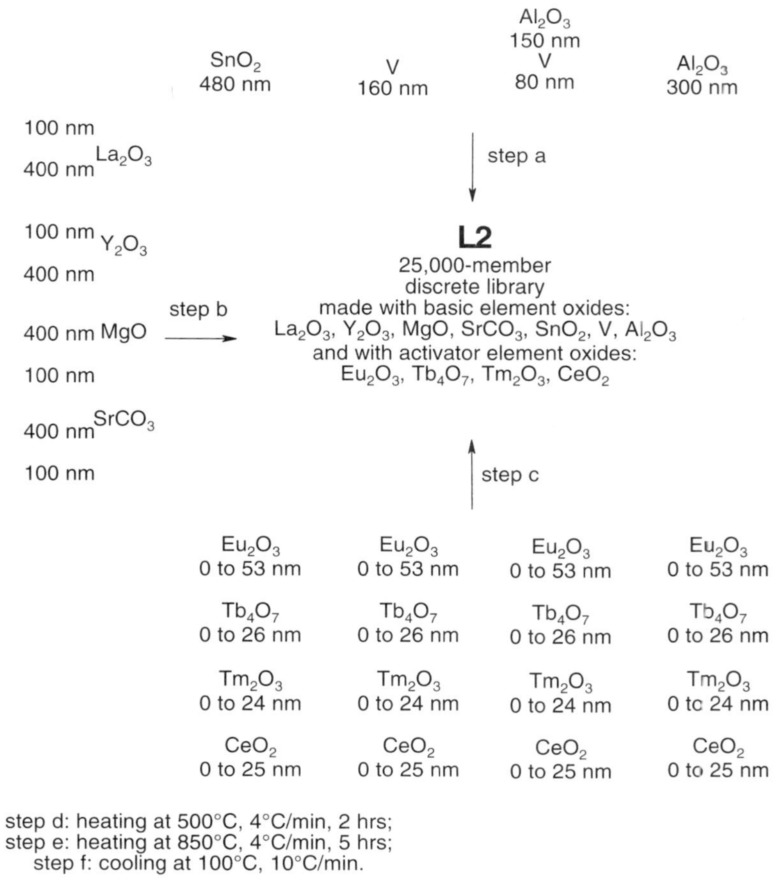

Figure 11.3 Structure and synthesis of the photoluminescent materials discrete library **L2**.

a library discrete with an empty space of 420 μm between each position to isolate the library elements. The thin films were deposited with an electron gun using solid, ultrapure pellets of the library components. At first four columns were produced (step a, Fig. 11.3) using a moving mask **M1** with a 19.1-mm-wide rectangular slit (one-quarter of the wafer width) and depositing, from left to right, SnO_2 (480 nm thickness), V (160 nm), Al_2O_3/V (150 + 80 nm), and Al_2O_3 (300 nm). This was followed by deposition of four rows using the same mask after a 90° rotation (step b), giving thin films of varying thicknesses, from top to bottom, of La_2O_3 (100 to 400 nm), Y_2O_3 (from 100 to 400 nm), MgO (from 400 to 100 nm), and $SrCO_3$ (from 400 to 100 nm). Each of the 16 subregions was then treated with thin films of activators using a moving mask **M2** with four 4.8-mm rectangular slits (one-sixteenth of the wafer width) to deposit layers of varying thicknesses (step c), from top to bottom, of Eu_2O_3 (from 0 to 53 nm), Tb_4O_7 (from 0 to 26 nm), Tm_2O_3 (from 0 to 24 nm), and CeO_2 (from 0 to 25 nm). Deposition was followed by slow heating to homogenize the deposited layers and to create the library individuals. **L2** was first heated to 500 °C using a 4 °C/min differential and maintained at 500 °C for 2 h (step d); then it was further heated to 850 °C using the same differential and kept at 850 °C for 5 h (step e) before cooling to 100 °C with a 10 °C/min differential (step f, Fig. 11.3). More than 2500 components showed red, green, or blue emission properties when excited at 254 nm with ultraviolet lamps. Some of the most active library individuals are reported in Table 11.1 on the basis of their chromaticity index (CIE, Table 11.1).

Red phosphors were the most active and numerous (>1700) library individuals found, with intensities comparable to known, commercially available luminescent materials. Their composition (Y, V, Al, and La with Eu as an activator) inspired the design of a focused library **L3** (Fig. 11.4), made by deposition of films of Eu_2O_3 (26.3 nm) and V (189.6 nm) on a triangular silicon wafer (step a), followed by the deposition

TABLE 11.1 Red, Green, and Blue Phosphors from L2

Composition	Ranking[a]	Relative Luminosity[b]	CIE (x, y)[c]
$Y_{0.34}V_{0.60}Eu_{0.06}$	1/1754	R, 1.00	0.65, 0.35
$Y_{0.35}V_{0.59}Eu_{0.06}$	4/1754	R, 0.95	0.65, 0.35
$Al_{0.34}La_{0.31}V_{0.28}Eu_{0.07}$	380/1754	R, 0.27	0.64, 0.34
$La_{0.51}V_{0.43}Eu_{0.06}$	516/1754	R, 0.21	0.64, 0.35
$Al_{0.21}Mg_{0.61}V_{0.17}Ce_{0.01}$	1/121	G, 1.00	0.40, 0.51
$Al_{0.22}Mg_{0.60}V_{0.17}Tb_{0.01}$	2/121	G, 1.00	0.38, 0.53
$Mg_{0.34}V_{0.63}Eu_{0.03}$	6/121	G, 0.73	0.37, 0.51
$Y_{0.41}V_{0.57}Tm_{0.02}$	1/728	B, 1.00	0.18, 0.09
$Y_{0.42}V_{0.56}Tm_{0.02}$	3/728	B, 0.94	0.18, 0.08
$Al_{0.38}Y_{0.31}V_{0.30}Tm_{0.01}$	29/728	B, 0.76	0.19, 0.11

[a]Relative luminosity ranking/population of the phosphor class.

[b]R = red, G = green, B = blue.

[c]Commercial standards: red, $Y_{1.9}O_2SEu_{0.1}$, $x = 0.629$, $y = 0.35$; green, ZnS–Cu, Al, $x = 0.266$, $y = 0.576$; blue: ZnS–Ag, Cl, $x = 0.15$, $y = 0.05$.

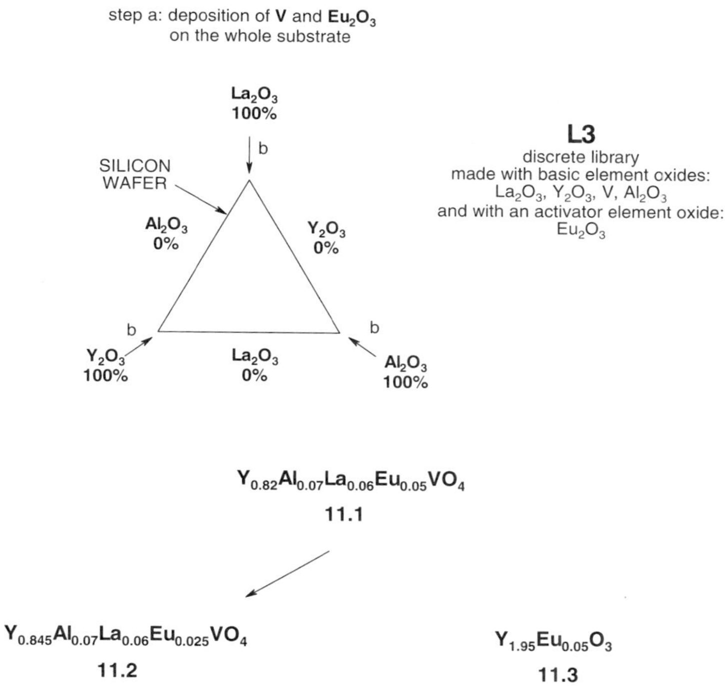

Figure 11.4 Structure of the photoluminescent materials discrete library **L3** and of the most active library individuals **11.1** and **11.2** from its screening.

of gradients of Y_2O_3, Al_2O_3, and La_2O_3 (step b, from 0 to 100%, Fig. 11.4). The composite **11.1** produced the maximum intensity as a red phosphor and was further optimized to the final compound **11.2** by variation of the activator concentration (Fig. 11.4).

Compound **11.2** was then prepared in large quantity and was fully characterized, showing similar if not better efficiency properties when compared with the commercial standard **11.3** (Fig. 11.4). Moreover, the quantities of both expensive elements Y and Eu were reduced in **11.2**, while V provided an increase of red chromaticity. A simple primary/focused library scheme thus allowed the identification of novel blue and green phosphor composites and the optimization of the properties and the reduction of the cost of efficient red phosphors.

The same library was used by the authors to identify a novel, blue-white emission composite identified as Sr_2CeO_4 (39, 40). Another prominent group identified magnetoresistant materials (41) and novel capacitors (42) from smaller discrete materials libraries, while joint efforts by a multi-lab group (43) proved the applicability of combinatorial technologies to the development of molecular plastic solar cells. Further examples will be described in the next section with greater focus on screening procedures.

11.1.4 Solution-Phase Synthesis of Materials Science Libraries

Solution-phase synthetic methods, as they were described for synthetic organic libraries, can also be applied to materials science and are devoid of the diffusion problems encountered in thin-film deposition. The reagent solutions are mixed and incubated following an appropriate procedure, and the final products are usually isolated by precipitation or crystallization. Automated liquid dispensing units with extreme precision and high reliability can be used in synthetic protocols. No major differences are presented in respect to solution-phase organic library synthesis (see Section 8.2.4). Several examples are briefly illustrated below to provide a quick overview of the currently reported synthetic methods in solution for materials libraries.

Hill and Damico Gall (44) reported the synthesis of a 39-member discrete library of homogeneous polyoxometalate (POM) catalysts for the oxidation of mustards and mustard analogues. The general synthesis of POMs (equation 1) and the derived structure of **L4** are reported in Fig. 11.5, together with the model catalyzed oxidation of the sulfide **11.4** to the sulfoxide **11.5** (equation 2). The total ratio of P versus (V+M), as expressed in Fig. 11.5, was kept constant at 1 versus 12, while the relative abundances of V (from 0 to 5), W, and Mo (from 0 to 12) were varied for each library individual. The POM library was prepared by simple addition of volumes of reagents, incubation/heating, and subsequent addition of the POM solutions to aliquots of the oxidation mixture. Automated instruments were not necessary and the determination of catalytic activity only required simple GC determinations of **11.4** and **11.5**. The whole 39-member set showed significant catalytic activity for the sulfide–sulfoxide transformation at pH 2.0 with no overoxidation. No significant SAR could be extracted from **L4**, and studies using either more demanding experimental protocols or more diverse libraries should be used in the future to differentiate more active and less active/inactive POM compositions.

$$PO_4^{3-} + xVO_3^{-} + (12 - x)MO_4^{2-} + HNO_3 \text{ (excess)} \xrightarrow{a} H_{(3+x)}[PV_xM_{(12-x)}O_{40}] + (12 - x)H_2O + NO_3^{-} \text{ (excess)} \qquad \textit{equation 1}$$

a: water, 24°C, 16 hrs.

M = **Mo** or **W**

$H_{(3+x)}[PV_xM_{(12-x)}O_{40}]$

L4

39-member
discrete library

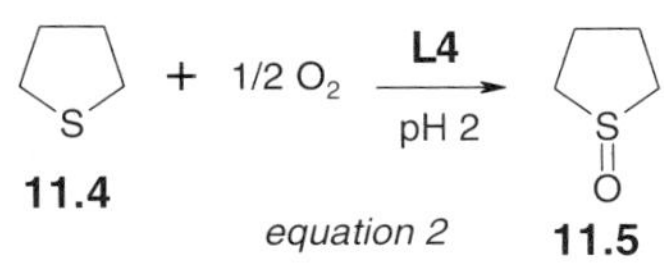

Figure 11.5 Structure and synthesis of the polyoxometalate (POMs) library **L4** and POM-catalyzed sulfide–sulfoxide oxidation of **11.4**.

More sophisticated and miniaturized liquid dispensing units, based on the inkjet technology (Section 6.4.1), were used by Sun et al. (45) to prepare a 100-member microscale discrete photoluminescent library **L5** made of 50 ternary (La, Al, Eu or Gd, Al, Eu) and 50 quaternary composites (La, Gd, Al, Eu; Fig. 11.6). The library was made in around 30 min on a scale of 0.1 mg of composite per well. The starting reagents were added as droplets from standard 0.5 *M* water/ethylene glycol solutions for La, Gd, and Al and a 0.1 *M* solution for Eu. The one hundred 1-mm-deep, 1-mm-diameter ceramic wells were treated, after evaporation of the solvents, at 900 °C for 1 h, and the phosphor activity was measured as reported in Section 11.1.3. A promising red phosphor **11.6**, with defined composition, was identified and structurally characterized (Fig. 11.6). The use of a sophisticated but "hand-made" inkjet delivery system allowed the miniaturization of volumes and quantities, the automation of the process, and its high reliability (<2% fluctuation in the dimensions and contents of each droplet). The forecasted scalability of up to thousands of library individuals makes this system extremely appealing for other applications.

Some interesting devices, both on classical (46) and miniaturized scale (47, 48) have been developed for the combinatorial hydrothermal synthesis of zeolites (49, 50). Solutions of suitable precursors were dispensed into the wells of the multireactor autoclaves, which were then sealed and heated following appropriate protocols. X-ray characterization of the individuals allowed the determination of the properties of the library individuals. Both protocols and instruments allowed the synthesis of hundreds of composites and their automated (47) or semiautomated (46, 48) work-up and characterization. Structural information was derived from both reports, proving the applicability of combinatorial technologies to hydrothermal synthesis procedures.

This and many other fields, using both solid-state and solution-phase synthetic procedures, are expected to benefit significantly from the use of combinatorial technologies. Unprecedented applications will surely appear due to the increasing popularity of materials science applications in the combinatorial chemistry scientific arena.

L5
100-member
discrete library
made with basic element nitrates:
$La(NO_3)_3$, $Gd(NO_3)_3$, $Al(NO_3)_3$
and with an activator element nitrate:
$Eu(NO_3)_3$

$Gd_{0.8}Al_{1.2}Eu^{3+}_{0.06}O_x$

11.6

Figure 11.6 Structure of the photoluminescent materials discrete library **L5** and the most active library individual **11.6** from its screening.

11.2 CHARACTERIZATION AND SCREENING OF MATERIALS SCIENCE LIBRARIES

11.2.1 General Considerations

Combinatorial technologies in pharmaceutical research have received a boost from the need to access chemical diversity with a similar throughput in respect to the available biological screening techniques, which can determine the activity of large libraries in short periods of time. High-throughput analytical techniques for the characterization of organic library individuals are also available, allowing reliable and fast quality control procedures. Even though the very first "combinatorial" report dealt with materials science (19), the scarce portfolio of high-throughput screening methods for composites has failed to stimulate the expansion of combinatorial materials science. The long and labor-intensive characterization process for novel composites has also contributed to prevent the growth of combinatorial applications in materials science.

The above-mentioned synthetic methods, either in solution or as solid-state protocols, have put additional pressure on the fast characterization and screening of materials libraries. The latter subject has been approached by various groups, and several high-technology methods have been reported. They will be briefly presented and discussed via several examples in this section. As of today, structural characterization of library individuals is still troublesome, nongeneralizable, and time consuming. Moreover, it has often proven to be essential to evaluate the quality of a library, and thus this process cannot be avoided. The tendency is to fully characterize a few library members and to assume that their composition and adherence to the planned composite in that library portion are indicative of the whole library purity and quality.

As a comment, solid-state thin-film deposition has significantly improved the throughput and the quality of prepared materials, but the resynthesis of active individuals in large quantities is often performed with more classical solid-state methods. The actual identity of the composition of the "combinatorial" and the "classical" material must be carefully controlled, because even small diffusion resistances during the nucleation/crystallization phases may produce highly different products with significant activity variations, leading to both false positives and negatives.

11.2.2 Screening Libraries for Heterogeneous Catalysts

Several libraries of heterogeneous catalysts have been screened using IR thermographic imaging (51), resonance-enhanced multiphoton ionization (52), and mass spectrometry (53), among others. Example of these screening methods are given below.

Holzwarth et al. (51) reported the synthesis and IR thermographic-imaging screening of a 37-member, focused discrete heterogeneous catalyst library **L6** for oxidations and reductions. The library was prepared using sol–gel solution synthetic protocols (47, 51) to produce the library individuals as amorphous microporous mixed oxides (AMMs), which have previously shown heterogeneous catalytic properties (54, 55). The scaffolding metal oxides contained either Ti (subset 1, Fig. 11.7) or Si (subset 2), and many active metal components were used. The complete structure of **L6** is reported

L6
37-member
discrete library
of Si- and Ti- based AMM oxides

subset 1:		*subset 2:*	
Ir_1Ti	Pd_1Ti	Ir_2Si	$Fe_{10}Si$
Pt_1Ti	Pd_5Ti	Ir_5Si	Pd_1Si
Pt_2Ti	Cr_5Ti	Pt_1Si	Pd_5Si
Pt_5Ti	Co_5Ti	Pt_2Si	Cr_5Si
Zn_5Ti	Ni_5Ti	Pt_5Si	Co_5Si
V_5Ti	Rh_5Ti	Zn_5Si	Ni_5Si
Mn_3Ti	Ru_5Ti	V_5Si	Rh_5Si
Mn_5Ti	Cu_5Ti	Mn_3Si	Ru_5Si
Fe_5Ti		Mn_5Si	Cu_5Si
		Fe_5Si	Ti_5Si

Figure 11.7 Structure of the amorphous microporous mixed (AMMs) oxides discrete catalytic library **L6**.

in Fig. 11.7 (Mn_3Ti signifies 3 mol % of manganese in 97% titania sol, and the same representation is used for the other library individuals). The Ti- or Si-based sols were prepared with the appropriate metal composition and added to the appropriate well of an inert reaction block (a few microliters per well, amounting to <200 μg). The sols were then concentrated, progressive heating was applied, and final reduction of the metal salts produced the AMM active library **L6**.

The library was tested as a source of heterogeneous catalysts for three processes: the reduction of 1-hexyne and the oxidation of iso-octane or toluene. IR thermographic imaging (56), a technique already applied to bead-based libraries (57) and to heterogeneous catalysis (58), was employed to screen **L6**. The reaction block was first heated to the desired temperature; then the appropriate reagent was passed over the block together with a stream of hydrogen (reduction) or air (oxidation). IR thermal images were taken with a steady flow of reagents and were used to discriminate catalytic efficiencies. The temperature of each library individual should only have been increased if catalytic activity was releasing a heat of reaction at its surface. The IR image of the library could be directly related to the temperature of each potential catalyst after adjusting the image of each well for its different composition by subtraction of a blank, that is, the heated reaction block in the absence of the reactants. The three reactions were analyzed, and the results clearly showed different SARs between all three reactions. For example, Pd_1Ti was extremely active for iso-octane oxidation but was inactive toward toluene oxidation, while Pt_5Ti showed the opposite behavior. Small differences could easily be spotted, even with <1 °C temperature differences and small quantities of catalysts, confirming the sensitivity and the reliability of the screening method.

Senkan and Ozturk (52) reported the synthesis and screening of a 66-member discrete heterogeneous catalyst library **L7**, containing Pt, Pd, and In, for the dehydrogenation of cyclohexane to benzene at 300 °C. The structure of **L7** and its synthesis are reported in Fig. 11.8. Sixty-six porous alumina pellets (30 mg each) were shaped into 0.3-cm-diameter, 0.1-cm-high cylinders (step a), then immersed in aqueous HCl solutions containing the 66 appropriate mixtures of $InCl_2$, $PdCl_2$, and H_2PtCl_6 (step

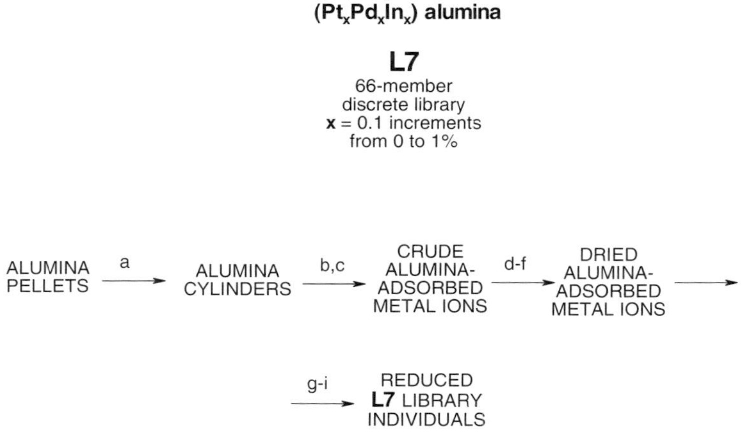

a: shaping of alumina pellets; b: treatment with aq. HCl solutions of H_2PtCl_6, $PdCl_2$ and $InCl_2$; c: incubation at rt for 24 hrs; d: solvent evaporation; e: drying at 90°C (4 hrs) and 120°C (4 hrs); f: calcination at 500°C (2 hrs); g: placement of the cylinders in the reaction block; h: heating at 350°C under helium; i: reduction with hydrogen (2 hrs).

Figure 11.8 Structure and synthesis of the sol–gel discrete library **L7**.

b). Adsorption was continued at rt for 24 h (step c); then the solvent was evaporated and the homogeneous metal–alumina pellets were dried (steps d and e). The library was calcinated for 2 h at 500 °C (step f) to give an average metal content and a total Cl^- content of around 1% in each library individual. The library was then placed in a particular microreactor that ensured the isolation of each pellet/library individual, the uniform exposure of the library to the same flow of reactants, and the uniform heating of the whole microreactor (step g). After heating to 350 °C under helium (step h), the gas source was switched to hydrogen for 2 h to reduce the metal cations to active metals, producing **L7** (step i, Fig. 11.8).

The screening protocol for catalytic activity consisted of lowering the temperature to 300 °C under helium (step a, Fig. 11.9), then introducing 10% cyclohexane in helium carrier gas to the microreactor (step b). Benzene formation was detected using resonance-enhanced multiphoton ionization (REMPI; 59, 60), a technique that detects molecules via their excitation when they are hit by a first photon (equation 1, Fig. 11.9), and their ionization caused by a second photon (equation 2, Fig. 11.9). The photons were generated by a suitable laser source tuned to 259.7 nm, a specific wavelength where benzene is excited and ionized while cyclohexane, argon, and hydrogen are not affected. The laser source was applied to a window in the microreactor allowing irradiation of the gas stream (step c), and a pair of moving microelectrodes were used to detect the relative amount of benzene-derived cations formed in the vicinity of each library pellet (step d, Fig. 11.9). The whole cycle was repeated five times as the microreactor could only host 17 separate reactions at once. The screening results showed how Pt and Pd were generally good catalysts, while In was inactive as

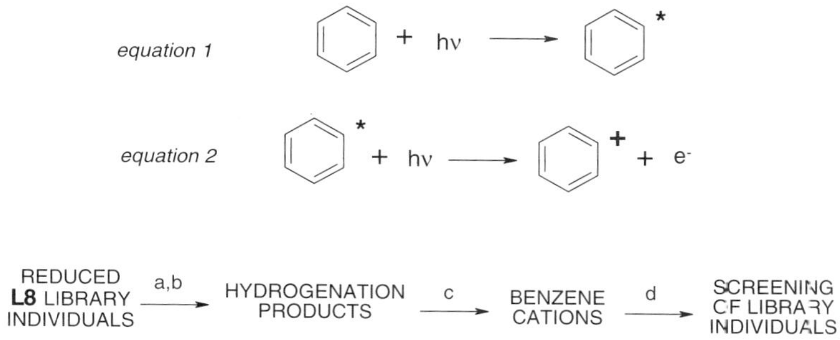

a: hydrogen removal under helium at 300°C; b: introduction of 10% cyclohexane gaseous reaction stream; c: laser source, ionization of benzene; d: REMPI quantitative determination.

BEST COMPOSITION:

($Pt_{0.8}Pd_{0.1}In_{0.1}$) alumina

11.7

Figure 11.9 Screening of the sol–gel discrete catalytic library **L7** with REMPI and structure of the most active catalyst **11.7** for the oxidation of cyclohexane.

expected. An a priori unpredictable ternary composition **11.7** (Fig. 11.9) was found to be the best catalytic library composite.

Cong et al. (53) reported the synthesis and screening of the 120-member discrete materials library **L8** as a source of heterogeneous catalysts for the oxidation of CO to CO_2, promoted either by O_2 (equation 1, Fig. 11.10) or by NO (equation 2, Fig. 11.10). The library were prepared on a circular, 7.5-cm-diameter quartz wafer using a triangular architecture. A 15 × 15 × 15 grid of Rh–Pd–Pt alloys was obtained by radiofrequency sputtering of the metals onto the substrate, using moving masks to obtain the desired composition for each library individual, and the desired metal gradient from an apex of the triangle (100% pure metal) to the facing side (0% of the same metal, binary mixtures of the other library components, Fig. 11.10). Ten deposition steps, each producing a thin metal film of around 10 nm, yielded an average of 2–4 μg of metal alloy with a thickness of around 100 nm for each library individual. The quality of the library and the complete intermetal mixing of the layers was confirmed by X-ray analysis.

The quartz substrate holding the library was then transferred to the screening apparatus, which is schematically depicted in Fig. 11.11 (61). A CO_2-heating laser was focused on a specific library individual, heating it to the desired temperature (300, 350, or 400 °C for O_2-promoted CO oxidation; 400, 500, or 600 °C for NO-promoted CO oxidation) without affecting the neighboring library individuals. Subsequently, a gas stream containing the appropriate reagents was delivered to the heated catalyst through a miniaturized system, with occasional sampling and delivery of the reaction gases to a mass detector to determine the quantity of the reaction products via the measured

$$CO + 1/2\,O_2 \xrightarrow{a} CO_2 \quad \textit{equation 1}$$

$$CO + NO \xrightarrow{b} CO_2 + 1/2\,N_2 \quad \textit{equation 2}$$

a: 300° - 400°C

b: 400° - 600°C

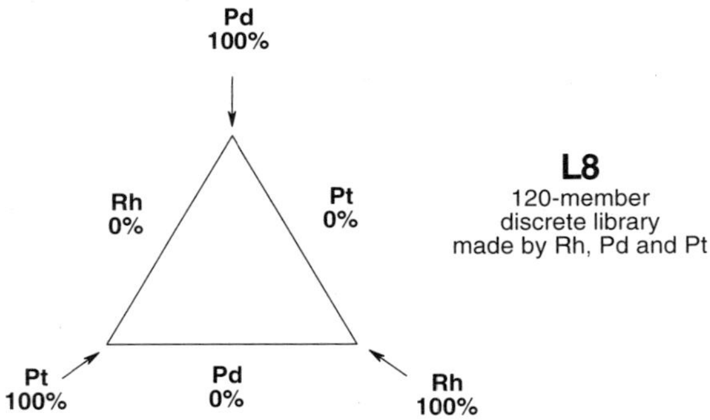

Figure 11.10 Structure of the Rh-Pt-Pd discrete catalytic library **L8** for the oxidation of CO.

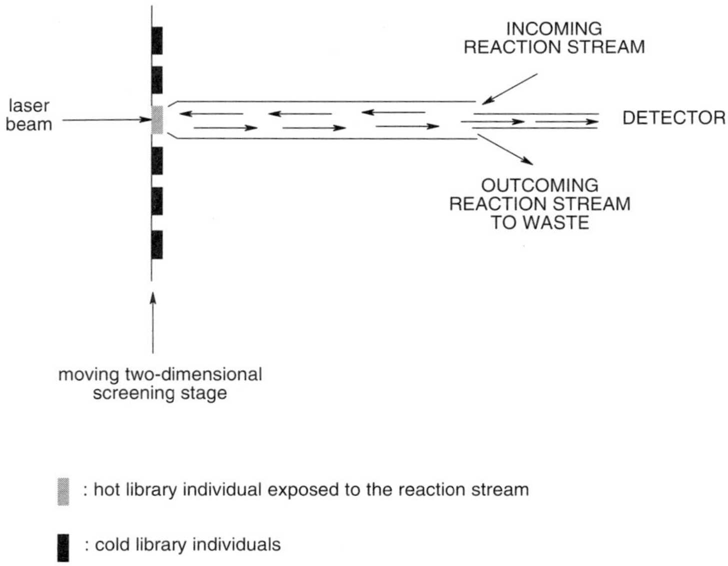

Figure 11.11 MS screening of the Rh-Pt-Pd discrete catalytic library schematic representation of the apparatus.

ion current. The whole process of measuring the activity of **L8** on a specific CO oxidation lasted only 2 h, representing a significant improvement in terms of throughput when compared to classical materials sciences procedures.

Screening results for **L8** in the O_2-promoted oxidation showed the expected Rh > Pd > Pt activity plus a temperature dependence for the most active alloy catalysts. Eighty percent to 85% of Rh was preferred in the alloys below 350 °C, while at higher temperatures the most active library individuals contained just 70–80% Rh. The NO-promoted oxidation could also lead to production of N_2O via an undesired secondary process that should preferably be minimized. The screening results (measured as N_2 and N_2O production) were often comparable in terms of metal reactivity, but N_2O formation at different temperatures was indeed observed and could be minimized, selecting Rh-rich alloys at 600 °C as the best catalysts from **L8**.

The same group reported the use of MS-based screening techniques for the dehydrogentation of ethane to ethylene, using also an optical detection scheme in combination with MS (62). A higher throughput was reported by Senkan et al. (63) using array microreactors (52) and capillary microprobe sampling with on-line Quadrupole Mass Spectrometry for the dehydrogenation of cyclohexane to benzene. More sensitive conditions to better discriminate catalyst selectivity were used by Orschel et al. (64), who employed spatially resolved MS for the oxidation of propene using air at normal pressure; the use of other spatially resolved techniques (GC, IR thermography, UV/VIS, and so on) was also hypothesized by authors. Hoffmann et al. (65) reported a system easily reconducible to conventional catalytic experiments for the oxidation of CO with Au-based catalysts at rt; MS, GC and non dispersive IR were used as screening methods.

The reported examples clearly show how accurate qualitative and/or quantitative screening protocols, based on more or less complex analytical techniques, allow the selection of heterogeneous catalysts and the establishment of SARs from hundreds of library individuals in an acceptable time. An increase in throughput via new technological developments, allowing materials libraries to reach the sizes obtained with synthetic organic libraries, will significantly increase the usefulness and the utilization of heterogeneous catalyst libraries and of materials libraries in general.

11.2.3 Screening Libraries for Electrochemical Catalysts

Reddington et al. (66) reported the synthesis and screening of a 645-member discrete materials library **L9** as a source of catalysts for the anode catalysis of direct methanol fuel cells (DMFCs), with the relevant goal of improving their properties as fuel cells for vehicles and other applications. The anode oxidation in DMFCs is reported in equation 1 (Fig. 11.12). At the time of the publication, state-of-the-art anode catalysts were either binary Pt–Ru alloys (67) or ternary Pt–Ru–Os alloys (68). A systematic exploration of ternary or higher order alloys as anode catalysts for DMFCs was not available, and predictive models to orient the efforts were also lacking.

$$CH_3OH + H_2O \longrightarrow CO_2 + 6H^+ + 6e^- \qquad \textit{equation 1}$$

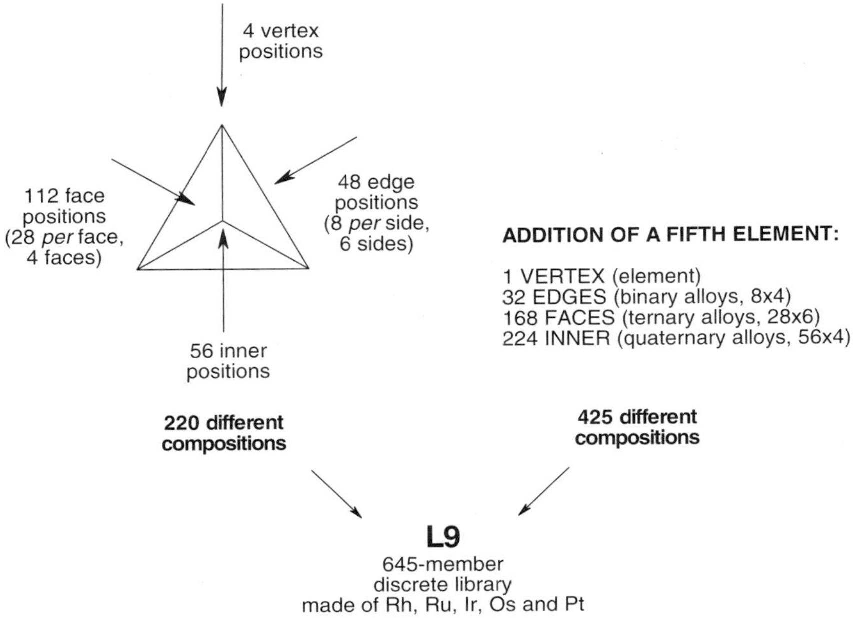

Figure 11.12 Structure of the Pt-Rh-Os-Ir-Ru discrete catalyst library **L9** for the oxidation of methanol.

L9 was built from the combination of five elements (Pt, Ru, Os, Ir, and Rh) in a quaternary phase diagram (combinations of up to four elements from the five possible; 57). The quaternary phase diagram can be represented by a $10 \times 10 \times 10 \times 10$ tetrahedron (Fig. 11.12), where the 4 vertices represent the pure elements, the 48 positions on the edges (6×8) represent the binary alloys, the 112 positions on the faces (4×28) represent the ternary alloys, and the 56 inner positions represent the quaternary alloys to give a total of 220 different compositions. The addition of a fifth element increased the library composition by 1 (element) + 32 (4×8, binary alloys) + 168 (6×28, ternary alloys) + 224 (4×56, quaternary alloys) = 425 different compositions, giving a total number of 645 library individuals for **L9** (Fig. 11.12).

The library was prepared by automated inkjet deposition of aqueous/glycerol solutions of metal salts with appropriate viscosity (step a, Fig. 11.13). The deposition surface was composed of carbon paper to ensure electric conduction and the desired chemical inertness. The deposited salts were reduced to metallic elements (step b);

acidic pH: **fluorescent**
neutral or basic pH: **non fluorescent**

11.8

DEPOSITION SURFACE —a→ PRINTED METAL ION ARRAYS IN SOLUTION —b,c→ **L9** 645-member discrete library —d,e→ SELECTED LIBRARY INDIVIDUALS

a: inkjet deposition of water/glycerol solutions of H_2PtCl_6, $RuCl_3$, $OsCl_3$, K_2IrCl_6 and $RhCl_3$; b: reduction with aqueous $NaBH_4$; c: washings and drying procedures; d: electric connection; e: fluorescence-based screening.

$\mathbf{Pt_{0.44}Ru_{0.41}Os_{0.10}Ir_{0.05}}$

11.9

$\mathbf{Pt_{0.50}Ru_{0.50}}$

11.10

Figure 11.13 Synthetic route to the Pt-Rh-O-Ir-Ru discrete catalyst library **L9** and fluorescence-based screening strategy based on **11.8** to discover the novel catalysts **11.9** and **11.10**.

then the substrate was washed and dried to give the final library **L9** (step c). The library was screened by measurement of the H^+ ions released in the anodic reaction, according to equation 1. First, **L9** was connected electrically to create an anode compartment (step d); then each library individual was tested (step e) using a moving electrode and an electrolyte solution containing methanol, water, and the fluorescent indicator **11.8** (Fig. 11.13). The Ni^{2+} complex **11.8** is not luminescent at neutral pH, but the release of protons makes it fluorescent at acidic pH and gives an optically readable, semiquantifiable signal.

Several active compositions were detected from screening. Among them, the quaternary alloy **11.9** (Fig. 11.13) was selected as the best candidate, reprepared in bulk quantity, and compared with the best commercially available binary alloy catalyst **11.10**. The performance of **11.9** was significantly superior to **11.10** in suitable DMFC conditions, even using a nonoptimized preparation of the quaternary alloy while **11.10** was tested as a high-surface-area catalyst. The structure of **11.9** could not easily have been predicted by rational means, especially with regard to the relative composition of the alloy, once more proving the usefulness of combinatorial technologies in this area.

11.2.4 Screening Libraries for Photoluminescent Materials

Wang et al. (69) reported the synthesis and the visible-light-imaging screening of a 1024-member discrete photoluminescent library **L10** (Fig. 11.14). The library was prepared on a Si surface (2.5 cm^2) containing 650-μm^2 individual sites spaced by 100-μm empty portions. Thin-film deposition was performed using five rotating masks **M1–M5** (Fig. 11.15) with the following scheme:

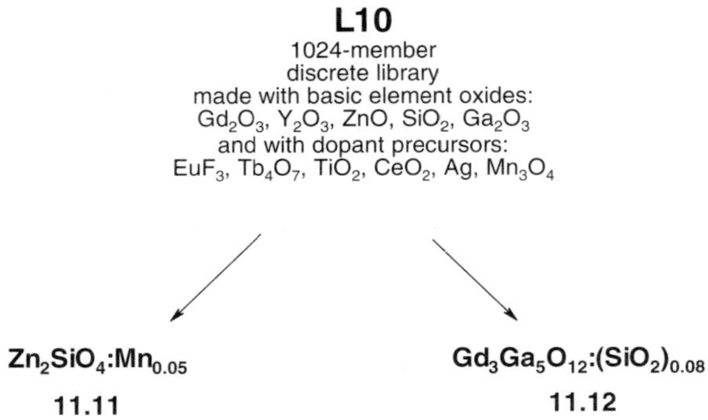

Figure 11.14 Structure of the photoluminescent materials discrete library **L10** and the most active library individual **11.11** and **11.12**.

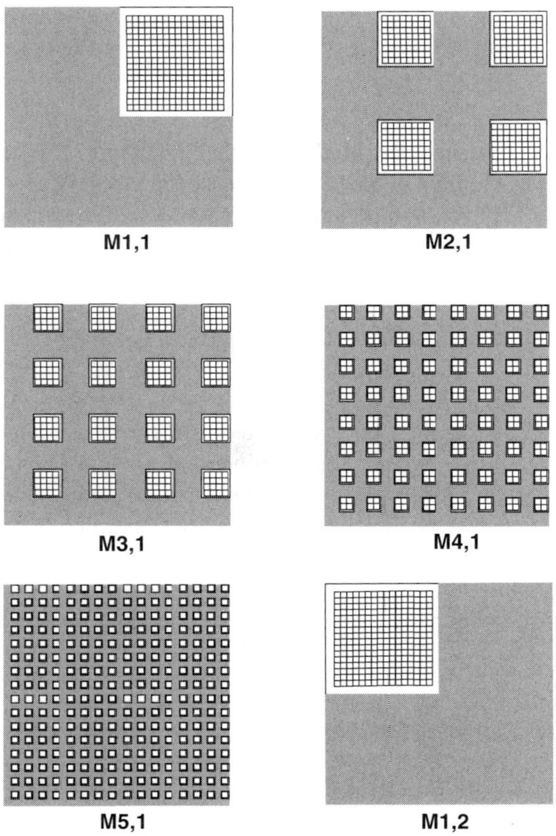

Figure 11.15 Moving masks used for the synthesis of the photluminescent materials discrete library **L10**.

1. **M1,1**: Ga_2O_3 (355 nm);
2. **M1,2**: Ga_2O_3 (426 nm);
3. **M1,3**: SiO_2 (200 nm);
4. **M1,4**: SiO_2 (400 nm);
5. **M4,1**: CeO_2 (3.5 nm);
6. **M4,2**: EuF_3 (11.3 nm);
7. **M4,3**: Tb_4O_7 (9.2 nm);
8. **M5,1**: Ag (3.8 nm);
9. **M5,2**: TiO_2 (6.9 nm);
10. **M5,3**: Mn_3O_4 (5.8 nm);
11. **M2,1**: Gd_2O_3 (577 nm);
12. **M2,2**: ZnO (105 nm);
13. **M2,3**: ZnO (210 nm);
14. **M3,1**: Gd_2O_3 (359 nm);
15. **M3,2**: Y_2O_3 (330 nm); and
16. **M3,3**: Y_2O_3 (82.5 nm).

The plates were rotated counterclockwise with 90° increments, considering their position in Fig. 11.15 as **Mx,1** and the positions **Mx,2**, **Mx,3**, and **Mx,4** respectively, as 90°, 180°, and 270° rotations (see **M1,2** in Fig. 11.15). The sequential deposition scheme allowed the combination of the library components (Ga_2O_3, SiO_2, Gd_2O_3, ZnO, Y_2O_3) and the dopants (EuF_3, CeO_2, Tb_4O_7, Ag, TiO_2, and Mn_3O_4) as 1024 different composites in a 32 × 32 grid. Extensive studies were performed to optimize the library processing to obtain high-quality, uniform materials in each position, and the adopted procedure afforded satisfactory overall results.

The library was screened under ultraviolet irradiation at 254 nm to excite the library individuals using several filters to determine the emission nature (green, red, or blue) of the library individuals. An image of the plate was taken using a charge-coupled detector (CCD) camera with high spatial resolution, and this was used to rapidly discriminate among the photoluminescent library composites. Quantitative determination of their properties was subsequently obtained via spectrophotometric acquisition of the emission and excitation spectra of the ≈100 luminescent samples. A standard phosphor (**11.11**, Zn_2SiO_4–Mn, Fig. 11.14) was included in the library and was used to measure the relative photon output Q for each sample. The most promising samples demonstrated a Q of 60–75% when compared to the standard **11.11**. They all contained Gd and Ga, with or without the addition of Ag as dopant. Resynthesis of pure individuals did not confirm the optical screening results, and the release of some Si from the surface was postulated. Additional studies and the synthesis of a focused library identified the promising Si-doped, blue photoluminescent material **11.12** (Fig. 11.14).

Similar studies (70, 71) were reported by the same authors on different composites using the same screening strategy. The ease and speed of this visual imaging screening, which has been used in many of the previously mentioned photoluminescent libraries, makes it suitable for processing even larger libraries in a short time. However, accurate characterization of the active samples still requires the use of classical analytic techniques. It is thus essential to carefully select the screening cutoffs so as to have only a small percentage of positives to be processed and confirmed.

Isaacs et al. (72) reported the synthesis and screening of an 128-member, discrete photoluminescent library **L11**. The library was prepared as 128-individual 1-mm-thick, 1-mm × 2-mm-rectangular films using radiofrequency sputtering to deposit thin-film layers on a polished $LaAlO_3$ single-crystal surface (Fig. 11.16). Conventional deposition/masking procedures using constitutive and excitatory metal salts were followed by reduction to the metals and heating/diffusion of the library individuals. The library was screened basically as described for **L10**, identifying some promising red, green, and blue phosphors. Its quality control was also assessed in a high-throughput fashion using a complex synchrotron X-ray microbeam source, which ensured the spatial resolution for each library individual while fully characterizing them from a structural point of view (homogeneity of components, morphology of the composites, etc.). This nondestructive, quantitative QC assessment represented a major breakthrough in the characterization of a structural materials library, which could theoretically be extended to other composites and applications. The authors claim a possible upgrade to the QC, which would allow processing of >1000 library individuals per hour using more complex synchrotron facilities and data software management system. The extreme sophistication of the technique and the instrumentation requirement will probably leave this approach largely unexplored, at least in the near future.

Other groups reported the solution synthesis of a cerium-doped inorganic phosphor library (73) to optimize their structure as luminescence down-converters in white Light Emitting Diodes (LEDs) and the use of cathodoluminescence (CL) (74) to identify

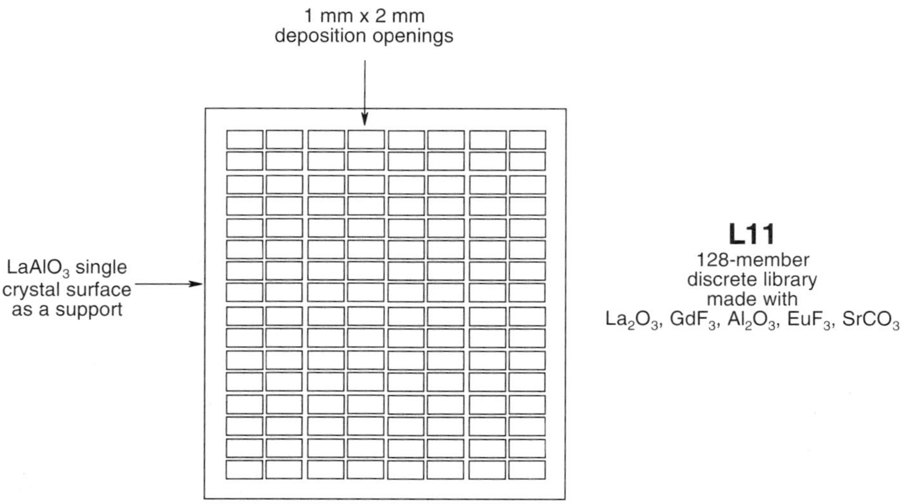

Figure 11.16 Structure of the photoluminescent materials discrete library **L11**.

luminescent properties of ion-implanted samples on a SiO_2 film. A recent review has specifically covered the examples of combinatorial libraries aimed to the discovery of novel luminescence materials (75).

11.2.5 Screening Libraries for Ferroelectric/Dielectric Materials

Chang et al. (76) reported the synthesis and screening of a 256-member discrete ferroelectric materials library **L12** to be used for microwave-related applications. The library was built on a $LaAlO_3$ single-crystal substrate using radiofrequency sputtering of known base elements (Ba, Sr, and Ti oxides) in different combinations, with additional dopants selected as a maximum of three out of nine metallic elements. The adopted synthetic sequence used moving masks **M1–M4** and 90° degree rotation increments as seen for **L10** (Fig. 11.15) in the following sequence:

1. **no mask**: TiO_2 (87 nm);
2. **M2,1**: Fe_2O_3 (0.7 nm);
3. **M2,2**: W (0.5 nm);
4. **M2,3**: CaF_2 (1.2 nm);
5. **M3,1**: Cr (0.4 nm);
6. **M3,2**: Mn_3O_4 (0.7 nm);
7. **M3,3**: CeO_2 (1.2 nm);
8. **M4,1**: MgO (0.7 nm);
9. **M4,2**: Y_2O_3 (1.0 nm);
10. **M4,3**: La_2O_3 (1.2 nm);
11. **M1,1**: BaF_2 (164 nm);
12. **M1,2**: SrF_2 (27 nm) + BaF_2 (132 nm);
13. **M1,3**: SrF_2 (41 nm) + BaF_2 (94 nm); and
14. **M1,4**: SrF_2 (68 nm) + BaF_2 (83 nm).

The library synthesis was completed by progressive heating under an oxygen atmosphere to thoroughly mix the library constituents and the dopants. Screening for ferroelectric/dielectric materials was performed using the previously reported scanning-tip microwave near-field microscope (STMNM) (77). This device allowed a low-micrometer resolution on the library substrate and a high-throughput characterization of the 256 individual composites. A microwave impedance image of the library was obtained, and the most active library composites were identified together with a reasonable SAR related to the dopants and the percentages of main library elements.

A recent report presented Continuous Composition Spread (CCS) approaches (37) to identify thin film dielectrics with varying composition among the system Zr-Sn-Ti-O (78). A structural-driven approach was applied to the discovery of a high dielectric perovskite polymorphic material (79).

Many other combinatorial technologies applications will appear in this emerging field, and their usefulness will also become more apparent for nonimmediate uses. A larger number of scientists working in this field should also be attracted by the potential benefits of a combinatorial approach, with obvious positive effects on the output of relevant materials science libraries.

11.3 POLYMERIC COMBINATORIAL LIBRARIES

11.3.1 General Considerations

Polymer science is a very popular subject in combinatorial technologies, as solid-phase chemistry is based on availability and flexibility of various solid supports made from different polymeric structures. This subject has been extensively covered in Section 1.1, and it remains the main interest of many groups that work to constantly improve the quality and the flexibility of available supports using classical polymer chemistry methods, and hence it will not be covered further here.

Polymer libraries have recently been the focus of several publications with the aim either to discover novel polymeric materials for a specific application (primary libraries) or to rapidly optimize the properties of known polymers (focused libraries). All the major areas where successful combinatorial approaches to polymer libraries have been reported are covered in the following sections with the help of a specific, recent example and up-to-date referencing. Several other papers dealing with polymer combinatorial libraries can be consulted by the interested reader (80–89).

11.3.2 Polymer Libraries as Sources of Soluble Copolymers

Soluble supports for solution-phase combinatorial synthesis were extensively covered in Section 8.5. A recent survey of available soluble supports, with respect to their use in the soluble supported synthesis of various classes of chemicals (90), highlights the wide range of physicochemical properties (especially regarding solubility, tendency to crystallize, and solubilization power) that are embedded in different polymers and copolymers. The assessment of a sort of SAR for the composition of copolymers versus their physicochemical properties would require the preparation of a large number of examples. Combinatorial approaches to soluble support libraries could be highly beneficial in this perspective.

An intriguing paper by Gravert et al. (91) reported the combinatorial synthesis of an array of copolymers **L13** using radical polymerization by means of the free-radical initiator **11.12** (92) and the vinylic monomers $\mathbf{M_1}$ and $\mathbf{M_2}$ (Fig. 11.17). The initiator contained a diazene moiety, known to initiate and propagate free-radical polymerization at 70 °C, and the TEMPO group, which requires 130 °C for initiation of the same process. The first monomer set (five representatives) was used as solvent for **11.12** and polymerization was conducted at 70 °C to give intermediates **11.13** (step a, 5 reactions, Fig. 11.17). Compounds **11.13** were then dissolved in the second monomer set (four from the five representatives excluding the five homopolymeric combinations) and polymerized at 130 °C (step b, 20 reactions, Fig. 11.17) to give the final array **L13**. The array was characterized for its composition using typical polymer chemistry methods, such as size exclusion chromatography (SEC) and transmission electron

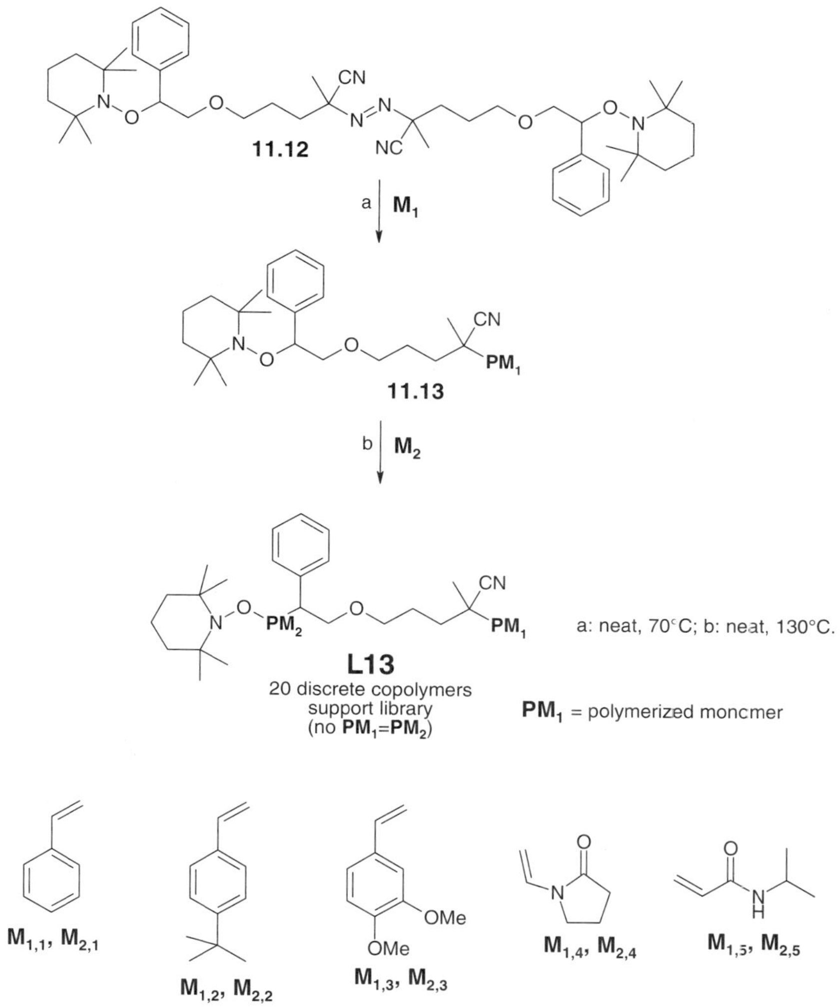

Figure 11.17 Synthesis of the solution-phase, discrete copolymer library of soluble supports **L13** and structures of monomers used for its synthesis.

microscopy (TEM). The excellent quality of each polymeric library component was verified.

Each **L13** individual was evaluated as a potential soluble support in liquid-phase combinatorial synthesis. A great variation in the solubility profiles was observed for the 20 copolymers, the most relevant of which are reported in Table 11.2. The copolymer $\mathbf{M_{1,2}}$–$\mathbf{M_{2,3}}$ was selected, being very soluble in apolar solvents such as THF and diethyl ether, insoluble in polar solvents such as water and alcohol, and non-swelling in any solvent. It thus resulted complementary to PEG-based supports and

TABLE 11.2 Solubility Properties of L13 Individuals[a]

Solubility	Toluene	Et_2O	THF	Acetone	MeCN	OCM	DMF	DMSO	MeOH	Water
$\mathbf{M_{1,1}\text{-}M_{2,2}}$	S	S	S	S	SW	S	SW	I	I	I
$\mathbf{M_{1,1}\text{-}M_{2,3}}$	S	SW	S	S	S	S	S	I	I	I
$\mathbf{M_{1,1}\text{-}M_{2,4}}$	S	SW	S	S	SW	S	S	I	I	I
$\mathbf{M_{1,1}\text{-}M_{2,5}}$	I	I	S	S	S	S	S	SW	S	I
$\mathbf{M_{1,2}\text{-}M_{2,1}}$	S	S	S	SW	I	S	I	I	I	I
$\mathbf{M_{1,2}\text{-}M_{2,3}}$	S	S	S	S	I	S	I	I	I	I
$\mathbf{M_{1,2}\text{-}M_{2,4}}$	I	I	S	I	S	I	I	I	I	I
$\mathbf{M_{1,2}\text{-}M_{2,5}}$	SW	I	S	SW	I	S	SW	I	I	I
$\mathbf{M_{1,3}\text{-}M_{2,1}}$	S	I	S	S	S	S	S	SW	I	I
$\mathbf{M_{1,3}\text{-}M_{2,2}}$	S	SW	S	S	SW	S	S	SW	I	I
$\mathbf{M_{1,3}\text{-}M_{2,4}}$	S	I	S	S	S	S	S	S	S	I
$\mathbf{M_{1,3}\text{-}M_{2,5}}$	SW	I	S	S	S	S	S	S	SW	SW
$\mathbf{M_{1,4}\text{-}M_{2,1}}$	SW	I	S	S	S	S	S	SW	SW	I
$\mathbf{M_{1,4}\text{-}M_{2,}}$	I	I	S	SW	I	S	I	I	I	I
$\mathbf{M_{1,4}\text{-}M_{2,3}}$	I	I	S	S	S	S	S	S	SW	I
$\mathbf{M_{1,4}\text{-}M_{2,5}}$	I	I	S	S	S	S	S	S	S	S
$\mathbf{M_{1,5}\text{-}M_{2,1}}$	I	I	S	S	I	S	S	I	I	I
$\mathbf{M_{1,5}\text{-}M_{2,2}}$	SW	I	S	SW	I	S	I	I	I	I
$\mathbf{M_{1,5}\text{-}M_{2,3}}$	SW	I	S	S	S	S	S	S	SW	I
$\mathbf{M_{1,5}\text{-}M_{2,4}}$	I	I	S	S	S	S	S	S	S	S

[a]S = soluble; SW = swelling, I = insoluble.

was considered a promising candidate for the transfer of PEG-incompatible synthetic routes onto soluble supports.

The combinatorial assembly of isocyanate-decorated dendrimers was reported by Newkome et al. (93) as a method to modulate the solubility, reactivity and viscosity properties of such popular materials. This modulation has an obvious impact for the discovery of high loading, soluble and flexible supports for high throughput organic synthesis.

The structures of five protected functionalized isocyanate building blocks **AP3**–**EP3** (94, 95) are reported in Fig. 11.18, top. The protected side chains for each building block allowed further increase of ramification and loading (n $\rightarrow$ 3n sites after an isocyanate coupling), but above all allowed a combinatorial deprotection/decoration strategy reported for a specific example in Fig. 11.18, bottom. The starting polyamine dendrimer **11.14** (32-PPI, schematically represented as in Fig. 11.18) (96) was treated with a stoichiometeric mixture of isocyanates **DP3** and **EP3** (0.5 eqs. each, step a, Fig. 11.18) to give the mixed, fully protected ureido dendrimer **11.15**. Treatment with formic acid (step b) selectively deprotected the amine functions of **DP3** to give **11.16** which was treated with a stoichiometric mixture of **AP3**, **BP3** and **DP3** (step c, Fig. 11.18). The resulting, hyperbranched **11.17** is an example of the obtainable multi-func-

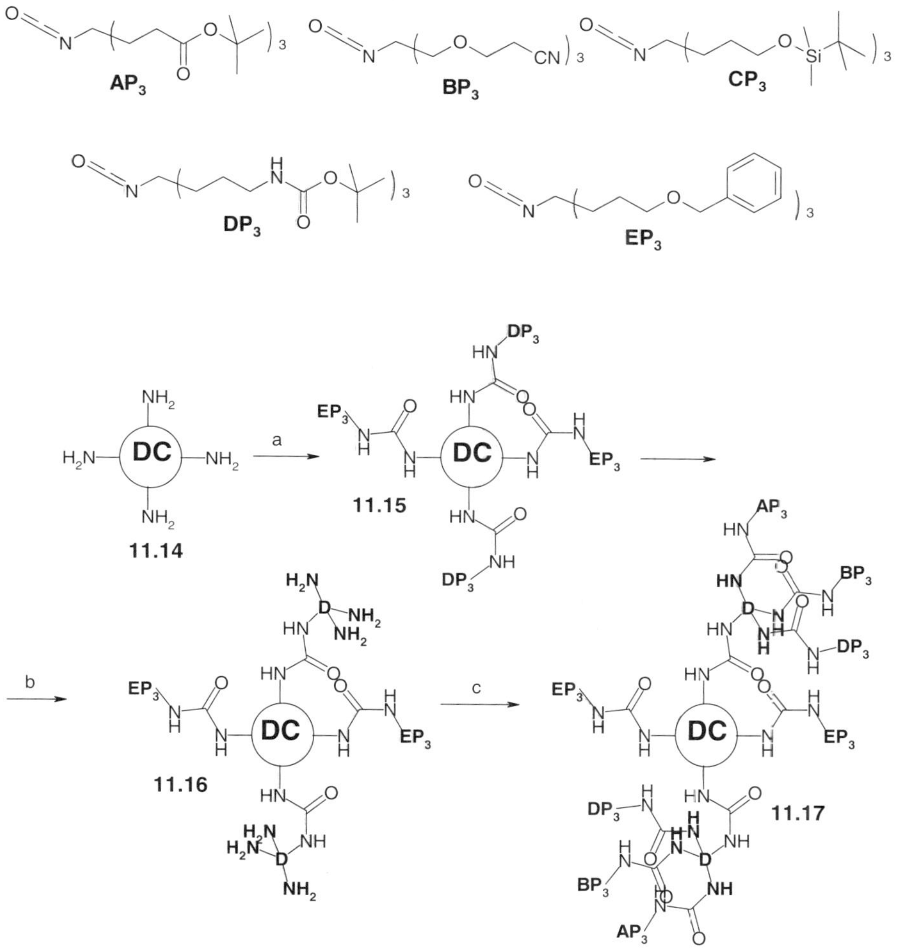

a: **DP_3/EP_3** (0.5/0.5 eqs.), DCM, reflux; b: HCOOH, rt; c: **AP_3/BP_3/DP_3** (0.375/0.25/0.375 eqs.), rt.

Figure 11.18 Structures of isocyanates AP3-EP3 and synthesis of the solution-phase hybrid dendrimers **11.15–11.17**.

tionalization, which may lead to great variability in physicochemical properties exploitable in disciplines such as catalysis and phase-transfer agents (93).

11.3.3 Polymer Libraries as Sources of Reagents

The use of polymeric reagents or catalysts is popular in organic synthesis. Usually, the reactive entity is attached to a solid support, and while its chemical properties are similar to its solution counterpart, the heterogeneicity of the solid supported version allows easier work-up/purification procedures, or stabilizes the supported reactant, or simplifies the recovery of precious catalysts. A general overview of supported reagents and catalysts, with particular attention to their application in solution-phase combinatorial library synthesis, was reported in Section 8.4.

Menger et al. (97) presented an alternative approach, where an 8198-member discrete polymer library **L14** was prepared and tested as a source of reducing agents for the ketone–alcohol transformation in aqueous media. The library features are reported in Fig. 11.19. Its synthesis was carried out using an automated liquid dispenser to deliver and withdraw the solutions of reagents. Two polymeric amines **11.18** and **11.19** were reacted with mixtures of three or four acids selected from the monomer set **M_1** (15 representatives, Fig. 11.19). The coupling was performed with a sub-stoichiometric mixture of acids (typically 20–60% of the polymer amine sites), leaving a substantial amount of free amino groups. The final structure of **L14** was obtained by functionalization of 5–10% of the loading sites with dihydropyridine moieties (DHP; 25–100 DHP units *per* polymer molecule) during the amide coupling. Known biological systems (98, 99), based on the oxidation of DHP moieties and the concomitant reduction of ketones to alcohols, suggested DHP as a potential polymer-supported ketone reducing agent in aqueous media.The presence of metal ions (Cu^{2+}, Mg^{2+}, or Zn^{2+}) was also required for the reducing activity. Two randomly selected library individuals with different polyamine scaffolds could be represented as **11.20** and **11.21** (Fig. 11.19), where the percentages refer to the quantity of each acid versus the number of amine loading sites. The exact determination of functionalized versus nonfunctionalized sites and the relative location of each acid molecule on the polymer backbone was obviously impossible. Each library individual, thus, was more precisely a mixture of structural isomers containing the same relative amounts of each monomer.

The selected test reduction of benzoyl formic acid **11.22** to racemic mandelic acid **11.23** was performed under carefully optimized reaction conditions (Fig. 11.20). The reaction was monitored by following the reduction of intensity of the DHP chromophore at 340 nm as it was oxidized, and the yields for each library individual were expressed as the percentage of oxidized DHP sites per polymer molecule. Yields varied from 0 to 50%, with average reaction times of 4 h. The accuracy of the correspondence between this value and the amount of recovered reaction product **11.23** was checked, performing the large-scale synthesis of a specific active library individual and purifying the obtained **11.23** by chromatography. The screening produced around 25 composites (0.3%) with a conversion yield of >40%. The structures and the reducing activity of several library individuals (**11.24–11.31**) are reported in Table 11.3.

The 10 best polymers were reprepared in larger amounts to confirm their properties, and each preparation was repeated twice to check if the random disposition of the acid-derived functional groups on the polyamine backbone was influencing the reducing efficiency. All the polymers were confirmed as active, and the two batches showed comparable activities, confirming the reproducibility of this synthetic method. Several crude indications in terms of SAR were identified, and a structural specificity of reducing efficiency was clearly present, albeit difficult to rationalize. For example, compare the activity of similar composites **11.27** (40% yield) and **11.29** (1.3% yield). Many modifications of the polymeric scaffold (different average MWs, increase of

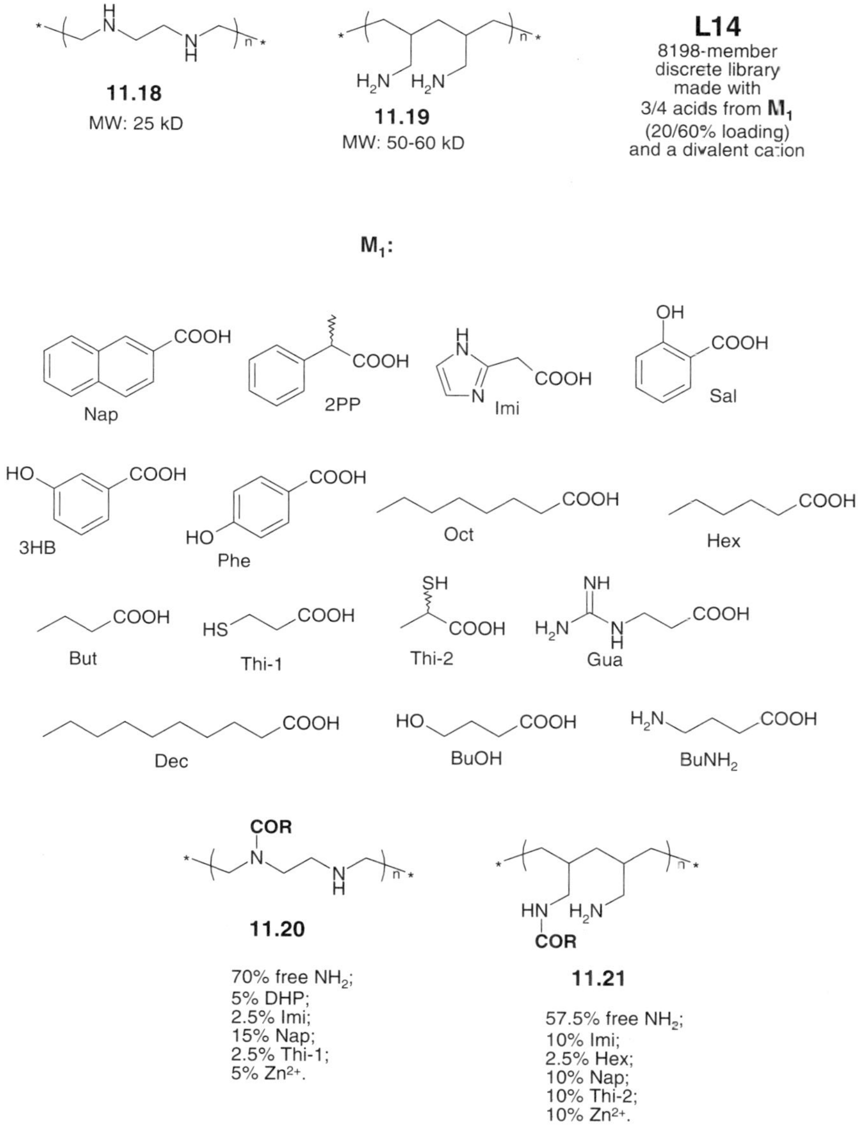

Figure 11.19 Structure of the monomer set **M1** used for the synthesis of the polyamine amide catalyst library **L14** and of the library individuals **11.20** and **11.21**.

DHP loading) or of the experimental reaction protocols (presence of cosolvents, absence of metal ions) completely canceled the reducing efficiency.

The cheap, commercially available precursors as well as the simplicity of the synthesis, characterization, and use of such a polymer as a reagent make this approach

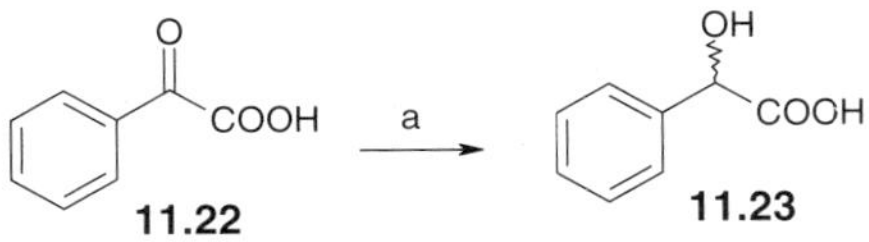

11.22 **11.23**

a: pH 7.2, aq. buffer, 24°C, divalent cation (0.125 μmoles), **L14** library individuals.

Figure 11.20 Screening of the polyamine amide catalyst library **L14** using the keto reduction of **11.22**.

appealing for other organic transformations. Recycling of the polymer by reduction of the oxidized DHPs could render this a true catalytic system, such as the ones are illustrated in the next section.

11.3.4 Polymer Libraries as Sources of Catalysts

Menger et al. (100) reported the synthesis and screening of a 1344-member discrete polymer library **L15** as a source of catalysts for the dehydration of the β-hydroxy ketone **11.33** to the enone **11.34** (equation 1, Fig. 11.21). The main features of **L15**, obtained from poly(acrylic anhydride) **11.32** (101) as a scaffold and the amine monomer set **M1** (11 representatives) are reported in Fig. 11.21. The protocols for library preparation followed the same principles seen for **L14** in the previous section. The presence of both acidic (the COOH backbone, to protonate the OH in **11.33** and promote its departure) and basic groups (side chains in some **M1** representatives, to promote proton abstraction) should fulfill the core functional requirements to exert the overall catalytic activity.

Screening of **L15** was carried out monitoring the appearance of the unsaturated ketone chromophore at 340 nm and calculating the reaction yields in reference to the

TABLE 11.3 Structure and Reducing Activity of Positives from L15

Compound	Polyamine[a]	Composition[b]	Activity (%)[c]
11.24	PEI	2.5Imi;5Hex;7.5Nap;2.5Thi-1;5Zn^{2+}	30
11.25	PEI	15Hex;2.5Gua;2.5BuOH;5Mg^{2+}	33
11.26	PEI	2.5Imi;15Nap;2.5Hex;5Mg^{2}	34
11.27	PEI	2.5Imi;15Nap;2.5Thi-1;10Mg^{2+}	40
11.28	PEI	2.5Imi;15Nap;2.5Thi-1;5Zn^{2+}	>25
11.29	PEI	10Imi;2.5Nap;7.5Thi-1;10Mg^{2+}	1.3
11.30	PAA	10Imi;2.5Hex;10Nap;10Thi-2;10Zn^{2+}	42
11.31	PAA	2.5Gua;7.5Hex;7.5Nap;2.5Thi-2;1Cu^{2+}	28

[a]PEI: poly(ethyleneirnine); PAA: poly(allylamine).

[b]The quantity of DHP was not reported.

[c]Calculated on the total number of DHP per polymer molecule.

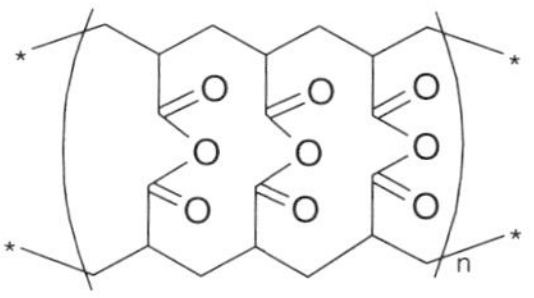

11.32

L15
1344-member
discrete library
made with
3/4 amines from M_1
(up to 20% loading)

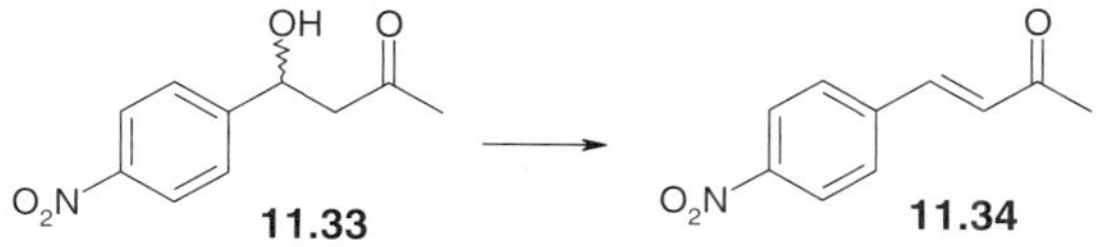

equation 1

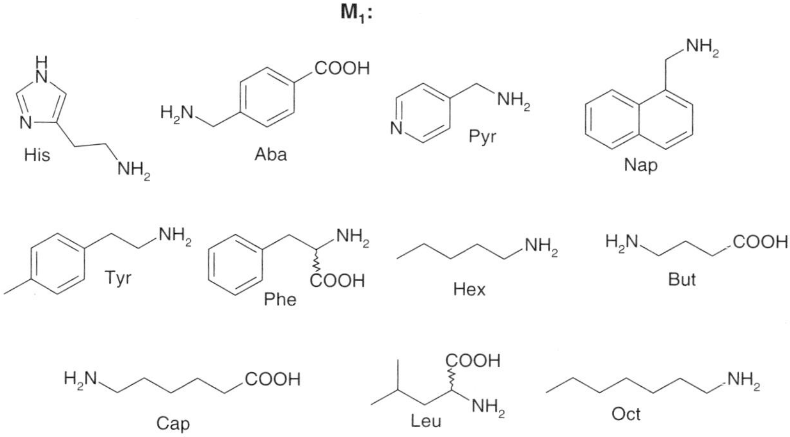

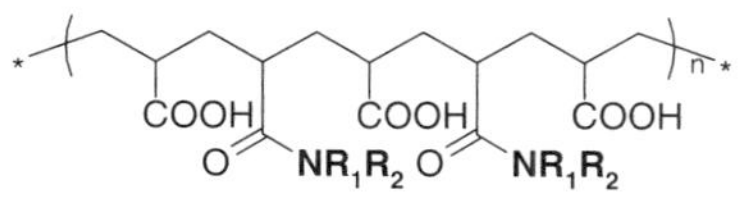

80% free COOH;
5% His;
5% Oct;
5% Phe;
5% Cap.

11.35

Figure 11.21 Structure of the monomer set **M1** used for the synthesis of the polyacid amide artificial enzyme library **L15** and the library individual **11.35**.

absorbance of a pure, independently prepared solution of the product **11.34**. Several library individuals (<1% of **L15**) showed an efficient catalysis of the dehydration reaction, as, for example, **11.35** (Fig. 11.21). An average 1000-fold enhancement of the uncatalyzed dehydration reaction rate was estimated. The reliability and reproducibility of the results was confirmed, as in the previous example, by larger repreparations of multiple batches of active library individuals.

An interesting finding was a common lag period of around 24 h for each catalyst, which was followed for the active library individuals by a fast and efficient catalytic cycle. This catalytic activity was eradicated by low (0 °C) or high (40 °C) reaction temperatures, by sonication of the reaction mixture, or by cross-linking and stiffening of the polymeric matrix. A hypothesis in keeping with all the experimental findings is the slow substrate-induced modification of the polymer into a catalytically active conformation, which could have been disrupted by sonication or higher temperature/energy, or prevented by the rigid structure of the cross-linked matrix, or by the lower temperature/energy.

Menger et al. (33) also reported a similar polymeric library to identify catalysts for the hydrolysis of a phosphate ester (102), where a 30,000-fold increase versus the uncatalyzed reaction rate was obtained. Miller and Ford (103) reported the synthesis of a 32-member discrete polymer library based on anion-exchange latexes and its screening for the alkaline hydrolysis of *p*-nitrophenylalkane carboxylates. The reported efforts may represent just the tip of an iceberg in terms of opportunities granted by polymeric catalysts, and an expansion of knowledge derived from further efforts is to be expected in the near future.

11.3.5 Polymer Libraries as Sources of Biodegradable Medical Implants

The use of combinatorial technologies in pharmaceutical research has been vastly limited to the early phases of the process of drug discovery (see Section 9.1), while later phases in the development of a drug and medical applications of library individuals have always been considered exempt from a combinatorial way of thinking. In fact, their synthetic throughput is usually low, because the efforts are extremely focused and parallel/combinatorial approaches are not considered necessary.

Brocchini et al. (104, 105) reported an intriguing application of parallel synthesis for the fast identification and characterization of polymeric materials suitable for biomedical applications. An 112-member focused polymer library **L16** was prepared and screened as a source of suitable materials for medical implants. Its features and the structure of the monomer sets $\mathbf{M_1}$ (14 diphenols) (106) and $\mathbf{M_2}$ (8 diacids) are reported in Fig. 11.22. The library was designed using degradable, biocompatible monomers to help stimulate the regrowth of functional tissue in the human body. An extensive amount of available information regarding similar structures was used to help select the building blocks and the properties to be monitored for each library individual (*vide infra*). The synthetic protocol was adapted from reported procedures (106, 107) and was carefully set up to accommodate a higher synthetic throughput together with a high control of the polymeric reaction products. The monomer representatives were selected among candidates that exhibited similar reactivity during monomer rehearsal, and the parallel experimental procedures (stirring, solvent delivery, etc.) were regulated to obtain an average quantity of 200 mg of each library component. Careful weighing of monomers was required to ensure their stoichiometric equivalence and thus the high quality of final library individuals.

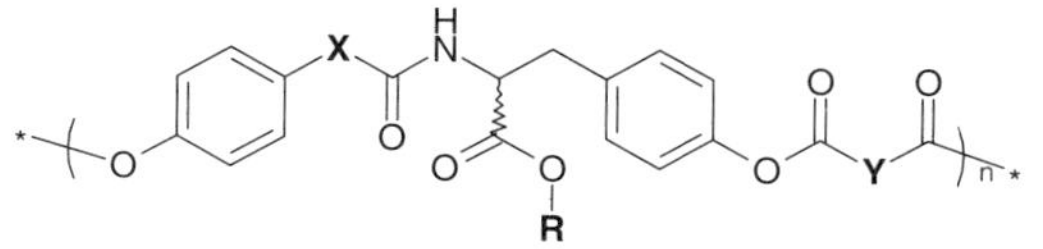

L16
112-member
discrete library
made with
monomer sets M_1 and M_2

X = CH_2 or CH_2CH_2 (see M_1)
Y = see M_2
R = see M_1

M_1:

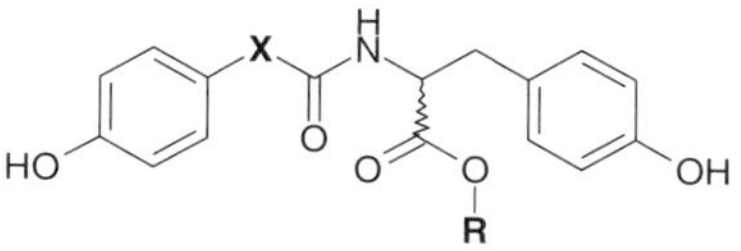

14 representatives

X = CH_2; R = Me, Et, i-Pr, n-Bu, i-Bu, s-Bu, n-Hex, n-Oct, n-Dodecyl, benzyl, EtOEtOEt
X = CH_2CH_2; R = Et, n-Hex, n-Oct

M_2:

8 representatives

Y = $(CH_2)_3$, $(CH_2)_2$, $(CH_2)_4$, $(CH_2)_6$, $(CH_2)_8$, $CH_2CH(CH_3)CH_2CH_2$, CH_2OCH_2, $CH_2OCH_2CH_2OCH_2$

Figure 11.22 Structure of the monomer sets **M1** and **M2** used to prepare the biocompatible polymer amide ester library **L16**.

The materials were characterized using classical polymer science techniques, with less than 40 mg of each sample being required for the measurement of their most relevant physicochemical properties. They all proved to be stable amorphous solids (decomposition above 300 °C), and all were suitable for larger scale production. Their high MW (50,000 < MW < 150,000) and their polymer polydispersity index (from 1.4 to 2) were also appropriate for use as medical implants. Despite their structural similarity, the different backbones and pendent chains produced both stiff and flexible individuals with low (down to 2 °C) or high (up to 91 °C) glass transition temperatures and with varying air–water contact angles (from 64° to 101°). The 112 library individuals were thus similar enough to biomedically relevant models to represent suitable potential candidates for medical applications, but they were also different enough to deserve testing and screening to build an SAR and to eventually select the best candidates for different applications.

A sublibrary **L17**, made of 42 diverse library individuals (Fig. 11.23), was screened in an in vivo model for the proliferation of fibroblasts, where a high value represented a candidate for the tissue reconstruction process around the medical implant. The screening protocol was performed in parallel, and a correlation was established between the fibroblast proliferation (biological property) and the air–water contact angle (physicochemical property). A crude SAR was assessed for further, more

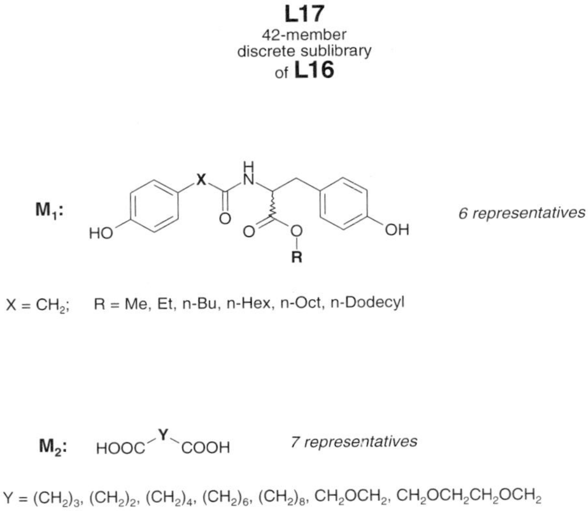

Figure 11.23 Structure of the monomer sets **M1** and **M2** used to prepare the **L16**-derived, biocompatible polymer amide ester sublibrary **L17**.

focused efforts, and several composites were considered appealing candidates for realizing novel medical implant devices. Several other significant findings were obtained from the analysis of **L16** and **L17**, and the potential applicability of other polymers in other fields of polymer science was mentioned by the authors.

The application of combinatorial technologies, or better combinatorial/parallel protocols, to more classical, downstream projects, including in vivo testing such as the example reported above, has the potential to significantly decrease the time needed for information gathering and to better rationalize largely untouched areas. The largely diffused belief that the usefulness of combinatorial technologies is limited to the early research phases should be disproven more and more by similar examples.

11.3.6 Molecularly Imprinted Polymer Libraries

Molecularly imprinted polymers (MIPs) (108, 109) have recently emerged as a class of artificial receptors with desirable properties. They are produced by copolymerization of functional monomers and cross-linking agents in the presence of a template molecule, which drives the polymerization so as to maximize the noncovalent interactions between the polymer backbone and the template (Fig. 11.24; compare the template-driven path b with the random process of path a). Simple removal of the soluble template from the cavities in the polymer structure leaves a template-specific imprint. The resulting MIPs may display high affinities for the template and, when compared to biological antibodies or receptors, possess a higher stability and can be handled safely. They have found applications in many disciplines, and it is easy to predict an exponential growth of interest in MIPs from the scientific community. There

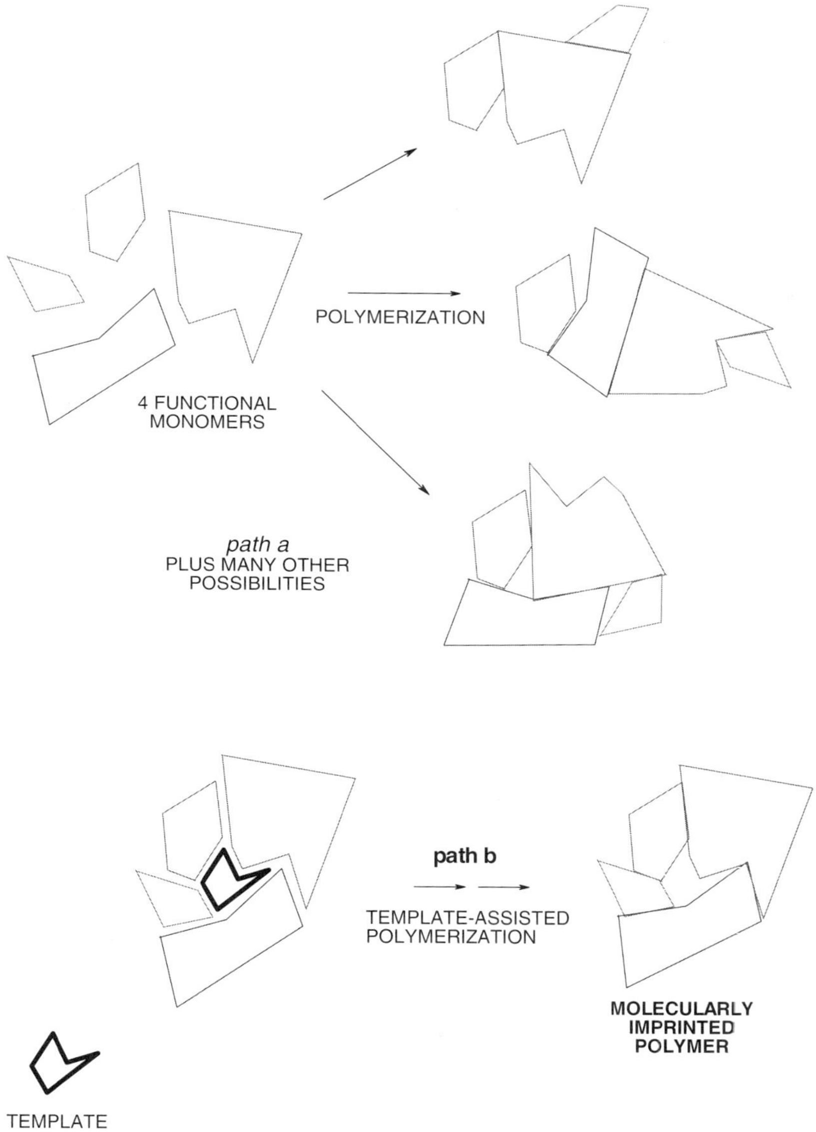

Figure 11.24 Molecularly imprinted polymers (MIP): random (top) and template-assisted (bottom) polymerization process.

are, however, a number of drawbacks that are limiting MIP-related research, such as the typically long process (up to several days) for their preparation and characterization and the absence of assessed SARs/established information databases able to drive the rational design of MIPs for a specific purpose.

Takeuchi et al. (110) recently reported the synthesis and screening of two 49-member parallel MIP libraries **L18** and **L19**, composed of varying percentages of the two acrylate monomers **11.36** and **11.37**, using two triazine herbicides, ametryn **11.38** for **L18** and atrazine **11.38** for **L19**, as imprinting templates. The structures of monomers and templates to produce **L18** and **L19** are reported in Fig. 11.25. The classical, lengthy polymerization protocol was replaced by an automated procedure comprised of the dispensation, under nitrogen, of the appropriate reagent solutions using an automated liquid handler and the polymerization of the reaction mixture to form a thin polymer film in each reaction vial. The whole parallel, simultaneous synthesis of 98 polymers (**L18** + **L19**) lasted less than one day, amounting to a >100-fold reduction of the time required in comparison with a classical, sequential synthesis.

The same automated instrument was used to determine the affinity of each library individual for its respective template. The faster, less accurate screening procedure (path a, Fig. 11.26) consisted of stirring the crude polymer thin film with acetonitrile and measuring the amount of released template by HPLC. Both solvent handling and autosampling into the HPLC were managed by the automated liquid handler. The more rigorous screening (path b, Fig. 11.26) consisted of a thorough washing of the polymer followed by incubation with carefully determined template solutions. Analytical

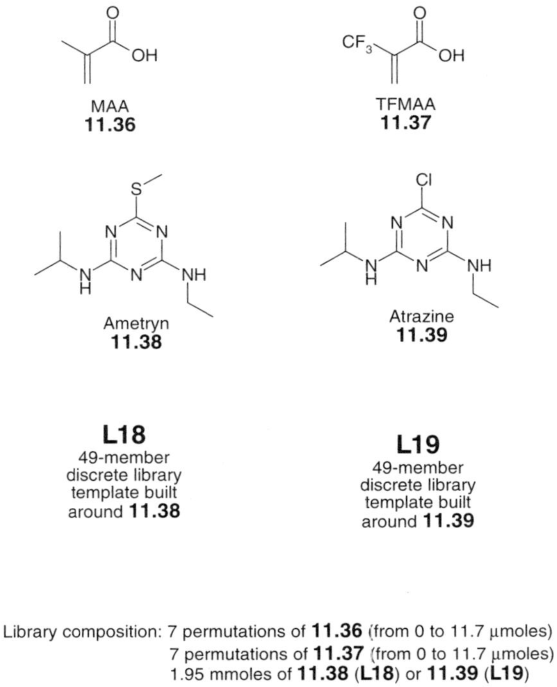

Figure 11.25 Structure of the monomeric units (**11.36** and **11.37**) and of the imprinting templates (**11.38** and **11.39**) used to prepare the MIP discrete libraries **L18** and **L19**.

CRUDE LIBRARIES —a→ EXTRACTION SOLUTIONS —b→ SELECTION OF BEST MIPs *path a*

a: extraction with acetonitrile under stirring, 24 hrs, rt;
b: HPLC quantitation of released **11.38** or **11.39**.

CRUDE LIBRARIES —a→ WASHED LIBRARIES —b→ EXTRACTION SOLUTIONS —c→ SELECTION OF BEST MIPs *path b*

a: 10 washings under stirring with MeOH-AcOH-water 7-1-2, 2 hrs each;
b: incubation with **11.38** or **11.39** solutions (500 μM);
c: HPLC quantitation of templates in solution and quantitation of bound **11.38** or **11.39**.

Figure 11.26 Double-screening procedure for the MIP discrete libraries **L18** and **L19**.

determination of the template left in the supernatant produced, by difference, the amount of adsorbed template.

The screening results for both procedures were coherent and showed some interesting trends. Ametryn was better adsorbed by **L18** individuals containing larger quantities of **11.37**, whereas atrazine was better absorbed by **11.36**-rich **L19** individuals. Even more importantly, a high preference for ametryn versus atrazine was observed for **L18**, showing that good selectivity is possible even in the presence of extremely similar template structures (see reference 91 for additional details). The semiautomated nature of the library synthesis and screening significantly helped the acquisition and utilization of the large amount of data generated.

The use of higher throughput, combinatorial synthetic/screening processes was extremely useful in observing these trends, which could eventually be used to further refine analogous MIPs. Several other reports have presented small MIP libraries (111) or reported their use in combinatorial technologies as artificial targets to select positives from combinatorial libraries (112–114). Reports on parallel copolymer libraries with varying amounts of functional monomers should become more frequent in the near future.

11.3.7 Sensor Libraries and Technologies

The use of chemical sensors to detect small quantities of a specific analyte, mostly for gas-phase sensing but also for solution substances, has gained importance in the last years in many disciplines (115, 116). Few of these sensors, though, have multianalyte specificity together with a high sensitivity, and research is ongoing to improve the characteristics of the materials that compose the sensors. A very active field is the so-called electronic nose sensor (117, 118), where an artificial sensor system allows both the identification and the quantification of complex vapor mixtures, thus mimicking the sense of smell. Complex computer algorithms then allow the pattern detected by the sensor to be recorded and the vapor components to be identified,

L20

4-member
discrete library
from 2 functional monomers

Monomer 1: **PS802**
Monomer 2: **MMA**

Library composition: **0, 6.7, 33.3 or 50%** MAA in PS802

Figure 11.27 Composition of the chemosensor polymer discrete library **L20**.

mimicking the neural processing of the olfactory stimulus in the brain. Foods and perfumes, environmental monitoring, medical diagnosis, and quality control are only several of the disciplines that can benefit from these sensors. Among the many approaches utilized, polymeric materials have been widely used to build electronic noses, and the access to novel, complementary polymers is very important.

Dickinson and Walt (119) recently reported the synthesis and characterization of a small four-member model library **L20** using two monomers in different relative amounts. The composition of **L20** is represented in Fig. 11.27. The library individuals were polymerized at the end of a fiber-optic sensor, and a solvatochromic dye was entrapped in the matrix during the polymerization. The exposure of such a sensor to vapors causes the swelling of the four polymers, depending on their nature and composition, and a spectral shift of the polarity-sensitive dye. The two effects combined lead to a sensing region–specific temporal change in the fluorescence emission

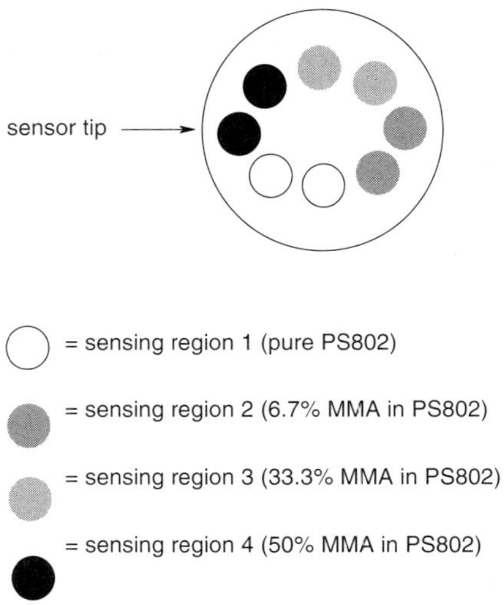

Figure 11.28 Structure of the fiber-optic sensor bearing the chemosensor polymer discrete library **L20**.

TABLE 11.4 Fluorescence Spectra of L20: Main Features[a]

Library Individual[b]Vapor (%)	Benzene	Hexane	2-Propanol	Ethyl Acetate
0	N	N	P	N
6.7	N	N	P	N
33.3	P	B	P	P
50	N	N	N	N

[a]N = negative, P = positive, B = biphasic fluorescence.
[b]Percent of MMA in PS802.

spectrum, and thus an optical screening method was adopted to detect the response for **L20** to various gas phases.

An array of two copies of each library individual (eight sensing regions) was prepared and exposed to four solvent vapors, measuring the fluorescent response of the sensor array for each of them. A 535-nm light source was used to illuminate the array, and emission spectra were monitored at 629 nm with a CCD camera using 1-s vapor pulses and 5-s recording periods. Forty different images were collected in the 4 s following the vapor pulse. A schematic representation of the sensor is reported in Fig. 11.28, highlighting the presence of the eight sensing regions on the same sensor. The four library components showed high sensitivity and reproducibility of results, but most importantly, the four compositions gave different temporal responses, sometimes varying from positive to negative or even biphasic fluorescence changes in response to the same vapor stimulus (see Table 11.4).

The creation of many effective combinations of artificial nose sensors with only a few functional monomers used in different relative amounts was proven here. The application of combinatorial technologies to the discovery of novel materials for more recent, miniaturized electronic nose systems based on small polymer beads (120), and to equally intriguing electronic tonguelike microsensors in solution to mimick the sense of taste for solution mixtures (121), should be highly beneficial and thus is to be expected in the near future.

The same prediction can easily be formulated for combinatorial polymerization of functional monomers to develop novel materials or to speed the optimization of the properties of existing composites. This is probably, together with materials science, the field where the potential of combinatorial technologies has only barely been perceived and thus is also a discipline for which major outcomes have to be expected in the near future.

REFERENCES

1. Xiang, X.-D., *Mat. Sci. Eng.* **B56**, 246–250 (1998).
2. Weinberg, W. H., Jandeleit, B., Self, K. and Turner, H., *Curr. Opin. Solid State Mater. Sci.* **3**, 104–110 (1998).

3. Xiang, X.-D., *Chem. Ind.,* 800–802 (1998).
4. Jandeleit, B., Turner, H. W., Uno, T., Van Beek, J. A. M. and Weinberg, W. H., *CATTECH2,* 101–123 (1998).
5. Schlogl, R., *Angew. Chem. Int. Ed.* **37**, 2333–2336 (1998).
6. McFarland, E. W. and Weinberg, W. H., *Trends Biotechnol.* **17**, 107–115 (1999).
7. Bein, T., *Angew. Chem. Int. Ed.* **38**, 323–326 (1999).
8. Maier, W. F., *Angew. Chem. Int. Ed.* **38**, 1216–1218 (1999).
9. Newsam, J. M. and Schuth, F., *Biotech. Bioeng.* **61**, 203–216 (1999).
10. Xiang, X.-D., *Biotech. Bioeng.* **61**, 227–241 (1999).
11. Xiang, X.-D., *Annul Rev. Mater. Sci.* **29**, 149–171 (1999).
12. Adachi, G., *Textbook in Chemistry: Fundamentals of Solid State Chemistry and Inorganic Materials.* Maruzen, Tokyo, Japan, 1995.
13. Poeppelmeier, K. R., *Chem. Mater.* **10**, 2577–2578 (1998).
14. Janek, J. and Ruck, M., *Nachr. Chem., Tech. Lab.* **47**, 143–153 (1999).
15. *Solid-State Chemistry of Inorganic Materials II,* S. M. Kauzlarich, E. M. McCarron, III, A. W. Sleight and H.-C. Zur Loye (Eds.). Mater. Res. Soc., Warrendale, PA, 1999.
16. Campbell, D. S., *Thin Solid Films* **32**, 3–10 (1976).
17. Dini, J. W., *Mater. Manuf. Processes* **12**, 437–472 (1997).
18. Pawlowski, L., *Galvano-Orgrano* **66**, 532–537 (1997).
19. Hanak, J. J., *J. Mater. Sci.* **5**, 964–971 (1970).
20. DaSilva, E. M. and Terhune, C. E., *IBM Tech. Disclosure Bull.* **22**, 2922 (1979).
21. Fodor, S. P. A., Read, J. L., Pirrung, M. C., Stryer, L. Lu, A. T. and Solas, D., *Science* **251**, 767–773 (1991).
22. Xiang, X.-D., Sun, X., Briceno, G., Lou, Y., Wang, K.-A., Chang, H., Wallace Freedman, W. G., Chen, S.-W. and Schultz, P. G., *Science* **268**, 1738–1740 (1995).
23. Kay, E., *Tech. Metal Res.* **1**, 1269–1309 (1968).
24. Rohde, S. L. and Muenz, W., *Adv. Surf. Coat.,* 92–126 (1991).
25. Raven, M. S., *J. Mater. Sci.: Mater. Electron.* **5**, 129–146 (1994).
26. Powell, C. F., *Vapor Deposition*, 221–248 (1966).
27. *Handbook of Physical Vapor Deposition (PVD) Processing: Film Formation, Adhesion, Surface Preparation and Contamination Control,* D. M. Mattox (Ed.). Noyes, Westwood, NJ, 1998.
28. Oettel, H., *Freiberg. Forschungsh. B* **B287**, 17–33 (1998).
29. Conway, B. E., *Prog. Surf. Sci.* **16**, 1–137 (1984).
30. *Electrochemical Phase Formation and Growth: An Introduction to the Initial Stages of Metal Deposition.* E. Budevski, G. Staikov, W. J. Lorenz (Eds.) VCH, 1996, Weinheim, Germany.
31. Oskam, G., Long, J. G., Natarajan, A. and Searson, P. C., *J. Phys. D: Appl. Phys.* **31**, 1927–1949 (1998).
32. Egermeier, J. and Hill, R. J., *Proc. Annul Tech. Conf.—30th Soc. Vac. Coaters,* 329–337 (1987).
33. Ray, A. K., *Trans. Met. Finish. Assoc. India* **4**, 49–55 (1995).

34. Dieleman, J., Van de Riet, E. and Kools, J. C. S., *Jpn. J. Appl. Phys., Part 1* **31**, 1964–1971 (1992).

35. Hanabusa, M., *Proc. Int. Conf: Lasers*, 55–62 (1994).

36. Stuke, M., *Vide: Sci., Tech. Appl.* **54**, 484–515 (1998).

37. van Dover, R. B., Schneemeyer, L. F. and Fleming, R. M., *Nature* **392**, 162–164 (1998).

38. Danielson, E., Golden, J. H., McFarland, E. W., Reaves, C. M., Weinberg, W. H. and Wu, X. D., *Nature* **389**, 944–948 (1997).

39. Danielson, E., Devenney, M., Giaquinta, D. M., Golden, J. H., Haushalter, R. C., McFarland, E. W., Poojary, D. M., Reaves, C. M., Weinberg, W. H. and Wu, X. D., *Science* **279**, 837–839 (1998).

40. Danielson, E., Devenney, M., Giaquinta, D. M., Golden, J. H., Haushalter, R. C., McFfarland, E. W., Poojary, D. M., Reaves, C. M., Weinberg, W. H. and Wu, X. D., *J. Mol. Struc.* **470**, 229–235 (1998).

41. Briceno, G., Chang, H., Sun, X., Schultz, P. G. and Xiang, X.-D., *Science* **270**, 273–275 (1995).

42. Takeuchi, I., Chang, H., Gao, C., Schultz, P. G., Xiang, X.-D., Sharma, R. P., Downes, M. J. and Venkatesan, T., *Appl. Phys. Lett.* **73**, 894–896 (1998).

43. Godovsky, D., Inganas, O., Brabec, C. J., Sariciftci, N. S., Hummelen, J. C., Janssen, R. A. I., Prato, M., Maggini, M., Segura, J. and Martin, N., *AIP Conf: Proc.* **486**, 483–486 (1999).

44. Hill, C. L. and Damico Gall, R., J. *Mol. Catal. A* **114**, 103–111 (1996).

45. Sun, X.-D., Wang, K. A., Yoo, Y., Wallace-Freedman, W. G., Gao, C., Xiang, X. D. and Schultz, P. G., *Adv. Mater.* **9**, 1046–1049 (1997).

46. Akporiaye, D. E., Dahl, I. M., Karlsson, A. and Wendelbo, R., *Angew. Chem. Int. Ed.* **37**, 609–611 (1998).

47. Klein, J., Lehmann, C. W., Schmidt, H.-W. and Maier, W. F., *Angew. Chem. Int. Ed.* **37**, 3369–3372 (1998).

48. Choi, K., Gardner, D., Hilbrandt, N. and Bein, T., *Angew. Chem. Int. Ed.* **38**, 2891–2894 (1999).

49. Feijen, E. P. and Martens, P. A., *Stud. Surf: Sci. Catal.* **84**, 3–35 (1994).

50. Corma, A., *Chem. Rev.* **97**, 2373–2420 (1997).

51. Holzwarth, A., Schmidt. H.-W. and Maier, W. F., *Angew. Chem. Int. Ed.* **37**, 2644–2647 (1998).

52. Senkan, S. M. and Ozturk, S., *Angew. Chem. Int. Ed.* **38**, 791–795 (1999).

53. Cong, P., Doolen, R. D., Fan, Q., Giaquinta, D. M., Guan, S., McFarland, E. W., Poojary, D. M., Self, K., Turner, H. W. and Weinberg, W. H., *Angew. Chem. Int. Ed.* **38**, 484–488 (1999).

54. Klein, S., Thorimbert, S. and Maier, W. F., *J. Catal.* **163**, 476–488 (1996).

55. Storck, S., Maier, W. F., Salvado, I. M. M., Ferreira, J. M. F., Guhl, D., Souverijns, W. and Martens, J. A., *J. Catal.* **172**, 414–426 (1997).

56. Georgiades, G., Self, V. A. and Sermon, P. A., *Angew. Chem. Int. Ed. Engl.* **26**, 1050–1052 (1987).

57. Taylor, S. J. and Morken, J. P., *Science* **280**, 267–270 (1998).

58. Moates, F. C., Somani, M., Annamalai, J., Richardson, J. T., Luss, D. and Willson, R. C., *Ind. Eng. Chem. Res.* **35**, 4801–4803 (1996).

59. Srinivasan, R., Hsing, I-M., Berger, P. E., Jensen, K. F., Firebaugh, S. L., Schmidt, M. A., Harold, M. P., Lerou, J. J. and Ryley, J. F., *AlChE J.* **43**, 3059–3069 (1997).

60. Senkan, S. M., *Nature* **394**, 350–353 (1998).

61. Weinberg, W. H., McFarland, E. W., Cong, P. and Guan, S., *WO Patent 9815969 A2* (1998).

62. Cong, P., Dehestani, A., Doolen, R., Giaquinta, D. M., Guan, S., Markov, V., Poojary, D., Self, K., Turner, H. and Weinberg, W. H., *Proc. Natl. Acad. Sci. USA* **96**, 11077–11080 (1999).

63. Senkan, S., Krantz, K., Ozturk, S., Zengin, V. and Onal, I., *Angew. Chem. Int. Ed.* **38**, 2794–2799 (1999).

64. Orschel, M., Klein, J., Schmidt, H.-W. and Maier, W. F., *Angew. Chem. Int. Ed.* **38**, 2791–2794 (1999).

65. Hoffmann, C., Wolf, A. and Schuth, F., *Angew. Chem. Int. Ed.* **38**, 2800–2803 (1999).

66. Reddington, E., Sapienza, A., Gurau, B., Viswanathan, R., Sarangapani, S., Smotkin, E. S. and Mallouk, T. E., *Science* **280**, 1735–1737 (1998).

67. Watanabe, M., Uchida, M. and Motoo, S., *J. Electroanal. Chem. Interfacial Electrochem.* **229**, 395–406 (1987).

68. Ley, K. L., Liu, R., Pu, C., Fan, Q., Leyarovska, N., Segre, C. and Smotkin, E. S., *J. Electrochem. Soc.* **144**, 1543–1548 (1997).

69. Wang, J., Yoo, Y., Gao, C., Takeuchi, I., Sun, X., Chang, H., Xiang, X. -D. and Schultz, P G., *Science* **279**, 1712–1714 (1998).

70. Sun, X.-D., Gao, C., Wang, J. and Xiang, X.-D., *Appl. Phys. Lett.* **70**, 3353–3355 (1997).

71. Sun, X.-D. and Xiang, X.-D., *Appl. Phys. Lett.* **72**, 525–527 (1998).

72. Isaacs, E. D., Marcus, M., Aeppli, G., Xiang, X.-D., Sun, X.-D., Schultz, P., Kao, H.-K., Cargill, G. S. III and Haushalter, R., *Appl. Phys. Lett.* **73**, 1820–1822 (1998).

73. Wu, J. L., Devenney, M., Danielson, E., Poojary, D. and Weinberg, W. H., *Mat. Res. Soc. Symp. Proc.* **560**, 65–70 (1999).

74. Chen, C.-M., Pan, H. C., Zhu, D. Z., Hu, J. and Li, M. Q., *Nucl. Instr. Meth. Phys. Res. B* **159**, 81–88 (1999).

75. Sun, T. X., *Biotechnol. Bioeng.* **61**, 193–201 (1999).

76. Chang, H., Gao, C., Takeuchi, I., Yoo, Y., Wang, J., Schultz, P. G., Xiang, X.-D., Sharma, R. P., Downes, M. and Venkatesan, T., *Appl. Phys. Lett.* **72**, 2185–2187 (1998).

77. Wei, T., Xiang, X.-D., Wallace-Freedman. W. G. and Schultz, P. G., *Appl. Phys. Lett.* **68**, 3506–3508 (1996).

78. van Dover, R. B., Schneemeyer, L. F., Fleming, R. M. and Huggins, H. A., *Biotechnol. Bioeng.* **61**, 217–225 (1999).

79. Park, J.-H., Parise, J. B., Woodward, P. M., Lubomirsky, I. and Stafsudd, O., *J. Mater. Res.* **14**, 3192–3195 (1999).

80. Klaerner, G., Safir, A. L., Chang, H.-T., Petro, M. and Nielsen, R. B., *Polym. Prepr.* **40**, 469 (1999).

81. Hawker, C. J., Benoil, D., Rivera, F. Jr., Chaplinski, V., Nilsen, A. and Braslau, R., *Polym. Mater. Sci. Eng.* **80**, 90–91 (1999).

82. Nielsen, R. B., Safir, A. L., Petro, M., Lee, T. S. and Huefner, P., *Polym. Mater. Sci. Eng.* **80**, 92 (1999).

83. Petro, M., Safr, A. L. and Nielsen, R. B., *Polym. Mater. Sci. Eng.* **80**, 702 (1999).

84. Frechet, J. M. J., *Polym. Mater. Sci. Eng.* **80**, 494 (1999).

85. Schmitz, C., Posch, P., Thelakkat, M. and Schmidt, H.-W., *Phys. Chem. Chem. Phys.* **1**, 1777–1781 (1999).

86. Wurthner, F., Sens, R., Etzbach, K.-H. and Seybold, G., *Angew. Chem. Int. Ed.* **38**, 1649–1652 (1999).

87. Wang, J. and Gonsalves, K. E., *J. Comb. Chem.* **1**, 216–222 (1999).

88. Reynolds, C. H., *J. Comb. Chem.* **1**, 297–306 (1999).

89. Newkome, G. R., *Pure Appl. Chem.* **70**, 2337–2343 (1999).

90. Gravert, D. J. and Janda, K. D., *Chem. Rev.* **97**, 489–509 (1997).

91. Gravert, D. J., Datta, A., Wentworth, P., Jr. and Janda, K. D., *J. Am. Chem. Soc.* **120**, 9481–9495 (1998).

92. Li, I. Q., Howell, B. A., Dineen, M. T., Kastl, P. E., Lyons, J. W., Meunier, D. M., Smith, P. B. and Priddy, D. B., *Macromolecules* **30**, 5195–5199 (1997).

93. Newkome, G. R., Weis, C. D., Moorefield, C. N., Baker, G. R., Childs, B. J. and Epperson, J., *Angew. Chem. Int. Ed.* **37**, 307–310 (1998).

94. Newkome, G. R., Weis, C. D., Moorefield, C. N. and Fronczek, F. R., *Tetrahedron Lett.* **38**, 7053–7056 (1997).

95. Newkome, G. R., Weis, C. D. and Childs, B. J., *Des. Monom. Polym.* **1**, 3–14 (1998).

96. Worner, C. and Mulhaupt, R., *Angew. Chem., Int. Ed. Engl.* **32**, 1306–1308 (1993).

97. Menger, F. M., West, C. A. and Ding, J., *Chem. Commun.*, 633–634 (1997).

98. Talma, A. G., Jouin, P., De Vries, J. G., Troostwijk, C. B., Werumeus Buning, G. H., Waininge, J. K., Visscher, J. and Kellogg, R. M., *J. Am. Chem. Soc.* **107**, 3981–3997 (1985).

99. Engbersen, J. F. J., Koudijs, A. and Van der Plas, H. C., *J. Org. Chem.* **55**, 3647–3654 (1990).

100. Menger, F. M., Ding, J. and Barragan, V., *J. Org. Chem.* **63**, 7578–7579 (1998).

101. Choi, S.-K., Mammen, M. and Whitesides, G. M., *J. Am. Chem. Soc.* **119**, 4103–4111 (1997).

102. Menger, F. M., Eliseev, A. V. and Migulin, V. A., *J. Org. Chem.* **60**, 6666–6667 (1995).

103. Miller, P. D. and Ford, W. T., *Chem. Commun.*, 1151–1152 (1998).

104. Brocchini, S., James, K., Tangpasuthadol, V. and Kohn, J., *J. Am. Chem. Soc.* **119**, 4553–4554 (1997).

105. Brocchini, S., James, K., Tangpasuthadol, V. and Kohn, J., *J. Biomed. Mater. Res.* **42**, 66–75 (1998).

106. Hooper, K. A. and Kohn, J., *J. Bioact. Compat. Polym.* **10**, 327–340 (1995).

107. Fiordeliso, J., Bron. S. and Kohn, J., *J. Biomater. Sci. Polym. Ed.* **5**, 497–510 (1994).

108. Wulff, G., *Angew. Chem. Int. Ed. Engl.* **34**, 1812–1832 (1995).

109. Ensing, K. and de Boer, T., *Trends Anal. Sci.* **18**, 138–145 (1999).

110. Takeuchi, T., Fukuma, D. and Matsui, J., *Anal. Chem.* **71**, 285–290 (1999).

111. Sabourin, L., Ansell, R. J., Mosbach, K. and Nicholls, I. A., *Anal. Commun.* **35**, 285–287 (1998).

112. Berglund, J., Lindbladh, C., Mosbach, K. and Nicholls, I. A., *Anal. Commun.* **35**, 37 (1998).

113. Ramstrom, O., Ye, L. and Mosbach, K., *Anal. Commun.* **35**, 9–11 (1998).

114. Ramstroem, O., Ye. L., Krook, M. and Mosbach, K., *Chromatographia* **47**, 465–469 (1998).

115. Ricco, A. J. and Crooks, R. M., *Acc. Chem. Res.* **31**, 200 (1998).

116. Crooks, R. J. and Ricco, A. J., *Acc. Chem. Res.* **31**, 219–227 (1998).

117. Walt, D. R., *Acc. Chem. Res.* **31**, 267–278 (1998).

118. Dickinson, T. A., White, J., Kauer, J. S. and Walt. D. R., *Trends Biotechnol.* **16**, 250–258 (1998).

119. Dickinson, T. A. and Walt, D. R., *Anal. Chem.* **69**, 3413–3418 (1997).

120. Dickinson, T. A., Michael, K. L., Kauer, J. S. and Walt, D. R., *Anal. Chem.* **71**, 2192–2198 (1999).

121. Lavigne, J. J., Metzger, A., Niikura, K., Cabell, L. A., Savoy, S. M., Yoo, J. S.-J., McDevitt, J. T., Neikirk, D. P., Shear, J. B. and Anslyn, E. V., *Proc. SPIE-Int. Soc. Opt. Eng.* **3602**, 17–26 (1999).

INDEX